AF410866

Herbaceous Field Crops Cultivation

Herbaceous Field Crops Cultivation

Editors

Sara Lombardo
Giovanni Mauromicale

MDPI • Basel • Beijing • Wuhan • Barcelona • Belgrade • Manchester • Tokyo • Cluj • Tianjin

Editors

Sara Lombardo
Department of Agriculture,
Food and Environment (Di3A)
University of Catania
Catania
Italy

Giovanni Mauromicale
Department of Agriculture,
Food and Environment (Di3A)
University of Catania
Catania
Italy

Editorial Office
MDPI
St. Alban-Anlage 66
4052 Basel, Switzerland

This is a reprint of articles from the Special Issue published online in the open access journal *Agronomy* (ISSN 2073-4395) (available at: www.mdpi.com/journal/agronomy/special_issues/Herbaceous_Crops_Cultivation).

For citation purposes, cite each article independently as indicated on the article page online and as indicated below:

LastName, A.A.; LastName, B.B.; LastName, C.C. Article Title. *Journal Name* **Year**, *Volume Number*, Page Range.

ISBN 978-3-0365-2535-8 (Hbk)
ISBN 978-3-0365-2534-1 (PDF)

Contents

About the Editors

Sara Lombardo

Sara Lombardo is a researcher within the "Agronomy and herbaceous field crops" scientific disciplinary sector, at the University of Catania (Italy). Her main research subjects are: i) yield and nutritional responses of different Mediterranean crops to environmental conditions; ii) effects of crop management on the quality of Mediterranean crops; iii) bioactive compounds and possible alternative utilization of *Cynara cardunculus* L.; iv) effects of organic farming on food quality; and v) interaction between pre-harvest factors and post-harvest treatments on quality of globe artichoke and potato.

Giovanni Mauromicale

Giovanni Mauromicale is full professor of Agronomy and Field Crops at the University of Catania (Italy). His main research interests are: a) crop ecophysiological and yield responses to environmental stresses and agronomic management; b) functional substances and possible alternative utilization of *Cynara cardunculus* L.; c) genetic and inbreeding aspects on globe artichoke; d) soil solarization and control of *Orobanche* spp.; e) role of cover crops in agroecosystems; and f) utilization of allelopathic mechanism for weed and microbial control.

Preface to "Herbaceous Field Crops Cultivation"

This book, entitled "Herbaceous Field Crops Cultivation", includes 24 articles from researchers worldwide and provides detailed research on several aspects of herbaceous field crops' cultivation. Particularly, it contains two reviews and 22 original research papers aimed at elucidating the influence of agricultural management factors (i.e., genetic selection, planting density and arrangement, fertilization, irrigation, weed control and harvest time) on the yield and qualitative performances of 11 field crops (wheat, cardoon, potato, clary sage, basil, sugarcane, canola, cotton, tomato, lettuce and hemp). An editorial is also included in order to provide an overview of current and future challenges in the perspective to boost yield and quality of the main herbaceous field crops, especially with a view of environmental sustainability of agricultural practices.

Gathering contributions by eminent experts in the field, to whom our personal acknowledgements are directed, this book is addressed to a wide range of readers, such as plant physiologists, environmental scientists, biotechnologists, botanists, soil chemists and agronomists.

Sara Lombardo, Giovanni Mauromicale
Editors

Editorial

Herbaceous Field Crops' Cultivation

Sara Lombardo * and Giovanni Mauromicale

Dipartimento di Agricoltura, Alimentazione e Ambiente (Di3A), University of Catania, via Valdisavoia 5, 95123 Catania, Italy; g.mauromicale@unict.it
* Correspondence: saralomb@unict.it; Tel.: +39-095-4783421

Citation: Lombardo, S.; Mauromicale, G. Herbaceous Field Crops' Cultivation. *Agronomy* **2021**, *11*, 742. https://doi.org/10.3390/agronomy11040742

Received: 10 March 2021
Accepted: 7 April 2021
Published: 11 April 2021

Publisher's Note: MDPI stays neutral with regard to jurisdictional claims in published maps and institutional affiliations.

Herbaceous field crops include several hundred plant species spread worldwide for different end-uses, from food to non-food applications. Among them are cereals, grain legumes, sugar beet, potato, cotton, tobacco, sunflower, safflower, rape, flax, soybean, alfalfa, clover *spp.* and other fodder crops. Only 15–20 species play a relevant role in the global economy, representing about 1600 Mha of harvested area in total. Herbaceous field crops can be grouped according to different standpoints, as follows:

- Taxonomy (division, subdivision, family, etc.);
- Life cycle (annual, biennial or perennial crops);
- Climate (tropical, sub-tropical or temperate crops);
- Growing season (spring-summer-autumn, autumn-winter-spring, indifferent);
- Primary end-use (cereals or grain crops, grain legumes, sugar crops, oil crops, fiber crops, rubber crops, fodder crops, aromatic crops, bioenergy crops);
- Used plant part (reproductive organs, subterranean organs, foliage, grass or foraged materials).

In recent decades, the rapid increase in global population and the parallel decrease in arable land has necessitated efforts to develop sustainable agricultural systems for the cultivation of herbaceous field crops. In light of this, the present special issue entitled "Herbaceous Field Crops' Cultivation" publishes articles from colleagues worldwide and provides detailed research involving several aspects of herbaceous field crops' cultivation. It contains two reviews and 22 original research papers devoted to elucidating the impacts of management factors (i.e., genetic background, planting density and arrangement, fertilization management, irrigation, weed control and harvest time) on the yield and qualitative performances of 11 field crops (wheat, cardoon, potato, clary sage, basil, sugarcane, canola, cotton, tomato, lettuce and hemp).

The current challenge of agriculture is to reconsider our production systems in search of the best agronomic practices that are able to reduce yield losses by enhancing the resilience and sustainability of crops. In this spirit, natural genetic variability within crop species gives plant breeders the opportunity to develop new and improved genotypes with desirable characteristics (yield potential, pest and disease resistance, etc.). As a result, nowadays, there is a need to take into account new breeding methods, given that several factors limited conventional breeding, based on phenotypic selection, for some crops. To this end, Yadav et al. [1] provide an overview of genomic selection, based on DNA marker profiles, in sugarcane breeding programs. Indeed, molecular markers are advantageous when compared to conventional phenotype-based alternatives since they are stable and detectable in all tissues regardless of the growth, differentiation, development or defense status of the cell. The potential to reduce the breeding cycle length, to increase the prediction accuracy for clonal performance and to increase the accuracy of breeding values for parent selection is also greatly documented by Yadav et al. [1], especially in comparison with other crops. However, in our opinion, a breeding strategy based on molecular markers may be implemented for any crop only through an integrated and collaborative approach with agronomists, engineers and farmers. Indeed, this allows us to select new and improved cultivars within crop species with desirable traits (yield potential, pest resistance, etc.), which must be confirmed in either multi-year or multi-site experimental field trials. From this perspective, most of the research articles included in the present special issue elucidate interactions between genotype and climatic conditions (due to different growing seasons or locations), as well as between genotype and other agronomic factors, such as planting density and fertilization rate. The exception is the study conducted in South Italy by Tuttolomondo et al. [2], which only focuses on the genotype's effect. In particular, the obtained results demonstrate as the biometric and production traits could be used for differentiating clary sage accessions with the aim of achieving a wider expansion of this medicinal and aromatic species. Indeed, an increase in crop genetic diversity, in general, may allow more flexibility for agricultural production. Taking that thought, as an instance, the 'Messina' accessions are of particular note for obtaining a higher essential oil yield performance per hectare.

The objective of increasing the productivity of herbaceous field crops may be rationally reached by modulating planting density and arrangement since the latter may affect plant architecture and growth, resource utilization, disease and pest tolerance, as well as carbohydrate production and partitioning. The proper planning density for any crop can vary considerably depending on many agronomic factors such as sowing time, fertilization, soil moisture and pest management for different geographic locations. Accordingly, Khan et al. [3] find that the adoption of a planting density of 8.7 plants m^{-2} enhances cotton yield and fiber quality, as compared to the conventional wider rows and lower plants ha^{-1} adopted in the studied area. Through further field trials, Khan et al. [4] examine over two growing seasons the effects of three planting densities (low, 3×10^4; medium, 6×10^4; high, 9×10^4 plant ha^{-1}) on lint yield, leaf structure, chlorophyll fluorescence and leaf gas exchange attributes in two cotton cultivars ('Zhongmian-16' and 'J-4B'). The results evidence that medium and low planting densities are able to improve the leaf structural and functional traits of cotton cultivars grown in subtropical regions. Once again, however, the crucial role of varietal choice is confirmed, as highlighted by the different canopy architectures and yield formations of the selected cultivars in relation to the planting densities. Conversely, Zaheer et al. [5] find that planting density (20 vs. 40 plants m^{-2}) does not have a significant effect on the grain yield of canola. Indeed, the effect of planting density on yield can vary with geographic location and cultivar. In addition to the possible effects on plant physiology and crop yield, planting density and arrangement can also influence crop quality. In this framework, according to Deng et al. [6], it is not appropriate to increase the planting density to over 32–37 plants m^{-2} in hemp production, if you are to ensure a high fiber yield per area. Indeed, as reported by the researchers, when the planting density reaches a certain level, hemp fiber yield decreases due to a self-thinning effect.

In a scenario characterized by declining natural resources, climatic changes, demographic increases in urban areas and depopulation in agricultural ones, the improvement of soil fertility is imperative for future food security, and can be achieved by sustainable agricultural production systems. From this perspective, multiple cropping systems offer undoubted agroecosystemic services, including soil preservation. Although there exists well-documented literature on intercropping approaches, a successful multicropping system necessitates specific consideration of the agronomic management practices that are able to overcome some disadvantages experienced in intercropping systems, such as yield reduction of the main crop and higher labor costs over monocultures. As such, the management of planting density and spatial arrangement has a crucial role to play in reducing intra- and inter-specific competition for natural resources and external inputs. Nadeem et al. [7] suggest the adoption of a 120 cm trench planting pattern, along with lentil intercropping for improved LER (land equivalent ratio), economic return and seed yield of sugarcane, as compared to other intercropping patterns and a control (sole cropping), likely due to the improved utilization of farming inputs and an increased lentil plant population. This positively influences variability in millable canes m^{-2} due to increased nutrient accessibility, better air circulation and interception of light, resulting in reduced shoot mortality and better growth of canes. In line with these principles, sugarcane planted via 120 cm trench planting also presents greater LAI (leaf area index) and plant height, as well as higher values of total sugar yield.

The maximum yield potential of herbaceous field crops can be successfully achieved with a balanced mineral nutrients supply. Among the primary nutrients required by plants, nitrogen (N) is one of the most limiting factors for plant growth and crop yield formation. Nevertheless, the current awareness of the environmental impacts of agricultural practices has enforced additional efforts to reduce the N losses from crop production systems. A study conducted in Southern Italy [8] on potato indicates that the adoption of cultivars characterized by high nitrogen use efficiency (NUE) at a low N fertilization rate and a soil nitrate test prior to planting are effective tools for achieving more sustainable and cost-effective N fertilization management. As reported by the researchers, however, N fertilization should be commensurate to climatic conditions. Indeed, only a small rate of N fertilizer applied in surplus to potato carries over to the succeeding crops, while most of this is probably lost over summer by volatilization (N_2O and NH_3) and in autumn, when rainfall exceeds evapotranspiration, by leaching of $NO_3{}^-$. Thanks to the research of Lombardo et al. [9] the effect of the N fertilization rate on both the agronomic and qualitative traits of potato is elucidated. Particularly, a higher nutritional profile of the tuber (i.e., high levels of dry matter, starch, total polyphenols and ascorbic acid, and low nitrate amount) is obtained by supplying 140 kg N ha^{-1}, as compared to the conventionally adopted 280 kg N ha^{-1}. This is relevant to reducing the N fertilization rate while enhancing the yield and quality of the product. However, it is important to underline that the proper N fertilization rate required to maximize potato yield may vary on the basis of soil traits, cultivar choice and the type of N fertilizer. The potential to reduce environmental N losses by increasing the NUE of crops is also explored by Conversa and Elia [10] on lettuce, the most important leafy vegetable worldwide. On the basis of the critical N curve, i.e., plotting at each time interval the minimum N concentration corresponding to the maximum aboveground dry weight, the authors suggest the use of a butterhead typology as compared to crisphead ones. As the authors speculate, the differences in growth between fertilized butterhead and crisphead typologies may strictly depend upon their light interception capacities, which in turn, are due to their specific head shapes. In particular, the greater root apparatus of the butterhead type, resulting in a larger uptake of soil-N, may explain its higher shoot dry weight accumulation. In addition, the values of NUE underline the poor ability of the crisphead type to absorb soil N and utilize the absorbed N to produce a foliar dry biomass concentration, in comparison to the butterhead typology. Considering the relevant roles of other agronomic factors on crop NUE, Zaheer et al. [5] explore the relationships among planting density, time of N fertilizer application and N fertilization rates, with the aim of enhancing canola yield and quality. Briefly, the researchers conclude that the application of N fertilizers in two splits at 120 kg N ha^{-1} combined with 20 plants m^{-2} could be a valuable strategy to achieve good qualitative attributes (especially in terms of glucosinolates and protein levels) and yields of canola. An optimal N supply at a proper time is a feasible strategy for mitigating N losses from any crop, as the split application of N fertilizers ensures the availability of this nutrient when it is required by plants. In addition, high N fertilization rates as a basal dose can be toxic to seeds and potentially expose the emerging crop to losses. A split N fertilization supply can be, therefore, beneficial from agronomic, economic and environmental standpoints.

In this scenario, the use of legumes within multicropping systems represents an excellent alternative to conventional N fertilization by providing multiple services in line with sustainability principles. Accordingly, Toukabri et al. [11] prove that the mixture of fenugreek and clover, as companion plants to durum wheat, may preserve soil moisture and, hence, help to mitigate the plant's water stress. Intercropping with legumes is a valuable option, especially under water-limiting field conditions, since these crops provide an extra canopy, which is able to minimize soil water evaporation losses. However, several criteria (e.g., pedoclimatic conditions, biological cycles of the main crop and companion ones, biomass and nutrients produced by the legume crop, etc.) must be considered as part of the choice of a single legume crop or a mix of species.

Phosphorus (P), along with N, plays a pivotal role in the reproductive growth and yield formation of herbaceous field crops. However, an excessive P fertilization rate increases the risk of P losses to surface and ground waters, impairing aquatic ecosystems through eutrophication. Although soil P reserves vary across the world, the estimated increase of the costs of high-quality P fertilizers and of the global demand for these supports the need to manage P fertilization on the basis of a suitable soil test and a prediction of crop requirement. In research conducted by Iqbal et al. [12] in Jiangsu (China), increasing the P rate up to 200 kg ha^{-1} improves the yield and reproductive organ biomass, as well as macronutrients' (N, P and K) accumulation, in cotton cultivars, especially those with low P-sensitivity. In other words, managing P fertilization according to cultivar P-sensitivity is essential to effectively boost cotton yield. In addition, Iqbal et al. [12] find that, because of the indeterminate growth of cotton, the nutrients' deposition varies with the advancement of growth stages. This knowledge may be useful for growers making management decisions to maximize seed cotton yield. In particular, it is unquestionable that a better equilibrium among vegetative and reproductive growth, achieved by supplying a proper P fertilization rate, is essential to establishing a good and balanced source–sink relationship for any crop species.

The importance of potassium (K) for the physiological processes vital to plant nutrient and water uptake, nutrient transport, plant growth, dry matter production and transportation is largely documented, especially under adverse conditions. Although K fertilization is the primary source of this macronutrient in agricultural production systems, the application rate of K fertilizer is often insufficient to overcome the severe soil K deficiencies that regularly occur in some geographic areas. Therefore, an evaluation of the optimum K fertilizer recommendations at a local scale is needed for several crops. Accordingly, Ma et al. [13] suggest that a K fertilizer dose of 210 kg ha^{-1} is able to enhance cotton biomass, fiber quality and economic profit in the Yangtze River Valley (China) and similar climatic regions. In particular, the researchers highlight, for both P and K fertilization, how a proper rate may positively influence the yield and qualitative traits of cotton, provided that specific attention is given to climatic conditions and genotype. In this context, while a major research effort has been devoted to demonstrating the yield and qualitative benefits to herbaceous field crops of the fertilizer application of individual macronutrients, there is little understanding to date about the impact of NPK fertilizers' application. Due to the low utilization efficiency of fertilizers, especially NPK ones, over 50% of the nutrients supplied are wasted and can contaminate soil and water resources. Thus, it is crucial to establish how to effectively utilize fertilizers or to increase the nutrient use efficiency of crops to obtain high yields in a sustainable way. Comprehensive research conducted by Deng et al. [7] elucidates the potential to improve hemp fiber yield by combining a proper NPK application rate with an adequate choice of planting density. Since increasing the P or N fertilization rate generates a positive effect on hemp yield, while increasing the K fertilization rate or planting density has a negative impact, the authors conclude that to obtain yields of hemp with high-quality fiber of greater than 2200 kg ha^{-1}, the optimal ranges for cultivation conditions are: 329,950–371,500 plants ha^{-1}, 251–273 kg N ha^{-1}, 85–95 kg P$_2$O$_5$ ha^{-1}, and 212–238 K$_2$O kg ha^{-1}. These indications are useful for reducing the environmental impacts of hemp production that result from the large amounts of fertilizers required during the growing period to achieve high biomasses and rapid plant development.

Fertilization management under organic farming deserves specific attention due to the limited range of fertilizers and plant protection products allowed. As a result of this, the search for alternative nutrient sources is constantly evolving under organic farming. Recently, biochar, produced through the pyrolysis of lignocellulosic biomasses, has been attracting interest for its ability to improve the water-holding capacity and organic matter content of soil. This is relevant in marginal lands, particularly those that are scarcely rainfed and where irrigation is difficult for several reasons. In a study conducted in northern Italy, Ronga et al. [14] suggest the use of digestate and biochar fertilizers for processing tomato, a globally important cash crop. Indeed, both liquid digestate and biochar ensure greater yields, allow a higher plant growth (expressed in terms of a higher fruit number per plant, fruit weight, main stem length and aboveground biomass), and also improve the °Brix and Bostwick viscosity, two important fruit quality parameters when processing the tomato crop. In particular, the highest values for fruit number per plant and fruit weight found when using liquid digestate and biochar fertilizers may be conducive to an increase in water (rainfall and irrigation) and nutrient retention (carried by liquid digestate) in the soil due to biochar supply. In addition, as reported by the authors, digestates present phytohormones and other bioactive compounds that are able to improve plant growth.

Recently the use of silicon(Si)-based fertilizers has been increasingly reconsidered by researchers on account of the numerous benefits of this element to plants. A main function of Si is to enhance plant growth and yield, especially under conditions of stress, by increasing resistance to diseases and pathogens, metal toxicities, salinity and drought stresses. In addition, Si fertilization has the potential to affect the absorption and translocation of several macro- and micronutrients. In this framework, Kowalska et al. [15] propose that Si fertilization is effective for minimizing the negative impact of drought stress on wheat that is organically grown. In addition, a tendency is noted for protein accumulation in the grain of the cv. Rusałka to be promoted when fertilized with Si, as a seed dressing and foliar spray. In our opinion, however, as the positive effects provided by Si fertilization are strictly related to its accumulation in plant tissue, further studies elucidating the possible mechanisms of Si uptake and transport in plants are needed.

The soil inoculum of arbuscular mycorrhizal fungi (AMF) may also play a key role in plant nutrition under organic farming, since plants may benefit from mycorrhizal symbiosis through a better uptake of mineralized soil nutrients present at low concentrations. Indeed, Lombardo et al. [16] find that the application of AMF improves the marketable yield of organic potato, especially when grown in low fertility soils such as calcareous ones. In addition, it is demonstrated that AMF application may enhance the marketable yield of potato plants, as well as the efficiency of photosynthesis rate and stomatal conductance, especially when adopting halved fertilizer doses and using locations with unfavorable soil conditions for potato growth. Accordingly, it should be emphasized once again how the use of AMF is in line with the principles of greater sustainability in modern agriculture. Another important issue of agricultural sustainability concerns competition for water, between agriculture and civil uses, industrial production and environmental needs. In addition, current climatic changes are leading to crop adaptation in stressful drought environments. The most efficient agricultural tool for achieving water saving is to improve water productivity, thereby producing more food per unit of water used. In this spirit, by comparing the effects of three irrigation treatments (100, 70 and 40% of the full irrigation requirements) on the water use efficiency (WUE) of five basil cultivars, Kalamartzis et al. [17] confirm that an appropriate cultivar choice (i.e., 'Mrs. Burns') is essential to achieving a higher WUE and may allow water resources to be saved, especially in drought areas, while also obtaining high dry weight accumulation and essential oil yield. Similarly, in an attempt to develop a water conservation strategy in drought lands, Gao et al. [18] find that the drip irrigation level of 540–600 m^3 ha^{-1}, combined with low mepiquat chloride application, may represent a good strategy in cotton to achieve higher water productivity and lint yield, thanks to improved leaf photosynthetic traits and reproductive organ biomass accumulation. According to the authors' perspective, this finding is relevant since mepiquat chloride, a growth regulator used in cotton production since 1975, affects plant structure in complex hormonal ways and, therefore, optimal mepiquat chloride schedules are difficult to identify. Interestingly, in this study, moderately reduced drip irrigation rates (540 and 480 m^3 ha^{-1}) do not significantly affect cotton fiber quality parameters, such as fiber length and uniformity, nor specific strength and micronaire values. Further to this, Chen et al. [19] highlight that pre-sowing irrigation combined with basal surface fertilization ensures the higher root morphological and physiological activity [i.e., greater root biomass, longer root length in the surface soil profile (0–30 cm) and higher root nitrate reductase activity in the surface or deep soil profile (60–80 cm) at the boll setting stage] and water-nitrogen productivity of cotton crops in arid regions. A future challenge should be how the irrigation method and scheduling can affect water-nutrient efficiency, considering that in the case of cotton, as reported by Chen et al. [19], a resource conservation strategy should balance growth and development between the aerial and underground parts of the cotton plant.

From the first steps of agriculture, a problematic aspect affecting the productivity of herbaceous field crops is certainly weed control management. Current awareness of the possible impact of chemical weeding on the environment has led to the development of integrated weed management (IWM). As described in a review from Scavo and Mauromicale [20], this represents a feasible approach, especially under organic farming and low-input agricultural systems. Scavo and Mauromicale [20], through a holistic view, present a literature and expert analysis of the different tactics (preventive, cultural, mechanical and chemical) to be adopted for effective IWM. Indeed, a single weed control measure is unlikely due to the presence of different weed species with highly diverse life cycles and survival strategies. In addition, the adoption of a singular control method enforces weed adaptability to this practice. Therefore, advancements in non-chemical weed control are inevitable if IWM is to be achieved. In particular, with a view to sustainability, the authors explore the possible integration of allelopathy for weed control. The use of cover crops for this purpose is still being studied, particularly in order to verify the best agronomic technique for exploiting their natural herbicide potential. In this sense, considering the different suppression indices of allelopathic plants, Carrubba et al. [21] study the herbicidal potential of five plant water extracts (from *Artemisia arborescens*, *Rhus coriaria*, *Lantana camara*, *Thymus vulgaris*, and *Euphorbia characias*) on durum wheat (cv. Valbelice). Although none of the tested treatments (including a chemical control) are able to eradicate weeds from the field, the lack of a significant difference in grain yield between chemically treated plots and untreated ones demonstrates that weed control with chemical herbicides does not necessarily result in a significant grain yield increase. Though not yet conclusive about which allelochemical extract exerts predictable effects on crop yield and development, Carrubba et al. [21] state the need to use a broader range of crops and allelochemicals. It is noteworthy that non-chemical weeding can minimize, but not necessarily eliminate, all weeds. The latter may even be welcomed due to their contribution of organic matter to soil during tillage. Always on wheat durum, the already cited research from Toukabri et al. [11] highlights that the mixture of fenugreek and clover, as companion plants, allows weed suppression that is comparable to herbicides in efficiency. So, intercropping with legumes can be considered effective to limit pesticide dependency and, hence, to mitigate food-related chemical hazards. Besides suppressing weeds, intercropping with legume crops is well-suited to the holistic approach of IWM, as it provides several ecosystemic services, such as improving soil organic matter content, reducing runoff and soil erosion, and minimizing dependency on external fertilizers, by fixing atmospheric nitrogen. However, the use of cover crops as a weed management tool needs to be carefully followed up throughout the growing period, unlike the use of herbicides. According to Scavo and Mauromicale [20], another important aspect of an IWM approach is the selection of genotypes able to tolerate weeds' competition while maintaining a high yield. Accordingly, Milan et al. [22] evaluate, at two experimental sites in Northern Italy, the difference between hybrid and conventional wheat cultivars in terms of response to weed pressure. This study presents interesting preliminary results on the adoption of hybrid cultivars, which needs a reconsideration of the production system. Indeed, the higher cost of seeds requires a reduction in seeding rate (by about one-third of that ordinarily adopted for conventional cultivars), which may cause delayed canopy development and, as a consequence, more bare soil that is potentially colonizable by weeds in the early stages of the growing season.

Interestingly, on fields characterized by reduced weed pressure and in the case of weed infestation mostly represented by early emerging weeds, hybrid cultivars may not be significantly affected by yield losses. However, the obtained results must be corroborated by further studies in other geographic locations and with a major number of genotypes.

A further aspect impairing the yield and quality of herbaceous field crops is harvest time, which is associated with both the product maturation stage at harvest and the climatic conditions before collection. In particular, timely harvesting is crucial to crop loss prevention, good quality and high market value for any crop. In the present special issue, two studies conducted in Greece find a variation in the chemical composition of cultivated cardoon bracts [23] and heads [24] in relation to the maturation stage. As an example, it is highlighted that the content of phenolic compounds decrease with increasing maturity as a consequence of the lignification of bracts tissues. By contrast, mature bracts present higher amounts of sugars than immature ones, due to inulin biosynthesis and carbohydrate translocation in other plant parts, such as the heads. Understanding the relationships between metabolite accumulation in the plant parts and harvest time provides useful information to increase the quality and added value of this crop for the possible extraction of phytochemical compounds. Despite ongoing progress in synthetic chemistry, natural products are more characterized by enormous scaffold diversity and structural complexity. Therefore, challenges in agricultural practices should be focused not only on yield increase but also on the maintenance and/or enrichment of the phytochemicals present in plants.

In conclusion, the current special issue includes several topics of research relative to herbaceous field crops, highlighting the importance of biodiversity and environment preservation while showing agronomic practices that are able to improve crop yield and quality. Furthermore, this special issue provides an overview of current and future challenges in the sustainable cultivation of our main herbaceous field crops.

Author Contributions: S.L. and G.M. wrote this editorial for the introduction of the Special Issue of Agronomy, entitled "Herbaceous Field Crops' Cultivation", and edited the Special Issue. All authors have read and agreed to the published version of the manuscript.

Funding: This work received no external funding.

Acknowledgments: We are grateful to all authors who submitted their valuable manuscripts to the Special Issue of Agronomy, entitled "Herbaceous Field Crops' Cultivation".

Conflicts of Interest: The authors declare no conflict of interest.

References

1. Yadav, S.; Jackson, P.; Wei, X.; Ross, E.; Aitken, K.; Deomano, E.; Atkin, F.; Hayes, B.; Voss-Fels, K. Accelerating Genetic Gain in Sugarcane Breeding Using Genomic Selection. *Agronomy* **2020**, *10*, 585. [CrossRef]
2. Tuttolomondo, T.; Iapichino, G.; Licata, M.; Virga, G.; Leto, C.; La Bella, S. Agronomic Evaluation and Chemical Characterization of Sicilian *Salvia sclarea* L. Accessions. *Agronomy* **2020**, *10*, 1114. [CrossRef]
3. Khan, N.; Xing, F.; Feng, L.; Wang, Z.; Xin, M.; Xiong, S.; Wang, G.; Chen, H.; Du, W.; Li, Y. Comparative Yield, Fiber Quality and Dry Matter Production of Cotton Planted at Various Densities under Equidistant Row Arrangement. *Agronomy* **2020**, *10*, 232. [CrossRef]
4. Khan, A.; Zheng, J.; Tan, D.; Khan, A.; Akhtar, K.; Kong, X.; Munsif, F.; Iqbal, A.; Afridi, M.; Ullah, A.; et al. Changes in Leaf Structural and Functional Characteristics when Changing Planting Density at Different Growth Stages Alters Cotton Lint Yield under a New Planting Model. *Agronomy* **2019**, *9*, 859. [CrossRef]
5. Zaheer, S.; Arif, M.; Akhtar, K.; Khan, A.; Khan, A.; Bibi, S.; Muhammad Aslam, M.; Ali, S.; Munsif, F.; Jalal, F.; et al. Grazing and Cutting under Different Nitrogen Rates, Application Methods and Planting Density Strongly Influence Qualitative Traits and Yield of Canola Crop. *Agronomy* **2020**, *10*, 404. [CrossRef]
6. Deng, G.; Du, G.; Yang, Y.; Bao, Y.; Liu, F. Planting Density and Fertilization Evidently Influence the Fiber Yield of Hemp (*Cannabis sativa* L.). *Agronomy* **2019**, *9*, 368. [CrossRef]
7. Nadeem, M.; Tanveer, A.; Sandhu, H.; Javed, S.; Safdar, M.; Ibrahim, M.; Shabir, M.; Sarwar, M.; Arshad, U. Agronomic and Economic Evaluation of Autumn Planted Sugarcane under Different Planting Patterns with Lentil Intercropping. *Agronomy* **2020**, *10*, 644. [CrossRef]
8. Ierna, A.; Mauromicale, G. Sustainable and Profitable Nitrogen Fertilization Management of Potato. *Agronomy* **2019**, *9*, 582. [CrossRef]

9.   Lombardo, S.; Pandino, G.; Mauromicale, G. Optimizing Nitrogen Fertilization to Improve Qualitative Performances and Physiological and Yield Responses of Potato (*Solanum tuberosum* L.). *Agronomy* **2020**, *10*, 352. [CrossRef]
10. Conversa, G.; Elia, A. Growth, Critical N Concentration and Crop N Demand in Butterhead and Crisphead Lettuce Grown under Mediterranean Conditions. *Agronomy* **2019**, *9*, 681. [CrossRef]
11. Toukabri, W.; Ferchichi, N.; Hlel, D.; Jadlaoui, M.; Kheriji, O.; Zribi, F.; Taamalli, W.; Mhamdi, R.; Trabelsi, D. Improvements of Durum Wheat Main Crop in Weed Control, Productivity and Grain Quality through the Inclusion of Fenugreek and Clover as Companion Plants: Effect of N Fertilization Regime. *Agronomy* **2021**, *11*, 78. [CrossRef]
12. Iqbal, B.; Kong, F.; Ullah, I.; Ali, S.; Li, H.; Wang, J.; Khattak, W.; Zhou, Z. Phosphorus Application Improves the Cotton Yield by Enhancing Reproductive Organ Biomass and Nutrient Accumulation in Two Cotton Cultivars with Different Phosphorus Sensitivity. *Agronomy* **2020**, *10*, 153. [CrossRef]
13. Ma, X.; Ali, S.; Hafeez, A.; Liu, A.; Liu, J.; Zhang, Z.; Luo, D.; Noor Shah, A.; Yang, G. Equal K Amounts to N Achieved Optimal Biomass and Better Fiber Quality of Late Sown Cotton in Yangtze River Valley. *Agronomy* **2020**, *10*, 112. [CrossRef]
14. Ronga, D.; Caradonia, F.; Parisi, M.; Bezzi, G.; Parisi, B.; Allesina, G.; Pedrazzi, S.; Francia, E. Using Digestate and Biochar as Fertilizers to Improve Processing Tomato Production Sustainability. *Agronomy* **2020**, *10*, 138. [CrossRef]
15. Kowalska, J.; Tyburski, J.; Bocianowski, J.; Krzymińska, J.; Matysiak, K. Methods of Silicon Application on Organic Spring Wheat (*Triticum aestivum* L. spp. *vulgare*) Cultivars Grown across Two Contrasting Precipitation Years. *Agronomy* **2020**, *10*, 1655. [CrossRef]
16. Lombardo, S.; Abbate, C.; Pandino, G.; Parisi, B.; Scavo, A.; Mauromicale, G. Productive and Physiological Response of Organic Potato Grown under Highly Calcareous Soils to Fertilization and Mycorrhization Management. *Agronomy* **2020**, *10*, 1200. [CrossRef]
17. Kalamartzis, I.; Dordas, C.; Georgiou, P.; Menexes, G. The Use of Appropriate Cultivar of Basil (*Ocimum basilicum*) Can Increase Water Use Efficiency under Water Stress. *Agronomy* **2020**, *10*, 70. [CrossRef]
18. Gao, H.; Ma, H.; Khan, A.; Xia, J.; Hao, X.; Wang, F.; Luo, H. Moderate Drip Irrigation Level with Low Mepiquat Chloride Application Increases Cotton Lint Yield by Improving Leaf Photosynthetic Rate and Reproductive Organ Biomass Accumulation in Arid Region. *Agronomy* **2019**, *9*, 834. [CrossRef]
19. Chen, Z.; Gao, H.; Hou, F.; Khan, A.; Luo, H. Pre-Sowing Irrigation Plus Surface Fertilization Improves Morpho-Physiological Traits and Sustaining Water-Nitrogen Productivity of Cotton. *Agronomy* **2019**, *9*, 772. [CrossRef]
20. Scavo, A.; Mauromicale, G. Integrated Weed Management in Herbaceous Field Crops. *Agronomy* **2020**, *10*, 466. [CrossRef]
21. Carrubba, A.; Labruzzo, A.; Comparato, A.; Muccilli, S.; Spina, A. Use of Plant Water Extracts for Weed Control in Durum Wheat (*Triticum turgidum* L. subsp. durum Desf.). *Agronomy* **2020**, *10*, 364. [CrossRef]
22. Milan, M.; Fogliatto, S.; Blandino, M.; Vidotto, F. Are Wheat Hybrids More Affected by Weed Competition than Conventional Cultivars? *Agronomy* **2020**, *10*, 526. [CrossRef]
23. Mandim, F.; Petropoulos, S.; Giannoulis, K.; Santos-Buelga, C.; Ferreira, I.; Barros, L. Chemical Composition of *Cynara cardunculus* L. var. *altilis* Bracts Cultivated in Central Greece: The Impact of Harvesting Time. *Agronomy* **2020**, *10*, 1976. [CrossRef]
24. Mandim, F.; Petropoulos, S.; Fernandes, Â.; Santos-Buelga, C.; Ferreira, I.; Barros, L. Chemical Composition of *Cynara cardunculus* L. var. *altilis* Heads: The Impact of Harvesting Time. *Agronomy* **2020**, *10*, 1088. [CrossRef]

 *agronomy*

Article

Improvements of Durum Wheat Main Crop in Weed Control, Productivity and Grain Quality through the Inclusion of FenuGreek and Clover as Companion Plants: Effect of N FertilizaTion Regime

Wael Toukabri [1,2,3], Nouha Ferchichi [1,2], Dorsaf Hlel [3], Mohamed Jadlaoui [3], Oussema Kheriji [3], Fathia Zribi [4], Wael Taamalli [4,5,6], Ridha Mhamdi [2] and Darine Trabelsi [2,*]

[1] Faculté des Sciences de Tunis, Université de Tunis El Manar, El Manar Tunis 2092, Tunisia; waeltoukebri@gmail.com (W.T.); nouha.f08@gmail.com (N.F.)

[2] Laboratory of Legumes, Centre of Biotechnology of Borj-Cedria (CBBC), BP 901, Hammam-Lif 2050, Tunisia; ridha.mhamdi@cbbc.rnrt.tn

[3] Institut National des Grandes Cultures, BP 120, Bou Salem 8170, Tunisia; dorsafhlel@gmail.com (D.H.); mohamed.jadlaoui@gmail.com (M.J.); oussama.kheriji@ingc.tn (O.K.)

[4] Centre of Biotechnology of Borj-Cedria (CBBC), Laboratory of Extremophile Plants, BP 901, Hammam-Lif 2050, Tunisia; zr.feth@gmail.com (F.Z.); wael.taamalli@cbbc.rnrt.tn (W.T.)

[5] Higher Institute of Biotechnology of Béja, University of Jendouba, BP 382, Béja 9000, Tunisia

[6] Laboratory of Olive Biotechnology, Centre of Biotechnology of Borj-Cedria (CBBC), BP 901, Hammam-Lif 2050, Tunisia

* Correspondence: trabelsi_darine2@yahoo.fr

check for **updates**

Citation: Toukabri, W.; Ferchichi, N.; Hlel, D.; Jadlaoui, M.; Kheriji, O.; Zribi, F.; Taamalli, W.; Mhamdi, R.; Trabelsi, D. Improvements of Durum Wheat Main Crop in Weed Control, Productivity and Grain Quality through the Inclusion of FenuGreek and Clover as Companion Plants: Effect of N FertilizaTion Regime. *Agronomy* **2021**, *11*, 78. https://doi.org/10.3390/agronomy11010078

Received: 9 December 2020
Accepted: 27 December 2020
Published: 31 December 2020

Publisher's Note: MDPI stays neutral with regard to jurisdictional claims in published maps and institutional affiliations.

Abstract: Assessing the performance of legume species as companion plants is a prerequisite for promoting a low chemical-input durum wheat production system. This study aims to evaluate fenugreek (IC-Fen), clover (IC-Clo) and their mixture (IC-Mix) performances on weed control, productivity, and grain quality of durum wheat main crop under different N fertilization regimes, as compared to durum wheat alone with (SC-H) and without (SC-NH) herbicide. On-field experimentations were carried out in humid and semi-arid conditions. Results showed that legumes offer significant advantages in terms of weed control, soil moisture conservation, productivity, and grain quality for durum wheat cash crops. Results explain that these benefits depend on the legume part and the adopted N fertilization regime. Most significant improvements occurred with the IC-Mix under unfertilized conditions (N0) and relatively low and late N regimes (N1 and N2) where, for example, the partial land equivalent ratio of durum wheat grain yield (PLER) reached 1.25 compared to the SC-NH, with no need to sort the raw grain product (legumes seeds not exceeding 4.3%). Our study illustrates that under low and late N-fertilization condition using promising legumes species combinations result in the improvement of N fertilizer land-use efficiency and hence help to reduce N-fertilization inputs.

Keywords: companion plants; N-fertilization; partial land equivalent ratio (PLER); weed control; grain quality; productivity

1. Introduction

Durum wheat (*Triticum turgidum subsp. durum* (Desf.) Husn) constitutes one of the most pivotal cereal crops for global food security. Worldwide, durum wheat crops cover nearly 30–35 million hectares [1]. In Tunisia, durum wheat represents more than 27% of the total cultivated area [2]. Farmers in these areas confront various natural constraints, including low and erratic rainfall and low fertility of most lands [3]. These constraints are reinforced by a limited land potential for most farmers (about 89% have less than 20 hectares) and the steadily rising chemical input prices, leading to a continuous increase in production costs [2]. Besides, chemical inputs performance has significantly reduced, given the increase in herbicide-resistant weeds [4] and the low N fertilizer efficiency not exceeding 50% [5]. On the other hand, chemical inputs are also increasingly recognized as majors factors driving global environmental change and food safety hazards [6]. There is a need to develop more sustainable and cost-efficient cereal cropping systems that can be readily adopted by smallholders so as not to impose new burdens on their poor resource.

Mixed crops systems appear as a potential alternative towards more sustainable and efficient production systems. These systems value complementarity and facilitation processes between plants leading to better use of soil resources [7–9]. There are different forms of mixed crops: the intercropping that involves simultaneous cropping of two or more species on the same land [10]; the cover crops that include crops as cover replacing bare fallow, which is grown as green manure prior sowing the main crop [11]; and the intercrops that involve the main cash crop with a cover crop also called companion plants that are sown not to be harvested but to provide agroecological services to the cash crop [12]. For this latter intercropping system, forage legumes appear a suitable candidate as companion plants in cereal crops. Their use as companion plants could contribute to the N requirements of cereal cash crops through their biological N fixation ability and the facilitation processes involving the N transfer from legumes to cereals [13,14]. Hence, this may improve cereal cash crops productivity [15] and well help to reduce N inputs [16]. Legumes as companion plants can also help to limit weed growth through several mechanisms [8]. Some legumes species, like fenugreek (*Trigonella foenum-graecum*) [17] and clover (*Trifolium alexandrinum*) [18] emit allelochemicals that are detrimental for weeds growth [19]. However, those services of legumes as companion plants may depend heavily on the adopted cropping system, including species combinations, and may also vary according to local environmental conditions and soil nutrient availability, mainly N [20,21]. The success of cereal-legume intercrops depends on promoting a niche of complementarity and facilitation between species, which improves the N use efficiency [7,21] and hence helps to reduce N inputs. Therefore, the selection of legume species as companion plants within cereal crops constitute a crucial factor, as they should be less susceptible to N inputs often required for cereal to ensure high yields.

Despite large intercropping literature, the study on companion plants is relatively scarce compared to other intercropping types. Therefore, further research is needed to identify optimal species combinations and N fertilization management to achieve high yields and high N use efficiency simultaneously. Also, to our knowledge, no research studies are available on the mixture of legumes as companion plants within cereal cash crops intended for foods production. The present study aims to evaluate the performance of two legume species (fenugreek and clover) added as companion plants to suppress weeds and improve productivity and grain quality of durum wheat as a cash crop. As well, we aim to evaluate the mixture of these legume species against their added separately as companion plants within durum wheat crop. We hypothesized that the N-fertilization regime could modify the competition between species and could influence the performance of legumes as companion plants within durum wheat cropping systems. Our study, therefore, assesses the effects of N-fertilization regimes (different doses and application times) on the effectiveness of the legumes as companion plants to enhance durum wheat productivity and grain quality.

2. Materials and Methods

2.1. Experimental Sites and Environmental Conditions

Field experiments were conducted from 2015 to 2018 at two sites in the North-west of Tunisia: the Beja El Gnadil site (denoted by BEG) (36°7258′ N, 9°3043′ E) and the Siliana Bourouis site (denoted by SBR) (36°2098′ N, 9°0665′ E) (Figure S1). Agronomic characteristics and soil physicochemical properties are given in the Supplementary Table S1. At both sites, farming practices were conventional based on cereal-legume rotation, with a predominance of cereal crops where the previous crops before the experimental trials was oats (*Avena sativa* L.). Generally, in the north-west of Tunisia, chemical input management in conventional cereal crops relies mostly on the use of herbicides and N-fertilizers, about 90 to 120 kg ha^{-1} of N-fertilizer for durum wheat [22]. The BEG site has a sub-humid climate while the SBR site has a semi-arid climate. Agricultural production at both sites depends only on natural precipitation. Figure 1 shows data on environmental conditions (monthly precipitation and temperature) for the three experimental seasons. Precipitation varied greatly between both sites but was generally similar between the three experimental seasons. At BEG site, the climate was rainy weather with an average annual precipitation of 621.1 mm, 646.3 mm and 595.3 mm respectively during 2015–2016, 2016–2017 and 2017–2018 seasons. However, at SBR, the climate was semi-arid with an average annual precipitation of 364 mm, 332.2 mm and 355.2 mm respectively during 2015–2016, 2016–2017 and 2017–2018 seasons. The minimum temperature seldom drops below 0 °C during winter, so no frost damage was observed in legumes which are less frost-resistant than cereals (Figure 1).

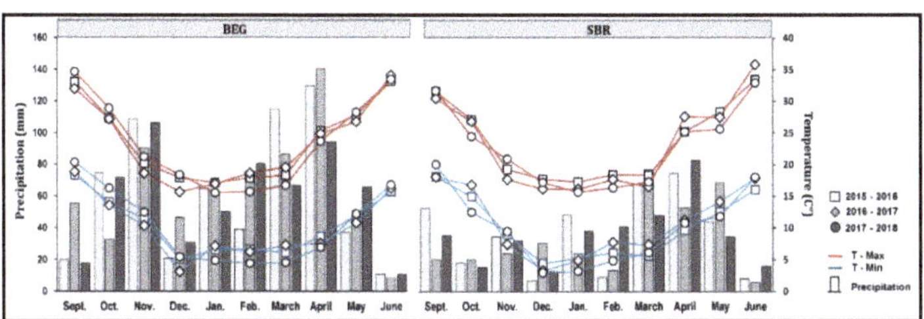

Figure 1. Average monthly precipitation, minimum and maximum temperature at both sites during the three experimental seasons (Data provided by the National institute of meteorology, Tunisia).

2.2. Experimental Design

Durum wheat-legume intercropping patterns were based on the additive principle: durum wheat, (Triticum turgidum subsp. durum (Desf.) Husn) as cash crop was sown at standard sowing density (320 plant m^{-2}) and, legumes were added as companion plants. The legumes compounds were fenugreek (50 plant·m^{-2}) or/and clover (100 plant m^{-2}). Durum wheat variety used was Maali. For legumes, a local cultivar was used for fenugreek (*Trigonella foenum graecum* L). and the variety Masri Baladi for clover (*Trifolium alexandrinum* L.).

The experiment was arranged in split-plot design with the main plot factor randomized according to a RCBD. Nitrogen treatments constituted the main plots and cropping patterns the sub-plots. The sub-plot size was 3.6 m × 5 m.

The following N-treatments were assigned to the main plots:

- N0: Unfertilized treatment.
- N1: Low and late N-fertilization treatment (receiving 30 kg ha^{-1} of N-fertilizer at durum wheat heading stage).
- N2: Moderate N-fertilization beginning at durum wheat stem elongation (receiving two equal N-fertilizer doses of 30 kg ha^{-1}, one at durum wheat stem elongation and the other at heading stage).
- N3: Medium N-fertilization beginning at durum wheat tillering (receiving three equal N-fertilizer doses of 30 kg ha^{-1}, the first at durum wheat tillering, the second at stem elongation and the third at heading).
- N4: High N-fertilization beginning at durum wheat tillering (receiving three equal N-fertilizer doses of 40 kg ha^{-1}, the first at durum wheat tillering, the second at stem elongation and the third at heading).

The following cropping patterns were assigned to the sub-plots:

- SC-H: Durum wheat sole cropping with a conventional weed control using herbicides.
- SC-NH: Durum wheat sole cropping without herbicides application.
- IC-Fen: Intercrops of Durum wheat-Fenugreek.
- IC-Clo: Intercrops of Durum wheat-Clover.

- IC-Mix: Intercrops of Durum wheat-Mixture (fenugreek+clover).

2.3. Sites Management

At both sites the false seedbed technique that consists of preparing a regular seedbed (early) before sowing the actual crops was adopted to better control weeds [23]. Before sowing, the soil was ploughed with disk harrow at a tillage depth of 10 cm, followed by one pass with a rotary disc to prepare the seedbed. All species were sown simultaneously in November with a precision seed drill (Wintersteiger Plotseed, Austria) at a depth of 4 cm. Durum wheat was sown in single rows 0.2 m apart in the sole crop. For intercrops, legumes were sown in the middle of wheat inter-row space. Plants emergence was satisfactory owing to high water availability during the 15 days after sowing (Table S2). To evaluate the effect of the experimental treatments on weed infestation, weeds in all plots were untreated, except for the sole-crop with herbicide (SC-H) plots where the Amilcar OD herbicide was applied at the recommended dose approximately five weeks after sowing (Figure S2). Nitrogen fertilizer was applied, using the Ammonitrate (33.5%), at durum wheat tillering (i), stem elongation (ii) and heading (iii) according to the different N treatments as described above (Figure S2). Durum wheat was harvested at maturity (Zadoks 9-ripening) [24]. Physiological maturity was generally observed in late June at the BEG site and mid-June at the SBR site.

2.4. Sampling, Measurements, and Calculation

2.4.1. Biomass

At durum wheat heading stage (Zadoks 45), weeds biomass was determined by collecting weeds from the central three-meter square in each plot. At durum wheat maturity (Zadoks 80), legumes above-ground biomass was determined by cutting all the plants at ground level from four linear meters in each plot. Then, after drying for 72 h at 75 °C, the weeds and legumes biomass were weighed, and recorded values were converted to kg ha^{-1}. For straw yield determination, at harvesting, only the central rows of plots were harvested (1.2 m) with a 10 cm cutting bar using experimental combine harvester (Wintersteiger Plot combine, Autriche). Straw was weighed, and values were converted to t ha^{-1}.

2.4.2. Nodule Weight

At legumes flowering (firstly the fenugreek and then the clover), ten plants from each legume's species were selected randomly from each sub-plot, and the roots were excavated using a spade. The soil was removed carefully from roots to ensure that roots and nodules were as much as possible recovered. Roots were washed carefully with distilled water, and absorbed residual water with absorbent paper, then the nodules were removed quickly. Pink nodules (representing a high efficiency in N fixation) were weighted. Mean values of nodules fresh weight derived from the ten plants were recorded and expressed as mg of nodules per plant.

2.4.3. Net Photosynthetic Rate (*Pn*)

The net photosynthetic rate (*Pn*) was measured for two growing seasons (2016 and 2017) using a portable gas-exchange system (Model Li-Cor 6200, Li-Cor, Lincoln, NE, USA). Measurements for durum wheat and both legumes were performed on sunny days (above 800 µmol m^{-2} s^{-1} PPFD) at durum wheat stem elongation stage (Zadoks 45), between 9.00 a.m. and 1 p.m. (solar time), by holding the chamber perpendicular to the incident light from the sun (flag leaf). The leaves were kept in the chamber until the photosynthesis values were observed as constant as possible, i.e., "steady state" ($\pm$ 2 min). The chambers were open most of the time, exposing the chamber interior to the ambient conditions.

2.4.4. Soil Moisture Analysis

The soil was sampled at legumes flowering for two growing seasons (2016 and 2017). From each plot, four random samples were collected with a soil corer at a depth of 25 cm. Soil moisture was determined gravimetrically by drying fresh soil samples for 48 h at 105 °C according to [25].

2.4.5. Durum Wheat Grain Yield and Quality

After harvesting, in the laboratory, grain samples of each plot were vigorously cleaned. The impurity content (legume seeds percentage) was determined. The moisture content of grain samples was determined using NIR Inframatic 9500 analyzer (Perten Instruments, Sweden) and grain yield in t ha^{-1} was expressed based on 12% moisture content. Thousand kernel weight (TKW) was determined on an analytical balance ($\pm$ 0.1mg) after counting 1000 grains by a seed counter (Numigral II Chopin seed counter, France). Grain N concentration was determined by the Micro-Kjeldahl method. Grain protein content (GPC) was expressed as crude protein by multiplying the value of grain N concentration by 5.7. Ash content was determined according to the AACC method 08-01 [26].

2.4.6. Durum Wheat Intercropping Patterns Efficiency

To evaluate the efficiency of the three intercropping patterns against durum wheat sole crops with no herbicide (SC-NH), the partial land equivalent ratio based on the grain yield (PLER) was calculated. The partial land equivalent ratio is defined as the relative yield of an intercropped species compared to its yield in a sole crop, which can be interpreted, in the present study, as a measure for the contribution of legumes species to the efficiency of land use by the durum wheat crop [10,27]. The PLER was calculated for each N treatment as:

$$PLER = Y_{(IN-L)}/Y_{(SC-NH)} \tag{1}$$

where Y was the grain yield of the intercropping patterns (IN-L) and of the durum wheat sole cropping pattern with no herbicide (SC-NH). PLER values above one, indicate that the intercropping pattern is more productive and more efficient in using N resources than the durum wheat sole cropping pattern (SC-NH) [10].

By analogy with the partial land equivalent ratio, the herbicide response ratio based on the grain yield (HRR) was calculated and compared with the different obtained PLER values. The HRR was calculated as the ratio between the durum wheat sole crops with herbicide (SC-H) and the durum wheat sole crops with no herbicide (SC-NH) regarding the grain yield for each N treatment as:

$$HRR = Y_{(SC-H)}/Y_{(SC-NH)} \tag{2}$$

where Y is the grain yield of the sole crops with herbicide (SC-H) and of the sole crops without herbicide (SC-NH).

2.4.7. Statistical Analyses

All statistical analyses were performed using R [28]. Mixed-effects models were used to analyze the data of each site to produce ANOVA p-values for main effects and all interactions using the using the lme function in the nlme package. Three-factor analyses with Season (S), N Treatment (NT), and Cropping Pattern (CP) as fixed effects were carried out. The hierarchical nature of the split plot design was reflected in the random error structures that were specified as S/block/mainplot, where mainplot is an ID for the main plots of a trial [29]. All models were visually checked for homogeneity of variance and normal distribution of residuals using the ggResidpanel package. Only for *Pn* and soil moisture measurements each season under each site were analyzed separately using a split-plot ANOVA model in the R package "Agricolae" [30] for randomized complete block design (RCBD) to assess the effects of NT and CS and their interaction. When the ANOVA indicated significant effects, Tukey's HSD test (α = 0.05) was used to determine significant differences among factor levels. Weed biomass data were log-transformed to meet model residuals requirements, using the ln (x + 1) transformation to account for zeros in the data. The relationships between weeds biomass and legumes biomass in each intercropping pattern were tested using the Spearman's correlation method and also, the linear regression with block within season as a random effect.

3. Results

3.1. Durum Wheat Grain Yield, PLER, and HRR

At both sites, durum wheat grain yield was highly affected by N treatment (NT) and cropping patterns (CP) ($p < 0.000$). High significant interaction between NT and CP was found at BEG site ($p = 0.004$) whereas not significant at SBR ($p = 0.068$) (Table S3).

According to NT, grain yield was significantly increased in response to the increase of N-input (Figure 2A). Average across cropping patterns, grain yield increased from 3068 (N0) to 4426 kg ha^{-1} (N4) at BEG and from 2057 (N0) to 2470 kg ha^{-1} (N4) at SBR. The response to N-input was shown higher at BEG compared to SBR.

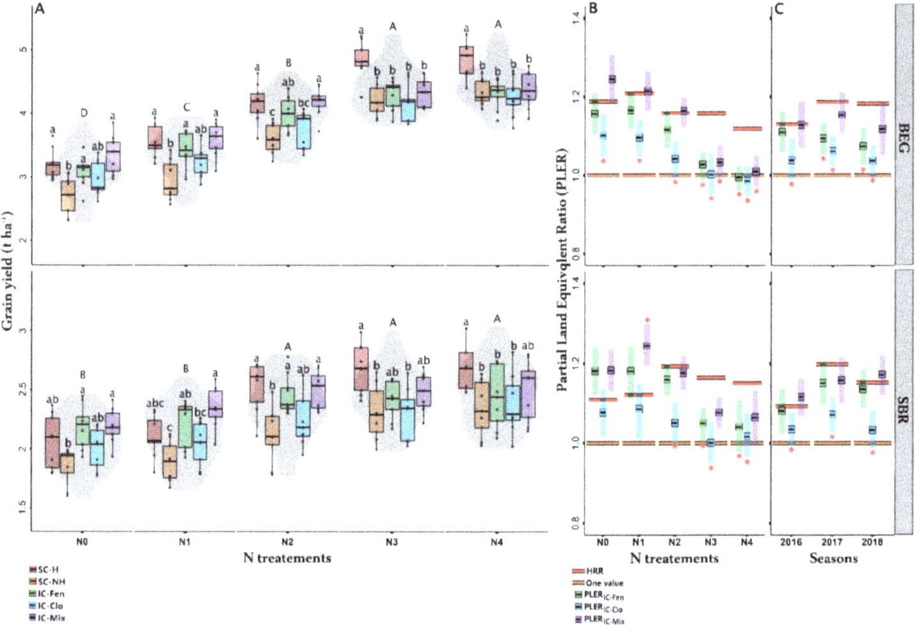

Figure 2. Durum wheat grain yield and PLER at both sites. (**A**) Grain yield affected by N treatment (Violin plot) and cropping pattern within each N treatment (Box plot) averaged over the three seasons; Violin plots headed by a common capital letter are not significantly different at $p < 0.05$ according to Tukey's HSD-test. Within each violin plot, box plots with a common lowercase letter are not significantly different at $p < 0.05$ according to Tukey's HSD-test (for statistical output see Table S3). (**B**): PLER of the three intercropping patterns compared to the herbicide response ratio (HRR) in each N treatment averaged over the three seasons; (**C**): PLER of the three intercropping patterns compared to the herbicide response ratio (HRR) in each season averaged over the five N treatments; In (**B**,**C**) panels, PLER: partial land equivalent ratio-yield in intercrop divided by yield in sole crop with no herbicide; HRR: herbicide response ratio-yield in sole crop with herbicide divided by yield in sole crop with no herbicide; Orange lines indicate a relative yield of one, i.e., with yield equal to that of a sole durum wheat with no herbicide; Red lines indicate the HRR; Tiles represent the means PLER and 95% confidence intervals of each intercropping pattern; Top asterisk (*) indicates a significant increase and bottom asterisk indicates a significant decrease compared to the HRR (Tukey-test, $\alpha = 0.05$).

Within the sole cropping patterns, grain yield was significantly higher in the herbicide-treated pattern (SC-H) compared to the no-treated pattern (SC-NH) under all N treatment, except under N0 and N1 treatments at SBR site (Figure 2A). The herbicide response ratio

(HRR) was between 1.11 and 1.21, that reflected a moderate weed infestation impact (Figure 2B).

The existing of legumes as companion plants showed significant grain yield advantage compared to the SC-NH, depending on the intercropping pattern and the NT (Figure 2A). The highest advantage was obtained with the IC-Mix pattern under N0, N1 and N2 treatments. PLERs in the IC-Mix were 1.25, 1.22 and 1.16 at BEG and were 1.18, 1.24 and 1.18 at SBR, respectively under N0, N1 and N2 treatments (Figure 2B). Under these N treatments, IC-Mix PLERs were globally very close to and even higher than HRRs, where for instance PLER was significantly higher than the HRR under N1at SBR (Figure 2B). Similar effects were obtained with the IC-Fen pattern, but with lower yield advantage compared to the IC-Mix (PLERs in the IC-Fen were between 1.12 and 1.18 under N0, N1 and N2). However, grain yield in the IC-Clo pattern was statistically similar to the SC-NH pattern under all NT. Globally under N3 and N4 treatments, no significant difference between the three intercropping patterns and also the SC-NH pattern with regard of grain yield (Figure 2A). For these N treatments and particularly at BEG, PLERs were between 0.99 and 1.03 which were significantly lower than the HRR (Figure 2B). Yet, at SBR, even though the lowest PLERs were shown with these N treatments, the IC-Mix pattern resulted in PLERs (averaged 1.07) statistically equivalent to the HRR (averaged 1.15) (Figure 2B).

Overall, grain yields of intercropping patterns were largely sustained over the cropping seasons (Figure 2C). The IC-Clo showed the lowest PLERs significantly lower than the HRR and the IC-Mix showed the highest PLERs statistically equivalent to the HRR (Figure 2C). Overall, averaged over seasons, sites and N treatments, grain yield was the lowest in the SC-NH (2.8 t ha^{-1}), very close in the IC-Clo (+4%), intermediate in the IC-Fen (+9%), and the highest in the IC-Mix (+13%) and in the SC-NH (+16%) (Figure 2C).

3.2. Legumes Biomass and Nodulation

Fenugreek biomass and nodules weight were greatly affected by NT ($p < 0.000$) (Table 1). As compared to N0, fenugreek biomass and nodules weight under N1 and N2 showed no significant difference; however, under N3 and N4 significant decreases of both biomass and nodules weight were shown. Average reduction rates under N3 and N4 were 39% and 26% for the biomass and 60% and 50% for the nodules weight, respectively at BEG and SBR sites (Table 1). Fenugreek biomass was, also, affected by the presence of clover in the same row (IC-Fen vs IC-Mix) ($p < 0.000$). As compared to the IC-Fen, fenugreek biomass was 22% and 17% lower in the IC-Mix, respectively at BEG and SBR sites (Table 1). However, fenugreek nodules weight was not affected by the presence of clover in the IC-Mix compared to the IC-Fen (Table 1).

Clover biomass and nodules weight at BEG site were significantly affected by NT and CP and their interaction (Table 1). Within the IC-Clo pattern, clover biomass and nodules weight were significantly reduced under N2, N3 and N4 treatments compared to N0 and N1. Average reduction rate was 53% for the biomass and 72% for the nodule weight (Table 1). However, with the presence of fenugreek in the same raw (IC-Mix), clover biomass and nodules weight was showed significantly reduced only under N3 and N4 treatments compared to N0. On the other hand, under N0 and N1 treatment, clover biomass was showed reduced by the presence of fenugreek (IC-Mix) compared to the IC-Clo (−25%) while the nodules weight was not affected. Interestingly, clover nodules weight was significantly higher with the IC-Mix compared to the IC-Clo in the N2 treatment (+141%) (Table 1). At SBR site clover biomass and nodules weight were significantly affected by NT, while only the nodules weight was significantly affected by CP (Table 1). Overall clover biomass was significantly reduced under N3 and N4 treatment compared to N0 (Averaged −33%). However, clover nodules weight at this site was greatly affected by the presence of fenugreek in the same row (IC-Clo vs. IC-Mix) ($p < 0.000$) (Table 1). Indeed, although the interaction between NT and CP was not significant ($p = 0.091$), clover nodules weight was significantly higher with the IC-Mix compared to the IC-Clo under N0 (+21%), N1 (+20%) and N$_2$ (+54%) treatments.

Table 1. Effects of cropping pattern and N-treatment on legumes biomass and nodule fresh weight at both sites averaged across the three experimental seasons.

		Legumes Biomass (kg ha^{-1})						Nodules Weight (mg Plant^{-1})					
		N0	N1	N2	N3	N4	Mean	N0	N1	N2	N3	N4	Mean
Fenugreek													
BEG	IC-Fen	976 aA	993 aA	979 aA	630 aB	582 aB	832 a	156 aA	163 aA	143 aA	61 aB	67 aB	118 a
	IC-Mix	768 bA	781 bA	745 bA	472 aB	468 aB	647 b	153 aA	148 aA	140 aA	64 aB	57 aB	112 a
	Mean	872 A	887 A	862 A	551 B	525 B		154 A	155 A	142 A	63 B	62 B	
SBR	IC-Fen	663 aAB	697 aA	629 aAB	521 aB	505 aB	603 a	122 aA	120 aA	112 aA	59 aB	58 aB	94 a
	IC-Mix	574 aA	582 aA	542 aAB	419 aBC	399 aC	503 b	112 aA	117 aA	111 aA	66 aB	53 aB	92 a
	Mean	618 A	639 A	586 A	470 B	452 B		117 A	118 A	111 A	62 B	55 B	
Clover													
BEG	IC-Clo	646 aA	679 aA	353 aB	267 aB	341 aB	457 a	103 aA	112 aA	32 bB	31 aB	28 aB	61 a
	IC-Mix	492 aA	499 bA	413 aA	232 aB	241 aB	375 b	105 aA	108 aA	78 aB	37 aC	29 aC	71 a
	Mean	569 A	589 A	383 B	250 C	291 C		104 A	110 A	55 B	34 C	28 C	
SBR	IC-Clo	318 aA	334 aA	202 aA	227 aA	229 aA	262 a	68 bA	65 bAB	43 bBC	26 aC	28 aC	46 b
	IC-Mix	307 aA	275 aAB	262 aABC	204 aBC	185 aC	247 a	82 aA	78 aA	67 aB	31 aC	28 aC	57 a
	Mean	312 A	305 A	232 AB	216 C	207 C		75 A	71 A	55 B	29 C	28 C	

ANOVA

Source of Variation	df	Fenugreek		Clover		Fenugreek		Clover	
		BEG	SBR	BEG	SBR	BEG	SBR	BEG	SBR
S	2	0.126	0.044 *	0.195	0.662	0.063 °	0.189	0.155	0.483
NT	4	0.000 ***	0.000 ***	0.000 ***	0.004 **	0.000 ***	0.000 ***	0.000 ***	0.000 ***
CP	1	0.000 ***	0.000 ***	0.001 **	0.221	0.080 °	0.206	0.012*	0.000 ***
S × NT	8	0.390	0.783	0.619	0.735	0.676	0.315	0.454	0.781
S × CP	2	0.420	0.423	0.723	0.345	0.246	0.608	0.765	0.741
N × CP	4	0.569	0.796	0.019 *	0.230	0.252	0.277	0.001 **	0.091 °
S × NT × CP	8	0.992	0.977	0.996	0.824	0.953	0.923	0.955	0.999

Within each column, means followed by a common lowercase letter are not significantly different at $p < 0.05$ according to Tukey's HSD-test. Within each line, means followed by a common capital letter are not significantly different at $p < 0.05$ according to Tukey's HSD-test. ANOVA shows p-values for the main effects and their interactions; S: Season, NT: N Treatment, and CP: Cropping Pattern. Note: °, *, ** and ***: $p \leq 0.1$, 0.05, 0.01 and 0.001, respectively.

Overall, results showed that fenugreek has produced higher biomass compared to clover. Globally, N fertilization applied three times beginning at durum wheat tillering (N3 and N4), greatly reduced both legumes biomass and nodulation in all intercropping patterns. However, N fertilization beginning at durum wheat stem elongation (N2) negatively affected only clover biomass and clover nodules weight in the C-IN pattern (Table 1). The effect of NT was showed more pronounced at BEG site compared to SBR site.

Although biomass of each legumes species was reduced in the IC-Mix pattern compared to their use separately, the nodules weight was not affected and in contrast it was enhanced particularly for the clover. Overall, total legume biomass (fenugreek biomass+clover biomass) and nodule occurrence are clearly above in the IC-Mix pattern compared to the IC-Fen and IC-Clo patterns.

3.3. Weed Biomass

Results showed that weed biomass was influenced exclusively by the adopted cropping pattern ($p < 0.000$) (Table S3). The SC-NH showed the highest weed biomass averaged 258 and 212 kg ha^{-1} at BEG and SBR, respectively. (Figure 3A). Herbicide application (SC-H) successfully controlled weeds, with an average reduction rate of weed biomass by 78% relative to the SC-NH (Figure 3A).

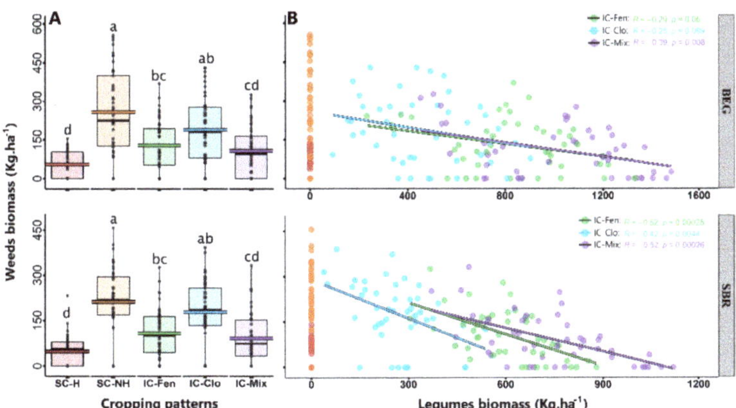

Figure 3. Effect of cropping patterns on weed biomass, averaged over the three seasons and the five N treatments (**A**) and the relationship between weed biomass and legume biomass in each intercropping pattern (**B**) at both sites.(**A**): Box plot indicating the cropping pattern effects on the weed biomass averaged over the three seasons and the five N treatments (Colored lines indicate the mean); Box plots with a common letter are not significantly different at $p < 0.05$ according to Tukey's HSD-test. (**B**): Relationship between weeds biomass and legumes biomass in each intercropping pattern (R-values are the correlation coefficient (Spearman's correlation); colored lines represent linear regression for each intercropping pattern).

Within the intercrop's patterns, the IC-Mix showed the highest weeds suppression performance. Weed biomass in the IC-Mix was significantly reduced by averaged 58% relative to the SC-NH which is statistically equivalent to the herbicide application effect (SC-H). The IC-Fen pattern showed, also, a significant reduction of weed biomass by average 49%. However, in the IC-Clo pattern weed biomass was slightly reduced compared to the SC-NH by 27% at BEG and 17% at SBR (not significant). Our results showed that weed suppression in intercrops was mostly due to fenugreek, along with the increase in the total legume sowing density in the IC-Mix pattern. Indeed, legumes biomass was showed significantly negatively correlated with weeds biomass. The higher the legumes biomass, the fewer weeds could establish (Figure 3B).

3.4. Net Photosynthetic Rate (Pn)

The Net Photosynthetic Rate (*Pn*) was measured at crops flowering stage in 2016 and 2017 seasons. Durum wheat *Pn* at BEG site was greatly affected by NT, CP and NT $\times$ CP ($p < 0.000$) (Table S3). Compared with N0, the application of N fertilizer significantly increased *Pn*. Within the sole cropping patterns, the *Pn* values under SC-H in all N treatments were significantly higher than those under SC-NH by 4–11% (Table 2). Within the intercrop's patterns, the *Pn* values under IC-Mix in the N0, N1and N2 treatments were significantly higher than those under SC-NH by 3–9%, which were statistically equivalent to those in SC-H pattern. While the *Pn* values under IC-Fen and IC-Clo only in the N0 and N1 treatments were higher than those under SC-NH by 4–7 and 2–3%, respectively (Table 2). However, in treatments N3 and N4, *Pn* values were largely the same in the intercropping patterns and the SC-NH pattern, which were in all patterns significantly lower than those in the SC-H pattern (Table 2). These results illustrate that only under unfertilized conditions (N0) and with a relatively low and late N-input (N1 and N2), the *Pn* of durum wheat shows an improvement especially with the mixture of fenugreek and clover (IC- Mix) compared to the sole crop. Overall, similar trends were also shown at the SBR site with regard to the effects of CP and NT on *Pn*. Yet, particularly at this site, *Pn* values under IC-Mix were statistically equivalent to those in SC-H pattern in all N treatments (Table 2). The not significant interaction term between NT and CP for *Pn* further showed that CP effects were not NT dependent at this site.

Table 2. Effect of different cropping patterns and N treatments on the net photosynthetic rate (Pn, μmol CO_2 m^{-2} s^{-1}) of durum wheat, fenugreek and clover at flowering stage in 2016 and 2017 seasons.

		BEG						SBR					
		N0	N1	N2	N3	N4	Mean	N0	N1	N2	N3	N4	Mean
Durum wheat													
2016	SC-H	15.7 abcD	17.1 abC	18.3 aB	19.1 aA	18.8 aAB	17.8 a	14.2 abC	14.9 abB	15.6 abA	16.1 aA	15.8 aA	15.3 a
	SC-NH	15 cC	16.3cB	17.2 cA	17.2 cA	17.4 bcA	16.6 c	13.9 bC	14.6 bB	14.8 bB	15.4 bA	15.3 cA	14.8 b
	IC-Fen	15.8 abC	16.9 abB	17.7 bA	17.9 bA	18 bA	17.3 b	14.7 aB	15.2 abAB	15.7 aA	15.6 bA	15.7 abcA	15.4 a
	IC-Clo	15.3 bcB	16.8 bA	17.1 cA	17.3 cA	17.3 cA	16.8 c	14.1 abB	15 abA	15.2 abA	15.5 bA	15.4 bcA	15 b
	IC-Mix	16.1 aC	17.3 aB	18.1 abA	17.9 bA	17.9 bcA	17.5 b	14.7 aB	15.4 aA	15.8 aA	15.8 abA	15.8 abA	15.5 a
	Mean	15.6 C	16.9 B	17.7 A	17.9 A	17.9 A		14.3 C	15 B	15.4 A	15.7A	15.6 A	
2017	SC-H	16 bC	17.1 abB	18.1 aA	18.4 aA	18.4 aA	17.6 a	15.5 bcC	16.1 abB	16.8 aA	17.3 aA	17.2 aA	16.6 a
	SC-NH	15.2 cD	16.3 cC	17.2 bcB	17.5 bAB	17.7 bA	16.8 c	15.2 cD	15.7 bCD	16b BC	16.4 bAB	16.5 bA	16 b
	IC-Fen	16.3 abC	17.2 abB	17.4 bcAB	17.6 bA	17.6 bA	17.2 b	15.9 aB	16.4 aAB	16.7 abA	16.9 abA	16.8 abA	16.5 a
	IC-Clo	15.7 bcB	16.8 bA	17 cA	17.4 bA	17.4 bA	16.9 c	15.5 bC	15.9 abBC	16.1b ABC	16.5 abA	16.4 bA	16.1 b
	IC-Mix	16.6 aB	17.3 aA	17.7 abA	17.7 bA	17.6 bA	17.4 b	16.1 aB	16.5 aAB	17 aA	17.1 abA	17ab A	16.7 a
	Mean	16 D	16.9 C	17.5 B	17.7 A	17.7 A		15.6 D	16.1 C	16.5 B	16.8 A	16.8 A	
Fenugreek													
2016	IC-Fen	18 aA	18.3 aA	18 aA	16.5 aB	16.3 aB	17.4 a	15.3 aA	15.3 aA	15.1 aAB	14.9 aB	14.9 aB	15.1 a
	IC-Mix	18.2 aA	18.4 aA	18.5 aA	16.5 aB	16.4 aB	17.6 a	15.2aA	15.2 aA	15 aA	14.9 aA	14.9 aA	15 a
	Mean	18.1 A	18.3 A	18.2 A	16.5 B	16.3 B		15.2A	15.3 A	15 A	14.9 A	14.9 A	

Table 2. Cont.

		BEG						SBR				
	N0	N1	N2	N3	N4	Mean	N0	N1	N2	N3	N4	Mean
Durum wheat												
2017 IC-Fen	17.3 aA	17.5 aA	17.7 aA	15.9 aB	15.7 aB	16.8 a	16 aA	16.2 aA	16.2 aA	15.7 aA	15.7aA	16 a
IC-Mix	17.3 aA	17.3 aA	17.4 aA	16 aB	15.9 aB	16.8 a	15.9 aAB	16.1 aA	16.2 aA	15.8 aAB	15.6 aB	15.9 a
Mean	17.3 aA	17.4 A	17.6 A	16 B	15.8 B		16A	16.2 A	16.2 A	15.7 A	15.7 A	
Clover												
2016 IC-Clo	16.9 aA	17 aA	15.3 aB	14.1 aC	14a C	15.5 a	14.7 aA	14.8 aA	13.1 bB	13.2 aB	13 aB	13.7 b
IC-Mix	16.9 aAB	17.1 aA	16.1 aB	14a C	14.1 aC	15.6 a	15 aA	15.2 aA	14.6 aAB	13.4 aC	13.6 aBC	14.3 a
Mean	16.9 A	17 A	15.7 B	14.1 C	14 C		14.8 A	15 A	13.8 B	13.3 B	13.3 B	
2017 IC-Clo	17 aA	17.2 aA	15.3 bB	14.8 aC	14.8 aC	15.8 b	14.9 aA	15 A	13.5b AB	13.6 aAB	13.1 aB	14 b
IC-Mix	16.9 aA	17.1 aA	16.3a A	14.8 aB	14.7 aB	16 a	15.3 aA	15.4 A	14.8a AB	13.6 aB	13.7 aB	14.5 a
Mean	16.9 A	17.1 A	15.8 B	14.8 C	14.8 C		15.1 A	15.2 A	14.2 B	13.6 B	13.4 B	

Within each column, means followed by a common lowercase letter are not significantly different at $p < 0.05$ according to Tukey's HSD-test. Within each line, means followed by a common capital letter are not significantly different at $p < 0.05$ according to Tukey's HSD-test. For statistical details regarding factors and their interaction significance, see Table S4.

With regard to fenugreek, *Pn* was only affected by NT (Table 2 & Table S4). The effect of NT was showed more pronounced at BEG site compared to SBR. Further, at BEG, the *Pn* values under N0, N1 and N2 treatments were maintained similar and were significantly higher than those under N3 and N4 by 9–12%. Likewise, for clover, Pn was greatly affected by NT ($p < 0.000$). However, the effect of NT on *Pn* was showed dependent on CP, since the interaction NT × CP was significant at BEG site for both seasons (Table 2 & Table S4). Indeed, within the IC-Mix pattern the Pn values of clover under N0, N1 and N2 treatments were similar and significantly higher than those under N3 and N4 by 19–23%, while within the IC-Clo the *Pn* values were significantly the highest only under N0 and N1 treatments. Further, with the presence of fenugreek (IC-Mix) the *Pn* values of clover were globally enhanced particularly under N2 treatment (Table 2). This effect was clearly showed at BEG site for 2017 season and at SBR site for both seasons, where Pn values under treatment N2 were significantly higher under IC-Mix than those under IC-Clo. However, under N3 and N4 treatments, the *Pn* values of clover were decreased in both IC-Clo and IC-Mix mostly at BEG site.

3.5. Soil Moisture

At BEG site, soil moisture at legumes flowering was generally homogeneous between all cropping patterns and N treatments over both seasons (Figure 4). However, at SBR site, soil moisture was showed greatly affected by the adopted cropping pattern ($p < 0.001$). The highest soil moisture values were recorded in the IC-Mix and also the IC-Fen. Soil moisture in these intercropping patterns was significantly higher than both SC-NH and SC-H patterns. The improvements of soil moisture in the IC-Fen and IC-Mix patterns were showed maintained in all NT (Figure 4). These results illustrate that the existing legumes as companion plants provide a better canopy that contributes to minimizes soil water evaporation and conserves soil moisture. Indeed, in the SBR site, which is characterized by a semi-arid climate, soil moisture was positively correlated with legumes biomass ($p < 0.001$).

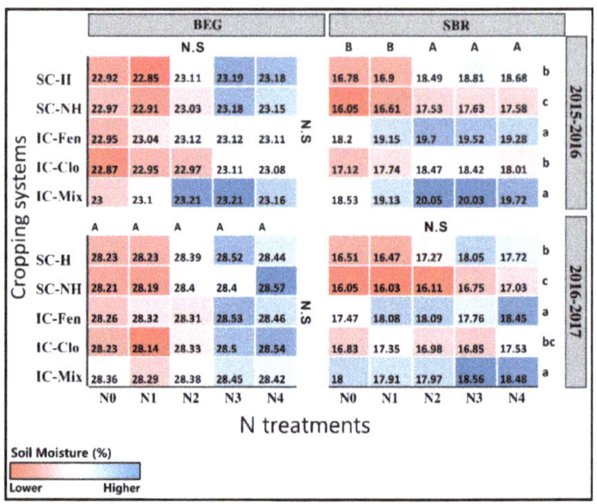

Figure 4. Effects of cropping pattern and N treatment on soil moisture at legumes flowering in 2016 and 2017 seasons, at both sites. Capital letters (A, B) indicate significant differences between N treatments (cropping patterns combined), according to Tukey's HSD-test. Lower-case letters (a, b, c) indicate significant differences between cropping patterns (N treatments combined), according to Tukey's HSD-test. N.S. not significant.

3.6. Grain Product Quality

Overall, durum wheat thousand kernel weight (TKW), grain protein content (GPC), and grain ash content (GAC) were showed improved in the intercropping patterns mainly in the IC-Mix pattern compared to both SC-NH and SC-H (Table 3). The improvement of durum wheat grain quality appears dependent on NT, particularly at BEG. Indeed, only under N0, N1 and N2 treatments there were a significant improvement of TKW, GPC and GAC mainly with the IC-Mix compared to the SC-NH. However, under N3 and N4 treatments, TKW, GPC and GAC were broadly similar whatever the cropping patterns (Table 3).

Table 3. Effects of cropping patten and N treatment on thousand kernel weights (TKW), grain protein content (GPC), grain ash content (GAC) and gross grain product impurity rate of durum wheat at both sites averaged over the three seasons.

	BEG						SBR					
	N0	N1	N2	N3	N4	Mean	N0	N1	N2	N3	N4	Mean
Thousand Kernel Weights (TKW g)												
SC-H	48.9 abB	49.3 abA	49.2 bAB	49.1 aAB	49 aAB	49.1 abc	40.2 bcA	40.4 bcA	40.6 aA	40.5 aA	40.5 aA	40.5 b
SC-NH	48.7 bB	49.1 bAB	49.2 bA	48.8 aAB	48.8 aAB	48.9 c	40.1 cA	40.3 cA	40.4 aA	40.1 aA	40.1 bA	40.2 c
IC-Fen	49 BabC	49.5 abA	49.4 abAB	48.9 aBC	48.8 aC	49.1 ab	40.5 abAB	40.8 abA	40.6 aAB	40.4 aB	40.3 abB	40.5 ab
IC-Clo	49 abAB	49.4 abA	49.1 bAB	48.9 aAB	48.8 aB	49 bc	40.3 bcA	40.5 bcA	40.4 aA	40.1 aA	40.1 abA	40.3 c
IC-Mix	49.2 aB	49.6 aA	49.7 aA	48.9 aB	48.8 aB	49.2 a	40.7 aA	40.9 aA	40.7 aA	40.5 aA	40.4 abA	40.7 a
Mean	48.9 B	49.4 A	49.3 A	48.9 B	48.8 B		40.4 BC	40.6 A	40.5 AB	40.3 C	40.3 C	

Table 3. *Cont.*

	BEG						SBR					
	N0	N1	N2	N3	N4	Mean	N0	N1	N2	N3	N4	Mean
Grain Protein Content (GPC%)												
SC-H	12.5 cB	13.2 aA	13.5 abA	13.2 aA	13.2 aA	13.1 ab	14.1 cB	14.5 bB	15.2 abA	15.4 aA	15.3 aA	14.9 bc
SC-NH	12.5 cB	12.8 bA	13 cA	13 aA	13.1 aA	12.9 c	14.3 bcC	14.6 bBC	14.7 bABC	15 aAB	15.2 aA	14.8 c
IC-Fen	12.8 abB	13.3 aA	13.4 abcA	13.1 aA	13.1 aA	13.1 ab	14.6 abB	14.9 abAB	15.3 abA	15.2 aA	15.3 aA	15.1 b
IC-Clo	12.7 bcB	13.1 abA	13.1 bcA	13.1 aA	13.1 aA	13 bc	14.5 abcB	14.8 abAB	14.9 bAB	15 aAB	15.2 aA	14.9 bc
IC-Mix	13 aB	13.4 aAB	13.5 aA	13.2 aAB	13.2 aAB	13.3 a	14.8 aB	15.2 aAB	15.6 aA	15.5 aA	15.5 aA	15.3 a
Mean	12.7 C	13.2 AB	13.3 A	13.1 B	13.2 AB		14.5 C	14.8 B	15.1 A	15.3 A	15.3 A	
Grain Ash Content (%)												
SC-H	1.92 aB	1.97 aA	2.01 aA	2.01 aA	2.01 aA	1.98 a	1.98 bcC	2.02 bcBC	2.05 abAB	2.07 aA	2.06 aA	2.04 a
SC-NH	1.88 bB	1.92 bB	1.97 bA	1.95 cA	1.96 bA	1.93 c	1.96 cB	1.99 cAB	2.01 bA	2.02 bA	2.02 bA	2 b
IC-Fen	1.94 aB	1.99 aA	2.01 aA	1.99 abA	1.99 abA	1.99 a	2.01 abB	2.05 abAB	2.06 aA	2.04 abAB	2.05 abAB	2.04 a
IC-Clo	1.93 aA	1.96 abA	1.97 bA	1.96 bcA	1.96 bA	1.96 b	1.99 abcA	2.02 bcA	2.02 bA	2.02 bA	2.02 abA	2.02 b
IC-Mix	1.94 aB	2 aA	2.02 aA	2 aA	2ab A	1.99 a	2.02 aB	2.07 aA	2.07 aA	2.05 abAB	2.05 abAB	2.05 a
Mean	1.92 C	1.97 B	1.99 A	1.98 AB	1.99 A		1.99 B	2.03 A	2.04 A	2.04 A	2.04 A	
Impurity rate (% of fenugreek seeds)												
IC-Fen	2.6	2.1	2	1.2	1.2	1.8	4	3.5	3.3	2.6	2.2	3.1
IC-Mix	3	2.6	2.2	1.4	1.5	2.1	4.3	3.6	2.6	2.4	2.6	3.1
Mean	2.8	2.4	2.1	1.3	1.4		4.1	3.6	3	2.5	2.4	

Capital letters indicate significant differences between N treatments (horizontal comparison) and lower-case significant differences between cropping patterns (vertical comparison) (Tukey-Test, $\alpha = 0.05$). For statistical details regarding factors and their interaction significance, see Table S3.

3.7. Straw Yield

At both sites, durum wheat straw yield significantly varied according to the CP and NT ($p < 0.05$) (Table S3). Straw yield was increased in all intercropping patterns as compared to the SC-NH (Figure 5A). The IC-Mix pattern showed the highest straw yield; which was statistically higher than those in the SC-NH and SC-H (Figure 5A). As compared to SC-NH, straw yield increase rate in the IC-Mix pattern was 13% and 16% respectively at BEG and SBR sites. Overall, regardless of the cropping patterns straw yield was significantly increased when N-fertilization was applied at durum wheat tillering (N3 and N4) or, at the latest, at durum wheat stem elongation (N2) (Figure 5B). Although there was no significant interaction between NT and CP, within the IC-Mix and IC-Fen, the highest additional straw yields relative to the SC-NH pattern were obtained in N0 and N1 treatments (18–27%), followed by N2 treatment (13–16%) over N3 and N4 treatments (4–11%) (Figure 5B).

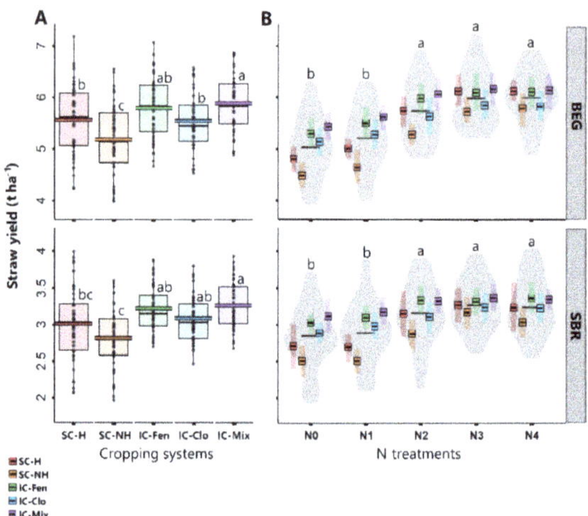

Figure 5. Effects of cropping pattern and N treatment on the straw yield at both sites averaged across the three experimental seasons.(**A**): Box plot indicating the cropping pattern effects on the straw yield averaged over the three seasons and the five N treatments;(**B**): Violin plot indicating the N treatment effects on the straw yield averaged over the three seasons and the five cropping patterns (black lines indicate mean grain yield; Tiles represent the means and 95% confidence intervals according to the cropping pattern in each N treatment); Box plots (**A**) and violin plots (**B**) with a common letter are not significantly different at $p < 0.05$ according to Tukey's HSD-test (for statistical output see Table S3).

4. Discussion

Results of the present study highlight that the performance of legumes to improve durum wheat main crop depend largely on the legume part itself, the adopted N fertilization regime and the interaction between them especially at the humid site (BEG). When added separately as companion plants, i.e., IC-Fen vs IC-Clo, fenugreek performs better than clover in terms of weed suppression and the improvement in durum wheat productivity and grain quality. Still, their mixture in the IC-Mix results in the greatest improvement of durum wheat crop. Compared to durum wheat sole crop with no-herbicide (SC-NH), only under unfertilized conditions (N0) and under relatively low and late N fertilization regimes (N1 and N2), IC-Mix resulted in a clear improvement in productivity (grain and straw yields) and grain quality (TKW, GPC and GAC) of durum wheat, where values were closely similar and even higher than those reported with the use of herbicides (SC-H). However, under the highest N fertilization regimes (N3 and N4), all intercrops had no significant advantages over the SC-NH, especially at the rainy site (BEG). Particularly at this site, durum wheat response to different cropping patterns (CP) was found significantly dependent on the N fertilization treatment (NT) regarding the grain yield, GPC and GAC. Further, the superiority of the conventional durum wheat cropping system, i.e., the sole durum wheat crop herbicide treated (SC-H) under N3 and N4, was clear in terms of grain yield. On the semi-arid site (SBR), even with these N fertilization regimes (N3 and N4), there was a slight improvement in durum wheat grain productivity and quality with the IC-Mix pattern compared to the SC-NH. This could be related to the greater responsiveness of the durum wheat crop to N fertilization at the humid site (BEG) compared to the semi-arid site (SBR) (Figure 1), as crop responses to N fertilization are dependent on soil water availability [31].

In the present study, globally, weeds infestation was not severe. This may be attributed to the sites long-term management for farming production and the adoption of the false seedbed technique [23]. Accordingly, weeds adverse effects regarding durum wheat sole crops productivity and grain quality (SC-NH vs SC-H) were showed globally moderate as compared for instance to other studies [32,33]. Also, this could be, partly, because weeds were removed at durum wheat heading stage (Zadoks 45), which could have minimized weed competition. Nevertheless, weed suppression under the IC-Mix pattern involving both the fenugreek and the clover as companion plants was very resilient, which reached the herbicide weed control efficiency (in the SC-H). Both legumes, the fenugreek [17] and the clover [18] emit allelochemicals that are detrimental for weeds growth [19]. Our study, therefore, suggests that the adoption of the false seedbed technique combined with the use of fenugreek and clover as companion plants with durum wheat main crop help to suppress weeds. This effect was shown consistent across seasons at both sites. Results show, also, that fenugreek performs better than clover in terms of weed suppression (IC-Fen vs IC-Clo). This may be due to the fast growth features and the high competitive ability of fenugreek compared to clover [34,35] which help to fast-close crop canopy over weeds and thereby limiting early their growth [36].

The benefits of legumes, as companion plants, were shown dependent on their biomass, nodulation, and photosynthetic rate (*Pn*), which were greatly influenced by the adopted N fertilization regime. One of the most important advantages of the cereal-legumes intercrops is the improvement of legumes nodulation and N_2 fixation capacity [37,38]. This is attributed to soil N depletion by cereals, which reduces nitrate inhibition of nodulation and nodule functioning, and also because cereals are more competitive for soil N, forcing legumes to rely on biological N fixation [39]. However, as shown in our study, both legumes biomass and nodulation were significantly reduced under N fertilization regimes beginning at durum wheat tillering stage (N3 and N4). Mostly at BEG (humid), this can be explained by the fact that early and high soil N availability (N3 and N4) leads to a competitive imbalance that favours durum wheat growth. High durum wheat growth at early stage causes shading effects on the under-sown legumes. Hence, resulting in the greatest decrease in their *Pn*. Reduced legumes *Pn* leads to a significant decrease in energy supply to nodules, resulting in reduced nodulation and N_2 fixation [40,41]. Particularly at BEG site, both legumes *Pn* and nodule weight were positively correlated ($p < 0.001$). Similar observations were also reported by [42] under cereal-pea intercropping condition. At SBR (semi-arid), the decrease of legumes nodulation under N3 and N4 treatments may be explained, also, by a nitrate inhibitory effect [43] given that durum wheat N-uptake and thus soil N depletion in the rhizosphere environment may be limited by water-limiting availability. Indeed, under water-limiting conditions such as SBR site, soil water availability has a direct effect on productivity and an indirect effect through its regulatory role in soil N availability [44,45]. Thus, the limited durum wheat biomass establishment (estimated via the straw yield), at SBR compared to BEG, could have minimized light competitiveness between durum wheat and legumes which explain the relatively lower adverse effect of high soil N-availability under N3 and N4 treatment on legumes biomass and *Pn* at SBR compared to BEG site. Our study suggests that competition for light under the humid site and high N availability in the rhizosphere under the semi-arid site are the main factors limiting legume nodulation. The reduced nodulation and N fixing capacity of legumes lead to the lack of N facilitation process, which consists in the transfer of N from legumes to cereals [46,47], hence, the non-benefit of durum wheat from N2 fixation and also the non-advantages of legumes as companion plants as shown in our study in terms of PLER and durum wheat grain quality (GPC, GAC), mostly at the humid site. Our study, therefore, suggests avoiding N-fertilization regimes beginning at tillering stage in cereals-legumes intercrops systems. Of interest to indicate that mainly with the IC-Mix pattern, under a relatively low and late N-fertilization regime (N1 and N2) there was no effect on both legumes' biomass, *Pn* and nodulation compared to the N0 treatment, yet significantly increased overall durum wheat productivity and grain quality. This suggests that the benefits of legumes companion plants for the durum wheat crop increases with the decrease in N availability mostly at the early growth stage.

The effect of N fertilization regime on legumes biomass, *Pn* and nodulation depend, also, on the legume's species. Thus, N fertilization regime beginning at cereal stem elongation (N2) negatively affected only the clover biomass, and nodulation weighs in the IC-Clo without any significant effect on fenugreek in IC-Fen. Our findings suggest that under conditions of increased soil N availability at durum wheat stem elongation, trait differences among fenugreek and clover influenced their corresponding growth through changes in resource availability. Indeed, clover is known as a low-growing species compared to fenugreek [34,35], so its growth may be more affected by the durum wheat dominance and also by the weeds when N availability increase at durum wheat stem elongation. Durum wheat -and weeds- may on the one hand push legumes to increases its N_2-fixation reliance via competition and depletion for soil N, and on the other hand, reduce growth and N_2 fixation in low-growing legumes like clover via strong competition for light. The same finding concerning clover species was reported by [48], where N-fertilization applied at cereal stem elongation stage significantly decreased both red and white clover biomass compared to the unfertilized treatment under clover-wheat intercrops conditions. When both legumes were added as a mixture (IC-Mix), albeit each legume biomass was relatively reduced compared to its addition alone (IC-Fen and IC-Clo), their nodules weight and *Pn* were maintained unaffected and, instead, improved in particular for clover under N2 treatment. Our results suggest that by mixing these legume species -and thus increasing the diversity of traits-, the interactions between plants into the cropping system i.e., between weeds-legumes (fenugreek+clover)-durum wheat can be modulated under certain condition of soil N availability. Thus, the better weed suppression performance of fenugreek due to its fast growth features could have minimized weed competition early and therefore giving a better growth condition for clover into the IC-Mix pattern. On the other hand, driven by trait differences between fenugreek and clover (i.e., fast growth—low growth and medium root depth—relatively deeper root depth), the frequency of legumes roots and nodules (fenugreek+clover) could be substantially higher in either time and space in the IC-Mix pattern which results in greater bioavailability and -reachability- for durum wheat to benefit from the N_2 fixation through the facilitation process [46,47] and also could further induce nodule function [46,47,49]. Accordingly, this may constitute an additional source of N, which resulted in the improvement of durum wheat *Pn*, productivity (PLER) and grain quality (TKW, GPC) with the IC-Mix under N0, N1 and N2 treatments. Our study, therefore, illustrates that under certain condition of soil N availability using promising legumes species combinations could result in the improvement of N fertilizer land-use efficiency and hence help to reduce N-fertilization inputs for an eco-friendly durum wheat production.

Particularly under water-limiting conditions such as SBR site, our study illustrate that the existing of legumes provides an extra canopy that confers the shading which minimizes soil water evaporation losses. Thus, within the intercrop's patterns, mostly the IC-Mix and also the IC-Fen, there was a significant improvement of soil moisture, which was positively correlated with the legume biomass ($p < 0.001$). This result agrees with previous studies, which demonstrated that intercropping decreased water evaporation and conserved soil moisture [21,50]. Our study provides that under low precipitation conditions, legumes as companion plants improve soil water availability for durum wheat as cash crops mainly at grain filling stage, which results in the increases of the TKW.

The use of legumes, as companion plants, can be considered a powerful strategy to limit pesticide dependency [51], as shown in our study for herbicides but also for insecticides [52], and for disease control [53], which can mitigate food-related chemicals hazards [54]. Thus, using legumes as companion plants could provide incentives to ensure food safety and provide better nutritional values with limited repercussions on the processing and acceptability of cereal-based products. As our study confirms, legumes as companion plants can improve cereal grain weight, protein, and ash content relative to cereals sole crops. Compared to other cereal-legume intercropping systems that require a sorting step of raw cereal products for acceptability when processed for human consumption [55], the use of legumes such as the fenugreek and clover as companion plants helps to avoid such sorting process. On the other hand, having fenugreek seeds in raw grain product at low levels, not exceeding 4.3% as in this study (Table 3), could improve foods nutritional value and support functional food concept without repercussions on the organoleptic properties [56].

5. Conclusions

In the present study, we evaluated the performance of two legume species (fenugreek, clover and a mixture of them) added as companion plants to enhance durum wheat crop under five N fertilization regimes compared to durum wheat sole crops with and without herbicide. Result reveled that the mixture of fenugreek and clover as companion plants (IC-Mix) may offer significant opportunities for developing sustainable durum wheat production. The mixture of fenugreek and clover as companion plants (combined with better seedbed preparation using for instance the false seedbed technique as the present study) resulted in a better weed suppression performance that reached herbicide efficiency. Mostly under rainy conditions, the highest's performance of the mixture of fenugreek and clover to improve durum wheat productivity and grain quality were found under the unfertilized conditions (N0) or the relatively low and late N-fertilization regimes (N1 and N2), suggesting that N fertilization regime requires attention to ensure the expected benefits of such intercropping systems. Particularly at the semi-arid site (SBR), findings proved that the mixture of fenugreek and clover as companion plants help to preserve soil moisture and hence help to mitigate the water stress. This study proves that the use of legumes, as companion plants, represents an excellent alternative to the conventional cereal cropping system by providing multiple services in line with the sustainability principles.

Supplementary Materials: The following are available online at https://www.mdpi.com/2073-439 5/11/1/78/s1, Figure S1: Map of Tunisia showing the experimental sites, Figure S2: Approximate crop management dates adopted at each site. Grey oval indicates cereal growth stage according to Zadoks; Red diamonds indicate the date of herbicide application; Orange pentagon indicate timing of N-fertilization; Green tiles indicate the date of hand weeding; Green triangles indicate the date of legumes biomass determining. Table S1: Agronomic characteristics and soil physicochemical properties of both sites., Table S2: Cereals and legumes plant density after 20 days of sowing date during the three seasons at both sites. Table S3: *p*-value of ANOVA's statistical output showing the influence of season (S), N treatment (NT), cropping pattern (CP), and their interactions, on the different studied parameters, in durum wheat crops at both sites. Table S4: P-value of ANOVA's statistical output showing the influence of N treatment (NT), cropping pattern (CP), and their interaction on durum wheat, fenugreek and clover net photosynthetic rate (Pn, μmol CO_2 m^{-2} s^{-1}) at crops flowering stage in 2016 and 2017 seasons at both sites.

Author Contributions: W.T. (Wael Toukabri) conceptualization. W.T. (Wael Toukabri) and D.T. designed the research methodology. W.T. (Wael Toukabri) conducted the experiments. D.H. and M.J. helps in fields experimentation. D.H. and N.F. contributed to sample collection and analysis. F.Z. contributed to sample collection. W.T. (Wael Toukabri) analyzed the data with support from W.T. (Wael Taamalli). N.F contributed to the interpretation of the results. W.T. (Wael Toukabri) and D.T. wrote the manuscript. O.K. and R.M. Funding acquisition and Resources. R.M. helped supervise the project. D.T. supervised the project. All authors have read and agreed to the published version of the manuscript.

Funding: This work is part of a research and development agreement between the Centre of Biotechnology of Borj-Cedria (CBBC) and the Institute National des Grandes Cultures (INGC).

Data Availability Statement: The data presented in this study are available on request from the corresponding author.

Acknowledgments: This work is part of a research and development agreement between the Centre of Biotechnology of Borj-Cedria (CBBC) and the Institute National des Grandes Cultures (INGC). The authors sincerely thank all field and laboratory staff for their support in field operations and their assistance in laboratory experiments.

Conflicts of Interest: The authors declare no conflict of interest.

References

1. Chairi, F.; Aparicio, N.; Serret, M.D.; Araus, J.L. Breeding effects on the genotype × environment interaction for yield of durum wheat grown after the Green Revolution: The case of Spain. *Crop J.* **2020**, *8*, 623–634. [CrossRef]
2. Jouili, M. Tunisian Agriculture: Are Small Farms Doomed to Disappear? 111 EAAE-IAAE Seminar 'Small Farms: Decline or Persistence', University of Kent, Canterbury, 26–27 June 2009. Available online: http://ageconsearch.tind.io/record/52816/files/051.pdf (accessed on 8 July 2020).
3. Latiri, K.; Lhomme, J.P.; Annabi, M.; Setter, T.L. Wheat production in Tunisia: Progress, inter-annual variability and relation to rainfall. *Eur. J. Agron.* **2010**, *33*, 33–42. [CrossRef]
4. Heap, I. Global perspective of herbicide-resistant weeds. *Pest Manag. Sci.* **2014**, *70*, 1306–1315. [CrossRef] [PubMed]
5. Sharma, L.; Bali, S. A Review of Methods to Improve Nitrogen Use Efficiency in Agriculture. *Sustainability* **2017**, *10*, 51. [CrossRef]
6. Bernhardt, E.S.; Rosi, E.J.; Gessner, M.O. Synthetic chemicals as agents of global change. *Front. Ecol. Environ.* **2017**, *15*, 84–90. [CrossRef]
7. Duchene, O.; Vian, J.-F.; Celette, F. Intercropping with legume for agroecological cropping systems: Complementarity and facilitation processes and the importance of soil microorganisms. A review. *Agric. Ecosyst. Environ.* **2017**, *240*, 148–161. [CrossRef]
8. Verret, V.; Gardarin, A.; Pelzer, E.; Médiène, S.; Makowski, D.; Valantin-Morison, M. Can legume companion plants control weeds without decreasing crop yield? A meta-analysis. *Field Crop. Res.* **2017**, *204*, 158–168. [CrossRef]
9. Hauggaard-Nielsen, H.; Jørnsgaard, B.; Kinane, J.; Jensen, E.S. Grain legume-Cereal intercropping: The practical application of diversity, competition and facilitation in arable and organic cropping systems. *Renew. Agric. Food Syst.* **2008**, *23*, 3–12. [CrossRef]
10. Mead, R.; Willey, R.W. The Concept of a 'Land Equivalent Ratio' and Advantages in Yields from Intercropping. *Exp. Agric.* **1980**, *16*, 217–228. [CrossRef]
11. Dabney, S.M.; Delgado, J.A.; Reeves, D.W. Using winter cover crops to improve soil and water quality. *Commun. Soil Sci. Plant Anal.* **2001**, *32*, 1221–1250. [CrossRef]
12. Hartwig, N.L.; Ammon, H.U. Cover crops and living mulches. *Weed Sci.* **2002**, *50*, 688–699. [CrossRef]
13. Pirhofer-Walzl, K.; Rasmussen, J.; Høgh-Jensen, H.; Eriksen, J.; Søegaard, K.; Rasmussen, J. Nitrogen transfer from forage legumes to nine neighbouring plants in a multi-species grassland. *Plant Soil* **2012**, *350*, 71–84. [CrossRef]
14. Frankow-Lindberg, B.E.; Dahlin, A.S. N2 fixation, N transfer, and yield in grassland communities including a deep-rooted legume or non-legume species. *Plant Soil* **2013**, *370*, 567–581. [CrossRef]
15. Pappa, V.A.; Rees, R.M.; Walker, R.L.; Baddeley, J.A.; Watson, C.A. Legumes intercropped with spring barley contribute to increased biomass production and carry-over effects. *J. Agric. Sci.* **2012**, *150*, 584–594. [CrossRef]
16. Xiao, J.; Yin, X.; Ren, J.; Zhang, M.; Tang, L.; Zheng, Y. Complementation drives higher growth rate and yield of wheat and saves nitrogen fertilizer in wheat and faba bean intercropping. *Field Crop. Res.* **2018**, *221*, 119–129. [CrossRef]
17. Fernández-Aparicio, M.; Emeran, A.A.; Rubiales, D. Control of Orobanche crenata in legumes intercropped with fenugreek (*Trigonella foenum-graecum*). *Crop Prot.* **2008**, *27*, 653–659. [CrossRef]
18. Fernández-Aparicio, M.; Emeran, A.A.; Rubiales, D. Inter-cropping with berseem clover (*Trifolium alexandrinum*) reduces infection by Orobanche crenata in legumes. *Crop Prot.* **2010**, *29*, 867–871. [CrossRef]
19. Jabran, K.; Mahajan, G.; Sardana, V.; Chauhan, B.S. Allelopathy for weed control in agricultural systems. *Crop Prot.* **2015**, *72*, 57–65. [CrossRef]
20. Gou, F.; Yin, W.; Hong, Y.; van der Werf, W.; Chai, Q.; Heerink, N.; van Ittersum, M.K. On yield gaps and yield gains in intercropping: Opportunities for increasing grain production in northwest China. *Agric. Syst.* **2017**, *151*, 96–105. [CrossRef]
21. Brooker, R.W.; Bennett, A.E.; Cong, W.F.; Daniell, T.J.; George, T.S.; Hallett, P.D.; Hawes, C.; Iannetta, P.P.; Jones, H.G.; Karley, A.J.; et al. Improving intercropping: A synthesis of research in agronomy, plant physiology and ecology. *New Phytol.* **2015**, *206*, 107–117. [CrossRef]
22. Cossani, C.M.; Thabet, C.; Mellouli, H.J.; Slafer, G.A. Improving wheat yields through N fertilization in mediterranean tunisia. *Exp. Agric.* **2011**, *47*, 459–475. [CrossRef]
23. Kanatas, P.J.; Travlos, I.S.; Gazoulis, J.; Antonopoulos, N.; Tsekoura, A.; Tataridas, A.; Zannopoulos, S. The combined effects of false seedbed technique, post-emergence chemical control and cultivar on weed management and yield of barley in Greece. *Phytoparasitica* **2020**, *48*, 131–143. [CrossRef]
24. Zadoks, J.C.; Chang, T.T.; Konzak, C.F. A decimal code for the growth stages of cereals. *Weed Res.* **1974**, *14*, 415–421. [CrossRef]
25. Diakhaté, S.; Gueye, M.; Chevallier, T.; Diallo, N.H.; Assigbetse, K.; Abadie, J.; Diouf, M.; Masse, D.; Sembène, M.; Ndour, Y.B.; et al. Soil microbial functional capacity and diversity in a millet-shrub intercropping system of semi-arid Senegal. *J. Arid. Environ.* **2016**, *129*, 71–79. [CrossRef]
26. *AACC Approved Methods of Analysis*, 11th ed.; Method 08-01.01, Ash–Basic Method; Cereals & Grains Association: St. Paul, MN, USA, 2009; Available online: http://dx.doi.org/10.1094/AACCIntMethod-08-01.01 (accessed on 8 July 2020).
27. Yu, Y.; Stomph, T.J.; Makowski, D.; Zhang, L.; van der Werf, W. A meta-analysis of relative crop yields in cereal/legume mixtures suggests options for management. *Field Crop. Res.* **2016**, *198*, 269–279. [CrossRef]
28. R Development Core Team R: A Language and Environment for Statistical Computing 2017. R Foundation for Statistical Computing, Vienna, Austria. Available online: https://www.R-project.org/ (accessed on 8 July 2020).
29. Crawley, M.J. *The R Book*; John Wiley & Sons, Ltd.: Chichester, UK, 2012.

30. Mendiburu, F. Agricolae: Statistical Procedures for Agricultural Research. *R Package*. 2017. Available online: http://CRAN.R-project.org/package=agricolae (accessed on 8 July 2020).
31. Albrizio, R.; Todorovic, M.; Matic, T.; Stellacci, A.M. Comparing the interactive effects of water and nitrogen on durum wheat and barley grown in a Mediterranean environment. *Field Crop. Res.* **2010**, *115*, 179–190. [CrossRef]
32. Gharde, Y.; Singh, P.K.; Dubey, R.P.; Gupta, P.K. Assessment of yield and economic losses in agriculture due to weeds in India. *Crop Prot.* **2018**, *107*, 12–18. [CrossRef]
33. Fahad, S.; Hussain, S.; Chauhan, B.S.; Saud, S.; Wu, C.; Hassan, S.; Tanveer, M.; Jan, A.; Huang, J. Weed growth and crop yield loss in wheat as influenced by row spacing and weed emergence times. *Crop Prot.* **2015**, *71*, 101–108. [CrossRef]
34. Petropoulos, G.A. *Fenugreek: The Genus Trigonella*; Taylor and Francis: Oxfordshire, UK, 2002.
35. Taylor, N.L.; Norman, L. *Clover Science and Technology*; American Society of Agronomy: Madison, WI, USA, 1985.
36. Pouryousef, M.; Yousefi, A.R.; Oveisi, M.; Asadi, F. Intercropping of fenugreek as living mulch at different densities for weed suppression in coriander. *Crop Prot.* **2015**, *69*, 60–64. [CrossRef]
37. Zhang, F.; Shen, J.; Zhang, J.; Zuo, Y.; Li, L.; Chen, X. Rhizosphere Processes and Management for Improving Nutrient Use Efficiency and Crop Productivity: Implications for China. *Adv. Agron.* **2010**, *107*, 1–32.
38. Nabel, M.; Schrey, S.D.; Temperton, V.M.; Harrison, L.; Jablonowski, N.D. Legume Intercropping with the Bioenergy Crop Sida hermaphrodita on Marginal Soil. *Front. Plant Sci.* **2018**, *9*, 905. [CrossRef] [PubMed]
39. Hauggaard-Nielsen, H.; Gooding, M.; Ambus, P.; Corre-Hellou, G.; Crozat, Y.; Dahlmann, C.; Dibet, A.; von Fragstein, P.; Pristeri, A.; Monti, M.; et al. Pea–barley intercropping for efficient symbiotic N2-fixation, soil N acquisition and use of other nutrients in European organic cropping systems. *Field Crop. Res.* **2009**, *113*, 64–71. [CrossRef]
40. Zhou, Y.; Zhang, Y.; Wang, X.; Cui, J.; Xia, X.; Shi, K.; Yu, J. Effects of nitrogen form on growth, CO_2 assimilation, chlorophyll fluorescence, and photosynthetic electron allocation in cucumber and rice plants. *J. Zhejiang Univ. Sci. B* **2011**, *12*, 126–134. [CrossRef] [PubMed]
41. Li, Y.-Y.; Yu, C.-B.; Cheng, X.; Li, C.-J.; Sun, J.-H.; Zhang, F.-S.; Lambers, H.; Li, L. Intercropping alleviates the inhibitory effect of N fertilization on nodulation and symbiotic N2 fixation of faba bean. *Plant Soil* **2009**, *323*, 295–308. [CrossRef]
42. Pellicanò, A.; Romeo, M.; Pristeri, A.; Preiti, G.; Monti, M. Cereal-pea intercrops to improve sustainability in bioethanol production. *Agron. Sustain. Dev.* **2015**, *35*, 827–835. [CrossRef]
43. Rose, T.J.; Julia, C.C.; Shepherd, M.; Rose, M.T.; Van Zwieten, L. Faba bean is less susceptible to fertiliser N impacts on biological N2 fixation than chickpea in monoculture and intercropping systems. *Biol. Fertil. Soils* **2016**, *52*, 271–276. [CrossRef]
44. Cossani, C.M.; Slafer, G.A.; Savin, R. Nitrogen and water use efficiencies of wheat and barley under a Mediterranean environment in Catalonia. *Field Crop. Res.* **2012**, *128*, 109–118. [CrossRef]
45. Nguyen, G.N.; Joshi, S.; Kant, S. Water availability and nitrogen use in plants: Effects, interaction, and underlying molecular mechanisms. *Plant Macronutr. Use Effic.* **2017**, 233–243. [CrossRef]
46. Thilakarathna, M.S.; McElroy, M.S.; Chapagain, T.; Papadopoulos, Y.A.; Raizada, M.N. Belowground nitrogen transfer from legumes to non-legumes under managed herbaceous cropping systems. A review. *Agron. Sustain. Dev.* **2016**, *36*, 58. [CrossRef]
47. Tsialtas, I.T.; Baxevanos, D.; Vlachostergios, D.N.; Dordas, C.; Lithourgidis, A. Cultivar complementarity for symbiotic nitrogen fixation and water use efficiency in pea-oat intercrops and its effect on forage yield and quality. *Field Crop. Res.* **2018**, *226*, 28–37. [CrossRef]
48. Bergkvist, G.; Stenberg, M.; Wetterlind, J.; Båth, B.; Elfstrand, S. Clover cover crops under-sown in winter wheat increase yield of subsequent spring barley—Effect of N dose and companion grass. *Field Crop. Res.* **2011**, *120*, 292–298. [CrossRef]
49. Makoi, J.H.J.R.; Chimphango, S.B.M.; Dakora, F.D. Effect of legume plant density and mixed culture on symbiotic N2 fixation in five cowpea (*Vigna unguiculata* L. Walp.) genotypes in South Africa. *Symbiosis* **2009**, *48*, 57–67. [CrossRef]
50. Nyawade, S.O.; Karanja, N.N.; Gachene, C.K.K.; Gitari, H.I.; Schulte-Geldermann, E.; Parker, M.L. Intercropping Optimizes Soil Temperature and Increases Crop Water Productivity and Radiation Use Efficiency of Rainfed Potato. *Am. J. Potato Res.* **2019**, *96*, 457–471. [CrossRef]
51. Zhang, L.; Li, X.; Yu, J.; Yao, X. Toward cleaner production: What drives farmers to adopt eco-friendly agricultural production? *J. Clean. Prod.* **2018**, *184*, 550–558. [CrossRef]
52. Letourneau, D.K.; Armbrecht, I.; Rivera, B.S.; Lerma, J.M.; Carmona, E.J.; Daza, M.C.; Escobar, S.; Galindo, V.; Gutiérrez, C.; López, S.D.; et al. Does plant diversity benefit agroecosystems? A synthetic review. *Ecol. Appl.* **2011**, *21*, 9–21. [CrossRef]
53. Uzokwe, V.N.E.; Mlay, D.P.; Masunga, H.R.; Kanju, E.; Odeh, I.O.A.; Onyeka, J. Combating viral mosaic disease of cassava in the Lake Zone of Tanzania by intercropping with legumes. *Crop Prot.* **2016**, *84*, 69–80. [CrossRef]
54. Alldrick, A.J. Food Safety Aspects of Grain and Cereal Product Quality. *Cereal Grains* **2017**, 393–424. [CrossRef]
55. Bedoussac, L.; Journet, E.P.; Hauggaard-Nielsen, H.; Naudin, C.; Corre-Hellou, G.; Jensen, E.S.; Prieur, L.; Justes, E. Ecological principles underlying the increase of productivity achieved by cereal-grain legume intercrops in organic farming. A review. *Agron. Sustain. Dev.* **2015**, *35*, 911–935. [CrossRef]
56. Wani, S.A.; Kumar, P. Fenugreek: A review on its nutraceutical properties and utilization in various food products. *J. Saudi Soc. Agric. Sci.* **2018**, *17*, 97–106. [CrossRef]

 *agronomy*

Article

Productive and Physiological Response of Organic Potato Grown under Highly Calcareous Soils to Fertilization and Mycorrhization Management

Sara Lombardo [1], Cristina Abbate [1], Gaetano Pandino [1], Bruno Parisi [2], Aurelio Scavo [1,*] and Giovanni Mauromicale [1]

1   Department of Agriculture, Food and Environment (Di3A), University of Catania, 95123 Catania, Italy; saralomb@unict.it (S.L.); cristina.abbate@unict.it (C.A.); g.pandino@unict.it (G.P.); g.mauromicale@unict.it (G.M.)
2   Research Centre for Cereal and Industrial Crops (CREA-CI), Via di Corticella, 133-40128 Bologna, Italy; bruno.parisi@crea.gov.it
*   Correspondence: aurelio.scavo@unict.it; Tel.: +39-0954783415

Received: 9 July 2020; Accepted: 12 August 2020; Published: 14 August 2020

Abstract: The enhancement of the actual low yields is the most important challenge regarding organic farming management. In this view, a valid tool may arise by the improvement of fertilization management and efficiency. In this regard, arbuscular mycorrhizal fungi (AMF) can play an important role, especially in low fertility soils such as calcareous ones, through a better nutrient uptake and by alleviating abiotic stresses. A replicated-space experiment was carried out to investigate the role of mycorrhizal-based inoculants combined with full or halved fertilizer doses on yield and physiological traits of three early potato cultivars organically grown in highly calcareous and alkaline soils. The results indicate that AMF symbiosis ameliorated, in comparison to the not-inoculated plants, the potato tolerance to limestone stress by enhancing the potential quantum efficiency of photosystem II (F_v/F_0) and plant gas-exchange parameters (photosynthesis rate and stomatal conductance). Moreover, a significant improvement of marketable yield (+25%) was observed, mainly due to an increase of the number of tubers plant^{-1} (+21%) and, to a lesser extent, of average tuber weight (+10%). The AMF efficiency was higher applying halved fertilizer doses and in the location where soil conditions were unfavourable for potato growth. Moreover, the qRT-PCR highlighted that AMF colonization was similar in each location, demonstrating their tolerance to limestone, alkalinity and P stresses. These findings outlined that AMF are good candidate to bio-ameliorate calcareous soils and are very useful for improving potato yields under organic farming, limiting external fertilizers supply and environmental pollution.

Keywords: arbuscular mycorrhizal fungi; potato; organic farming; fertilization; calcareous soils; crop physiology; tuber yield; sustainability

---

1. Introduction

In the latest years, an increasing public interest for environmental safety and food quality has driven a major expansion in the organic farming sector all over the world [1]. In Europe, Spain, Italy and France are the countries with the largest organic agricultural land (2.1, 1.9 and 1.7 Mha, respectively [2]). The entire European Union organic food market is composed of 37% cereal products, 11% dairy products, 11% meat and 41% all other products [3]. Among the arable crops, potato (*Solanum tuberosum* L.) is successfully organically grown both in Italy [4] and elsewhere [5,6]. The production of organic early potato tubers is of particular economic relevance. They can be sold around 200–250 euros € per tonne in Italy [7]. The high content of minerals such as potassium, phosphorus and calcium [8,9], as well as that of polyphenols and carotenoids [10,11], is an attractive feature of the early crop potato (winter–spring cycle; planted from November to January and harvested from March to early-June). In addition, early potato tubers are also useful as feedstock for industrial products [12,13]. Therefore, crop conventional producers often adopt inorganic fertilizers and pesticides [14], which can result in the build-up of undesirable residues in both tubers and soil [15–17]. As a result, the share of organic production of early potato tubers has been increasing. However, yield levels are typically lower in organic systems than in conventional high-input ones [7,18,19]. Indeed, organic restrictions on fertilization mainly cause a reduced N availability [20–22], resulting in a detrimental effect on potato plant growth and tuber development. In addition, the early potato cycle is often characterized by a relatively low temperature, a short photoperiod and limited solar radiation, which are conditions with an appreciable effect on plant growth, substantially modifying the morphological and phenological characteristics of the plants (for example, most potato cultivars do not flower) compared to those cultivated in the common spring–summer cycle [23,24].

In the coastal agricultural areas of the Mediterranean basin, the early crop potato is commonly cultivated under calcareous soils, so containing a high concentration (>15%) of calcium carbonate ($CaCO_3$) and HCO_3^- in soil solution, and a reaction included in neutral–alkaline range, always <8.5 [25]. Calcareous soils have been estimated covering more than 30% of the world's land surface area, resulting in ~800 Mha according to FAO [26] and especially widespread in the arid and semi-arid regions because of the low leaching process. These soils are characterised by crust formation due to an inadequate quality of irrigation water, a high degree of P-fixation and Fe-precipitation, a low availability of nitrogen, magnesium and zinc [25], thus impairing the plant's mineral nutrition and worsening yield performances.

A reasonable agronomic measure for enhancing early potato organic production is represented by an adequate nutrient management [27] and an improvement of fertilization use efficiency [28]. Particularly, this may be achievable by applying arbuscular mycorrhizal fungi (AMF), obligate symbionts of the vast majority of land plants [29]. The main advantages arising from the application of AMF as inoculum for agricultural purposes are: (i) the increase of root system extension by more than 100-fold; (ii) the enhanced uptake of the soil immobile mineral nutrients; (iii) the reduction of abiotic stresses such as water scarcity and thermal imbalances; (iv) a better soil aggregation, which is important in improving soil structure and preventing soil erosion [29,30]. Therefore, the AMF application may play a key role under organic farming, since plants may particularly benefit by mycorrhizal symbiosis through a better uptake of mineralized soil nutrients present at low concentrations [31]. Black and Tinker [32] firstly reported the interaction between AMF symbiosis and potato in field conditions. After them, other researches have been carried out with different results considering the potato cultivar and the mycorrhizal fungus isolate. In several scientific studies reviewed by Wu et al. [33], potato crop was chosen as a case study for evaluating the impact of AMF on crop production, due to its worldwide diffusion and recognized nutritional value in the human diet. However, to our knowledge, few literature data [34–36] are available concerning the influence of AMF on potato crop performances in a large-scale production system under organic farming. The role of AMF on organic early potato grown in highly calcareous soils is still unknown. Moreover, specific attention must be directed to the cultivar choice, since this has a relevant role in the crop productive and qualitative performances under organic farming [8,37]. In particular, adaptable cultivars for organic production need to show a reliably high yield under low input production system, efficiency in nutrient uptake, fast early ground cover, good level of resistance/tolerance to the common biotic and abiotic stresses and a high suitability to low temperatures of storage [38]. In addition, it is recognized that AMF isolates may show a host genotype-specificity [35]. For example, the symbiosis with *Glomus fasciculatum* was found to increase the potato yield, contrariwise to *G. mosseae*, which had not effect [39].

Taking into account all these considerations, the present research was designed (1) to evaluate the influence of AMF application on the yield performances and plant physiology profile of three early potato genotypes organically grown in open-field conditions; (2) to investigate whether it is possible by using AMF inoculants to halve the organic fertilization rate, while keeping yield reduction to a minimum and having positive effects on crop physiology; (3) to observe the aforementioned effects of AMF application in different highly calcareous soils; (4) to verify the exploitation of the inherent advantages of the quantitative real-time PCR (qRT-PCR) technique in the detection and quantification of AMF *Glomus* spp. and *Gigaspora* spp. in soil.

2. Materials and Methods

2.1. Site, Soil and Climate

To consider the influence of soil type and the between-site variability, the research was replicated in space in accordance with Johnstone et al. [40]. The trials were conducted during the 2017 growing season in three different experimental fields (hereafter referred to as location I, II and III) placed on the coastal plain of South Siracusa (36°49′ N, 14°57′ E, 130 m a.s.l., south-eastern Sicily, Italy), a typical area for 'early' potato cultivation in the southern Italy. The soil, moderately deep, was Calcixerollic Xerochrepts on the basis of the USDA Soil Taxonomy Classification [41]. A layer, 0.25 m thick (from −0.05 to −0.30 m), where about 90% of active potato roots were located, was considered for the soil analysis. All soil analyses were carried out using the procedures approved by the Italian Society of Soil Science [42]. The three locations were characterized by various soil types, whose characteristics are reported in Table 1.

Table 1. Soil physio-chemical characteristics (−30 cm depth) of the three locations under study.

Soil Characteristic	Location I	Location II	Location III
Sand (%)	42.6	54.1	51.8
Silt (%)	38.7	24.8	22.0
Clay (%)	18.7	21.1	26.2
Total limestone (%)	68.0	44.2	65.6
Active limestone (%)	27.9	15.5	18.0
Organic matter (%)	2.18	1.7	2.6
Organic carbon (%)	1.27	1.0	1.5
C/N ratio	7.0	7.5	7.5
Total N (g kg^{-1})	1.8	1.3	2.0
Assimilable P_2O_5 (mg kg^{-1})	28.5	66	135
Exchangeable K_2O (mg kg^{-1})	197	455	612
pH	8	7.8	7.5
Electrical conductivity (dS m^{-1})	1.7	1.32	1.14
Cation exchange capacity (meq 100 g^{-1})	17.2	22.8	26.0

The active limestone level was high in all the soils, with the highest amount (~28%) in location I, +80% and +55% with respect to location II and III, respectively (Table 1). On the contrary, the lowest amount of P_2O_5 was detected in location I (28.5 mg kg^{-1}), with increasing values in location II (+132%) and III (+374%). Also, the K_2O concentration followed this trend. The three soils showed medium organic matter contents, but are characterised by a high level of organic matter mineralization. The climate of the area including the three locations under study, which are 4–5 km apart, is semi-arid Mediterranean with mild wet winters and common rainless springs. A meteorological station (Mod. Multirecorder 2.40; ETG, Firenze, Italy), located on the experimental field of location I, was used to daily record the air temperatures (minima, maxima and mean) and rainfall during the growing season. Both maximum and minimum monthly average temperature and total monthly rainfall during the early potato crop production (January–May) were calculated (Figure 1). The total rainfall of the growing season (115 mm) was lower compared to 179 mm of 30-year period. February experienced 45% of the rainfall, while April was particularly dry (only 2 mm). Minima temperatures never fell below 7.8 °C during the growing season, while the mean maximum temperature was above 16.4 °C at the plants' emergence (February) and 20.6 °C at the tuberification stage (April). The mean maximum temperature

(18.8 °C) and the mean minimum temperature (10.4 °C) were consistent with the long-term average (18.7 °C and 9.8 °C, respectively).

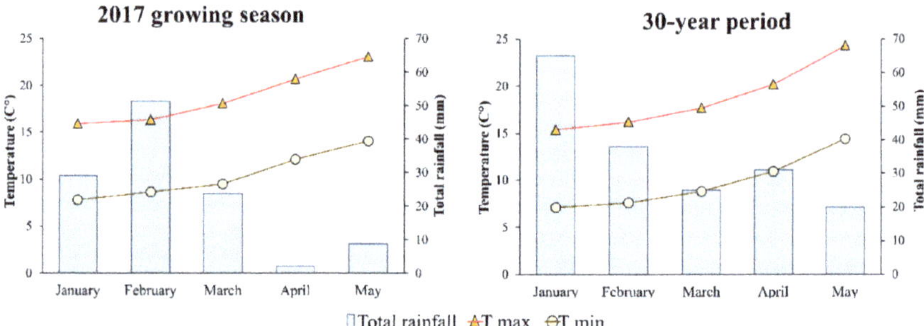

Figure 1. Total rainfall and average monthly maxima and minima temperatures for the 2017 growing season and long-term period (1977–2006).

2.2. Field Experimental Design, Plant Material and Management Practices

In each location, the experiment was arranged in a randomized split-plot design with three replications including three potato cultivars (i.e., Arizona, Mondial and Universa) as the main plots, and three fertilization management treatments as the sub-plots. The fertilization management treatments were summarized in Table 2 and, in particular, they included: (a) plots optimally fertilized and not inoculated, as a control (F100); (b) plots optimally fertilized and mycorrhizal inoculated (F100+M); (c) plots sub-optimally fertilized (with halved fertilizer doses respect to the other treatments) and mycorrhizal inoculated (F50+M).

Table 2. Agronomic management treatments of 'early' crop potato under organic farming.

Fertilization Management Treatment (F)	Phenological Stage of Application	Commercial Product	No. of Applications	Dose Rate per Application
F100	At sowing	Ricin-Xed®	1	1.2 t ha^{-1}
	"	Xedaneem Pel®	1	1.2 t ha^{-1}
	"	Kalisop®	1	0.6 t ha^{-1}
	"	Fosfonature 26®	1	0.4 t ha^{-1}
	After emergence	Biosin®	3	150 cc hL
F100+M	At sowing	Ricin-Xed®	1	1.2 t ha^{-1}
	"	Xedaneem Pel®	1	1.2 t ha^{-1}
	"	Kalisop®	1	0.6 t ha^{-1}
	"	Fosfonature 26®	1	0.4 t ha^{-1}
	"	Xedaopen®		40 kg ha^{-1}
	After emergence	Biosin®	3	150 cc hL
F50+M	At sowing	Ricin-Xed®	1	0.6 t ha^{-1}
	"	Xedaneem Pel®	1	0.6 t ha^{-1}
	"	Kalisop®	1	0.3 t ha^{-1}
	"	Fosfonature 26®	1	0.2 t ha^{-1}
	"	Xedaopen®		40 kg ha^{-1}
	After emergence	Biosin®	3	75 cc hL

The optimal fertilization was formulated on the basis of the recommendations provided by Research Institute of Organic Agriculture (FiBL) [43] and Sicily Department of Agriculture (www.regionesicilia.it), while considering both the NPK uptake by potato crop in Sicily with target yield of 20 t ha^{-1} and average NPK availability of experimental soils. At sowing, N was soil-applied by commercial organic sources derived from castor seeds (4% of N, Ricin-Xed®, XEDA Italia s.r.l., Forlì, Italy) and Neem seeds (3% of N, Xedaneem Pel®, XEDA Italia s.r.l., Forlì, Italy) after oil extraction, K_2O by applying a commercial granular product allowed in organic farming (50% of K_2O and 45% of SO_3, Kalisop®, K+S KALI GmbH, Verona, Italy) and P_2O_5 by a complex of 'Pheoflore' algal origin (26% of P_2O_5 and 41% of CaO, Fosfonature 26®, TIMAC Agro, Milan, Italy). After potato plants' emergence, a further N organic application was provided in three times by using a commercial liquid product (Biosin®, XEDA Italia s.r.l., Forlì, Italy) with 7.7% of N. In F100+M and F50+M treatments, the mycorrhizal inoculation (40 kg ha^{-1}) was also provided by using a commercially available inoculant (Xedaopen®, Xeda s.r.l., Forlì, Italy), containing 7 active propagules g^{-1} of the genus *Glomus* spp. and *Gigaspora* spp., as guaranteed by the manufacturer. The inoculation was manually carried out by placing the microgranules of 1.5 mm directly beneath the tuber seed at sowing. The cultivars utilized in this research differ for their morphological, biological, physiological and productive characteristics. 'Mondial' (Spunta × VE66-295) is a Dutch B cooking type (by EAPR classification) cultivar with high tuberification speed, medium to high vigour and late cycle. 'Arizona' (UK 150-19D22 × Mascotte) is a new Dutch AB cooking type cultivar with medium to late cycle, high vigour and low resistance to common scab. 'Universa' (Agata × 88F164.1) is a French AB cooking type cultivar, very common in Sicily, with high potential yield, medium vigour and early to medium cycle. All cultivars are skin and flesh yellow coloured, and rather used for production of early' potato crop.

The experimental fields have been cultivated in a potato–lettuce–carrot rotation over the last twenty years, as commonly in the cultivation area. Obliviously, the previous crop was carrot and the three locations were fertilized with the same dose of NPK (120, 80 and 130 kg ha^{-1}). In the three locations, tillage consists of a 30 cm depth ploughing followed by harrowing in October. Disease-free, no-pre-sprouted "seed" tubers, from a single seed lot, were manually planted on January 6th 2017 in the three experimental fields. Whole tubers were planted at 0.24 m intervals in rows and 0.75 m apart, corresponding to a planting density of 5.55 plants m^{-2}. Each sub-plot size was 4.2 × 4.2 m and consisted of six rows. The two external rows and two plants on each row-end were used as border to minimize contamination from adjacent treatments. The two middle rows per plot were harvested to assess the yield when about 70% of leaves were dry (121 days after planting, DAP). Drip irrigation was provided once the accumulated daily evaporation rate (derived from measurements of an unscreened class A-Pan evaporimeter) had reached about 30 mm. Over the crop cycle, about 180 mm irrigation water was provided by five applications. Weed and pest control followed current EU regulations (Regulation CE 834/2007, 889/2008, 967/2008, 1235/2008 and 1254/2008) for organic farming.

2.3. Crop Physiology, Measurements and Calculations

The physiological variables detected in the present study were the photosynthetic rate (P_r), the stomatal conductance (g), the chlorophyll (Chl) content and the Chl fluorescence parameters F_v/F_m and F_v/F_0. Concerning these ratios, F_0 is the initial fluorescence (the basal emission of Chl fluorescence when redox components of photosystems are fully oxidised), F_m is the maximum fluorescence (the situation under fully saturated irradiance, when the electron acceptor Q_A is fully reduced) and F_v is the variable fluorescence (the reduction at a given time of the primary electron acceptor, which, in the oxidised state, quenches fluorescence) [44,45], calculated as F_m-F_0. The F_v/F_m ratio is considered a measure of the photochemical efficiency of the electron transport in photosystem II (PSII) and it is well correlated with the quantum yield of net photosynthesis [46]. The F_v/F_0 ratio is a more sensitive parameter than F_v/F_m since exhibits a higher dynamic range, given that both components are considered at any time and thus it is very fast in response [47].

In each location, three physiological measurements (each one performed over three consecutive days in the 3 locations) during the potato plant growth were made from the youngest fully expanded leaf (usually the 3rd or 4th leaf from the apex) and at the same hours (10:00–12:00, local solar time). In particular, they were determined at 81, 91 and 99 DAP in location I, at 82, 92 and 100 DAP in location II and at 83, 93 and 101 DAP in location III. For simplicity, measurement date (M) will be indicated as M1, M2 and M3 for all the locations. At each time point, measurements per each fertilization management treatment and cultivar were taken in duplicate on the same leaves of five plants (10 readings per sub-plot), previously marked for the purpose. Chl fluorescence parameters were detected with a with a portable fluorescence induction monitor (Fi_m 1500; Alma Group Company, Hoddesdon, Herts, UK) by applying a clip on the terminal of full sun-exposed leaflets after a 20 min dark adaptation period. Chl fluorescence measurements were carried out with saturation irradiance up to 3000 µmol m^{-2} s^{-1}. Leaf SPAD absorbance readings (correlated to Chl content) were detected by using a portable absorbance-based Chl meter (SPAD-502 model, Konica Minolta, Sakai, Osaka, Japan). P_r and g were measured by a LI-6200 closed gas-exchange system (LI-Cor Inc., Lincoln, NE, USA) previously calibrated according to manufacturer's instructions. Instantaneous gas-exchange measurements were taken on the same leaves previously used for Chl fluorescence measurements inside a 250 cm^3 chamber in the closed-circuit mode. Days on which P_r was measured were typically clear sunny and characterized by a PAR $\leq$ 1800 µmol photons m^{-2} s^{-1}. Air temperatures varied only slightly during each measuring hour, but ranged between 18 and 24 °C during the period of measurements.

2.4. Crop Yield and Its Components

In each location, for the determination of yield and its components, tubers (from each sub-plot and replicate) were harvested manually when about 70% of leaves and haulms were fully desiccated (i.e., at 120, 122 and 124 DAP in location I, II and III, respectively), and the number and weight of both marketable and unmarketable tubers per plant were determined. Tubers, which were greened, misshapen or displayed pathological damage were classed as unmarketable, as well as those with weight lower than 20 g. This allowed the calculation of the number of marketable tubers per plant (NMTP), average marketable tuber weight (AMTW) and marketable yield (MY). The yield of unmarketable tubers was very low (below 2.0%) and hence excluded from the data.

2.5. Tuber Dry Matter Determination

In laboratory, for each location, a sub-sample of 15 marketable tubers per replicate was washed with tap water, dried with tissue paper, diced and immediately oven-dried at 65 °C (Binder, Milan, Italy), until a constant weight was reached, in order to determine the tuber dry matter percentage (TDMP).

2.6. Soil Sampling and DNA Extraction

In this study, the qRT-PCR conjugated with the fluorescent SYBR Green I dye was used to quantify the AMF *Glomus* spp. and *Gigaspora* spp. in soil. In each location, three soil samples for each sub-plot [each adjacent (±5 cm distance) to a standing plant and weighting 500 g] were collected from the first 20 cm layer, excluding the outer 3 m of each plot and the non-homogeneous areas, by taking care not to include weeds. Then they were sieved through 2 mm pores and kept frozen at −20 °C for DNA extraction. Each soil sample derived from the composition of three soil cores, giving a total of 81 cores (3 plots × 3 cultivars × 3 locations × 3 cores). The extraction of soil DNA was carried out following Scavo et al. [48]. The purified DNA was stored at −20 °C until RT-PCR amplification. Purified DNA was quantified spectrophotometrically (all with 260:280 ratios above 1.7).

2.7. Real-Time Quantitative PCR Assay of Soil DNA Extracts

The qRT-PCR is a very powerful and sensitive technique to determine the amount of PCR product. The absolute quantification method was used to analyze data from RT-quantitative PCR experiments. Absolute quantification determines the input copy number of the gene of interest, usually by relating the PCR signal to a standard curve [49]. A DNA-binding dye, such as SYBR Green, binds to all double-stranded DNA in PCR, causing fluorescence of the dye. An increase in DNA product during PCR therefore leads to an increase in fluorescence intensity and is measured at each cycle, thus allowing DNA concentrations to be quantified. In qRT-PCR assay, a positive reaction is detected by accumulation of a fluorescent signal. The Ct (cycle threshold) is defined as the number of cycles required for the fluorescent signal to cross the threshold (i.e., exceeds background level). Ct levels are inversely proportional to the amount of target nucleic acid in the sample.

In this study, a iCycler iQ^{TM}5 (BIORAD) detection system was used. Reactions were 25-µL volumes using Platinum Quantitative PCR Supermix-UDG (Invitrogen, Carlsbad, CA, USA). Two sets of fungal 28S rDNA primers were used to amplify *Glomus* spp. and *Gigaspora* spp. For *Glomus* spp. Glofor (5′-GAAGTCAGTCATACCAACGGGAA-3′) and Glorev (5′-CTCGCGAATCCGAAGGC-3′) oligonucleotides, flanking a 101 bp DNA fragment, were used (Alkan et al., 2006). For *Gigaspora* spp. the primer pair Gigfor (5′-CTTTGAAAAGAGAGTTAAATAG-3′) and Gigrev (5′-GTCCATAACCCAACACC-3′) was used to generate a DNA product of 272 bp [50].

The conditions for *Glomus* spp. DNA template amplification were initial denaturation at 95 °C for 10 min followed by 40 cycles at 95 °C for 15 s, 62 °C for 30 s, 72 °C for 30 s. For testing the primers, *Glomus mosseae* (BEG12) was directly used as a source of DNA template in a 25-µL reaction. The optimized cycling conditions for *Gigaspora* spp. were established as follows: initial DNA denaturation at 95 °C for 15 min, then 45 cycles each with denaturation at 95 °C for 10 s, annealing at 48 °C for 30 s and elongation at 72 °C for 1 s. For testing the primers, *Gigaspora margarita* (BEG34) was used. The same strains were used as standard for calibration curves and the subsequent calculation of their amount. Threshold cycle (Ct) values were determined, in triplicate, using 2-µL samples of each soil DNA extract per PCR reaction. The concentrations of fungal genomic DNA in soil experimental treatments were calculated by comparing the Ct values to the crossing point values of the linear regression line of the standard curve.

2.8. Statistical Analysis

Productive, physiological and microbiological data were analysed statistically through analysis of variance (ANOVA) by using the CoStat® computer package version 6.003 (CoHort Software, Monterey, CA, USA). Untransformed data are reported and presented as means ± standard deviation.

Concerning yield data, a three-way ANOVA 'fertilization management × cultivar × location' was used and, when needed, two-way ANOVAs were performed at each location. To remedy deviations from the ANOVA basic assumptions, data NMTP were square-root transformed, while an arcsine-square root transformation was applied to tuber dry matter percentage. Then, homoscedasticity was verified with the Bartlett's test and normality through a graphical inspection of the residuals, which showed not significant deviations. Pairwise mean comparisons were carried out with the Fisher's protected Least Significant Difference (LSD) test at $\alpha = 0.05$. Data about potato plant physiology were initially analysed according to a four-way ANOVA factorial model with '3 fertilization managements', '3 cultivars', '3 locations' and '3 measurement dates' as the main factors. Since the four-way ANOVA showed a high significance ($p \leq 0.001$) of location for all the variables under study, data were processed according to a generalised mixed model with 'fertilization management', 'cultivar' and 'measurement date' as the main factors and 'location' as random factor [51]. In some cases, two-way ANOVAs for each measurement date were conducted. Since these data did not show any violation of basic assumptions, they were not transformed before ANOVA. Following the procedure of yield, a three-way factorial ANOVA model 'fertilization management × cultivar × location' was applied to the statistical analysis of microbiological data. In accordance with Scavo et al. [52], they were log-transformed prior to ANOVA for the homogeneity of variances.

3. Results

3.1. Mycorrhizal Colonization

ANOVA results showed that mycorrhizal colonization was significantly influenced ($p \leq 0.001$) by main factors and even their interactions (Table 3).

Table 3. *F*-values as absolute value of main factors and their interactions resulting from ANOVA of qRT-PCR analysis, yield and its components.

Source of Variation	df	qRT-PCR Analysis		Yield and Its Components			
		Gigaspora spp.	*Glomus* spp.	MY	AMTW	NMTP	TDMP
Main factors							
Fertilization management (F)	2	4641.5 ***	3240.5 ***	22.1 ***	9.3 ***	12.7 ***	5.5 **
Cultivar (C)	2	573.1 ***	342.7 ***	53.4 ***	199.1 ***	11.7 ***	19.9 ***
Location (L)	2	39.1 ***	122.5 ***	219.9 ***	141.9 ***	80.2 ***	13.1 ***
Interactions							
(F) × (C)	4	171.9 ***	30.5 ***	2.4 [NS]	9.3 ***	5.8 ***	1.2 [NS]
(F) × (L)	4	37.1 ***	61.5 ***	4.5 **	25.2 ***	2.3 [NS]	0.6 [NS]
(C) × (L)	4	18.4 ***	10.4 ***	24.4 ***	23.9 ***	6.5 ***	2.3 [NS]
(F) × (C) × (L)	8	13.8 ***	11.0 ***	2.3 [NS]	3.5 **	1.1 [NS]	0.4 [NS]

Values are given as F of Fisher. *** and ** indicate significant at $p \leq 0.001$ and $p \leq 0.01$, respectively, and [NS], not significant. MY: Marketable yield; AMTW: Average marketable tuber weight; NMTP: Number of marketable tubers plant^{-1}; TDMP: Tuber dry matter percentage.

Overall, the effect of fertilization management accounted for 84.5 and 84.8% of the variance for *Gigaspora* spp. and *Glomus* spp., respectively. In all the locations, DNA extraction by qRT-PCR pointed out that soil samples of F100+M and F50+M were efficiently colonized by both mycorrhiza, which showed a very similar trend. Keeping in mind that Ct levels are inversely proportional to the amount of target nucleic acids in the sample, a decrease of 31.9% and 28.1% of F100+M and F50+M as compared to F100 was observed in location I, of 27.7% and 26.4% in location II, and of 30.7% and 30.9% in location III (Figure 2). Therefore, the trend was also constant in relation to the soil type. It is interesting how the cultivar, explaining 10.3% and 8.9% of the total variance for *Gigaspora* spp. and *Glomus* spp., also significantly affected the mycorrhizal colonization, showing the highest Ct levels (and thus the lowest amount of nucleic acids) for both genera in 'Universa'.

3.2. Marketable Yield and Its Components

ANOVA demonstrated that the productive response of early crop potato, evaluated by MY and yield components, varied in relation to fertilization management, location's soil characteristics and cultivar (Table 3). The use of AMF with half fertilizer doses (F50+M) has led to an increase of 25.5 and 15.1% of MY as compared to F100 and F100+M, respectively (Figure 3). The MY increase highlighted by inoculated (F100+M and F50+M) sub-plots than not inoculated ones (F100) was more marked in location I (on average 217%) than in location II (87%) and III (72.8%), as demonstrated by the significance of the 'fertilization management × location' interaction ($F = 4.5$). Such MY differences can be attributed to the higher NMTP, observed in inoculated sub-plots (on average 7.4) than not inoculated ones (6.3), and, in location I, also to the higher AMTW (Figure 3). The F50+M also caused a reduction of unmarketable yield in location I and III (data not shown). Location provided the largest contribution (66.7%) to variance, followed by cultivar (16%) (Table 3). Overall, the mean MY values were 2.0 and 2.2 fold higher in location II and III than in location I, respectively (39.8 and 42.3 t ha^{-1} vs. 19.7 t ha^{-1}). Although the lowest mycorrhizal colonization detected, 'Universa' interestingly had the highest mean MY (40 t ha^{-1}), followed by 'Arizona' (33 t ha^{-1}) and 'Mondial' (28 t ha^{-1}).

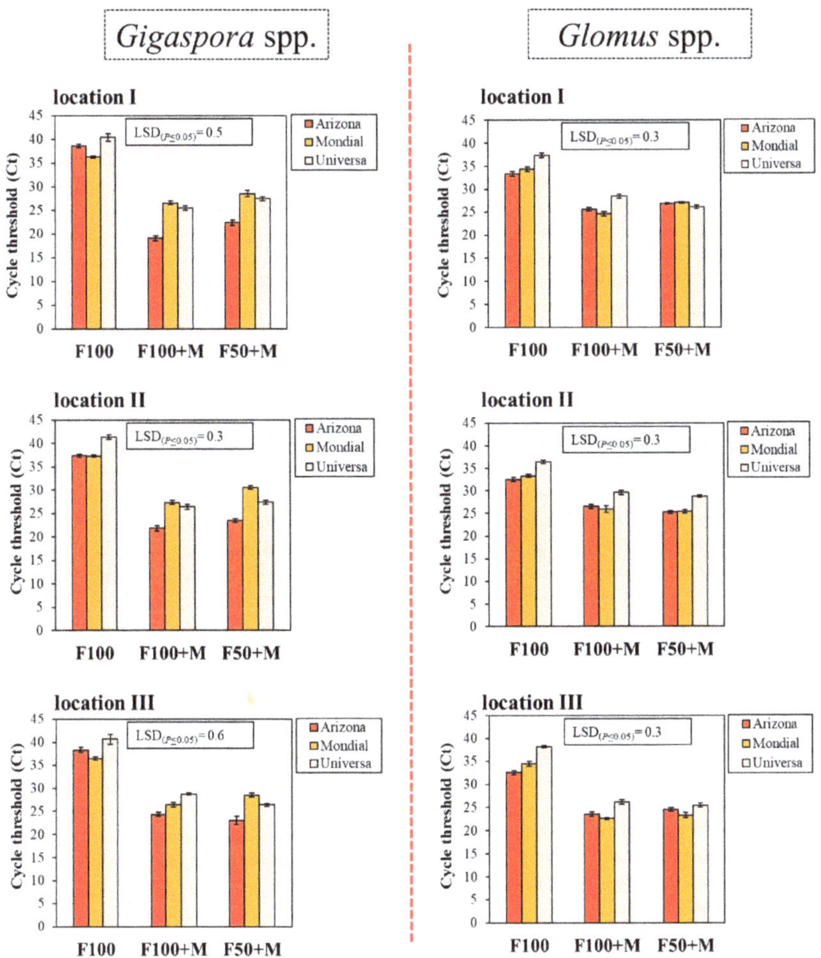

Figure 2. Amount of AMF *Glomus* spp. and *Gigaspora* spp., detected by qRT-PCR, in three different highly calcareous soils cultivated with organic early potato. The Least Significant Difference (LSD) interaction was calculated with the Fisher's protected LSD test at $\alpha = 0.05$. Each bar indicates means ± standard deviation ($n = 3$). F100: optimal fertilization without mycorrhizal inoculation (control); F100+M: optimal fertilization with mychorrhizal inoculation; F50+M: sub-optimal fertilization with mycorrhizal inoculation; 'Arizona', 'Mondial' and 'Universa': potato cultivars. Ct levels are inversely proportional to the amount of target nucleic acids in the soil sample.

Following the trend of MY, higher AMTWs were observed in F50+M (90 g) than in F100+M (86 g) and F100 (82 g), and mainly in location I with an increase of 66% (Figure 3). Similarly, F50+M showed higher mean NMTP (7.8) than F100 (6.3). Regardless of fertilization management and cultivar, the highest AMTW was observed in location III (100 g), while location II expressed the highest NMTP (8.2). Among cultivars, 'Universa' showed the highest AMTW (108 g), while 'Arizona' had the highest NMTP (7.6) and 'Mondial' the highest TDMP (19.3%). As observed for MY and AMTW, TDMP was higher in F50+M than in F100 (18.8% vs. 17.7%, $p \leq 0.005$) and increased by 10% from location I to location III ($p \leq 0.005$) (data not shown).

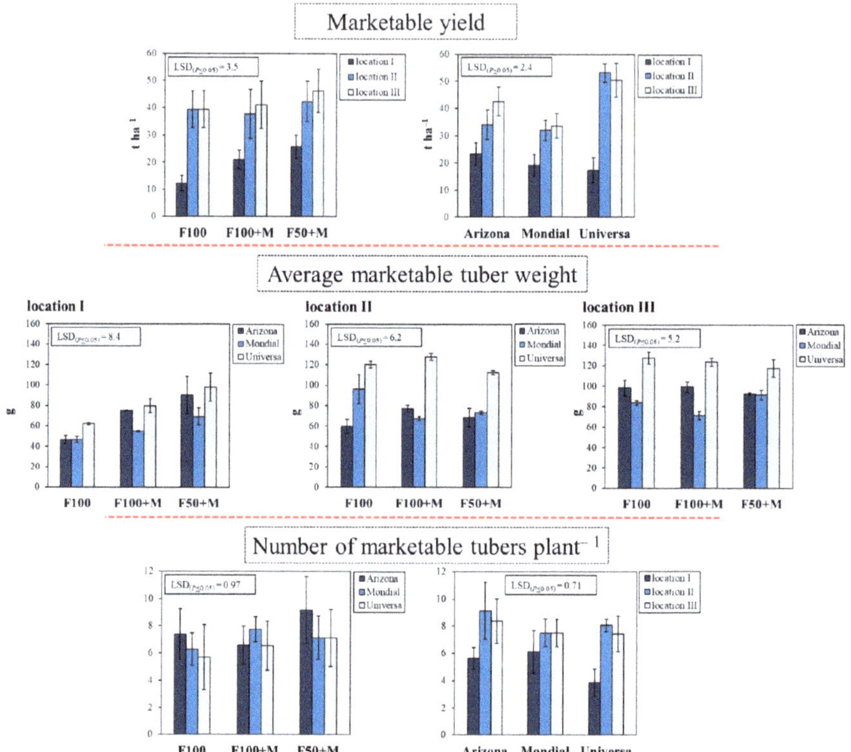

Figure 3. Marketable yield (t ha^{-1}), average marketable tuber weight (g) and number of marketable tubers plant^{-1} of early crop potato organically grown in highly calcareous soils as affected by fertilization management, soil type and potato cultivar. The Least Significant Difference (LSD) interaction was calculated with the Fisher's protected LSD test at α = 0.05. Each bar means ± standard deviation (n = 3). F100: optimal fertilization without mycorrhizal inoculation (control); F100+M: optimal fertilization with mycorrhizal inoculation; F50+M: sub-optimal fertilization with mycorrhizal inoculation; 'Arizona', 'Mondial' and 'Universa': potato cultivars.

3.3. Photosynthesis Rate (P$_r$)

The three-way interaction for P_r was only significant ($p \leq 0.001$) in location I and II (Table 4). In particular, the main factors (namely fertilization management, cultivar and measurement date) significantly affected P_r in the three locations, except for cultivar in location II and fertilization management in location III, with measurement date providing always the major source of variation. The effect of fertilization management and cultivar on P_r was clearly influenced by soil characteristics of each location (Figure 4). In both location I and III, the mycorrhizal inoculation was not consistent. On the contrary, in location II, F50+M determined a significantly higher P_r than F100 at each measurement date. In this location the increase, averaged over all measurement date, was equivalent to 8.3% (11.7 vs. 10.8 µmol CO_2 m^{-2} s^{-1}). The mean P_r was highest for 'Mondial' in location I (11.2 µmol CO_2 m^{-2} s^{-1}), for 'Universa' in location II (11.3 µmol CO_2 m^{-2} s^{-1}) and for 'Arizona' in location III (10.1 µmol CO_2 m^{-2} s^{-1}). Moreover, regardless of fertilization management, P_r significantly decreased with plant age in each location. Indeed, P_r declined by 32.6–10.2 and 21.3% respectively in location I, II and III passing from M1 to M3.

Table 4. *F*-values as absolute value of main factors and their interactions resulting from ANOVA of photosynthesis rate (Pr), chlorophyll content (Chl), Chl fluorescence parameters and stomatal conductance (*g*).

	Location I						
	Main Factors				**Interactions**		
	Fertilization Management (F)	Cultivar (C)	Measurement Date (M)	(F) × (C)	(F) × (M)	(C) × (M)	(F) × (C) × (M)
degrees of freedom	2	2	2	4	4	4	8
P_r	3.3 *	8.7 ***	208.0 ***	8.6 ***	3.6 **	14.6 ***	6.8 ***
Chl content	66.1 ***	0.3 NS	32.9 ***	4.9 **	18.7 ***	12.1 ***	1.6 NS
F_v/F_m	0.1 NS	7.3 **	0.5 NS	6.7 ***	1.5 NS	2.5 *	1.1 NS
F_v/F_0	2.9 NS	7.4 **	0.9 NS	14.8 ***	5.7 ***	3.3 *	4.4 ***
g	35.8 ***	40.3 ***	94.7 ***	3.1 *	3.1 *	2.7 *	2.3 *
	Location II						
P_r	21.0 ***	2.4 NS	64.2 ***	6.6 ***	1.3 NS	4.1 **	12.3 ***
Chl content	11.8 ***	44.0 ***	144.2 ***	29.5 ***	2.6 *	10.6 ***	1.0 NS
F_v/F_m	1.2 NS	8.2 ***	1.0 NS	1.4 NS	0.6 NS	1.3 NS	3.6 **
F_v/F_0	0.4 NS	36.2 **	5.8 **	9.6 ***	8.9 ***	12.6 ***	13.1 ***
g	21.2 ***	30.3 ***	33.1 ***	10.2 ***	4.3 **	10.9 ***	3.1 **
	Location III						
P_r	0.2 NS	6.3 **	120.6 ***	0.8 NS	2.5 NS	1.3 NS	0.8 NS
Chl content	15.1 ***	96.6 ***	317.5 ***	19.9 ***	6.0 ***	9.7 ***	1.7 NS
F_v/F_m	0.3 NS	2.3 NS	4.3 *	0.5 NS	1.5 NS	0.7 NS	0.8 NS
F_v/F_0	4.1 *	5.1 **	9.1 ***	3.1 *	4.9 **	8.7 ***	5.4 ***
g	24.7 ***	26.4 ***	32.7 ***	6.3 ***	5.8 ***	7.2 ***	4.4 ***

Values are given as *F* of Fisher. ***, ** and * indicate significant at $p \leq 0.001$, $p \leq 0.01$ and $p \leq 0.05$, respectively, and NS, not significant.

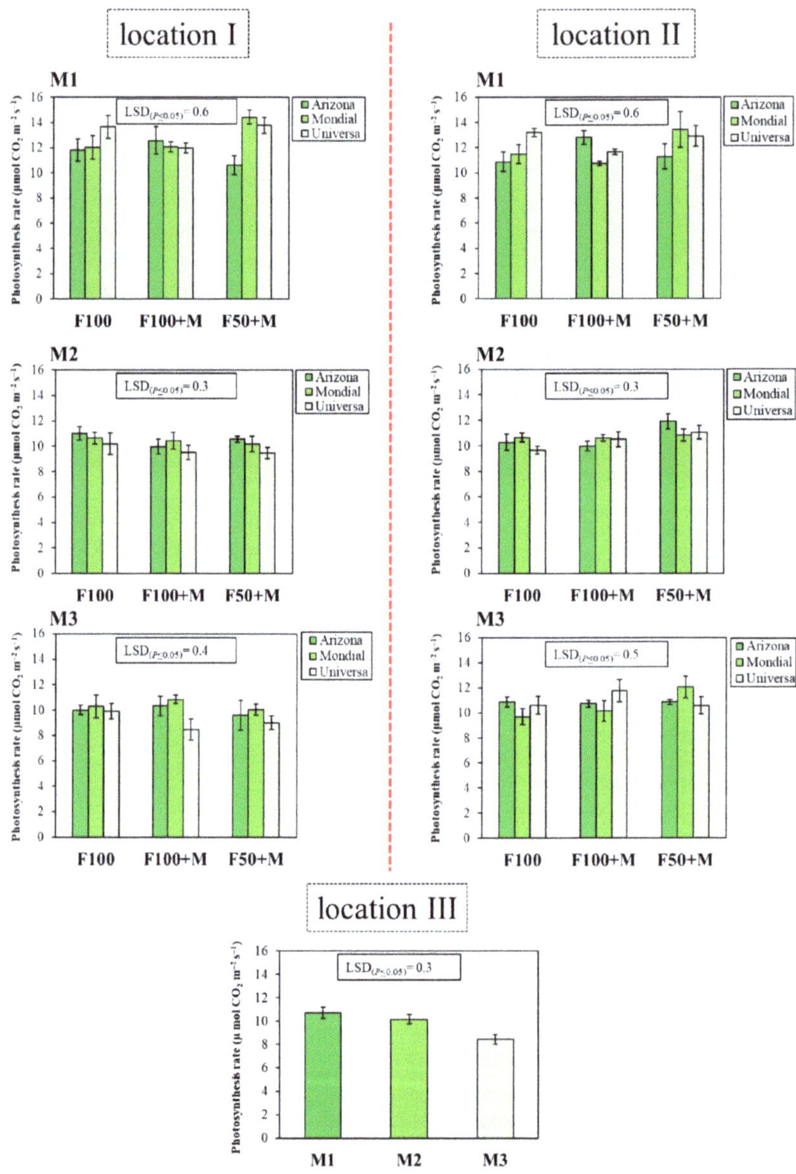

Figure 4. Photosynthetic rate (μmol CO$_2$ m^{-2} s^{-1}) of leaves of early crop potato organically grown in highly calcareous soils as affected by fertilization management, soil type and potato cultivar. The Least Significant Difference (LSD) interaction was calculated with the Fisher's protected LSD test at α = 0.05. Each bar indicates means ± standard deviation (n = 5). F100: optimal fertilization without mycorrhizal inoculation (control); F100+M: optimal fertilization with mycorrhizal inoculation; F50+M: sub-optimal fertilization with mycorrhizal inoculation; 'Arizona', 'Mondial' and 'Universa': potato cultivars. M1, M2 and M3: first, second and third measurement date.

3.4. Chl Content and Chl Fluorescence

The effect of fertilization management provided the largest source of variation for Chl content in location I, while becoming less significance for the other two locations in favour of the measurement date as index of plant age (Table 4). At each measurement date, on the average of cultivars, F50+M plants had the lowest Chl content in the three locations (Figure 5). The differences among the studied fertilization management treatments were mainly evident in location I, given that the Chl content, averaged over the measurement dates, was 47.3 for F100, 45.8 for F100+M and 43.1 for F50+M. In accordance with P_r, Chl content declined with increasing plant age in each location. This was particularly highlighted in location III, where from M1 to M3 the Chl values decreased by 20%. The effect of cultivar was not significant in location I, while in the other two locations 'Universa' showed the highest Chl content (39.6 and 43.0 in location II and III), followed by 'Mondial' and 'Arizona' (37.2).

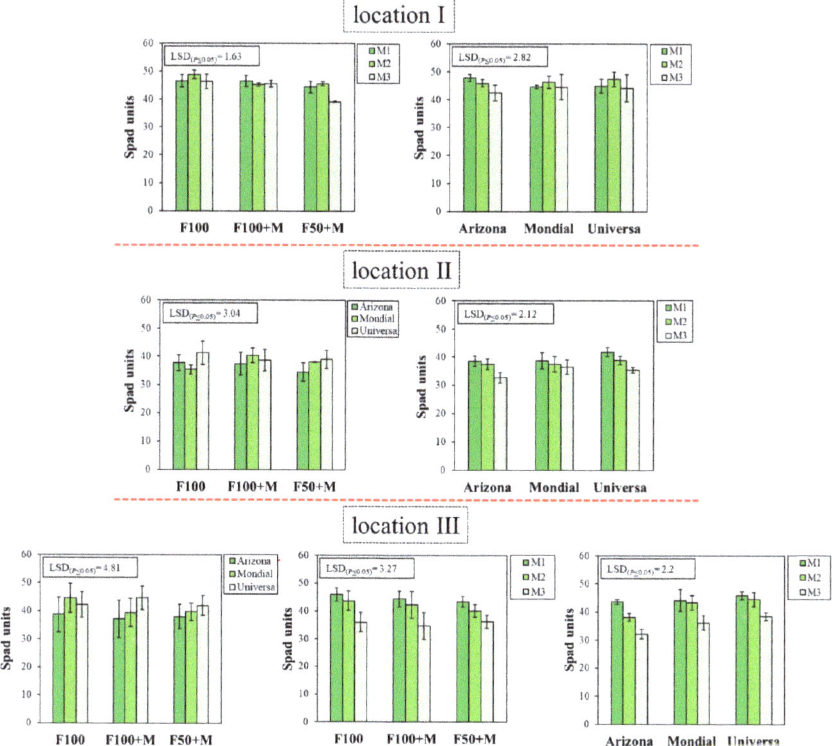

Figure 5. Chlorophyll content (spad units) of leaves of early crop potato organically grown in highly calcareous soils as affected by fertilization management, soil type and potato cultivar. The Least Significant Difference (LSD) interaction was calculated with the Fisher's protected LSD test at $\alpha = 0.05$. Each bar indicates means ± standard deviation ($n = 5$). F100: optimal fertilization without mycorrhizal inoculation (control); F100+M: optimal fertilization with mycorrhizal inoculation; F50+M: sub-optimal fertilization with mycorrhizal inoculation; 'Arizona', 'Mondial' and 'Universa': potato cultivars. M1, M2 and M3: first, second and third measurement date.

The F_v/F_m ratio was not relevant in location I and III, while showing significant results in location II. In particular, it was higher in the mycorrhizal inoculated plots for each measurement date and particularly for 'Mondial'. More clear results were observed for the F_v/F_0 ratio, for which the three-way interaction was significant at $p \leq 0.001$ in each location (Table 4). Except for location II, it was significantly higher in F50+M plants respect to F100 ones (+20% in location I and +8% in location III) (Figure 6). Furthermore, location II and III exhibited more than 4-fold higher values of the F_v/F_0 ratio than location I, thus pointing out how soil characteristics closely affected this parameter. Cultivar was the main cause of variance in location I (19%) and II (42%), and the second one in location III (12.5%) after the measurement date (22.4%) (Table 4). The performances of the three cultivars were markedly different in relation to soil characteristics for which, averaged over the other factors, 'Arizona' had the highest F_v/F_0 in location I (0.64), 'Mondial' in location II (2.49) and 'Universa' in location III (2.12), even with not statistical differences with 'Arizona' (Figure 6).

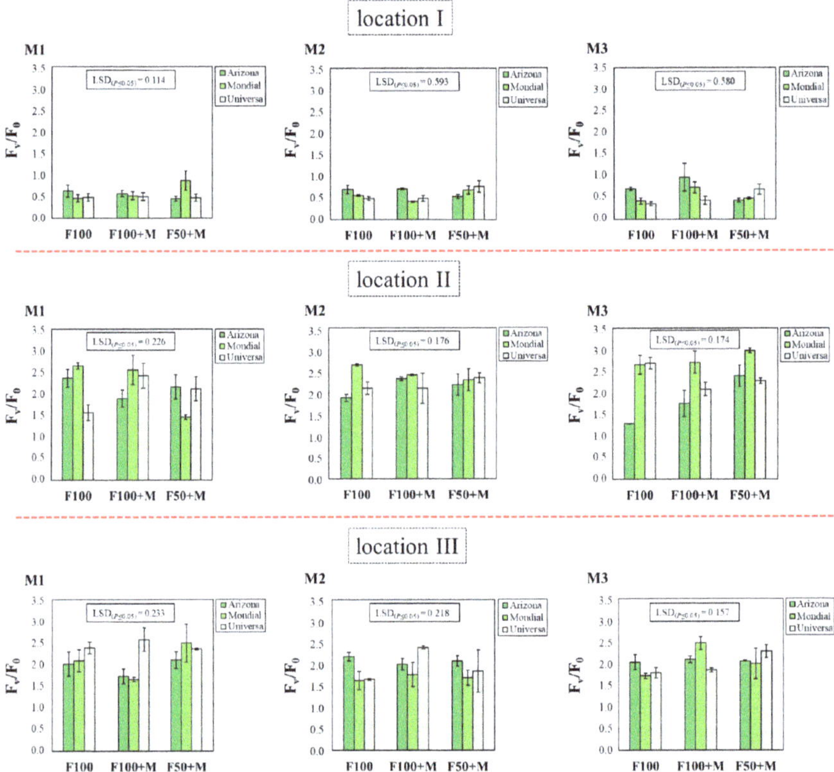

Figure 6. F_v/F_m ratio of leaves of early crop potato organically grown in highly calcareous soils as affected by fertilization management, soil type and potato cultivar. The Least Significant Difference (LSD) interaction was calculated with the Fisher's protected LSD test at $\alpha = 0.05$. Each bar indicates means ± standard deviation ($n = 5$). F100: optimal fertilization without mycorrhizal inoculation (control); F100+M: optimal fertilization with mycorrhizal inoculation; F50+M: sub-optimal fertilization with mycorrhizal inoculation; 'Arizona', 'Mondial' and 'Universa': potato cultivars. M1, M2 and M3: first, second and third measurement date.

3.5. Stomatal Conductance (g)

ANOVA showed a high significance ($p \leq 0.001$) of main factors for each location (Table 4). As observed for P_r, the measurement date contributed to the largest part of the overall variance (by 52.0–29.3 and 30.3% in location I, II and III, respectively), followed by cultivar (22.1%, 26.8% and 24.6%, respectively). With reference to the effect of fertilization management, the mycorrhizal inoculated plots (i.e., F100+M and F50+M) significantly lowered g in location I (0.34 and 0.29 vs. 0.40 mol H_2O m^{-2} s^{-1}

of F100) (Figure 7). On the contrary, F100+M and F50+M caused a significantly higher g than F100 both in location II and III. Concerning the effect of measurement date, g changed based on location. In location I, the highest g was observed at M1, while in location II and III at M2 and M3, respectively. In each location, regardless of fertilization management and cultivar, g showed an opposite trend than P_r, with increasing values at declining plant age. This was particularly marked in location III, which for g reported an increase by 53% from M1 to M3. Moreover, g was higher for 'Mondial' in each location (on average 0.44 mol H_2O m^{-2} s^{-1}), while 'Arizona' and 'Universa' recorded values not statistically different.

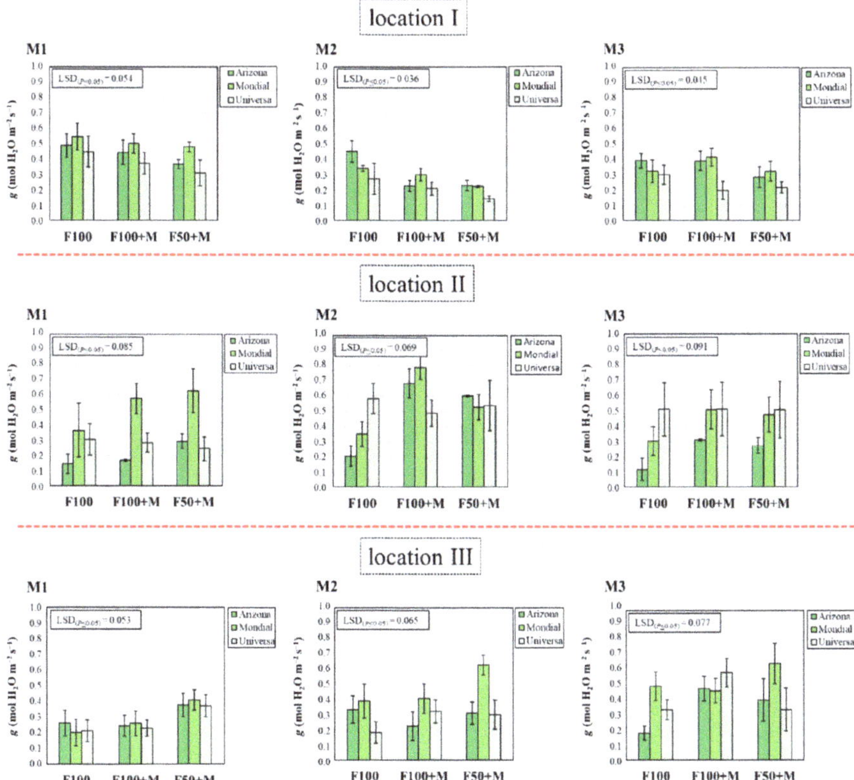

Figure 7. Stomatal conductance (mol H_2O m^{-2} s^{-1}) of leaves of early crop potato organically grown in highly calcareous soils as affected by fertilization management, soil type and potato cultivar. The Least Significant Difference (LSD) interaction was calculated with the Fisher's protected LSD test at α = 0.05. Each bar indicates means ± standard deviation (n = 5). F100: optimal fertilization without mycorrhizal inoculation (control); F100+M: optimal fertilization with mycorrhizal inoculation; F50+M: sub-optimal fertilization with mycorrhizal inoculation; 'Arizona', 'Mondial' and 'Universa': potato cultivars. M1, M2 and M3: first, second and third measurement date.

4. Discussion

The changes in terms of yield, including its components, and physiological traits in early potato crop organically grown in relation to mycorrhizal colonization and cultivar were studied in three different locations. The latter (namely location I, II and III) were characterized by highly calcareous and alkaline soils, as common in the coastal-areas of the Mediterranean basin. Such conditions are usually referred to affect negatively crop yield by impairing the availability and uptake of minerals, especially

P, also when additionally added by inorganic based-P fertilizers due to the rapid transformation into stable minerals relatively unavailable to crops [53]. Location I was characterised by the highest active limestone content (28%), the highest pH (8.0) and the lowest P_2O_5 (28.5 mg kg^{-1}) and K_2O (197 mg kg^{-1}) levels, thus offering the most detrimental conditions for potato growth. Location III showed opposite characteristics to location I, i.e., lower active limestone content (18%), the lowest pH (7.5) and the highest P_2O_5 (135 mg kg^{-1}) and K_2O (612 mg kg^{-1}) levels, while location II had intermediate conditions.

To our knowledge, few information can be found in literature regarding the AMF effect on potato crop in highly calcareous soil. In our research, the qRT-PCR, carried out on DNA extracted by soil samples, revealed that mycorrhizal colonization occurred with similar percentages in each location, independently from the soil characteristics, demonstrating that in the specific conditions of this research AMF were able to occur in highly calcareous soils, alkaline and with high P levels. Our results are in contrast with those commonly reported in literature indicating a negative relationship between soil P level and mycorrhizal colonization [29], but at the same time, are consistent with Sylvia and Schenc [54] and Alkan et al. [55], who observed that AMF differ in their ability to P-tolerance, with *G. mosseae* showing a high sensitivity to increasing levels of P. Interestingly, in our study AMF were also detected in not-inoculated sub-plots in accordance to Hijri [36], reporting a dataset of 231 field trials with AMF-potato associations. This should not be surprising since AMF naturally occur in field soils but their abundance, diversity and time needed for the establishment can be negatively affected by crop management practices both directly or indirectly [56]. Generally, mycorrhizal colonization is favoured under organic farming. For instance, in a long-term field trial Mäder et al. [31] found that AM root colonization of vetch-rye, winter wheat and grass-clover crops was 30–60% higher in low-input farming systems than in the conventional ones. However, in some cases (e.g., due to soil with P concentrations too high caused by high P fertilizers doses, excessive tillage for weed control or plant diseases) the performance of AMF is low, likely because the benefit received by modern cultivars from mycorrhizal association is poor [56]. Regarding potato cultivar differences, the lowest level of colonization of both mycorrhizal genera was detected in 'Universa' plots due to its high potential yield, while 'Arizona' was more colonized by *Gigaspora* spp. and 'Mondial' by *Glomus* spp.

In accordance with previous findings [34–36], this research outlined a significant increase of potato MY, AMTW, NMTP and TDMP due to mycorrhizal colonization. The mycorrhizal inoculation with halved fertilizers dose (F50+M) showed the best results in terms of yield and its components in all the locations. However, the application of full fertilizer doses (F100+M) reported worse results than halved-inoculated sub-plots, likely due to a negative impact of the full dosage on AMF. The high amounts of organic amendments fertilizers, which are generally high in P, is reported to negatively affect the AMF symbiosis with crops [56], and probably this was amplified in the high P soils of the present research. The average yields obtained here with F50+M (38 t ha^{-1}) were higher than those obtained by Lombardo et al. [7] (~20 t ha^{-1}) and Maggio et al. [4] (16 t ha^{-1}) under organic farming management. Douds et al. [34] reported higher yields and larger tubers with commercial inoculants of *G. intraradices* than by using conventional chemical fertilizers. The enhancement in potato tuber production by AMF inoculation could be attributable to many reasons, but most of researches indicate the increased nutrient uptake, mainly P due to the ability of mycorrhizal fungal hyphae to acquire P well beyond the limits of the rhizosphere, and the disease resistance to *Fusarium* spp. [33,57] as the most reasonable ones. In our specific field conditions, we also hypothesize an increased tolerance to active limestone and alkalinity, as found by Romero-Munar et al. [58] for salinity in *Arundo donax* L., through ion homeostasis, vacuoles-compartmentalization and Na$^+$ translocation, as suggested by Ruiz-Lozano and Azcón [59]. Moreover, as originally supposed, the highest MY and AMTW were found in location III, which offered better soil conditions for potato growth and AMF, but the highest yield increase, caused by F50+M, was found in location I (109%), compared to an increase of 8% and 17% in location II and III, respectively. The lower AMF efficiency in location III may be ascribed to its high soil P_2O_5 level, since the potential for a mycorrhiza-mediated growth benefit decreases as soil P increases [32]. According to our data, AMF showed the highest efficiency applying halved fertilizer doses and in the location where soil conditions were unfavourable for potato growth. It should also be noted that 'Universa' reported the highest MY and AMTW, despite the mycorrhizal colonization was the lowest. Cultivar differences in response to AMF inoculation, that in field are attributable to a number of factors, were observed both for potato [34] and other crops such as wheat [60], barley [61], white clover [62], globe artichoke [63], etc.

The beneficial effects of AMF on potato yield and its components were also consistent in terms of physiological traits. Indeed, under the specific conditions in which the experiment was conducted, AMF increased the stomatal conductance in location II and III, enhanced the photochemical efficiency and improved the photosynthesis rate, even if statistical significance was only recorded in location II. The poor response in P_r could be attributable to imbalance in the energetic status of the plant, since the light energy absorbed by chlorophylls can be used for photosynthesis, re-emitted as light-chlorophyll fluorescence or dissipated by heat, and these three processes are competitive to each other's [64]. Soil abiotic factors are reported to inhibit the photosynthetic processes by over-reducing the reaction centres in PSII or inhibiting specific enzymes involved for the synthesis of photosynthetic pigments, thus causing a reduction in plant chlorophyll content [65]. Among soil abiotic factors, salinity is the most common and discussed in literature [64], but also calcareous soils are reported to be negatively correlated to the plant's photosynthetic machine [66]. AMF are able to ameliorate salt stress by improving the photosynthetic activity, the photochemical properties, the source-sink ratio or the water use efficiency [64,65]. Mycorrhizal-inoculated rice plants in saline soils were found to present a higher photochemical efficiency for CO_2 fixation and solar energy utilization than not-inoculated plants through an increase in actual quantum yield of PSII photochemistry, net photosynthetic rate, stomatal conductance and transpiration rate as well as by stimulating carbohydrate transport and metabolism between source and sink tissues [64,67]. A similar behaviour was found by Hajiboland et al. [68] in tomato. Studying the influence of AMF symbiosis in *A. donax* grown under low P availability, Romero-Munar et al. [58] indicated that AMF conferred salt tolerance by enhancing nutrient use efficiency rather than nutrient uptake: worse Na^+ uptake, Na^+ root-to-shoot translocation and Na^+/K^+ ratio, and better P and K use efficiencies. The authors also reported that the mycorrhizal symbiosis ameliorated the response of *A. donax* to combine low P and mild salinization conditions, and that the plant growth was driven by salinity rather than P availability. Under our experimental conditions, there was also an effect of concurrent abiotic stresses on plant growth caused by the high active limestone content, alkalinity, low nutrient efficiency (mainly for P and Fe) and high organic matter mineralization. Since soil colonization by *Gigaspora* spp. and *Glomus* spp. was found in the three locations with very similar results, we hypothesize a tolerance of AMF inoculates to the above-mentioned abiotic factors, at least in terms of primary colonization, while the effects on secondary colonizations are unknown. To better understand the photosynthetic ability and energy conversion efficiency to abiotic stress, the PSII photochemical efficiency has been studied through the F_v/F_m and F_v/F_0 ratios. According to Pinior et al. [69], under disturbance of biotic or abiotic stresses, the plant dissipates its redundant energy to avoid damage of tissues, and such dissipation can occur via heat or chlorophyll fluorescence. The potential quantum efficiency of PSII is widely reflected with the F_v/F_m ratio. It did not response well in this research, probably due to high variability of field conditions. For this reason, we calculated the F_v/F_0 ratio which is a more sensitive and dynamic parameter to better investigate the PSII efficiency [70]. Indeed, since F_m represents the sum of F_v and F_0, the F_v/F_m ratio is slow when F_v slightly decreases and F_0 slightly increases [47]. Its mean values were very low in location I (~0.3–0.8), when the stress conditions were higher and impaired the PSII electron transport, and increased in the other two locations to optimal values. Except for location II, AMF increased F_v/F_0 by 11.7% in location I and 8.2% in locations III, indicating a better performance in more detrimental soil conditions, as demonstrated by the higher yield increase in the same location.

5. Conclusions

To summarize, the results obtained in the present research demonstrated that AMF inoculation is a useful tool for enhancing early potato yield and physiological traits in highly calcareous and alkaline soils. These results are particularly noticeable by providing halved fertilization doses to the potato crop. Despite the detrimental soil conditions, the qRT-PCR highlighted that AMF colonized all the experimental soils, showing good tolerance to high active limestone, pH and P levels. Furthermore, AMF ameliorated the early potato tolerance to such abiotic stresses by increasing the plant's gas-exchange capacity and the PSII photochemical efficiency. These findings are of key importance not only for improving the yield of early potato under organic farming, but also for the sustainable management of fertilization by halving the doses with better results in terms of production and environmental impact, as well as for the possibility of a profitable potato cultivation in the coastal agricultural areas of the Mediterranean basin and other arid or semi-arid regions. To this end, further investigations will be necessary to clarify the mycorrhizal association with potato, particularly the role of indigenous AMF communities in this process, as well as to investigate other possible integrations of mycorrhizal-based inoculants with other agronomic practices.

Author Contributions: Conceptualization, G.M., B.P. and S.L.; methodology, A.S., S.L., C.A., G.P., B.P. and G.M.; investigation, S.L., C.A., A.S., G.P. and G.M.; data curation and statistical analysis, A.S. and S.L.; writing—original draft preparation, A.S., S.L. and C.A.; writing—review and editing, A.S., S.L., G.P. and C.A.; supervision, G.M. All authors have read and agreed to the published version of the manuscript.

Funding: This research was funded by "Fondi di ateneo 2020–2022, Università di Catania, linea Open Access".

Acknowledgments: The authors gratefully acknowledge the BIOVERDE S.S. farm (Rosolini, Italy) for hosting the field experimental trials, as well as XEDA Italia s.r.l. (Forlì, Italy) for kindly providing the agrochemical products. Thanks are also due to Umberto Burdieri and Angelo Litrico for their excellent technical assistance in conducting the field trials.

Conflicts of Interest: The authors declare no conflict of interest.

References

1. Yiridoe, E.K.; Bonti-Ankomah, S.; Martin, R.C. Comparison of consumer perceptions and preference toward organic versus conventionally produced foods: A review and update of the literature. *Renew. Agric. Food Syst.* **2005**, *20*, 193–205. [CrossRef]
2. FiBL-AMI. Statistics on Organic Agriculture. 2018. Available online: http://www.fibl.org/en/themes/organic-farmingstatistics (accessed on 20 April 2020).
3. Willer, H.; Schaack, D.; Lernoud, J. Organic farming and market development in Europe and the European Union. In *The World of Organic Agriculture. Statistics and Emerging Trends*; Willer, H., Lernoud, J., Eds.; FiBL & IFOAM–Organics International: Nürnberg, Germany, 2019; pp. 217–254.
4. Maggio, A.; Carillo, P.; Bulmetti, G.S.; Fuggi, A.; Barbieri, G.; De Pascale, S. Potato yield and metabolic profiling under conventional and organic farming. *Eur. J. Agron.* **2008**, *28*, 343–350. [CrossRef]
5. Bacchi, M.A.; De Nadai Fernandes, E.A.; Tsai, S.M.; Santos, L.G.C. Conventional and organic potatoes: Assessment of elemental composition using k_0-INAA. *J. Radioanal. Nucl. Chem.* **2004**, *259*, 421–424. [CrossRef]
6. Warman, P.R.; Havard, K.A. Yield, vitamin and mineral contents of organically and conventionally grown potatoes and sweet corn. *Agric. Ecosyst. Environ.* **1998**, *68*, 207–216. [CrossRef]
7. Lombardo, S.; Lo Monaco, A.; Pandino, G.; Parisi, B.; Mauromicale, G. The phenology, yield and tuber composition of 'early' crop potatoes: A comparison between organic and conventional cultivation systems. *Renew. Agric. Food Syst.* **2013**, *28*, 50–58. [CrossRef]
8. Lombardo, S.; Pandino, G.; Mauromicale, G. The influence of growing environment on the antioxidant and mineral content of "early" crop potato. *J. Food Compos. Anal.* **2013**, *32*, 28–35. [CrossRef]
9. Lombardo, S.; Pandino, G.; Mauromicale, G. The mineral profile in organically and conventionally grown "early" crop potato tubers. *Sci. Hortic.* **2014**, *167*, 169–173. [CrossRef]
10. Buono, V.; Paradiso, A.; Serio, F.; Gonnella, M.; De Gara, L.; Santamaria, P. Tuber quality and nutritional components of early potato subjected to chemical haulm desiccation. *J. Food Compos. Anal.* **2009**, *22*, 556–562. [CrossRef]

11. Lombardo, S.; Pandino, G.; Mauromicale, G. The effect on tuber quality of an organic versus a conventional cultivation system in the early crop potato. *J. Food Compos. Anal.* **2017**, *62*, 189–196. [CrossRef]

12. Izmirlioglu, G.; Demirci, A. Enhanced bio-ethanol production from industrial potato waste by statistical medium optimization. *Int. J. Mol. Sci.* **2015**, *16*, 24490–24505. [CrossRef]

13. Jagatee, S.; Behera, S.; Dash, P.K.; Sahoo, S.; Mohanty, R.C. Bioprospecting starchy feedstocks for bioethanol production: A future perspective. *JMRR* **2015**, *3*, 24–42.

14. Ierna, A.; Mauromicale, G. Potato growth, yield and water productivity response to different irrigation and fertilization regimes. *Agric. Water Manag.* **2018**, *201*, 21–26. [CrossRef]

15. Canali, S.; Ciaccia, C.; Antichi, D.; Barberi, P.; Montemurro, F.; Tittarelli, F. Interactions between green manure and amendment type and rate: Effects on organic potato and soil mineral N dynamic. *J. Food Agric. Environ.* **2010**, *8*, 537–543.

16. Hepperly, P.; Lotter, D.; Ziegler, C.; Seidel, R.; Reider, C. Compost, manure and synthetic fertilizer influences crop yields, soil properties: Nitrate leaching and crop nutrient content. *Compos. Sci. Util.* **2009**, *17*, 117–126. [CrossRef]

17. Rosen, C.J.; Allan, D.L. Exploring the benefits of organic nutrient sources for crop production and soil quality. *HortTechnology* **2007**, *17*, 422–430. [CrossRef]

18. Caliskan, M.E.; Kilic, S.; Gunel, E.; Mert, M. Effect of farmyard manure and mineral fertilization on growth and yield of early potato (*Solanum tuberosum* L.) under the Mediterranean conditions in Turkey. *Indian J. Agron.* **2004**, *49*, 198–200.

19. Ierna, A.; Parisi, B. Crop growth and tuber yield of "early" potato crop under organic and conventional farming. *Sci. Hortic.* **2014**, *165*, 260–265. [CrossRef]

20. Clark, M.S.; Horwath, W.R.; Shennan, C.; Scow, K.M.; Lantni, W.T.; Ferris, H. Nitrogen, weeds and water as yield-limiting factors in conventional, low-input, and organic tomato systems. *Agric. Ecosyst. Environ.* **1999**, *73*, 257–270. [CrossRef]

21. Lynch, D.H.; Sharifi, M.; Hammermeister, A.; Burton, D. Nitrogen management in organic potato production. In *Sustainable Potato Production: Global Case Studies*; He, Z., Larkin, R., Honeycutt, W., Eds.; Springer: Dordrecht, The Netherlands, 2012; pp. 209–231. [CrossRef]

22. Van Delden, A.; Schroder, J.J.; Kropff, M.J.; Grashoff, C.; Booij, R. Simulated potato yield and crop and soil nitrogen dynamics under different organic nitrogen management strategies in the Netherlands. *Agric. Ecosyst. Environ.* **2003**, *96*, 77–95. [CrossRef]

23. Ierna, A.; Mauromicale, G. Tuber yield and irrigation water productivity in early potatoes as affected by irrigation regime. *Agric. Water Manag.* **2012**, *115*, 276–284. [CrossRef]

24. Mauromicale, G.; Signorelli, P.; Ierna, A.; Foti, S. Effects of intraspecific competition on yield of early potato grown in Mediterranean environment. *Am. J. Potato Res.* **2003**, *80*, 281–288. [CrossRef]

25. Taalab, A.S.; Ageeb, G.W.; Siam, H.S.; Mahmoud, S.A. Some characteristics of calcareous soils. A review. *Middle East J. Agric. Res.* **2019**, *8*, 96–105.

26. Di Gregorio, A. *Land Cover Classification System, Classification Concepts and User Manual*, 3rd ed.; FAO: Rome, Italy, 2016.

27. Koch, M.; Naumann, M.; Pawelzik, E.; Gransee, A.; Thiel, H. The importance of nutrient management for potato production Part I: Plant nutrition and yield. *Potato Res.* **2019**, *63*, 97–119. [CrossRef]

28. Lombardo, S.; Pandino, G.; Mauromicale, G. Optimizing nitrogen fertilization to improve qualitative performances and physiological and yield responses of potato (*Solanum tuberosum* L.). *Agronomy* **2020**, *10*, 352. [CrossRef]

29. Smith, S.E.; Read, D.J. *Mycorrhizal Symbiosis*, 3rd ed.; Academic Press: London, UK, 2008.

30. Scavo, A.; Abbate, C.; Mauromicale, G. Plant allelochemicals: Agronomic, nutritional and ecological relevance in the soil system. *Plant Soil* **2019**, *442*, 23–48. [CrossRef]

31. Mäder, P.; Edenhofer, S.; Boller, T.; Wiemken, A.; Niggli, U. Arbuscular mycorrhizae in a long-term field trial comparing low-input (organic, biological) and high-input (conventional) farming systems in a crop rotation. *Biol. Fertil. Soils* **2000**, *31*, 150–156. [CrossRef]

32. Black, R.L.B.; Tinker, P.B. Interaction between effects of vesicular-arbuscular mycorrhizae and fertilizer phosphorus on yields of potatoes in the field. *Nature* **1977**, *267*, 510–511. [CrossRef]

33. Wu, F.; Wang, W.; Ma, Y.; Liu, Y.; Ma, X.; An, L.; Feng, H. Prospect of beneficial microorganisms applied in potato cultivation for sustainable agriculture. *Afr. J. Microbiol. Res.* **2013**, *7*, 2150–2158. [CrossRef]

34. Douds, D.D., Jr.; Nagahashi, G.; Reider, C.; Hepperly, P.R. Inoculation with arbuscular mycorrhizal fungi increases the yield of potatoes in a high P soil. *Biol. Agric. Hortic.* **2007**, *25*, 67–78. [CrossRef]

35. Duffy, E.M.; Cassells, A.C. The effect of inoculation of potato (*Solanum tuberosum* L.) microplants with arbuscular mycorrhizal fungi on tuber yield and tuber size distribution. *Appl. Soil Ecol.* **2000**, *15*, 137–144. [CrossRef]

36. Hijri, M. Analysis of a large dataset of mycorrhiza inoculation field trials on potato shows highly significant increases in yield. *Mycorrhiza* **2016**, *26*, 209–214. [CrossRef] [PubMed]

37. Lombardo, S.; Pandino, G.; Mauromicale, G. Nutritional and sensory characteristics of "early" potato cultivars under organic and conventional cultivation systems. *Food Chem.* **2012**, *133*, 1249–1254. [CrossRef]

38. Parisi, B.; Govoni, F.; Mainolfi, A.; Baschieri, T.; Ranalli, P. Nuovi cloni italiani per la pataticoltura nazionale. *L'Informatore Agrar.* **2002**, *46*, 41–45.

39. Graham, S.O.; Green, N.E.; Hendrix, J.W. The influence of vesicular-arbuscular mycorrhizal fungi on growth and tuberization of potatoes. *Mycologia* **1976**, *68*, 925–929. [CrossRef]

40. Johnstone, P.D.; Lowther, W.L.; Keoghan, J.M. Design and analysis of multi-site agronomic evaluation trials. *N. Z. J. Agric. Res.* **1993**, *36*, 323–326. [CrossRef]

41. Soil Survey Staff. *Soil Taxonomy: A Basic System of Soil Classification for Making and Interpreting Soil Surveys*, 2nd ed.; US. Gov. Print. Office: Washington, DC, USA, 1999.

42. Italian Society of Soil Science. *Metodi Normalizzati Di Analisi Del Suolo*; Edagricole: Bologna, Italy, 1985.

43. Dierauer, H.; Siegrist, F.; Weidmann, G.; Commercial Organic Fertiliser as Supplementary Fertilisers in Potato Crop Production. Research Institute of Organic Agriculture-FiBL, Practice Abstract. 2017. Available online: https://www.orgprints.org/31027/ (accessed on 20 April 2020).

44. Mauromicale, G.; Ierna, A.; Marchese, M. Chlorophyll fluorescence and chlorophyll content in field-grown potato as affected by nitrogen supply, genotype, and plant age. *Photosynthetica* **2006**, *44*, 76–82. [CrossRef]

45. Mauromicale, G.; Lo Monaco, A.; Longo, A.M.G. Effect of branched broomrape (*Orobanche ramosa*) infection on the growth and photosynthesis of tomato. *Weed Sci.* **2008**, *56*, 574–581. [CrossRef]

46. Maxwell, K.; Johnson, G.N. Chlorophyll fluorescence—A practical guide. *J. Exp. Bot.* **2000**, *51*, 659–668. [CrossRef]

47. Babani, F.; Lichtenthaler, H.K. Light-induced and age-dependent development of chloroplasts in etiolated barley leaves as visualized by determination of photosynthetic pigments, CO_2 assimilation rates and different kinds of chlorophyll fluorescence ratios. *J. Plant Physiol.* **1996**, *148*, 555–566. [CrossRef]

48. Scavo, A.; Restuccia, A.; Abbate, C.; Mauromicale, G. Seeming field allelopathic activity of *Cynara cardunculus* L. reduces the soil weed seed bank. *Agron. Sustain. Develop.* **2019**, *39*, 41. [CrossRef]

49. Livak, K.J.; Schmittgen, T.D. Analysis of relative gene expression data using real-time quantitative PCR and the 2(-Delta Delta C(T)). *Methods* **2001**, *25*, 402–408. [CrossRef] [PubMed]

50. Thonar, C.; Erb, A.; Jansa, J. Real-time PCR to quantify composition of arbuscular mycorrhizal fungal communities-marker design, verification, calibration and field validation. *Mol. Ecol. Resour.* **2012**, *12*, 219–232. [CrossRef] [PubMed]

51. Gomez, K.A.; Gomez, A.A. *Statistical Procedures for Agricultural Research*; John Wiley & Sons: New York, NY, USA, 1984.

52. Scavo, A.; Restuccia, A.; Lombardo, S.; Fontanazza, S.; Abbate, C.; Pandino, G.; Anastasi, U.; Onofri, A.; Mauromicale, G. Improving soil health, weed management and nitrogen dynamics by *Trifolium subterraneum* cover cropping. *Agron. Sustain. Develop.* **2020**, *40*, 18. [CrossRef]

53. Khademi, Z.; Jones, D.L.; Malakouti, M.J.; Asadi, F. Organic acids differ in enhancing phosphorus uptake by *Triticum aestivum* L.—Effects of rhizosphere concentration and counterion. *Plant Soil* **2010**, *334*, 151–159. [CrossRef]

54. Sylvia, D.M.; Schenck, N.C. Application of superphosphate to mycorrhizal plants stimulates sporulation of phosphorus-tolerate vesicular-arbuscular mycorrhizal fungi. *New Phytol.* **1983**, *95*, 655–661. [CrossRef]

55. Alkan, N.; Gadkar, V.; Yarden, O.; Kapulnik, Y. Analysis of quantitative interactions between two species of arbuscular mycorrhizal fungi, *Glomus mosseae* and *G. intraradices*, by Real-Time PCR. *Appl. Environ. Microb.* **2006**, *72*, 4192–4199. [CrossRef]

56. Gosling, P.; Hodge, A.; Goodlass, G.; Bending, G.D. Arbuscular mycorrhizal fungi and organic farming. *Agr. Ecosyst. Environ.* **2006**, *113*, 17–35. [CrossRef]

57. Ismail, Y.; McCormick, S.; Hijri, M. A fungal symbiont of plant-roots modulates mycotoxin gene expression in the pathogen *Fusarium sambucinum*. *PLoS ONE* **2011**, *6*, e17990. [CrossRef]
58. Romero-Munar, A.; Baraza, E.; Gulías, J.; Cabot, C. Arbuscular mycorrhizal fungi confer salt tolerance in giant reed (*Arundo donax* L.) plants grown under low phosphorus by reducing leaf Na^+ concentration and improving phosphorus use efficiency. *Front. Plant Sci.* **2019**, *10*, 843. [CrossRef]
59. Ruiz-Lozano, J.M.; Porcel, R.; Azcón, C.; Aroca, R. Regulation by arbuscular mycorrhizae of the integrated physiological response to salinity in plants: New challenges in physiological and molecular studies. *J. Exp. Bot.* **2012**, *63*, 4033–4044. [CrossRef]
60. Al-Karaki, G.; McMichael, B.; Zak, J. Field response of wheat to arbuscular mycorrhizal fungi and drought stress. *Mycorrhiza* **2004**, *14*, 263–269. [CrossRef] [PubMed]
61. Zhu, Y.G.; Smith, F.A.; Smith, S.E. Phosphorus efficiencies and responses of barley (*Hordeum vulgare* L.) to arbuscular mycorrhizal fungi grown in highly calcareous soil. *Mycorrhiza* **2003**, *13*, 93–100. [CrossRef] [PubMed]
62. Eason, W.R.; Webb, K.J.; Michaelson-Yeates, T.P.T.; Abberton, M.T.; Griffith, G.W.; Culshaw, C.M.; Hooker, J.E.; Dhanoa, M.S. Effect of genotype of *Trifolium repens* on myconhizal symbiosis with *Glomus mosseae*. *J. Agric. Sci.* **2001**, *137*, 27–36. [CrossRef]
63. Pandino, G.; Lombardo, S.; Lo Monaco, A.; Ruta, C.; Mauromicale, G. In vitro micropropagation and mycorrhizal treatment influences the polyphenols content profile of globe artichoke under field conditions. *Food Res. Int.* **2017**, *99*, 385–392. [CrossRef] [PubMed]
64. Porcel, R.; Redondo-Gómez, S.; Mateos-Naranjo, E.; Aroca, R.; Garcia, R.; Ruiz-Lozano, J.M. Arbuscular mycorrhizal symbiosis ameliorates the optimum quantum yield of photosystem II and reduces non-photochemical quenching in rice plants subjected to salt stress. *J. Plant Physiol.* **2015**, *185*, 75–83. [CrossRef]
65. Sheng, M.; Tang, M.; Chen, H.; Yang, B.; Zhang, F.; Huang, Y. Influence of arbuscular mycorrhizae on photosynthesis and water status of maize plants under salt stress. *Mycorrhiza* **2008**, *18*, 287–296. [CrossRef]
66. Bavaresco, L.; Poni, S. Effect of calcareous soil on photosynthesis rate, mineral nutrition, and Source-Sink ratio of table grape. *J. Plant Nutr.* **2007**, *26*, 2123–2135. [CrossRef]
67. Tisarum, R.; Theerawitaya, C.; Samphumphuang, T.; Polispitak, K.; Thongpoem, P.; Singh, H.P.; Cha-um, S. Alleviation of salt stress in upland rice (*Oryza sativa* L. ssp. *indica* cv. Leum Pua) using arbuscular mycorrhizal fungi inoculation. *Front. Plant Sci.* **2020**, *11*, 348. [CrossRef]
68. Hajiboland, R.; Aliasgharzadeh, N.; Laiegh, S.F.; Poschenrieder, C. Colonization with arbuscular mycorrhizal fungi improves salinity tolerance of tomato (*Solanum lycopersicum* L.) plants. *Plant Soil* **2010**, *331*, 313–327. [CrossRef]
69. Pinior, A.; Grunewaldt-Stöcker, G.; von Alten, H.; Strasser, R.T. Mycorrhizal impact on drought stress tolerance of rose plants probed by chlorophyll a fluorescence, proline content and visual scoring. *Mycorrhiza* **2005**, *15*, 596–605. [CrossRef]
70. Lima, J.D.; Mosquim, P.R.; Da Matta, F.M. Leaf gas exchange and chlorophyll fluorescence parameters in Phaseolus vulgaris as affected by nitrogen and phosphorus deficiency. *Photosynthetica* **1999**, *37*, 113–121. [CrossRef]

 *agronomy*

Article

Chemical Composition of *Cynara cardunculus* L. var. *altilis* Bracts Cultivated in Central Greece: The Impact of Harvesting Time

Filipa Mandim [1,2], Spyridon A. Petropoulos [3,*], Kyriakos D. Giannoulis [3], Celestino Santos-Buelga [2], Isabel C. F. R. Ferreira [1] and Lillian Barros [1,*]

[1] Centro de Investigação de Montanha (CIMO), Instituto Politécnico de Bragança, Campus de Santa Apolónia, 5300-253 Bragança, Portugal; filipamandim@ipb.pt (F.M.); iferreira@ipb.pt (I.C.F.R.F.)

[2] Grupo de Investigación en Polifenoles (GIP-USAL), Facultad de Farmacia, Universidad de Salamanca, Campus Miguel de Unamuno s/n, 37002 Salamanca, Spain; csb@usal.es

[3] Department of Agriculture, Crop Production and Rural Environment, University of Thessaly, 38446 N. Ionia, Magnissia, Greece; kgiannoulis@uth.gr

* Correspondence: spetropoulos@uth.gr (S.A.P.); lillian@ipb.pt (L.B.)

Received: 23 November 2020; Accepted: 14 December 2020; Published: 16 December 2020

Abstract: The present study evaluated the effect of maturity stage on the chemical composition of cardoon bracts. Plant material was collected in Greece at eight different maturation stages (C1–C8) and the chemical composition was analyzed in regard to lipidic fraction and the content in fatty acids, tocopherols, organic acids, and free sugars. Samples of late maturity (C6–C8) revealed the lowest lipidic content, while a total of 29 fatty acids was identified in all the samples, with palmitic, stearic, oleic, and eicosatrienoic acids present in the highest levels depending on harvesting time. Immature (C1) and mature (C8) bracts were more abundant in saturated fatty acids (SFA) than bracts of medium-to-late maturity (C5, C6), where the monounsaturated fatty acids (MUFA) were the prevalent class. The α- and γ-tocopherols were the only identified isoforms of vitamin E, while the highest content was observed in sample C8 (199 µg/100 g dry weight (dw). The detected organic acids were oxalic, quinic, malic, citric, and fumaric acids, while fructose, glucose, sucrose, trehalose, and raffinose were the main detected sugars. The results of the present study allowed us to reveal the effect of maturity stage on cardoon bracts chemical composition and further valorize this byproduct by improving its bioactive compounds content.

Keywords: seasonal variation; chemical composition; free sugars; tocopherols; *Cynara cardunculus* L.; lipidic fraction; fatty acids; organic acids

1. Introduction

Cynara cardunculus L. is a species widely distributed throughout the world, especially in the circum-Mediterranean sea area where it was domesticated for the first time [1]. Belonging to one of the largest families of the plant kingdom, the *Asteraceae* family, this species includes three botanical varieties, namely: the cultivated cardoon (*Cynara cardunculus* var. *altilis* DC), the globe or head artichoke (*Cynara cardunculus* var. *scolymus* (L.) Fiori), and the wild cardoon (*Cynara cardunculus* L. var. *sylvestris* Lamk Fiori). Commonly known as cardoon or artichoke thistle, it is widely used due to its multifaceted properties not only as a food ingredient but also in various industrial applications [2–4].

This species has high nutritional, pharmacological, and industrial value, and, although it has been used since ancient times, it was only in the last decades that cardoon gained attention [5]. In addition to having multiple applications, it is a plant highly resistant to variations in climatic conditions and abiotic stressors, characteristic of the Mediterranean regions [6–8]. Widely consumed in typical Mediterranean countries as a source of fibers, minerals, and inulin, cardoon is a species that contains a great variety of compounds with important bioactive and nutritional properties. Literature reports refer to the presence of various phenolic compounds, mostly derived from caffeoylquinic and dicaffeoylquinic acids, as well as apigenin and luteolin derivatives, while the presence of sesquiterpenes, lignans, and anthocyanins has also been described [4,9–13].

This actual wealth of compounds with bioactive potential has boosted their exploitation in several sectors of the industry [14]. One of its best-known applications is its use as vegetable rennet for the production of protected designation of origin cheeses (PDO) [3,15]. It is also used for biomass and bioenergy production, as well in the papermaking industry due to its high content of cellulose and hemicellulose [5,16–20]. Its application as a food additive or in nutraceutical and cosmetic products has been also explored [21–23]. There are several studies described in the literature regarding the various industrial applications and biochemical potential associated with cardoon vegetable tissues [5,20,24]. However, about 60% to 85% of the plant material resulting from industrial processes is discarded, increasing the environmental burden and the footprint of the crop and necessitating the channeling of these byproducts in alternative sectors that could increase the added value of the crop. The wasted material consists of bracts, stems, and leaves, which can be a source of important bioactive compounds such as phenolic acids, fibers, minerals, and inulin, which could be used for medicinal and nutraceutical purposes [18,20,21,25–27]. Therefore, the exploitation of these plant tissues can be an important contribution to their economic recovery and reuse, thus reducing waste and sources of environmental contamination and reinforcing the circular economy [5,10,22,28]. In addition, during the whole growth cycle, plants are subjected to variable conditions and cultivation practices. Environmental factors such as the water availability, temperature, light intensity and quality, soil type, and nutritional status, reveal a significant impact on plant metabolism and consequently on the chemical composition of the species throughout the growing season [29–31].

Considering the great amount of waste generated from the cardoon crop, alternative uses of byproducts are essential to increase the added value of this important crop. So far, the studies regarding the influence that the different plant growth stages may have on quality properties such as the chemical composition and the bioactive potential of vegetative tissues are very scarce. Moreover, the valorization of byproducts focuses mostly on biomass and energy production and scarce reports are available regarding the recovering of bioactive compounds from discarded plant parts. Therefore, the aim of this study was to report for the first time the influence that the maturation stage may have on the chemical composition of cardoon bracts collected in central Greece at eight different growth stages, focusing on the lipidic fraction, tocopherols, organic acids, and free sugars composition and content. The presented results could be helpful for the identification of growth stages where the content of specific bioactive contents may increase, as well as for the valorization of such byproducts and the improvement of the overall crop added value.

2. Materials and Methods

2.1. Plant Material

Bracts samples of *Cynara cardunculus* var. *altilis* DC cv. *Bianco Avorio* (Fratelli Ingegnoli Spa, Milano, Italy) were collected during the cultivating period of 2017–2018 at the experimental farm of the University of Thessaly in Velestino, Greece (22.756 E, 39.396 N). Bracts were collected from 15 individual heads for eight harvesting dates according to the principal growth stages (PGS) defined by the Biologische Bundesanstalt, Bundessortenamt und Chemische Industrie (BBCH) scale, comprising stages between PGS 5 and PGS 8/9 [32]. Sample C1 was collected at the end of April (PGS 5), sample C2 at the beginning of May (PGS 5/6), sample C3 at the end of May (PGS 6), sample C4 at the beginning of June (PGS 6/7), sample C5 at the beginning of July (PGS 7), sample C6 at the end of July (PGS 7/8), sample C7 at the beginning of August (PGS 8), and finally, sample C8 was collected at the end of August (PGS 8/9).

The climate conditions and the procedure used for the collection and sampling treatments of plant material were previously described by Mandim et al. [33].

2.2. Chemical Composition Analysis

2.2.1. Fatty Acids

The lipidic fraction of cardoon bracts was extracted through a Soxhlet extraction apparatus with petroleum ether at 120 °C, as recommended by Association of Official Agricultural Chemists (AOAC) procedures [34]. Subsequently, the fat content was subjected to a transesterification process and the fatty acids content was analyzed by Gas-liquid Chromatography (GC), coupled to a Flame Ionization Detector (FID) according to the conditions previously described in Reference [34]. The identification and quantification of fatty acids was performed with the Clarity DataApex 4.0 software (DataApex, Prague, Czech Republic). The identification was based on the comparison of the retention times of the Fatty Acid Methyl Ester (FAME) peaks from samples with commercial standards (reference standard mixture 47885-U; Sigma-Aldrich, St. Louis, MO, USA), and a quantification was made from the area of the peaks. Final results were expressed as relative percentages and in mg of each identified fatty acid per 100 g of dry weight (dw) of plant material.

2.2.2. Tocopherols

The tocopherols content was determined by high-performance liquid chromatography (HPLC, Knauer, Smartline system 1000, Berlin, Germany) coupled to a fluorescence detector (FP-2020, Jasco, Easton, PA, USA) programmed for excitation at 290 nm and emission at 330 nm, according to the procedure described by Barros et al. [35]. The identification and quantification were performed using the Clarity 2.4 software (DataApex, Prague, Czech Republic) and the internal standard (IS) method, through comparison of the retention times and spectra with tocopherols' standards (α-, β-, γ-, and δ-isophorms). Results were expressed in µg per 100 g of dw.

2.2.3. Organic Acids

For organic acids identification, cardoon samples were analyzed by Ultrafast Liquid Chromatography (UPLC, Shimadzu 20A series, Kyoto, Japan) coupled to a Diode Array Detector (UFLC-PDA, Shimadzu Corporation, Kyoto, Japan), according to the chromatographic conditions previously described by Mandim et al. [36]. The identification was performed using the LabSolutions Multi LC-PDA software (Shimadzu Corporation, Kyoto, Japan) and through the comparison of the chromatographic data (retention times and spectra) with commercial standards (oxalic, quinic, malic, ascorbic, citric, and fumaric acids), while their respective calibration curves were used for the quantification based on the area of each peak. Results were presented in g per 100 g of dw.

2.2.4. Free Sugars

The content in free sugars was determined by High-Performance Liquid Chromatography (HPLC, Knauer Smartline 2300, Knauer, Berlin, Germany), coupled to a refractive index detector (RI detector, Knauer Smartline 2300, Knauer, Berlin, Germany), according to the procedure previously described by Dias et al. [37]. The identification and quantification of free sugars were performed using the Clarity 2.4 software (DataApex, Prague, Czech Republic) and through the comparison with commercial standards, namely D-(−)-fructose, D-(+)-sucrose, D-(+)-glucose, D-(+)-trehalose, and D-(+)-raffinose pentahydrate (Sigma-Aldrich, St. Louis, MO, USA).

2.3. Statistical Analysis

The performed assays were carried out in triplicate. The obtained results were presented as mean values ± standard deviation (SD). Means and standard deviations were calculated using Microsoft Excel. SPSS Statistics software (IBM SPSS Statistics for Mac OS, Version 26.0; IBM Corp., Armonk, NY, USA) was used to determine differences between samples. The results were subject to an analysis of variance (ANOVA), while the Tukey's honest significance test (HSD) test ($p = 0.05$) was used to determine the significant differences among samples.

3. Results and Discussion

3.1. Lipidic Fraction and Fatty Acids Composition

The results related to the lipidic fraction and the fatty acids composition (relative percentage and concentration) are shown in Tables 1 and 2, as well as the proportions of saturated fatty acids (SFA), monounsaturated fatty acids (MUFA), and polyunsaturated fatty acids (PUFA), and the PUFA/SFA and n-6/n-3 ratios. Samples collected at late stages of maturity, namely C6, C7, and C8, presented the lowest lipidic content (2.2–2.9 g/100 g dw). In contrast, immature bracts (C2) revealed the highest lipidic levels, being 5.95 times higher than those in the sample of late maturity (C8). Twenty-nine individual fatty acids were identified in cardoon bracts collected at different maturation stages, with palmitic (C16:0, 0.95–44%), stearic (C18:0, 6.64–44.37%), oleic (C18:1n9c, 4.16–29.0%), and eicosatrienoic (C18:2n6c, 2.27–16.852%) acids being present in the highest concentrations. A representative chromatogram of the fatty acids profile is presented in Figure 1 where the retention times and the peaks of the individual detected fatty acids are illustrated. Regarding the effect of maturity stage, palmitic and eicosatrienoic acids were detected in higher levels in immature bracts (samples C1 and C2), while stearic acid revealed higher abundance in bracts of mid-maturity (samples C4, C5, and C6), and oleic acid in bracts of late maturity stages (samples C7 and C8). Saturated fatty acids (SFAs) were the most abundant class of fatty acids in immature bracts (samples C1–C4) and samples of late maturity (samples C7 and C8) due to the high content in palmitic and stearic acids. In turn, monounsaturated fatty acids (MUFAs) were the class with the highest abundance in samples C5 and C6, due to the high content of pentadecanoic acid (C15:1; 31.01–31.2%). Our results also revealed that the tested cardoon bracts did not present an analogous composition and abundance of fatty acids over time, which was also reflected to the recorded PUFA/SFA and n-6/n-3 ratios. The PUFA/SFA ratio was higher than 0.45 in samples C4 and C6, whereas the values of the n-6/n-3 ratio were below 4.0 in all samples, except sample C8. These results verify that the state of maturity influences the composition and abundance of fatty acids in bracts, a finding which is in agreement with previous studies of our team where other cardoon parts were examined [33,38]. In particular, Mandim et al. [33] also reported that lipidic content in cardoon heads decreased with the maturation process and suggested that the differences in the environmental conditions during the growing period could be responsible for the observed differences. Considering that this study was carried out under the same conditions and with the same plant material as in our study, it could be suggested that the environmental factors are also the key drivers for the observed differences in the present study. Similarly, Curt et al. [24], who tested different locations and growing years, highlighted the significant effect of environmental conditions on fatty acid composition of cardoon seeds. To the best of our knowledge, this is the first report that analyzes the influence of the growth cycle on these parameters of chemical composition of bracts, where according to our results, immature bracts (sample C2) presented the highest contents in lipidic components.

Table 1. Lipidic fraction and fatty acids composition of *Cynara cardunculus* bracts in relation to maturity stage (C1–C8).

	C1	C2	C3	C4	C5	C6	C7	C8
	Total lipidic fraction (g/100 g dw)							
	4.7 ± 0.1 [c]	13.1 ± 0.2 [a]	6.1 ± 0.2 [b]	4.0 ± 0.1 [d]	4.9 ± 0.1 [c]	2.9 ± 0.1 [e]	2.4 ± 0.2 [f]	2.2 ± 0.2 [f]
	Fatty acids (relative percentage, %)							
C6:0	0.20 ± 0.01 [g]	0.26 ± 0.02 [f]	0.27 ± 0.03 [f]	0.488 ± 0.005 [e]	0.56 ± 0.01 [d]	0.87 ± 0.01 [c]	1.12 ± 0.03 [b]	2.28 ± 0.04 [a]
C8:0	0.29 ± 0.03 [f]	0.33 ± 0.01 [e]	0.29 ± 0.03 [f]	0.56 ± 0.03 [c]	0.615 ± 0.004 [b]	0.78 ± 0.02 [a]	0.36 n ± 0.01 [d]	0.61 ± 0.02 [b]
C10:0	0.261 ± 0.002 [f]	0.22 ± 0.02 [f]	0.7 ± 0.1 [c]	0.75 ± 0.06 [bc]	0.784 ± 0.001 [b]	1.124 ± 0.002 [a]	0.52 ± 0.02 [e]	0.608 ± 0.001 [d]
C11:0	1.666 ± 0.001 [b]	0.70 ± 0.02 [c]	0.60 ± 0.05 [d]	0.62 ± 0.02 [d]	0.65 ± 0.01 [d]	1.04 ± 0.02 [b]	1.03 ± 0.05 [b]	1.14 ± 0.05 [a]
C12:0	0.65 ± 0.06 [e]	1.911 ± 0.004 [b]	2.5 ± 0.3 [a]	1.46 ± 0.02 [c]	1.057 ± 0.004 [d]	1.04 ± 0.02 [d]	0.62 ± 0.04 [e]	0.330 ± 0.003 [f]
C13:0	0.095 ± 0.009 [a]	0.081 ± 0.001 [b]	0.052 ± 0.001 [d]	0.092 ± 0.003 [a]	0.064 ± 0.001 [c]	n.d.	n.d.	n.d.
C14:0	1.9 ± 0.1 [d]	2.56 ± 0.01 [b]	2.8 ± 0.2 [a]	2.63 ± 0.01 [ab]	2.19 ± 0.02 [c]	1.90 ± 0.02 [d]	1.8 ± 0.1 [de]	1.7 ± 0.1 [e]
C14:1	0.39 ± 0.01 [a]	0.064 ± 0.004 [b]	0.041 ± 0.001 [c]	0.062 ± 0.004 [b]	n.d.	n.d.	n.d.	n.d.
C15:0	0.33 ± 0.03 [b]	0.70 ± 0.01 [d]	0.6 ± 0.1 [e]	0.71 ± 0.02 [d]	0.866 ± 0.001 [b]	1.02 ± 0.01 [a]	0.78 ± 0.01 [c]	0.74 ± 0.04 [cd]
C15:1	0.066 ± 0.003 [d]	0.062 ± 0.002 [d]	0.043 ± 0.001 [d]	0.057 ± 0.003 [d]	31.01 ± 0.01 [b]	31.2 ± 0.2 [a]	0.15 ± 0.01 [cd]	0.19 ± 0.01 [c]
C16:0	47.3 ± 0.6 [a]	47.2 ± 0.1 [a]	44 ± 2 [b]	0.95 ± 0.01 [e]	0.98 ± 0.02 [e]	1.27 ± 0.01 [e]	36.7 ± 0.4 [d]	41 ± 1 [c]
C16:1	0.16 ± 0.01 [f]	0.257 ± 0.002 [e]	0.29 ± 0.03 [e]	0.894 ± 0.001 [b]	0.79 ± 0.01 [c]	1.3 ± 0.1 [a]	0.84 ± 0.01 [c]	0.49 ± 0.02 [e]
C17:0	0.83 ± 0.04 [c]	0.699 ± 0.001 [d]	0.62 ± 0.03 [e]	0.57 ± 0.01 [f]	0.545 ± 0.004 [f]	0.64 ± 0.02 [e]	1.18 ± 0.05 [b]	1.3 ± 0.1 [a]
C18:0	6.6 ± 0.1 [g]	7.58 ± 0.01 [f]	8.7 ± 0.5 [e]	44.34 ± 0.1 [a]	29.80 ± 0.04 [b]	24.57 ± 0.01 [c]	8.71 ± 0.03 [e]	10.4 ± 0.2 [d]
C18:1n9c	4.2 ± 0.1 [h]	6.90 ± 0.01 [g]	14.6 ± 0.3 [c]	13.0 ± 0.4 [d]	8.14 ± 0.01 [f]	10.0 ± 0.1 [e]	29.0 ± 0.4 [a]	19 ± 1 [b]
C18:2n6c	12.59 ± 0.03 [b]	16.852 ± 0.001 [a]	10.5 ± 0.3 [c]	4.96 ± 0.01 [d]	2.69 ± 0.03 [f]	2.8 ± 0.2 [f]	2.3 ± 0.1 [g]	3.8 ± 0.2 [e]
C18:3n3	5.60 ± 0.02 [a]	4.15 ± 0.01 [b]	2.7 ± 0.3 [c]	0.64 ± 0.01 [d]	0.354 ± 0.004 [e]	0.51 ± 0.02 [de]	0.48 ± 0.02 [de]	0.35 ± 0.01 [e]
C20:0	4.3 ± 0.1 [a]	2.819 ± 0.004 [d]	2.43 ± 0.04 [e]	0.12 ± 0.01 [g]	0.55 ± 0.02 [f]	0.63 ± 0.02 [f]	3.32 ± 0.02 [c]	3.6 ± 0.2 [b]
C20:1	0.089 ± 0.002 [f]	0.08 ± 0.01 [f]	0.52 ± 0.05 [d]	0.69 ± 0.01 [c]	0.369 ± 0.004 [e]	1.06 ± 0.03 [a]	1.11 ± 0.05 [a]	0.8 ± 0.1 [b]
C20:2	0.03 ± 0.01 [d]	0.17 ± 0.02 [c]	0.35 ± 0.03 [a]	0.21 ± 0.01 [b]	n.d.	n.d.	n.d.	n.d.
C21:0	0.853 ± 0.002 [a]	0.54 ± 0.03 [d]	0.17 ± 0.02 [e]	0.13 ± 0.02 [e]	0.16 ± 0.01 [e]	n.d.	0.67 ± 0.05 [c]	0.79 ± 0.03 [b]
C20:3n6	0.279 ± 0.003 [d]	0.16 ± 0.01 [e]	n.d.	1.76 ± 0.03 [c]	2.078 ± 0.001 [a]	1.80 ± 0.02 [b]	n.d.	n.d.
C20:3n3	1.6 ± 0.1 [c]	1.19 ± 0.03 [d]	2.2 ± 0.2 [b]	3.4 ± 0.1 [a]	0.35 ± 0.01 [e]	1.65 ± 0.02 [c]	n.d.	n.d.
C22:0	4.58 ± 0.03 [a]	3.41 ± 0.04 [b]	3.2 ± 0.3 [c]	1.9 ± 0.1 [f]	1.87 ± 0.02 [f]	2.18 ± 0.04 [e]	2.9 ± 0.1 [d]	4.41 ± 0.03 [a]
C22:1	0.5c ± 0.02 [d]	0.11 ± 0.01 [g]	0.37 ± 0.02 [e]	0.80 ± 0.02 [c]	1.2 ± 0.1 [b]	1.44 ± 0.04 [a]	0.405 ± 0.001 [e]	0.16 ± 0.01 [f]
C20:5n3	0.5 ± 0.1 [d]	0.031 ± 0.001 [h]	0.20 ± 0.02 [g]	1.50 ± 0.05 [a]	1.1 ± 0.1 [c]	1.20 ± 0.03 [b]	0.42 ± 0.01 [e]	0.33 ± 0.01 [f]
C22:2	0.437 ± 0.001 [d]	n.d.	n.d.	16.7 ± 0.5 [a]	11.3 ± 0.1 [b]	10.052 ± 0.005 [c]	n.d.	0.37 ± 0.01 [d]
C23:0	1.28 ± 0.02 [b]	0.078 ± 0.001 [e]	0.74 ± 0.04 [d]	n.d.	n.d.	n.d.	1.20 ± 0.01 [c]	1.50 ± 0.05 [a]
C24:0	2.7135 ± 0.1 [c]	0.90 ± 0.04 [d]	0.8 ± 0.1 [d]	n.d.	n.d.	n.d.	4.4 ± 0.2 [a]	4.2 ± 0.2 [b]
SFA	73.8 ± 0.1 [a]	69.98 ± 0.02 [b]	68.3 ± 0.5 [c]	55.3 ± 0.1 [e]	40.7 ± 0.1 [f]	37.05 ± 0.08 [g]	65.3 ± 0.4 [d]	75 ± 1 [a]
MUFA	5.2 ± 0.1 [g]	7.47 ± 0.02 [f]	15.8 ± 0.3 [e]	15.5 ± 0.4 [e]	41.5 ± 0.1 [b]	44.9 ± 0.1 [a]	31.5 ± 0.5 [c]	21 ± 1 [d]
PUFA	21.1 ± 0.3 [c]	22.55 ± 0.01 [b]	15.9 ± 0.8 [e]	29.2 ± 0.3 [a]	17.82 ± 0.03 [d]	18.0 ± 0.2 [d]	3.2 ± 0.1 [f]	4.8 ± 0.2 [g]
PUFA/SFA	0.286 ± 0.001 [e]	0.322 ± 0.001 [d]	0.23 ± 0.01 [f]	0.527 ± 0.005 [a]	0.4381 ± 0.0001 [c]	0.486 ± 0.005 [b]	0.048 ± 0.001 [h]	0.065 ± 0.004 [g]
n-6/n-3	2.32 ± 0.01 [e]	3.82 ± 0.01 [b]	2.7 ± 0.2 [d]	3.1 ± 0.1 [c]	3.7 ± 0.1 [b]	2.59 ± 0.03 [de]	3.1 ± 0.1 [c]	6.6 ± 0.5 [a]

Results are presented as mean ± standard deviation. Different letters correspond to significant differences ($p < 0.05$). Fatty acids are expressed as relative percentage of each fatty acid. dw—dry weight; n.d.—not detected; C6:0—caproic acid; C8:0—caprylic acid; C10:0—capric acid; C11:0—undecanoic acid; C12:0—lauric acid; C13:0—tridecanoic acid; C14:0—myristic acid; C14:1—tetradecanoic acid; C15:0—pentadecanoic acid; C15:1—pentadecenoic acid; C16:0—palmitic acid; C16:1—palmitoleic acid; C17:0—heptadecanoic acid; C18:0—stearic acid; C18:1n9c—oleic acid; C18:2n6c—linoleic acid; C18:3n3—linolenic acid; C20:0—arachidic acid; C20:1—gadoleic acid; C20:2—eicosadieoic acid; C21:0—heneicosanoic acid; C20:3n6—eicosatrienoic acid; C20:3n3—11,14,17-eicosatrienoic acid; C22:0—behenic acid; C22:1—eicosenoic acid; C20:5n3—eicosapentaenoic acid; C22:2—behenic acid; C23:0—tricosanoic acid; C24:0—lignoceric acid; SFA—saturated fatty acids; MUFA—monounsaturated fatty acids; PUFA—polyunsaturated fatty acids; n-6/n-3: ratio of omega 6/omega 3 fatty acids.

Table 2. Composition of fatty acids (mg/100 g dw) of *Cynara cardunculus* L. bracts in relation to maturity stage (C1–C8; mean ± standard deviation (SD); n = 3).

	C1	C2	C3	C4	C5	C6	C7	C8
				Fatty acids (mg/100 g dw)				
C6:0	9.4 ± 0.5 [g]	33 ± 3 [b]	17 ± 2 [f]	19.5 ± 0.2 [e]	27.2 ± 0.4 [c]	25.2 ± 0.3 [d]	27 ± 1 [cd]	50 ± 1 [a]
C8:0	13 ± 1 [e]	42.9 ± 0.8 [a]	18 ± 2 [d]	22 ± 1 [c]	30.1 ± 0.2 [b]	22.6 ± 0.5 [c]	8.6 ± 0.2 [f]	13.4 ± 0.4 [e]
C10:0	12.2 ± 0.1 [e]	29 ± 2 [d]	44 ± 4 [a]	30 ± 3 [cd]	38.4 ± 0.1 [b]	32.6 ± 0.1 [c]	12.5 ± 0.5 [e]	13.39 ± 0.03 [e]
C11:0	50.13 ± 0.03 [b]	92 ± 2 [a]	37 ± 3 [c]	25 ± 1 [e]	32 ± 1 [d]	30.2 ± 0.5 [d]	25 ± 1 [e]	25 ± 1 [e]
C12:0	30 ± 3 [d]	250.2 ± 0.5 [a]	150 ± 16 [b]	59 ± 1 [c]	51.8 ± 0.2 [c]	30.0 ± 0.5 [d]	15 ± 1 [e]	7.3 ± 0.1 [e]
C13:0	4.4 ± 0.4 [b]	10.5 ± 0.1 [a]	3.2 ± 0.1 [d]	3.7 ± 0.1 [c]	3 ± 1 [d]	n.d.	n.d.	n.d.
C14:0	89 ± 5 [d]	335.4 ± 0.9 [a]	170 ± 14 [b]	105.4 ± 0.4 [c]	107 ± 1 [c]	55 ± 1 [e]	43 ± 2 [f]	37 ± 2 [f]
C14:1	4.4 ± 0.4 [b]	8.3 ± 0.5 [a]	2.5 ± 0.1 [c]	2.5 ± 0.1 [c]	n.d.	n.d.	n.d.	n.d.
C15:0	39 ± 2 [c]	92 ± 1 [a]	37 ± 4 [c]	28 ± 1 [d]	42.4 ± 0.1 [b]	29.6 ± 0.3 [d]	18.6 ± 0.3 [e]	16 ± 1 [e]
C15:1	3.1 ± 0.1 [d]	8.1 ± 0.3 [c]	2.6 ± 0.1 [d]	2.3 ± 0.1 [d]	1519.7 ± 0.5 [a]	905 ± 5 [b]	3.5 ± 0.2 [d]	4.1 ± 0.1 [d]
C16:0	2223 ± 30 [c]	6183 ± 10 [a]	2672 ± 128 [b]	38.1 ± 0.5 [e]	48 ± 1 [e]	36.7 ± 0.2 [e]	881 ± 10 [d]	901 ± 36 [d]
C16:1	7.6 ± 0.3 [g]	33.6 ± 0.3 [c]	18 ± 2 [e]	35.74 ± 0.02 [b]	39 ± 1 [a]	36 ± 2 [b]	20.1 ± 0.2 [d]	10.7 ± 0.5 [f]
C17:0	39 ± 2 [b]	91.5 ± 0.1 [a]	38 ± 2 [b]	22.5 ± 0.4 [e]	26.7 ± 0.2 [d]	18.4 ± 0.5 [f]	28 ± 1 [cd]	29 ± 1 [c]
C18:0	312 ± 4 [f]	993.0 ± 0.6 [c]	531 ± 29 [e]	1775 ± 4 [a]	1460 ± 2 [b]	712.4 ± 0.2 [d]	209 ± 1 [h]	230 ± 4 [g]
C18:1n9c	195 ± 3 [f]	903 ± 2 [a]	890 ± 17 [a]	520 ± 16 [c]	398.8 ± 0.4 [d]	290 ± 2 [e]	697 ± 11 [b]	417 ± 24 [d]
C18:2n6c	592 ± 2 [c]	2208 ± 1 [a]	641 ± 20 [b]	198.5 ± 0.4 [d]	132 ± 2 [e]	82 ± 5 [f]	54 ± 2 [g]	83 ± 5 [f]
C18:3n3	263 ± 1 [b]	543 ± 1 [a]	163 ± 16 [c]	25.6 ± 0.3 [d]	17.3 ± 0.2 [de]	14.9 ± 0.5 [e]	11.6 ± 0.4 [e]	7.6 ± 0.3 [e]
C20:0	202 ± 6 [b]	369.3 ± 0.6 [a]	148 ± 3 [c]	4.9 ± 0.4 [g]	27 ± 1 [e]	18.4 ± 0.5 [f]	80 ± 1 [d]	80 ± 4 [d]
C20:1	4.2 ± 0.1 [e]	11 ± 1 [d]	32 ± 3 [a]	27.5 ± 0.6 [b]	18.1 ± 0.2 [c]	31 ± 1 [a]	27 ± 1 [b]	17 ± 1 [c]
C20:2	1.5 ± 0.1 [c]	22 ± 2 [a]	21 ± 2 [a]	8 ± 1 [b]	n.d.	n.d.	n.d.	n.d.
C21:0	40 ± 4 [b]	70 ± 3 [a]	11 ± 1 [d]	5 ± 1 [e]	7.8 ± 0.3 [de]	n.d.	16 ± 1 [c]	17.5 ± 0.5 [c]
C20:3n6	13 ± 1 [e]	21.6 ± 0.7 [d]	n.d.	71 ± 1 [b]	101.80 ± 0.03 [a]	52 ± 1 [c]	n.d.	n.d.
C20:3n3	76 ± 1 [c]	156 ± 3 [a]	131 ± 13 [b]	136 ± 3 [b]	17.2 ± 0.3 [e]	48 ± 1 [d]	n.d.	n.d.
C22:0	215 ± 6 [b]	447 ± 5 [a]	194 ± 19 [c]	75 ± 3 [e]	92 ± 1 [d]	63 ± 1 [e]	79 ± 1 [e]	97.0 ± 0.5 [d]
C22:1	27.7 ± 0.5 [d]	14 ± 1 [f]	22 ± 1 [e]	32 ± 1 [c]	58 ± 3 [a]	42 ± 1 [b]	9.71 ± 0.02 [g]	3.6 ± 0.4 [h]
C20:5n3	25.2 ± 0.4 [d]	3.9 ± 0.1 [g]	12 ± 1 [e]	60 ± 2 [a]	53 ± 4 [b]	35 ± 1 [c]	9.9 ± 0.2 [ef]	7.3 ± 0.2 [f]
C22:2	21 ± 2 [d]	n.d.	n.d.	669 ± 19 [a]	553 ± 7 [b]	291.5 ± 0.1 [c]	n.d.	8.1 ± 0.2 [e]
C23:0	60 ± 1 [a]	10.3 ± 0.1 [e]	45 ± 2 [b]	n.d.	n.d.	n.d.	28.8 ± 0.3 [d]	33 ± 1 [c]
C24:0	128 ± 6 [a]	118 ± 5 [b]	50 ± 4 [e]	n.d.	n.d.	n.d.	106 ± 4 [c]	92 ± 5 [d]
SFA	3466 ± 5 [c]	9167 ± 3 [a]	4164 ± 29 [b]	2213 ± 4 [d]	1993 ± 3 [e]	1074 ± 2 [h]	1568 ± 10 [g]	1641 ± 31 [f]
MUFA	242 ± 4 [g]	978 ± 2 [c]	967 ± 20 [c]	619 ± 16 [e]	2034 ± 5 [a]	1303 ± 2 [b]	756 ± 12 [d]	453 ± 26 [f]
PUFA	991 ± 1 [c]	2955 ± 1 [a]	970 ± 49 [c]	1167 ± 13 [b]	873 ± 1 [d]	523 ± 4 [e]	76 ± 2 [g]	106 ± 4 [f]

Results are presented as mean ± standard deviation. Different letters correspond to significant differences ($p < 0.05$). Fatty acids are expressed as mg per 100 g of dw of each fatty acid. dw—dry weight; n.d.—not detected; C6:0—caproic acid; C8:0—caprylic acid; C10:0—capric acid; C11:0—undecanoic acid; C12:0—lauric acid; C13:0—tridecanoic acid; C14:0—myristic acid; C14:1—tetradecanoic acid; C15:0—pentadecanoic acid; C15:1—pentadecenoic acid; C16:0—palmitic acid; C16:1—palmitoleic acid; C17:0—heptadecanoic acid; C18:0—stearic acid; C18:1n9c—oleic acid; C18:2n6c—linoleic acid; C18:3n3—linolenic acid; C20:0—arachidic acid; C20:1—gadoleic acid; C20:2—eicosadieoic acid; C21:0—heneicosanoic acid; C20:3n6—eicosatrienoic acid; C20:3n3—11,14,17-eicosatrienoic acid; C22:0—behenic acid; C22:1—eicosenoic acid; C20:5n3—eicosapentaenoic acid; C22:2—behenic acid; C23:0—tricosanoic acid; C24:0—lignoceric acid; SFA—saturated fatty acids; MUFA—monounsaturated fatty acids; PUFA—polyunsaturated fatty acids.

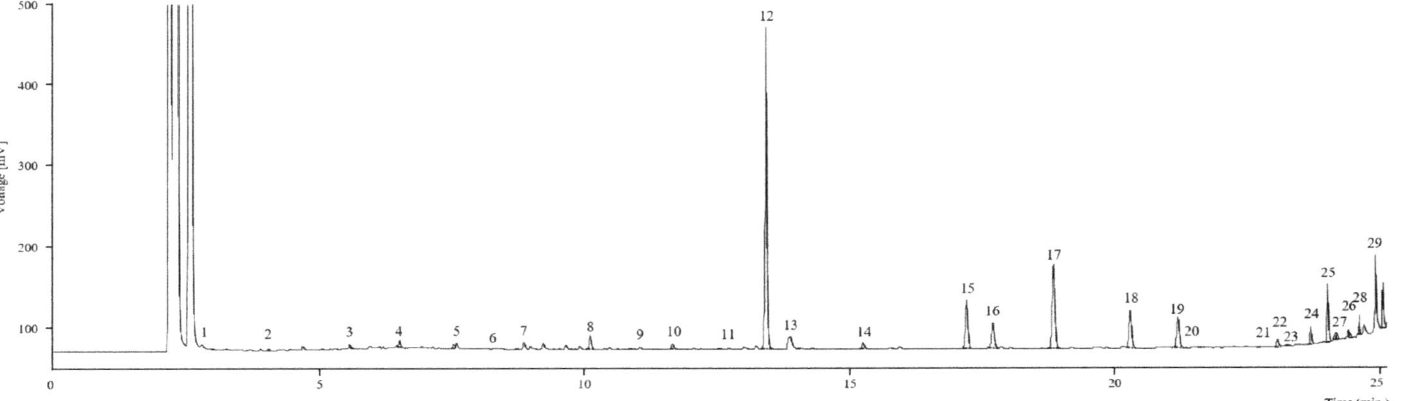

Figure 1. Chromatogram of fatty acids profile of *Cynara cardunculus* bracts (sample C1; collected at the end of April). 1. C6:0—caproic acid; 2. C8:0 caprylic acid; 3. C10:0—capric acid; 4. C11:0—undecanoic acid; 5. C12:0—lauric acid; 6. C13:0—tridecanoic acid; 7. C14:0—myristic acid; 8. C14:1—tetradecanoic acid; 9. C15:0—pentadecanoic acid; 10. C15:1—pentadecenoic acid; 11. C16:0—palmitic acid; 12. C16:1—palmitoleic acid; 13. C17:0—heptadecanoic acid; 14. C18:0—stearic acid; 15. C18:1n9—oleic acid; 16. C18:2n6c—linoleic acid; 17. C18:3n3—alpha-linolenic acid; 18. C20:0—arachidic acid; 19. C20:1—gadoleic acid; 20. C20:2—eicosadieoic acid; 21. C21:0—heneicosanoic acid; 22. C20:3n6—eicosatrienoic acid; 23. C20:3n3—11,14,17-eicosatrienoic acid; 24. C22:0—behenic acid; 25. C22:1—erucic acid; 26. C20:5n3—eicosapentaenoic acid; 27. C22:2—docosadienoic acid; 28. C23:0—tricosanoic acid; 29. C24:0—lignoceric acid.

3.2. Tocopherols, Organic Acids, and Free Sugars Content

In Table 3, the qualitative and quantitative information regarding tocopherols, organic acids, and free sugars identified in the cardoon bracts harvested at different maturation stages are presented. The α- and γ-tocopherols were the only vitamin E isoforms detected in the studied cardoon bracts. Isoform γ-tocopherol was detected in only three maturity stages (samples C2, C4, and C6) in higher concentration than α-tocopherol. The highest abundance of total tocopherols was detected in bracts of late maturity (sample C8; 199 µg/100 g dw), a finding that could be associated with environmental conditions such as solar radiation (light quality and increasing light intensity), as well as the increasing average air temperature. On the contrary, the lowest content of tocopherols was detected in sample C5 (11.7 µg/100 g dw). Due to the fact that tocopherols are antioxidant molecules, they are susceptible to oxidation reactions; thus, they can be strongly influenced by the various environmental conditions to which the plant is subjected throughout its growth cycle [30]. This fact could justify the variations in the content of tocopherols in bracts, as well as the fact that in our previous study [39], phenolic compounds content showed a decrease with increasing maturity, explained by the lignification of bracts tissues [12]. Considering the protective role of tocopherols and polyphenols in the overall antioxidant mechanism of the plant, the observed increase of tocopherols at late maturity could compensate the decreased content of polyphenols and provide defense against abiotic stress. Moreover, although the sample C8 had a higher content in tocopherols and was more efficient to inhibit oxidative hemolysis (OxHLIA), the same was not true for the inhibition of lipid peroxidation (thiobarbituric acid reactive substances; TBARS), where sample C1 revealed a superior antioxidant potential [32]. The same observation was also made in samples of cardoon heads [40], suggesting that other classes of compounds are involved in the antioxidant capacity demonstrated by the analyzed samples. Moreover, it is very common in natural matrices to exhibit variable effectiveness in various antioxidant activity assays, since different compounds and mechanisms are involved in different assays [41,42]. Similarly to our study, the reduced variety of tocopherols found in cardoon bracts has also been reported for other plant parts such as heads and seeds, indicating that α-tocopherol is the main vitamin E isoform detected in the species [30,40,43].

Regarding the organic acids' composition (Table 3), oxalic, quinic, malic, citric, and fumaric acids were the detected compounds. As verified for the other studied parameters, the organic acids composition showed a variation along the maturation process. Bracts collected at the eighth principal growth stage (PSG 8; sample C7) revealed the highest abundance in organic acids (15.6 g/100 g dw), whereas sample C1 (PSG 5) had the lowest abundance (1.96 g/100 g dw). Malic acid was the most relevant organic acid (0.81–1.87 g/100 g dw) at early- to mid-maturation stages (samples C1–C5), whereas in later stages, quinic (samples C6 and C8) and oxalic acids (sample C7) reached the highest concentrations (0.92–4.82 and 9.5 g/100 g dw, respectively). The tested cardoon bracts reveal a similar organic acids profile to that observed for cardoon heads of the same genetic material, with malic acid being present in higher levels in immature heads, while oxalic and quinic acids were more abundant in samples collected at more advanced states of maturity [40].

Table 3. Tocopherols, organic acids, and free sugars of *Cynara cardunculus* L. bracts in relation to maturity stage (C1–C8).

	C1	C2	C3	C4	C5	C6	C7	C8
Tocopherols (µg/100 g dw)								
α-Tocopherol	36.2 ± 0.1 [b]	62 ± 2 [a]	19.8 ± 0.8 [c]	9.4 ± 0.3 [e]	11.7 ± 0.5 [d]	8.1 ± 0.3 [f]	n.d.	199 ± 7 [g]
γ-Tocopherol	n.d.	87 ± 3 [b]	n.d.	82 ± 3 [c]	n.d.	120 ± 2 [a]	n.d.	n.d.
Total tocopherols	36.2 ± 0.1 [d]	149 ± 1 [a]	19.8 ± 0.8 [e]	91 ± 5 [c]	11.7 ± 0.5 [f]	128 ± 2 [b]	n.d.	199 ± 7 [g]
Organic acids (g/100 g dw)								
Oxalic acid	0.320 ± 0.002 [b]	0.328 ± 0.002 [b]	0.093 ± 0.002 [d]	0.181 ± 0.001 [cd]	0.206 ± 0.003 [c]	0.129 ± 0.001 [cd]	9.5 ± 0.2 [a]	0.31 ± 0.01 [b]
Quinic acid	0.43 ± 0.01 [d]	0.29 ± 0.02 [e]	tr	tr	0.056 ± 0.001 [f]	0.92 ± 0.01 [c]	4.2 ± 0.1 [b]	4.82 ± 0.06 [a]
Malic acid	0.81 ± 0.02 [e]	1.87 ± 0.01 [a]	1.42 ± 0.02 [d]	1.62 ± 0.02 [b]	1.51 ± 0.01 [c]	0.40 ± 0.02 [f]	0.008 ± 0.001 [g]	tr
Citric acid	0.39 ± 0.02 [f]	0.55 ± 0.01 [d]	0.49 ± 0.02 [e]	0.75 ± 0.04 [c]	1.15 ± 0.04 [b]	0.77 ± 0.03 [c]	1.9 ± 0.1 [a]	n.d.
Fumaric acid	0.0076 ± 0.0004 [a]	0.0049 ± 0.0002 [b]	tr	tr	tr	tr	0.0019 ± 0.0001 [c]	n.d.
Total organic acids	1.96 ± 0.05 [f]	3.042 ± 0.003 [c]	2.002 ± 0.004 [f]	2.55 ± 0.02 [d]	2.92 ± 0.03 [c]	2.22 ± 0.04 [e]	15.6 ± 0.3 [a]	4.95 ± 0.06 [b]
Free Sugars (g/100 g dw)								
Fructose	0.41 ± 0.07 [e]	1.41 ± 0.09 [a]	1.1 ± 0.1 [b]	0.91 ± 0.05 [c]	0.53 ± 0.08 [d]	0.21 ± 0.06 [f]	0.14 ± 0.01 [f]	0.15 ± 0.02 [f]
Glucose	0.144 ± 0.003 [d]	0.19 ± 0.01 [c]	0.29 ± 0.03 [b]	0.29 ± 0.03 [b]	0.30 ± 0.01 [b]	0.10 ± 0.07 [e]	0.27 ± 0.01 [b]	0.557 ± 0.004 [a]
Sucrose	1.73 ± 0.07 [d]	4.97 ± 0.07 [a]	2.97 ± 0.03 [b]	2.82 ± 0.04 [c]	1.3 ± 0.1 [e]	0.33 ± 0.07 [f]	0.28 ± 0.02 [g]	0.12 ± 0.01 [h]
Trehalose	0.32 ± 0.06 [e]	0.30 ± 0.05 [ef]	0.75 ± 0.04 [c]	1.16 ± 0.03 [a]	0.90 ± 0.02 [b]	0.24 ± 0.05 [f]	0.34 ± 0.02 [e]	0.57 ± 0.02 [d]
Raffinose	1.77 ± 0.08 [b]	2.13 ± 0.04 [a]	1.76 ± 0.04 [b]	1.72 ± 0.06 [b]	n.d.	n.d.	n.d.	n.d.
Total free sugars	4.4 ± 0.1 [c]	9.0 ± 0.2 [a]	6.8 ± 0.2 [b]	6.9 ± 0.2 [b]	3.0 ± 0.2 [d]	0.9 ± 0.2 [f]	1.03 ± 0.02 [f]	1.40 ± 0.03 [e]

Results are presented as mean ± standard deviation. Different letters correspond to significant differences ($p < 0.05$). dw—dry weight; tr—traces; n.d.—not detected. Calibration curves for organic acids: oxalic acid ($y = 9106x + 45.973$, $R^2 = 0.9901$); Quinic acid ($y = 610{,}607x + 46.061$, $R^2 = 0.9995$); Citric acid ($y = 1106 x + 45.682$, $R^2 = 0.9997$).

The free sugars composition of cardoon bracts is presented in Table 3. A great variation in sugar composition was observed when considering the effect of maturation stage, which suggests that the variation in environmental conditions could be the reason for the observed oscillations. Bracts of early- to mid-maturity stages (samples C1–C6) presented higher concentrations of sucrose and raffinose (0.12–4.97 and 1.72–2.13 g/100 g dw, respectively), whereas in the remaining samples (C7 and C8), the total free sugars content decreased significantly, and trehalose was the sugar present in higher abundance (0.34–0.57 g/100 g dw). Moreover, immature bracts presented higher levels of sugars than the mature ones, a trend that could be associated with the increase of organic acids at late maturity stages. According to Mandim et al. [33], the decrease of free sugars at late maturity stages could be attributed firstly to inulin formation and carbohydrate translocation in other plant parts such as heads, and secondly to the increased needs of osmolytes that help plants to overcome the developing stressful conditions over time, as already reported in other wild species grown under stress conditions [41,44,45]. To the best of our knowledge, this is the first report that analyzes the influence of the growth cycle on these parameters of chemical composition, where according to our results, immature bracts (sample C2) presented the highest contents and total free sugars, while samples with higher grade of maturation (samples C7 and C8) presented the highest content in organic acids and tocopherols.

4. Conclusions

The climatic conditions and the physiological changes to which cardoon plants are subjected throughout their growth cycle have a high impact on their chemical composition. In this work, it was found that the state of maturation has a high influence on the chemical composition of cardoon bracts in regard to lipidic fraction and the content in fatty acids, tocopherols, organic acids, and free sugars. This study is an important contribution to a more complete characterization of the chemical composition of cardoon bracts and reveals how the different phases of growth cycle can influence bioactive compounds content. The obtained results can be used for the sustainable use of bracts through the extraction of compounds with high biochemical value and consequently for the valorization of this species and the increase of the added value of this multifaceted crop.

Author Contributions: Methodology, Investigation, Writing—Original Draft: F.M.; Writing—Reviewing and Editing: S.A.P.; Methodology: K.D.G.; Conceptualization, Methodology, Writing—Reviewing and Editing: C.S.-B., S.A.P., I.C.F.R.F. and L.B. All authors have read and agreed to the published version of the manuscript.

Funding: This work was financially support by national funds FCT/MCTES to CIMO (UIDB/00690/2020). We also thank the FCT for the PhD grant (SFRH/BD/146614/2019) of F. Mandim. The authors are also grateful to the financial support though the TRANSCoLAB (0612_TRANS_CO_LAB_2_P).

Acknowledgments: The authors wish to thank the Foundation for Science and Technology (FCT, Portugal) for the contract of L. Barros through the institutional scientific employment program.

Conflicts of Interest: The authors declare no conflict of interest.

References

1. Sonnante, G.; Pignone, D.; Hammer, K. The domestication of artichoke and cardoon: From Roman times to the genomic age. *Ann. Bot.* **2007**, *100*, 1095–1100. [CrossRef]
2. Pappalardo, H.D.; Toscano, V.; Puglia, G.D.; Genovese, C.; Raccuia, S.A. *Cynara cardunculus* L. as a Multipurpose Crop for Plant Secondary Metabolites Production in Marginal Stressed Lands. *Front. Plant. Sci.* **2020**, *11*, 1–14. [CrossRef] [PubMed]
3. Conceição, C.; Martins, P.; Alvarenga, N.; Dias, J.; Lamy, E.; Garrido, L.; Gomes, S.; Freitas, S.; Belo, A.; Brás, T.; et al. *Cynara cardunculus*: Use in Cheesemaking and pharmaceutical applications. In *Technological Approaches for Novel Applications in Dairy Processing*; IntechOpen: London, UK, 2012; pp. 73–107.
4. Zayed, A.; Serag, A.; Farag, M.A. *Cynara cardunculus* L.: Outgoing and potential trends of phytochemical, industrial, nutritive and medicinal merits. *J. Funct. Foods* **2020**, *69*, 103937. [CrossRef]
5. Gominho, J.; Curt, M.D.; Lourenço, A.; Fernández, J.; Pereira, H. *Cynara cardunculus* L. as a biomass and multi-purpose crop: A review of 30 years of research. *Biomass Bioenergy* **2018**, *109*, 257–275. [CrossRef]
6. Fernández, J.; Curt, M.D.; Aguado, P.L. Industrial applications of *Cynara cardunculus* L. for energy and other uses. *Ind. Crops Prod.* **2006**, *24*, 222–229. [CrossRef]

7. Ierna, A.; Sortino, O.; Mauromicale, G. Biomass, seed and energy yield of *Cynara cardunculus* L. as affected by environment and season. *Agronomy* **2020**, *10*, 1548. [CrossRef]
8. Docimo, T.; De Stefano, R.; Cappetta, E.; Lisa Piccinelli, A.; Celano, R.; De Palma, M.; Tucci, M. Physiological, biochemical, and metabolic responses to short and prolonged saline stress in two cultivated cardoon genotypes. *Plants* **2020**, *9*, 554. [CrossRef]
9. Gostin, A.I.; Waisundara, V.Y. Edible flowers as functional food: A review on artichoke (*Cynara cardunculus* L.). *Trends Food Sci. Technol.* **2019**, *86*, 381–391. [CrossRef]
10. Barracosa, P.; Barracosa, M.; Pires, E. Cardoon as a Sustainable Crop for Biomass and Bioactive Compounds Production. *Chem. Biodivers.* **2019**, *16*, e1900498. [CrossRef]
11. Abu-Reidah, I.M.; Arraez-Roman, D.; Segura-Carretero, A.; Fernandez-Gutierrez, A. Extensive characterisation of bioactive phenolic constituents from globe artichoke (*Cynara scolymus* L.) by HPLC-DAD-ESI-QTOF-MS. *Food Chem.* **2013**, *141*, 2269–2277. [CrossRef]
12. Pandino, G.; Lombardo, S.; Mauromicale, G.; Williamson, G. Phenolic acids and flavonoids in leaf and floral stem of cultivated and wild *Cynara cardunculus* L. genotypes. *Food Chem.* **2011**, *126*, 417–422. [CrossRef]
13. Eljounaidi, K.; Comino, C.; Moglia, A.; Cankar, K.; Genre, A.; Hehn, A.; Bourgaud, F.; Beekwilder, J.; Lanteri, S. Accumulation of cynaropicrin in globe artichoke and localization of enzymes involved in its biosynthesis. *Plant. Sci.* **2015**, *239*, 128–136. [CrossRef] [PubMed]
14. Capozzi, F.; Sorrentino, M.C.; Caporale, A.G.; Fiorentino, N.; Giordano, S.; Spagnuolo, V. Exploring the phytoremediation potential of *Cynara cardunculus*: A trial on an industrial soil highly contaminated by heavy metals. *Environ. Sci. Pollut. Res.* **2020**, *27*, 9075–9084. [CrossRef] [PubMed]
15. Almeida, C.M.; Simões, I. Cardoon-based rennets for cheese production. *Appl. Microbiol. Biotechnol.* **2018**, *102*, 4675–4686. [CrossRef]
16. Bousios, S.; Worrell, E. Towards a Multiple Input-Multiple Output paper mill: Opportunities for alternative raw materials and sidestream valorisation in the paper and board industry. *Resour. Conserv. Recycl.* **2017**, *125*, 218–232. [CrossRef]
17. Francaviglia, R.; Bruno, A.; Falcucci, M.; Farina, R.; Renzi, G.; Russo, D.E.; Sepe, L.; Neri, U. Yields and quality of *Cynara cardunculus* L. wild and cultivated cardoon genotypes. A case study from a marginal land in Central Italy. *Eur. J. Agron.* **2016**, *72*, 10–19. [CrossRef]
18. Piluzza, G.; Molinu, M.G.; Re, G.A.; Sulas, L. Phenolic compounds content and antioxidant capacity in cardoon achenes from different head orders. *Nat. Prod. Res.* **2019**, *5*, 1–5. [CrossRef]
19. Gominho, J.; Lourenço, A.; Palma, P.; Lourenço, M.E.; Curt, M.D.; Fernández, J.; Pereira, H. Large scale cultivation of *Cynara cardunculus* L. for biomass production—A case study. *Ind. Crops Prod.* **2011**, *33*, 1–6. [CrossRef]
20. Chihoub, W.; Dias, M.I.; Barros, L.; Calhelha, R.C.; Alves, M.J.; Harzallah-Skhiri, F.; Ferreira, I.C.F.R. Valorisation of the green waste parts from turnip, radish and wild cardoon: Nutritional value, phenolic profile and bioactivity evaluation. *Food Res. Int.* **2019**, *126*, 108651. [CrossRef]
21. Pagano, I.; Piccinelli, A.L.; Celano, R.; Campone, L.; Gazzerro, P.; Falco, E.; Rastrelli, L. Chemical profile and cellular antioxidant activity of artichoke by-products. *Food Funct.* **2016**, *7*, 4841–4850. [CrossRef]
22. D'Antuono, I.; Carola, A.; Sena, L.M.; Linsalata, V.; Cardinali, A.; Logrieco, A.F.; Colucci, M.G.; Apone, F. Artichoke polyphenols produce skin anti-age effects by improving endothelial cell integrity and functionality. *Molecules* **2018**, *23*, 2729. [CrossRef] [PubMed]
23. Faria-Silva, C.; Ascenso, A.; Costa, A.M.; Marto, J.; Carvalheiro, M.; Ribeiro, H.M.; Simões, S. Feeding the skin: A new trend in food and cosmetics convergence. *Trends Food Sci. Technol.* **2020**, *95*, 21–32. [CrossRef]
24. Curt, M.D.; Sánchez, G.; Fernández, J. The potential of Cynara cardunculus L. for seed oil production in a perennial cultivation system. *Biomass Bioenergy* **2002**, *23*, 33–46. [CrossRef]
25. Jiménez-Moreno, N.; Cimminelli, M.J.; Volpe, F.; Ansó, R.; Esparza, I.; Mármol, I.; Rodríguez-Yoldi, M.J.; Ancín-Azpilicueta, C. Phenolic composition of Artichoke waste and its antioxidant capacity on differentiated Caco-2 cells. *Nutrients* **2019**, *11*, 1723. [CrossRef]
26. Petropoulos, S.A.; Pereira, C.; Barros, L.; Ferreira, I.C.F.R. Leaf parts from Greek artichoke genotypes as a good source of bioactive compounds and antioxidants. *Food Funct.* **2017**, *8*, 2022–2029. [CrossRef]
27. Petropoulos, S.A.; Pereira, C.; Tzortzakis, N.; Barros, L.; Ferreira, I.C.F.R. Nutritional value and bioactive compounds characterization of plant parts from *Cynara cardunculus* L. (Asteraceae) cultivated in central Greece. *Front. Plant. Sci.* **2018**, *9*, 1–12. [CrossRef]
28. Barbulova, A.; Colucci, G.; Apone, F. New trends in cosmetics: By-products of plant origin and their potential use as cosmetic active ingredients. *Cosmetics* **2015**, *2*, 82–92. [CrossRef]

29. Li, Y.; Kong, D.; Fu, Y.; Sussman, M.R.; Wu, H. The effect of developmental and environmental factors on secondary metabolites in medicinal plants. *Plant. Physiol. Biochem.* **2020**, *148*, 80–89. [CrossRef]

30. Mandim, F.; Dias, M.I.; Pinela, J.; Barracosa, P.; Ivanov, M.; Ferreira, I.C.F.R. Chemical composition and *in vitro* biological activities of cardoon (*Cynara cardunculus* L. var. *altilis* DC.) seeds as influenced by viability. *Food Chem.* **2020**, *323*, 126838. [CrossRef]

31. Pandino, G.; Lombardo, S.; Mauromicale, G. Globe artichoke leaves and floral stems as a source of bioactive compounds. *Ind. Crops Prod.* **2013**, *44*, 44–49. [CrossRef]

32. Archontoulis, S.V.; Struik, P.C.; Vos, J.; Danalatos, N.G. Phenological growth stages of *Cynara cardunculus*: Codification and description according to the BBCH scale. *Ann. Appl. B* **2013**, *156*, 253–270. [CrossRef]

33. Mandim, F.; Petropoulos, S.A.; Fernandes, Â.; Santos-Buelga, C.; Ferreira, I.C.F.R.; Barros, L. Chemical composition of *Cynara cardunculus* L. var. *altilis* heads: The impact of harvesting time. *Agronomy* **2020**, *10*, 1088. [CrossRef]

34. Petropoulos, S.A.; Pereira, C.; Ntatsi, G.; Danalatos, N.; Barros, L.; Ferreira, I.C.F.R. Nutritional value and chemical composition of Greek artichoke genotypes. *Food Chem.* **2018**, *267*, 296–302. [CrossRef] [PubMed]

35. Barros, L.; Pereira, E.; Calhelha, R.C.; Dueñas, M.; Carvalho, A.M.; Santos-Buelga, C.; Ferreira, I.C.F.R. Bioactivity and chemical characterization in hydrophilic and lipophilic compounds of *Chenopodium ambrosioides* L. *J. Funct. Foods* **2013**, *5*, 1732–1740. [CrossRef]

36. Mandim, F.; Barros, L.; Calhelha, R.C.; Abreu, R.M.V.; Pinela, J.; Alves, M.J.; Heleno, S.; Santos, P.F.; Ferreira, I.C.F.R. *Calluna vulgaris* (L.) Hull: Chemical characterization, evaluation of its bioactive properties and effect on the vaginal microbiota. *Food Funct.* **2019**, *10*, 78–89. [CrossRef]

37. Dias, M.I.; Barros, L.; Morales, P.; Sánchez-Mata, M.C.; Oliveira, M.B.P.P.; Ferreira, I.C.F.R. Nutritional parameters of infusions and decoctions obtained from *Fragaria vesca* L. roots and vegetative parts. *LWT Food Sci. Technol.* **2015**, *62*, 32–38. [CrossRef]

38. Petropoulos, S.A.; Fernandes, Â.; Calhelha, R.C.; Danalatos, N.; Barros, L.; Ferreira, I.C.F.R. How extraction method affects yield, fatty acids composition and bioactive properties of cardoon seed oil? *Ind. Crops Prod.* **2018**, *124*, 459–465. [CrossRef]

39. Mandim, F.; Petropoulos, S.A.; Dias, M.I.; Pinela, J.; Kostic, M.; Soković, M.; Santos-Buelga, C.; Ferreira, I.C.F.R.; Barros, L. Seasonal variation in bioactive properties and phenolic composition of cardoon (*Cynara cardunculus* var. *altilis*) bracts. *Food Chem.* **2020**, *336*, 127744. [CrossRef]

40. Mandim, F.; Petropoulos, S.A.; Giannoulis, K.D.; Dias, M.I.; Fernandes, Â.; Pinela, J.; Kostic, M.; Soković, M.; Barros, L.; Santos-Buelga, C.; et al. Seasonal variation of bioactive properties and phenolic composition of *Cynara cardunculus* var. *altilis*. *Food Res. Int.* **2020**, *134*, 109281. [CrossRef]

41. Petropoulos, S.A.; Fernandes, Â.; Dias, M.I.; Pereira, C.; Calhelha, R.; Di Gioia, F.; Tzortzakis, N.; Ivanov, M.; Sokovic, M.; Barros, L.; et al. Wild and cultivated *Centaurea raphanina* subsp. *mixta*: A valuable source of bioactive compounds. *Antioxidants* **2020**, *9*, 314. [CrossRef]

42. Georgieva, E.; Karamalakova, Y.; Nikolova, G.; Grigorov, B.; Pavlov, D.; Gadjeva, V.; Zheleva, A. Radical scavenging capacity of seeds and leaves ethanol extracts of *Cynara scolymus* L.—A comparative study. *Biotechnol. Biotechnol. Equip.* **2014**, *26*, 151–155. [CrossRef]

43. Petropoulos, S.; Fernandes, Â.; Pereira, C.; Tzortzakis, N.; Vaz, J.; Soković, M.; Barros, L.; Ferreira, I.C.F.R. Bioactivities, chemical composition and nutritional value of *Cynara cardunculus* L. seeds. *Food Chem.* **2019**, *289*, 404–412. [CrossRef] [PubMed]

44. Petropoulos, S.A.; Fernandes, Â.; Dias, M.I.; Pereira, C.; Calhelha, R.C.; Chrysargyris, A.; Tzortzakis, N.; Ivanov, M.; Sokovic, M.D.; Barros, L.; et al. Chemical composition and plant growth of *Centaurea raphanina* subsp. *mixta* plants cultivated under saline conditions. *Molecules* **2020**, *25*, 2204. [CrossRef] [PubMed]

45. Petropoulos, S.; Levizou, E.; Ntatsi, G.; Fernandes, Â.; Petrotos, K.; Akoumianakis, K.; Barros, L.; Ferreira, I. Salinity effect on nutritional value, chemical composition and bioactive compounds content of *Cichorium spinosum* L. *Food Chem.* **2017**, *214*, 129–136. [CrossRef]

Publisher's Note: MDPI stays neutral with regard to jurisdictional claims in published maps and institutional affiliations.

 *agronomy*

Article

Methods of Silicon Application on Organic Spring Wheat (*Triticum aestivum* L. spp. vulgare) Cultivars Grown across Two Contrasting Precipitation Years

Jolanta Kowalska [1,*], Józef Tyburski [2], Jan Bocianowski [3], Joanna Krzymińska [1] and Kinga Matysiak [4]

[1] Department of Organic Agriculture and Environmental Protection, Institute of Plant Protection—National Research Institute, Władysława Węgorka 20, 60–318 Poznań, Poland; J.Krzyminska@iorpib.poznan.pl
[2] Department of Agroecosystems, University of Warmia and Mazury in Olsztyn, Michała Oczapowskiego 2, 10-719 Olsztyn, Poland; jozef.tyburski@uwm.edu.pl
[3] Department of Mathematical and Statistical Methods, Poznań University of Life Sciences, Wojska Polskiego 28, 60–637 Poznań, Poland; jan.bocianowski@up.poznan.pl
[4] Department of Weed Science and Plant Protection Techniques, Institute of Plant Protection—National Research Institute, Władysława Węgorka 20, 60–318 Poznań, Poland; K.Matysiak@iorpib.poznan.pl
* Correspondence: j.kowalska@iorpib.poznan.pl; Tel.: +48-61864-9077

Received: 30 September 2020; Accepted: 20 October 2020; Published: 27 October 2020

Abstract: The potential of silicon used in two forms, two methods and three cultivars of spring wheat cultivated under organic farming conditions is high, as it helps plants to alleviate abiotic stresses. The research hypotheses of paper were the assumptions that the effectiveness of silicon may differ not only by the form of silicon and the method of its application, but also by the variety of common wheat and different water conditions in the soil during the growing season. These hypotheses were confirmed. The aim of this study was to determine the effectiveness of liquid and powder silicon forms and different methods of application in three cultivars (Harenda, Serenada and Rusałka) of spring wheat organically grown under a specific field experiment in water stress vs. no stress conditions. The water stress of plants was assessed on the basis of the sum of precipitation in the winter–spring and vegetation season in each year. The differences in water availability for the plants in the experimental years were confirmed. Silicon (Si) was used for seed dressing and/or for leaf spraying. In the first case, the powdered form of Si was used at a dose of 0.5 kg/100 kg of seeds; it was used together with the liquid form at a dose of 0.5 L/100 kg of seeds, and in the second, the liquid form of Si was used at a dose of 0.5 L per 200 L of water per hectare; spraying was carried out at the following plant development stages: three tillers detectable, the first node and the flag leaf. The application of Si positively influenced the wheat yield depending on the method of Si application, wheat variety and severity of water stress. The cultivar Harenda was more susceptible to lower water content in the soil than the cultivars Rusałka and Serenada. Under conditions of water stress, the use of Si slowed the development of young Harenda plants, but ultimately, the variety increased its grain yield to a greater extent than the other two varieties. The lowest weight of a thousand grains (TGW) was found in the Harenda variety; however, Si treatment improved this parameter. Si increased the yields of the three wheat varieties, and the highest were harvested in plots with combined Si treatments. The yields of the Rusałka and Serenada cultivars on these plots were 14 to 28% higher compared to the control. Harenda was the least fertile variety, but it increased its yield more than the other two varieties. This variety increased its yield in 2018 (year of average rainfall) by 26% from 2.92 to 3.94 tons per hectare, and in 2019 (a year of drought) by 42% from 1.66 to 2.87 tons per hectare. It can be concluded that Si improves the wheat yield, and its efficiency depends on the scale of water stress, the method of application and the variety. The simplest and most adaptable method of Si application is seed dressing and has prospects for wider application, especially in organic farming.

Keywords: diatomaceous earth; monosilicic acid; Si application method; soil water conditions; wheat cultivar

---

1. Introduction

Climate change, including rising temperatures and increasingly severe droughts, has hampered crop development and yields. Plants mitigate the effects of less soil water content using physiological mechanisms and produce bioactive compounds such as antioxidants and osmolytes. In the literature, the influence of ascorbic acid, glutathione and proline on the alleviation of the harmful effects of drought stress for plants is noted. Studies on the use of egozogenic antioxidants and proline are described; the synergistic effect of these substances is noted [1]. Proline has possible options for consideration as an indicator and a potential marker of clinical damage by osmotic stress; however, degradation and toxicity are potential threats posed by proline. Mycorrhizal plants and their non-mycorrhizal counterparts show varied expression patterns regarding proline [2].

Another of the alternative methods for alleviating negative stress effects might be application of silicon as a fertilizer (root or foliar application). An abundant mineral element in plant tissues, silicon (Si) provides structural support and improves tolerance to disease, drought, metal toxicity [3,4] and biotic stresses including plant pathogen and insect pests [5]. The excessive use of Si does not harm or pollute plants or corrode machinery [6]. Silicon is present in plants in amounts equivalent to those of macronutrient elements such as calcium, magnesium and phosphorus, and in grasses, often at higher levels than any other inorganic constituent [7]. Plants take up Si from soil solution both passively and actively and are unable to accumulate Si. Studies conducted about fifty years ago showed that species of Poaceae contained up to ten to twenty times more Si than non-monocotyledonous species [8]. More recent research on specimens from botanical gardens indicates that high Si accumulation is restricted to primitive plants and to some monocot clades, namely the Poaceae, Cyperaceae and Commelinaceae [9]. According to Sacała [10], many plants, particularly monocotyledonous species, contain large amounts of Si (up to 10% of dry mass). The role of Si in plants is not restricted to the formation of a physical or mechanical barrier (as precipitated amorphous silica) in cell walls, lumens and intercellular voids; silicon can also modulate plants' metabolism and alter physiological activities, particularly in plants subjected to stress conditions. However, in some plants, increased silicon application does not improve plant growth; a better understanding of the interactions between silicon application and plant responses will contribute to more efficient Si use, especially under stress conditions [10].

Usually, plants in natural conditions do not demonstrate Si shortages. Nevertheless, in crop production, fertilizers containing Si are often applied to crops such as rice and sugar cane to increase their yield and quality [11–13], but the positive effect of Si application is not restricted to monocotyledonous plants. In tests with collard, silicon suppressed the harmful effects of drought on leaf and root length. This suggests that monocots (e.g., *Triticum aestivum* L.) react to Si application similarly to non-monocot plant species (e.g., *Brassica oleracea*) [14]. Another advantage is that Si application increases the respiration of soils deficient in phosphorous. A major component in regulating P mobilization in Arctic soils, Si is assumed to play a role in the management of P availability in all types of soils [15].

Silicon has been widely reported to improve the growth, biomass, yield and quality of a wide variety of crops including monocotyledonous crops such as wheat, rice, maize and barley. The observed increases in grain yield, however, may be due not only to the beneficial effects of Si fertilization (such as growth promotion, lodging resistance, and biotic and abiotic stress resistance), but also to certain indirect effects such slight pH changes and the uptake of macro- and micronutrients contained in the Si-based fertilizers [16]. Early studies on the effect of Si on plant growth were inconclusive, but studies that are more recent indicate that Si can have a beneficial effect on many aspects of plant growth, most notably in rice. Tibbitts [17] reported the effect of supplemental silicon (Si) on wheat (*T. aestivum* cv. "USU-Apogee") from studies in a mini-lysimeter system imposing drought and salinity stress. There was no effect of Si on the harvest index, TGW or grains per spike.

Increased Si concentrations in plants not only maintain the water status but also improve drought resistance by regulating the leaf water potential, helping in CO_2 assimilation and decreasing transpiration through the adjustment of the leaf area [18]. A study by Ming et al. [19] suggests that Si application increases the water and osmotic potential in roots and leaves. Xu et al. [20] reported that Si-mediated changes result in a new balance of endogenous hormones and enhance the tolerance of the wheat plants to drought stress. Based on the results of previous studies, it can be concluded that foliar nutrition should be introduced as a standard treatment in the crop management of many species of agricultural plants. It can help farmers to increase crop yields [21].

Si should be used in organic crops where problems sometimes occur with plant nutrition and pest pressure, and only a limited range of fertilizers and plant protection products is permitted for use in organic farming. This limited range contributes to water stress that is especially hard for inadequately nourished plants to tolerate. In another paper was assessed how Si influenced the growth parameters and yield of spring wheat, both in powder and liquid form, applied to soil and leaves, respectively, and in combined methods of application. Si stimulated the growth of organic spring wheat and increased grain yields. Liquid Si was more effective than powdered Si, and the combined application of Si to soil and leaves was more effective than only soil or only foliar [22]. The research hypothesis of this work was based on the assumption that the reaction of wheat to the use of silicon may vary depending on the variety, also. Moreover, the work took into account an additional factor determining the effective use of silicon, concerning the varied water conditions in the course of wheat vegetation. Since the effectiveness of silicon application depends on the structure of silicon compounds and the way they are used, the question arises whether the usefulness of silicon may differ by genotype within a given crop species.

The potential of Si should be used in organic crops where there are problems with plant nutrition and pests due to the strict rules about the application of fertilizers and conservation measures in organic farming. In this growing system, water stress tolerance is particularly difficult for plants that are often less well fed. Therefore, an effective method of silicon delivery, including seed treatment, needs to be developed to minimize the harmful effects of abiotic and biotic stress factors. Chemical dressing is forbidden in organic farming; therefore, natural products are tested for this purpose [23,24].

The aim of the study was to determine the effect of powder and liquid silicon used as seed dressing and foliar treatments on the growth and quality and quantity of yield of three cultivars of wheat grown organically under various conditions of water availability in soil.

2. Materials and Methods

The field experiments under an organic regime were conducted in the years of 2018–2019 at an experimental agricultural station (52°2′ N; 17°4′ E) of the Institute of Plant Protection, National Research Institute (IPP-NRI), in Poland. The experiment was performed with Harenda, Rusałka and Serenada cultivars of *T. aestivum* L., which were grown in medium-heavy soil and followed potatoes in crop rotation. The three selected spring wheat cultivars were recommended in Poland to be grown in organic farming [25]. Soil samples were taken at the level of 0–20 cm (as a spade test), and the soil chemical properties of the experimental fields were analyzed. The average values of soil fertility from the two experimental years were as follows: the soil pH was slightly acidic (6.1), the organic matter content low (1.3%), P (112 mg kg^{-1}) and K (129 mg kg^{-1}) medium, and Mg (64 mg kg^{-1}) content high, which made the soil appropriate for growing spring wheat.

2.1. Summary of the Varieties Used in the Experiments

Rusałka—Qualitative variety (group A). Yield good or very good. Poor disease resistance. Medium TGW, high to very-high bulk density. High protein content, high amount of gluten. Large to very-high SDS sedimentation rate. The flour yield is quite low.

Serenade—Qualitative variety (group A). Good yield. Moderate disease resistance. TGW very high, bulk density high to very-high. High to very-high protein content, very high amount of gluten. Very high SDS sedimentation rate. Flour yield is average.

Harenda—Bread variety (group B). High disease resistance. Average TGW, high bulk density. Protein and gluten content are quite high. Very high SDS sedimentation rate. Average flour yield.

2.2. Meteorological Conditions during Tests

The meteorological conditions differed markedly in the growing seasons of 2018 and 2019. According to data from the Agricultural Meteorological Station of the Institute of Plant Protection, National Research Institute, located in Winna Góra, precipitation in the 2018 growing season was very similar to the average for the period of 1998–2017, but the temperature was higher by 2.5 °C (Table 1). The water available in the soil was assessed on the basis of both the precipitation/snowfall in the winter time until sowing time (November–April) and precipitation during the growing season in both experimental years (Table 1). In 2019, the rainfall for the growing season (April–August) dropped from 245.6 mm to 101.1 mm and was ca. 2.5-times lower than in 2018, causing high water stress. To sum up, the growing seasons of 2018 and 2019 had contrasting weather conditions characterized by a typical rainfall rate in 2018 and a very dry season in 2019 (Table 1). In the 2017/2018 winter–spring time, the total rainfall was 195.7 mm, and in the 2018/2019 season, a much lower snow/rain total of only 128.7 mm was recorded.

Table 1. Mean air temperatures and rainfall during spring wheat vegetation; data for the period 1998–2017 and for 2018 and 2019, Meteorological Station in Winna Góra.

Month/year	Means for 1998–2017		2018		2019	
	Temp., °C	Rainfall, mm	Temp., °C	Rainfall, mm	Temp., °C	Rainfall, mm
April	9.5	30.5	13.7	28.9	10.7	6.1
May	14.2	52.7	17.5	37.2	12.4	83.4
June	17.4	52.4	19.0	48.0	22.7	2.1
July	19.7	85.5	20.9	112.8	19.5	4.7
August	19.1	63.1	21.6	18.7	21.2	4.8
Mean/sum April–August	16.0	284.2	18.5	245.6	17.3	101.1

2.3. Field Experiment Design

Two silicon products were used—AdeSil® as a powder formulation and ZumSil® as a liquid trade formulation. ZumSil™ is a 24% solution of monosilicic acid. AdeSil® is amorphous diatomaceous earth with a flour texture and contains 89–95% amorphous silica (SiO_2). The studies concerned seed dressings carried out as a simple treatment or combined with three foliar treatments. Different combinations were used: (1) untreated plot, (2) only seed dressing, (3) three foliar treatments, and (4) seed dressing combined with three foliar treatments. For the seed dressing, a powdered form of silicon was used at a dose of 0.5 kg/100 kg of seed and then mixed with a liquid form of silicon at a dose of 0.5 L/100 kg of seed. The foliar treatments were performed with a liquid form of silicon at a dose of 0.5 L with 200 L of water per hectare. Three foliar applications were performed at the following stages of plant development: BBCH 23 (3 tillers detectable), BBCH 31 (first node) and BBCH 39 (flag leaf); the time of the intervals between the foliar sprays was 7–10 days. Each combination was used on plots of 24 m^2 and repeated three times. The experiment was carried out using the random plot system. Plots devoid of silicon application were used as the control. Wheat was sown on the 8th and 2nd of April of 2018 and 2019, respectively, with standard row spacing (12.5 cm), with a standard sowing ratio of 200 kg of grain per hectare. Due to the prohibition of synthetic herbicides in organic farming, only mechanical weed control was performed. No mineral fertilizers were used. Ten young plants were collected from each plot, and two growth parameters (the lengths of their leaves and roots) after the first foliar treatments (at the BBCH 29 stage—end of tillering) were measured manually using graph paper. After the harvest, the TGW and the yield were established. In 2018, spring wheat was harvested on 12 August, while in 2019, when there was a severe drought, harvest was performed on 29 July.

2.4. Laboratory Analysis Grain Quality Parameters

The quality of the harvested grain was evaluated using multifunctional equipment available in the laboratory. A qualitative analysis was performed using a FOSS Infratec™ 1241 Grain Analyzer (FOSS, Hilleroed, Denmark). For the analysis of crude protein density, wet gluten and bulk density, a cleaned and dry 0.5 kg grain sample was collected from each combination of experiments in 2018 and 2019. Each analyzed grain sample (0.5 kg) consisted of sub-samples taken from each plot in one combination. The device is calibrated and accredited once a year.

2.5. Statistical Analysis

The normality of the distribution of the observed traits was tested. Three-way (year, cultivar and method of silicon supply) analysis of variance (ANOVA) was performed to verify the hypotheses of the lack of effects of the year, cultivar and method of silicon supply as well as the interactions year × cultivar, year × method of silicon supply, cultivar × method of silicon supply and year × cultivar × method of silicon supply on the variability of the thousand-grain weight, yield, length of leaves per plant and length of roots per plant. The means values and standard deviations were calculated for all the observed traits. The significance of the differences between the mean values was verified with Tukey's test at a level of $p < 0.05$. The GenStat v. 18 statistical software package was used for all the analyses.

3. Results

The results of analysis of variance indicated that all four observed traits (the thousand-grain weight, yield, length of leaves per plant and length of roots per plant) were influenced by the cultivar, method of silicon application and water availability in the soil (Tables 3–6), as well as the interaction of year × cultivar × method of silicon supply being confirmed (Table 2).

Table 2. Mean squares (m.s.) from three-way analysis of variances for four observed traits: thousand-grain weight (TGW), yield, length of leaves per plant and length of roots per plant.

Source of Variation	Thousand-Grain Weight		Yield		Length of Leaves per Plant		Length of Roots per Plant	
	d.f.	m.s.	d.f.	m.s.	d.f.	m.s.	d.f.	m.s.
Year	1	3701.2 ***	1	12.47 ***	1	7.427	1	7.001
Method	3	39.08 ***	3	2.84 ***	2	235.3 ***	2	24.87 **
Cultivar	2	115.75 ***	2	6.466 ***	3	26.09 ***	3	10.99 *
Year × Method	3	2.705 ***	3	0.013	2	24.14 **	2	2.618
Year × Cultivar	2	12.64 ***	2	0.198 ***	3	3.967	3	1.296
Method × Cultivar	6	1.800 ***	6	0.100 ***	6	159.7 ***	6	73.69 ***
Year × Method × Cultivar	6	1.008 ***	6	0.060 ***	6	13.4 **	6	7.157
Residual	36	0.031	36	0.005	372	3.631	372	4.076

$* p < 0.05; ** p < 0.01; *** p < 0.001$; d.f.—number of degrees of freedom.

The main effects of the year as well as the year × cultivar interaction were significant for the thousand-grain weight and yield (Table 2). The year × method of silicon supply interaction was statistically significant for the thousand-grain weight and length of leaves per plant; however, the year × cultivar × method of silicon supply interaction was significant for the thousand-grain weight, yield and length of leaves per plant (Table 2). The effect of Si application on the growth parameters of young wheat development, the grain yield and its quality was significant and related to water soil content; the results are presented in Table 3, Table 4, Table 5, and Table 6.

Table 3. Growth parameters of common wheat plants evaluated in the BBCH 29 growth stage depending on different methods of Si application.

Cultivar	Year	Si Application Method			
		Untreated	Seed Dressing	Foliar Treatments	Seed and Foliar Treatments
		Mean length of leaves per plant, cm			
Harenda	2018	20.60 ± 0.34 [a]	21.05 ± 0.12 [a]	20.00 ± 0.32 [a]	20.40 ± 0.09 [a]
	2019	21.88 ± 0.06 [a]	19.94 ± 0.26 [b]	18.92 ± 0.63 [bc]	17.73 ± 0.42 [c]
Rusałka	2018	20.36 ± 0.21 [a]	20.93 ± 0.11 [a]	20.24 ± 0.08 [a]	20.81 ± 0.23 [a]
	2019	15.25 ± 0.50 [b]	20.20 ± 0.78 a	19.57 ± 0.38 [a]	20.49 ± 0.62 [a]
Serenada	2018	20.75 ± 0.11 [a]	20.68 ± 0.32 [a]	20.27 ± 0.21 [a]	20.44 ± 0.15 [a]
	2019	19.26 ± 0.21 [b]	23.79 ± 1.04 [a]	23.24 ± 0.42 [a]	21.55 ± 1.07 [a]
		Mean length of roots per plant, cm			
Harenda	2018	13.03 ± 0.10 [a]	12.96 ± 0.21 [a]	12.82 ± 0.09 [a]	12.84 ± 0.51 [a]
	2019	14.44 ± 0.21 [a]	12.41 ± 0.43 [b]	11.89 ± 0.69 [b]	10.19 ± 0.22 [c]
Rusałka	2018	12.89 ± 0.34 [a]	12.43 ± 0.56 [a]	12.84 ± 0.23 [a]	12.68 ± 0.19 [a]
	2019	11.02 ± 0.72 [c]	12.51 ± 1.04 [bc]	14.15 ± 0.41 [a]	13.55 ± 0.45 [ab]
Serenada	2018	13.98 ± 0.05 [a]	13.85 ± 0.14 [a]	13.73 ± 0.45 [a]	14.01 ± 0.28 [a]
	2019	11.74 ± 1.05 [b]	13.29 ± 0.81 [a]	13.56 ± 0.22 [a]	13.76 ± 0.37 [a]

Values followed by the same letter are not statistically different at $p < 0.05$.

Table 4. Thousand-grain weight (TGW) (g) of wheat in relation to method of Si application and wheat cultivar.

Cultivar	Year	Si Application Method			
		Untreated	Seed Dressing	Foliar Treatments	Seed and Foliar Treatments
Harenda	2018	26.32 ± 0.22 [c]	29.01 ± 0.31 [b]	28.78 ± 0.32 [b]	31.38 ± 0.50 [a]
	2019	26.48 ± 0.77 [c]	27.85 ± 0.70 [b]	26.62 ± 0.48 [b]	29.37 ± 0.47 [a]
Rusałka	2018	30.09 ± 0.85 [a]	32.09 ± 0.50 [a]	30.34 ± 0.57 [a]	33.00 ± 0.27 [a]
	2019	28.85 ± 0.82 [a]	29.05 ± 0.11 [a]	28.98 ± 0.86 [a]	29.80 ± 0.30 [a]
Serenada	2018	31.89 ± 0.41 [b]	32.09 ± 0.78 [a]	32.02 ± 0.23 [a]	33.03 ± 0.51 [a]
	2019	30.80 ± 0.48 [a]	30.20 ± 0.64 [ca]	30.67 ± 0.59 [a]	31.18 ± 0.15 [a]

Values followed by the same letter are not statistically different at $p < 0.05$.

Table 5. Yield (t) in relation to method of Si application and wheat cultivar.

Cultivar	Year	Si Application Method			
		Untreated	Seed Dressing	Foliar Treatments	Seed and Foliar Treatments
Harenda	2018	2.92 ± 0.18 [b]	3.62 ± 0.26 [a]	3.23 ± 0.20 [a]	3.94 ± 0.19 [a]
	2019	1.66 ± 0.16 [c]	2.48 ± 0.24 [b]	2.04 ± 0.03 [bc]	2.87 ± 0.12 [a]
Rusałka	2018	3.60 ± 0.28 [b]	3.93 ± 0.19 [b]	3.78 ± 0.21 [b]	4.98 ± 0.33 [a]
	2019	2.89 ± 0.05 [c]	3.28 ± 0.13 [b]	3.09 ± 0.07 [bc]	3.74 ± 0.18 [a]
Serenada	2018	4.06 ± 0.32 [b]	4.47 ± 0.34 [ab]	4.32 ± 0.07 [b]	4.71 ± 0.19 [a]
	2019	3.12 ± 0.13 [c]	3.74 ± 0.06 [ab]	3.33 ± 0.18 [bc]	4.10 ± 0.22 [a]

Values followed by the same letter are not statistically different at $p < 0.05$.

Table 6. Selected parameters of wheat grain quality in relation to Si application method and wheat cultivar.

Cultivar	Year	Si Application Method			
		Untreated	Seed Dressing	Foliar Treatments	Seed and Foliar Treatments
		Crude Protein (%)			
Harenda	2018	13.7 ± 0.43 [a]	14.1 ± 0.36 [a]	13.9 ± 0.36 [a]	14.4 ± 0.51 [a]
	2019	14.3 ± 0.14 [b]	15.9 ± 0.34 [a]	14.8 ± 0.37 [b]	16.5 ± 0.41 [a]
Rusałka	2018	13.7 ± 0.35 [a]	14.0 ± 0.22 [a]	13.8 ± 0.10 [a]	14.2 ± 0.11 [a]
	2019	14.5 ± 0.26 [a]	14.9 ± 0.18 [a]	14.6 ± 0.14 [a]	15.0 ± 0.19 [a]
Serenada	2018	13.2 ± 0.27 [a]	13.5 ± 0.46 [a]	13.4 ± 0.42 [a]	13.7 ± 0.11 [a]
	2019	13.9 ± 0.15 [a]	14.5 ± 0.37 [a]	14.4 ± 0.41 [a]	14.6 ± 0.14 [a]
		Wet Gluten (cm^3)			
Harenda	2018	30.1 ± 0.76 [a]	31.8 ± 0.76 [a]	30.9 ± 0.45 [a]	32.1 ± 0.98 [a]
	2019	31.4 ± 0.98 [b]	34.9 ± 0.87 [a]	32.7 ± 0.98 [b]	36.0 ± 1.80 [a]
Rusałka	2018	29.4 ± 0.41 [a]	30.1 ± 0.98 [a]	29.7 ± 0.87 [a]	30.2 ± 0.65 [a]
	2019	32.2 ± 0.43 [a]	32.6 ± 0.78 [a]	32.6 ± 0.76 [a]	33.9 ± 0.65 [a]
Serenada	2018	33.7 ± 0.56 [a]	34.8 ± 0.54 [a]	34.2 ± 0.62 [a]	35.5 ± 0.78 [a]
	2019	38.2 ± 0.71 [a]	40.2 ± 0.91 [a]	39.4 ± 0.75 [a]	40.4 ± 0.86 [a]

For each parameter of wheat grain quality, values followed by the same letter are not statistically different at $p < 0.05$.

3.1. Growth Parameters

In 2018, when no problems with the availability of soil water for plants were observed, no differences in the development of the leaves and roots of young wheat plants were noticed (Table 3). The measurements were made after the first Si foliar application. Neither the wheat cultivar nor the Si form and method of application influenced the wheat.

In 2019, a dramatic water shortage changed the growth pattern of young wheat plants. The rainfall in April was five-times lower than usual (Table 1), but sufficient rain in May saved the plants from wilting. The wheat cultivars showed a different pattern of reaction to these spring water conditions. Cultivar Harenda developed the longest leaves without Si. Seed treatment significantly reduced leaf development. Treatment with foliar Si made the situation worse, and the combination of seed and foliar Si produced the worst results (Table 3). The same pattern of reaction to Si application was noted in the root development of this cultivar. Contrasting results were noted in 2019 in the development of young wheat plants of the Rusałka and Serenda cultivars. The mean length of Rusałka leaves was the shortest on the control object, and the leaves were shorter by ca. 25% than the leaves of the plants treated with Si, regardless of the form and the method of application (Table 3). The Rusałka roots were also the shortest in the control plots; however, the differences were not statistically significant compared to plants developed from seed dressing. This cultivar developed longer roots in response to Si seed dressing and foliar application, and the longest when only foliar applications were used. The Serenada cultivar showed almost the same pattern of development as Rusałka. It had the shortest leaves and roots when no Si treatment was applied; however, the length of the leaves and roots increased to the same degree as those of Rusałka, regardless of the form and method of Si application.

3.2. Impact of Si on Grain and Yield Development

The thousand-grain weight (TGW) is significantly affected by water stress and wheat cultivar [26]. A water deficit can affect plant growth and development in all stages; in early stages, the rate of tiller appearance, leaf appearance and leaf area are reduced; later on, the length of the stems is reduced together with the number of grains per spike, and stress after anthesis shortens the duration of grain filling, thus reducing the grain size [27]. In the presented results, the wheat cultivar also had a marked impact on the grain parameter, with the lowest TGW noted in the Harenda cultivar not treated with Si. This effect was found in both 2018 and 2019, and the TGWs stood at 26.32 and 26.48 g, respectively (Table 4).

The reduction of the grain weight as a basic parameter resulted in a very low wheat yield in the control plots, as the plant density (data not shown) was the same in all combinations of the study. The Si treatment had a positive effect on grain development, and the highest values of TGW were associated with the combined (seed and foliar) Si treatment. The larger grain size of the cultivar Rusałka was more frequent than in the cultivar Harenda, but in this case, no statistical effect of Si treatment on the TGW was observed. The grain development of the Serenada variety was similar to that of the Rusałka cutivar, although in 2018, the Serenada grains harvested from untreated plots were smaller than in the other combinations (Table 4).

In 2018, the weather during the growing season was typical, especially for rainfall; the average grain yield of the three spring wheat cultivars was almost four tons per hectare (3.90 t per hectare). In 2019, which was very dry, the average yields of wheat dropped by almost 1 t per hectare (to 3.03 t per hectare), that is, by ca. 25%. The cultivar choice, also an important factor, influenced the grain yields; the highest grain yields were obtained from the Serenada cultivar (3.98 t per hectare), somewhat lower yields were obtained from the Rusałka cultivar (3.66), and the lowest were from the Harenda cultivar (2.85 t per hectare) (data not shown).

The strongest response to Si treatment was found in the Harenda cultivar, both in normal 2018 and dry 2019 (Table 5). In both years, the lowest yields were harvested from the control plots. In 2018, the difference in grain yield between the control and foliar-treated wheat was 10%, and in dry 2019, it was 19%. The yield of wheat harvested in the control plots in 2019 was only 1.66 t per hectare, which is probably not profitable for organic-wheat growers. In the same year, the yields from the control plots of the Rusałka and Serenada cultivars were 2.89 and 3.12 t per hectare, respectively. Thus, the yield of the Serenada cultivar was almost double that of the Harenda cultivar. The most efficient way of Si application was the combined seed and foliar treatment, which resulted in a yield increase of the Harenda cultivar by 26% in 2018 and 42% in 2019. Although the increases in yields were high in both years, the result obtained in dry 2019 was exceptionally high.

The most effective method of Si application was the combined treatment of seed dressing and three sprayings. This application method increased the grain yields of the Rusałka cultivar by 28% in 2018 and 23% in 2019. In the case of the Serenada variety, the increase in yields was particularly noticeable in 2018 (18.8%), when no water-limiting drought was recorded, and much higher in dry 2019, when the yield increased by 24%.

3.3. Grain Quality

A few differences in the grain quality parameters in relation to Si application and cultivar choice were noted. In the case of the Rusałka and Serenada wheat cultivars, Si application had a statistically insignificant effect on the protein content in grain, but a tendency to promote protein accumulation in the grain of the Rusałka cultivar when the wheat was treated with Si as a seed dressing and foliar spraying was noted (Table 6). In 2019, for the Harenda cultivar, a higher protein content was the result of Si seed dressing (15.9%) and the combined Si treatment (seed dressing and three foliar sprayings) (16.5%). A lower protein content was found in the smallest grains from the control plots (14.3%) and foliar treatments (14.8%).

Gluten, another basic parameter of wheat grain quality, usually correlates with protein content. In general, the gluten content was high, showing a high baking quality for the grain. In a comparison of the results from 2018 and 2019 (normal and dry growing seasons), a tendency towards higher gluten content was found in the latter year. The Harenda cultivar grain harvested in 2018 had no difference in quality; in the dry 2019, the highest gluten content was in grain harvested from the plants treated with Si, seed dressing and combined application methods; the lowest content of gluten was noted for untreated crops (Table 6). The gluten contents of the Rusałka and Serenada cultivars were the same in all combinations, although Serenada showed a tendency towards greater gluten accumulation as a result of the combined seed dressing and foliar Si treatments. Another basic parameter used to assess the baking quality of the wheat grain was its bulk density, and a minimum value of 73 kg per hectoliter is expected for the best quality grain. This value was not always met in our research (Figure 1). This was the case with the Harenda variety in all experimental variants during the 2019 dry season, so this variety not only produced lower yields but also, in some respects, lower quality.

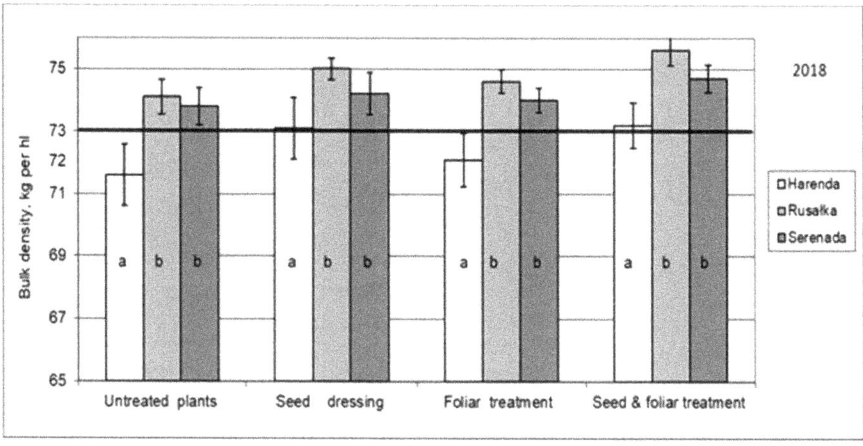

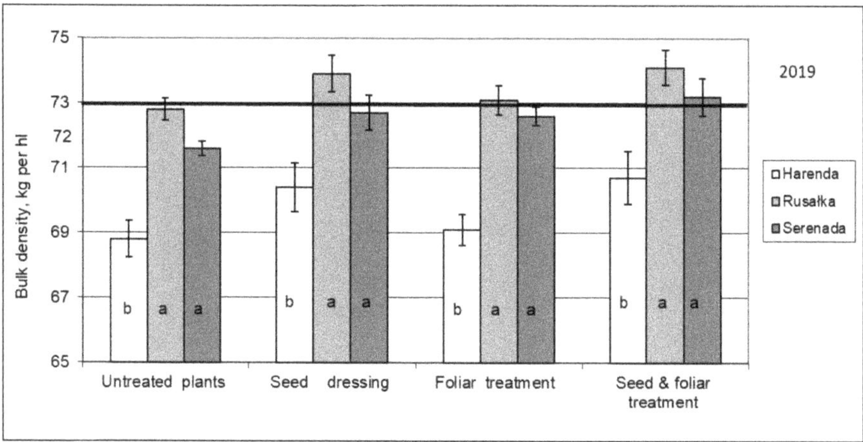

Figure 1. Effect of Si application method on bulk density of wheat cultivars in 2018 and 2019. Values followed by the same letter are not statistically different at $p < 0.05$. The bold line indicates the bulk density guide value (73 kg per hL) for wheat grain.

4. Discussion

In our research, it was shown that the use of silicon had a positive effect on the development and yielding of wheat grown organically, and the scale of this effect depended on both rainfall deficiencies and the variety. There are limited studies available in the literature on the effectiveness of Si in wheat, especially in organic farming and under drought stress. Ahmad et al. [28] investigated the role of silicon in the fertilization of wheat (*T. aestivum*) under various soil moisture conditions and found that Si application significantly improved plant biomass, growth and spike weight. This statement is in line with our observations. The use of monosilicic acid—as in the presented paper—absorbed by plants in almost every crop (compared to the control) resulted in an increase in the root mass, the development of thicker shoots, a larger leaf area and a higher chlorophyll content. Thus, the use of Si may have an indirect effect on the improvement of plant growth parameters by increasing the mass of roots and is associated with a higher uptake of nutrients (PCa, K, Si and Bo) from the soil [29]. The synergistic effects of silicon (Si) and salicylic acid (SA) applied at 6 mM Si, 1 mM SA and 6 mM Si + 1 mM SA on the grain yield and some key physiological characteristics of the wheat cultivars Shiraz (drought-sensitive) and Sirvan (drought-tolerant) were investigated [30]. Water-stress alleviation and yield improvement in the wheat cultivars by Si and SA application were attributed partly to improved osmotic adjustment and antioxidant activity, as well as to a more favourable water status under stress conditions. Generally, it was concluded that Si and SA application proved to have a great potential in advancing the grain yield of wheat in drought-prone areas [30]. This statement is confirmed by our results and can be a recommendation for farmers fighting with growing problems with water limitation during the growing season. The studies conducted by Guevel et al. [31] also confirm the beneficial effect of silicon application on wheat plants; however, effect Si amendment, either through the roots or the leaves had a biostimulating effect and did not increase plant growth. Their results lead to the conclusion that Si is primarily, if not exclusively, absorbed by the root system and that such absorption by the roots is necessary for an optimal prophylactic effect against powdery mildew. Although less effective than root applications, foliar treatments with both Si and nutrient salt solutions led to a significant reduction in powdery mildew on wheat plants. This suggests a direct effect of the products on powdery mildew rather than one mediated by the plant as in the case of root amendments. In our experiments were also made observations on the healthiness of the plants, and the frequency of spike diseases caused by *Septoria nodorum* and Fusarium spp. were noted. The severity of these diseases varied depending on the variety and use of silicon—this issue will be discussed in the next manuscript. The foliar application of silicon has a biostimulative effect, and the best results are observed in conditions stressful for plants such as salinity, a deficiency or excess of water, high and low temperatures, and the pressure of diseases and pests, etc. [21].

The plant response to Si application is greatly influenced by the genotype, and this phenomenon was noted by Dufey et al. [32] in rice crops. Our study also confirmed the impact of the genotype of wheat on the efficacy of silicon, similar to other studies [30]. The yielding of wheat depends on the wheat cultivar and conditions of growing. We grew the Harenda, Rusałka and Serenada cultivars, as they are recommended for organic farming by the Institute of Soil Science and Plant Cultivation for organic farming [27]. In that study, yields of 5.58, 4.90 and 4.89 t per hectare were obtained for the Harenda, Serenada and Rusałka cultivars, respectively [33]. In our research, the yields of these cultivars were much lower and were 3.98, 3.66 and 2.85 t per hectare for the Serenada, Rusałka and Harenda cultivars, respectively. Our field studies were carried out in the region of Wielkopolska, the Polish region with the greatest rainfall shortages, so the yields are much lower than in the other regions of Poland. Among the cultivars grown, Harenda gave the lowest yields, indicating that it did not adapt to such environmental pressure. At the same time, Harenda reacted the best to Si application, especially in the growing season of dry 2019. Dufey et al. [32] stated that the choice of stronger Si-accumulating varieties could be valuable in the improvement of wheat resistance to drought stress. The same cultivars grown in a conventional farming system showed almost no difference in yield, and, according to data from the Research Centre for Cultivar Testing in Poland (COBORU), the three wheat cultivars yielded 10.3, 10.0 and 9.8 t per hectare, respectively, in 2018 [34]. This proves that the genotype of the wheat cultivars was a factor determining the development and yielding of the plants depending on the agricultural cultivation system. In this way, treatments with silicon can minimize the negative impact of a stress factor, e.g., a lack of water, especially in organic farming.

The most effective method of silicon application was the combined method, which increased the wheat yield by ca. 25% in both study years (normal and dry), but in the case of the Harenda cultivar, during the dry year, the yields almost doubled, increasing by 42%. The combined Si application produced the best results; the yield response to Si may be related to an improved uptake of this element and the methods of delivery to plants, and it was confirmed for sugarcanes [35]. Guevel et al. [31] also concluded that the combined foliar and soil application was the most effective for wheat health. Segalin et al. [36] revealed that the foliar application of silicon affected neither the yield nor quality of the wheat grain of different cultivars. Walsh et al. [16] also could not confirm any beneficial effects on the plant growth, grain yield and grain protein of irrigated winter wheat grown in non-stressed conditions; a Si product (sourced from a high-energy amorphous, non-crystalized volcanic tuff) was applied twice at rates of 140, 280 and 560 kg Si ha^{-1}, once at planting and once at tilling time. Korunic et al. [37] evaluated the effect of diatomaceous earth (DE) on grain quality and noted that it reduced the bulk density of durum wheat. In our observation, this fact was not confirmed; lower values of bulk density were noted only in 2019 (dry year) compared to 2018 (Figure 1). The grain quality depends not only on the cultivar but also on the management system (organic vs. conventional). Spring wheat cultivars grown under organic and conventional management systems were found to have different quality yield parameters (bread-making) and phospholipid fatty acid (PLFA) profiles. In the organic system, the wheat yields were roughly half of the conventional yields, but the protein content was higher in the organic system. In general, high protein and gluten contents were obtained in our study, although this is not in agreement with findings by Nelson et al. [38], in which research from different parts of the world reported lower quality parameters (contents of protein and gluten, the TGW and the sedimentation index described by Zeleny) of wheat grown under an organic farming regime [39–42]. In our study, the bulk density of the Harenda cultivar in the control plots in 2018 and all the study variants in the dry year of 2019 did not meet the EU standards for bread-making wheat [43]. Moreover, our results show no influence of silicon on the protein and gluten contents, with the exception of for the Harenda cultivar in the dry 2019, when all the quality parameters (protein and gluten contents and bulk density) were the lowest in the control area. In the case of the Harenda cultivar, a positive effect of Si application on the TGW was also found.

5. Conclusions

The increasing incidences of different biotic and abiotic stresses, especially drought, throughout the world have restrained the growth of wheat. Si application has a positive effect on wheat yields, and the scale of the effect depends on the application method, the wheat cultivar and the severity of the different water conditions in the soil. The Harenda cultivar was more prone to less soil water content than the Rusałka and Serenada cultivars. Under severe water stress, young Harenda plants slowed down their development after Si application but eventually increased their grain yield to a greater degree than in the case of the other two cultivars. Silicon increased the yields of the three wheat cultivars, and the highest yields were harvested in the plots with combined silicon treatments. The values of the thousand-grain weight, yield, length of leaves per plant and length of roots per plant were statistically significantly determined by the cultivar and method of silicon supply as well as the interaction of the cultivar × method of silicon supply. Confirming the interaction leads to the conclusion that the effectiveness of using silicon may vary depending on the wheat variety. The value of the presented research is the confirmation of the possibility of mitigating plant stress due to limited water availability in the soil, especially in the case of cultivars sensitive to this stress factor. An extremely important conclusion is also demonstrating the effectiveness of the methods of silicon application, with an indication of the combined method of silicon application and the seed dressing method dedicated especially to organic farming.

Author Contributions: J.K. (Jolanta Kowalska) and J.T. designed and drafted the work. K.M and J.K. (Joanna Krzymińska) contributed to the revision of the manuscript. J.B. contributed to the statistical analysis. All authors have read and agreed to the published version of the manuscript.

Funding: The funding body had no role in the design of the study; in the collection, analyses or interpretation of data; in the writing of the manuscript; or in the decision to publish the results.

Acknowledgments: This study was part of the statutory activity of the Institute of Plant Protection—National Research Institute in Poland. The project "Development of a strategy for the use of microorganisms and natural products in organic farming" (No. BIOEKO 02) was supported by the Polish Ministry of Science and Higher Education.

Conflicts of Interest: The authors declare no conflict of interest.

References

1. El-Beltagi, H.S.; Mohamed, H.I.; Sofy, M.R. Role of ascorbic acid, glutathione and proline applied as singly or in sequence combination in improving chickpea plant through physiological change and antioxidant defense under different levels of irrigation intervals. *Molecules* **2020**, *25*, 1702. [CrossRef] [PubMed]
2. Chun, S.C.; Paramasivan, M.; Chandrasekaran, M. Proline accumulation influenced by osmotic stress in arbuscular mycorrhizal symbiotic plants. *Front. Microbiol.* **2018**, *9*, 2525. [CrossRef] [PubMed]
3. Epstein, E. Silicon: Its manifold roles in plants. *Ann. Appl. Biol.* **2009**, *155*, 155–160. [CrossRef]
4. Tubana, B.S.; Babu, T.; Datnoff, L.E.A. Review of Silicon in Soils and Plants and Its Role in US Agriculture History and Future Perspectives. *Soil Sci.* **2016**, *181*, 393–411. [CrossRef]
5. Haynes, R.A. Contemporary overview of silicon availability in agricultural soils. *J. Soil Sci. Plant Nutr.* **2014**, *177*, 831–844. [CrossRef]
6. Etesami, H.; Jeong, B.R. Silicon (Si): Review and future prospects on the action mechanisms in alleviating biotic and abiotic stresses in plants. *Ecotoxicol. Environ. Saf.* **2018**, *147*, 881–896. [CrossRef]
7. Epstein, E. SILICON. *Annu. Rev. Plant Biol.* **1999**, *50*, 641–664. [CrossRef]
8. Jones, L.H.P.; Handreck, K. Silica in soils, plants and animals. *Adv. Agron.* **1967**, *19*, 107–149.
9. Ma, J.F.; Takahashi, E. *Soil, Fertilizer, and Silicon Research in Japan*; Elsevier: Amsterdam, The Netherlands, 2002; p. 294. ISBN 9780444511669.
10. Sacała, E. Role of silicon in plant resistance to water stress. *J. Elem.* **2009**, *14*, 619–630. [CrossRef]
11. Pereira, H.S.; Korndörfer, G.H.; Vidal, A.D.; de Camargo, M.S. Silicon sources for rice crop. *Sci. Agric.* **2004**, *61*, 522–528. [CrossRef]
12. Savant, N.K.; Korndörfer, G.H.; Datnoff, L.E.; Snyder, G.H. Silicon nutrition and sugarcane production: A review. *J. Plant Nutr.* **1999**, *22*, 1853–1903. [CrossRef]
13. Verma, K.K.; Anas, M.; Chen, Z.; Rajput, V.D.; Malviya, M.K.; Verma, C.L.; Singh, R.K.; Singh, P.; Song, X.P.; Li, Y.R. Silicon supply improves leaf gas exchange, antioxidant defense system and growth in sugarcane responsive to water limitation. *Plants* **2020**, *9*, 1032. [CrossRef] [PubMed]
14. Teixeira, N.C.; Oliveira, J.S.; Goreti, V.M.; Wellington, O.A.; Campos, G. Combined effects of soil silicon and drought stress on host plant chemical and ultrastructural quality for leaf-chewing and sap-sucking insects. *J. Agron. Crop Sci.* **2020**, *206*, 187–201. [CrossRef]
15. Schaller, J.; Faucherre, S.; Joss, H.; Obs, M.; Goeckede, M.; Planer-Friedrich, B.; Peiffer, S.; Gilfedder, B.; Elberling, B. Silicon increases the phosphorus availability of arctic soils. *Sci. Rep.* **2019**, *9*, 1–11. [CrossRef]
16. Walsh, O.S.; Shafian, S.; McClintick-Chess, J.R.; Belmont, K.M.; Blanscet, S.M. Potential of Silicon Amendment for Improved Wheat Production. *Plants* **2018**, *7*, 26. [CrossRef]
17. Tibbitts, S.A. Effect of Silicon on Wheat Growth and Development in Drought and Salinity Stress. Master's Thesis, 2018; p. 6925. Available online: https://digitalcommons.usu.edu/etd/6925 (accessed on 15 October 2020).
18. Zhu, Y.; Haijun, G. Beneficial effects of silicon on salt and drought tolerance in plants. *Agron. Sustain. Dev.* **2014**, *34*, 455–472. [CrossRef]
19. Ming, D.; Pei, Z.; Naeem, M.; Gong, H.; Zhou, W. Silicon alleviates PEG-induced water-deficit stress in upland rice seedlings by enhancing osmotic adjustment. *J. Agron. Crop Sci.* **2012**, *198*, 14–26. [CrossRef]
20. Xu, L.; Islam, F.; Ali, B.; Pei, Z.; Li, J.; Ghani, M.A.; Zhou, W. Silicon and water-deficit stress differentially modulate physiology and ultrastructure in wheat (*Triticum aestivum* L.). *Biotechnology* **2017**, *7*, 273. [CrossRef]
21. Artyszak, A. Effect of silicon fertilization on crop yield quantity and quality—A Literature Review in Europe. *Plants* **2018**, *7*, 54. [CrossRef]
22. Kowalska, J.; Tyburski, J.; Jakubowska, M.; Krzyminska, J. Effect of different forms of silicon on growth of spring wheat cultivated in organic farming system. *Silicon* **2020**, *2*. [CrossRef]
23. Borgen, A.; Kristensen, L. Use of mustard flour and milk powder to control common bunt (*Tilletia tritici*) in wheat and stem smut (*Urocystis occulta*) in rye in organic agriculture. *BCPC Symp. Proc. Seed Treat. Chall. Oppor.* **2001**, *75*, 141–148.

24. Plakholm, G.; Sollinger, J. Seed treatment for common wheat-bunt (*Tilletia caries* (DC) Tul.) according to organic farming principles. In Proceedings of the 13th International IFOAM Scientic Conference 2000, Basel, Switzerland, 28–31 August 2000; p. 139.

25. Ekologiczne doświadczalnictwo Odmianowe, Institute of Soil Science and Plant Cultivation. Zalecenia, Odmiany Pszenicy Jarej do Rolnictwa Ekologicznego. Instrukcja Upowszechnieniowa nr 237. Puławy 2019. 2019. Available online: http://iung.pulawy.pl/edo/zalecenia.html (accessed on 15 October 2020). (In Polish).

26. Moussavi-Nik, M.; Mobasser, H.R.; Mheraban, A. Effect of Water Stress and Potassium Chloride on Biological and Grain Yield of Different Wheat Cultivars. In *Wheat Production in Stressed Environments. Developments in Plant Breeding*; Buck, H.T., Nisi, J.E., Salomón, N., Eds.; Springer: Dordrecht, The Netherlands, 2007; Volume 12. [CrossRef]

27. Smutná, P.; Elzner, P.; Středa, T. The effect of water deficit on yield and yield component variation in winter wheat. *Agric. Conspec. Sci.* **2018**, *83*, 105–111.

28. Ahmad, F.; Aziz, T.; Maqsood, M.A.; Tahir, M.A.; Kanwal, S. Effect of silicon application on wheat (*Triticum aestivum* L.) growth under water deficiency stress. *Emir. J. Food Agric.* **2007**, *19*, 1–7. [CrossRef]

29. Laane, H.M. The efficacy of the Silicic Acid Agro Technology: The use of stabilized silicic acid. In Proceedings of the 6th Int. Conf on Silicon in Agriculture, Stockholm, Sweden, 26–30 August 2014; p. 110.

30. Maghsoudi, K.; Yahya, A.E.; Emam, B.; Muhammad, A.C.; Mohammad, J.; Arvin, D. Alleviation of field water stress in wheat cultivars by using silicon and salicylic acid applied separately or in combination. *Crop Pasture Sci.* **2019**, *70*, 36–43. [CrossRef]

31. Guevel, M.H.; Menzies, J.G.; Belanger, R.R. Effect of root and foliar applications of soluble silicon on powdery mildew control and growth of wheat plants. *Eur. J. Plant Pathol.* **2007**, *119*, 429–436. [CrossRef]

32. Dufey, I.; Gheysens, S.; Ingabire, A.; Lutts, S.; Bertin, P. Silicon application in cultivated rices (*Oryza sativa* L and *Oryza glaberrima* Steud) alleviates iron toxicity symptoms through the reduction in iron concentration in the leaf tissue. *J. Agron. Crop Sci.* **2013**, *200*, 132–142. [CrossRef]

33. Institute of Soil Science and Plant Cultivation, Badania nad Przydatnością Odmian Zbóż Jarych do Uprawy w Rolnictwie Ekologicznym w Ramach Ekologicznego Doświadczalnictwa Odmianowego—EDO dla zbóż Jarych. 2019. Available online: http://www.iung.pulawy.pl/index.php?option=com_content&view=article&id=175&Itemid=155 (accessed on 15 October 2020). (In Polish).

34. Publikacje COBORU. Wstępne Wyniki Plonowania Odmian w Doświadczeniach Porejestrowych. Zboża Jare; COBO 46/2020; 2019; p. 14. Available online: https://coboru.gov.pl/Polska/Publikacje/publikacje.aspx (accessed on 15 October 2020). (In Polish)

35. Huang, X.; Zhang, Z.; Ke, Y.; Xiao, C.; Peng, Z.; Wu, L.; Zhong, S. Effects of silicate fertilizer on nutrition of leaves. yield and sugar of sugarcanes. *J. Trop. Subtrop. Soil Sci.* **1997**, *6*, 242–246.

36. Segalin, S.R.; Huth, C.; Rosa, T.A.; Pahins, D.B.; Mertz, L.M.; Nunes, U.R. Foliar application of silicon and the effect on wheat seed yield and quality. *J. Seed Sci.* **2013**, *35*, 86–91. [CrossRef]

37. Korunic, Z.; Fields, P.G.; Kovacs, M.I.P.; Noll, J.S.; Lukow, O.M.; Demianyk, C.J.; Shibley, K.J. The effect of diatomaceous earth on grain quality. *Postharvest Biol. Tec.* **1996**, *9*, 373–387. [CrossRef]

38. Nelson, A.G.; Quideau, S.; Frick, B.; Niziol, D.; Clapperton, J.; Spaner, D. Spring wheat genotypes differentially alter soil microbial communities and wheat bread making quality in organic and conventional systems. *Can. J. Plant Sci.* **2011**, *91*, 485–495. [CrossRef]

39. Langenkämper, G.; Zörb, C.; Seifert, M.; Mäder, P.; Fretzdorff, B.; Betsche, T. Nutritional quality of organic and conventional wheat. *J. Appl. Bot. Food Qual.* **2006**, *80*, 150–154.

40. Krejčířová, L.; Capouchová, I.; Petr, J.; Bicanová, E.; Faměra, O. The effect of organic and conventional growing systems on quality and storage protein composition of winter wheat. *Plant. Soil Environ.* **2007**, *53*, 499–505. [CrossRef]

41. Casagrande, M.; David, C.; Valantin-Morison, M.; Makowski, D.; Jeuffroy, M.-H. Factors limiting the grain protein content of organic winter wheat in south-eastern France: A mixed-model approach. *Agron. Sustain. Dev.* **2009**, *29*, 565–574. [CrossRef]

42. Park, E.Y.; Fuerst, E.P.; Miller, P.R.; Machado, S.; Burke, I.C.; Baik, B.K. Functional and Nutritional Characteristics of Wheat Grown in Organic, No-Till and Conventional Cropping Systems. 2014. Available online: http://smallgrains.wsu.edu/wp-content/uploads/2014/01/Wheat-Quality-in-Organic-No-till-and-Conventional-Croppping-Systems.pdf (accessed on 15 October 2020).
43. Commission Regulation (EU) No 742/2010 of 17 August 2010 amending Regulation (EU) No 1272/2009 Laying down Common Detailed Rules for the Implementation of Council Regulation (EC) No 1234/2007 as Regards Buying-in and Selling of Agricultural Products under Public Intervention. Available online: http://data.europa.eu/eli/reg/2010/742/oj (accessed on 15 October 2020).

Publisher's Note: MDPI stays neutral with regard to jurisdictional claims in published maps and institutional affiliations.

 *agronomy*

Article

Agronomic Evaluation and Chemical Characterization of Sicilian *Salvia sclarea* L. Accessions

Teresa Tuttolomondo [1], Giovanni Iapichino [1], Mario Licata [1,*], Giuseppe Virga [2,*], Claudio Leto [1,2] and Salvatore La Bella [1]

[1] Department of Agricultural, Food and Forest Sciences, Università degli Studi di Palermo, 90128 Palermo, Italy; teresa.tuttolomondo@unipa.it (T.T.); giovanni.iapichino@unipa.it (G.I.); claudio.leto@unipa.it (C.L.); salvatore.labella@unipa.it (S.L.B.)

[2] Consorzio di Ricerca per lo Sviluppo di Sistemi Innovativi Agroambientali, 90143 Palermo, Italy

* Correspondence: mario.licata@unipa.it (M.L.); giuseppe251@hotmail.it (G.V.)

Received: 9 June 2020; Accepted: 29 July 2020; Published: 1 August 2020

Abstract: Clary sage (*Salvia sclarea* L.), known for its aromatic and medicinal properties, belongs to the *Lamiaceae* family. Although the species grows wild throughout Sicily, knowledge of its production and qualitative properties is limited. The aim of this study was to evaluate the agronomic behavior of the species over two years of testing and to characterize the chemical properties of its wild counterparts in order to identify the most promising accessions for cropping or for use in breeding programs. Tests were carried out during 2008, 2009, and 2010. During the first year, the plot was established. Subsequently, the main parameters for bio-agronomic evaluation were taken in 2009 and 2010. Regarding qualitative characterization, essential oils (EO) were extracted from flowering samples of clary sage. The accessions in the study showed satisfactory adaptation capacity to cropping. The accessions examined belong to the "linalyl acetate" (range 36–43%) chemotype. Test results show good potential for Mediterranean cropping systems, helping to increase the range of medicinal and aromatic species in cultivation.

Keywords: clary sage; essential oil; aromatic plant species; biometric and agronomic characteristics

---

1. Introduction

Clary sage (*Salvia sclarea* L.) is a biennial or perennial, heliophilous, and xerophytic herbaceous plant belonging to the *Lamiaceae* [1] family. It is found in the north of the Mediterranean, central Asia, and some areas of North Africa. It grows throughout Italy, thriving both on dry hilly slopes and on scrubland [2]. In Sicily, it is mostly found growing in mountainous and hilly areas [3].

Clary sage has been highly valued for its aromatics and medicinal properties since ancient times and is one of the most important species for the production of essential oils, together with *Salvia officinalis* and *Salvia lavandulifolia*, with an estimated production of between 50 and 100 tons a year [4]. The whole plant is highly aromatic, the inflorescences in particular, and the essential oils possess a fresh, floral fragrance [5]. Sage essential oils are used as an aromatic agent in food products, as an ingredient in liqueurs and tobacco and as a scent component in perfumes and cosmetic formulas [6,7]. It is currently cultivated in Bulgaria, France, Russian and Morocco for essential oils used by the perfume industry [8]. It is also well known in traditional medicine for the treatment of several common ailments. In Turkey, for example, the leaves and flowers are used in infusions for the treatment of sore throats, coughs, gynecological disturbances, ulcers and intestinal cramps [9,10]. Several scientific studies have also demonstrated the antioxidant, neuroprotective, anti-depressive, anti-inflammatory, antifungal, antiviral and antimicrobial [11–22] activity of the essential oil of clary sage.

The essential oils of various species from the *Lamiaceae* family demonstrate a certain degree of chemical variability due to a range of factors [23–28]. Regarding the chemical composition of clary sage, various authors note that in most cases, the principal volatile compounds are terpenoids [29,30], among which linalyl acetate and linalool. These components are central to good quality oil for use as an aromatizing agent [6,12,14,31]. Another principal component of clary sageis sclareol. Sclareol is used as a base for the chemical synthesis of Ambrox, a central component in perfume production and an alternative to the more naturally obtained ambergris [32,33]. Linalool, linalyl acetate, and sclareol are the essential oil components predominant in the flowers, while germacrene D, bicyclogermacrene, beta-caryophyllene and spathulenol are most abundant in the leaves [34,35]. The species grows in the wild in Sicily (Italy); however, little is known of its production and qualitative properties.

The agronomic characteristics of clary sage found in scientific literature relate to locations with different environmental conditions compared to those of Italy. Studies have been carried out in Brazil, with Mossi [36] reporting data on biomass production, plant size and leaf size, on the color and characteristics of the inflorescences, and on essential oil production. Other studies, which highlight the variability found in biomass and essential oil production, were carried out in various sites throughout Spain over medium to long test periods, and in India [37,38]. In Italy, however, very little research data is available on the agronomic characteristics of germplasm of this species.

The aim of the study was to evaluate the agronomic behavior of clary sage over two years of tests and to characterize the chemical properties of its wild counterparts to determine the most promising accessions for cropping in the Mediterranean or to use in breeding programs.

2. Materials and Methods

2.1. Site of Experiments and Treatments

The 3-year study (2008, 2009 and 2010) was carried out at the Orleans Experimental Station, University of Palermo (Italy) (38°06′26.2″ N, 13°20′56.0″ E, 31 m a.s.l.). The plot was established during the first year following sowing and measurements of the main parameters for bio-agronomic evaluation were taken in 2009 and 2010. Soils in the test area were sandy clay loam (Aric Regosol, 54% sand, 23% clay, 21% silt) with a pH of 7.6, 14 g kg^{-1} organic matter, 3.70% active carbonates, 1.32% total nitrogen, 18.1 ppm available phosphorus and 320 ppm exchangeable potassium. The climate in the area is Mediterranean with mild, humid winters and hot, dry summers. Seeds from local accessions of clary sage from the island were sown in March 2008. The plants of these populations were characterized taxonomically using analytical keys and compared to exsiccatae stored at the Botanical Gardens of the University of Palermo. In total, 9 accessions of clary sage were used, gathered from 3 sites in Sicily located in the Province of Agrigento (AG), Palermo (PA) and Messina (ME) (Figure 1).

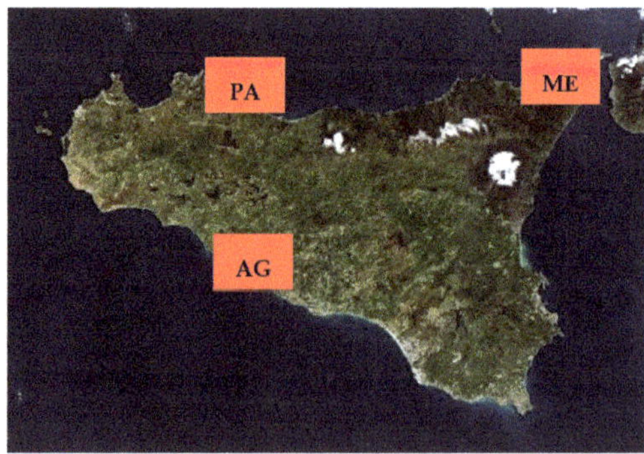

Figure 1. Sampling sites of clary sage in Sicily. Abbreviations of the provinces of origin: AG = Agrigento; ME = Messina; PA = Palermo.

Three accessions per province were identified using an initial followed by a numerical code. The initials SS indicate accessions from Agrigento, PR from Palermo and AF from Messina (Table 1).

Table 1. List of accessions in the test.

Accessions	Provenance Initials	Province
SS4		
SS7	SS	AG
SS9		
PR1		
PR5	PR	PA
PR4		
AF2		
AF3	AF	ME
AF8		

Seeds from each of the accessions were placed in 84-hole seed trays and set in a cold frame. Following emergence 10 days after sowing, the seedlings were transferred to 10 cm pots. During the second 10-days of May, the plantlings were planted in the open field. Each plot measured 30 m^2 (5 m × 6 m). The test plot was created using a density of 20,000 plants per hectare and a randomized plot design with 3 replications (Figure 2). This plant density was chosen to evaluate better the growth of each plant, limiting the competition levels for the main environmental factors between the plants. Tests were carried out in dry conditions, this being a traditional practice used for cultivation of aromatic and medicinal plants in the Mediterranean region. Agronomic management included, however, 2 supplementary irrigation events applied during the summer months, immediately after planting, to foster establishment, and manual weed control. During the test, no additional chemical fertilization applications were given. Tests were carried out under organic farming conditions; the residual mineral soil fertility of the previous crop was, then, exploited to allow the plants to grow. The previous crop was *Hedysarum coronarium* L., a perennial legume. Finally, no pathologies or insects were observed.

Figure 2. Plants of clary sage at full flowering stage.

2.2. Plant Measurements

Biometric and production observations were made in 2009 and 2010 on a sample plot of 10 plants, excluding the border rows. Harvesting of accessions was carried out during the second 10-day period of May for both years. The samples were gathered (through reaping of the whole area), when 70% of the plants were in full flowering stage. The following parameters were recorded during harvesting: plant height (cm), plant fresh weight (g), plant dry weight (g), number of branches (no.), number of stems (no.), floral spike length (cm), inflorescence dry weight (g), leaf dry weight (g) and stem dry weight (g). Inflorescence, leaf and stem ratios (as a percentage of the total dry weight of the plant) were also measured. Dry matter weight was calculated when constant sample weight was reached (dried in a shaded and well aerated environment at a temperature of approx. 30 °C). Inflorescence yields (d.m.) per hectare (Mg ha^{-1}) were also estimated.

2.3. Essential Oil Extraction and Oil Yield Calculation

On a sample of 500 g of dried inflorescences, the total essential oil content was determined, expressed as a % *v/w* (oil volume/sample weight in g) and extracted using steam distillation. Oil yields were calculated by multiplying inflorescence yields by oil content and 0.90 (approximate specific gravity of oil) [8]. Clary sage inflorescences were then divided into inflorescences from the main stem (ISP) and inflorescences from the secondary stem (ISS) to evaluate both the content and composition of the essential oils. The length of the ISP spike and the ISS spike was also measured.

2.4. GC- and GC/MS Analyses of Essential Oils

Gas chromatographic (GC) analyses were run on a Shimadzu gas chromatograph, Model 17-A (Shimadzu Corporation, Duisburg, Germany) equipped with a flame ionization detector (FID) and operating software Class VP Chromatography Date System version 4.3 (Shimadzu). Analytical conditions: SPB-5 capillary column(15 m × 0.10 mm × 0.15 μm), helium as carrier gas (1 mL/min). Injection in split mode (1:200), injected volume 1 ll (4% essential oil/CH_2Cl_2 *v/v*), injector and detector temp. 250–280 °C, resp. Linear velocity in column 19 cm/s. The oven temperature was held at 60 °C for 1 min, then programmed from 60 to 280 °C at 10 °C min^{-1}, then 280 °C for 1 min. Percentages of compounds were determined from their peak areas in the GC/FID profiles.

Gas-chromatography-mass spectrometry (GC/MS) was carried out in the fast mode on a Shimadzu GC/MS mod. GCMS-QP5050A, with the same column and the same operative conditions used for analytical GC/FID, operating software GC/MS solution version 1.02 (Shimadzu). Ionization voltage 70 eV, electron multiplier 900 V, ion source temp. 180 °C. Mass spectra data were acquired in the scan mode in *m/z* range 40–400. The same oil solutions (1 ll) were injected with the split mode (1:96).

2.5. Identification of Components of Essential Oils

The identity of components was based on GC retention index (relative to C9–C22 n-alkanes on the SPB-5 column), computer matching of spectral MS data with those from NIST MS libraries, [39] comparison of the fragmentation patterns with those reported in the literature [40] and, where possible, co-injections with authentic samples.

2.6. Statistical Analyses

Data of all biometric and production parameters were processed using analysis of variance. The difference between means was carried out using the Tukey test. In addition, a correlation matrix was determined for the main parameters recorded for each year, based on standardization of data. To carry out an overall analysis of the structure of agronomic variability and to determine the weight of each parameter on the total variance [41], principle component analysis (PCA) and cluster analysis (UPGMA) were carried out by grouping data from the two years. This latter analysis is shown graphically on the principal components plot (PC1 and PC2), where test accessions were projected according to test year using factor scores. The software "Past" V. 3.16 for Windows was used for data analysis [42].

3. Results

3.1. Analyses of Rainfall and Temperature Trends in the Test Site

Rainfall and temperature trends during 2008, 2009, and 2010 are shown in Figure 3. Rainfall levels during the three test years were quite unalike. In 2008, the year when the plot was established, there was a marked lack of rainfall (365 mm), although minimum and maximum temperatures were consistent with the test environment. In 2009, rainfall levels were high with over 1200 mm for the year. Rainfall events were concentrated above all between January and May and between September and December, with an absence of rainfall between June and August. In 2010, although with the same rainfall distribution was the same, rainfall levels were lower, at an annual level of 700 mm, typical of the test environment. Average minimum temperatures (2009: 14.10 °C–2010: 14.70 °C) and average maximum temperatures (2009: 23.40 °C–2010: 23.00 °C) were not found to be different from average temperatures for the area during the test period.

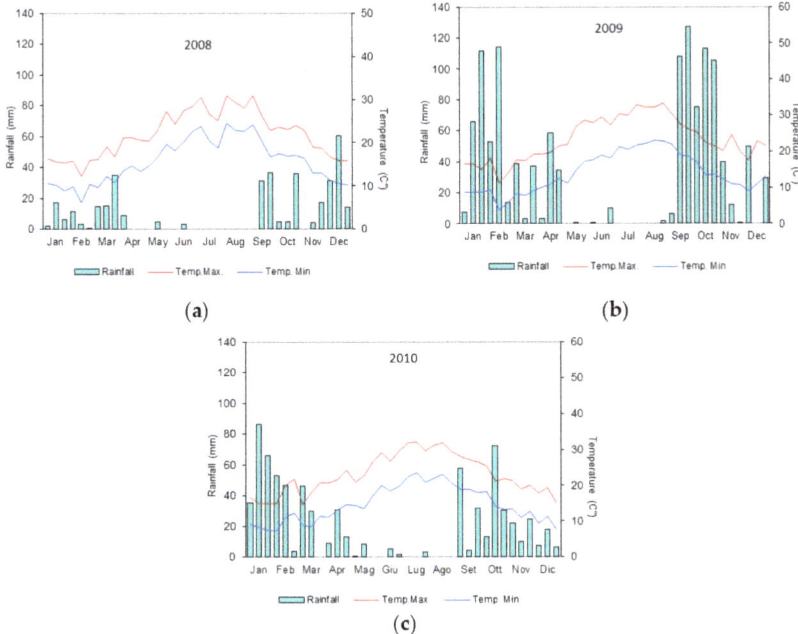

Figure 3. Orleans (Palermo, PA). Rainfall and temperature trends during the test period. Graph (**a**) refers to 2008, graph (**b**) refers to 2009 and graph (**c**) refers to 2010.

3.2. Analyses of Biometric and Production Parameters in the Study

The factors "year" and "accession" and the year-by-accession interaction determined highly significant differences for all the parameters in the study. The differences found on the tested parameters over the years (Table 2) highlight the influence of the environmental factors on the biometric and productive characteristics.

Table 2. Orleans (Palermo, PA). Yearly averages for biometric and production parameters.

Year	Plant Height (cm)	Plant Fresh Weight (g)	Plant Dry Weight (g)	Floral Spike Dry Weight (g)	Leaf Dry Weight (g)	Stem Dry Weight (g)
2009	183.86 **	1190.41 **	412.08 **	142.92 **	95.33 **	173.83 **
2010	142.97 **	886.21 **	193.08 **	81.68 **	39.56 **	71.42 **

Year	Branches Plant^{-1} (no.)	Stems Plant^{-1} (no.)	Spike Length (cm)	ISP Spike Length (cm)	ISS Spike Length (cm)
2009	8.83 **	3.07 **	49.34 **	57.83 **	46.41 **
2010	15.77 **	5.44 **	42.59 **	58.36 **	50.76 **

** = significant at $p \leq 0.01$.

It is also worth noting variations in the percentage ratio of inflorescences, leaves, and stems for the two years under consideration (Table 3). The table shows that in the drier year (2010), inflorescence incidence was 8 percentage points higher than the year with greater rainfall. In contrast, leaf incidence and stem incidence were 3 and 5 percentage points higher, respectively, in the rainier year than in 2010.

Table 3. Percentage distribution of biomass components (d.m.).

Year	Floral Spike (%)	Leaves (%)	Stems (%)
2009	35 **	23 **	42 **
2010	43 **	20 **	37 **

** = significant at $p \leq 0.01$.

Average plant height (Table 2) was found to be below 150 cm in both years. Branch number, stem number per plant, and ISS floral spike length were higher in the less rainy year. Floral spike yields were higher in 2009 (2.84 Mg ha^{-1}) than in 2010 (1.60 Mg ha^{-1}) with a difference of 1.24 Mg ha^{-1}. Essential oil content in percentage terms was slightly higher in 2010, while in terms of EO yield, it was found to be 9.10 kg ha^{-1} higher in 2009 (23.51 kg ha^{-1}) (Table 4).

Table 4. Year averages for floral spike and EO yields.

Year	Spike Yield (Mg ha^{-1})	Essential Oil Content (% *v/w*)	Essential Oil Yield (kg ha^{-1})
2009	2.84 **	0.92 **	23.50 **
2010	1.60 **	1.00 **	14.40 **

** = significant at $p \leq 0.01$.

Relative to the 2 types of floral spike (ISP and ISS), average ISS spike lengths ranged between 46.41 cm in 2009 and 50.76 cm in 2010, while ISP spike lengths were found to be around 58 cm in both years Furthermore, they showed significant differences over the two years, highlighting a greater oil percentage content in 2010 (Table 5).

Table 5. Average annual production parameters–floral spike principal stem (ISP) and secondary stem (ISS).

Year	ISP Spike Length (cm)	Essential Oil Yield % ISP	ISS Spike Length (cm)	EO % Yield ISS
2009	57.83 **	0.81 **	46.41 **	0.98 **
2010	58.36 **	0.88 **	50.76 **	1.12 **

** = significant at $p \leq 0.01$.

Between the accessions (Table 6), SS4 (3.21 Mg ha^{-1}), SS7 (2.60 Mg ha^{-1}), SS9 (2.52 Mg ha^{-1}) and AF2 (2.53 Mg ha^{-1}) were of high interest regarding to the floral spike yield. These accessions also recorded the higher values of the number of stems, number of branches, and ISS spike length. PR5 (1.52 Mg ha^{-1}) and AF8 (1.63 Mg ha^{-1}) were, instead, the less productive accessions. The highest oil percentage values were obtained by AF8 (1.36%), AF3 (1.28%) and AF2 (1.25%), while the lowest values of oil percentage were found in the SS accessions and, in particular, in SS7 (0.65%). The higher oil yield values were recorded by AF2 (28.05 Mg ha^{-1}) and AF3 (25.10 Mg ha^{-1}) with respect to other accessions. AF8 had the highest oil percentage value in the ISS (1.68%), while PR4 obtained the highest oil percentage value in the ISP (1.12%).

Table 6. Mean values of biometric and production parameters of the accessions over the years.

Accession	SY	EOP	EOY	PH	PFW	PDW	FSDW	LDW	SDW	NB	NS	SL
SS4	3.21 a	0.71 f	18.41 bc	145.05 b	1189.79 d	423.92 b	163.09 a	80.28 b	180.54 b	4.38 bc	13.52 b	48.45 c
SS7	2.60 ab	0.65 g	14.71 c	137.91 e	953.09 g	320.45 e	127.65 b	57.49 e	135.31 c	3.85 cd	13.75 b	43.58 f
SS9	2.52 abc	0.69 fg	14.85 c	143.02 c	1198.35 c	323.82 c	123.70 d	71.86 c	128.27 e	4.63 b	14.53 a	46.27 d
PR1	2.13 bcd	0.77 e	13.95 c	148.12 a	976.27 e	254.24 g	105.82 f	59.28 d	89.14 g	4.38 bc	8.02 e	53.77 a
PR5	1.52 d	0.4 d	12.55 c	132.31 f	579.31 i	174.38 i	78.38 i	27.98 g	65.52 i	3.88 bcd	9.64 d	45.18 e
PR4	1.83 cd	1.00 c	15.65 c	139.12 d	750.98 h	214.91 h	90.87 g	40.31 f	83.74 h	3.38 d	11.77 c	50.49 b
AF2	2.53 abc	1.25 b	28.05 a	143.51 c	1523.66 a	429.85 a	125.85 c	119.10 a	184.91 a	6.00 a	11.66 c	40.94 g
AF3	2.24 bcd	1.28 b	25.1 ab	139.60 d	1217.88 b	322.43 d	111.93 e	80.08 b	130.42 d	3.75 cd	13.89 ab	43.56 f
AF8	1.63 d	1.36 a	19.45 bc	130.52 g	955.51 f	259.25 f	83.45 h	72.48 c	103.32 f	4.13 bcd	14.11 ab	41.54 g

Accession	ISPL	ISPEOP	ISSL	ISSEOP
SS4	59.01 d	0.53 f	52.81 a	0.87 d
SS7	55.75 f	0.61 e	52.54 a	0.69 f
SS9	60.11 c	0.61 e	51.64 b	0.77 e
PR1	65.13 a	0.78 d	43.12 g	0.75 e
PR5	58.12 e	1.05 b	50.19 c	0.83 d
5PR4	61.75 b	1.12 a	49.10 d	0.88 d
AF2	53.88 g	0.88 c	45.8 f	1.61 b
AF3	56.38 f	1.02 b	45.25 f	1.53 c
AF8	53.21 h	1.03 b	46.84 e	1.68 a

SY = spike yield; EOP = essential oil percentage; EOY = essential oil yield; PH = plant height; PFW = plant fresh weight; PDW = plant dry weight; FSDW = floral spike dry weight; LDW = leaf dry weight; SDW = stem dry weight; NB = no. branches; NS = no. stems; SL = spike length; ISPL = ISPL spike length; ISPEOP = ISP essential oil percentage content; ISSL = ISS spike length; ISSEOP = ISS essential oil percentage content. Means followed by the same letter are not significantly different for $p \leq 0.05$ according to Test of Tukey.

Considering the results of accessions in the study, during the years 2009 and 2010 (Table 7a,b), the best production results for 2009 were found in the Agrigento accessions (SS), followed by those of Messina (AF).

The best accessions, as regards inflorescence yields, were found to be SS4 and SS7 with 4.81 and 4.20 Mg ha^{-1}, respectively. Worthy of note among the Messina accessions was AF2 which, compared to other accessions from the same area, produced higher yields of 3.41 Mg ha^{-1}. This value was the minimum yield obtained by the Agrigento accession SS9 and above the average for the field (2.84 Mg ha^{-1}). Regarding the Palermo (PR) accessions, with the exception of PR4 with a yield of 2.43 Mg ha^{-1}, the remaining accessions produced inflorescence yields of approx. 1.50 Mg ha^{-1}, thus appearing the least productive of the accessions.

The greatest plant height was obtained by SS4 (151.50 cm), higher than all the others in that year. This was followed by PR4 (142.00 cm) and SS9 (141.52 cm), while the shortest plant size was that of PR5 and AF3 at a height of below 130 cm. Relating to the fresh and dry weight of the plants, it is worth noting that all of the Palermo accessions (PR) were found to be well below the average for the field (2.84 Mg ha^{-1}), together with AF8 from Messina, albeit to a far lesser extent. The number of branches, number of stems and secondary floral spike length were found to be higher in the SS and AF accessions, in particular in the more productive accessions. SS4 (12.01) and SS7 (12.51) were of interest regarding the number of branches, AF2 (4.52) regarding the number of stems and, concerning the secondary floral spike length (ISS), accession SS4 (60.06) was worthy of note. Average floral spike length together with that of the ISP was greater in the SS and PR accessions, with PR4 obtaining the greatest values.

The highest percentage content in oil was found in the AF accessions followed by those from Palermo and the lowest content was found in the SS accessions SS. In terms of oil yields, however, the most productive were the Messina accessions (AF2 38.41 kg ha^{-1}), followed by the Agrigento accessions.

In the second year, there was a fall in production which altered the ranking of the accessions based on results in the study. More specifically, in 2010, similar to the first year, the Messina accessions were ranked as the most productive. The Palermo accessions, in contrast to the previous year, were also ranked at the top of the list. The above-mentioned areas of provenance were found to perform better than the Agrigento accessions not only in terms of average inflorescence yields but of all the other parameters examined. However, if we look at the Agrigento data more closely, regarding inflorescence yields, we can see that two of the three accessions, SS4 and SS9, produced values equal to the average for the field (1.60 Mg ha^{-1}), only SS7 producing far lower yields (1.00 Mg ha^{-1}). The AF accessions were ranked second. Of the three accessions, AF2 (1.61 Mg ha^{-1}) and AF3 (1.83 Mg ha^{-1}) produced inflorescence yields which were respectively equal or slightly higher than the average for the fields, while AF8 (1.41 Mg ha^{-1}) was a little lower. Accession PR1 bolstered the results for the PR accessions with inflorescence yields of 2.81 Mg ha^{-1}; 1.20 Mg ha^{-1} higher than the average for the field. In contrast, the other Palermo accession, PR4 and PR5 produced yields slightly below the average for the field.

Once again, PR1 obtained the highest plant fresh weight (1346.88 g) and dry weight (318.72 g), in addition to the highest inflorescence dry weight (139.64 g) and leaf dry weight (80.56 g). The greatest plant size was also obtained by PR1 (159.75 cm), statistically different from all the other plants. PR1 was followed by AF3 (152 cm) and AF2 (150.5 cm), while AF8 produced the smallest size plant (129.75 cm). In 2010, the greatest number of branches, number of stems, and floral spike lengths were obtained from the most productive accessions in the various areas. Regarding the number of branches, AF8 (19.75), AF3 (18.77) and SS9 (19.55) were worthy of note, as was AF2 (7.52) concerning the number of stems. The greatest main floral spike (ISP) and secondary spike (ISS) lengths were obtained from the Palermo accession PR1 at 72.75 cm and 57.40 cm, respectively. Concerning the percentage content of oil, in the second year, a general increase in production was evident in all of the accessions belonging to the Messina group and the Agrigento group, whereas a slight decrease was recorded in the remaining accessions; the ranking list remained unchanged, however, compared to the previous year.

Table 7. (**a**) Mean values of biometric and productive parameters of clary sage accessions during 2009 and 2010. (**b**) Mean values of biometric and productive parameters of clary sage accessions during 2009 and 2010.

(a)

2009

Accession	SY	EOP	EOY	PH	PFW	PDW	FSDW	LDW	SDW	NB	NS	SL
SS4	4.81 a	0.58 c	24.81 bc	151.50 a	1604.85 c	653.01 a	244.51 a	126.51	282.01a	12.01 a	3.51 bc	54.48 b
SS7	4.20 ab	0.61 c	23.10 bc	140.75 c	1327.31 e	526.25 c	207.25 b	91.01 f	228.01 c	12.51 a	2.95 bcd	51.13 d
SS9	3.42 b	0.62 c	18.81 bcd	141.52 bc	1623.20 b	462.75 d	167.25 d	110.25 d	185.25 d	9.53 b	3.25 bc	52.94 c
PR1	1.40 c	0.86 b	10.81 d	136.50 d	605.67 h	189.75 i	72.01 i	38.01 h	79.75 h	3.75 g	2.02 d	52.74 c
PR5	1.61 c	0.89 b	12.71 cs	124.51 f	555.56 i	198.51 h	82.25 h	37.51 h	78.75 h	6.75 f	3.52 bc	45.95 e
PR4	2.43 bc	0.91 b	19.52 bc	142.00 b	828.08 g	285.25 g	115.25 f	50.25 g	119.75 g	9.75 bc	2.02 d	58.69 a
AF2	3.41 b	1.26 a	38.41 a	136.51 d	1734.32 a	622.51 b	173.75 c	178.75 a	270.01 b	8.02 d	4.52 a	42.91 f
AF3	2.63 bc	1.25 a	29.11 ab	127.25 f	1386.63 d	414.75 e	128.75 e	118.75 c	167.25 e	9.01 bcd	2.75 cd	42.62 f
AF8	1.82 c	1.30 a	21.10 bcd	131.25 e	1048.13 f	356.01 f	95.25 g	107.01 e	153.75 f	8.25 d	3.25 c	42.67 f

2010

Accession	SY	EOP	EOY	PH	PFW	PDW	FSDW	LDW	SDW	NB	NS	SL
SS4	1.62 ab	0.84 d	12.01 ab	139.01 e	774.75 e	194.81 d	81.67 c	34.06 e	79.07 d	15.03 c	5.25 cd	42.42 c
SS7	1.02 bc	0.69 e	6.32 b	135.01 g	578.88 i	114.64 i	48.04 i	23.98 g	42.62 i	15.01 cd	4.75 d	36.03 g
SS9	1.62 ab	0.76 de	10.91 ab	144.51 d	773.51 f	184.89 e	80.14 d	33.47 e	71.28 e	19.55 a	6.04 c	39.60 df
PR1	2.81 a	0.68 de	17.11 ab	159.75 a	1346.88 a	318.72 a	139.64 a	80.56 a	98.52 b	12.25 e	6.75 ab	54.80 a
PR5	1.40 b	0.99 c	12.40 ab	140.02 e	603.06 h	150.25 g	74.51 f	18.46 h	57.29 f	12.53 e	4.25 d	44.41 b
5PR4	1.22 b	1.09 b	11.81 ab	136.25 f	673.88 g	144.47 h	66.48 h	30.37 f	47.72 h	13.78 d	4.75 d	42.28 c
AF2	1.61 ab	1.23 a	17.72 ab	150.51 c	1313.02 b	237.21 b	77.94 e	59.44 b	99.82 a	15.32 bc	7.52 a	38.95 d
AF3	1.83 ab	1.31 a	21.10 a	152.01 b	1049.13 c	230.11 c	95.11 b	41.40 c	93.59 c	18.77 a	4.75 d	44.52 b
AF8	1.41 b	1.41 a	17.82 ab	129.75 h	862.88 d	162.51 f	71.65 g	37.96 d	52.89 g	19.75 a	5.02 cd	40.40 f

Table 7. *Cont.*

(b)

2009

Accession	ISPL	ISPEOP	ISSL	ISSEOP
SS4	64.25 b	0.51 b	60.06 a	0.65 c
SS7	60.25 c	0.54 b	58.17 b	0.68 c
SS9	63.75 b	0.53 b	51.78 c	0.71 c
PR1	57.52 d	0.92 a	28.83 h	0.79 b
PR5	53.75 e	0.94 a	44.38 f	0.83 b
PR4	66.25 a	0.98 a	45.33 ef	0.83 b
AF2	52.75 e	0.92 a	45.92 e	1.59 a
AF3	50.55 f	0.96 a	35.69 g	1.53 a
AF8	51.57 f	0.99 a	47.54 d	1.61 a

2010

Accession	ISPL	ISPEOP	ISSL	ISSEOP
SS4	53.75 d	0.56 e	45.54 f	1.09 g
SS7	51.25 e	0.67 d	46.92 e	0.71 f
SS9	56.25 c	0.68 d	51.52 d	0.84 e
PR1	72.75 a	0.64 de	57.42 a	0.72 f
PR5	62.25 b	1.15 b	56.01 b	0.83 e
PR4	57.25 c	1.25 a	52.86 c	0.94 d
AF2	55.07 d	0.83 c	45.68 efg	1.63 b
AF3	62.25 b	1.08 b	54.81 b	1.53 c
AF8	54.54 d	1.07 b	46.13 efg	1.76 a

SY = spike yield; EOP = essential oil percentage; EOY = essential oil yield; PH = plant height; PFW = plant fresh weight; PDW = plant dry weight; FSDW = floral spike. dry weight; LDW = leaf dry weight; SDW = stem dry weight; NB = no. branches; NS = no. stems; SL = spike length. Means followed by the same letter are not significantly different for $p \leq 0.05$ according to Test of Tukey. ISPL = ISPL spike length; ISPEOP = ISP essential oil percentage content; ISSL = ISS spike length; ISSEOP = ISS essential oil percentage content. Means followed by the same letter are not significantly different for $p \leq 0.05$ according to Test of Tukey.

In addition to the AF accessions, having a range between 1.23% (AF2) and 1.42% (AF8), other productive accessions were PR4 (1.10%), PR5 (0.99%) and SS4 (0.83%). In terms of oil yield, however, the most productive accessions were those belonging to the Messina group (17.70–21.10 kg ha^{-1}) followed by PR1 (17.10 kg ha^{-1}). The lowest values were, instead, achieved by SS7 (6.80 kg ha^{-1}). Furthermore, values slightly higher than 10 kg ha^{-1} were found in other accessions.

Of interest here are average EO yields of the Agrigento accessions. EO yields in 2009 were 22.20 kg ha^{-1}, decreasing to approx. 10.00 kg ha^{-1} the following year. This year witnessed a general fall in accession production. The Palermo accessions, however, maintained average EO yields of approx. 14.00 kg ha^{-1} during both years, due to the production performance of PR4 in 2009 and PR1 in 2010. The above-mentioned variations, as is generally known, can be attributed to several factors, including environmental and genetic differences.

3.3. Correlation Matrix

Correlation analysis (Table 8) for both years showed a positive and highly significant correlation between floral spike yields and parameters linked to vigor and plant development, such as height, fresh weight and dry weight of the plant and components (inflorescences, leaves, and stems). In addition to these clear relationships, however, several rather different relationships were observed over the two years which are worthy of note.

In 2009, spike yield was positively affected by ISS spike length (r = 0.80) and branch number (r = 0.84). Furthermore, branch number was positively correlated with most of the parameters in the study (some with statistical significance). In 2010, however, branch number was inversely correlated, although not markedly, with all biometric and production parameters (except for EO%). In the same year, a positive correlation was found between floral spike yield and ISP spike length (r = 0.85), floral spike length (r = 0.88) and ISS spike length (r = 0.52).

Table 8. Correlation matrix of the main biometric and production parameters.

	Characters	SY	EOP	EOY	PH	PFW	PDW	FSDW	LDW	SDW	NB	NS	SL	ISPL	ISSL
								2009							
	SY		−0.512	0.535	**0.735** *	**0.807** **	**0.910** **	**0.998** **	**0.600** *	**0.904** **	**0.843** **	0.398	0.259	0.464	**0.800** **
	EOP	−0.234		0.424	−0.654	−0.057	−0.152	−0.493	0.298	−0.143	−0.407	0.192	**−0.759** *	**−0.801** **	−0.523
	EOY	0.545	**0.673** *		0.121	**0.792** *	0.787 *	0.545	**0.912** **	**0.786** *	0.412	0.616	−0.433	−0.267	0.253
	PH	**0.838** **	−0.205	0.507		0.453	0.572	**0.731** *	0.222	0.583	0.556	−0.043	**0.736** *	**0.830** **	0.596
	PFW	**0.788** *	0.174	**0.749** *	**0.803** *		**0.936** **	**0.812** **	**0.915** **	**0.912** **	0.604	0.596	−0.148	0.092	0.536
	PDW	**0.951** **	−0.078	0.663	**0.899** **	**0.918** **		0.924 **	**0.871** **	**0.997** **	**0.702** *	0.630	−0.073	0.154	**0.681** *
2010	FSDW	**0.993** **	−0.187	0.576	**0.825** **	**0.745** *	**0.935** **		0.622	**0.919** **	**0.827** **	0.432	0.299	0.436	**0.802** **
	LDW	**0.861** **	−0.054	0.586	**0.778** *	**0.946** **	**0.929** **	**0.823** **		**0.862** **	0.395	**0.736** *	−0.442	−0.299	0.364
	SDW	**0.768** *	0.042	**0.678** *	**0.883** **	**0.882** **	0.909 **	**0.748** *	**0.799** *		**0.695** *	0.629	−0.070	0.150	**0.685** *
	NB	−0.212	0.452	0.273	−0.266	−0.028	−0.146	−0.220	−0.168	−0.012		0.197	0.252	0.454	**0.861** **
	NS	0.556	−0.124	0.315	0.634	0.815 **	**0.707** *	0.481	**0.811** **	**0.713** *	−0.054		−0.556	−0.344	0.433
	SL	**0.884** **	−0.204	0.448	0.661	0.518	**0.57** *	**0.921** **	0.641	0.503	−0.452	0.197		**0.956** **	0.244
	ISPL	**0.854** **	−0.174	0.467	**0.741** *	0.530	0.737 *	**0.889** **	0.615	0.507	−0.411	0.207	**0.954** **		0.476
	ISSL	0.520	−0.185	0.216	0.531	0.128	0.366	0.586	0.193	0.185	−0.351	−0.149	**0.740** *	**0.859** **	

SY = spike yield; EOP = essential oil percentage; EOY = essential oil yield; PH = plant height; PFW = plant fresh weight; PDW = plant dry weight; FSDW = floral spike dry weight; LDW = leaf dry weight; SDW = stem dry weight; NB = no. branches; NS = no. stems; SL = spike length; ISPL = ISP spike length; ISSL = ISS spike length. **: correlation is significant at the 0.01 level. *: correlation is significant at the 0.05 level.

3.4. PCA Analysis

PCA analysis was carried out to provide an overall assessment of the accessions over the 2 years and showed that the 2 biggest principal components accounted for 71.00% of total variability, rising to over 86.00% with the 3rd component (Table 9). For analytical purposes, however, only the first two were considered of interest.

Table 9. Variance in principal components and cumulative contribution to total variation.

	PC1	PC2	PC3
Eigenvalues	6891	3030	2171
% Variance	49,218	21,642	15,505
% Cumulative variance	49,218	70,860	86,365

In Table 10, it is clear that the biggest principal component (PC1), accounting for 49.20%, was strongly and directly correlated with as many as 7 characteristics out of the 14. In particular, it is associated with floral spike yield, essential oil yield, fresh and dry weight of the plant, and the dry weight of the inflorescences, leaves, and stems.

Table 10. Factor weights of properties on the two PC.

	PC1	PC2
Spike yield	0.972	0.110
Essential oil percentage	−0.278	−0.519
Essential oil yield	0.781	−0.390
Plant height	0.313	0.822
Plant fresh weight	0.879	−0.023
Plant dry weight	0.977	−0.153
Floral spike dry weight	0.973	0.104
Leaf dry weight	0.873	−0.356
Stem dry weight	0.955	−0.203
No. branches	−0.354	0.527
No. stems	−0.295	0.491
Spike length	0.591	0.290
ISP spike length	0.380	0.766
ISS spike length	0.329	0.727

The second component, accounting for 22.00% of total variation, is strongly linked to the biometric characteristics, such as plant size, main floral spike length (ISP) and secondary floral spike length (ISS). Less marked, however, was the correlation found with two other biometric properties: branch no. and stem no., and inversely correlated, with the same intensity, with % content of EO. Figure 4 shows a loading plot of the factor weights pertaining to the main two principal components.

From Figure 5, which projects the distribution and clustering of the accessions on the plot for the two principal components, statistical data can be extracted.

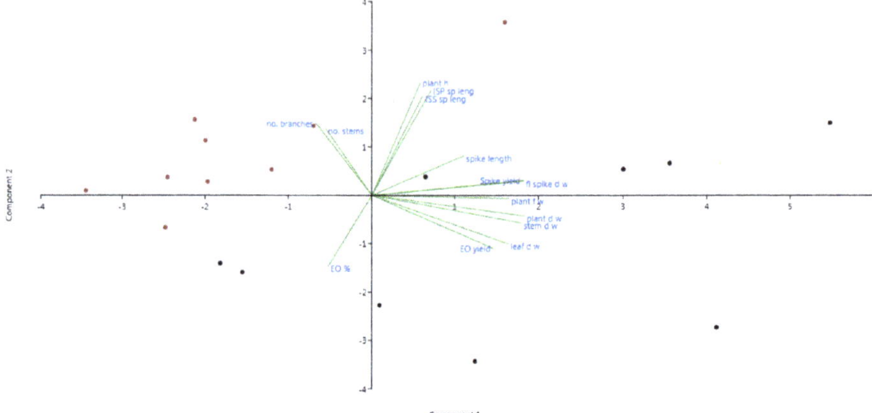

Figure 4. Loading plot of the factor weights pertaining to the main two Principal Component. In the graph, black points refer to clary sage accessions cultivated in the first year while red points refer to accessions cultivated in the second year.

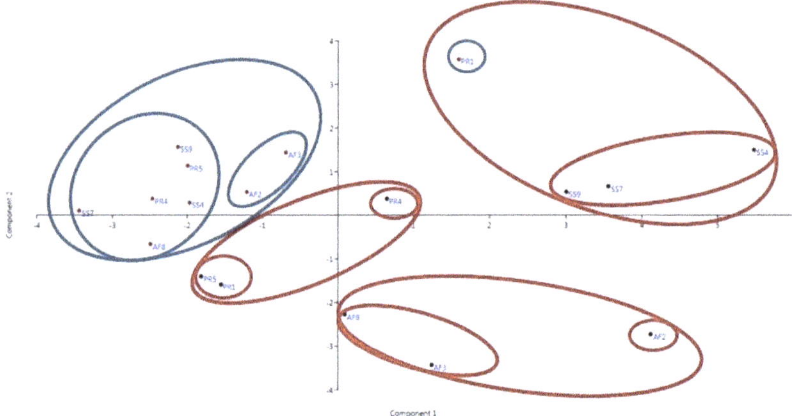

Figure 5. Clustering of the accessions on the score plot for the main two principal components. In the graph, black points refer to clary sage accessions cultivated in the first year while red points refer to accessions cultivated in the second year.

In general, allocation of the accessions led to the identification of 2 clusters which corresponded to the two years. To the right of the origin are the accessions grown during the first year and to the left those grown during the second year. Hence, variations along PC1 can represent environmental variability during the two test years. Characteristics linked to this variability are those which are most affected by the factor "year". However, several heterogeneities emerge which can be attributed to the Palermo accessions. More specifically, PR1 from 2010, allocated in the upper right quadrant, and PR1 and PR5 from 2009, in the lower left quadrant, diverge from this trend. Cluster analysis carried out led to the creation of 4 main groups: 3 for the first year and 1 for the second year. Relative to the first year (2009), each of the 3 main groups was formed by accessions from the same area of provenance. The exception to this concerned the Agrigento accessions, which also included the Palermo accession PR1 from 2010. These Agrigento accessions, however, when associated with the various production parameters, formed a subgroup which is in the upper right quadrant. This is an area which corresponds to high values for the first principal component; accession SS4 being of particular note. PR4, belonging to the other main group from the first year, is located in the same quadrant, though slightly more to the left. PR4, in contrast to other accessions from the same area (PR1 and PR5) managed to ensure a positive production outcome, although in a somewhat more contained manner. PR1 and PR5 formed a subgroup in the lower left quadrant in an area with negative values from both PC2 and PC1. The remaining group from this year, the Messina accessions (AF), are all located in the lower right quadrant, thus possessing negative values from the second principal component. As this is strongly correlated with plant height, ISP lengths and ISS length directly, and moderately correlated with percentage content of EO inversely, these accessions are small in size with a short floral spike but have a high % content of EO. Accession AF3 and, in particular, AF8 (lying near the axis of the second principal component) are more defined by PC2 compared to AF2. This latter is separate from the first two, forming a separate subgroup. In 2010, cluster analysis highlighted one main group with 2 subgroups: 1 of which containing 70.00% of the accessions and the other subgroup containing only AF3 and AF2. These latter were located to the left of the origin, and, therefore, with negative PC1 values. In general, relating to 2010, a more uniform distribution is apparent from the new positions occupied by the accessions, except for PR1. The position of this latter, as previously mentioned, is peculiar, given its position in a cluster located in the upper right quadrant and, thus, far from the main distribution for that year. For most of the accessions, small variations in position concern those along the PC1 axis, except for SS7. This accession was most affected, in production terms, by the environment. However, greater variations were found along the PC2 axis; in particular, in the area with high values concerning morphological characteristics. Several accessions, although occupying similar positions along the PC1 axis, obtained different yields and were mainly defined by those parameters directly linked to PC2. This was also true of accession AF8. Although it located in the lower left quadrant (the only accession in the subgroup located below PC1), it was the one with a greater percentage content in EO, as this parameter was inversely correlated with PC2. Nearer to the origin but positioned in the upper left quadrant, are AF2 and AF3. In this subgroup, AF3, which lies above AF2 along the PC2 axis, obtained production values which were even higher than the average for the field.

3.5. Essential Oil Composition

GC-MS analysis of the essential oil from the clary sage accessions, grouped by area of provenance, led to the identification of 76 chemical compounds, representing 98.00% of the chemical profile (Table 11).

Table 11. Chemical compound main classes of clary sage essential oil.

	ISP						ISS					
	SS		PR		AF		SS		PR		AF	
Class/Compound	m	sd	m	sd	m	sd	m	sd	m	sd	m	sd
Monoterpene hydrocarbons	5.14	0.57	5.16	0.12	4.42	1.06	5.20	0.11	6.06	0.69	5.30	0.24
Oxygenated monoterpenes	77.87	5.30	79.35	2.13	67.30	11.45	79.22	3.88	78.10	2.58	78.54	3.33
Sesquiterpenes	11.18	3.90	11.27	1.54	16.61	5.92	9.70	2.12	10.77	1.88	10.94	2.09
Diterpenes	4.06	1.20	3.32	0.40	9.23	5.34	3.90	1.00	3.87	1.20	3.81	1.00

m = mean; sd = standard deviation.

The most abundant class was that of oxygenated monoterpenes (67–79.00%), followed by sesquiterpenes at levels above 10.00%. Monoterpenes, hydrocarbons, and diterpenes remained below 10.00%. The compounds linalyl acetate and linalool, in a ratio of 2:1, accounted for approx. 60.00% of the total, with α-terpineol as the third most abundant component (Table 12). This pattern was found to be common to all the samples tested, including those on the SP and the SS floral spike, and was also reflected in the profile of the minor components. The chemical composition of the two types of inflorescence did not show intraspecific chemical differences, and the content of the principal components was highly uniform. The accessions in the study are of chemotype "linalyl acetate" (range 36.00–43.00%).

Table 12. Chemical composition of clary sage essential oils.

				ISP						ISS					
				SS		PR		AF		SS		PR		AF	
Peak [a]	RI Lit. [b]	RI Exp. [c]	Class/Compound	m	sd	m	sd	m	sd	m	sd	m	sd	m	sd
			Monoterpene Hydrocarbons												
6	991	994	ß-Mircene	1.39	0.42	1.72	0.09	1.67	0.26	1.74	0.16	2.10	0.32	1.73	0.03
10	1050	1041	*trans*-Ocimene	1.07	0.25	1.35	0.08	1.28	0.10	1.33	0.07	1.59	0.09	1.32	0.04
			Oxygenates Monoterpenes												
14	1097	1094	Linalool	21.42	3.50	23.79	0.37	24.40	3.24	23.43	1.22	27.31	2.63	23.59	1.92
22	1189	1200	a-terpineol	4.83	2.12	6.40	0.23	6.38	1.49	6.69	0.74	7.90	1.20	6.58	0.20
25	1130	1240	Nerol	1.07	0.36	1.27	0.04	1.23	0.23	1.32	0.10	1.52	0.21	1.28	0.01
26	1257	1271	Lynalin-acetate	35.80	4.78	42.46	2.63	40.85	2.11	41.71	1.30	35.46	2.18	42.42	2.79
34	1362	1369	Neryl acetate	1.43	0.33	1.62	0.09	1.53	0.18	1.66	0.11	1.81	0.18	1.56	0.08
36	1381	1389	Geranyl acetate	2.26	0.69	2.76	0.13	2.62	0.42	3.02	0.23	3.32	0.33	2.85	0.07
			sesquiterpenes												
41	1419	1432	b-caryophyllene	3.28	1.73	2.03	0.14	2.39	1.04	2.03	0.64	1.97	0.53	2.07	0.28
49	1485	1494	Germacrene D	5.08	1.09	4.55	0.91	3.95	1.41	3.76	0.49	4.42	0.90	3.27	0.92
50	1496	1499	Valencene	3.17	1.20	1.93	0.43	1.88	0.63	1.94	0.31	1.60	0.26	1.72	0.51
			Diterpenes												
76	2223	2168	Sclareol	7.76	4.48	2.71	0.34	3.29	0.95	3.25	0.96	3.15	1.07	3.14	0.87

m = mean; sd = standard deviation. [a] the numbering refers to elution order, and values (relative peak area percent) represent averages of 3 determinations and standard deviation; [b] Literature Retention Index; [c] Experimental Retention Index relative to standard mixture of n-alkanes on SPB-5 column.

4. Discussion

For cropping systems wishing to increase levels of diversification while ensuring low input, wild species, including many medicinal and aromatic plants, are considered to be a strategic choice due to their rusticity, adaptability, and sustainability, often expression of the combination of environment and ecotype. However, compared to the great number of aromatic plants known to us, few are cultivated [43]. To domesticate and ensure diffusion of wild species, it is necessary to understand how they behave during cultivation, both in agronomic terms and in terms of quality [44]. As noted by Mossi et al. [36]; there is a lack of information on clary sage regarding biometric and agronomic aspects, in contrast to studies on other species of the genus Salvia, such as *S. officinalis* and *Salvia triloba*.

All the accessions grew in a regular although controlled manner, maintaining rosette stage. During new growth stage, plant stems developed and flowering stage was reached the following year, confirmation of their biennial habitus, as noted by other authors [6]. Averages obtained for the field during the two years of tests showed far higher values in 2009 than in 2010 for the biometric and production parameters being examined. Environmental factors undoubtedly played a key role in the different production results. More specifically, the heterogeneous rainfall levels throughout the test period were considered highly influential, with 2009 experiencing exceptionally high rainfall levels and 2010 much lower and consistent with the test environment.

Considering fresh biomass inflorescence yields, results obtained in this test (regarding both productive and less productive accessions) were consistent with those reported by Yessen et al. [8,37] in tests carried out in India in a subtropical environment. Dry biomass yields, however, appeared to be slightly lower than those reported in tests carried out in Brazil by Mossi et al. [36]. It is worth noting that estimates of this parameter were made using a greater plant density per hectare compared to this study, while production results per plant were similar. Considering the three classifications (low, medium and high) determined by Yessen et al., [37] based on plant height, we can classify all of the Sicilian accessions in the study as medium-sized (100–150 cm). This contrasts with other studies carried out in Brazil, Spain, and Sicily, where plants of almost 1 m can be classified as small [6,36,38,45]. As regards no. of branches (9.17), fresh plant biomass production (1098 g) and dry plant biomass production (361 g), results for 2009 were consistent with those found in Mossi et al. [36]. Results from 2010, albeit lower than the previous year, were, in some cases, far higher, than those obtained by other researchers in Mediterranean environments. In the Aragona region in Spain, for example, Alquezar [38] obtained a plant fresh weight in the first year of 201 g and 830 g in the second year. This latter was similar to the weight obtained by us in the drier year. Once again in Sicily, Carrubba et al. [6], in highly arid conditions, obtained very low values for plant dry weight (75.80 g), except for number of branches (23 branches). In agreement with this, it is worth noting that in our study, as regards number of branches, this parameter was higher in the year with lower rainfall, with nearly a two-fold increase compared to the previous year (16 branches).

In addition to the above-mentioned agro-morphological parameters, the biometric and production characteristics of the floral spike are of particular interest in the scientific literature [37] on medicinal and aromatic plants. These characteristics are considered to be a reliable factor when selecting species with a high EO content.

Furthermore, in this study, we considered it important to analyze the floral spike by distinguishing between the main stem floral spike (ISP) and the floral spike of the secondary stem (ISS), an aspect not discussed in the literature. Of interest is the fact that spike length was longer in the drier year and EO was slightly higher in ISS compared to ISP. This would suggest the importance of bearing this peculiar characteristic in mind in the selection of accessions for production purposes, above all in Mediterranean areas. In the two years of tests, average ISP spike length did not vary and was relatively high. The great variability found, above all in some of the accessions, in production and biometric parameters was revealed by analysis of the principal components, with production parameters accounting for approx. 49.00% and biometric parameters 22.00% of the total variation.

Oil percentage content in this study, over the two years, was reasonably high in all the accessions (0.58–1.8%). Values for this parameter found in the literature are mostly much lower. Research carried out in various regions in the north of Iran (0.31–0.65%) and in Leskovac, Serbia (0.78–0.83%) [31,46] reported similar results to some of the values in this study. In this study, the accessions from Messina obtained satisfactory results not only for the above-mentioned parameter, but also for EO yields per hectare, as these accessions were also among the most productive accessions. In fact, the 3 Messina accessions produced, on average, an EO yield of 19.00 kg ha^{-1} in 2010, consistent with Mossi et al. [36]. In 2009, however, despite a lower percentage content of EO, the same accessions produced 29.50 kg ha^{-1} of EO, as the yield was obviously connected to spike yield. The above-mentioned variations, as is generally known, can be attributed to several factors, among which the various environmental and genetic differences.

As regards the chemical composition, there are various chemotypes known for clary sage. In this study, it is clear from the chemical analyses of the essential oils extracted from the inflorescences that the accessions were chemotypes rich in linalyl acetate and linalool. This agrees with results from other authors [1,6,14,34,37]. The chemical composition of the two inflorescence types, ISP and ISS, from the three areas (an average of the 3 accessions from each area of provenance) did not show differences, presenting a highly uniform content of principal components. Nevertheless, further research is needed to increase knowledge of this, especially in relation to the geographic origin.

Furthermore, regarding the biometric characteristics, PCA also revealed distinctive features which enabled a series of clustering. This clustering highlighted, in general, the effects of the different years and the subdivision of the accessions, mainly based on ISP and ISS spike length. In 2009, abundant rainfall led to improved biometric results and yields, allowing us to identify accessions with a tendency towards medium high production levels. Furthermore, within the macro-groups, we were able to identify subgroups linked to the area of provenance. The lower rainfall levels in 2010, although consistent with the test environment, not only limited the biometric and production characteristics, but also annulled the link to provenance, which, however, was evident in the first year. In this year, accessions were found to be less evenly distributed along the PC axes compared to 2010 (mostly concentrated along the PC2 axis). The accessions which maintained or exceeded averages for the field were those accessions which modified their production parameters regarding the second principal component, clearly actuating an adaptation strategy triggered by adverse environmental conditions. Of the accessions, PR1 is worthy of note while SS7 showed poor results, despite being one of the best performers in the more favorable year.

5. Conclusions

Results obtained in this study represent a valid contribution to the acquisition of knowledge of the adaptability and production potential of a medicinal and aromatic species of interest to industry in the Mediterranean area. The production levels obtained are interesting and promising, even though variable over the two years. Most of the clary sage wild accessions showed a satisfactory production response, reaching higher or around averages for the field in the two years. In general, the rainy year led to more vigorous plants with a higher percentage incidence of stems and leaves, while in 2010, the plants were sparser and with a greater incidence of inflorescences compared to other plant components. On this note, regarding to the determination of spike yield, analysis of variance and multivariate analysis (PCA) highlighted the considerable importance of several biometric properties, among which number of stems, number of branches and ISP spike length and, in particular, ISS spike length. The accessions which maintained or exceeded the averages for the field, above all in the drier year, were the ones with a longer ISS and ISP spike, demonstrating this as a production adaptation strategy to adverse environmental conditions. From an agronomic point of view, PR1 was found to be worth of note. Accessions SS7 and PR4 obtained interesting results only in the first year, as they were found to be highly adversely affected by environmental conditions in the second year. The remaining accessions; however, were consistent with the average for the field. Relative to the essential oil content, all the accessions produced high EO content; the Messina and Palermo accessions being of particular note. The accessions with the best EO yield performance per hectare were the Messina accessions in both years; the Agrigento accessions and PR1 in 2009; and PR1 in 2010. All the accessions in the study were "linalyl acetate" chemotype (range 36.00–42.60%). The chemical composition did not vary between the two types of inflorescence, ISP and ISS. The different biometric and production properties of the accessions in the study could be of use for the selection of biotypes for future use. In conclusion, the results show, in addition to good adaptability to the environment, good potential for introduction into Mediterranean cropping systems, fostering the expansion of medicinal and aromatic crops.

Author Contributions: Conceptualization, T.T. and S.L.B.; methodology, T.T. and S.L.B.; software, T.T. and G.V.; validation, C.L., G.I. and M.L.; formal analysis, T.T., M.L., G.V. and G.I.; investigation, G.V., M.L., S.L.B. and C.L.; resources, T.T. and S.L.B.; data curation, G.I., M.L., G.V. and C.L.; writing—original draft preparation, G.I., M.L., G.V. and CL.; writing—review and editing, M.L. and G.V.; visualization, G.I., M.L. and C.L.; supervision, T.T. and S.L.B.; project administration, C.L.; funding acquisition, C.L. All authors have read and agreed to the published version of the manuscript.

Funding: This research was funded by the Sicilian Regional Ministry of Agriculture and Food Resources (Italy), project "Environmental and plant resources in the Mediterranean: study, valorisation and defence", grant number 2309/2005.

Acknowledgments: The authors would like to thank Giuseppe Ruberto, from Istituto di Chimica Biomolecolare (CN) of Catania (Italy), for his contribution in the chemical analyses of the EOs. A special thanks goes to Lucie Branwen Hornsby for her linguistic assistance.

Conflicts of Interest: The authors declare no conflict of interest. The funders had no role in the design of the study; in the collection, analyses, or interpretation of data; in the writing of the manuscript, or in the decision to publish the results.

References

1. Kumar, R.; Kaundal, M.; Sharma, S.; Thakur, M.; Kumar, N.; Kaur, T.; Vyas, D.; Kumar, S. Effect of elevated [CO$_2$] and temperature on growth, physiology and essential oil composition of *Salvia sclarea* L. in the western Himalayas. *J. Appl. Res. Med. Aroma.* **2017**, *6*, 22–30. [CrossRef]

2. Pignatti, S. *Flora d'Italia*; Edagricole: Bologna, Italy, 2003.

3. Lojacono-Pojero, M. Flora Sicula o descrizione delle piante spontanee o indigenate in Sicilia. In *Monocotyledones-Cryptogamae Vasculares—Scuola Tip*; Boccone del Povero: Palermo, Sicily, Italy, 2013; Volume 3, pp. 483–495.

4. Lubbea, A.; Verpoorte, R. Cultivation of medicinal and aromatic plants for speciality industrial materials. *Ind. Crop. Prod.* **2011**, *34*, 785–801. [CrossRef]

5. Cai, J.; Lin, P.; Zhu, X.; Su, Q. Comparative analysis of clary sage (*S. sclarea* L.) oil volatiles by GC-FTIR and GC-MS. *Food Chem.* **2006**, *99*, 401–407. [CrossRef]

6. Carrubba, A.; Torre, R.; Piccaglia, R.; Marotti, M. Characterization pf an Italian biotype of clary sage (*Salvia sclarea* L) grown in semi-arid Mediterranean environmenr. *Flavour Fragr. J.* **2002**, *17*, 191–194. [CrossRef]

7. Lawrence, B.M. Progress in essential oils. Clary sage oil. *Perfum. Flavorist* **1986**, *11*, 111.

8. Yaseen, M.; Singh, M.; Ram, D.; Singh, K. Production potential, nitrogen use efficiency and economics of clary sage (*Salvia sclarea* L.) varieties as influenced by nitrogen levels under different locations. *Ind. Crop. Prod.* **2014**, *54*, 86–91. [CrossRef]

9. Fakir, H.; Korkmaz, M.; Güller, B. Medicinal plant diversity of western Mediterranean region in Turkey. *J. Appl. Biol. Sci.* **2009**, *3*, 30–40.

10. Özdemir, E.; Alpınar, K. An ethnobotanical survey of medicinal plants in western part of central Taurus Mountains, Aladaglar (Nigde–Turkey). *J. Ethnopharmacol.* **2015**, *166*, 53–65. [CrossRef]

11. Peana, A.T.; Moretti, M.D.; Juliano, C. Chemical composition and antimicrobial action of the essential oils of Salvia desoleana and *S. Sclarea*. *Planta Med.* **1999**, *65*, 752–754. [CrossRef]

12. Pitarokili, D.; Couladis, M.; Petsikos-Panayotarou, N.; Tzakou, O. Composition and antifungal activity on soil-borne pathogens of the essential oil of *Salvia sclarea* from Greece. *J. Agric. Food Chem.* **2002**, *50*, 6688–6691. [CrossRef]

13. Gülçin, I. Evaluation of the antioxidant and antimicrobial activities of clary sage (*Salvia sclarea* L.). *Turk. J. Agric. For.* **2004**, *28*, 25–33.

14. Fraternale, D.; Giamperi, L.; Bucchini, A.; Ricci, D.; Epifano, F.; Genovese, S.; Curini, M. Composition and antifungal activity of essential oil of *Salvia sclarea* from Italy. *Chem. Nat. Comp.* **2005**, *41*, 604–606. [CrossRef]

15. Georgiev, E.; Stoyanova, A. *A Guide for the Specialist in Aromatic Industry*; UFT Academic Publishing House: Plovdiv, Bulgaria, 2006.

16. Jirovetz, L.; Buchbauer, G.; Denkova, Z.; Slavchev, A.; Stoyanova, A.; Schmidt, E. Chemical composition, antimicrobial activities and odor descriptions of various *Salvia* sp. and *Thuja* sp. essential oils. *Nutrition* **2006**, *90*, 152–159.

17. Jirovetz, L.; Wlcek, K.; Buchbauer, G.; Gochev, V.; Girova, T.; Stoyanova, A.; Schmidt, E.; Geissler, M. Antifungal activities of essential oils of Salvia lavandulifolia, Salvia officinalis and *Salvia sclarea* against various pathogenic Candida species. *J. Essent. Oil-Bear. Plants* **2007**, *10*, 430–439. [CrossRef]

18. Kuźma, L.; Kalemba, D.; Różalski, M.; Różalski, F.; Więckowaska-Szakiel, M.; Krajewska, U.; Wysokińska, H. Chemical composition and biological activities of essential oil from *Salvia sclarea* plants regenerated in vitro. *Molecules* **2009**, *14*, 1438–1447. [CrossRef]

19. Geun, H.S.; Hyun, S.S.; Pill-Joo, K.; Hean, K.M.; Ki, H.L.; Insop, S.; Suk, S.M. Antidepressant-like effect of *Salvia sclarea* is explained by modulation of dopamine activities in rats. *J. Ethnopharmacol.* **2010**, *130*, 187–190. [CrossRef]

20. Kostić, M.; Kitić, D.; Petrović, M.B.; Jevtović-Stoimenov, T.; Jović, M.; Petrović, A.; Živanović, S. Anti-inflammatory effect of the *Salvia sclarea* L. ethanolic extract on lipopolysaccharide-induced periodontitis in rats. *J. Ethnopharmacol.* **2017**, *199*, 2–59. [CrossRef]

21. Hao, D.C.; Chen, S.L.; Osbourn, A.; Kontogianni, V.G.; Liu, L.W.; Jordán, M.J. Temporal transcriptome changes induced by methyl jasmonate in *Salvia sclarea*. *Gene* **2015**, *558*, 41–53. [CrossRef]

22. Kuźma, Ł.; Różalski, M.; Walencka, E.; Różalska, B.; Wysokińska, H. Antimicrobial activity of diterpenoids from hairy roots of *Salvia sclarea* L.: Salvipisone as a potential anti-biofilm agent active against antibiotic resistant Staphylococci. *Phytomedicine* **2007**, *14*, 31–35. [CrossRef]

23. Tuttolomondo, T.; Leto, C.; Leone, R.; Licata, M.; Virga, G.; Ruberto, G.; Napoli, E.M.; La Bella, S. Essential oil characteristics of wild Sicilian oregano populations in relation to environmental conditions. *J. Essent. Oil Res.* **2014**, *26*, 210–220. [CrossRef]

24. Tuttolomondo, T.; Dugo, G.; Leto, C.; Cicero, N.; Tropea, A.; Virga, G.; Leone, R.; Licata, M.; La Bella, S. Agronomical and chemical characterisation of *Thymbra capitata* (L.) Cav. biotypes from Sicily, Italy. *Nat. Prod. Res.* **2015**, *29*, 1289–1299. [CrossRef] [PubMed]

25. Militello, M.; Carrubba, A.; Blázquez, M.A. Artemisia arborescens L.: Essential oil composition and effects of plant growth stage in some genotypes from Sicily. *J. Essent. Oil Res.* **2012**, *24*, 229–235. [CrossRef]

26. Napoli, E.M.; Siracusa, L.; Saija, A.; Speciale, A.; Trombetta, D.; Tuttolomondo, T.; La Bella, S.; Licata, M.; Virga, G.; Leone, R.; et al. Wild Sicilian rosemary: Phytochemical and morphological screening and antioxidant activity evaluation of extracts and essential oils. *Chem. Biodivers.* **2015**, *12*, 1075–1094. [CrossRef] [PubMed]

27. Saija, A.; Speciale, A.; Trombetta, D.; Leto, C.; Tuttolomondo, T.; La Bella, S.; Licata, M.; Virga, G.; Bonsangue, G.; Gennaro, M.C.; et al. Phytochemical, ecological and antioxidant evaluation of wild Sicilian thyme: Thymbra capitata (L.) Cav. *Chem. Biodivers.* **2016**, *13*, 1641–1655. [CrossRef]

28. Tuttolomondo, T.; Dugo, G.; Ruberto, G.; Leto, C.; Napoli, E.M.; Potortì, A.G.; Fede, M.R.; Virga, G.; Leone, R.; D'Anna, E.; et al. Agronomical evaluation of Sicilian biotypes of Lavandula stoechas L. spp. Stoechas and analysis of the essential oils. *J. Essent. Oil Res.* **2015**, *27*, 115–124. [CrossRef]

29. Yuce, E.; Yildirim, N.; Yildirim, N.C.; Paksoy, M.Y.; Bagci, E. Essential oil composition, antioxidant and antifungal activities of *Salvia sclarea* L. from Munzur Valley in Tunceli, Turkey. *Cell. Mol. Biol.* **2014**, *60*, 1–5. [CrossRef] [PubMed]

30. Saeidnia, S.; Gohari, A.R.; Haddadi, A.; Amin, G.; Nikan, M.; Hadjiakhoondi, A. Presence of monoterpene synthase in four *Labiatae* species and solid-phase microextraction–gas chromatography–mass spectroscopy analysis of their aroma profiles. *Pharmacognosy Res.* **2014**, *6*, 138–142. [CrossRef]

31. Pesic, P.Z.; Bankovi, V.M. Investigation on the essential oil of cultivated *Salvia sclarea* L. *Flavour Fragr. J.* **2003**, *18*, 228–230. [CrossRef]

32. Ohloff, G. *Riechstoffe und Geruchssinn: Die molekulare Welt der Düfte*; Springer: Berlin/Heidelberg, Germany; New York, NY, USA, 1990; pp. 208–214.

33. Moulines, J.; Bats, J.P.; Lamidey, A.M.; Da Silva, N. About a practical synthesis of Ambrox from sclareol: A new preparation of a ketone key intermediate and a close look at its Baeyer-Villiger oxidation. *Helv. Chim. Acta* **2004**, *87*, 2695–2705. [CrossRef]
34. Farkas, P.; Holla, M.; Jozef, T.; Mellen, S. Composition of the essential oils from the flowers and leaves of *Salvia sclarea* L. (Lamiaceae) cultivated in Slovak Republic. *J. Essent. Oil Res.* **2005**, *17*, 141–144. [CrossRef]
35. Schmiderer, C.; Grassi, P.; Novak, J.; Weber, M.; Franz, C. Diversity of essential oil glands of clary sage (*Salvia sclarea*, L., Lamiaceae). *Plant Biol.* **2008**, *10*, 433–440. [CrossRef] [PubMed]
36. Mossi, A.J.; Cansian, R.L.; Paroul, N.; Toniazzo, G.; Oliveira, J.V.; Pierozan, M.K.; Pauletti, G.; Rota, L.; Santos, A.C.A.; Serafini, L.A. Morphological characterization and agronomical parameters of different species of *Salvia* sp. (Lamiaceae). *Braz. J. Biol.* **2011**, *71*, 121–129. [CrossRef] [PubMed]
37. Yaseen, M.; Kumar, B.; Ram, D.; Singh, M.; Anand, S.; Yadav, H.K.; Samad, A. Agro morphological, chemical and genetic variability studies for yield assessment in clary sage (*Salvia sclarea* L.). *Ind. Crop. Prod.* **2015**, *77*, 640–647. [CrossRef]
38. Alquezar, J.B. *Investigación y Experimentación de Plantas Aromàticas e Medicinalesen Aragón*; Cultivo, Trasformación y Analìtica, Espanha: Governo de Aragón, Zaragoza, Spain, 2003; p. 262.
39. *Nist, 'National Institute of Standard and Technology'*; Mass Spectral Library: Gaithersburg, MD, USA, 1998.
40. Adam, R.M. *Identification of Essential Oil Components by Gas Chromatography/Mass Spectrometry*, 4th ed.; Allured Publisghing Corporation: Carol Stream, IL, USA, 2007.
41. Lee, J.; Kaltsikes, P.J. Multivariate statistical analysis of grain yield and agronomic characters in durum wheat. *Theor. Appl. Genet.* **1973**, *43*, 226–231. [CrossRef]
42. Harper, H.H.; David, A.T.; Ryan, P.D. Past: Paleontological Statistics Software Package for Education and Data Analysis. *Palaeontol. Electron.* **2001**, *4*, 9.
43. Catizone, P.; Marotti, M.; Toderi, G.; Tetenyi, P. *Coltivazione Delle Piante Medicinali e Aromatiche*; Patron: Bologna, Italy, 1986.
44. Carrubba, A.; Militello, M. Nonchemical weeding of medicinal and aromatic plants. *Agron. Sustain. Dev.* **2013**, *33*, 551–561. [CrossRef]
45. Quer, P.F. *Plantas Medicinales: El Dioscórides Rinovado*, 1st ed.; Penìnsula: Barcelona, Spain, 1999.
46. Rajabi, Z.; Ebrahimi, M.; Farajpour, M.; Mirza, M.; Ramshini, H. Compositions and yield variation of essential oils among and within nine Salvia species from various areas of Iran. *Ind. Crop. Prod.* **2014**, *61*, 233–239. [CrossRef]

 *agronomy*

Article

Chemical Composition of *Cynara Cardunculus* L. var. *altilis* Heads: The Impact of Harvesting Time

Filipa Mandim [1,2], Spyridon A. Petropoulos [3,*], Ângela Fernandes [1],
Celestino Santos-Buelga [2], Isabel C. F. R. Ferreira [1] and Lillian Barros [1,*]

[1] Mountain Research Center (CIMO), Polytechnic Institute of Bragança, Campus de Santa Apolónia,
5300-253 Bragança, Portugal; filipamandim@ipb.pt (F.M.); afeitor@ipb.pt (Â.F.); iferreira@ipb.pt (I.C.F.R.F.)
[2] The Polyphenols Research Group (GIP-USAL), Faculty of pharmacy, University of Salamanca, Campus
Miguel de Umaamuno s/n, 37007 Salamanca, Spain; csb@usal.es
[3] Department of Agriculture, Crop Production and Rural Environment, University of Thessaly, N. Iona,
38446 Volos, Greece
* Correspondence: spetropoulos@uth.gr (S.A.P.); lillian@ipb.pt (L.B.)

Received: 2 July 2020; Accepted: 23 July 2020; Published: 27 July 2020

Abstract: Cardoon is a multi-purpose crop with several industrial applications, while the heads (capitula) are edible and commonly used in various dishes of the Mediterranean diet. Several reports in the literature study the chemical composition of the various plants parts (leaves, flower stalks, bracts, seeds) aiming to industrial applications of crop bio-waste, whereas for the heads, most of the studies are limited to the chemical composition and bioactive properties at the edible stage. In the present study, cardoon heads were collected at six different maturation stages and their chemical composition was evaluated in order to determine the effect of harvesting stage and examine the potential of alternative uses in the food and nutraceutical industries. Lipidic fraction and the content in fatty acids, tocopherols, organic acids, and free sugars were determined. Lipidic content decreases with the maturation process, while 22 fatty acids were detected in total, with palmitic, oleic, and linoleic acids being those with the highest abundance depending on harvesting time. In particular, immature heads have a higher abundance in saturated fatty acids (SFA), whereas the samples of mature heads were the richest in monounsaturated fatty acids (MUFA). The α-tocopherol was the only isoform detected being present in higher amounts in sample Car B (619 µg/100 g dw). Oxalic, quinic, malic, citric and fumaric acids were the detected organic acids, and the higher content was observed in sample Car E (15.7 g/100 g dw). The detected sugars were fructose, glucose, sucrose, trehalose and raffinose, while the highest content (7.4 g/100 g dw) was recorded in sample Car C. In conclusion, the maturation stage of cardoon heads influences their chemical composition and harvesting time could be a useful means to increase the quality and the added value of the final product by introducing this material in the food and nutraceutical industries.

Keywords: seasonal variation; fatty acids; free sugars; chemical composition; *Cynara cardunculus* L.; cardoon; organic acids

---

1. Introduction

Cynara cardunculus L., or commonly known as cardoon, belongs to the *Asteraceae* family which is one of the largest families of the plant kingdom with more than 2000 species. *Cynara cardunculus* comprises three botanical varieties, all native to the Mediterranean basin, the wild cardoon (var. *sylvestris*), the domesticated cardoon (var. *altilis* DC), and the globe artichoke (var. *scolymus*) [1,2]. This crop has been gaining attention due to the high biological and industrial potential that it has shown in several studies described in the literature [1–4]. Despite being present all over the world, countries like Spain, France, and Italy are responsible for almost 80% of its production worldwide [1].

Cardoon is a species highly resistant to the fluctuation of weather conditions with low precipitation and hot and dry summers, characteristic of the Mediterranean basin climate. Its high resistance against adverse conditions and weather extremities, together with its multifaceted applicability, favor its exploration and the multiple uses in different industrial applications [4,5]. The industrial applications of cardoon are diverse, since it is used as plant rennet in the food industry to produce cheeses of protected designation of origin (PDO) [6,7]; it is also used for the production of paper pulp, due to its high content in cellulose and hemicellulose [8,9], as well as for bioenergy and biomass production [1,10,11].

Cardoon is also used in traditional medicine since ancient times due to its health-promoting benefits. This species is widely consumed as a result of its antidiabetic, antihemorrhodial, cardiotonic, choleretic, and lipid-lowering actions. Furthermore, several studies have demonstrated other health-promoting properties, such as antioxidant, anti-HIV, anti-inflammatory, cytotoxic, antifungal, and antibacterial properties [4,12–15]. Several studies suggested that the miscellaneous medicinal properties confirmed so far are related to the presence of a high variety of bioactive compounds and phytochemicals. For this purpose, cardoon tissues are being widely explored as a result of its high concentration and variety in compounds with important biological effects and industrial purposes [4,5,16,17]. Literature reports also indicate cardoon as an important source of dietary components such as fibers, inulin, and minerals, but also of phenolic acids, mostly caffeoylquinic and dicafeoylquinic acids derivatives, flavonoids such luteolin and apigenin derivatives, anthocyanins and sesquiterpene lactones [8,18–21]. The presence and abundance of the detected biological compounds could be influenced by several factors, namely by the genotype, the pre and post-harvest conditions, the parts considered for chemical analysis (heads, leaves, bracts, flowers, pappi, receptacle, and petioles), the growing conditions, and the physiological stage at harvest [4,19,22–25], while the choice of harvesting time has been suggested to affect the polyphenols content and composition [20]. Although the chemical composition of cardoon is widely described in the literature, further studies are needed to evaluate the influence of the abovementioned parameters on its chemical composition, thus allowing a more complete knowledge and the adequate use of the species based on specific bioactive compounds content.

Considering the lack of information in the scientific literature, the aim of this study was to determine the impact of the maturation stage on the lipidic content and on the profile of fatty acids, tocopherols, organic acids, and free sugars present in cardoon heads collected in central Greece and to evaluate alternative uses of cardoon heads that are overripe, e.g., they have passed the edible stage. The results of this study will contribute to better understanding the influence of the seasonal changes on the chemical composition of cardoon, resulting in an in-depth knowledge of the species and possible alternative uses in different areas of application that would add economic importance in this multi-purpose crop.

2. Materials and Methods

2.1. Standards and Reagents

All solvents were of analytical grade and were purchased from Fisher Scientific (Lisbon, Portugal). The fatty acids methyl esters (FAME) mixture (standard 47885-U) and standards of organic acids and sugars were acquired from Sigma-Aldrich (St Louis, MO, USA). Tocopherol standards were acquired from Matreya (Pleasant Gap, PA, USA). Other reagents and solvents of analytical grade were purchased from common sources. Water treatment was performed using a Milli-Q water purification system (TGI Pure Water Systems, Greenville, SC, USA).

2.2. Plant Material

Head (capitulum) samples of *Cynara cardunculus* var. *altilis* DC cv. *Bianco Avorio* (Fratelli Ingegnoli Spa, Milano, Italy) were collected from the experimental farm of the University of Thessaly in Velestino, in central Greece (22.756 E, 39.396 N), during the growing period of 2017–2018 (Figure 1). Heads were collected from 15 individual plants (n = 15) at the beginning of the flower's development, in full maturity, and in seed ripening stages. The climate conditions during the growing period and the procedure used for the collection and treatment of plant material were described by Mandim et al. [19]. Briefly, Car A corresponds to principal growth stage (PGS) 5 (harvest: 26 April 2018), Car B corresponds to PGS 5/6 (harvest: 10 May 2018), Car C corresponds to PGS 6 (harvest: 24 May 2018), Car D corresponds to PGS 6/7 (harvest: 04 July 2018), and finally the samples Car E and Car F correspond to PGS 7 and PGS8, respectively (harvest: 09 August 2018 and 29 August 2018, respectively) [19,26].

Figure 1. Evolution of maturity (Car A to F from left to right) of *Cynara cardunculus* L. heads during the growing period.

The plant material and the growing condition have been described in previous reports by our team. Briefly, bract samples were collected from 8-year-old plants sexually propagated from seeds in 2010. Soil parameters were the following, as previously described by the authors [27]: loam texture (48% Sand; 29% Silt; 23% Clay); Organic matter: 1.3%; pH: 7.9; EC: 1.4 dS/m; NO^{-3}: 9.49 mg/kg; P: 74.53 mg/kg; K_{exch}: 0.98 cmol$_c$/kg; Ca_{exch}: 13.96 cmol$_c$/kg; Mg: 4.32 cmol$_c$/kg. Prior to crop establishment a base dressing was applied by using 50 kg/ha N, 90 kg/ha P_2O_5 and 40 kg/ha K_2O. After crop establishment, nitrogen fertilizers were applied with side dressing at each growing period and before plant regrowth (100 kg/ha N). Plant density was 40,000/ha with distances of approximately 0.6 m between rows and 0.4 m within rows. Irrigation was applied monthly during the first growing period (staring on April and until July) with water cannons, whereas in the following years irrigation was applied only twice in each growing period (on April and May) due to the extensive root system that plants form after the second year of establishment. Weed control was applied with hoeing after plant regrowth at each growing period, since at later growth stages plant is very competitive against weeds. No pesticides and fungicides were applied. Climate conditions during the experimental period (shoot emergence until senescence) are presented in Figure 2.

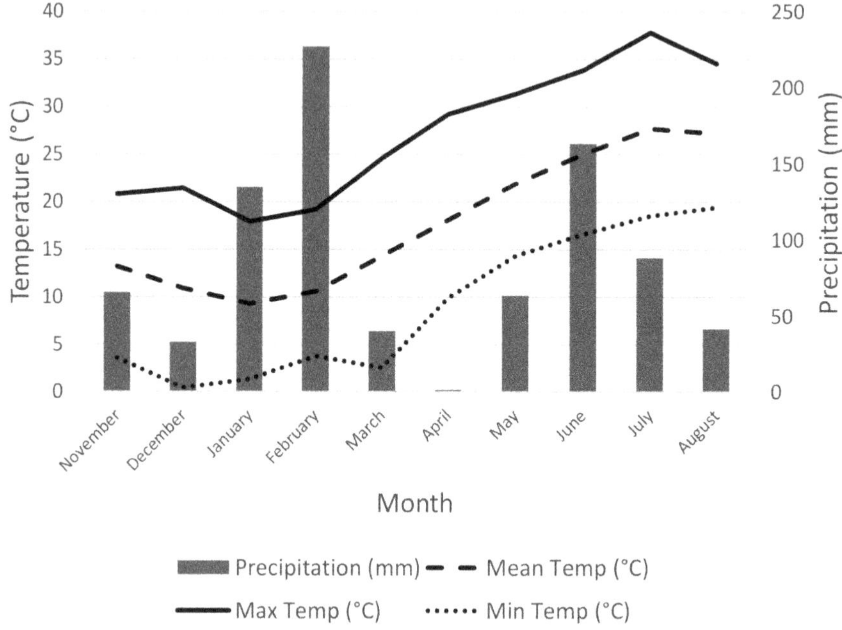

Figure 2. Climate conditions during the experimental period (Mean temperature (°C); Max temperature (°C); Min temperature (°C); Precipitation (mm).

2.3. Chemical Composition Analysis

2.3.1. Fatty Acids

For the analysis of fatty acids composition, the lipidic fraction was extracted with petroleum ether through a Soxhlet extraction system at 120 °C. After a transesterification process, the fatty acids content was analyzed by Gas-liquid Chromatography (GC), coupled to a Flame Ionization Detector (FID) at 260 °C and according to the analytical conditions previously described [27]. The identification and quantification of fatty acids were performed by comparing the relative retention times of FAME peaks from samples with standards (reference standard mixture 47,885-U), using Clarity DataApex 4.0 software (Prague, Czech Republic). The results were expressed as relative percentages and in mg per 100 g dw of each detected fatty acid.

2.3.2. Tocopherols

Tocopherols composition was analyzed using a high-performance liquid chromatography system (HPLC, Knauer, Smartline system 1000, Berlin, Germany) coupled to a fluorescence detector (FP-2020; Jasco, Easton, USA) programmed for excitation at 290 nm and emission at 330 nm, according to the procedure previously described [28]. The qualitative and quantitative analysis were performed using the Clarity 2.4 software (DataApex, Prague, Czech Republic) and was achieved through comparison of the chromatographic data (retention times and spectra) with commercial standards, using the internal standard method. Tocopherols standards (α-, β-, γ-, and δ-isophorms,) were used for compounds identification and quantification by the internal standard method. Results were expressed in mg per 100 g of dry weight (dw) and were processed using the Clarity 2.4 software (DataApex, Prague, Czech Republic).

2.3.3. Organic Acids

Organic acids were determined following the procedure previously described [29]. The samples were analyzed by Ultrafast Liquid Chromatography (UPLC, Shimadzu 20A series, Kyoto, Japan) coupled to a Diode Array Detector (UFLC-PDA, Shimadzu Corporation, Kyoto, Japan). Data were analyzed using the LabSolutions Multi LC-PDA software (Shimadzu Corporation, Kyoto, Japan). The identification was accomplished through the comparison of the retention times and the spectra obtained to the commercial standards (oxalic, quinic, malic, ascorbic, citric and fumaric acids), and their respective calibration curves were used to determine the quantity based on the area of the peaks. Results were expressed in g per 100 g of dw.

2.3.4. Free Sugars

Free sugars content was analyzed by High Performance Liquid Chromatography (HPLC, Knauer, Smartline system 1000) coupled to a refractive index detector (RI detector, Knauer Smartilne 2300, Knauer, Berlin, Germany), according to the analytical conditions previously described [30]. Data was analyzed using the Clarity 2.4 software (DataApex) and the identification was performed through the comparison with standards (D (-)-fructose, D (+)-sucrose, D (+)-glucose, D (+)-trehalose and D (+)-raffinose pentahydrate (Sigma-Aldrich, St. Louis, MO, USA). The quantification was performed using melezitose (Matreya, PA, USA) as internal standard (IS). The results were processed through Clarity 2.4 software (DataApex, Prague, Czech Republic) through the comparison of retention times, UV-Vis and mass spectra of the sample compounds with those obtained from the available standards and the literature information available and presented in g per 100 g of dry weight (dw).

2.4. Statistical Analysis

All the performed experiments were executed in triplicate. Results were presented as mean value ± standard deviation. Means and standard deviations were calculated using Microsoft Excel. Differences among samples were analyzed using SPSS Statistics software (IBM SPSS Statistics for Mac OS, Version 26.0. Armonk, NY: IBM Corp.). The results were subject to an analysis of variance (ANOVA), while the Tukey's HSD test ($\alpha = 0.05$) was used to assess the significant differences between the samples. For the comparison between two samples, a two-tailed paired Student's t-test was applied to assess the statistical differences ($\alpha = 0.05$).

Moreover, a Principal Component Analysis (PCA) was performed in order to examine the contribution of each variable to the total diversity and classify the studied maturation stages according to their chemical composition and nutritional value by using the statistical software Statgraphics 5.1.plus (Statpoint Technologies, Inc., VA, USA).

3. Results

3.1. Lipid Fraction and Fatty Acids Composition

In Table 1 are presented the lipid and tocopherols content of cardoon heads collected at different maturation stages. The highest total lipidic fraction and a-tocopherol content was observed at early maturity stages (Car A and Car B for total lipidic fraction and α-tocopherol content, respectively) and immature heads (Car A) had 10.9 times higher lipidic content than the sample of late maturity (Car F). In contrast, the progress of maturation process resulted in a gradual decrease of total lipid fraction, whereas α-tocopherol content showed fluctuating trends with the lowest content being observed in Car C samples (mid-maturity stages). In particular, the highest amount was detected in the sample Car B (PSG 5/6) (619 µg/100 g dw), while sample Car C (PSG 6) presented the lowest abundance (25 µg/100 g dw).

Table 1. Lipidic fraction and tocopherols of *Cynara cardunculus* heads at various stages of maturity (mean ± SD; n = 3).

Sample	Total Lipidic Fraction (g/100 g dw)	α-Tocopherol (µg/100 g dw)
Car A	17.5 ± 0.1 [a]	264 ± 1 [b]
Car B	5.1 ± 0.1 [b]	619 ± 4 [a]
Car C	3.5 ± 0.2 [d]	25 ± 2 [f]
Car D	4.31 ± 0.04 [c]	107 ± 1 [e]
Car E	1.9 ± 0.2 [e]	162 ± 1 [c]
Car F	1.6 ± 0.1 [f]	117 ± 5 [d]

Results are presented as mean ± standard deviation. Different letters correspond to significant differences according to Tukey's HSD test ($p < 0.05$). dw–dry weight. Tocopherols calibration curves: α-tocopherol (y = 1.295 x; R^2 = 0.991; LOD = 18.06 ng/mL; LOQ = 60.20 ng/mL).

Regarding the fatty acid composition of the tested cardoon heads in relation to maturity stage, the results are presented in Table 2, as well as the total saturated fatty acids, (SFA), monounsaturated fatty acids (MUFA), and polyunsaturated fatty acids (PUFA) content and PUFA/SFA and n-6/n-3 ratios. The values of the concentrations of all the referred parameters are provided as Supplementary Materials (Table S1). Thirty individual fatty acids were identified, with the palmitic (C16:0; 14.62–43.8%), oleic (C18:1n9c; 4.48–46.6%), and linoleic (C18:2n6c; 0.748–30.6%) acids being detected in higher abundance. A typical chromatogram of fatty acids profile is presented in Figure 3. The highest content of palmitic acid was observed at early maturity stages (sample Car A), while a gradual decrease was observed as a maturation progress evolved. In contrast, oleic acid content showed increasing trends until mid-maturity (sample Car C), followed by a slight decrease at the following maturity stages (samples Car D–F). Finally, linoleic acids exhibited fluctuating trends with the highest and lowest content being observed in samples B and E, respectively. SFAs and MUFAs were the most abundant class of fatty acids due to the high content of palmitic and linoleic acids, respectively. Moreover, the recorded values of PUFA/SFA ratio were higher than 0.45 in samples B, D and F, whereas the values of n6/n3 ratio were below 4.0 in samples A, B and E.

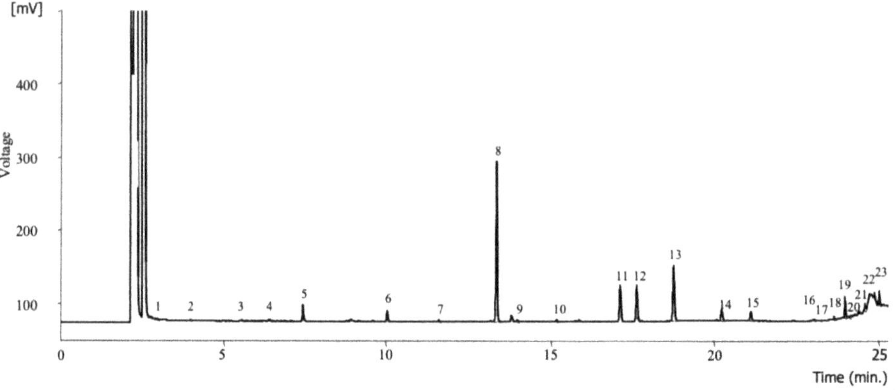

Figure 3. Chromatogram of fatty acids profile of *Cynara cardunculus* heads (sample Car A). 1. C6:0–caproic acid; 2. C8:0–caprylic acid; 3. C10:0–capric acid; 4. C11:0-undecanoic acid; 5. C12:0-lauric acid; 6. C14:0–myristic acid; 7. C15:0–pentadecanoic acid; 8. C16:0–palmitic acid; 9. C16:1–palmitoleic acid; 10. C17:0–heptadecanoic acid; 11. C18:0–stearic acid; 12. C18:1n9–oleic acid; 13. C18:2n6c–linoleic acid; 14. C18:3n3–alpha-linolenic acid; 15. C20:0–arachidic acid; 16. C21:0–heneicosanoic acid; 17. C20:3n6–eicosatrienoic acid; 18. C20:3n3–11,14,17-eicosatrienoic acid; 19. C22:1–erucic acid; 20. C20:5n3–eicosapentaenoic acid; 21. C22:2–docosadienoic acid; 22. C23:0–tricosanoic acid; 23. C24:0–lignoceric acid.

Table 2. Fatty acids composition (relative %) of *Cynara cardunculus* heads in relation to maturity stage (mean ± SD; n = 3).

	Car A	Car B	Car C	Car D	Car E	Car F
Fatty Acids (Relative Percentage, %)						
C6:0	0.49 ± 0.01 [c]	0.082 ± 0.006 [f]	0.123 ± 0.001 [d]	0.094 ± 0.002 [e]	3.71 ± 0.01 [a]	0.84 ± 0.02 [b]
C8:0	0.250 ± 0.002 [c]	0.19 ± 0.01 [d]	0.057 ± 0.003 [e]	0.059 ± 0.006 [e]	1.314 ± 0.004 [a]	0.29 ± 0.01 [b]
C10:0	0.205 ± 0.008 [cd]	0.198 ± 0.001 [d]	0.254 ±0.006 [b]	0.21 ± 0.02 [c]	0.473 ± 0.006 [a]	0.186 ± 0.002 [e]
C11:0	0.72 ± 0.02 [a]	0.335 ± 0.001 [c]	0.237 ± 0.001 [d]	0.16 ± 0.02 [e]	0.579 ± 0.005 [b]	0.17 ± 0.01 [e]
C12:0	1.8 ± 0.1 [b]	2.57 ± 0.04 [a]	0.406 ± 0.002 [e]	0.82 ± 0.07 [d]	1.53 ± 0.07 [c]	0.326 ± 0.004 [f]
C13:0	n.d.	0.028 ± 0.001 [d]	0.0375 ± 0.0007 [b]	0.084 ± 0.004 [a]	0.030 ± 0.001 [c]	0.027 ± 0.003 [d]
C14:0	1.90 ± 0.02 [b]	0.58 ± 0.02 [f]	1.27 ± 0.02 [d]	1.450 ± 0.002 [c]	2.69 ± 0.01 [a]	1.0 ± 0.1 [e]
C14:1	n.d.	0.53 ± 0.01 [b]	n.d.	n.d.	0.54 ± 0.01 [a]	0.23 ± 0.01 [c]
C15:0	0.48 ± 0.01 [a]	n.d.	0.193 ± 0.001 [d]	0.176 ± 0.009 [e]	0.427 ± 0.006 [b]	0.28 ± 0.02 [c]
C15:1	n.d.	n.d.	n.d.	n.d.	1.36 ± 0.02*	0.122 ± 0.003 *
C16:0	43.8 ± 0.1 [a]	30.4 ± 0.8 [b]	22.577 ± 0.003 [d]	14.62 ± 0.03 [f]	25.60 ± 0.08 [c]	17.8 ± 0.2 [e]
C16:1	0.43 ± 0.01 [e]	0.317 ± 0.003 [f]	0.827 ± 0.002 [d]	12.62 ± 0.03 [b]	12.76 ± 0.03 [a]	6.69 ± 0.04 [c]
C17:0	0.779 ± 0.001 [a]	0.666 ± 0.001 [b]	0.313 ± 0.003 [e]	0.239 ± 0.001 [f]	0.462 ± 0.001 [d]	0.579 ± 0.004 [c]
C18:0	6.0 ± 0.1 [a]	2.96 ± 0.05 [e]	3.236 ± 0.008 [d]	2.687 ± 0.004 [f]	5.68 ± 0.01 [b]	4.599 ± 0.001 [c]
C18:1n9	7.7 ± 0.1 [e]	4.48 ± 0.04 [f]	46.6 ± 0.1 [a]	32.8 ± 0.1 [c]	32.47 ± 0.08 [d]	33.7 ± 0.8 [b]
C18:2n6c	20.1 ± 0.1 [d]	30.6 ± 0.4 [a]	6.23 ± 0.03 [e]	25.82 ± 0.08 [c]	0.748 ± 0.002 [f]	27.2 ± 0.4 [b]
C18:3n6	n.d.	0.176 ± 0.006 [a]	0.049 ± 0.004 [d]	n.d.	0.067 ± 0.001 [c]	0.145 ± 0.005 [b]
C18:3n3	5.55 ± 0.05 [b]	7.5 ± 0.1 [a]	1.02 ± 0.01 [e]	2.705 ± 0.008 [c]	0.3675 ± 0.0007 [f]	1.38 ± 0.06 [d]
C20:0	2.18 ± 0.02 [b]	3.225 ± 0.005 [a]	0.655 ± 0.007 [e]	0.377 ± 0.001 [f]	0.882 ± 0.009 [c]	0.672 ± 0.005 [d]
C20:1	n.d.	0.159 ± 0.002 [c]	0.196 ± 0.004 [b]	0.114 ± 0.006 [e]	4.52 ± 0.01 [a]	0.138 ± 0.001 [d]
C20:2	n.d.	0.223 ± 0.001 [b]	0.107 ± 0.001 [d]	0.182 ± 0.005 [c]	0.0845 ± 0.0007 [e]	0.31 ± 0.02 [a]
C21:0	0.276 ± 0.002 [b]	0.324 ± 0.004 [a]	0.092 ± 0.004 [e]	0.070 ± 0.001 [f]	0.2695 ± 0.0007 [c]	0.169 ± 0.005 [d]
C20:3n6	0.23 ± 0.02 [b]	8.9 ± 0.3 [a]	0.103 ± 0.009 [c]	0.101 ± 0.006 [c]	n.d.	n.d.
C20:3n3	1.14 ± 0.04 [b]	0.142 ± 0.001 [d]	0.12 ± 0.01 [d]	1.38 ± 0.08 [a]	n.d.	0.22 ± 0.02 [c]
C22:0	n.d.	0.81 ± 0.01 [d]	2.6365 ± 0.0007 [a]	1.56 ± 0.08 [c]	1.645 ± 0.004 [b]	n.d.
C22:1	2.249 ± 0.008 [b]	0.12 ± 0.01 [f]	4.9505 ± 0.0007 [a]	1.22 ± 0.02 [c]	0.44 ± 0.01 [e]	0.82 ± 0.06 [d]
C20:5n3	0.38 ± 0.04 [c]	n.d.	0.036 ± 0.001 [e]	0.32 ± 0.03 [d]	0.6285 ± 0.002 [a]	0.51 ± 0.01 [b]
C22:2	0.30 ± 0.01*	0.184 ± 0.001 *	n.d.	n.d.	n.d.	n.d.
C23:0	1.61 ± 0.09 [a]	1.47 ± 0.09 [b]	0.26 ± 0.02 [e]	0.2615 ± 0.0007 [e]	0.308 ± 0.001 [d]	0.52 ± 0.02 [c]
C24:0	1.4 ± 0.1 [c]	2.88 ± 0.05 [b]	7.411 ± 0.001 [a]	n.d.	0.413 ± 0.001 [e]	1.1 ± 0.1 [d]
SFA	61.9 ± 0.4 [a]	46.7 ± 0.7 [b]	39.75 ± 0.06 [d]	22.9 ± 0.2 [e]	46.0 ± 0.1 [c]	28.5 ± 0.5 [f]
MUFA	10.4 ± 0.1 [e]	5.61 ± 0.04 [f]	52.6 ± 0.1 [a]	46.73 ± 0.07 [c]	52.1 ± 0.1 [b]	41.7 ± 0.9 [d]
PUFA	27.7 ± 0.3 [d]	47.7 ± 0.8 [a]	7.66 ± 0.05 [e]	30.4 ± 0.1 [b]	1.895 ± 0.006 [f]	29.8 ± 0.4 [c]
PUFA/SFA	0.45 ± 0.01 [d]	1.02 ± 0.03 [c]	0.193 ± 0.001 [e]	1.33 ± 0.01 [a]	0.0412 ± 0.0002 [f]	1.043 ± 0.002 [b]
n-6/n-3	2.76 ± 0.04 [d]	1.88 ± 0.02 [e]	5.02 ± 0.04 [c]	5.8 ± 0.1 [b]	0.902 ± 0.002 [f]	13.1 ± 0.3 [a]

Results are presented as mean ± standard deviation. Concentration values are given as Supplementary Materials (Table S1). Different letters in the same line correspond to significant differences according to Tukey's honest significance test (HSD) test ($p < 0.05$). Fatty acids are expressed as relative percentage of each fatty acid. dw–dry weight; n.d.–not detected; C6:0–caproic acid; C8:0–caprylic acid; C10:0–capric acid; C11:0-undecanoic acid; C12:0–lauric acid; C13:0–tridecanoic acid; C14:0–myristic acid; C14:1–tetradecanoic acid; C15:0–pentadecanoic acid; C15:1–pentadecenoic acid; C16:0–palmitic acid; C16:1–palmitoleic acid; C17:0–heptadecanoic acid; C18:0–stearic acid; C18:1n9–oleic acid; C18:2n6c–linoleic acid; C18:3n6-gamma-linolenic acid; C18:3n3–alpha-linolenic acid; C20:0–arachidic acid; C20:1–gondoic acid; C20:2–eicosadienoic acid; C21:0–heneicosanoic acid; C20:3n6–eicosatrienoic acid; C20:3n3–11,14,17-eicosatrienoic acid; C22:0–behenic acid; C22:1–erucic acid; C20:5n3–eicosapentaenoic acid; C22:2–docosadienoic acid; C23:0–tricosanoic acid; C24:0–lignoceric acid; SFA–saturated fatty acids; MUFA–monounsaturated fatty acids; PUFA–polyunsaturated fatty acids; n-6/n-3: ratio of omega 6/omega 3 fatty acids. * Means statistical differences obtained by the t-student test, p value < 0.01.

3.2. Organic Acids and Free Sugars

In Table 3 are presented the results regarding the content in organic acids of *Cynara cardunculus* heads, in relation to maturity stage. The main detected organic acids were oxalic, quinic and malic acid, followed by citric and fumaric acids which were detected in lower amounts. Moreover, a great variation in individual organic acids content was observed at different maturity stages. In particular, at early stages (samples Car A and B) malic acid was the most abundant organic acid (1.45 and 2.31 g/100 g dw, respectively), while at late stages (sample F) oxalic acid was the richest organic acid (12.1 g/100 g dw) followed by quinic acid (3.3 g/100 g dw).

Table 3. Organic acids composition (g/100 g dw) of *Cynara cardunculus* heads in relation to maturity stage (mean ± SD; n = 3).

	Organic Acids (g/100 g dw)					
	Car A	Car B	Car C	Car D	Car E	Car F
Oxalic acid	0.324 ± 0.002 [f]	0.98 ± 0.01 [b]	0.3994 ± 0.0001 [e]	0.650 ± 0.001 [c]	12.1± 0.1 [a]	0.44 ± 0.01 [d]
Quinic acid	0.87 ± 0.04 [b]	0.46 ± 0.01 [c]	0.17 ± 0.01 [e]	0.017 ± 0.001 [f]	3.3 ± 0.1 [a]	0.45 ± 0.02 [d]
Malic acid	1.45 ± 0.06 [c]	2.31 ± 0.02 [a]	1.495 ± 0.001 [b]	0.0149 ± 0.0005 [e]	0.36 ± 0.02 [d]	n.d.
Citric acid	0.70 ± 0.04 [b]	0.86 ± 0.05 [a]	0.66 ± 0.02 [c]	0.84 ± 0.01 [a]	tr	n.d.
Fumaric acid	0.046 ± 0.001 [b]	0.0542 ± 0.0003 [a]	0.0110 ± 0.0003 [c]	tr	0.0045 ± 0.0001 [d]	n.d.
Total	3.39 ± 0.06 [c]	4.67 ± 0.06 [b]	2.74 ± 0.03 [d]	1.52 ± 0.01 [e]	15.7 ± 0.2 [a]	0.89 ± 0.01 [f]

Results are presented as mean ± standard deviation. Different letters in the same line correspond to significant differences according to Tukey's HSD test ($p < 0.05$). dw–dry weight; tr–traces (below limit of quantification (LOQ) values); n.d.–not detected (below limit of detection (LOD) values). Calibration curves for organic acids: oxalic acid ($y = 1 \times 10^6 x + 231891$, $R^2 = 0.9999$; LOD = 12.55 µg/mL; LOQ = 41.82 µg/mL); quinic acid ($y = 671557x + 14583$, $R^2 = 0.9998$; LOD = 24.18 µg/mL; LOQ = 80.61 µg/mL); malic acid ($y = 950041x + 6255.6$, $R^2 = 0.9999$; LOD = 35.76 µg/mL; LOQ = 119.18 µg/mL); citric acid ($y = 1 \times 10^5 x + 10277$, $R^2 = 0.9997$; LOD = 10.47 µg/mL; LOQ = 34.91 µg/mL) and fumaric acid ($y = 1 \times 10^7 x + 614399$, $R^2 = 0.9986$; LOD = 0.08 µg/mL; LOQ = 0.26 µg/mL).

The free sugars composition of cardoon heads in relation to maturity stage is presented in Table 4. Sucrose, glucose and raffinose were the main detected sugars, followed by fructose and trehalose, while a great variation in sugar composition was observed in response to maturity stage. In particular, sample Car C had the highest content in total sugars (7.4 g/100 g dw), with raffinose and fructose (1.8 and 1.64 mg/100 g dw, respectively) being present in higher amounts; the same sample was also the only one containing all the detected sugars. Samples Car D and Car E revealed the lowest content of total sugars (1.03 mg/100 g dw). Moreover, in the most advanced maturation stages (samples Car D-F) trehalose was the free sugar being recorded in higher amounts (0.34–0.96 mg/100 g dw), whereas raffinose was not detected. In the remaining samples car B and C, the presence of raffinose and sucrose stands out, except for sample Car B where sucrose was not detected.

Table 4. Free sugars composition (g/100 g dw) of *Cynara cardunculus* heads in relation to maturity stage (mean ± SD; n = 3).

	Free Sugars (g/100 g dw)					
	Car A	Car B	Car C	Car D	Car E	Car F
Fructose	0.13 ± 0.03 [d]	0.51 ± 0.04 [b]	1.64 ± 0.06 [a]	0.013 ± 0.004 [e]	0.14 ± 0.02 [d]	0.184 ± 0.001 [c]
Glucose	n.d.	2.02 ± 008 [a]	0.68 ± 0.03 [b]	0.06 ± 0.01 [e]	0.27 ± 0.01 [d]	0.26 ± 0.02 [c]
Sucrose	2.39 ± 0.06 [b]	n.d.	3.0 ± 0.1 [a]	n.d.	0.28 ± 0.02 [c]	0.11 ± 0.01 [d]
Trehalose	0.23 ± 0.04 [d]	0.98 ± 0.02 [a]	0.26 ± 0.02 [d]	0.96 ± 0.09 [a]	0.34 ± 0.02 [b]	0.797 ± 0.005 [c]
Raffinose	2.24 ± 0.07 [b]	2.62 ± 0.06 [a]	1.8 ± 0.1 [c]	n.d.	n.d.	n.d.
Total	5.0 ± 0.2 [c]	6.12 ± 0.08 [b]	7.4 ± 0.1 [a]	1.03 ± 0.09 [e]	1.03 ± 0.02 [d]	1.35 ± 0.03 [d]

Results are presented as mean ± standard deviation. Different letters correspond to significant differences according to Tukey's. HSD test ($p < 0.05$). dw–dry weight; n.d.–not detected (bellow limit of detection (LOD) values). Free sugars calibration curves: fructose ($y = 1.04\,x$, $R^2 = 0.999$; LOD = 0.05 mg/mL), glucose ($y = 0.935\,x$, $R^2 = 0.999$; LOD = 0.08 mg/mL; limit of quantification (LOQ) = 0.25 mg/mL) and trehalose ($y = 0.991\,x$, $R^2 = 0.999$; LOD = 0.07 mg/mL, LOQ = 0.24 mg/mL).

3.3. Principal Component Analysis (PCA)

Principal component analysis (PCA) is used to reduce multivariate data complexity as a method of identifying patterns and expressing data in ways that highlight similarities and differences, and further identify groups of samples according to their maturation stage [31,32]. The first four principal components (PCs) were associated with Eigen values higher than 1 and explained 97.2% of the cumulative variance, with PC1 accounting for 43.8%, PC2 for 22.9%, PC3 for 19.4% and finally PC4 for 11.1%. PC1 was positively correlated to lipidic fraction, raffinose and total sugars, malic, citric and fumaric acid, linoleic and α-linolenic acid and PUFA content, whereas it was negatively correlated to oxalic and quinic acid, total organic acids and MUFA content. PC2 was positively correlated to sucrose, quinic acid, palmitic and oleic acid and SFA content, whereas it was negatively correlated to trehalose, linoleic and α-linolenic acid and PUFA. Similarly, PC3 was positively correlated to α-tocopherol, tregalose and oxalic acid content, whereas it was negatively correlated to fructose, sucrose, raffinose, total sugars, malic and citric acid and oleic acid. Finally, PC4 was positively correlated to fructose, sucrose, total organic acids, oleic acid and MUFA content, whereas it was negatively correlated to palmitic and stearic acid and SFA content. These results indicating a correct application of the PCA allowing differentiation between the tested maturity stages, as shown in the corresponding scatterplot (Figure 4). Moreover, the plot suggests that the differences in the chemical composition of the tested samples are correlated with the maturation stage. The early (sampes Car A and B) and mid and late stages (samples Car D and F) are closely positioned, whereas samples Car C and Car E are clearly distinct due to the very low α-tocopherol and very high organic acids content, respectively, compared to the rest of the tested maturity stages. The loading plot (Figure 5) of the first two components revealed groups of positively correlated variables, namely the upper right quadrant comprising fructose, sucrose, raffinose, total sugars, lipidic fraction, SFA, palmitic acid, malic acid and fumaric acid; the lower right quadrant comprising tocopherols, glucose, citric acid, PUFA and and linoleic acid; the upper left quadrant comprising quinic acid, oxalic acid, total organic acids and oleic acid; the lower left quadant comprising trehalose and α-linolenic acid

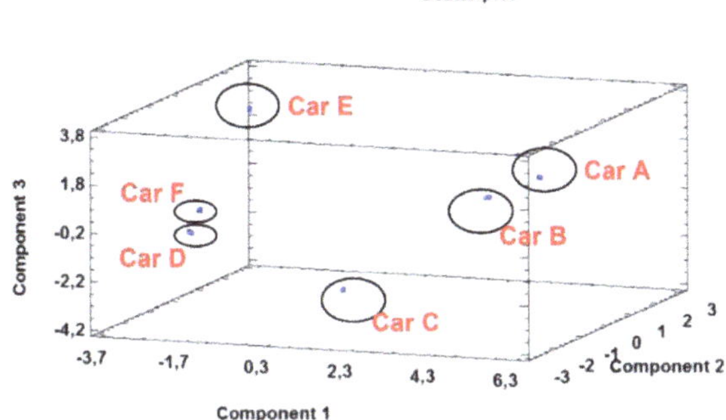

Figure 4. Three dimensional principal components scatterplot of the tested variables at different maturation stages of cardoon heads (samples Car A–F).

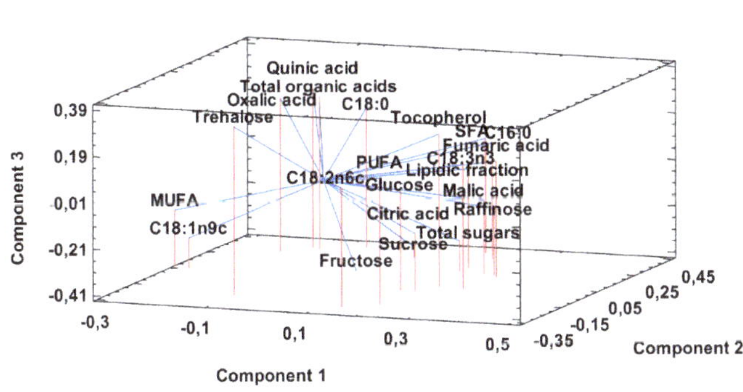

Figure 5. The principal components loading plot of the tested variables at different maturation stages of cardoon heads. SFA–saturated fatty acids, MUFA–Monounsaturated fatty acids, PUFA–polyunsaturated fatty acids.

4. Discussion

It could be suggested that the maturation stage of cardoon heads affects the fatty acids content which do not present a similar qualitative and quantitative profile over the maturation process. For example, pentadecanoic acid (C15:1) was detected only in senescent head samples (Car E and F), whereas docosadienoic acid (C22:2) was only detected in immature heads (samples Car A and B). The fatty acid profile of cardoon has already been studied by our team who studied cardoon seeds collected at full maturity and when heads were dry and senesced and the influence of harvesting time was suggested [33]. Moreover, harvesting time is also associated with variable climate conditions as shown in Figure 2, which according to the literature may also affect the compositional profile of globe artichoke [20]. Therefore, the dry conditions during the period of March-May in our study could be implicated in the observed differences among the tested maturation stages. Analogous fatty acids composition in cardoon heads has also been previously described by Petropoulos et al. [34], who studied different cardoon genotypes and verified that palmitic and linoleic acids were present in higher abundance in all the studied heads. Palmitic and linoleic acids were also suggested as the main fatty acid in globe artichoke heads by Dosi et al. [35] although they reported significantly higher amounts of linoleic than palmitic acid (55.20 and 34.80 mg/100 g fw, respectively) compared to our study. The studies of the lipidic fraction in cardoon tissues mostly refer to its seeds [36–38], as a result of the great interest for its industrial potential for the biodiesel production, although fatty acids composition of stalks, capitula and leaves has been also reported [39]. The high content in fatty acids present in cardoon heads, particularly the essential linoleic and oleic acids, is an added value to this multi-purpose crop that could be used for the production of these acids in industrial scale [40]. To the best of our knowledge, this is the first report that evaluated the influence of the physiological stages of cardoon heads collected during all the flowering stage.

Regarding fatty acids classification, our results showed that the saturated fatty acids (SFA) were the most abundant class of fatty acids in heads of early maturity (Car A–C), whereas monounsaturated fatty acids (MUFA) were the predominant ones in the remaining samples (samples D–F). Furthermore, the maturation stage of the samples reveals a strong influence both on the fatty acids profile, as well as on the proportion of PUFA/SFA and n-6/n-3, parameters which are associated with the nutritional value and the functional properties of food products [33,41]. Sample Car B was the only sample with a PUFA/SFA value higher than 0.45 and an n-6/n-3 ratio value lower than 4.0, both characteristics being associated with good nutritional properties. This evidence agrees with the popular medicine, since in the Mediterranean cousin cardoon heads are used when immature [19,42,43]. With the results obtained in this study, we corroborate the consumption of immature cardoon heads, preferably harvested in the beginning of May, based on the conditions of the growing location.

The tocopherols content detected in cardoon heads is presented in Table 1. The α-tocopherol was the only isoform identified and was detected in all the studied samples. Tocopherols are antioxidant compounds with high capacity to undergo oxidation reactions, therefore fluctuations of environmental conditions that may induce plant stress throughout the growth cycle of the heads could be a justification for the observed fluctuations in tocopherols content [23]. As shown in Figure 2, variable climatic conditions prevailed during and prior the harvesting period in our study, especially a dry period during March–May which could be associated with stressful conditions that resulted in an increase of tocopherols content. In a previous study, we studied the bioactive properties of cardoon heads also analyzed in the present work and despite the antioxidant capacity associated with tocopherols, the anticipation that the sample Car B with the highest tocopherols content would show the greater antioxidant activity was not verified. This fact suggests that other classes of compounds, such as phenolic compounds, could be also related to the demonstrated antioxidant potential [19], while Kukić et al. [44] suggested that β-sitosterol possessed a strong antioxidant capacity in extracts obtained from cardoon bracts. Despite that, the highest content of α-tocopherol in this sample could be associated with the highest content in PUFAs (see Table 2) highlighting the protective effects of tocopherols against lipid peroxidation [45,46]. Similarly to tocopherols composition of heads, other plant tissues of cardoon do not present a wide variety of tocopherols isoforms and the α-tocopherol was the only isoform detected in cardoon seeds [23,47], contrary to the cardoon seed oil which was rich in α- and δ-tocopherols [47–50] and leaf blades that contained all tocopherol isoforms [51].

The organic acids profile (Table 3) was also different between samples with a great variation during the maturation process of the detected organic acids: oxalic, quinic, malic, citric, and fumaric. The sample collected at the principal growth stage (PGS) 7 (Car E) revealed the highest content in organic acids (15.7 mg/100 g dw), and sample Car F (PGS 8) the lowest abundance (0.89 mg/100 g dw). Immature to mid maturity heads (samples Car A-C) had malic acid as the most abundant organic acid (1.45–2.31 mg/100 g dw). In sample D, the organic acid present in higher quantities was citric acid (0.84 g/100 g dw), while in advanced maturation stages, oxalic and quinic acids were the organic acids present in higher amounts. Moreover, considering the concomitant decrease in total sugars content at late maturity stages, the increase of organic acids especially in sample Car E could be explained by the increased requirements in osmolytes as a mechanism of cardoon plants to overcome stress conditions induced by the high mean and max temperatures and the concomitant low precipitation (see Figure 2) [52,53]. Compared to other reports, cardoon heads revealed greater variety and lower abundance of organic acids than seeds [23,47], while organic acids profile was similar to aerial parts [54]. Finally, the high quinic acid in sample Car E could be valorized for its confirmed high antioxidant potential [55,56].

The qualitative and quantitative information regarding the free sugars composition of cardoon heads are presented in Table 4. The samples analyzed revealed significant differences in free sugars content throughout the maturation process. Therefore, in early stages (sample car A) sucrose and raffinose were the main detected sugars, while sugar composition was altered in the following stage (sample car B) and raffinose and glucose were the most abundant sugars. Similarly, in sample C sucrose, fructose and raffinose were detected in the highest amounts, whereas in the following stages (samples D–F) total sugars content was significantly reduced consisting mainly of trehalose. Considering that the tested samples were collected under environmental conditions [19] could justify the observed differences in sugar composition between the various maturity stages. According to Petropoulos et al. [34] who studied the influence of the genotype on different parameters of cardoon heads, significant differences between the genotypes tested were also suggested. Previous studies identified sucrose [31,34,57] and glucose [58,59] as the most abundant sugars. These differences with the present work could be justified by the effect of factors such as the geographic location and growing conditions [58] or the stage of maturity of the analyzed samples [60], especially at early maturity stages where heads are edible and sweet in taste. In the same context and considering that artichoke is a rich source of inulin [61,62], free sugars content at late maturity stages (samples car D–E) exhibits a significant reduction probably due to the inulin formation and the accumulation of storage carbohydrates [63,64]. This argument could also explain the high content of trehalose at late maturity stages where the amounts of other free sugars is reduced, while the evolving environmental conditions (increasing temperatures, water shortage) could pose plants under stress and increase the needs of osmolytes for stress tolerance [52,65]. Finally, the decrease of total free sugars at late maturity stages should be linked with the lignification process and the hard texture of cardoon heads, since lignin biosynthesis involves the binding of non-structural carbohydrates for the formation of lignin carbohydrate complexes (LCC) [65].

5. Conclusions

Cardoon heads are widely consumed in several dishes of the Mediterranean countries, such as salad and soups, due to their richness in health-promoting compounds and to their well-recognized nutritional value. However, the stage when heads are edible is relatively short and is highly affected by environmental conditions during harvesting that may result in heads of hard texture due to high fibers content which make them inedible. Therefore, a considerable number of heads are considered as waste with potential suggested uses the energy and biomass production. With the present study we aimed to analyze cardoon heads chemical composition and evaluate the potential of alternative uses that will increase the added value of the crop. Our results allowed a complete characterization of the chemical composition of cardoon heads and the evaluation of the effect that the maturity stage has on the lipid, fatty acids, tocopherols, organic acids, and free sugars content. To the best of our knowledge, this is the first report that studied the influence of the maturity stage on the composition in lipidic content, fatty acids, tocopherols, organic acids, and free sugars content. In particular, immature heads (samples Car A and B) exhibited the highest content in lipidic fraction and α-tocopherol, respectively. Moreover, samples of mid- (sample Car C) and late maturity (sample Car E) had the highest total sugars and total organic acids content, respectively. With the present results we verified that the maturation stage of cardoon heads had a significant influence on the chemical composition. The obtained information should be implemented by different application areas, e.g., nutraceuticals and food supplements, in order to obtain the maximum potential of cardoon heads utilization, as well as to improve the production and commercialization techniques.

Supplementary Materials: The following are available online at http://www.mdpi.com/2073-4395/10/8/1088/s1, Table S1: Composition of fatty acids (mg/100 g dw) of *Cynara cardunculus* heads in relation to maturity 16 stage (mean ± SD; n = 3).

Author Contributions: Methodology, Investigation, Writing–Original Draft, F.M. and Â.F.; Methodology, Writing–Reviewing and Editing, S.A.P.; Conceptualization, Methodology, Writing-Reviewing and Editing, C.S.-B., I.C.F.R.F. and L.B. All authors have read and agreed to the published version of the manuscript.

Funding: This work was financially support by national funds FCT/MCTES to CIMO (UIDB/00690/2020). We also thank the FCT for the PhD grant (SFRH/BD/146614/2019) of F. Mandim. The authors are also grateful to the financial support though the TRANSCoLAB (0612_TRANS_CO_LAB_2_P).

Acknowledgments: To the Foundation for Science and Technology (FCT, Portugal) for the contracts of Â. Fernandes and L. Barros through the institutional scientific employment program.

Conflicts of Interest: The authors declare no conflict of interest.

References

1. Gominho, J.; Curt, M.D.; Lourenço, A.; Fernández, J.; Pereira, H. *Cynara cardunculus* L. as a biomass and multi-purpose crop: A review of 30 years of research. *Biomass Bioenergy* **2018**, *109*, 257–275. [CrossRef]
2. Zayed, A.; Serag, A.; Farag, M.A. *Cynara cardunculus* L.: Outgoing and potential trends of phytochemical, industrial, nutritive and medicinal merits. *J. Funct. Foods* **2020**, *69*, 103937. [CrossRef]
3. De Falco, B.; Incerti, G.; Amato, M.; Lanzotti, V. Artichoke: Botanical, agronomical, phytochemical, and pharmacological overview. *Phytochem. Rev.* **2015**, *14*, 993–1018. [CrossRef]
4. Gostin, A.I.; Waisundara, V.Y. Edible flowers as functional food: A review on artichoke (*Cynara cardunculus* L.). *Trends Food Sci. Technol.* **2019**, *86*, 381–391. [CrossRef]
5. Brás, T.; Paulino, A.F.C.; Neves, L.A.; Crespo, J.G.; Duarte, M.F. Ultrasound assisted extraction of cynaropicrin from *Cynara cardunculus* leaves: Optimization using the response surface methodology and the effect of pulse mode. *Ind. Crops Prod.* **2020**, *150*, 112395. [CrossRef]
6. Amira, A.B.; Besbes, S.; Attia, H.; Blecker, C. Milk-clotting properties of plant rennets and their enzymatic rheological, and sensory role in cheese making: A review. *Int. J. Food Prop.* **2017**, *28*, 576–593.
7. Gomes, S.; Belo, A.T.; Alvarenga, N.; Dias, J.; Lage, P.; Pinheiro, C.; Pinto-Cruz, C.; Brás, T.; Duarte, M.F.; Martins, A.P.L. Characterization of *Cynara cardunculus* L. flower from Alentejo as a coagulant agent for cheesemaking. *Int. Dairy J.* **2019**, *91*, 178–184. [CrossRef]
8. Jiménez-Moreno, N.; Cimminelli, M.J.; Volpe, F.; Ansó, R.; Esparza, I.; Mármol, I.; Rodríguez-Yoldi, M.J.; Ancín-Azpilicueta, C. Phenolic composition of Artichoke waste and its antioxidant capacity on differentiated Caco-2 cells. *Nutrients* **2019**, *11*, 1723. [CrossRef]
9. Pari, L.; Alfano, V.; Mattei, P.; Santangelo, E. Pappi of cardoon (*Cynara cardunculus* L.): The use of wetting during the harvesting aimed at recovering for the biorefinery. *Ind. Crops Prod.* **2017**, *108*, 722–728. [CrossRef]
10. Francaviglia, R.; Bruno, A.; Falcucci, M.; Farina, R.; Renzi, G.; Russo, D.E.; Sepe, L.; Neri, U. Yields and quality of *Cynara cardunculus* L. wild and cultivated cardoon genotypes. A case study from a marginal land in Central Italy. *Eur. J. Agron.* **2016**, *72*, 10–19. [CrossRef]

11. Lourenço, A.; Neiva, D.M.; Gominho, J.; Curt, M.D.; Fernández, J.; Marques, A.V.; Pereira, H. Biomass production of four *Cynara cardunculus* clones and lignin composition analysis. *Biomass Bioenergy* **2015**, *76*, 86–95. [CrossRef]

12. Conceição, C.; Martins, P.; Alvarenga, N.; Dias, J.; Lamy, E.; Garrido, L.; Gomes, S.; Freiras, S.; Belo, A.; Brás, T.; et al. Cynara cardunculus: Use in cheesemaking and pharmaceutical applications. In *Technological Approaches for Novel Applications in Dairy Processing*; Intech Open: London, UK, 2016; Volume 1, pp. 73–107.

13. D'Antuono, I.; Carola, A.; Sena, L.M.; Linsalata, V.; Cardinali, A.; Logrieco, A.F.; Colucci, M.G.; Apone, F. Artichoke polyphenols produce skin anti-age effects by improving endothelial cell integrity and functionality. *Molecules* **2018**, *23*, 2729. [CrossRef] [PubMed]

14. Salem, M.B.; Affes, H.; Athmouni, K.; Ksouda, K.; Dhouibi, R.; Sahnoun, Z.; Hammami, S.; Zeghal, K.M. Chemicals compositions, antioxidant and anti-inflammatory activity of *Cynara scolymus* leaves extracts, and analysis of major bioactive polyphenols by HPLC. *Evid. Based Complement. Altern. Med.* **2017**, *2017*, 4951937.

15. Marques, P.; Marto, J.; Gonçalves, L.M.; Pacheco, R.; Fitas, M.; Pinto, P.; Serralheiro, M.L.M.; Ribeiro, H. *Cynara scolymus* L.: A promising Mediterranean extract for topical anti-aging prevention. *Ind. Crops Prod.* **2017**, *109*, 699–706. [CrossRef]

16. Almeida, C.M.; Simões, I. Cardoon-based rennets for cheese production. *Appl. Microbiol. Biotechnol.* **2018**, *102*, 4675–4686. [CrossRef] [PubMed]

17. Faria-Silva, C.; Ascenso, A.; Costa, A.M.; Marto, J.; Carvalheiro, M.; Ribeiro, H.M.; Simões, S. Feeding the skin: A new trend in food and cosmetics convergence. *Trends Food Sci. Technol.* **2020**, *95*, 21–32. [CrossRef]

18. Dias, M.I.; Barros, L.; Barreira, J.C.M.; Alves, M.J.; Barracosa, P.; Ferreira, I.C.F.R. Phenolic profile and bioactivity of cardoon (*Cynara cardunculus* L.) inflorescence parts: Selecting the best genotype for food applications. *Food Chem.* **2018**, *268*, 196–202. [CrossRef]

19. Mandim, F.; Petropoulos, S.A.; Giannoulis, K.D.; Dias, M.I.; Fernandes, Â.; Pinela, J.; Kostic, M.; Soković, M.; Barros, L.; Santos-Buelga, C.; et al. Seasonal variation of bioactive properties and phenolic composition of *Cynara cardunculus* var. *altilis*. *Food Res. Int.* **2020**, *134*, 109281. [CrossRef]

20. Pandino, G.; Lombardo, S.; Lo Monaco, A.; Mauromicale, G. Choice of time of harvest influences the polyphenol profile of globe artichoke. *J. Funct. Foods* **2013**, *5*, 1822–1828. [CrossRef]

21. Scavo, A.; Pandino, G.; Restuccia, C.; Parafati, L.; Cirvilleri, G.; Mauromicale, G. Antimicrobial activity of cultivated cardoon (*Cynara cardunculus* L. var. *altilis* DC.) leaf extracts against bacterial species of agricultural and food interest. *Ind. Crops Prod.* **2019**, *129*, 206–211. [CrossRef]

22. Li, Y.; Kong, D.; Fu, Y.; Sussman, M.R.; Wu, H. The effect of developmental and environmental factors on secondary metabolites in medicinal plants. *Plant. Physiol. Biochem.* **2020**, *148*, 80–89. [CrossRef]

23. Mandim, F.; Dias, M.I.; Pinela, J.; Barracosa, P.; Ivanov, M.; Ferreira, I.C.F.R. Chemical composition and in vitro biological activities of cardoon (*Cynara cardunculus* L. var. *altilis* DC.) seeds as influenced by viability. *Food Chem.* **2020**, *323*, 126838. [CrossRef] [PubMed]

24. Ben Amira, A.; Blecker, C.; Richel, A.; Arias, A.A.; Fickers, P.; Francis, F.; Besbes, S.; Attia, H. Influence of the ripening stage and the lyophilization of wild cardoon flowers on their chemical composition, enzymatic activities of extracts and technological properties of cheese curds. *Food Chem.* **2018**, *245*, 919–925. [CrossRef] [PubMed]

25. Claus, T.; Maruyama, S.A.; Palombini, S.V.; Montanher, P.F.; Bonafé, E.G.; de Oliveira Santos Junior, O.; Matsushita, M.; Visentainer, J.V. Chemical characterization and use of artichoke parts for protection from oxidative stress in canola oil. *LWT Food Sci. Technol.* **2015**, *61*, 346–351. [CrossRef]

26. Archontoulis, S.V.; Struik, P.C.; Vos, J.; Danalatos, N.G. Phenological growth stages of *Cynara cardunculus*: Codification and description according to the BBCH scale. *Ann. Appl. B* **2013**, *156*, 253–270. [CrossRef]

27. Petropoulos, S.A.; Pereira, C.; Tzortzakis, N.; Barros, L.; Ferreira, I.C.F.R. Nutritional value and bioactive compounds characterization of plant parts from *Cynara cardunculus* L. (Asteraceae) cultivated in central Greece. *Front. Plant. Sci.* **2018**, *9*, 1–12. [CrossRef]

28. Barros, L.; Pereira, E.; Calhelha, R.C.; Dueñas, M.; Carvalho, A.M.; Santos-Buelga, C.; Ferreira, I.C.F.R. Bioactivity and chemical characterization in hydrophilic and lipophilic compounds of *Chenopodium ambrosioides* L. *J. Funct. Foods* **2013**, *5*, 1732–1740. [CrossRef]

29. Mandim, F.; Barros, L.; Calhelha, R.C.; Abreu, R.M.V.; Pinela, J.; Alves, M.J.; Heleno, S.; Santos, P.F.; Ferreira, I.C.F.R. *Calluna vulgaris* (L.) Hull: Chemical characterization, evaluation of its bioactive properties and effect on the vaginal microbiota. *Food Funct.* **2019**, *10*, 78–89. [CrossRef]

30. Dias, M.I.; Barros, L.; Morales, P.; Sánchez-Mata, M.C.; Oliveira, M.B.P.P.; Ferreira, I.C.F.R. Nutritional parameters of infusions and decoctions obtained from *Fragaria vesca* L. roots and vegetative parts. *LWT Food Sci. Technol.* **2015**, *62*, 32–38. [CrossRef]

31. Pereira, E.; Barros, L.; Calhelha, R.C.; Dueñas, M.; Carvalho, A.M.; Santos-Buelga, C.; Ferreira, I.C.F.R. Bioactivity and phytochemical characterization of *Arenaria Montana* L. *Food Funct.* **2014**, *5*, 1848–1855. [CrossRef]

32. El-Nakhel, C.; Petropoulos, S.A.; Pannico, A.; Kyriacou, M.C.; Giordano, M.; Colla, G.; Troise, A.D.; Vitaglione, P.; De Pascale, S.; Rouphael, Y. The bioactive profile of lettuce produced in a closed soilless system as configured by combinatorial effects of genotype and macrocation supply composition. *Food Chem.* **2020**, *309*, 125713. [CrossRef] [PubMed]

33. Petropoulos, S.A.; Fernandes, Â.; Calhelha, R.C.; Danalatos, N.; Barros, L.; Ferreira, I.C.F.R. How extraction method affects yield, fatty acids composition and bioactive properties of cardoon seed oil? *Ind. Crops Prod.* **2018**, *124*, 459–465. [CrossRef]

34. Petropoulos, S.A.; Pereira, C.; Ntatsi, G.; Danalatos, N.; Barros, L.; Ferreira, I.C.F.R. Nutritional value and chemical composition of Greek artichoke genotypes. *Food Chem.* **2018**, *267*, 296–302. [CrossRef] [PubMed]

35. Dosi, R.; Daniele, A.; Guida, V.; Ferrara, L.; Severino, V.; Di Maro, A. Nutritional and metabolic profiling of the globe artichoke (*Cynara scolymus* L. *"Capuanella"* heads) in province of caserta, Italy. *Aust. J. Crop. Sci.* **2013**, *7*, 1927–1934.

36. Petropoulos, S.; Fernandes, Â.; Pereira, C.; Tzortzakis, N.; Vaz, J.; Soković, M.; Barros, L.; Ferreira, I.C.F.R. Bioactivities, chemical composition and nutritional value of *Cynara cardunculus* L. seeds. *Food Chem.* **2019**, *289*, 404–412. [CrossRef] [PubMed]

37. Alexandre, A.M.R.C.; Dias, A.M.A.; Seabra, I.J.; Portugal, A.A.T.G.; De Sousa, H.C.; Braga, M.E.M. Biodiesel obtained from supercritical carbon dioxide oil of *Cynara cardunculus* L. *J. Supercrit. Fluids* **2012**, *68*, 52–63. [CrossRef]

38. Raccuia, S.A.; Piscioneri, I.; Sharma, N.; Melilli, M.G. Genetic variability in *Cynara cardunculus* L. domestic and wild types for grain oil production and fatty acids composition. *Biomass Bioenergy* **2011**, *35*, 3167–3173. [CrossRef]

39. Ramos, P.A.B.; Guerra, A.R.; Guerreiro, O.; Freire, C.S.R.; Silva, A.M.S.; Duarte, M.F.; Silvestre, A.J.D. Lipophilic extracts of *Cynara cardunculus* L. var. *altilis* (DC): A source of valuable bioactive terpenic compounds. *J. Agric. Food Chem.* **2013**, *61*, 8420–8429. [CrossRef]

40. Petropoulos, S.A.; Perreira, C.; Barros, L.; Ferreira, I.C.F.R. Leaf parts from Greek artichoke genotypes as a good source of bioactive compounds and antioxidants. *Food Funct.* **2017**, *8*, 2022–2029. [CrossRef]

41. Liu, L.; Hu, Q.; Wu, H.; Xue, Y.; Cai, L.; Fang, M.; Liu, Z.; Yao, P.; Wu, Y.; Gong, Z. Protective role of n6/n3 PUFA supplementation with varying DHA/EPA ratios against atherosclerosis in mice. *J. Nutr. Biochem.* **2016**, *32*, 171–180. [CrossRef]

42. Ierna, A.; Mauromicale, G. *Cynara cardunculus* L. genotypes as a crop for energy purposes in a Mediterranean environment. *Biomass Bioenergy* **2010**, *34*, 754–760. [CrossRef]

43. Pandino, G.; Lombardo, S.; Mauromicale, G.; Williamson, G. Profile of polyphenols and phenolic acids in bracts and receptacles of globe artichoke (*Cynara cardunculus* var. *scolymus*) germplasm. *J. Food Compos. Anal.* **2011**, *24*, 148–153. [CrossRef]

44. Kukić, J.; Popovic, V.; Petrović, S.; Mucaji, P.; Ćirić, A.; Stojković, D.; Soković, M. Antioxidant and antimicrobial activity of *Cynara cardunculus* extracts. *Food Chem.* **2008**, *107*, 861–868. [CrossRef]

45. Petropoulos, S.A.; Fernandes, Â.; Dias, M.I.; Vasilakoglou, I.B.; Petrotos, K.; Barros, L.; Ferreira, I.C.F.R. Nutritional value, chemical composition and cytotoxic properties of common purslane (*Portulaca oleracea* L.) in relation to harvesting stage and plant part. *Antioxidants* **2019**, *8*, 293. [CrossRef] [PubMed]

46. Da Silva, L.P.; Pereira, E.; Pires, T.C.S.P.; Alves, M.J.; Pereira, O.R.; Barros, L.; Ferreira, I.C.F.R. *Rubus ulmifolius* Schott fruits: A detailed study of its nutritional, chemical and bioactive properties. *Food Res. Int.* **2019**, *119*, 34–43. [CrossRef] [PubMed]

47. Ferreira-Dias, S.; Gominho, J.; Baptista, I.; Pereira, H. Pattern recognition of cardoon oil from different large-scale field trials. *Ind. Crops Prod.* **2018**, *118*, 236–245. [CrossRef]

48. Raccuia, S.A.; Melilli, M.G. Biomass and grain oil yields in *Cynara cardunculus* L. genotypes grown in a Mediterranean environment. *Field Crops Res.* **2007**, *101*, 187–197. [CrossRef]

49. Maccarone, E.; Fallico, B.; Fanella, F.; Mauromicale, G.; Raccuia, S.A. Possible alternative utilization of *Cynara* spp. II. Chemical characterization of their grain oil. *Ind. Crops Prod.* **1999**, *10*, 229–237. [CrossRef]

50. Chihoub, W.; Dias, M.I.; Barros, L.; Calhelha, R.C.; Alves, M.J.; Harzallah-Skhiri, F.; Ferreira, I.C.F.R. Valorisation of the green waste parts from turnip, radish and wild cardoon: Nutritional value, phenolic profile and bioactivity evaluation. *Food Res. Int.* **2019**, *126*, 108651. [CrossRef]

51. Petropoulos, S.A.A.; Karkanis, A.; Martins, N.; Ferreira, I.C.F.R. Edible halophytes of the Mediterranean basin: Potential candidates for novel food products. *Trends Food Sci. Technol.* **2018**, *74*, 69–84. [CrossRef]

52. Petropoulos, S.A.; Fernandes, Â.; Dias, M.I.; Pereira, C.; Calhelha, R.C.; Chrysargyris, A.; Tzortzakis, N.; Ivanov, M.; Sokovic, M.D.; Barros, L.; et al. Chemical composition and plant growth of *Centaurea raphanina* subsp. *mixta* plants cultivated under saline conditions. *Molecules* **2020**, *25*, 2204. [CrossRef] [PubMed]

53. Pereira, C.; Barros, L.; Ferreira, I.C.F.R. Analytical tools used to distinguish chemical profiles of plants widely consumed as infusions and dietary supplements: Artichoke, milk thistle, and borututu. *Food Anal. Methods* **2014**, *7*, 1604–1611. [CrossRef]

54. Šušaníková, I.; Balleková, J.; Štefek, M.; Hošek, J.; Mučaji, P. Artichoke leaf extract, as AKR1B1 inhibitor, decreases sorbitol level in the rat eye lenses under high glucose conditions ex vivo. *Phyther. Res.* **2018**, *32*, 2389–2395. [CrossRef]

55. Miláčková, I.; Kapustová, K.; Mučaji, P.; Hošek, J. Artichoke Leaf Extract Inhibits AKR1B1 and Reduces NF-κB Activity in Human Leukemic Cells. *Phyther. Res.* **2017**, *31*, 488–496. [CrossRef] [PubMed]

56. Hernández-Hernández, O.; Ruiz-Aceituno, L.; Sanz, M.L.; Martínez-Castro, I. Determination of free inositols and other low molecular weight carbohydrates in vegetables. *J. Agric. Food Chem.* **2011**, *59*, 2451–2455. [CrossRef] [PubMed]

57. Nicoletto, C.; Santagata, S.; Tosini, F.; Sambo, P. Qualitative and healthy traits of different Italian typical artichoke genotypes. *CyTA J. Food* **2013**, *11*, 108–113. [CrossRef]

58. Lombardo, S.; Restuccia, C.; Muratore, G.; Barbagallo, R.N.; Licciardello, F.; Pandino, G.; Scifò, O.; Mazzaglia, A.; Ragonese, F.; Mauromicale, G. Effect of nitrogen fertilization on the overall quality of minimally processed globe artichoke heads. *J. Sci. Food Agric.* **2017**, *97*, 650–658. [CrossRef]

59. Di Salvo, R.; Fadda, C.; Sanguinetti, A.M.; Naes, T.; Del Caro, A. Effect of harvest time and geographical area on sensory and instrumental texture profile of a PDO artichoke. *Int. J. Food Sci. Technol.* **2014**, *49*, 1231–1237. [CrossRef]

60. Leroy, G.; Grongnet, J.F.; Mabeau, S.; le Corre, D.; Baty-Julien, C. Changes in inulin and soluble sugar concentration in artichokes (*Cynara scolymus* L.) during storage. *J. Sci. Food Agric.* **2010**, *90*, 1203–1209. [CrossRef]

61. Barracosa, P.; Barracosa, M.; Pires, E. Cardoon as a Sustainable Crop for Biomass and Bioactive Compounds Production. *Chem. Biodivers.* **2019**, *16*, e1900498. [CrossRef]

62. Vergauwen, R.; Van Laere, A.; Van Den Ende, W. Properties of fructan:fructan 1-fructosyltransferases from chicory and globe thistle, two asteracean plants storing greatly different types of inulin. *Plant. Physiol.* **2003**, *133*, 391–401. [CrossRef]

63. Roberfroid, M.B. Introducing inulin-type fructans. *Br. J. Nutr.* **2005**, *93*, S13–S25. [CrossRef] [PubMed]

64. Petropoulos, S.A.; Fernandes, Â.; Dias, M.I.; Pereira, C.; Calhelha, R.; Gioia, F.D.; Tzortzakis, N.; Ivanov, M.; Sokovic, M.; Barros, L.; et al. Wild and cultivated *Centaurea raphanina* subsp. *mixta*: A valuable source of bioactive compounds. *Antioxidants* **2020**, *9*, 314. [CrossRef] [PubMed]

65. Tarasov, D.; Leitch, M.; Fatehi, P. Lignin-carbohydrate complexes: Properties, applications, analyses, and methods of extraction: A review. *Biotechnol. Biofuels* **2018**, *11*, 1–28.

agronomy

Article

Agronomic and Economic Evaluation of Autumn Planted Sugarcane under Different Planting Patterns with Lentil Intercropping

Mubashar Nadeem [1,*], Asif Tanveer [1], Hardev Sandhu [2,*], Saba Javed [3], Muhammad Ehsan Safdar [4], Muhammad Ibrahim [5], Muhammad Atif Shabir [1], Muhammad Sarwar [1] and Usman Arshad [6]

[1] Department of Agronomy, University of Agriculture, Faisalabad 38040, Pakistan;
     drasiftanveeruaf@hotmail.com (A.T.); aatif.shabir@yahoo.com (M.A.S.); sarwar1406@gmail.com (M.S.)
[2] Everglades Research and Education Center, Institute of Food and Agricultural Sciences,
     University of Florida, 3200 East Palm Beach Road, Belle Glade, FL 32611, USA
[3] Institute of Agriculture and Resource Economics, University of Agriculture, Faisalabad 38040, Pakistan;
     sabasipra13@gmail.com
[4] Department of Agronomy, College of Agriculture, University of Sargodha, Sargodha 40100, Pakistan;
     ehsan_safdar2002@yahoo.com
[5] Department of Agronomy, Faculty of Agricultural Sciences, Ghazi University, Dera Ghazi Khan 32200,
     Pakistan; chibrahim@gudgk.edu.pk
[6] Department of Plant Pathology, University of Agriculture, Faisalabad 38040, Pakistan;
     usman.jajja909@gmail.com
*   Correspondence: mubasharuaf@hotmail.com (M.N.); hsandhu@ufl.edu (H.S.)

Received: 1 March 2020; Accepted: 21 April 2020; Published: 1 May 2020

Abstract: Proper sowing orientation and spacing are important factors for best crop growth. A field experiment was conducted to study the effect of different planting patterns with and without lentil intercropping on sugarcane growth and yield and farm economics. Each of these treatments were planted as sole crop and intercropped with lentil. Data were collected on plant cane and first ratoon crop. The maximum stripped cane yields (154.36 t/ha and 130.28 t/ha in plant and ratoon crop, respectively) were obtained from sugarcane planted at 120 cm trench planting both as sole as well as lentil intercropped. This treatment also attained 61% and 43% higher total sugar yields compared to traditional 60 cm single rows planting in plant and ratoon crops, respectively. Lentil intercropping did not have any significant effect on sugarcane yield, but trench planting at 120 cm with lentil intercropping had the highest lentil seed yield (598.0 in 2013–2014 and 629.8 kg ha^{-1} in 2014–2015) along with maximum land equivalent ratio (1.40 and 1.37), net return (Rs.321254/ha), net field benefit (Rs.491703/ha) and benefit cost ratio (2.01). Sugarcane at 120 cm trench planting with lentil intercropping also outperformed other planting patterns in improving economic returns.

Keywords: pit plantation; planting patterns; ratoon crop; sowing techniques; sugarcane yield; quality

1. Introduction

Sugarcane plays a significant role in Pakistan's agriculture by producing sugar and other byproducts such as biofuel, fiber and press mud. It has a pivotal role in elevating the economic power of the farming community, as it contributes 3.6% value addition in agriculture and 0.7% in gross domestic products of Pakistan [1]. The harvestable yield potential of the current gene pool of sugarcane cultivars is more than 150 t ha^{-1}; however, the national average yield is 60 t ha^{-1}, which is below the genetic potential as well as the national average (80–100 t ha^{-1}) of numerous advanced countries [2]. This yield gap must be bridged to match the sugar demands of an ever-increasing population. To increase national sugar production, sugarcane yield needs to be increased by making use of the latest production technology [3]. Low profitability resulting from high production cost, low gross income, and delayed payment to growers is also responsible for low sugarcane production in the country as farmers' interest in attaining higher yields is lost [4]. Therefore, production technology should be of such type that it may be able to increase the net income of sugarcane besides getting higher yield.

Planting geometry refers to the spatial distribution of plants over a specified field area. Appropriate planting geometry ensures the efficient and judicious use of the irrigation water. Desired level of seed germination, vigorous and healthy root and shoot growth and equal distribution in time and space are some of the key benefits realized by ensuring appropriate planting geometry. Altering the planting geometries not only influences the morphological characteristics; rather, all the physiological and quantitative parameters are influenced significantly as well [5]. Different planting geometries are being practiced by sugarcane growers of Pakistan, which attains same plant population [6]. Conventionally, sugarcane planting is planted at 60 to 90 cm row spacing, which increases initial plant population per unit area but it increases plant competition for sunlight, nutrients and water. Narrow spacing also obstructs several management operations necessary for good crop production, which resulted in crop lodging, and hence, the yield is considerably reduced [7]. On the other hand, wider spacing in the pit and trench planting method proved more suitable and effective compared to a conventional planting system because it maintains a high plant population throughout the growth, saves water (up to 20%), prevents sugarcane lodging, eases earthing-up and inter-culture practices, and produces more net returns [7,8]. Observed higher stripped-cane yield (120.5 t/ha) in 120 cm row spacing than 60 cm row spacing (68.42 t ha^{-1}). Similarly, an increased cane length (3.0 m) was observed in wider row spacing than narrow row spacing [9]. Trench planting is likely to provide enough space for post-planting management operations and reduce plant damage while maintaining optimum plant population.

The main aim of imposing such types of planting patterns is to overcome the limitations of older ones such as facilitation of inter-tillage practices, avoidance from lodging, maintaining optimum plant population and ease of irrigation, fertilizer, and plant protection [10]. Pit planting technique of sugarcane is a method with one of the highest potentials among methods used with space limitations. It was introduced in Pakistan as an efficient sowing method which promotes better germination, attains desirable plant population, and enhances sugar recovery [11]. It was pointed out that significantly higher cane yield of sugarcane could be achieved by planting sugarcane in 100 cm × 100 cm pits that were 50 cm apart than the conventional method. The enhanced cane yield in pit planting is attributed to increased germination, leaf area index, and crop growth rate [8,12,13].

The ratooning potential is the ability of sugarcane to re-sprout from left-over plant parts in the field after harvesting. From the farmer's point of view, it is considered the most desirable character of sugarcane cultivar [14]. This is due to fact that in case of ratoon crop, the cost of production is reduced up to 25–30% compared to fresh planted crop. In addition, there is considerable reduction in costs of land preparation, labor, irrigation quantity, and seed [15]. Ratoon crop also ensures an early supply of the sugarcane to the market; thus, benefitting the farming community economically as compared with the plant crop [16]. The area of Pakistan under ratoon crop of sugarcane is nearly 50% of its total cropped area, while about 25–30% of total sugarcane production is derived from ratoon crop [17]. Although ratoon crop attains 10–30% less cane yield compared to that achieved by the freshly planted crop, a yield gap of more than 35% still exists between its potential and realized yield [18]. The major factors responsible for the low yield of the ratoon crop are its inappropriate planting technique and poor management.

There are many ways to assure near future food safety and to boost the per unit crop yield in different farming systems; inter-cropping can be a viable option. It is a more efficient and eco-friendly method that results in enhanced production. Intercropping ensures efficient utilization of natural assets and harmonizes the effect of two or more crops grown simultaneously on same unit of land;

thus, it is a very good option in the development of sustainable food production systems [19–21]. For developing countries such as Pakistan, intercropping can be introduced in the existing system of monocropping because it results in an increased farm income and better utilization of resources. The farming community can get great inspiration and higher net returns from this system [22].

Sugarcane is a relatively long-duration crop sown on wider spaced rows with slow initial growth. After emergence (5–6 weeks), it remains dormant for a period of 3–4 months due to low temperature. In order to drive benefits from its slow growth and make better use of resources, intercropping of some short duration crop (leguminous crop) can be explored. Lentil is one potentially viable option because it is a short-duration crop (3–4 months), fixes atmospheric nitrogen via symbiotic rhizobia in root nodules, and consequently, has, in rotation, the potential for maintaining soil fertility, and helps in controlling weeds.

More cane yield of the autumn planted sugarcane with lentil intercrop than sugarcane alone was reported by [23]. Other authors [24] observed a higher cane equivalent yield and the heaviest cane growth in a sugarcane + lentil system compared to sugarcane alone.

A comprehensive study was planned and executed for getting higher sugarcane yield. The specific objective of the study was to compare the different planting patterns, namely, single row, double and trench planting, pit plantation techniques, and ratoon crop with and without lentil intercropping aimed at getting higher cane yield, quality, and net economic returns under agro-ecological conditions of Punjab, Pakistan. We hypothesized trench planting of sugarcane and its intercropping with lentil will improve the growth and yield of both the crops.

2. Materials and Methods

2.1. Site and Soil

The field studies were carried out consecutively for two years 2013–2015 and 2015–2016 at Agronomy farm, University of Agriculture, Faisalabad, Pakistan. The soil is alluvial in nature and the area is canal irrigated. The geographic location of Faisalabad is 31.5° N latitude and 73° E longitude, with 184.4 m altitude above sea level. The weather is considered semi-arid with very hot and humid summers and cool dry winters. The summer season starts from the month of April, which lasts up to October, whereas the hottest months of the years are May to July. December to February are known as the coldest months in Pakistan.

Prior to the start of each experiment, a composite soil sample to 0–30 cm depth from the experimental site was analyzed for various physicochemical characteristics of the soil. For analysis of nitrogen (N), phosphorous (P) and potash (K) soil sampling was done after the harvest of the crop. The analysis was carried out in the Institute of Soil and Environmental Sciences (ISES) and results are given in Table 1. The soil analysis showed that the soil of the trial site was sandy loam, slightly alkaline, and highly deficient in nitrogen and phosphorus.

Table 1. Soil analysis of the experimental soil.

Soil Characteristics	2013–2015	2015–2016
A. Physical characteristics		
Sand %	63.12	61.85
Silt %	19.75	16.39
Clay %	19.29	20.95
B. Chemical analysis		
pH	7.80	7.85
EC_e (dSm^{-1})	1.21	1.19
Organic matter (%)	0.79	0.80
Available N (%)	0.041	0.043
Available Phosphorus P_2O_5 (ppm)	6.99	6.88
Available Potassium K_2O (ppm)	140	135

2.2. Meteorological Data

Agricultural Meteorology Cell, Agriculture University, provided the measured metrological data. Data are presented in Figure 1. Total rainfall during the growing period of the freshly-planted crop was 378 mm, while that of ratoon crop was 413 mm.

(a)

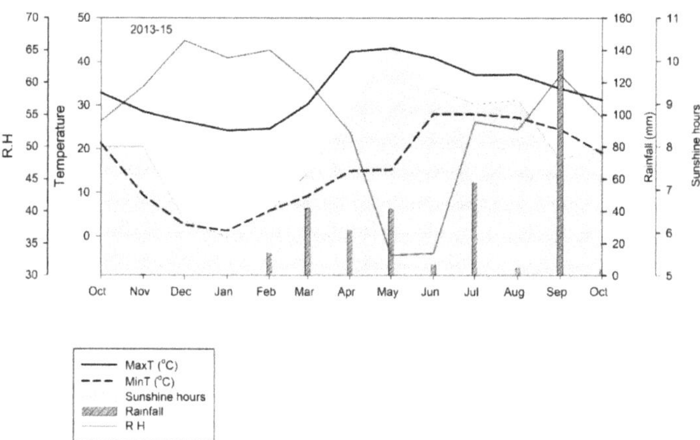

(b)

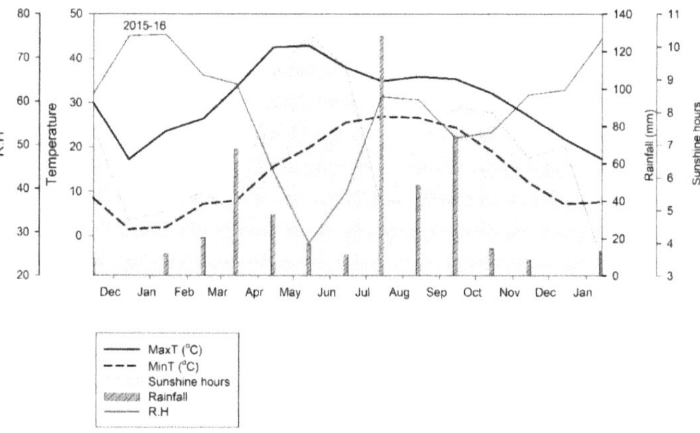

Figure 1. Meteorological data during. (**a**) Plant crop and (**b**) ratoon crop seasons of sugarcane.

2.3. Experimental Design and Treatments

The experiment was laid out in randomized complete block design with split plot arrangement having four replications. Treatments included seven planting patterns of sugarcane allotted to sub-plots in both plant cane and ratoon crops. In pit planting, the 90 cm diameter pits were established at 45, 60, 75, and 90 cm row-to-row as well as plant-to-plant spacing in treatments P1, P2, P3, and P4, respectively. In treatment P5, sugarcane was planted in single rows with 60 cm row-to-row spacing. In treatment P6, there were twin rows or double rows with 90 cm row-to-row spacing. Treatment P7 consisted of trench planting with 120 cm row to row distance with double rows of double budded setts placed at 20–25 cm depth. The plot size was 4.2 m × 9 m in P1, 4.5 m × 9 m in P2, 4.95 m × 9 m in P3, 4.95 m × 9 m in P4, 4.80 m × 9 m in P5, 4.80 m × 9 m in P6 and 7.2 m × 9 m in P7. Each of these treatments was planted as sole as well as intercropped with lentil that acted as the main plots. Lentil was intercropped in between the empty spaces of sugarcane rows/pits of each treatment at 30 cm distance (Figure 2).

2.4. Crop Husbandry

The field preparation in case of single, double, and trench planting methods was carried out using the standard procedure of ploughing, cultivation and levelling. However, in the case of pit plantation, a tractor-mounted post-hole digger was used to dig round pits of 60 cm depth in zero tilled soil. The 90 cm double rows planting and 120 cm trenches were made with the help of a sugarcane ridger. Afterwards, digging pits were again filled with the same soil to a depth of 45 cm. Sugarcane's variety CPF-247 was selected as a test variety for both years of experiment. Sowing was done on 14th of September 2013 and lentil was intercropped on 25th October 2013. However, in case of ratoon crop, lentil was sown just after the harvesting of plant crop. Lentil variety (NIAB Masoor-2006) was used as an intercrop. In different planting methods according to the space available, different number of lentil rows was sown, i.e., in 45 cm-spaced round pits with one row of lentil, 60 cm-spaced round pits with two rows of lentil, 75 cm-spaced round pits with two rows of lentil, 90 cm-spaced round pits with three rows of lentil, 60 cm-single row planting (conventional method) with one row of lentil, 90 cm-double row planting with two rows of lentil, and 120 cm-trench planting with four rows of lentil that were in between the empty spaces of sugarcane, while Lentil alone was planted in 30 cm-spaced single rows with 12 rows.

The 120 cm trenches, 90 cm double rows planting, and 60 cm single rows plantation of sugarcane were given two manual hoeings and one earthing-up. The first two hoeings were done on 28th November and 7th February, respectively. Earthing-up was done in the middle of March. In ratoon crop, earthing up was done by the end of May. The pit-planted sugarcane crop was not earthed-up at any stage. However, one hoeing was done two months after planting to control weeds growing in the space between the pits. The pits were inter-connected with one another through small water channels to design a basin irrigation system. Irrigation and fertilizer application were restricted only to pits. A total of 20 irrigations each of 10 cm depth were applied to sugarcane, while lentil crop was given only one–two irrigations during the whole growing period. The total amount of irrigation water applied to the plant crop was 1977 mm, while to ratoon crop it was 1950 mm. The N, P, and K were applied at a rate of 165, 110, and 110 kg ha^{-1} in the form of urea, di-ammonium phosphate, and sulfate of potash, respectively. However, ratoon crop was fertilized at 30% higher amounts compared to the first-year fresh planted crop. Insect-pests were controlled through chlorpyriphos at 5 L/ha with first irrigation. Plant crop of sugarcane was harvested on 30th November 2014, while ratoon crop of sugarcane was harvested on January 15th 2016. Lentil harvesting was done on 21st of March 2014 and 2nd April 2015.

2.5. Observations

Sugarcane growth, i.e., leaf area index (LAI), leaf area duration (LAD), crop growth rate (CGR), net assimilation rate (NAR), and total dry matter (TDM), yield and yield related traits, i.e., number of millable canes m^{-2}, plant height, cane length, cane diameter, and stripped cane yield, and total sugar yield, lentil yield and yield components (1000-seed weight, biological yield, seed yield, harvest index), and land equivalent ratio were recorded through their standard procedures.

LAI: With the help of leaf area meter (ΔT area meter MK2) leaf area of green laminae was recorded. Following formula was used for its calculation:

$$LAI = \frac{Laef\ area\ of\ crop\ plants}{Land\ area\ of\ crop\ plants}. \tag{1}$$

LAD (Days): The authors in [23] proposed the following formula for computing LAD:

$$LAD = \left[(LAI_1 + LAI_2) \times \frac{(T_2 - T_1)}{2}\right] days \tag{2}$$

where LAI_1 = Leaf area index at t_1, LAI_2 = Leaf area index at t_2, T_1 = Time of first observation, T_2 = Time of second observation, T_1 and T_2 are with 30 days interval, while the following 60 days of planting were recorded.

CGR (g m^{-2} day^{-1}): The authors in [25] projected the CGR formula as follows:

$$CGR = \frac{(W_2 - W_1)}{(T_2 - T_1)} \tag{3}$$

where W_1 is considered the plant dry weight at time t_1, W_2 is known as plant dry weight during the time t_2. T_1 and T_2 are the harvest time for the 1st and 2nd time, respectively. W_2 and W_1 are the total dry weights harvested at time T_1 and T_2, respectively; first data were collected at 60 days after planting.

NAR (g m^{-2} day^{-1}): NAR was determined by using the formula proposed by the authors in [25]:

$$NAR = \frac{TDM}{LAD} \tag{4}$$

where TDM = Total dry matter, LAD = Leaf area duration

TDM (t/ha): After a 30-day interval, plant sampling was carried out on a random basis from each experimental unit. To estimate the fresh weight, plants were separated into leaves, stem, and trash. To determine the dry weight, a 10 g plant sample was taken from each portion after drying at 65 °C. TDM was ascertained in (g) and converted into t/ha after adding the dry weights of the leaves, stem, and trash.

Number of millable canes/m^2: From each experimental unit at harvest, the number of millable canes was tallied and then calculated in m^{-2} area.

Plant height (cm): At harvest, 10 stripped canes from each treatment were collected. From the plant base to the base of top visible dewlap (TVD), the leaf plant height of each plant was calculated. Then, their average was calculated.

Cane length (cm): Ten randomly selected stripped canes were measured and their average values were recorded.

Cane diameter (cm): Ten randomly stripped canes were collected at the time of harvesting. The top, middle, and base of the cane was used for the diameter determination with a vernier caliper and then averaged.

Stripped cane yield (t/ha): All stripped canes of each plot were weighed, and then the number was transformed to t/ha.

Total sugar yield (t/ha): Sugar yield (t/ha) was calculated using the following formula:

$$Sugar\ yield\ (t/ha) = \frac{Stripped\ cane\ yield\ (t/ha)}{100} \times CCS\ \% \tag{5}$$

The commercial cane sugar (CCS%) was calculated through formula of [26]:

$$CCS = \frac{3P}{2}\left[1 - \frac{(F+5)}{100}\right] - \frac{1}{2}B\left[1 - \frac{(F+3)}{100}\right] \tag{6}$$

where P is the percentage of pol in juice, B is the percentage of brix in juice, and F is the percentage of fiber in juice (12.5%)

Land equivalent ratio: The authors in [27] proposed the formula for computing the term land equivalent ratio (LER):

$$LER = (Yab/Yaa) + (Yba/Ybb), \tag{7}$$

where Yaa = pure stand yield of crop a (sugarcane), Ybb = pure stand yield of crop b (lentil), Yab = intercrop yield of crop a (sugarcane), Yba = intercrop yield of crop b (lentil).

2.6. Economic Analysis

After deduction of the gross investment from the gross field benefits, the net filed benefit (NFB) was calculated [28]. Gross benefit refers to the gross income generated from the main and by-products from component crops in an intercropping system. The total variable cost (PKR Rs./ha) was attained by computing the total variable cost of the production of sugarcane and intercrop in each treatment. The benefit–cost ratio (BCR) was determined through dividing the gross income with the total cost of production. The marginal analysis comprises the dominance analysis (DA) and the marginal rate of return (MRR). In DA, the treatments were arranged in increasing variable cost order. A treatment was dominant (D) when its variable costs was more than the previous treatment, but its NFB was lower or equal [28]. MRR % is the marginal net field benefits (MNB) of the variation in NFB divided by the marginal costs (MC), i.e., the variation in costs expressed as a Percentage. MRR was calculated using the formula given by [26]:

$$\mathrm{MRR}\ (\%) = \frac{\mathrm{MNB}}{\mathrm{MC}} \times 100 \tag{8}$$

2.7. Statistical Analysis

Fisher's analysis of variance was used for statistically analysis of collected data and for comparison of differences among treatment means; a least significant difference (LSD) test was used at 5% probability [29]. Statistics 10 (Tallahassee, FL 32317) was used for the determination of statistical difference.

Figure 2. *Cont.*

Figure 2. Lentil intercropping in sugarcane.

3. Results

3.1. Sugarcane Growth

Different planting patterns of sugarcane showed a significant effect on growth parameters of sugarcane planted alone and intercropped with lentil presented in (Figures 3–7). In plant crop, sugarcane sown in 120 cm trench planting (sugarcane alone) attained the highest LAI (8.97), LAD (1346 days), CGR (12.42 g m^{-2} day^{-1}), NAR (3.15 g m^{-2} day^{-1}), and TDM (34.64 t ha^{-1}). The same planting geometry with lentil intercropping as well as 45 cm pits with 90 cm diameter, 60 cm pits with 90 cm diameter, and 75 cm pits with 90 cm diameter both in sugarcane planted alone and intercropped with lentil gave similar results regarding the growth parameters of sugarcane. In contrast, the minimum values of these parameters were recorded from the treatment where sugarcane was planted in 60 cm single rows with and without lentil as an intercrop.

In ratoon crop, the maximum LAI (7.88), LAD (1160 days), CGR (11.11 g m^{-2} day^{-1}), NAR (2.75 g m^{-2} day^{-1}), and TDM (29.62 t ha^{-1}) were recorded in trench planting at 120 cm (sugarcane alone), followed by 45 cm pits with 90 cm diameter, 120 cm trench planting with lentil intercrop, and 60 cm pits with 90 cm diameter. Sugarcane planted in 60 cm single rows planting + lentil as an intercrop and sugarcane sown in 60 cm apart as sole crop showed the minimum values of these parameters.

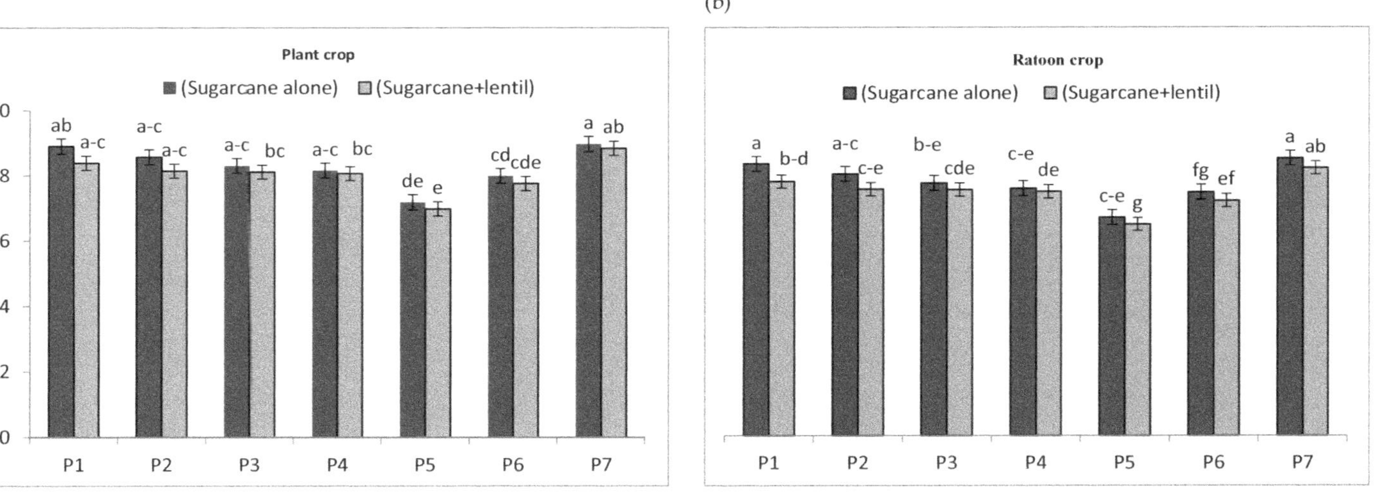

Figure 3. Leaf area index of sugarcane as influenced by various planting patterns and lentil intercropping in (**a**) Plant crop and (**b**) ratoon crop. P_1: 45 cm pits with 90 cm diameter, P_2: 60 cm pits with 90 cm diameter, P_3: 75 cm pits with 90 cm diameter, P_4: 90 cm pits with 90 cm diameter, P_5: 60 cm single rows planting, P_6: 90 cm double rows planting, P_7: Trench planting at 120 cm. Any two means not sharing a common letter(s) differ significantly at 5% probability.

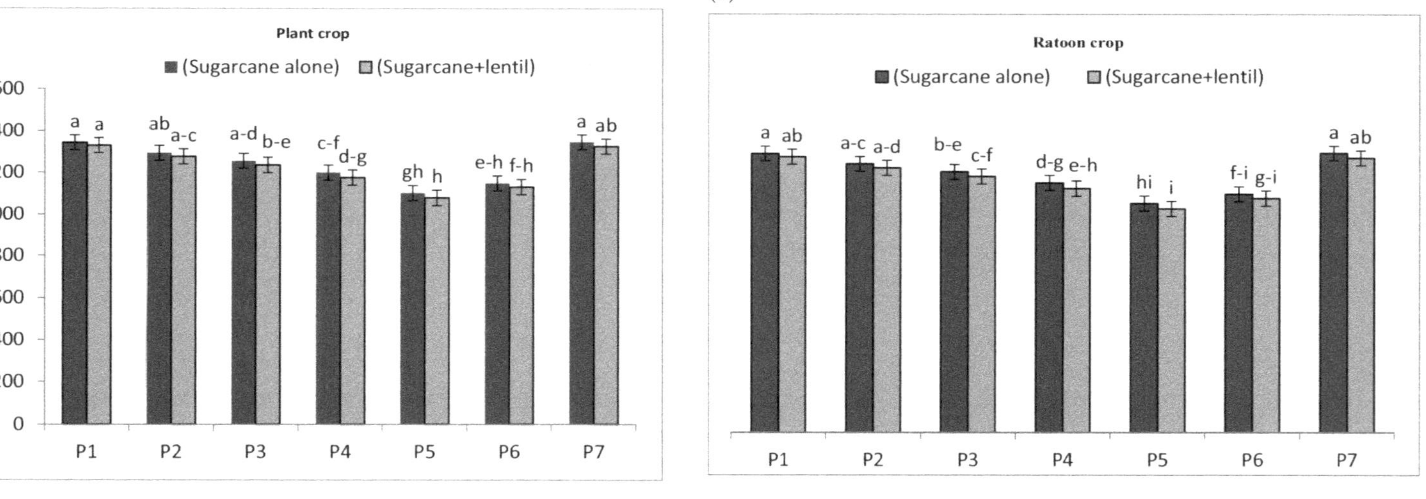

Figure 4. Leaf Area Duration (Days) of sugarcane as influenced by various planting patterns and lentil intercropping in (**a**) Plant crop and (**b**) ratoon crop. P_1: 45 cm pits with 90 cm diameter, P_2: 60 cm pits with 90 cm diameter, P_3: 75 cm pits with 90 cm diameter, P_4: 90 cm pits with 90 cm diameter, P_5: 60 cm single rows planting, P_6: 90 cm double rows planting, P_7: Trench planting at 120 cm. Any two means not sharing a common letter(s) differ significantly at 5% probability.

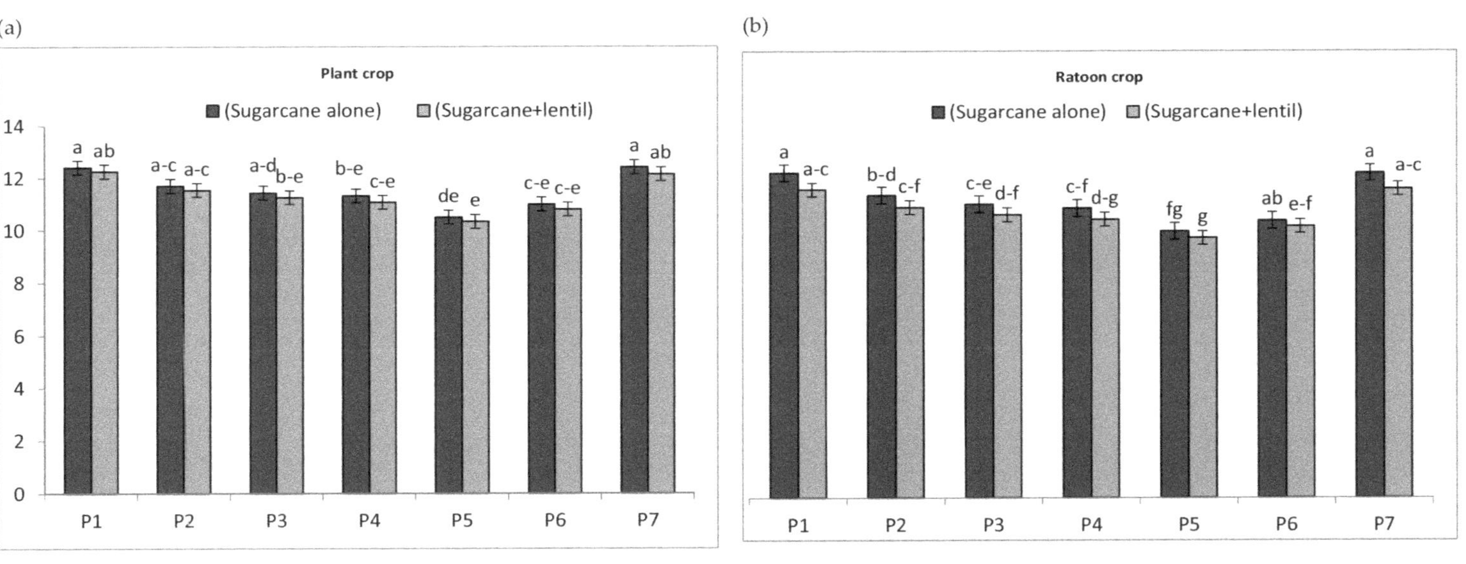

Figure 5. Crop growth rate (g m^{-2} day $^{-1}$) of sugarcane as influenced by various planting patterns and lentil intercropping in (**a**) Plant crop and (**b**) ratoon crop. P_1: 45 cm pits with 90 cm diameter, P_2: 60 cm pits with 90 cm diameter, P_3: 75 cm pits with 90 cm diameter, P_4: 90 cm pits with 90 cm diameter, P_5: 60 cm single rows planting, P_6: 90 cm double rows planting, P_7: Trench planting at 120 cm. Any two means not sharing a common letter(s) differ significantly at 5% probability.

(a)

(b)

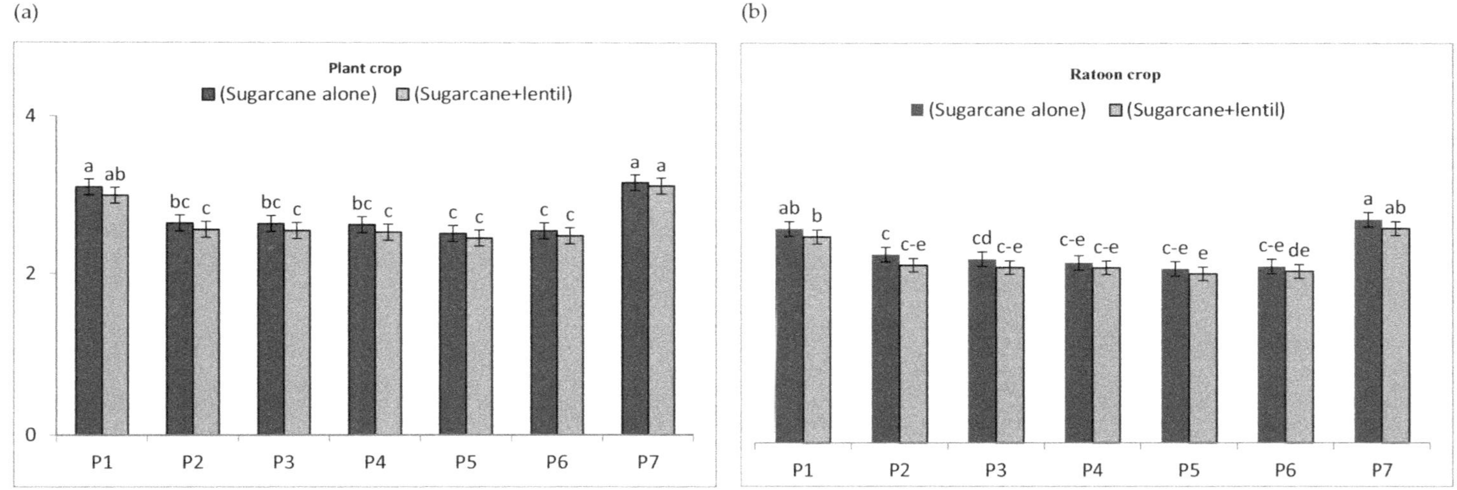

Figure 6. Net assimilation rate (g m^{-2} day^{-1}) of sugarcane as influenced by various planting patterns and lentil intercropping in (**a**) Plant crop and (**b**) ratoon crop. P_1: 45 cm pits with 90 cm diameter, P_2: 60 cm pits with 90 cm diameter, P_3: 75 cm pits with 90 cm diameter, P_4: 90 cm pits with 90 cm diameter, P_5: 60 cm single rows planting, P_6: 90 cm double rows planting, P_7: Trench planting at 120 cm. Any two means not sharing a common letter(s) differ significantly at 5% probability.

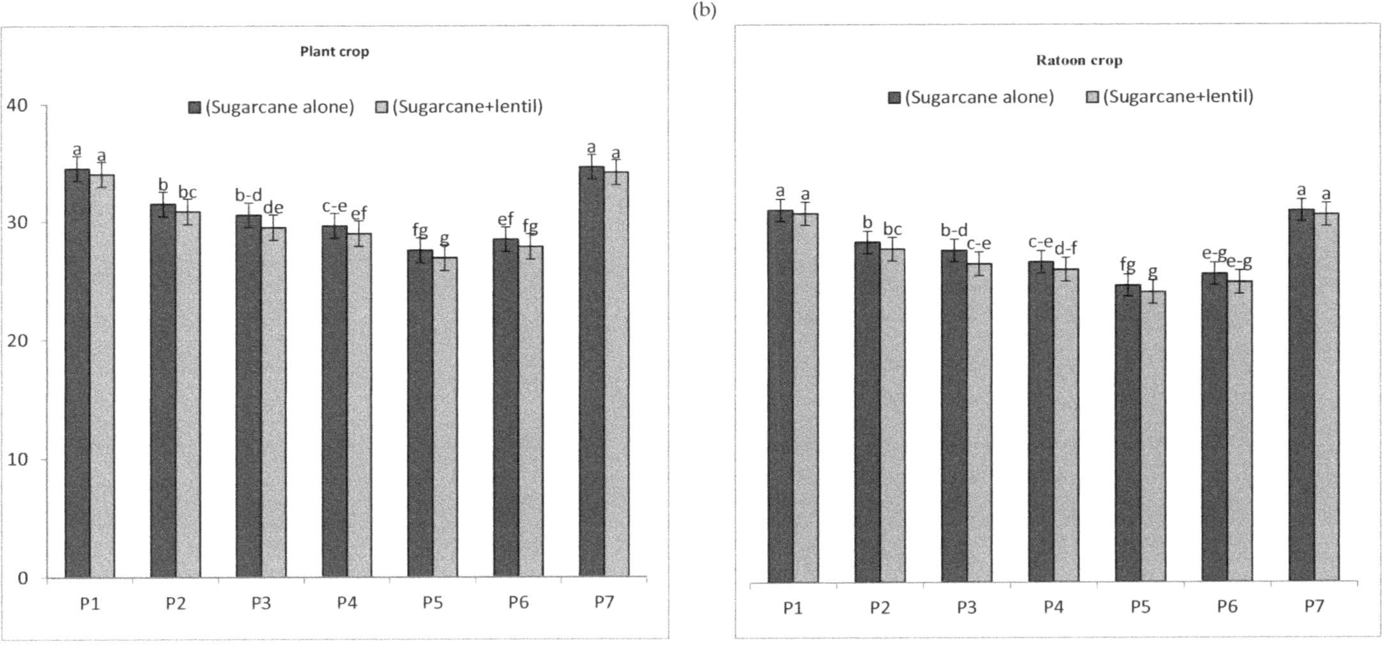

Figure 7. Total dry matter (t/ha) of sugarcane as influenced by various planting patterns and lentil intercropping in (**a**) Plant crop and (**b**) ratoon crop. P_1: 45 cm pits with 90 cm diameter, P_2: 60 cm pits with 90 cm diameter, P_3: 75 cm pits with 90 cm diameter, P_4: 90 cm pits with 90 cm diameter, P_5: 60 cm single rows planting, P_6: 90 cm double rows planting, P_7: Trench planting at 120 cm. Any two means not sharing a common letter(s) differ significantly at 5% probability.

3.2. Sugarcane Yield and Yield Components

Results presented in Table 2; Table 3 reveals that various planting patterns significantly affected the parameters related to sugarcane yield and total sugar yield of sugarcane either grown alone or intercropped with lentil in both plant crop and ratoon crop. Among different planting patterns in plant cane, the number of millable canes in 45 cm pit planting was similar to 120 cm trench planting, but it was significantly greater than other planting patterns (Table 2). However, plant height was greater in 90 cm pit planting than other planting patterns except trench planting. Cane diameter was mostly similar in all planting patterns (Table 3). Stripped cane yield (SCY) and total sugar yield (TSY), with and without lentil intercropping, were significantly greater in 120 cm trench planting than all other planting patterns (Table 3). Similar results were seen in the ratoon crop with greater cane and sugar yields in 120 cm trench planting compared to other planting patterns. In comparison to all planting patterns, 90 cm apart single row planting remained at the bottom, as it gained the lowest values of all these parameters in both plant and ratoon crops. A similar trend of achieving higher plant height, cane diameter, cane length, stripped cane yield (SCY), and total sugar yield (TSY) was shown by the 120 m trench planting in ratoon crop. Compared to the traditional 60 cm apart single row planting pattern, 120 cm trench planting showed up to 4.34, 12.12, 8.63, 43.18, and 42.81% increase in plant height, cane diameter, cane length, SCY, and TSY, respectively.

3.3. Lentil Yield

Biological yields, seed yields, and harvest index's means of lentil intercropped with sugarcane under different planting patterns showed significant differences among themselves (Table 4). Lentil crop sown alone in 30 cm apart rows produced significantly higher biological and seed yields than other planting patterns in both 2013–2014 and 2014–2015 years. However, lentil intercrop performed better when sown along with 120 cm trench planted sugarcane, as it produced significantly higher biological and seed yields, which were 296% and 319% higher, respectively, than the lowest yields recorded with the lentil intercropped in 45 cm pit planting. Similarly, the harvest index (HI) of lentil was significantly greater in 120 cm trench planting compared to others, including the lentil sole planting method in 2013–2014. In 2014–2015, 120 cm trench planting had a similar HI as lentil alone, but both of them were significantly greater than other planting patterns. The comparison of land equivalent ratios (LER) calculated from different sugarcane-lentil intercropping geometries (Table 4) revealed that 120 cm trench planting with four rows of lentil intercrop attained significantly higher LER values (1.40 and 1.37 during cropping seasons 2013–2014 and 2014–2015, respectively) than all other planting patterns.

Table 2. Yield contributing traits of sugarcane under different planting patterns with lentil intercropping.

Planting Methods (P)	Number of Millable Canes/m^2		Plant Height (cm)		Cane Length (cm)	
	Sugarcane Alone (I_0)	Sugarcane + Lentil (I_1)	Sugarcane Alone (I_0)	Sugarcane + Lentil (I_1)	Sugarcane Alone (I_0)	Sugarcane + Lentil (I_1)
Plant crop						
P_1: 45 cm pits with 90 cm diameter	19.00 a	17.50 ab	367.17 cde	367.92 b–e	240.28 b–e	236.86 c–f
P_2: 60 cm pits with 90 cm diameter	16.00 b–e	14.75 e	370.78 bcd	364.99 c–f	243.67 bcd	237.62 b–f
P_3: 75 cm pits with 90 cm diameter	15.25 cde	14.25 ef	369.16 bcd	366.28 cde	246.45 bc	242.42 bcd
P_4: 90 cm pits with 90 cm diameter	15.00 de	13.75 efg	381.11 a	369.65 bcd	259.51 a	250.60 ab
P_5: 60 cm single row planting	11.75 gh	11.25 h	356.19 f	359.94 ef	225.75 fg	222.00 g
P_6: 90 cm double row planting	12.00 fgh	11.50 h	364.49 def	361.74 def	232.15 d–g	228.15 efg
P_7: Trench planting at 120 cm	17.50 ab	17.00 a–d	376.52 ab	374.44 abc	251.46 ab	247.26 abc
Mean	15.21 A	14.28 B	370.10 A	365.67 B	242.75 A	287.84 B
Ratoon crop						
P_1: 45 cm pits with 90 cm diameter	15.75 a	14.50 ab	322.39 b–e	320.64 b–e	216.31 a–e	208.55 def
P_2: 60 cm pits with 90 cm diameter	13.00 b–e	12.00 c–g	324.34 a–d	321.84 b–e	217.37 a–d	209.37 c–f
P_3: 75 cm pits with 90 cm diameter	12.75 b–f	11.50 d–i	325.88 abc	322.63 b–e	219.67 abc	214.17 b–e
P_4: 90 cm pits with 90 cm diameter	12.50 c–g	11.00 f–i	326.10 ab	323.02 a–e	225.76 a	222.69 ab
P_5: 60 cm single row planting	10.50 ghi	9.75 h	318.44 de	316.44 e	202.00 f	200.50 f
P_6: 90 cm double row planting	11.00 e–i	10.25 hi	320.49 b–e	318.74 cde	207.07 ef	205.90 ef
P_7: Trench planting at 120 cm	13.50 bc	13.50 bcd	330.20 a	323.41 a–e	224.80 a	221.51 ab
Mean	12.75 A	11.78 B	323.98 A	320.96 B	216.14 A	211.81 B

Any two means in a column not sharing a common letter(s) differ significantly at 5% probability.

Table 3. Yield and quality of sugarcane under different planting patterns with lentil intercropping.

Planting Methods (P)	Cane Diameter (cm)		Stripped Cane Yield (t/ha)		Total Sugar Yield (t/ha)	
	Sugarcane Alone (I_0)	Sugarcane + Lentil (I_1)	Sugarcane Alone (I_0)	Sugarcane + Lentil (I_1)	Sugarcane Alone (I_0)	Sugarcane + Lentil (I_1)
			Plant crop			
P_1: 45 cm pits with 90 cm diameter	2.40 ab	2.36 ab	142.88 b	140.50 bc	19.11 b	18.83 bc
P_2: 60 cm pits with 90 cm diameter	2.44 ab	2.39 ab	137.00 bc	136.61 c	18.44 c	18.27 c
P_3: 75 cm pits with 90 cm diameter	2.45 ab	2.41 ab	124.49 d	123.03 d	17.64 d	16.28 e
P_4: 90 cm pits with 90 cm diameter	2.51 ab	2.49 ab	115.97 e	115.50 e	15.64 f	15.35 f
P_5: 60 cm single rows planting	2.32 ab	2.28 b	99.49 g	95.32 g	13.40 h	12.70 i
P_6: 90 cm double rows planting	2.37 ab	2.34 ab	108.45 f	98.05 g	14.54 g	13.19 hi
P_7: Trench planting at 120 cm	2.54 a	2.52 a	154.36 a	153.10 a	20.62 a	20.50 a
Mean	2.43	2.39	126.09 A	123.09 B	17.06 A	16.44 B
			Ratoon crop			
P_1: 45 cm pits with 90 cm diameter	2.29 a–e	2.23 de	122.03 b	121.85 b	16.89 b	16.84 b
P_2: 60 cm pits with 90 cm diameter	2.32 a–d	2.24 cde	116.28 c	115.98 cd	16.02 c	16.00 c
P_3: 75 cm pits with 90 cm diameter	2.33 abc	2.25 b–e	111.73 de	110.78 ef	15.24 de	15.28 d
P_4: 90 cm pits with 90 cm diameter	2.35 ab	2.27 b–e	107.84 ef	106.84 f	14.79 f	14.80 ef
P_5: 60 cm single row planting	2.26 b–e	2.20 e	92.84 h	90.99 h	12.65 h	12.52 h
P_6: 90 cm double row planting	2.28 b–e	2.21 e	101.28 g	100.98 g	13.88 g	14.07 g
P_7: Trench planting at 120 cm	2.39 a(8.63)	2.29 a–e	130.28 a	130.10 a	17.88 a	17.86 a
Mean	2.31 A	2.24 B	111.75	111.07	15.30 A	14.36 B

Any two means in a column not sharing a common letter(s) differ significantly at 5% probability.

Table 4. Yields, harvest index of lentil, and land equivalent ratio under different planting patterns with lentil intercropping.

Planting Methods (P)	Biological Yield (kg/ha)		Seed Yield (kg/ha)		Harvest Index (HI)		Land Equivalent Ratio	
	2013–2014	2014–2015	2013–2014	2014–2015	2013–2014	2014–2015	2013–2014	2014–2015
P_1: 45 cm pits with 90 cm diameter + one row of lentil	517.50 e	587.50 e	150.30 f	165.30 e	29.03 ab	28.08 b	1.09 d	1.10 d
P_2: 60 cm pits with 90 cm diameter + two rows of lentil	1022.50 d	1102.50 d	265.30 de	275.00 d	25.94 c	24.93 c	1.18 c	1.16 c
P_3: 75 cm pits with 90 cm diameter + two rows of lentil	1032.00 d	1117.50 d	275.00 d	285.00 d	26.66 c	25.49 c	1.18 c	1.16 c
P_4: 90 cm pits with 90 cm diameter + three rows of lentil	1510.00 c	1570.50 c	394.50 c	402.00 c	26.13 c	25.60 c	1.26 b	1.23 b
P_5: 60 cm single row planting + one row of lentil	530.00 e	595.00 e	129.80 f	145.00 e	26.48 c	24.37 c	1.05 d	1.07 d
P_6: 90 cm double row planting + two rows of lentil	1027.50 d	1080.00 d	240.30 e	265.30 d	23.37 d	24.53 c	1.07 d	1.15 c
P_7: Trench planting at 120 cm + four rows of lentil	1987.00 b	2052.50 b	598.00 b	629.80 b	29.96 a	30.68 a	1.40 a	1.37 a
P_8: Lentil alone 30 cm spaced single rows (twelve rows)	5215.00 a	5390.00 a	1476.00 a	1700.00 a	28.30 b	31.54 a	-	-
LSD	45.20	61.25	27.40	29.87	1.54	1.28	0.049	0.054

Any two means in a column not sharing common letter(s) differ significantly at 5% probability.

3.4. Economic Analysis

Possibly, productivity as well as adoptability of an intercropping system are ultimately determined by their net monetary gain. Based on average net field benefits from both plant and ratoon crop, the highest NFB of Rs. 491703/ha was found in 120 cm trench planting + lentil intercropping, whereas the minimum NFB of Rs. 302559/ha was recorded in 60 cm single rows planting + lentil. Overall, higher net field benefits were obtained when lentil was intercropped in sugarcane. The better result of the 120 cm trench with lentil intercropping might be due to more lentil plant population when sown in between the empty spaces as compared to pit planting: 60 cm single row planting or 90 cm double row planting. Average data on net return of both plant and ratoon crops of sugarcane are presented in Table 5. Maximum net return of Rs. 321254/ha was recorded in 120 cm trench planting with lentil as an intercrop; however, the minimum net return of Rs. 132110/ha was noted in 60 cm single rows planting + lentil.

The results of BCR are presented in Table 5. Average data on BCR of both plant and ratoon crops of sugarcane are presented in Table 5. Maximum BCR (2.01) was recorded in 120 cm trench planting with lentil as an intercrop; however, the least BCR (1.50) was noted in 60 cm apart single row planting + lentil. Overall, the maximum BCR values were examined in all treatments of ratoon crop, because no sowing and tillage operations were carried out that ultimately resulted in a reduction of the overall cost of production.

Since the net field benefits are not the final criteria for recommendation to the farmers, the marginal analysis was done to conclude the most profitable sugarcane intercropping system. The dominance analysis is given in Table 5. Based on two-year average data, sugarcane planting methods (P_{12}, P_{13}, P_3, P_{10}, P_2, P_9, P_1, and P_8) underwent dominance analysis. Those treatments that were signed as "D" in the dominance analysis were not selected for marginal analysis. The results of sugarcane-lentil intercropping systems have been described in Table 5. Based on an average MRR, the maximum (16,902%) value was obtained from 120 cm trench planting.

Table 5. Effect of different sugarcane-lentil based intercropping systems on average net return, average net field benefits, and average benefit-cost ratio 2013–2016 (each value is the average total experimental duration). MRR: marginal rate of return.

Planting Methods (P)	Variable Cost (Rs. ha^{-1})	Total Cost (Rs. ha^{-1})	Gross Income (Rs. ha^{-1})	Net Return (Rs. ha^{-1})	Net Field Benefit (Rs. ha^{-1})	Benefit Cost Ratio	Cost That Vary (Rs.)	Net Field Benefits (Rs.)	Marginal Cost (Rs.)	Marginal Net Profit (Rs.)	MRR (%)
P$_1$: 45 cm pits with 90 cm diameter	152,705	323,154	553,993	230,839	401,288	1.71	92,245	310,269	-	-	-
P$_2$: 60 cm pits with 90 cm diameter	142,147	312,596	529,658	217,061	387,510	1.69	94,379	D	2133	−7710	D
P$_3$: 75 cm pits with 90 cm diameter	131,754	302,203	494,202	191,999	362,448	1.64	99,068	339,825	4689	37,266	795
P$_4$: 90 cm pits with 90 cm diameter	116,822	287,271	468,359	181,087	351,536	1.63	103,194	D	4126	−9257	D
P$_5$: 60 cm single row planting	92,245	262,694	402,514	139,820	310,269	1.53	116,822	351,536	13,628	20,969	154
P$_6$: 90 cm double row planting	99,068	269,517	438,893	169,376	339,825	1.63	129,127	365,253	12,305	13,716	111
P$_7$: Trench planting at 120 cm	129,720	300,169	595,207	295,038	465,487	1.98	129,720	465,487	593	100,234	16,902
P$_8$: 45 cm pits with 90 cm diameter + lentil	156,801	327,250	555,604	228,354	398,803	1.70	131,754	D	2034	−103,039	D
P$_9$: 60 cm pits with 90 cm diameter + lentil	150,725	321,174	545,894	224,720	395,169	1.70	139,953	D	8199	5684	D
P$_{10}$: 75 cm pits with 90 cm diameter + lentil	139,953	310,402	508,085	197,683	368,132	1.64	142,147	D	2194	19,378	D
P$_{11}$: 90 cm pits with 90 cm diameter + lentil	129,127	299,576	494,380	194,804	365,253	1.65	149,077	491,703	6930	104,193	1504
P$_{12}$: 60 cm single row planting + lentil	94,379	264,828	396,937	132,110	302,559	1.50	150,725	D	1648	−96,534	D
P$_{13}$: 90 cm double row planting + lentil	103,194	273,643	433,762	160,119	330,568	1.59	152,705	D	1981	6119	D
P$_{14}$: Trench planting at 120 cm + lentil	149,077	319,526	640,780	321,254	491,703	2.01	156,801	D	4096	−2485	D

Letter 'D' represent dominant treatments.

4. Discussion

Intercropping of a legume crop in sugarcane is advantageous as it improves the growth and yield of both crops by enhancing the soil nutrient contents and benefits the microbial population [30]. However, the choice of suitable sugarcane planting pattern for adjusting intercrop to draw maximum economic benefit is important [31]. The present study showed significant variation in growth, sugarcane yield parameters, and total sugar yield attributes of sugarcane sown along with lentil intercrop. Variability in LAI of sugarcane in different planting patterns was ascribed to variable plant population and availability of moisture, nutrients, optimal temperature, and aeration. The greater LAI of sugarcane sown in 120 cm trench planting/45 cm pit planting compared to other methods could be due to more plant population and better leaf development. These findings are in agreement with those of [32], who reported significant changes in LAI due to intercropping in sugarcane. The difference in LAD among various planting patterns was due to the variation in LAI achieved as a result of utilization of resources in different planting patterns; in some papers [33,34], it was also noted that different planting patterns had significant effects on LAD of sugarcane crop.

Significantly higher CGR in sole sugarcane (SC) might be due to less competition for free environment, and more availability of nutrients and space that ultimately led to a well-developed root system. A deep root system provides nutrient and moisture availability for plants. The above conclusions are matched with the those of [35], who reported more CGR of sole sugarcane planted at triple row strips than intercrops. The variation in NAR under different treatments might be ascribed to variation in LAI (Figure 6), LAD (Figure 4), production of total dry matter (Figure 7), and CGR (Figure 5). Variability in the production of total dry matter of sugarcane in different planting patterns was ascribed to variability in plant population, intercrop competition, and availability of different farm resources.

Variability in millable canes/m^2 was possibly due to appreciative effect of increased nutrient accessibility, better air circulation and interception of light that resulted in reduced shoot mortality and better growth of cane due to better utilization of farm inputs. An increase in millable canes/m^2 for sole sugarcane could be possibly due to a higher number of tillers/m^2. Similar results confirmed by [36–38], who reported a significant destructive effect of linseed, mustard, alfalfa, and sunflower on millable canes/m^2. These results are also in line with those of [39]. The variation in plant height under different treatments might be due to variation in utilization of farm resources by crop plants. Moreover, better penetration of sunlight to the crop plant might improve the availability of photosynthates to cane sown in trench planting at 120 cm and 90 cm pits that promoted growth, resulting in increased values of plant height. These results are in line with those of [40], who reported that sugarcane planted in paired rows accommodating two rows of grain *Amaranth* sp., as an intercrop produced plants of maximum height. These results were also confirmed by [41], who noted an increased plant height in trench planting at 120 cm as compared to 60 cm single rows. Additionally, the authors of [42] documented more height of sugarcane plants in 0.5 m-spaced rows compared to 1.5 m-spaced rows. The maximum reduction in cane length was noted when sugarcane was intercropped with lentil both in plants as well as in ratoon crop. The dissimilarity in cane length under different treatments might be due to a higher number of internodes per cane when sugarcane was planted as sole crop compared to sugarcane + lentil. The authors in [43] reported that intercropping in sugarcane significantly affected cane length.

Variability in cane diameter in pit and trench planting with 60 cm single row and 90 cm double row strips of sole sugarcane and sugarcane with lentil may be attributed to better cane growth in pits and trench planting because of adequate supply of various agricultural inputs, such as water, fertilizer, etc., which were applied directly to the pits and within the trench instead of mixing them with the surface soil over the entire plant area. These outcomes are in close conformity with the conclusions of [44], who determined that cane diameter was increased as a result of increasing row spacing, while [45] stated that the diameter of cane was not influenced through different sowing geometry. Significantly more SCY from trench planting in sugarcane treatments over other treatments was possibly due to maximum LAI (Figure 3), CGR (Figure 5), TDM (Figure 7), and NAR (Figure 6). The more cane yield of the SC might be due to less competition for resources that enhanced the cane potential during developmental stage. The above findings are in line with [46], who observed a 21.8% reduction in yield of sugarcane when it was intercropped with sarsoon (*Brassica compestris*). Likewise, a better yield of sugarcane was obtained when planted as a sole crop compared to the intercropped method [47]. The above results are also supported by [48], who observed that different intercrops except Sesbania reduced the yield of cane significantly. Higher values of total sugar yield by 120 cm trench planting might be due to higher SCY (Table 2). These results supported by [39,49], who noted that sugarcane sown at 120 spaced trenches and 90 cm double rows produced significantly more sugar yield (total sucrose) than that sown at 60 cm-spaced single rows. Similarly, [50] hypothesized that wider spaced rows gave more sugar yield than narrow spaced rows.

The results of the present study indicated that biological and seed yields as well as HI of lentil inter-crop were also varied significantly under different planting patterns of sugarcane. The variability in biological yield of lentil in intercropping treatments might be due to variable plant density. Diminution in the biological yield of lentil in intercropped treatments might be due to competition for light, water, nutrients, etc. between the sugarcane and lentil, while the increased biological yield in lentil alone can be attributed to competition free environment. These results are in conformity to those of [51], who found increased biological yield when mung bean (*vigna radiate*) was sown alone in an intercropping study. Less grain yield in the intercropping system is possibly due to low plant density establishment, seed per pods (data not shown), and 1000-grain weight in comparison with sole lentil. These results are supported by [52], who observed that lentil yield was reduced to a greater degree by the intercropping systems. The reduction in HI of lentil occurred due to its intercropping with sugarcane planted in 60 cm single row planting and 90 cm double row planting as well as 60 and 75 cm pits of 90 cm diameter. The reduction in HI of lentil with these intercropping patterns might be attributed to its reduced reproductive growth at the expense of higher vegetative growth. These findings are in line with those of [53], who observed the reduction in harvest index of lentil crop when it was intercropped in linseed (*Linum usitatissimum*), methra (*Trigonella foenum-graecum*), and wheat (*Triticum aestivum*). Regarding LER, the significantly highest land equivalent ratio was achieved in 120 cm trench planted sugarcane intercropping systems, likely due to better utilization of inputs. Sugarcane planted in 120 cm trench planting showed the highest LER due to the increased lentil plant population. The benefits of intercropping of sugarcane with lentil were also reported by [54], who found LERs of 1.43 and 1.38 during the two years of study.

In conclusion, 120 cm trench planting pattern of sugarcane along with lentil intercropping outperformed in improving the LER and gave maximum economic return as compared to other intercropping patterns and sole planting of sugarcane.

Author Contributions: M.N.: methodology, writing the original draft preparation, and collecting experimental data. A.T.: supervision. H.S.: Investigations and improving the first draft. S.J.: reviewing and editing the manuscript. M.E.S., M.S., and M.I.: reviewed the manuscript. M.A.S. and U.A.: data analysis. All authors have read and agreed to the published version of the manuscript.

Conflicts of Interest: The authors declare no conflict of interest.

References

1.   GOP. *Economic Survey of Pakistan 2017–2018*; Ministry of Food Agriculture and Livestock, Federal Bureau of Statistics: Islamabad, Pakistan, 2018.
2.   FAO. Statistical Database. Food and Agriculture Organization (FAO). 2012. Available online: www.faostat.fao.org (accessed on 15 December 2019).
3.   Majid, A. Sugarcane variety composition in Pakistan. *Pak. Sugar J.* **2007**, *22*, 2–21.
4.   Nazir, A.; Jariko, G.A.; Junejo, M.A. Factors affecting sugarcane production in Pakistan. *Pak. J. Commer. Soc. Sci.* **2013**, *7*, 89–95.

5.   Bashir, S.; Saeed, M. Biomass production and partitioning in sugarcane genotype SPSG 394 as influenced by planting pattern and seeding density. *J. Agric. Res.* **2000**, *36*, 129–137.

6.   Mahmood, A.; Ishfaq, M.; Iqbal, J.; Nazir, M.S. Agronomic performance and juice quality of autumn planted sugarcane (*Saccharum officinarum* L.) as affected by flat, ditch and pit planting under different spatial arrangements. *Int. J. Agric. Biol.* **2007**, *9*, 167–169.

7.   Ehsanullah; Jabran, K.; Jamil, M.; Ghaffar, A. Optimizing the sugarcane row spacing and seeding density to improve its yield and quality. *Crop Environ.* **2011**, *2*, 1–5.

8.   Maqsood, M.; Iqbal, M.; Tayyab, M. Comparative productivity performance of sugarcane (*Saccharum officinarum* L.) sown in different planting patterns at farmer's field. *Pak. J. Agric. Sci.* **2005**, *42*, 25–28.

9.   Zafar, M.; Tanveer, A.; Cheema, Z.A.; Ashraf, M. Weed-crop competition effects on growth and yield of sugarcane planted using two methods. *Pak. J. Bot.* **2010**, *42*, 815–823.

10.   Ghaffar, A.; Ehsanullah, N.A.; Khan, S.H. Influence of zinc and iron on yield and quality of sugarcane planted under various trench spacing. *Pak J. Agric. Sci.* **2011**, *48*, 25–33.

11.   Singh, N.; Jain, J.L.; Singh, D.K. Impact of planting techniques on sugarcane and sugar productivity at Harinagar, Bihar. *Indian Sugar Technol.* **2009**, *59*, 19–22.

12.   Sureka, B.K.; Shahi, V.P.; Singh, I.S.; Singh, N. Ring pit planting technique of sugarcane adopted to increase cane productivity at Bharat sugar mills, Sidhwalia, Gopalganj Bihar. *Indian Sugar* **2009**, *59*, 69–74.

13.   Chand, M.; Khippal, A.; Singh, S.; Lal, R.; Singh, R. Effect of planting material and seed rate in pit planted sugarcane (*Saccharum* spp. hybrid complex) in sub-tropical India. *Indian J. Agron.* **2011**, *56*, 78–82.

14.   Chattha, M.U.; Ehsanullah. Agro-quantitative and qualitative performance of different cultivars and strains of sugarcane (*Saccharum officinarum* L.). *Pak. Sugar J.* **2003**, *18*, 2–5.

15.   Akhtar, M.; Ashraf, M.; Akhtar, M.E. Sugarcane yield gap analysis: Future options for Pakistan. *Sci. Technol. Dev.* **2003**, *1*, 38–48.

16.   Yadav, R.L. *Sugarcane Production Technology*; Constraints and Potentialities; Oxford and IBH Publishing Co. (Pvt.) Ltd.: Bombay, India, 1991; p. 204.

17.   Rehman, A.; Ehsanullah. Increasing Yield of Ratoon Sugarcane. 2008. Available online: http://www.dawn.com/news/296976/increasing-yield-of-ratoon-sugarcane (accessed on 22 December 2019).

18.   Malik, K.B.; Ali, F.G.; Khaliq, A. Effect of plant population and row spacing on cane yield of spring-planted cane. *J. Agric. Res.* **1996**, *34*, 389–395.

19.   Eskandari, H.; Ghanbari, A. Environmental resource consumption in wheat and bean intercropping: Comparison of nutrient uptake and light interception. *Not. Sci. Biol.* **2010**, *2*, 100–103. [CrossRef]

20.   Imran, M.; Ali, A.; Waseem, M.; Tahir, M.; Mohsin, A.U.; Shehzad, M.; Ghaffari, A.; Rehman, H. Bio-economic assessment of sunflower mungbean intercropping system at different planting geometry. *Int. Res. J. Agric. Sci.* **2011**, *1*, 126–136.

21.   Alizadeh, K.; Silva, J.A.T. Mix cropping of annual feed legumes with barley improves feed quantity and crude protein content under dry land conditions in Iran. *Maejo Int. J. Sci. Technol.* **2013**, *7*, 42–47.

22.   Akhtar, M.; Yaqub, M.; Iqbal, Z.; Ashraf, M.Y.; Akhter, J.; Hussain, F. Improvement in yield and nutrient uptake by co-cropping of wheat and chickpea. *Pak. J. Bot.* **2010**, *42*, 4043–4049.

23.   Pawar, M.W.; More, D.B.; Amodkar, V.T.; Joshi, S. Effect of inter-settling spacing on sugarcane yield and quality. *Sugar Technol.* **2005**, *7*, 87–89. [CrossRef]

24.   Rana, N.S.; Kumar, S.; Saini, S.K.; Panwar, G.S. Production potential of autumn sugarcane-based intercropping systems as influenced by intercrops and row spacing. *Indian J. Agron.* **2006**, *51*, 31–33.

25.   Hunt, R. *Plant Growth Analysis*; Academic Division of Unwin Hyman Ltd.: Edward Arnold, UK, 1978; pp. 26–38.

26.   Spancer, G.L.; Meade, G.P. *Cane Sugar Hand Book*, 9th ed.; John Wiley and Sons, Inc.: New York, NY, USA, 1963; p. 17.

27.   Willey, R.W. Intercropping, its importance and research needs. *Agron. J.* **1979**, *71*, 115–119.

28.   CIMMYT. *From Agronomic Data to Farmers Recommendations: An Economics Training Manual*; CIMMYT: Mexico City, Mexico, 1988.

29.   Steel, R.G.D.; Torrie, J.H.; Dicky, D.A. *Principles and Procedures of Statistics, A Biometrical Approach*, 3rd ed.; McGraw Hill, Inc. Book Co.: New York, NY, USA, 1996.

30. Li, X.; Mu, Y.; Cheng, Y.; Liu, X.; Nian, H. Effects of intercropping sugarcane and soybean on growth, rhizosphere soil microbes, nitrogen and phosphorus availability. *Acta Physiol. Plant.* **2013**, *35*, 1113–1119. [CrossRef]

31. Yang, W.; Li, Z.; Wang, J.; Wu, P.; Zhang, Y. Crop yield, nitrogen acquisition and sugarcane quality as affected by interspecific competition and nitrogen application. *Field Crops Res.* **2013**, *146*, 44–50. [CrossRef]

32. Mendoza, T.C. Light interception and total biomass productivity in sugarcane intercropping. *Philipp. J. Crop Sci.* **1986**, *11*, 181–187.

33. Gill, M.B. Physio-Agronomic Studies on Flat Versus Pit Plantation of Autumn and Spring Sugarcane (*Saccharum officinarum* L.). Ph.D. Thesis, Department of Agronomy, University of Agriculture, Faisalabad, Pakistan, 1995; pp. 41–89.

34. Ahmad, I. Bio-Economic Efficiency of Spring-Planted Sugarcane as Influenced by Spatial Arrangement and Nutrient Management. Ph.D. Thesis, Department of Agronomy, University of Agriculture, Faisalabad, Pakistan, 2002.

35. Pammenter, N.W.; Allison, J.C.S. Effects of treatments potentially influencing the supply of assimilate on its partitioning in sugarcane. *Exp. Bot.* **2002**, *53*, 123–129. [CrossRef]

36. Ahmad, R. Studies on Geometry of Planting Autumn Sugarcane Facilitating Intercropping of Wheat and Berseem. Master's Thesis, Department of Agronomy, University of Agriculture, Faisalabad, Pakistan, 1982.

37. Bukhtiar, B.A.; Muhammad, G. Feasibility of companion cropping with autumn planted cane. *Pak. J. Agric. Res.* **1988**, *9*, 294–299.

38. Khan, A.A.; Khan, M.A.; Khan, Q. Economic analysis of sugarcane (*Saccharum officinarum* L.) intercropping with canola (*Brassica napus* L.). *Pak. J. Agric. Sci.* **2012**, *49*, 589–592.

39. Roodagi, L.I.; Itnal, C.J.; Karabantanal, S.S.; Rachappa, V. Influence of planting systems and intercross on sugarcane tillering and yield. *Indian Sugar* **2001**, *51*, 159–163.

40. Chogatapur, S.V.; Deepa, G.S.; Chandranath, H.T. Review on Intercropping in Sugarcane (*Saccharum officinarum* L.). *Int. J. Pure Appl. Biosci.* **2017**, *5*, 319–323. [CrossRef]

41. Chattha, M.U. Studies on Growth, Yield and Quality of Sugarcane (*Saccharum officinarum* L.) under Different Planting Techniques, Irrigation Methods, Water Levels and Mulch Types. Ph.D. Thesis, Department of Agronomy, University of Agriculture, Faisalabad, Pakistan, 2007.

42. Gascho, G.T.; Shih, S.F. Cultural methods to increase sucrose and energy yield of sugarcane. *Agron. J.* **1981**, *73*, 999–1003. [CrossRef]

43. Domini, M.E.; Plana, R. Effect of planting density on sugarcane stalk growth and yield. *Cultiaos Trop.* **1989**, *11*, 67–73.

44. Raskar, B.S.; Bhoi, P.G. Response of sugarcane to planting materials, intra-row spacings and fertilizer levels under drip irrigation. *Indian Sugar J.* **2003**, *53*, 685–690.

45. Aziz, K. Ratooning Potential of Autumn Sugarcane as Affected by Pit and Flat Planting. Master's Thesis, Department of Agronomy, University of Agriculture, Faisalabad, Pakistan, 1991.

46. Nazir, M.S.; Jabbar, A.; Ahmad, I.; Nawaz, S.; Bhatti, I.H. Production Potential and Economics of Intercropping in Autumn-Planted Sugarcane. *Int. J. Agric. Biol.* **2001**, *4*, 140–142.

47. Fareed, G. Effect of Associated Culture and Planting Geometry on the Yield and Quality of Autumn Sugarcane. Master's Thesis, Department of Agronomy, University of Agriculture, Faisalabad, Pakistan, 1990.

48. Kumar, S.; Angiras, N.N.; Singh, R. Effect of planting and weed control methods on weed growth and seed yield of Black gram. *Indian J. Weed Sci.* **2006**, *38*, 73–76.

49. Chattha, M.U.; Ali, A.; Bilal, M. Influence of planting techniques on growth and yield of spring planted sugarcane (*Saccharum officinarum* L.). *Pak. J. Agric. Sci.* **2007**, *44*, 452–455.

50. Devi, C.; Rao, K.L.; Raju, D.V.M. The effect of row spacing and nitrogen on yield and quality of early maturing sugarcane cultivars. *Indian Sugar* **1990**, *40*, 541–544.

51. Bhatti, I.H. Agro-Physiological Studies on Sesame-Legume Intercropping Systems in Different Geometric Arrangements. Ph.D. Thesis, Department of Agronomy, University of Agriculture, Faisalabad, Pakistan, 2005.

52. Patrick, M.C.; Gardner, J.C.; Schatz, B.G.; Zwinger, S.W.; Guldan, S.J. Grain yield and weed biomass of wheat-lentil intercrop. *Agron. J.* **1995**, *87*, 574–579.

53. Saleem, K. Effect on Some Legume and Non-Legume Associated Cultures on Growth and Yield of Lentil. Master's Thesis, Department of Agronomy, University of Agriculture, Faisalabad, Pakistan, 1991.

54. Ehsanullah. Agro-Economic Studies on Various Sugarcane Based Intercropping Systems. Ph.D. Thesis, Department of Agronomy, University of Agriculture, Faisalabad, Pakistan, 1995.

 *agronomy*

Review

Accelerating Genetic Gain in Sugarcane Breeding Using Genomic Selection

Seema Yadav [1], **Phillip Jackson** [2], **Xianming Wei** [3], **Elizabeth M. Ross** [1], **Karen Aitken** [4], **Emily Deomano** [5], **Felicity Atkin** [6], **Ben J. Hayes** [1] and **Kai P. Voss-Fels** [1,*]

[1] Queensland Alliance for Agriculture and Food Innovation, The University of Queensland, St. Lucia, QLD 4072, Australia; seema.yadav@uq.edu.au (S.Y.); e.ross@uq.edu.au (E.M.R.); b.hayes@uq.edu.au (B.J.H.)
[2] Agriculture and Food, CSIRO, Townsville, QLD 4811, Australia; Phillip.Jackson@csiro.au
[3] Sugar Research Australia, Mackay, QLD 4741, Australia; XWei@sugarresearch.com.au
[4] Agriculture and Food, CSIRO, QBP, St. Lucia, QLD 4067, Australia; Karen.Aitken@csiro.au
[5] Sugar Research Australia, P.O. Box 86, Indooroopilly, QLD 4068, Australia; EDeomano@sugarresearch.com.au
[6] Sugar Research Australia, Meringa Gordonvale, QLD 4865, Australia; FAtkin@sugarresearch.com.au
* Correspondence: k.vossfels@uq.edu.au; Tel.: +61-7-334-62288

Received: 24 March 2020; Accepted: 15 April 2020; Published: 19 April 2020

Abstract: Sugarcane is a major industrial crop cultivated in tropical and subtropical regions of the world. It is the primary source of sugar worldwide, accounting for more than 70% of world sugar consumption. Additionally, sugarcane is emerging as a source of sustainable bioenergy. However, the increase in productivity from sugarcane has been small compared to other major crops, and the rate of genetic gains from current breeding programs tends to be plateauing. In this review, some of the main contributors for the relatively slow rates of genetic gain are discussed, including (i) breeding cycle length and (ii) low narrow-sense heritability for major commercial traits, possibly reflecting strong non-additive genetic effects involved in quantitative trait expression. A general overview of genomic selection (GS), a modern breeding tool that has been very successfully applied in animal and plant breeding, is given. This review discusses key elements of GS and its potential to significantly increase the rate of genetic gain in sugarcane, mainly by (i) reducing the breeding cycle length, (ii) increasing the prediction accuracy for clonal performance, and (iii) increasing the accuracy of breeding values for parent selection. GS approaches that can accurately capture non-additive genetic effects and potentially improve the accuracy of genomic estimated breeding values are particularly promising for the adoption of GS in sugarcane breeding. Finally, different strategies for the efficient incorporation of GS in a practical sugarcane breeding context are presented. These proposed strategies hold the potential to substantially increase the rate of genetic gain in future sugarcane breeding.

Keywords: genetic gain; genomic selection; quantitative genetics; sugarcane breeding

---

1. The Commercial Importance of Sugarcane and Production Trends

Sugarcane (*Saccharum spp, Poaceae*) is a perennial C4 [1,2] grass, which is commercially grown in tropical and subtropical production regions worldwide [3]. Sugarcane is an industrial crop and is one of the oldest cultivated plants in the world. Sugarcane accounts for more than 70% of the total sugar produced globally, mostly consumed as refined sugar. Recently, sugarcane has received attention as an energy crop [4]; in many countries, including Australia [5], and Brazil [6], bagasse (the fibrous part after juice extraction) is burnt by sugar mills to produce electricity to power the mills' operations. Among C4 plants, sugarcane is highly efficient in solar energy conversion and accumulates the highest biomass yield [7,8]. Bioethanol, a source of renewable energy that can help to meet the world's growing demand for energy while reducing greenhouse-gas emissions [9] at the same time, has awakened a wider interest in this crop. Sugarcane is also used for animal feed (green leaves and top portion), alcoholic beverages, and as a fertilizer (trash) in crop production across the globe [10].

Sugarcane is the world's most produced crop (total production) and ranks among the ten most widely grown crops worldwide. The total global production of sugarcane in 2016–2017 was 1.9 billion tons, and it was grown in approximately 100 countries, covering an area of ~26 million hectares [11]. The largest sugarcane producer is Brazil (40% of the total production), followed by India, China, and Thailand. Other major sugarcane producing countries are Mexico, Pakistan, the United States, Colombia, Australia, Cuba, and the Philippines [11].

In the past 50 years, world sugarcane production increased almost three-fold, mainly because of the rising demand for sugar and ethanol. Production gains are partly attributed to the genetic improvement of sugarcane varieties that are adapted to particular target environments. Concurrently, improvements in management techniques, fertilization, and irrigation have all played a role in increasing sugarcane productivity [12]. The main driver to the total increase in production is the dramatic increase in cultivated land area. The cultivation area in Brazil, India, China, and Thailand has increased by nearly 500, 94, 237, and 286%, respectively, from 1973 to 2013. Increased yield per hectare in the same locations was only moderate (60%, 38%, 59%, and 11% respectively) at the same time [13].

Over the past 60 years, total sugarcane production has increased across the globe, but the rate of sugar yield improvement appears to have plateaued (Figure 1). Between 1970 to 1989, sugar yield plateaued in Australia, then increased by 12 tons/ha-yr at the end of 1995. The total cane yield in Australia nearly reached 95 tons/ha at the end of 1999. This increment can be attributed mainly to the genetic improvements of cultivars [3].

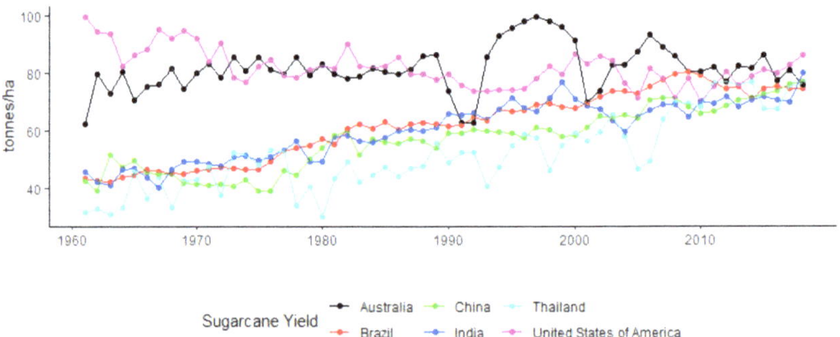

Figure 1. Sugarcane yields in Brazil, India, China, Thailand, Australia, and the United States of America from 1960–2018.

There were no substantial gains in cane yield in the top sugarcane producing countries in the last two decades (Figure 1). Several countries have been facing yield plateaus, and there are several different potential explanations for these production trends. Pests and diseases, the production potential of the land being used, and climatic conditions are all likely to contribute substantially to the observed reduced yield increases.

The occurrence of new diseases and pests could cause increased losses. Continuing monoculture cropping can build up soil pathogens and nematode pressure, which might be partly responsible for a lack of sugarcane yield increase worldwide [14]. Additionally, diseases have been observed to substantially impact sugarcane yield . Ratoon stunting disease (RSD) is one of the most economically important sugarcane diseases worldwide. Reported yield losses due to RSD are 15–50% in irrigated and rainfed trials in South Africa [15] and 29% in Fiji [16]. RSD primarily affects yield, while key quality characteristics like sugar content are only minimally affected. In 2000, a relatively new pathogenic race of orange rust destroyed the high yielding sugarcane variety Q124, which accounted for approximately 45% of the crop in Australia. It amounted to a loss of nearly $200M for the Australian sugarcane industry [17]. In this case, a considerable reduction in sugar content was also reported. Another major disease that affects sugarcane crops worldwide is sugarcane smut, which can have devastating impacts on yield. The estimated average potential losses due to sugarcane smut in the Herbert region in Australia was 26% [18]. Nearly 70% of the Australian sugarcane cultivars were susceptible to smut before 1998 [19]; sugarcane smut resistance is now one of the primary breeding objectives for Australian sugarcane. There was a significant increase in smut-resistance crosses in Australian breeding programs from 0.4 to 52% between 2000 to 2007 [20], nearly doubling the smut-resistant clones by the end of 2011 [21]. Many successful smut-resistant varieties are now bred in many sugarcane breeding programs worldwide.

The expansion of the sugarcane industry onto marginal land could be another possible reason that yield per hectare has plateaued. Regions that require significantly more inputs, such as irrigation, fertilizers, and high transportation costs, are now used to grow sugarcane. The adoption of mechanical harvesting in some countries and long-term degradation of soil fertility associated with cultivation might also have limiting effects on the productivity trends [3].

Extreme weather can also have significant impacts on sugarcane yield. In Fiji, favorable growing conditions in 1994 resulted in 5.2M tons of national production. In subsequent years, sugarcane production was reported to be reduced by half in the same region because of extreme climatic fluctuation [22]. Similar observations were reported in China in 2003–2004, where drought decreased average cane yields by around 18% [23]. However, as there is no evidence that these negative impacts have increased over the periods of low productivity improvement, the impact of environment-management is not sufficient to explain the continuous slow rate of improvement in sugarcane yield over time.

In addition to improving management practices, the genetic improvement of modern cultivars is a main avenue to enhance productivity in sugarcane. To overcome static yield trends, intensified breeding efforts are needed to develop new, improved varieties. However, there are several factors inherent to sugarcane biology, management and breeding practices that impose difficulties on the realization of genetic improvement through breeding.

2. Development of Modern Cultivars and Inherent Challenges

Sugarcane (*S. officinarum*) has been cultivated in India, China, and Papua New Guinea for sugar production for 10,000 years. The first sugarcane breeding programs were established in Java and Barbados in the late 1800s after the discovery that sugarcane can produce viable seeds [3,24]. Until the first quarter of the 20th century, sugarcane varieties used in industrial-scale production of sugar were *S. officinarum* clones, also known as a noble cane, originating from New Guinea. It is reported that *S. officinarum* species were domesticated from wild *S. robustum* in New Guinea around 8,000 years ago [3]. Unlike *S. officinarum* Indian cane (*S. barberi*) and Chinese cane (*S. sinense*) are derived from interspecific hybridization between octoploid *S. officinarum* (2n = 80) and *S. spontaneum* (2n = 40–128) with varying ploidy levels [25].

Historically, *S. officinarum* species had good commercial milling characteristics such as high sugar content, low impurity levels, and low fiber. However, this species lacked vigor, ratooning performance, and was susceptible to several diseases [24]. *S. spontaneum* is a genetically diverse wild species that is characterized by a lower commercial merit than *S. officinarum*, because of thin stalks and low sucrose content. Conversely, compared to *S. officinarum*, *S. spontaneum* has an increased ratooning capacity, a higher fiber level, and an overall superior adaptive capacity, characterized by an ability to perform better in unfavorable environmental conditions, such as drought, flood, or high salinity [26].

The genetic improvement of sugarcane can be divided in three main phases [27]. The first phase began with screening and intercrossing among *S. officinarum* clones. The major limitation of this approach was that noble canes, and hence progeny created from intercrossing, were susceptible to biotic and abiotic stresses. This led to the second phase, which involved the development of cultivars derived from interspecific hybridization between *S. officinarum* and *S. spontaneum*, and continuous backcrossing efforts with *S. officinarum* clones. Interspecific hybrids between *S. officinarum* and *S. spontaneum* were able to combine a high cane yield potential with increased disease resistance and improved ratooning ability.

An example is the cultivar "POJ2878," which led to a significant increment in productivity [28]. Many commercial cultivars used around the world today can be traced back to this cultivar [29]. In the third phase of modern genetic improvement of sugarcane, interspecific hybrids that were created in phase two were intensively exploited, through intercrossing among selected hybrids and recurrent selection among newly created progeny. This practice initially led to significant increases in genetic gain and still represents the main breeding strategy today.

Improved sugarcane varieties have played a pivotal role in the development of sugar industries throughout the world. There was a significant change in Hawaii's sugar yield from 1915–2003 by continuously updating sugarcane varieties. At the end of 2003, annual sugarcane production was around 15t/ha in Hawaii. Approximately 50% of Hawaii's sugar yield gains resulted from the genetic improvement of varieties [30]. The sugar yield in Colombia increased from 5t sugar/ha-year at the end of the 1950s to 8 t sugar/ha-year in the 1970s and recorded 12 t/ha-year at the end of 2000 [31]. Sugarcane production in Brazil and India increased throughout the same period and reached nearly 64–70 t/ha by the end of 2000. Results of a long-term study investigating productivity trends from 1968 to 2000 in Florida demonstrated significant improvements in cane and sucrose yield across the plant cane in first and second-ratoon crops. The positive impacts of genetic gain increases on Florida's sugarcane industry played a significant role in the country's economy across those years [32].

However, the observed increases in sucrose yield for the most recent varieties in Florida (unpublished data from a 2011 study) were associated with an increase in total cane yield, rather than improvements in CCS [13]. Similar results were reported from three small scale studies conducted in Australia where no significant differences for CCS could be found between older and new varieties [33]. Thus, genetic gain for key traits, particularly sucrose content and, to some extent, cane yield, has been stagnating in the past ten years in some countries. Conversely, genetic improvements for disease resistance achieved through traditional breeding programs have been very substantial.

On a global scale, most modern sugarcane cultivars are the product of only a few interspecific crosses between approximately 15–20 genotypes that can be traced back to ancestral sugarcane clones developed in Java and India [27]. In modern breeding programs, relatively old genetic material (>50 years old) is still widely used in crossing designs to create new varieties [34]. Thus, there have been few opportunities (~7–9 breeding generations) for chromosome recombination from the original founders. One consequence of the foundation bottleneck is strong genome-wide linkage disequilibrium (LD) patterns observed in elite germplasm [35] and a narrow genetic base in modern sugarcane germplasm [36].

Commercial hybrids originate from the initial hybrid (*S. officinarum* × *S. spontaneum*), which would have 2n transmission from the *S. officinarum* parent and n transmission from the *S. spontaneum* [37,38]. The hybrid is then crossed back to other hybrids to recover the high sugar phenotype, which breaks down the hybrid into n + n transmission [38]. Because of the narrow genetic base of important traits, genetic diversity could be reintroduced in sugarcane by utilizing the potential of wild relatives that are considered reservoirs of potentially useful alleles for important economic traits that might have been lost during domestication and breeding. Such practices of continual introgression of wild material into commercial breeding programs are used intensively in some breeding programs, e.g., in Louisiana.

New commercial hybrid cultivars have a complicated chromosome set, ranging between 2n = 100–130; 80% of the chromosomes are of *S. officinarum* origin, 10–15% of the chromosomes are of *S. spontaneum* origin, and the rest of the chromosomes are a combination of the two species [39–43]. Eight to 14 homo(eo)logus copies of alleles at a given locus in the hybrid genome are reported in the literature [44,45]. While the haploid genome of sugarcane is estimated at 1 Gb, the total size of the sugarcane nuclear genome is approximately 10Gb [46,47], making it ten times larger than the closest related genome sequenced species, which is sorghum [48].

The extreme polyploid genome of interspecific hybrids possesses irregular genetic characteristics that are passed from both parental species, making it more complicated than that of its precursors [40]. This phenomenon contributes substantially to the high level of heterozygosity observed between sugarcane cultivars [49]. Because of the random sorting of chromosomes in each crossing, the number of chromosomes varies between genotypes. The complex genetic composition of modern hybrids which are referred to as poly-aneuploids also results in inherent polygenic control of important agronomic traits. This complex genetic structure potentially makes the selection procedure slower and more complicated than in other major crop species.

3. Identifying and Overcoming Bottlenecks in Breeding Programs Using the Breeder's Equation

The overarching objective of any breeding program is to create new germplasm with improved genetic merit. The rate at which this improvement is realized in a given timeframe is referred to as "genetic gain." Increasing genetic gain in crop breeding has been identified as one of the key steps towards meeting the increasing future demand for plant-based products. In the context of the breeder's equation (Equation (1)), genetic gain (ΔG) [50] can be understood as the improvement in the mean genetic value of a trait of interest for a population over a defined time period, e.g., one breeding cycle. Following this equation, the expected rate of genetic gain that can be achieved in a given breeding cycle can be calculated as

$$\Delta G = \frac{i\,h^2\,\sigma_P}{L} \tag{1}$$

where ΔG is the rate of genetic gain, i represents the selection intensity, h^2 represents the narrow-sense heritability of the desired trait, σ_P is the observed phenotypic variation, and L is the total interval in time units to complete one cycle of selection. Selection intensity is related to the proportion of selected individuals as parents of the next breeding cycle, usually expressed in standard deviation units from the mean (assuming that most phenotypes are normally distributed). The narrow-sense heritability is a measure of the heritable (additive) genetic variation in the population relative to the total observed phenotypic variation (σ_P) in the population. This equation shows that increasing the heritability, selection intensity, and phenotypic variation increase the rate of genetic gain, while decreasing the breeding cycle length has the same effect. The breeder's equation (Equation (1)) provides a useful quantitative framework for the identification of potential bottlenecks in breeding programs that limit the rate of genetic gain, and for developing strategies to address these bottlenecks to accelerate gain in optimized breeding schemes.

4. Practices and Limitations of Conventional Sugarcane Breeding

A typical sugarcane breeding scheme (Figure 2) follows four key steps which include (i) the generation of a large progeny population generated from targeted crosses, (ii) the evaluation of those progeny in different phenotyping stages, (iii) the selection of clones with superior characteristics, and (iv) the recombination of selected clones to initiate the next breeding cycle [51].

To initiate a breeding cycle, parental clones are selected from a source population that has been characterized for key major commercial and agronomical traits, such as tones cane per hectare (TCH), sugar content measured as commercial cane sugar (CCS), fiber content and resistance to important diseases in the target environments. These parental clones may be sourced from intra- or inter-national breeding programs. The selected clones are crossed to create a large number of seedlings that are tested as families and then clonally propagated and selected throughout the remaining phenotypic testing phases [51] (Figure 2). Finally, the clones used as crossing parents are assessed based on the performance of their progeny. If the progeny performs relatively well, the corresponding crossing parents and the cross will be identified as "proven parents" and "proven cross," respectively, and may be used repeatedly to produce thousands of offspring that undergo selection.

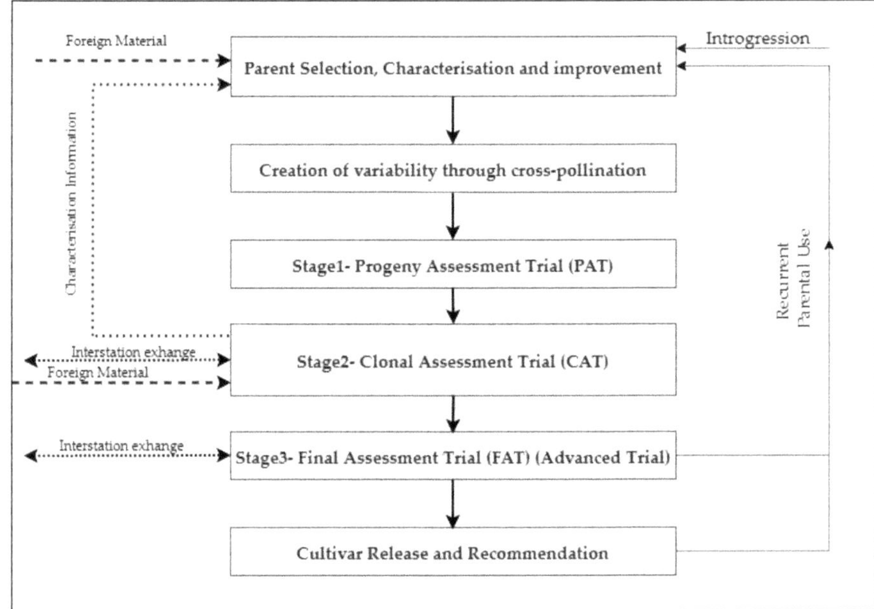

Figure 2. The conventional sugarcane breeding scheme. Adapted from [51]. A large number of progenies are generated from the targeted crosses, which are evaluated in three different stages of selection, best clones are screened in the advanced stage of selection which is intermated to initiate the new breeding cycle. PAT = progeny assessment trial; CAT = clonal assessment trial; FAT = final assessment trial.

For clonal improvement, the selection procedure depends on the crop-cycle length and number of ratoon cycles, which typically varies among breeding stations. In Australia, the conventional breeding program involves three stages that include seven years of selection, and three years of propagation [52]. At each stage, the top 5–10% of clones are progressed to the next stage of selection, and finally, a cultivar is released. In the process of releasing a variety, breeders test selected candidates in replicated multi-location trials to screen elite clones with high agronomic performances across a range of different environments. Because of the biology and management of sugarcane and the extensive phenotypic testing system, it can take more than ten years to complete a breeding cycle and even longer to commercially release a new cultivar.

In the context of increasing genetic gain using the breeder's equation framework (Equation (1)), it is widely reported that significant favorable genetic variation exists among the clones of *Saccharum* species. Since it is the additive genetic variance that selection acts on, this could potentially improve genetic gain in sugarcane. Increasing the selection intensity could also potentially lead to an improvement in genetic gain. However, simulation studies have shown that increasing the selection intensity can diminish the long-term selection response, decrease genetic diversity over time and increase inbreeding [53,54].

The relatively long breeding cycle (>10 years) in sugarcane, which consists of several resource-consuming selection stages, is one of the main constraints for improving genetic gain. Likewise, as important is the limited, narrow-sense heritability of economically important traits. Most commercial traits in sugarcane are likely significantly affected by non-additive gene-action [34,55]. This could explain the low levels of narrow-sense heritability estimates in empirical studies, e.g., for h^2_{TCH} = 0.13 [56], 0.003–0.032 [57], 0.03–0.40 [58]; h^2_{Brix} 0.034–0.101 [57], 0.21–0.67 [58] or fiber content h^2 = 0.629–0.813 [59].

Along with long breeding cycles and large proportions of non-additive genetic variance for key traits, breeders typically deal with other practical problems, for instance, the synchronization of flowering, which has been the focus of many studies [60–62]. Other factors, such as insufficient replication of new breeding materials in early generation trials, experimental errors, competition between adjacent plots [63,64], and G × E interaction effects [52,65], can also negatively affect the selection response for the target traits.

The most important traits in sugarcane are under quantitative genetic control, meaning that they are controlled by multiple genes along with environmental effects [66]. G × E interaction is an important source of phenotypic variation in sugarcane, especially for CCS and fiber content. G × E is difficult to account for in a breeding program, and therefore G × E interactions can reduce the rate of genetic improvement in sugarcane [52,67,68]. In the breeder's equation framework, this is due to the negative impact of G × E on the trait heritability. The genetic variance of a given trait can be biased by the variation caused by G × E interaction effects. Improved estimates of genetic variance can be obtained by partitioning the variation of G × E effects from the genetic variance [69].

To deal with G × E interaction, breeders typically test their breeding germplasm in multi-environment trials (MET), which ideally are a representative sample of the target production environment (also referred to as the target population of environments, TPE) and cover several locations and years. Breeders can significantly minimize the risk associated with fluctuating environmental conditions and improve the efficiency of their breeding program by understanding G × E interaction for their specific genotype-environment system. Several statistical methods have been developed specifically to explore and account for G × E interaction in plant breeding, essentially aiming to minimize its negative impact on the selection accuracy [70]. The development of methodologies and strategies that enable performance prediction under G × E interaction, especially for situations in which the aim is to predict the performance of novel genotypes in new (i.e., untested) environments, is a wide and active field of research [71].

Today, the vast majority of sugarcane breeding programs (outlined in Figure 2), which are based on phenotypic selection, are very cost- and time-consuming. Strategies that could enable the reduction of cycle length, as well as approaches that are more adequate for performance prediction and breeding value estimation, would be a major step forward for improving sugarcane breeding programs in the future.

5. Genomic Selection: A Powerful New Breeding Tool

Genomic selection (GS) is a relatively new breeding method in which individuals are selected based on their predicted breeding values that are calculated from genome-wide DNA marker profiles [72]. Decreasing costs of DNA marker screening methods such as high-density SNP arrays and genotyping by sequencing (GBS) approaches, and the development of statistical methods that can accurately predict marker effects are the main reasons why GS has increasingly been implemented in modern animal and plant breeding programs [73,74]. Two main avenues by which GS can accelerate the rate of genetic gain is by improving the accuracy at which individuals are selected and by reducing the length of the breeding cycle. However, the incorporation of GS into a breeding program is not a trivial task. It highly depends on several factors, such as the mating type, the genetic architecture and heritability of the target traits, the availability of genotyping platforms, and the total financial budget of the program to build large reference populations that are necessary to accurately estimate the typically small effects of DNA markers that are associated with the underlying causal mutations that affect the traits [53,75].

Conceptually, GS involves two main steps (Figure 3). The first step is to develop a prediction equation based on a training population (TP) that consists of individuals for which both high-quality phenotypes and genome-wide DNA marker profiles have been obtained.

The fundamental requirement for GS to work is that quantitative trait loci (QTL, the actual mutations) that are affecting the expression of the target trait are in LD with the DNA markers that are used for genotyping [72,75]. If this requirement is met, trait effects for DNA markers can be estimated and used in the prediction equation. In the second step, these marker effects are used to calculate the genomic estimated breeding values (GEBVs) of selection candidates (prediction population; PP) for which only genome-wide marker data (but no phenotypic data) are available. Genotypes can then be ranked based on their GEBVs to support selection decisions in a breeding program.

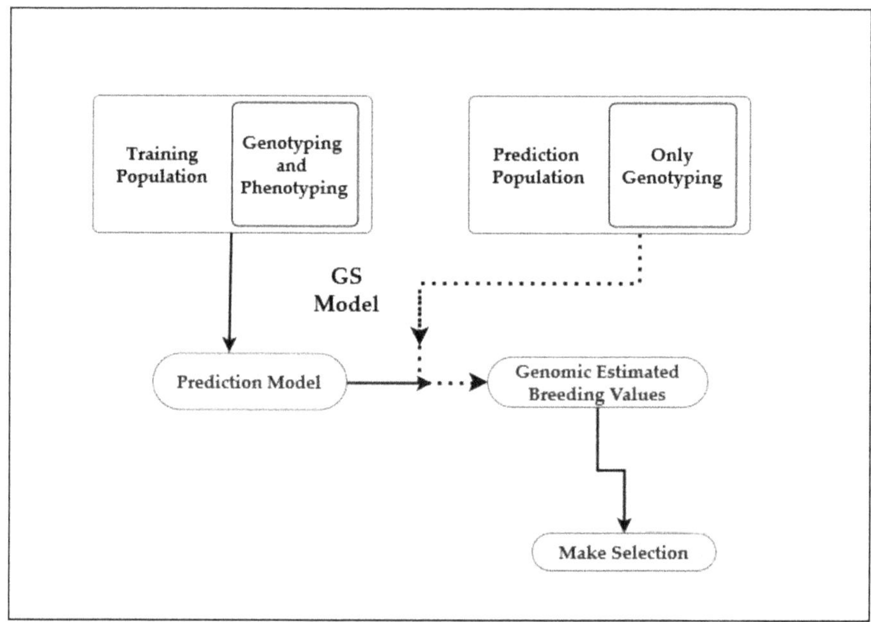

Figure 3. General overview of genomic selection (GS). A GS scheme starts with the training population (TP) that is used to estimate marker effects. These effects are used to calculate genomic estimated breeding values (GEBV) of clones in the prediction population (PP).

A number of statistical models and algorithms have been developed to deal with the problem that, in most situations, the number of DNA markers for which effects are to be estimated strongly exceeds the number of phenotypic observations, including parametric, Bayesian, and non-parametric methods. The most commonly used statistical methods RR-BLUP and GBLUP (which yield mathematically equivalent results) assume a normal distribution of SNP effects, while Bayesian approaches like BayesA, BayesB, BayesC(pi), and BayesR consider different variance distributions to allow for differences in marker effect sizes [76–78]. Kernel methods [79] utilize the distance (similarity) matrix, which is particularly useful for predicting non-additive effects. They also allow handling of complex multi-environment and/or multi-trait data and are, therefore becoming very popular in plant breeding [80].

A fundamental step for the implementation of GS is the development of the training population (TP). Numerous studies have demonstrated that in order to obtain high prediction accuracies, the TP has to be large and should include individuals with varying degrees of relationship [81–83]. Daetwyler et al. [84] reported an improvement in prediction accuracy of 50% by increasing the TP size from 500 to 2000. For wheat, Cericola et al. [85] observed an increment in prediction accuracy with an increase of the size of the TP, which included full-sibs, half-sibs, and less related lines from three continuous breeding cycles. This trend reached a plateau at around 700 breeding lines.

The expected prediction accuracy can be calculated as $r = \sqrt{\dfrac{N\,h^2}{N\,h^2 + M_e}}$ [86,87], in which r (the expected prediction accuracy) is affected by the size of the TP (N), the heritability of the trait (h^2), and the effective number of independent chromosomes segments in a given population (M_e) which is calculated as $2 \times N_e$ (effective population size) $\times$ L (the genome size in Morgan). To maximize GEBV accuracy, the TP should be related to the PP [88]. M_e can be estimated empirically by the mean LD (r^2) between all pairwise SNPs [89] or by using specific family structures [90].

To maintain a high prediction accuracy in GS-based breeding programs, the TP must be frequently updated with new phenotyped and genotyped accessions [91,92]. This is mainly due to the decrease in marker-QTL LD because of recombination events over time. For example, Auinger et al. [93] trained a prediction model for a rye breeding program by using multiple breeding cycles and demonstrated that prediction accuracies were significantly increased when the prediction model was constantly updated as the breeding program advanced.

Good quality phenotypic and genotypic data are the key factors to take full advantage of GS [73]. Because of inevitable constraints in operating budgets, breeders are always interested in finding the minimum number of markers needed to obtain to get useful GEBVs. The extent of LD (affected by N_e, and population structure) helps to determine the number of markers required for GS. High marker densities are desired for the prediction of far related individuals [53], because of reduced LD.

GS was implemented in animal breeding prior to its introduction to plant breeding. The implementation of GS in dairy cattle breeding programs have resulted in significant improvements compared to traditional phenotypic selection [94]. The reduction in total generation interval from 7 years to 1 year (young bulls are being ranked based on their GEBVs, and selected for artificial insemination) has almost doubled the rate of genetic gain. Furthermore, there has been a reduction in costs for progeny testing [94,95]. Interestingly, genetic gains were also reported for low-heritability traits such as disease resistance and fertility [73]. Consequently, GS has been implemented on a very large scale in other animal species such as beef cattle, pigs, sheep, and chicken [96,97].

In plant breeding, the potential of GS was first evaluated in corn (*Zea mays* L.) using simulations [98]. A range of simulation studies in different crop species such as wheat [92], barley [99], rice [100], and sorghum [101] have shown that implementing GS could result in a significant increase in genetic gain. However, only limited reports are available in crops on the realized genetic gain that were achieved as an outcome of implementing GS. One example is given by the drought-tolerant "AQUAmax" hybrid corn variety, which was created by integrating GS with enhanced phenotyping and crop growth modelling in a commercial maize breeding program [102]. Significantly higher yields were reported in the United States when growing "AQUAmax" maize hybrids under both drought and favorable conditions, with considerably improved yield stability underwater limitation [103].

6. Implementation of Genomic Selection in Sugarcane Breeding

Increasing the rate of genetic gain is a big challenge in sugarcane breeding, as implied by the static or slowly increasing yield trends in most countries. Several reasons for the observed yield plateaus have been proposed, such as a narrow genetic base of modern elite germplasm [36], highly complex genetic architectures for agronomically important quantitative traits for which non-additive gene action is likely playing a significant role, and very long breeding cycle lengths [34].

The use of molecular markers has become a standard practice in most important crop species. Traditionally, plant breeders have incorporated molecular markers in phenotypic selection for mono- or oligogenic traits to increase the efficiency of the breeding program. For instance, marker-assisted selection (MAS) has proven to be a practical approach for single gene introgression or pyramiding multiple genes in elite cultivars, to improve disease resistance or grain quality [104]. Despite the fact that a range of QTL mapping studies has been undertaken in sugarcane [105], the size and complexity of the sugarcane genome have limited DNA marker-based selection in this crop [44]. Generally, MAS has been largely ineffective for the improvement of highly quantitative traits because of several technical reasons that have been discussed extensively in the literature [106,107]. Polygenic traits are typically controlled by a huge number of QTL, each having infinitesimal small effects, or possibly with interactions among them as well as with environmental factors [108].

GS can be a promising tool for improving the rate of genetic gain for quantitative traits in sugarcane breeding. Since GS has not extensively been investigated in sugarcane and other highly polyploid crops, increased evaluation and validation efforts are needed to better understand the challenges associated with the implementation of the technology in breeding programs. A recent study investigated the potential use of GS in tetraploid potato and octoploid strawberry by the use of SNPs markers and partial sequence data, respectively. The authors concluded that the actual advantage of GS depends on the underlying genetic architecture of the trait [109]. For genetic improvement of quantitative traits in octoploid strawberry (e.g., yield and fruit quality), GS has been strongly recommended in practical breeding programs because of high prediction accuracies found in true validation trials [110].

Gouy et al. evaluated the potential of GS for sugarcane breeding in two different panels from a commercial breeding program in Reunion Island and Guadaloupe consisting of 167 clones each [111]. All 334 clones were genotyped with 1499 DArT markers and phenotyped for ten agronomically important traits. By comparing four genomic prediction models (Ridge Regression, Bayesian Lasso, Partial Least Square Regression, Reproducing Kernel Hilbert Space), prediction accuracies ranged from 0.11–0.62 within the panels and 0.13–0.55 between panels across the ten investigated traits which included morphological trait (stalk diameter, and millable stalk number), technological traits (bagasse content, brix), lignocellulosic traits (acid detergent fiber, invitro neutral detergent fiber digestibility of the bagasse, acid detergent lignin), and resistances to different diseases (yellow leaf disease, smut, and brown rust) [111]. These prediction accuracies seem promising, particularly when considering the relatively small size of the TP that was used in the study.

In another study, three different populations of clones from early and advanced selection stage of an established sugarcane breeding program were used to estimate the prediction accuracy of cane yield and sugar content. Different genomic prediction models (GBLUP, BayesA, BayesB, Bayesian LASSO, and RKHS) were compared with or without the use of pedigree information. The prediction accuracy for sugar content was highest in advanced stage trials while it was lower for cane yield. The prediction accuracies ranged from 0.25–0.45 in most data sets, which is promising and strongly supports the potential usefulness of GS for sugarcane breeding [112].

In sugarcane, modern germplasm can be traced back to only a small number of founder clones, which suggests that the effective population size N_e in elite germplasm is small. This is consistent with the high levels of LD reported in modern sugarcane breeding populations [35]. However, a considerable number of SNP markers still needs to be used to achieve accurate predictions due to the large size and complexity of the sugarcane genome.

Unlike major crops such as corn, wheat, or rice, high throughput genotyping is still relatively expensive in sugarcane (~AUD 95 per sample using the 50k Axiome SNP array). The cost associated with genotyping is still a major limiting factor for large scale genomic evaluation in commercial breeding programs. In addition to genotyping, high-throughput, and precision phenotyping, e.g., in multi-environment or managed trials, should be considered more seriously when GS is implemented because of potential negative effects of $G \times E$ interactions on genomic prediction accuracy [69]. Parameters that quantify critical environmental conditions could also be included in genomic prediction models to increase the heritability and hence the prediction accuracy for the target trait [113].

The use of advanced phenotyping methods might be helpful for improving the prediction accuracy in sugarcane. One main consideration is how to effectively use available information from modern high-throughput phenotyping in genomic prediction models [114]. An extensive review is given by Van Eeuwijk et al. [113] regarding a range of genotype-to-phenotype (G2P) modelling methods for the use of high-throughput phenotypes measured in field trials. The main idea is to collect data on secondary traits, e.g., time series traits such as dynamic measurement of canopy architecture or biomass, and include these data as covariates in genomic prediction models. Since this could allow to specifically target component traits that are important for performance under specific environmental conditions, approaches like this have the potential to better account for variation caused by environmental factors.

Therefore, the accuracy of estimated genetic merit of breeding germplasm in a given environmental context could ultimately be improved, which would directly translate into an increase in genetic gain.

Allelic and non-allelic (dominance and epistasis) interactions for target traits can create potential challenges for the implementation of GS in sugarcane breeding. The presence of dominance and epistasis genetic effects can change the average effects of allele substitution among populations that are targeted for selection in a breeding program because of the changes in allele frequencies that selection causes [115]. This results in a complicated situation when the ranks of the genotypes change as a consequence of changes in marker effect estimates [73]. Thus it is particularly necessary to update the training data set in the presence of strong epistatic effects [91]. This makes GS more expensive to implement for crop breeding.

The underlying assumption of most common genomic prediction approaches is that quantitative traits are determined by many additively acting genes. While approaches based on this assumption have been applied very successfully in plant and animal breeding, there is ample biological evidence that gene–gene interactions (epistasis) are important for agronomic traits. Because sugarcane cultivars are deployed as clones, all genetic effects could be utilized, and the accurate prediction of additive and non-additive genetic effects would be of great value in the future for predicting clonal performance and selecting parents for the next breeding cycle. Cheverud and Routman [116] proposed a new quantitative genetic parametrization for the analysis of physiological epistasis (i.e., on the genotype level) to understand the effect of gene-by-gene interaction on variance components that are important for quantitative genetics and breeding (additive, dominance and epistasis). They concluded that epistasis could be a source of increased additive genetic variance in populations that have undergone selection [117]. The use of extended statistical models that consider non-additive effects could be beneficial to derive precise marker effects and, ultimately, high prediction accuracies in crop breeding [118].

One challenge in polyploid species is to correctly distinguish between different types of heterozygotes. In polyploidy species, pseudo-diploid models are commonly used to account for heterozygosity. Polyploidy can create phenotypic variation through allele dosage. For instance, significant phenotypic differences in fruit size in tomato and plant architecture in corn were associated with allele dosage [119]. Therefore, the inclusion of allele dosage information has become a matter of high interest for genetic studies in polyploidy species. The explicit consideration of allele dosage in genomic prediction models might improve the prediction accuracy by providing a more realistic representation of genotypic class effects. For potato, an autotetraploid species, Endelman et al. showed significantly higher prediction accuracies by including digenic effects as well as accounting for allelic dosage using data from a SNP array [120]. Conclusively, the adequate treatment of non-additive effects and allele dosage in GS models could be very beneficial for sugarcane.

7. Recurrent Genomic Selection and Reciprocal Recurrent Genomic Selection: Two Strategies for the Incorporation of Genomic Selection in Sugarcane Breeding

Regarding the implementation of GS in sugarcane breeding, a key question is how to incorporate the technology into an existing breeding program. The first critical step in any breeding program is to create new genetic variation. In conventional sugarcane breeding, a large number of seedlings is created through targeted crossing, followed by several selection stages that aim to determine the relative genetic merit of the new germplasm in designed field trials. From the perspective of increasing genetic gain, a key bottleneck with this conventional approach is that alleles are only recombined in the crossing stage at the beginning of the breeding cycle. This could potentially be overcome by a breeding strategy called recurrent genomic selection (RGS) (Figure 4) which aims to rapidly improve the genetic merit of a population of heterozygous genotypes through rapid, recurrent selection and crossing of elite germplasm, and to simultaneously channel selected clones into advanced testing stages that ultimately develop commercial products.

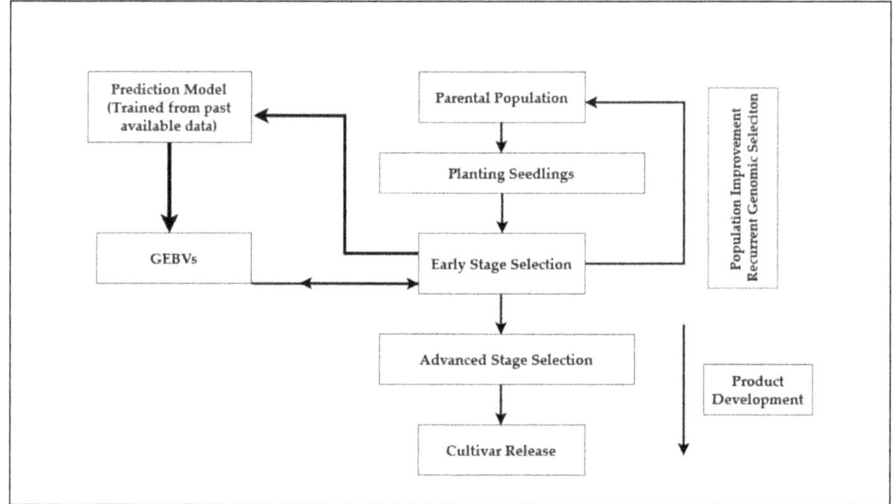

Figure 4. Flow diagram of a recurrent genomic selection breeding program for sugarcane. Conceptually, this can be divided into a population improvement component that uses recurrent genomic selection and a product development component in which clones with high GEBV enter advanced selection stages for variety development. The genomic prediction model is trained using data from previous trials. GEBV = genomic estimated breeding value.

Heffner et al. [53] first proposed the idea to separate population improvement from line development in a genomics-assisted plant breeding program. Later, Gaynor et al. [121] investigated RGS for a line breeding program using simulations by splitting the breeding program into a population improvement component and a product development component (cultivar release). They showed that a RGS-based program could generate up to 2.5 times more genetic than a conventional phenotypic selection scheme, and up to 1.5 times more genetic gain than the best-performing standard GS strategy in which GS is used to improve selection within the breeding cycle. A key role of phenotyping in a genomics-assisted breeding program is to (re)estimate marker effects. Changes in allele frequencies in populations under selection and epistatic gene-action result in changes in marker effect estimates that might reduce selection accuracy and hence realized genetic gains from GS-based breeding strategies [73]. Thus, there is a need for constant updating of the prediction model in each selection cycle, especially in an RGS system in which generation turnover and hence the number of recombination events is accelerated.

RGS breeding schemes that prioritize parents with high general combining ability typically capture and improve additive genetic effects in each generation cycle. The use of RGS for inter-population improvement may boost long-term selection gain in hybrid sugarcane breeding.

To maximize the response in crossbred populations, reciprocal recurrent selection (RRS) was proposed by Comstock et al. [122]. The RRS breeding scheme aims to simultaneously improve two genetically diverse, purebred populations that are used for targeted crossbreeding, ultimately aiming to maximally explore both general and specific combining ability. Individuals from purebred populations are selected based on their crossbred progeny performance. For instance, RRS was very successfully applied to improve general combining ability and specific combining ability for root yield and sucrose-content in sugarbeet [123], and grain yield and prolificacy in maize [124]. The main practical drawback of RRS is that generation intervals need to increase substantially, which can lead to a reduction in the overall genetic response to selection. An increase in generation intervals is necessary for RRS because selection decisions are made based on the performance of the crossbred progeny [122]. In the RRS scheme, GS can be used to predict crossbred performance and prioritize certain combinations of accessions from the distinct purebred pools. This practice is widely used in modern maize breeding [125].

In oil palm, Cros et al. [126] concluded that reciprocal recurrent genomic selection (RRGS) could increase annual gains by reducing the breeding cycle from 20 to six years compared to conventional RRS. Hence, RRGS seems to be a promising method to achieve long-term genetic gain under situations where traits are affected by heterosis, and when the breeding cycle is very long, as in the oil palm example. Rembe et al. [127] suggested that using RRGS-based breeding strategies that integrate product development and population improvement can increase long-term genetic gain in hybrid wheat breeding. Similar trends could be achieved in sugarcane.

Considering the importance of specific along with general combining ability effects in determining the performance of crosses, implementation of a modified version of RRGS, as shown in Figure 5, might improve long-term genetic gain in hybrid sugarcane breeding. Such a breeding scheme could begin with developing a genomic prediction model by using a reference population comprising a large number of progeny generated from a proven cross, say A × B where parent A and parent B are unrelated. One of the parents (e.g., parent B) and its derived self-progeny would be selected based on the predicted breeding value using a previously developed genomic prediction equation. If selfing is not feasible, very closely related clones (e.g., from one family) could be used instead. Several self-clones derived from parent B with high predicted breeding value would then be crossed with the opposite parent (Parent A). Potentially, the new crosses from the selected self-clones are better than the original high-value cross because of the improved genetic merit of the B-derived clones. The selected self-clones could also be crossed together and undergo further ongoing improvement cycles via rapid RGS. A similar breeding system could be initiated with a small number (2 or 3) parents on one, or both A and B sides (rather than single parents as in Figure 5), and progeny derived from crossing parents on one side would be selected for high predicted breeding values before crossing them with the opposite side. Extending the theory from Cheverud and Routman (1996) to a situation in which a quantitative trait is controlled by many epistatic QTL, in a modified RRGS breeding scheme, the QTL alleles in the opposite heterotic group could be fixed (remain unchanged). This could result in a genetic model with increased additive genetic variance and reduced statistical epistasis. This could contribute to an increase in predictability, leading to improved selection efficiency and higher genetic gain.

The proposed GS-based breeding schemes can be advantageous when the desired alleles for the traits of interest are available in the breeding germplasm. However, it could be the case that genetic variation for the trait of interest is limited in the primary gene-pool. In that situation, genetic variation in cultivated hybrid pools could be replenished by introgressing novel alleles from wild gene pools. This approach is time-consuming and cumbersome when a trait is affected by a large number of small-effect QTL. A well-designed pre-breeding program in which landraces and wild materials are exploited could be promising in maintaining and managing genetic diversity and long-term genetic gain in breeding programs. Pre-breeding programs could significantly benefit from GS approaches because they could help to prioritize accessions and track introgressions on the molecular level [128]. Incorporating GS without specific knowledge of the target QTL into a gene introgression program in fish was useful in preserving QTL. It sped up the process of introgression of a gene while increasing genetic gain compared to the classical selection, especially for disease resistance [129]. Integration of GS with genome-wide association studies (GWAS) can prevent the loss of target genes and sustain increased genetic gain through an appropriate capture of large- and small- effect QTL underlying a trait of interest [130].

One main drawback of genomic selection is that it can increase the rate of inbreeding per generation. However, Daetwlyer et al. [131] suggested that Mendelian sampling variation can be estimated more accurately using DNA markers, compared to traditional BLUP, and GS could reduce the probability of selecting siblings. Consequently, the inbreeding rate per generation can be reduced when DNA markers are used in the selection process. However, several simulation studies have shown that selection that is purely based on GEBVs can lead to a loss of genetic variance and hence an increase in the rate of inbreeding [75,126].

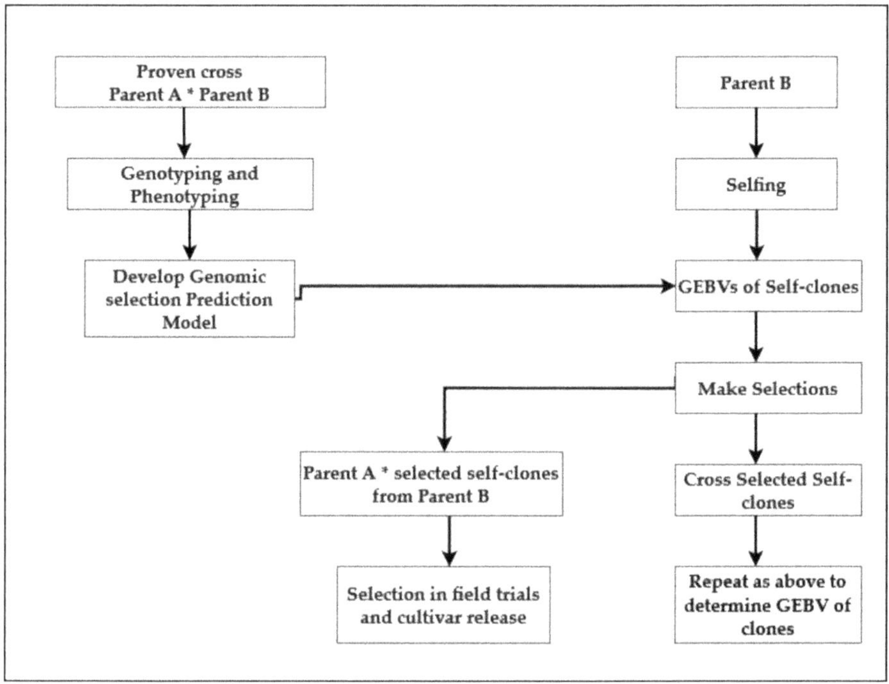

Figure 5. Flow diagram of a modified reciprocal recurrent genomic selection breeding scheme for sugarcane. The prediction model is trained by generating hundreds of offsprings from a proven cross of unrelated parents that are known to combine well. Either one or both clones in the cross are selfed, and offspring are selected based on their genomic estimated breeding values. If selfing is not feasible, closely related clones (e.g., from the same family) can be used instead. The selfed offspring is crossed with the opposite parent. GEBV = genomic estimated breeding value.

To avoid inbreeding depression in parental populations, the maintenance of genetic variation is necessary. Increasing the number of selected individuals could slow down the inbreeding rate, but at risk of a reduction in selection response [132]. Many modified selection criteria have been proposed to allow balancing genetic gain and maintaining genetic diversity while applying GS [133–137]. The main idea behind these selection criteria is to determine the exact contribution of an individual to the following generation based on its genetic merit and its genetic relationship with other individuals. Expanding on that principle, Toro and Varona [138] highlighted the potential of mate-allocation within a population. They used genomic prediction models, including dominance effects, to predict the performance of offspring generated through mating pairs of individuals. This was followed by an optimization procedure in which a set of mate pairs that can maximize performance in the subsequent generation was selected. In this example, selection and mating were simultaneously performed for improving the management of inbreeding. The advantage of an adequate mate allocation strategy is particularly relevant for improving complex traits with a high amount of non-additive genetic variance [118].

There are only a few studies that have investigated GS for sugarcane, and the empirical evaluation of different implementation strategies is impractical. Breeding simulations are an elegant way to assess the potential impacts that GS can have on sugarcane breeding efficiency because they require only a few physical resources. Furthermore, simulations can accommodate different genetic models with varying numbers of genes/alleles, dominance, epistatic gene effects, and also handle genotype-environment interaction effects [139]. A breeding program typically operates on fixed budgets, making the optimal allocation of resources very critical for breeders. Simulations allow one to investigate and compare breeding methods in terms of genetic gain and cost-effectiveness. Extensive simulation studies are needed in sugarcane to identify the best potential GS-based breeding scheme designs, e.g., RGS or RRGS, as discussed above, that can generate the highest rate of genetic gain per unit cost and time. Empirical validation experiments are then critical to test the most promising strategy in a practical breeding context. Thus, increased simulation efforts could provide valuable information and decision support for the design of empirical validation experiments, and ultimately for the efficient implementation of GS in practical sugarcane breeding.

While GS has the potential to tackle fundamental challenges associated with improving important traits in sugarcane, increased research efforts are needed to enable the implementation of the technology. The RGS or RRGS breeding schemes proposed in this paper hold the potential to increase long-term genetic gain for complex quantitative traits in sugarcane, but further investigations are needed.

Author Contributions: S.Y., K.P.V.-F., B.J.H., and P.J. jointly conceived the review. S.Y. conducted the literature review. S.Y. and K.P.V.-F. wrote the manuscript. B.J.H., P.J., X.W., E.M.R., K.A., E.D., and F.A. contributed critical input to different sections. All authors reviewed and edited the manuscript. All authors have read and agreed to the published version of the manuscript.

Funding: This research received was funded by Sugar Research Australia (project number: 2017/02).

Conflicts of Interest: The authors declare no conflict of interest.

References

1. Hatch, M.D. C4 photosynthesis: Discovery and resolution. In *Discoveries in Photosynthesis*; Govindjee, B.J.T., Gest, H., Allen, J.F., Eds.; Springer Netherlands: Dordrecht, The Netherlands, 2005; Volume 20, pp. 875–880. ISBN 978-1-4020-3324-7.

2. Hatch, M.D.; Slack, C.R.; Johnson, H.S. Further studies on a new pathway of photosynthetic carbon dioxide fixation in sugar-cane and its occurrence in other plant species. *Biochem. J.* **1967**, *102*, 417–422. [CrossRef] [PubMed]

3. Ming, R.; Moore, P.H.; Wu, K.K.; D'Hont, A.; Glaszmann, J.C.; Tew, T.L.; Mirkov, T.E.; da Silva, J.; Jifon, J.; Rai, M.; et al. *Plant Breeding Reviews*; Wiley Blackwell: Hoboken, NJ, USA, 2010; Volume 27, pp. 15–118. ISBN 978-0-4706-5034-9.

4. Souza, A.; Grandis, A.; Leite, D.; Buckeridge, M. Sugarcane as a bioenergy source: History, performance, and perspectives for second-generation bioethanol. *Bioenergy Res.* **2014**, *7*, 24–35. [CrossRef]

5. Mackintosh, D. Sugar Milling. Available online: http://hdl.handle.net/11079/15541 (accessed on 10 January 2020).

6. Cheavegatti-Gianotto, A.; Abreu, H.; Arruda, P.; Bespalhok Filho, J.; Burnquist, W.; Creste, S.; Ciero, L.; Ferro, J.; Oliveira Figueira, A.; Sousa Filgueiras, T.; et al. Sugarcane (*Saccharum X officinarum*): A reference study for the regulation of genetically modified cultivars in Brazil. *Trop. Plant Biol.* **2011**, *4*, 62–89. [CrossRef] [PubMed]

7. Henry, R.J. Evaluation of plant biomass resources available for replacement of fossil oil. *Plant Biotechnol. J.* **2010**, *8*, 288–293. [CrossRef] [PubMed]

8. Byrt, C.S.; Grof, C.; Furbank, R. C4 Plants as biofuel feedstocks: optimising biomass production and feedstock quality from a lignocellulosic perspective. *J. Integr. Plant Biol.* **2011**, *53*, 120–135. [CrossRef] [PubMed]

9. Goldemberg, J.; Coelho, S.T.; Guardabassi, P. The sustainability of ethanol production from sugarcane. *Energy Policy* **2008**, *36*, 2086–2097. [CrossRef]

10. Moore, P.H.; Paterson, A.H.; Tew, T. Sugarcane: The crop, the plant, and domestication. In *Sugarcane: Physiology, Biochemistry, and Functional Biology*; Moore, P.H., Botha, F.C., Eds.; John Wiley & Sons, Inc: Hoboken, NJ, USA, 2013; pp. 1–17. ISBN 978-1-1187-7128-0.

11. FAOSTAT. Food and Agricultural Organisation : Sugarcane Production Countries. 2018. Available online: http://www.fao.org/faostat (accessed on 10 January 2020).

12. De Morais, L.K.; de Aguiar, M.S.; de Albuquerque e Silva, P.; Câmara, T.M.M.; Cursi, D.E.; Júnior, A.R.F.; Chapola, R.G.; Carneiro, M.S.; Bespalhok Filho, J.C. Breeding of sugarcane. In *Industrial Crops: Breeding for BioEnergy and Bioproducts*; Cruz, V.M.V., Dierig, D.A., Eds.; Springer: New York, NY, USA, 2015; pp. 29–42. ISBN 978-1-4939-1447-0. [CrossRef]

13. Zhao, D.; Yang-Rui, L. Climate change and sugarcane production: Potential impact and mitigation strategies. *Int. J. Agron.* **2015**, *2015*, 1–10. [CrossRef]

14. Stirling, G.; Blair, B.; Pattemore, J.; Garside, A.; Bell, M. Changes in nematode populations on sugarcane following fallow, fumigation and crop rotation, and implications for the role of nematodes in yield decline. *Australas. Plant Pathol.* **2001**, *30*, 323–335. [CrossRef]

15. Bailey, R.; Bechet, G. The effect of ratoon stunting disease on the yield of some South African sugarcane varieties under irrigated and rainfed conditions. In Proceedings of the South African Sugar Technologists' Association: Annual conference, Durban, South Africa, 5–8 June 1995; pp. 74–78.

16. Johnson, S.S.; Tyagi, A.P. Effect of ratoon stunting disease (RSD) on sugarcane yield in Fiji. *South Pac. J. Nat. Appl. Sci.* **2010**, *28*, 69–73. [CrossRef]

17. Magarey, R.; Croft, B.; Willcox, T. An epidemic of orange rust on sugarcane in Australia. In Proceedings of the International Society of Sugar Cane Technologists: XXIV Congress, Brisbane, Australia, 17–21 September 2001; pp. 410–416.

18. Magarey, R.; Bull, J.; Sheahan, T.; Denney, D.; Bruce, R. Yield losses caused by sugarcane smut in several crops in Queensland. In Proceedings of the Australian Society of Sugar Cane Technologists: 32nd Annual Conference, Bundaberg, Australia, 11–14 May 2010; pp. 347–354.

19. Sundar, A.R.; Barnabas, E.L.; Malathi, P.; Viswanathan, R.; Sundar, A.; Barnabas, E.J.B. A mini-review on smut disease of sugarcane caused by *Sporisorium Scitamineum*. In *Botany*; Mworia, J.K., Ed.; InTech: Rijeka, Croatia, 2012; Volume 16, pp. 107–128. ISBN 978-953-51-0355-4.

20. Croft, B.J.; Berding, N.; Cox, M.C.; Bhuiyan, S. Breeding smut-resistant sugarcane varieties in Australia: Progress and future directions. In Proceedings of the Australian Society of Sugar Cane Technologists: XXX Annual Conference, Townsville, Australia, 29 April–2 May 2008; pp. 125–134.

21. Bhuiyan, S.; Croft, B.; Cox, M. Breeding for sugarcane smut resistance in Australia and industry response: 2006–2011. In Proceedings of the Australian Society of Sugar Cane Technologists : XXXV Annual Confernce, Townsville, QLD, Australia, 16–18 April 2013; pp. 1–9.

22. Gawander, J. The impact of climate change on sugar cane production in Fiji. *Wmo Bull.* **2007**, *56*, 34–39.

23. Li, Y.-R.; Wei, Y.-A. Sugar industry in China: R & D and policy initiatives to meet sugar and biofuel demand of future. *Sugar Tech* **2006**, *8*, 203–216. [CrossRef]

24. Stevenson, G.C. *Genetics and Breeding of Sugar Cane*; Longmans: London, UK, 1965.

25. D'Hont, A.; Paulet, F.; Glaszmann, J. Oligoclonal interspecific origin of 'North Indian' and 'Chinese' sugarcanes. *Chromosome Res.* **2002**, *10*, 253–262. [CrossRef] [PubMed]

26. Mohan Naidu, K.; Sreenivasan, T. Conservation of sugarcane germoplasm. In Proceedings of the Copersucar International Sugarcane Breeding Workshop, São Paulo, Brazil, May–June 1987; pp. 33–53.

27. Roach, B. Origin and improvement of the genetic base of sugarcane. In Proceedings of the Australian Society of Sugar Cane Technologists: Annual Conference, Tweed Heads, Australia, 2–5 May 1989; pp. 34–47.

28. Jeswiet, J. The development of selection and breeding of the sugar cane in Java. In Proceedings of the International Society of Sugarcane Technologists, Soerabaia, Indonesia, 7–19 June 1929; pp. 44–57.

29. Mangelsdorf, A.J. Sugar-Cane breeding methods. In Proceedings of the International Society of Sugar Cane Technologists: X Congress, Honolulu, HI, USA, 3–22 May 1959.

30. Hogarth, D.M. New varieties lift sugar production. *Prod. Rev.* **1976**, *66*, 21–22.

31. Cock, J.H. Sugarcane growth and development. *Int. Sugar J.* **2001**, *105*, 5–15.

32. Edme, S.J.; Miller, J.D.; Glaz, B.; Tai, P.Y.P.; Comstock, J.C. Genetic contribution to yield gains in the Florida sugarcane industry across 33 years. *Crop Sci.* **2005**, *45*, 92–97. [CrossRef]

33. Jackson, P.A. Breeding for improved sugar content in sugarcane. *Field Crop. Res.* **2005**, *92*, 277–290. [CrossRef]

34. Wei, X.; Jackson, P. Addressing slow rates of long-term genetic gain in sugarcane. In Proceedings of the International Society of Sugar Cane Technologists: XXIX Congress, Chiang Mai, Thailand, 5–9 December 2016; pp. 480–484.

35. Aitken, K.; Jackson, P.; McIntyre, C. Quantitative trait loci identified for sugar related traits in a sugarcane (Saccharum spp.) cultivar× Saccharum officinarum population. *Appl. Genet.* **2006**, *112*, 1306–1317. [CrossRef]

36. Raboin, L.-M.; Pauquet, J.; Butterfield, M.; D'Hont, A.; Glaszmann, J.-C. Analysis of genome-wide linkage disequilibrium in the highly polyploid sugarcane. *Appl. Genet.* **2008**, *116*, 701–714. [CrossRef]

37. Price, S. Cytogenetics of modern sugar canes. *Econ. Bot.* **1963**, *17*, 97–106. [CrossRef]

38. Bremer, G. Problems in breeding and cytology of sugar cane. *Euphytica* **1961**, *10*, 59–78. [CrossRef]

39. Sreenivasan, T.; Ahloowalia, B. Cytogenetics. In *Sugarcane Improvement Through Breeding*; Heinz, D.J., Ed.; Elsevier: Amsterdam, The Netherlands, 1987; pp. 211–256. ISBN 978-1-4832-8998-4.

40. D'Hont, A.; Grivet, L.; Feldmann, P.; Glaszmann, J.C.; Rao, S.; Berding, N. Characterisation of the double genome structure of modern sugarcane cultivars (*Saccharum spp.*) by molecular cytogenetics. *Mol. Genet. Genom.* **1996**, *250*, 405–413. [CrossRef]

41. D'Hont, A.; Ison, D.; Alix, K.; Roux, C.; Glaszmann, J. Determination of basic chromosome numbers in the genus *Saccharum* by physical mapping of ribosomal RNA genes. *Genome* **1998**, *41*, 221–225. [CrossRef]

42. Aitken, K.S.; McNeil, M.D.; Hermann, S.; Bundock, P.C.; Kilian, A.; Heller-Uszynska, K.; Henry, R.J.; Li, J. A comprehensive genetic map of sugarcane that provides enhanced map coverage and integrates high-throughput Diversity Array Technology (DArT) markers. *BMC Genom.* **2014**, *15*, 1–152. [CrossRef] [PubMed]

43. Garsmeur, O.; Droc, G.; Antonise, R.; Grimwood, J.; Potier, B.; Aitken, K.; Jenkins, J.; Martin, G.; Charron, C.; Hervouet, C.; et al. A mosaic monoploid reference sequence for the highly complex genome of sugarcane. *Nat. Commun.* **2018**, *9*, 1–10. [CrossRef] [PubMed]

44. Grivet, L.; Arruda, P. Sugarcane genomics: Depicting the complex genome of an important tropical crop. *Curr. Opin. Plant Biol.* **2002**, *5*, 122–127. [CrossRef]

45. Souza, G.; Berges, H.; Bocs, S.; Casu, R.; D'Hont, A.; Ferreira, J.; Henry, R.; Ming, R.; Potier, B.; Sluys, M.-A.; et al. The sugarcane genome challenge: Strategies for sequencing a highly complex genome. *Trop. Plant Biol.* **2011**, *4*, 145–156. [CrossRef]

46. D'Hont, A.; Glaszmann, J.C. Sugarcane genome analysis with molecular markers—A first decade of research. In Proceedings of the International Society of Sugar Cane Technologists: XXIV Congress, Brisbane, Australia, 17–21 September 2001; pp. 556–559.

47. Le Cunff, L.; Garsmeur, O.; Raboin, L.M.; Pauquet, J.; Telismart, H.; Selvi, A.; Grivet, L.; Philippe, R.; Begum, D.; Deu, M.J.G. Diploid/polyploid syntenic shuttle mapping and haplotype-specific chromosome walking toward a rust resistance gene (Bru1) in highly polyploid sugarcane (2n ~12x ~115). *Genetics* **2008**, *180*, 649–660. [CrossRef]

48. Aitken, K.; Berkman, P.; Rae, A. The first sugarcane genome assembly: How can we use it. In Proceedings of the Australian Society of Sugar Cane Technologists: 38th Annual Conference, Mackay, Australia, 27–29 April 2016.

49. Ming, R.; Liu, S.C.; Lin, Y.; Da Silva, J.; Wilson, W.; Braga, D.; van Deynze, A.; Wenslaff, T.; Wu, K.; Moore, P.; et al. Detailed alignment of *Saccharum* and sorghum chromosomes: Comparative organization of closely related diploid and polyploid genomes. *Genetics* **1998**, *150*, 1663–1682.

50. Lush, J.L. *Animal Breeding Plans. Ames*; Collegiate Press: Ames, IA, USA, 1937; ISBN 978-0-8138-2345-4.

51. Park, S.; Jackson, P.; Berding, N.; Inman-Bamber, G. Conventional breeding practices within the Australian sugarcane breeding program. In Proceedings of the Australian Society of Sugar Cane Technologists: 29th Annual Conference, Cairns, Australia, 8–11 May 2007; pp. 113–121.

52. Jackson, P.A.; Hogarth, D.M. Genotype x environment interactions in sugarcane, 1. Patterns of response across sites and crop-years in north Queensland. *Aust. J. Agric. Res.* **1992**, *43*, 1447–1459. [CrossRef]

53. Heffner, E.; Sorrells, M.E.; Jannink, J. Genomic selection for crop improvement. *Crop Sci.* **2009**, *49*, 1–12. [CrossRef]

54. Quinton, M.; Smith, C.; Goddard, M. Comparison of selection methods at the same level of inbreeding. *J. Anim. Sci.* **1992**, *70*, 1060–1067. [CrossRef] [PubMed]

55. Hogarth, D. Quantitative inheritance studies in sugar-cane. I. Estimation of variance components. *Aust. J. Agric. Res.* **1971**, *22*, 93–102. [CrossRef]

56. Barbosa, M.H.P.; Resende, M.D.V.; Bressiani, J.A.; Silveira, L.C.I.; Peternelli, L.A. Selection of sugarcane families and parents by Reml/Blup. *Crop Breed. Appl. Biotechnol.* **2005**, *5*, 443–450. [CrossRef]

57. Pisaroglo De Carvalho, M.; Gezan, S.A.; Peternelli, L.A.; Pereira Barbosa, M.H. Estimation of additive and nonadditive genetic components of sugarcane families using multitrait analysis. *Agron. J.* **2014**, *106*, 800–808. [CrossRef]

58. Hogarth, D.M.; Wu, K.K.; Heinz, D.J. Estimating genetic variance in sugarcane using a factorial cross design. *Crop Sci.* **1981**, *21*, 21–25. [CrossRef]

59. O'Reilly, K. An Investigation Into the Heritability of Commercially Important Traits in a Sugarcane Population Under Dryland Conditions. Master's Thesis of Science in Agriculture, University of Natal, Pietermaritzburg, South Africa, 1995.

60. Moore, P.; Nuss, K. Flowering and flower synchronization. In *Sugarcane Improvement Through Breeding*; Heinz, D.J., Ed.; Elsevier Science: Amsterdam, The Netherlands, 1987; pp. 273–311. ISBN 978-1-4832-8998-4.

61. Gosnell, J. Some factors affecting flowering in sugarcane. In Proceedings of the South African Sugar Technologists' Association, Durban and Mount Edgecombe, South Africa, 6–9 June 1983; pp. 144–147.

62. Berding, N.; Hogarth, D. Poor and variable flowering in tropical sugarcane improvement programs: Diagnosis and resolution of a major breeding impediment. In Proceedings of the International Society Sugar Cane Technologists: XXV congress, Guatemala City, Guatemala, 30 January–4 February 2005.

63. Jackson, P.; McRae, T. Selection of sugarcane clones in small plots effects of plot size and selection criteria. *Crop Sci.* **2001**, *41*, 315–322. [CrossRef]

64. Milligan, S.B.; Balzarini, M.; Gravois, K.A.; Bischoff, K.P. Early stage sugarcane selection using different plot sizes. *Crop Sci.* **2007**, *47*, 1859–1864. [CrossRef]

65. Tai, P.Y.P.; Rice, E.R.; Chew, V.; Miller, J.D. Phenotypic stability analyses of sugarcane cultivar performance tests1. *Crop Sci.* **1982**, *22*, 1179–1184. [CrossRef]

66. Casu, R.E.; Manners, J.M.; Bonnett, G.D.; Jackson, P.A.; McIntyre, C.L.; Dunne, R.; Chapman, S.C.; Rae, A.L.; Grof, C.P.L. Genomics approaches for the identification of genes determining important traits in sugarcane. *Field Crop. Res.* **2005**, *92*, 137–147. [CrossRef]

67. Kang, M.; Miller, J. Genotype × environment interactions for cane and sugar yield and their implications in sugarcane breeding 1. *Crop Sci.* **1984**, *24*, 435–440. [CrossRef]

68. Ramburan, S.; Zhou, M.; Labuschagne, M. Interpretation of genotype × environment interactions of sugarcane· Identifying significant environmental factors. *Field Crop. Res.* **2011**, *124*, 392–399. [CrossRef]

69. Hill, J. Genotype-environment interaction—A challenge for plant breeding. *J. Agric. Sci.* **1975**, *85*, 477–493. [CrossRef]

70. Freeman, G. Statistical methods for the analysis of genotype-environment interactions 2. *Heredity* **1973**, *31*, 339–354. [CrossRef] [PubMed]

71. Hammer, G.; Messina, C.; Wu, A.; Cooper, M. Biological reality and parsimony in crop models—Why we need both in crop improvement! *Silico Plants* **2019**, *1*, 1–21. [CrossRef]

72. Meuwissen, T.; Hayes, B.; Goddard, M. Prediction of total genetic value using genome-wide dense marker maps. *Genetics* **2001**, *157*, 1819–1829. [PubMed]

73. Voss-Fels, K.; Cooper, M.; Hayes, B. Accelerating crop genetic gains with genomic selection. *Appl. Genet.* **2019**, *132*, 669–686. [CrossRef]

74. Xu, Y.; Liu, X.; Fu, J.; Wang, H.; Wang, J.; Huang, C.; Prasanna, B.M.; Olsen, M.S.; Wang, G.; Zhang, A. Enhancing genetic gain through genomic selection: From livestock to plants. *Plant Commun.* **2020**, *1*, 2641–2666. [CrossRef]

75. Jannink, J.-L.; Lorenz, A.J.; Iwata, H. Genomic selection in plant breeding: From theory to practice. *Brief. Funct. Genom.* **2010**, *9*, 166–177. [CrossRef]

76. Heslot, N.; Yang, H.-P.; Sorrells, M.E.; Jannink, J.-L. Genomic selection in plant breeding: A comparison of models. *Crop Sci.* **2012**, *52*, 146–160. [CrossRef]

77. Campos, G.D.L.; Hickey, J.M.; Pong-Wong, R.; Daetwyler, H.D.; Calus, M.P.L. Whole-genome regression and prediction methods applied to plant and animal breeding. *Genetics* **2013**, *193*, 327–345. [CrossRef]

78. Erbe, M.; Hayes, B.J.; Matukumalli, L.K.; Goswami, S.; Bowman, P.J.; Reich, C.M.; Mason, B.A.; Goddard, M.E. Improving accuracy of genomic predictions within and between dairy cattle breeds with imputed high-density single nucleotide polymorphism panels. *J. Dairy Sci.* **2012**, *95*, 4114–4129. [CrossRef] [PubMed]

79. Gianola, D.; van Kaam, J.B.C.H.M. Reproducing kernel Hilbert spaces regression methods for genomic assisted prediction of quantitative traits. *Genetics* **2008**, *178*, 2289–2303. [CrossRef] [PubMed]

80. Sousa, M.; Cuevas, J.; Couto, E.; Perez-Rodriguez, P.; Jarquin, D.; Fritsche-Neto, R.; Burgueno, J.; Crossa, J. Genomic-enabled prediction in maize using kernel models with genotype x environment interaction. *G3 Bethesda* **2017**, *7*, 1995–2014. [CrossRef] [PubMed]

81. Calus, M.P.L. Genomic breeding value prediction: Methods and procedures. *Animal* **2010**, *4*, 157–164. [CrossRef] [PubMed]

82. Jenko, J.; Wiggans, G.R.; Cooper, T.A.; Eaglen, S.A.E.; Luff, W.G.d.L.; Bichard, M.; Pong-Wong, R.; Woolliams, J.A. Cow genotyping strategies for genomic selection in a small dairy cattle population. *J. Dairy Sci.* **2017**, *100*, 439–452. [CrossRef] [PubMed]

83. Zhang, A.; Wang, H.; Beyene, Y.; Semagn, K.; Liu, Y.; Cao, S.; Cui, Z.; Ruan, Y.; Burgueno, J.; Vicente, F.S.; et al. Effect of trait heritability, training population size and marker density on genomic prediction accuracy estimation in 22 bi-parental tropical maize populations. *Front. Plant Sci.* **2017**, *8*, 1–12. [CrossRef]

84. Daetwyler, H.D.; Pong-Wong, R.; Villanueva, B.; Woolliams, J.A. The impact of genetic architecture on genome-wide evaluation methods. *Genetics* **2010**, *185*, 1021–1031. [CrossRef]

85. Cericola, F.; Jahoor, A.; Orabi, J.; Andersen, J.R.; Janss, L.L.; Jensen, J. Optimizing training population size and genotyping strategy for genomic prediction using association study results and pedigree information. A case of study in advanced wheat breeding lines. *PLoS ONE* **2017**, *12*, e0169606. [CrossRef]

86. Daetwyler, H.D.; Villanueva, B.; Woolliams, J.A. Accuracy of predicting the genetic risk of disease using a genome-wide approach. *PLoS ONE* **2008**, *3*, e3395. [CrossRef]

87. Hayes Ben, J.; Bowman Phillip, J.; Chamberlain Amanda, C.; Verbyla, K.; Goddard Mike, E. Accuracy of genomic breeding values in multi-breed dairy cattle populations. *Genet. Sel. Evol.* **2009**, *41*, 1–9. [CrossRef]

88. Pszczola, M.; Strabel, T.; Mulder, H.A.; Calus, M.P.L. Reliability of direct genomic values for animals with different relationships within and to the reference population. *J. Dairy Sci.* **2012**, *95*, 389–400. [CrossRef] [PubMed]

89. Goddard, M.E.; Hayes, B.J.; Meuwissen, T.H.E. Using the genomic relationship matrix to predict the accuracy of genomic selection. *J. Anim. Breed. Genet.* **2011**, *128*, 409–421. [CrossRef] [PubMed]

90. Hayes, B.J.; Visscher, P.M.; Goddard, M.E. Increased accuracy of artificial selection by using the realized relationship matrix. *Genet. Res.* **2009**, *91*, 47–60. [CrossRef] [PubMed]

91. Podlich, D.W.; Winkler, C.R.; Cooper, M. Mapping as you go: An effective approach for marker-assisted selection of complex traits. *Crop Sci.* **2004**, *44*, 1560–1571. [CrossRef]

92. Elliot, L.H.; Jean-Luc, J.; Mark, E.S. Genomic selection accuracy using multifamily prediction models in a wheat breeding program. *Plant Genome* **2011**, *4*, 65–75. [CrossRef]

93. Auinger, H.-J.; Schönleben, M.; Lehermeier, C.; Schmidt, M.; Korzun, V.; Geiger, H.; Piepho, H.-P.; Gordillo, A.; Wilde, P.; Bauer, E.; et al. Model training across multiple breeding cycles significantly improves genomic prediction accuracy in rye (*Secale cereale* L.). *Appl. Genet.* **2016**, *129*, 2043–2053. [CrossRef]

94. García-Ruiz, A.; Cole, J.B.; Vanraden, P.M.; Wiggans, G.R.; Ruiz-López, F.J.; Van Tassell, C.P. Changes in genetic selection differentials and generation intervals in US Holstein dairy cattle as a result of genomic selection. *Proc. Natl. Acad. Sci. USA* **2016**, *113*, E3995–E4004. [CrossRef]

95. Hickey, J.M.; Chiurugwi, T.; Mackay, I.; Powell, W. Genomic prediction unifies animal and plant breeding programs to form platforms for biological discovery. *Nat. Genet.* **2017**, *49*, 1297–1303. [CrossRef]

96. Meuwissen, T.; Hayes, B.; Goddard, M. Genomic selection: A paradigm shift in animal breeding. *Anim. Front.* **2016**, *6*, 6–14. [CrossRef]

97. Mrode, R.; Ojango, J.; Okeyo, A.; Mwacharo, J. Genomic selection and use of molecular tools in breeding programs for indigenous and crossbred cattle in developing countries: Current status and future prospects. *Front. Genet.* **2019**, *9*, 1–11. [CrossRef]

98. Bernardo, R.; Yu, J. Prospects for genomewide selection for quantitative traits in maize. *Crop Sci.* **2007**, *47*, 1082–1090. [CrossRef]

99. Lorenz, A.J.; Smith, K.P.; Jannink, J.L. Potential and optimization of genomic selection for fusarium head blight resistance in six-row barley. *Crop Sci.* **2012**, *52*, 1609–1621. [CrossRef]

100. Spindel, J.; Begum, H.; Akdemir, D.; Virk, P.; Collard, B.; Redoña, E.; Atlin, G.; Jannink, J.-L.; McCouch, S.R. Genomic selection and association mapping in rice (Oryza Sativa): Effect of trait genetic architecture, training population composition, marker number and statistical model on accuracy of rice genomic selection in elite, tropical rice breeding lines. *PLoS Genet.* **2015**, *11*, 1–25. [CrossRef]

101. Muleta, K.T.; Pressoir, G.; Morris, G.P. Optimizing genomic selection for a sorghum breeding program in Haiti: A simulation study. *G3 Bethesda* **2019**, *9*, 391–401. [CrossRef] [PubMed]

102. Cooper, M.; Gho, C.; Leafgren, R.; Tang, T.; Messina, C. Breeding drought-tolerant maize hybrids for the US corn-belt: Discovery to product. *J. Exp. Bot.* **2014**, *65*, 6191–6204. [CrossRef]

103. Gaffney, J.; Schussler, J.; Löffler, C.; Cai, W.; Paszkiewicz, S.; Messina, C.; Groeteke, J.; Keaschall, J.; Cooper, M. Industry-scale evaluation of maize hybrids selected for increased yield in drought-stress conditions of the US corn belt. *Crop Sci.* **2015**, *55*, 1608–1618. [CrossRef]

104. Lee, M. DNA markers and plant breeding programs. *Adv. Agron.* **1995**, *55*, 265–344. [CrossRef]

105. Pastina, M.M.; Pinto, L.R.; Oliveira, K.M.; De Souza, A.P.; Garcia, A.A.F. Molecular mapping of complex traits. In *Genetics, Genomics and Breeding of Sugarcane*; Henry, R., Kole, C., Eds.; CRC Press: Boca Raton, FL, USA, 2010; pp. 117–148. ISBN 978-1-4398-4860-9. [CrossRef]

106. Beavis, W.D. QTL analyses: Power, precision, and accuracy. In *Molecular dissection of complex traits*; Patterson, A.H., Ed.; CRC Press: Boca Raton, FL, USA, 1998; Volume 1998, pp. 145–162. ISBN 978-0-8493-7686-3.

107. Xu, Y.; Crouch, J.H. Marker-assisted selection in plant breeding: From publications to practice. *Crop Sci.* **2008**, *48*, 391–407. [CrossRef]

108. Dekkers, J.C.M.; Hospital, F. The use of molecular genetics in the improvement of agricultural populations. *Nat. Rev. Genet.* **2002**, *3*, 22–32. [CrossRef]

109. Zingaretti, M.L.; Monfort, A.; Pérez-Enciso, M. pSBVB: A versatile simulation tool to evaluate genomic selection in polyploid species. *G3 Bethesda* **2019**, *9*, 327–334. [CrossRef]

110. Salvador, A.G.; Luis, F.O.; Verma, S.; Whitaker, V.M. An experimental validation of genomic selection in octoploid strawberry. *Hortic. Res.* **2017**, *4*, 1–9. [CrossRef]

111. Gouy, M.; Rousselle, Y.; Bastianelli, D.; Lecomte, P.; Bonnal, L.; Roques, D.; Efile, J.C.; Rocher, S.; Daugrois, J.; Toubi, L.; et al. Experimental assessment of the accuracy of genomic selection in sugarcane. *Appl. Genet.* **2013**, *126*, 2575–2586. [CrossRef] [PubMed]

112. Deomano, E.; Jakson, P.; Wei, X.; Aitken, K.; Kota, R.; Perez-Rodriguez, P. Genomic Prediction of sugar content an cane yield in sugar cane clones in different stages of selection in a breeding program, with and without pedigree information. *Mol. Breed.* **2020**, *40*. [CrossRef]

113. Van Eeuwijk, F.A.; Bustos-Korts, D.; Millet, E.J.; Boer, M.P.; Kruijer, W.; Thompson, A.; Malosetti, M.; Iwata, H.; Quiroz, R.; Kuppe, C.; et al. Modelling strategies for assessing and increasing the effectiveness of new phenotyping techniques in plant breeding. *Plant Sci.* **2019**, *282*, 23–39. [CrossRef] [PubMed]

114. Cooper, M.; Technow, F.; Messina, C.; Gho, C.; Totir, L.R. Use of crop growth models with whole-genome prediction: Application to a maize multienvironment trial. *Crop Sci.* **2016**, *56*, 2141–2156. [CrossRef]

115. Hill, W.G.; Goddard, M.E.; Visscher, P.M. Data and theory point to mainly additive genetic variance for complex traits. *PLoS Genet.* **2005**, *4*, 1–10. [CrossRef]

116. Cheverud, J.M.; Routman, E.J. Epistasis and its contribution to genetic variance components. *Genetics* **1995**, *139*, 1455–1461.

117. Cheverud, J.M.; Routman, E.J. Epistasis as a source of increased additive genetic variance at population bottlenecks. *Evolution* **1996**, *50*, 1042–1051. [CrossRef]

118. Varona, L.; Legarra, A.; Toro, M.A.; Vitezica, Z.G. Non-additive effects in genomic selection. *Front. Genet.* **2018**, *9*, 1–12. [CrossRef]

119. Osborn, T.C.; Chris Pires, J.; Birchler, J.A.; Auger, D.L.; Jeffery Chen, Z.; Lee, H.-S.; Comai, L.; Madlung, A.; Doerge, R.W.; Colot, V.; et al. Understanding mechanisms of novel gene expression in polyploids. *Trends Genet.* **2003**, *19*, 141–147. [CrossRef]

120. Endelman, J.B.; Carley, C.A.S.; Bethke, P.C.; Coombs, J.J.; Clough, M.E.; Da Silva, W.L.; De Jong, W.S.; Douches, D.S.; Frederick, C.M.; Haynes, K.G.; et al. Genetic variance partitioning and genome-wide prediction with allele dosage information in autotetraploid potato. *Genetics* **2018**, *209*, 77–87. [CrossRef] [PubMed]

121. Gaynor, R.C.; Gorjanc, G.; Bentley, A.R.; Ober, E.S.; Howell, P.; Jackson, R.; Mackay, I.J.; Hickey, J.M. A two-part strategy for using genomic selection to develop inbred lines. *Crop Sci.* **2017**, *57*, 2372–2386. [CrossRef]

122. Comstock, R.E.; Robinson, H.F.; Harvey, P.H. A breeding procedure designed to make maximum use of both general and specific combining ability1. *Agron. J.* **1949**, *41*, 360–367. [CrossRef]

123. Hecker, R.J. Recurrent and reciprocal recurrent selection in sugarbeet1. *Crop Sci.* **1978**, *18*, 805–809. [CrossRef]

124. Santos, M.F.; Moro, G.V.; Aguiar, A.M.; Souza, C.L.D., Jr. Responses to reciprocal recurrent selection and changes in genetic variability in IG-1 and IG-2 maize populations. *Genet. Mol. Biol.* **2005**, *28*, 781–788. [CrossRef]

125. Cooper, M.; Messina, C.D.; Podlich, D.; Totir, L.R.; Baumgarten, A.; Hausmann, N.J.; Wright, D.; Graham, G. Predicting the future of plant breeding: complementing empirical evaluation with genetic prediction. *Crop Pasture Sci.* **2014**, *65*, 311–336. [CrossRef]

126. Cros, D.; Denis, M.; Bouvet, J.-M.; Sánchez, L. Long-term genomic selection for heterosis without dominance in multiplicative traits: Case study of bunch production in oil palm. *BMC Genom.* **2015**, *16*, 1–17. [CrossRef]

127. Rembe, M.; Zhao, Y.; Jiang, Y.; Reif, J. Reciprocal recurrent genomic selection: An attractive tool to leverage hybrid wheat breeding. *Appl. Genet.* **2019**, *132*, 687–698. [CrossRef]

128. Crossa, J.; Pérez-Rodríguez, P.; Cuevas, J.; Montesinos-López, O.; Jarquín, D.; de Los Campos, G.; Burgueño, J.; González-Camacho, J.M.; Pérez-Elizalde, S.; Beyene, Y.; et al. Genomic selection in plant breeding: Methods, models, and perspectives. *Trends Plant Sci.* **2017**, *22*, 961–975. [CrossRef]

129. Ødegård, J.; Sonesson Anna, K.; Yazdi, M.H.; Meuwissen Theo, H. Introgression of a major QTL from an inferior into a superior population using genomic selection. *Genet. Sel. Evol.* **2009**, *41*, 1–10. [CrossRef]

130. Olatoye, M.; Clark, L.V.; Wang, J.P.; Yang, X.; Yamada, T.; Sacks, E.; Lipka, A.E. Evaluation of genomic selection and marker-assisted selection in *Miscanthus* and energycane. *Mol. Breed.* **2019**, *39*, 1–16. [CrossRef]

131. Daetwyler, H.D.; Villanueva, B.; Bijma, P.; Woolliams, J.A. Inbreeding in genome-wide selection. *J. Anim. Breed. Genet.* **2007**, *124*, 369–376. [CrossRef] [PubMed]

132. Bouquet, A.; Juga, J. Integrating genomic selection into dairy cattle breeding programmes: A review. *Animal* **2013**, *7*, 705–713. [CrossRef] [PubMed]

133. Meuwissen, T.H.E. Maximizing the response of selection with a predefined rate of inbreeding. *J. Anim. Sci.* **1997**, *75*, 934–940. [CrossRef] [PubMed]

134. Goddard, M. Genomic selection: prediction of accuracy and maximisation of long term response. *Genetica* **2009**, *136*, 245–257. [CrossRef] [PubMed]

135. Sonesson, A.K.; Woolliams, J.A.; Meuwissen, T.H.E. Genomic selection requires genomic control of inbreeding. *Genet. Sel. Evol.* **2012**, *44*, 1–10. [CrossRef]

136. Daetwyler, H.D.; Hayden, M.J.; Spangenberg, G.C.; Hayes, B.J. Selection on optimal haploid value increases genetic gain and preserves more genetic diversity relative to genomic selection. *Genetics* **2015**, *200*, 1341–1348. [CrossRef]

137. Müller, D.; Schopp, P.; Melchinger, A.E. Selection on expected maximum haploid breeding values can increase genetic gain in recurrent genomic selection. *G3 Bethesda* **2018**, *8*, 1173–1181. [CrossRef]

138. Toro, M.A.; Varona, L. A note on mate allocation for dominance handling in genomic selection. *Genet. Sel. Evol.* **2010**, *42*, 1–9. [CrossRef]

139. Li, X.; Zhu, C.; Wang, J.; Yu, J. Computer Simulation in plant breeding. In *Advances in Agronomy*; Sparks, D.L., Ed.; Elsevier Science & Technology: San Diego, CA, USA, 2012; Volume 116, pp. 219–264. ISBN 978-0-12-394277-7.

 **agronomy**

Article

Are Wheat Hybrids More Affected by Weed Competition than Conventional Cultivars?

Marco Milan *[ID], **Silvia Fogliatto**[ID], **Massimo Blandino**[ID] and **Francesco Vidotto**

Dipartimento di Scienze Agrarie, Forestali e Alimentari, Università di Torino, Largo Paolo Braccini 2, 10095 Grugliasco (TO), Italy; silvia.fogliatto@unito.it (S.F.); massimo.blandino@unito.it (M.B.); francesco.vidotto@unito.it (F.V.)
* Correspondence: marco.milan@unito.it

Received: 27 February 2020; Accepted: 4 April 2020; Published: 7 April 2020

Abstract: Seeding rates of hybrid wheat varieties are typically much lower than conventional varieties due to their higher seed costs, which could potentially delay canopy development leading to greater weed pressures. To test whether hybrid wheat crops are more affected by weed pressure than conventional cultivars, a conventional variety ("Illico") and a hybrid ("Hystar"), were compared in a three-year (2012–2016) field study at two sites in Northern Italy. Weed infestation was mainly characterized by weeds with an early growth pattern, and in only a few seasons did the hybrid crops show a higher weed density than the conventional cultivar. Despite the lower sowing rate, hybrids were able to achieve a similar crop density to the conventional cultivar even in years of delayed sowing or dry weather conditions. Normalized Difference Vegetation Index values were generally similar between cultivars across the years, regardless of the presence of weeds, except during the springtime. Occasionally, the test weight was significantly higher in weeded plots than un-weeded plots. Overall, the two cultivars showed similar yields within the same year. These results indicate that on fields with a low weed burden, and where these weeds emerge early, cultivars may not be significantly affected by productivity losses.

Keywords: hybrids; wheat; weeds; competition

1. Introduction

Modern agriculture relies on the use of herbicides which still represent the most cost-effective tool to control weeds [1,2]. However, in the last decades the sustainability of herbicide use is increasingly threatened by a continued rise in the number of herbicide-resistant weed species [2–5]. As a result, in the last decades weed management has moved towards more integrated weed control strategies, which include a variety of agronomic, mechanical, ecological, physical, and biological practices [6,7]. In particular, the ability of crops to tolerate weed competition while maintaining high yields can be usefully exploited. Crop competitiveness can be distinguished in weed tolerance and weed suppression [8,9]. Both abilities may help to reduce the development of weeds, diminishing the seed dispersal across the season, replenishing their weed seed bank [10].

Crop competitiveness is linked to different plant characteristics, but in particular it is directly related to the leaf area and the rapidity of canopy closure [11]. The possibility of exploiting crop competitiveness is of particular importance in organic farming, where herbicides are not allowed and the remaining control tools are not as effective as the chemical applications [12,13]. Cereals are generally considered more competitive crops than broadleaved species. However, not all cereals show the same competitive ability against weeds. Barley is more competitive than rye and wheat due to its early season growth vigour and to its more expansive root system [14,15].

Crops that rapidly shade the soil surface with their canopy show a more pronounced competitive ability against weeds [16]. This trait is influenced both by species and varieties, but also by technical-agronomic options. For example, cereals may allow high seeding rates and/or reduced row spacing that can eventually result in a significant decrease of weed density and weed competition [14,17]. Other studies reported a significant effect of row orientation on weed development in cereals such as wheat and barley [18].

Crop and livestock demand will increase worldwide according to the FAO prospects [19]. This represents a great challenge both for developed and developing countries. As cereals are still the main sources of food supply, stakeholders should make great efforts to achieve these objectives. The cereal production increased almost unabated over the last 60 years thanks mainly to breeding technologies, the enlargement of irrigated land area, and the wide use of chemical fertilizers, in particular those based on nitrogen and phosphorus [20]. While in developing countries we can expect an increase in the area under cereal cultivation, in developed countries we will see a further reduction in cultivated area [21]. Breeding technologies allowed very high yields to be gained for certain cereals such as maize and rice, while in other cereals the yield improvement has not been so dramatic. The potential advantages of hybrid wheats are higher biomass production, higher yields, and wider adaptation to different soil and climate conditions [22].

Despite hybrid wheat programs having been carried out over many decades, the impact of hybrid wheat varieties on the total sown area still remains limited [20]. Up to 2019, more than 140,000 ha have been cultivated with hybrid wheat in Europe, half of them in France. A more limited area is nowadays being cultivated in Hungary and Italy (30,000 ha) [23]. Overall, obtaining hybrids from autogamous species is not as easy as in allogamous species and it results in a considerably higher cost of seeds relative to conventional varieties. The systems used in the past to obtain hybrids in other cereal crops were less successful in wheat due to a variety of problems (fertility restoration, toxicity effect of hybridising agents, etc). Recently new hybridising systems have been proposed and followed thus refreshing the interest of certain companies in wheat hybrids [20].

Furthermore, we have also to consider that with wheat, yield benefits are controlled by dispersed dominant alleles [20].

The introduction of hybrid cultivar requires the reconsideration of the cropping system in order to enhance the strength of these genotypes and to minimize their weakness points. In particular, the higher cost of seeds determines the need to reduce the seeding rate, which rate isapproximately one third of the ordinary amount of conventional cultivars. This may lead to delayed canopy development, and a longer period of an open canopy, potentially allows a higher ingress of weeds and, as a consequence, a greater dependence on herbicides.

On the other hand, hybrids could recover the initial disadvantage by means of a higher tillering capacity. Despite the weed suppression ability of certain cereals, no information is available regarding the competitive behaviour of hybrid wheats compared to conventional varieties. The reduced seeding rate adopted in hybrid varieties leaves more bare soil free to be colonized by weeds in the early part of the growing season. This condition could potentially determine some yield reduction due to the high weed pressure. Up to now, these speculations have not yet been thoroughly investigated.

The aim of the present study was to verify whether wheat hybrids are potentially more affected by weed pressure. These hypotheses have not yet been investigated experimentally. For this scope, a conventional and a hybrid variety were compared in a three-year field study carried out in two pedo-climatic conditions. Specific crop and weed assessments were done in order to highlight the differences between the two varieties both in terms of yield performances and weed density.

2. Materials and Methods

2.1. Study Sites

Field experiments were carried out from 2012 to 2016 at two locations in Piemonte (Northern Italy): Cigliano (45°18′54.8″ N, 8 02′53.9″ E and 237 m a.s.l), and Grugliasco (45°04′00.5″ N, 7°35′35.3″ E and 293 m a.s.l.) (Figure 1). At Cigliano, the study was carried out in a private farm with a long history of cereal cultivation, while in Grugliasco the study was conducted at the experimental fields of the University of Turin. Cigliano site was characterised by a sandy-loam soil (sand 50.7%, silt 38.9%, clay 10.4%), Typic Hapludalfs (USDA classification), with sub-acidic reaction (pH 6.2), a very low cation exchange capacity (0.92 meq 100 g^{-1}) and a C/N rate of 10.7. The Grugliasco soil is classified as Typic Hapludalf, silty-loam (sand 41.0%, silt 48.1%, clay 10.9%), sub-acid (pH 6.1), mesic soil, with a low cation exchange capacity (8.9 meq 100 g^{-1}) and a C/N rate of 9.0. At Cigliano the study was carried out in the seasons 2012–13, 2013–14, and 2015–16, while at Grugliasco it was carried out in the seasons 2013–14, 2014–15, and 2015–16.

Figure 1. Geographical localization of the study area. The two red icons on the left image identify the two experimental locations.

2.2. Experimental Layout

Two wheat cultivars were compared: the conventional "Illico" (Syngenta Italia S.p.A.) and the hybrid "Hystar" (Venturoli Sementi s.r.l, Italy). In all seasons, "Illico" was sown at a rate of 220 kg ha^{-1}, while "Hystar" at a rate of 75 kg ha^{-1}. In both years and locations, fields were sown at 15 cm row spacing using a conventional seeder, in a north–south orientation. Both cultivars were sown on November 7, October 22 and October 21, in 2012, 2013, and 2015, respectively, at Cigliano site, and October 22, November 2 and October 21 in 2013, 2014, and 2015, at Grugliasco. Fields were managed according to the local agricultural practices. Seed-bed preparation consisted of an autumn ploughing at 30 cm, followed by a disk-harrowing. A total of 120 kg N ha^{-1} was applied as a granular ammonium nitrate fertilizer, split into 60 kg N ha^{-1} at tillering (growth stage 23 on the BBCH scale of Lancashire et al. [24] and 60 kg N ha^{-1} at stem elongation (BBCH 32). Phosphorus and potassium were applied in each site according to the ordinary management of the farms. All the plots were sprayed at the flowering stage with prothioconazole and tebuconazole [Prosaro®, Bayer, emulsifiable concentrate formulation (EC), applied at 0.125 kg of active ingredient (AI) ha^{-1}] to control fungal disease. Plots were characterized by homogenous weed infestations. Fields were routinely treated only in post-emergence to control weeds by means of a mixture of different post-emergence herbicides. As pre-emergence application of herbicide is an uncommon practice in the region, no pre-emergence treatments were done on the field. At Cigliano, weed control was undertaken by applying the herbicide Granstar Trio® at 50 g ha^{-1} [florasulam (5.25 g ha^{-1}) + metsulfuron-methyl (4.15 g ha^{-1}) + tribenuron (4.15 g ha^{-1}); DuPont de Nemours Italiana s.r.l.]. At Grugliasco weed control was done applying a mixture of Manta® Gold at 2.5 L ha^{-1} [fluroxypyr (150 g ha $^{-1}$) + clopyralid (58.25 g ha^{-1}) + MCPA (665 g ha^{-1}); Dow AgroSciences Italia s.r.l.] and Axial® at 0.4 L ha $^{-1}$ [(pinoxaden (40 g ha^{-1}) + cloquintocet-mexyl (5 g ha^{-1}); Syngenta Italia s.p.a.].

The experimental plots were located within contiguous fields cultivated respectively with the conventional and hybrid wheat cultivar. In each location, the compared treatments were represented by the following: absence of weed control (NOT-WEEDED treatment) and presence of weed control (WEEDED treatment). In the NOT-WEEDED treatment, plots were covered with a plastic film during herbicide spraying in order to avoid any drift from the adjacent treated plots. The experimental fields were previously managed under a continuous two-year rotation maize–wheat system (Cigliano) or maintained as meadow for 5 years (Grugliasco). The study was conducted every year on the same location but in different fields. At each location, experimental plots of 12 m^{-2} each were arranged in a complete randomized design with four replications for each cultivar and treatment.

2.3. Weed Assessments

Weed assessments were carried out by counting weed density (number of plants of each weed species per m^2) and visually evaluating weed cover on the ground (%). Weed density was assessed by counting the number of individuals of each species present within a metal quadrat frame of known area (0.625 m^2). Weed cover was evaluated by estimating the percentage of the area included in the metal frame covered by the weeds. At each assessment date, weed density and cover were assessed on three quadrat locations in each plot selected by randomly launching the metal frame. A total of 12 measurements were taken at each assessment (3 times per plot, 4 plots per field).

Two weed assessments were carried out, the first when the wheat crop was at early stem elongation, and the second between the start of heading and the late milk stage (Table 1). In Cigliano, the most abundant weed species were *Stellaria media* (L.) Vill., *Veronica persica* Poiret, *Matricaria chamomilla* L., *Panicum dichotomiflorum* (L.) Michaux and *Viola arvensis* Murray, while at Grugliasco the most detected weed species were *Poa* spp., *Stellaria media* (L.) Vill., *Veronica persica* Poiret and *Ranunculus repens* L.

Table 1. Wheat growth stage according to the BBCH scale [25] at the time of weed assessments carried out in the period 2012–2016 at the two locations.

Location	Season	First Assessment		Second Assessment	
		Crop BBCH Stage		*Crop BBCH Stage*	
		"Hystar"	"Illico"	"Hystar"	"Illico"
	2012–2013	31/32	31/32	50/51	50/51
Cigliano	2013–2014	32/33	32/33	62/63	64/65
	2015–2016	30/31	30/31	64/65	64/65
	2013–2014	30/31	30/31	39/40	39/40
Grugliasco	2014–2015	32/33	32/33	80/81	80/81
	2015–2016	30/31	30/31	77/78	77/78

2.4. Crop Assessments

2.4.1. Yield, Test Weight, and Moisture

Grain yield was measured by harvesting the whole plot using a plot combine (Wintersteiger Seedmech Ried im Innkreis, Austria). A sample taken from the bulk production harvested in each plot was used to determine the grain moisture and test weight (TW) (4 plots per treatment, 4 replicates). The TW was measured using a GAC® 2000 Grain Analyzer (Dickey-John Auburn, IL, USA) using the supplied program and after a validation with reference materials. The grain yield results were adjusted to a 13% moisture content.

2.4.2. Spike Density

At the end of flowering, the crop density was recorded by counting the number of spikes in a sampling surface of 0.04 m^2 per plot. The crop density was expressed as number of spikes per m^2. In each plot the number of spikes was counted twice and the mean value was used as the average number of spikes of the plot (4 plots per treatment, 4 replicates).

2.4.3. Crop Vigour after Winter Dormancy

A hand-held optical sensing device, GreenSeeker™ (Trimble, Sunnyvale, California, USA), was used to measure every 7 days the relative photosynthetically active biomass from the spring tillering stage after winter dormancy to the heading stage. In each plot the Normalized Difference Vegetation Index (NDVI) data acquisition was repeated twice and the mean value was used as the value of the whole plot (4 plot per treatment, 4 replicates). The measurement of NDVI helps to quantify the development of the crop canopy across the season. This device has its own consistent light emission source, photodiode detectors, and interference filters for red [Red] and near infrared [NIR] wavelengths at the 671 ± 6 nm and 780 ± 6 nm spectral bands, respectively; it provides the NDVI, which is calculated as follows [24]:

$$NDVI = \frac{RNIR - RRed}{RNIR + RRed} \tag{1}$$

where RNIR is the NIR radiation reflectance and RRed is the visible red radiation reflectance. The instrument was held approximately 80 cm above the canopy and its effective spatial resolution was 2 m^2. The NDVI values are proportional to the crop biomass and the greenness.

2.4.4. Statistical Analysis

Weed density, yield and the main yield-related parameters were compared by performing *t*-tests on differences between cultivars within the same treatment (WEEDED/NOT-WEEDED) and between cultivars in different treatments (WEEDED vs NOT-WEEDED). SPSS, version 25.00, (SPSS, IBM Corporation, 2008), was used for the statistical analysis.

2.5. Weather Conditions

The meteorological trend observed in the period 2012–2016 at the two locations under study is reported in Tables 2 and 3. The Grugliasco site showed the highest total rainfall as well as the greatest values of growing degree days (GDD). From November to March, the growing season 2015–2016 was the one with lowest rainfall, with less than 195 mm of rain fallen in Cigliano and 232 mm in Grugliasco. In terms of GDD, relevant variations were observed between years at Cigliano, while they were less pronounced at Grugliasco.

Table 2. Monthly rainfall and growing degree days (GDD) observed at Cigliano from sowing to the end of ripening in the period 2012–2016 [¥].

	2012–2013		2013–2014		2015–2016	
Month	**Rainfall (mm)**	**GDD (°C-Day)**	**Rainfall (mm)**	**GDD (°C-Day)**	**Rainfall (mm)**	**GDD (°C-Day)**
November	182.4	272.4	67.6	251.3	4.6	270.2
December	10.6	112.7	139.2	168.7	3.8	163.5
January	16.8	142.2	117.2	148.6	13.6	119.4
February	40.4	110.1	128.8	178.7	120.2	157.5
March	118.0	208.2	71.0	335.1	52.4	251.9
April	164.8	398.0	137.6	428.0	36.6	404.8
May	160.8	479.6	84.4	514.6	171.4	489.6
June	16.4	619.8	143.8	636.8	83.4	624.0
Nov.-June	710.2	2342.9	889.6	2661.7	486.0	2480.7
Nov.-Mar	368.2	845.5	523.8	1082.3	194.6	962.4
April-June	342.0	1497.4	365.8	1579.4	291.4	1518.3

[¥] Data from agrometeorological service of Regione Piemonte. GDD: Accumulated growing degree days for each month using a 0 °C base.

Table 3. Monthly rainfall and growing degree days (GDD) observed at Grugliasco from sowing to the end of ripening in the period 2013–2016 [¥].

	2013–2014		2014–2015		2015–2016	
Month	Rainfall (mm)	GDD (°C-Day)	Rainfall (mm)	GDD (°C-Day)	Rainfall (mm)	GDD (°C-Day)
November	118.8	272.25	255.2	308.1	2.6	301.8
December	82.4	168.5	61.4	200.7	2.8	189.55
January	83.4	166.4	21	166.95	9	163.25
February	119.8	188.3	103.4	144.05	136.2	193.7
March	89.2	346.9	119.6	320.75	80.8	303.4
April	79.2	450.05	81	435.55	66.4	441.55
May	59.4	533.8	46.8	584.2	119.2	518.45
June	102.4	648.85	141.6	674.3	41.2	644.55
Nov.-June	734.6	2755.0	830	2834.5	458.2	2756.2
Nov.-Mar	493.6	1142.3	560.6	1140.5	231.4	1151.7
April-June	241.0	1632.7	269.4	1694.0	226.8	1604.5

[¥] Data from agrometeorological service of Regione Piemonte. GDD: Accumulated growing degree days for each month using a 0 °C base.

3. Results

3.1. Weed Density and Weed Cover

During the 2012–2013 growing season the study was carried out only at the Cigliano site. Weed density and weed cover data are reported in Tables 4 and 5. At the first assessment, conducted almost at the beginning of the spring season (April 10), the average weed density was 109 plants m^{-2} in "Hystar" plots (hybrid variety) and 102 plants m^{-2} in "Illico" plots (conventional variety), without significant differences between the two cultivars. In both cultivars the most abundant weeds were *Stellaria media*, *Matricaria chamomilla* and *Polygonum aviculare*. The highest average weed cover was around 10%. At the second assessment a noticeable increase in weed density was observed in both cultivars, particularly on "Hystar" plots (198 plants m^{-2}). Weed flora was more diversified than the previous assessment as other annual summer weeds appeared (*Echinochloa crus-galli*, *Panicum dichotomiflorum*).

In 2013–2014 the study was carried out both at Cigliano and Grugliasco. At Cigliano, at the time of the first assessment (March 12), weed density was not significantly higher in hybrid variety ("Hystar") plots (50.7 plants m^{-2}) than in the conventional ("Illico") plots (51 plants m^{-2}). Less than 20% of the soil was covered by weeds. At the second assessment, weed infestation did not change greatly (Table 4), and *Stellaria media* was the most abundant species in both cases. At Grugliasco the most abundant weeds were *Veronica persica*, *Polygonum aviculare*, *Papaver rhoeas*, *Ranunculus repens* and *Poa* spp. On average, at the first assessment (March 14), weed infestation was significantly highest in "Illico" plots (414.7 plants m^{-2}). Despite the high weed density values, weed cover did not exceed 15%. Infestation was mainly composed by *Poa annua*, *Veronica persica* and *Ranunculus repens*. On the second assessment (April 17), weed infestation was on lower values compared to the previous assessment; a significantly high weed pressure was recorded in "Hystar" plots (242.9 plants m^{-2}) while an important decrease in weed density was observed on "Illico" plots (Table 4).

In 2014–2015 the study was carried out only on the Grugliasco site. At the first assessment (April 8) a high weed pressure was observed in "Illico" plots and it was mostly represented by *Poa* spp. and *Veronica persica*. During the last season under investigation (2015–2016), the study was carried out at both sites. In Cigliano at the assessment carried out in March, weed density showed low values at both sites (less than 14 plants/m^{-2}). In Grugliasco weed infestation was relevant: on the first assessment (March 24), weed density was 142.7 plants m^{-2} on "Hystar" plots and 126.5 plants m^{-2} on "Illico" plots, without significant differences between cultivars.

Table 4. Weed density measured at the two locations in the period 2012–2016 on WEEDED plots.

Location		Weed Density (Plants m^{-2})			
		First Assessment		Second Assessment	
		"Hystar"	"Illico"	"Hystar"	"Illico"
	2012/2013	109.0 ψ	102.0	198.0 * ψ	128.0 *
Cigliano	2013/2014	50.7	51.0	56.0	60.8
	2015/2016	9.3 ψ	13.3 ψ	46.7 ψ	38.7 ψ
	2013/2014	312.0	414.7 ψ	242.9 *	161.3 * ψ
Grugliasco	2014/2015	172.0 * ψ	350.7 * ψ	248.3 ψ	190.7 ψ
	2015/2016	142.7	126.5	128.0	149.3

Notes: * Statistical differences between cultivars within the same assessment ("Hystar" vs "Illico"); ψ Statistical differences between assessment within each cultivar (I° assessment vs II° assessment). "Histar" (hybrid); "Illico" (conventional).

Table 5. Weed cover (percent of ground cover) measured at the two locations in the period 2012–2016 on WEEDED plots.

Location		Weed Cover (%)			
		First Assessment		Second Assessment	
		"Hystar"	"Illico"	"Hystar"	"Illico"
	2012/2013	10.1 ψ	8.1 ψ	37.5 * ψ	16.7 * ψ
Cigliano	2013/2014	13.6 ψ	19.9 ψ	55.0 ψ	48.0 ψ
	2015/2016	3.6	3.8	4.4	4.9
	2013/2014	11.5 ψ	14.7 ψ	52.9 * ψ	36.7 * ψ
Grugliasco	2014/2015	15.0 ψ	16.7 ψ	37.9 * ψ	25.5 * ψ
	2015/2016	16.2 * ψ	10.8 ψ	47.5 * ψ	27.9 * ψ

Notes: * Statistical differences between cultivars within the same assessment ("Hystar" vs "Illico"); ψ Statistical differences between assessment within each cultivar (I° assessment vs II° assessment). "Histar" (hybrid); "Illico" (conventional).

3.2. Crop Assessments

3.2.1. Grain Yield

In all the seasons, in general no significant differences were observed in terms of grain yields between the two cultivars. The hybrid cultivar achieved the same yield performances of the conventional cultivar. The only exception was observed on WEEDED plots at Cigliano site in 2015–2016 (Table 6) when the yield observed in Hystar plots (8.4 t/ha) was significantly highest compared to the yield recorded on the conventional cultivar (7.2 t/ha). Even in NOT-WEEDED plots no significant differences were detected in both sites during the seasons under study. The presence of weeds seemed not to have a great influence on the yield performances of the two cultivars as the comparison of the yields achieved in presence or absence of weed pressure generally did not show significant yield differences. At Cigliano, in 2012–2013 "Hystar" showed a significantly higher yield in NOT WEEDED plots, while in the last season (2015–2016) despite the scarcity of rainfall during the winter time, high yields were reached by the two varieties in WEEDED plots. The highest yield monitored in "Hystar" plots may be related to the minor weed infestation observed, particularly at the first assessment. At Grugliasco, the analysis did not show any statistical differences between the yields measured in WEEDED and NOT-WEEDED plots, regardless of the cultivar. In this location, the highest grain yields were reached in Illico cultivar (≥ 9.6 t/ha) in 2013–2014 (Table 6), while the lowest was measured in 2015–2016 season in Illico plots. The reduced yields recorded in the 2015–2016 growing season at Grugliasco are likely attributable to the negative impact of the winter drought on crop growth.

Table 6. Grain yields recorded at the two locations in the studied period.

Location	Year	Grain Yield (t ha^{-1})			
		WEEDED		NOT-WEEDED	
		"Hystar"	"Illico"	"Hystar"	"Illico"
Cigliano	2012–2013	6.12 ψ	6.33	7.48 ψ	7.21
	2013–2014	8.41	8.03	8.18	8.08
	2015–2016	8.37 *	7.23 *	8.23	7.61
Grugliasco	2013–2014	8.74	9.49	8.20	9.59
	2014–2015	7.10	7.44	7.00	7.60
	2015–2016	6.96	7.52	6.03	6.80

Notes: * Statistical differences between cultivars within the same treatment (WEEDED/NOT WEEDED); ψ Statistical differences between treatments within each cultivar (WEEDED vs NOT WEEDED).

3.2.2. Test Weight and Grain Moisture

The statistical analysis showed a significant effect of the cultivar on these parameters. Overall, the highest test weight values were measured on "Illico" cultivar, both in WEEDED and NOT-WEEDED plots. The lowest test weights values were generally found in NOT-WEEDED plots, regardless of the cultivar (Table 7). Only at Cigliano site, in 2013–2014 on NOT-WEEDED plots, and in 2015–2016 on WEEDED plots, were no statistical differences observed between the two cultivars. At harvest time, grains of "Hystar" cultivar often reported the highest moisture values, in particular on weeded plots. At both sites, regardless of the presence or not of weeds, the highest grain moisture was measured during the 2013–2014 growing season that was characterized by high rainfall. On NOT-WEEDED plots, only in 2013–2014 were the grain moisture values statistically different between the two cultivars, in all the other years grain showed similar humidity (Table 8).

Table 7. Test weight measured at the harvest in the two locations.

Location	Year	Test Weight (kg hl^{-1})			
		WEEDED		NOT-WEEDED	
		"Hystar"	"Illico"	"Hystar"	"Illico"
Cigliano	2012–2013	77.66 *	81.84 *	78.05 *	80.94 *
	2013–2014	73.85 *	76.01 *	73.61	75.42
	2015–2016	77.09 ψ	80.76	77.76 ψ	81.04
Grugliasco	2013–2014	72.02 * ψ	75.17 * ψ	69.79 * ψ	73.74 * ψ
	2014–2015	73.89 *	78.89 *	72.37 *	78.89 *
	2015–2016	77.15 *	81.15* ψ	75.95 *	78.97 * ψ

Notes: * Statistical differences between cultivars within the same treatment (WEEDED/NOT WEEDED); ψ Statistical differences between treatments within each cultivar (WEEDED vs NOT- WEEDED).

Table 8. Grain moisture measured at the harvest in the two locations.

Location	Year	Grain Moisture (%)			
		WEEDED		NOT-WEEDED	
		"Hystar"	"Illico"	"Hystar"	"Illico"
Cigliano	2012–2013	12.31* ψ	12.14* ψ	13.56 ψ	13.14 ψ
	2013–2014	15.34 *	14.92 *	15.04	15.04
	2015–2016	13.91 ψ	13.82 ψ	13.36 ψ	13.77 ψ
Grugliasco	2013–2014	15.51* ψ	14.84* ψ	17.50* ψ	15.61* ψ
	2014–2015	12.90 *	13.49 *	13.89	13.41
	2015–2016	12.66* ψ	12.61* ψ	13.79 ψ	14.01 ψ

Notes: * Statistical differences between cultivars within the same treatment (WEEDED/NOT WEEDED); ψ Statistical differences between treatments within each cultivar (WEEDED vs NOT- WEEDED).

3.2.3. Spike Density

Spike density was not significantly affected by cultivar and weed competition. In particular, at the Grugliasco site, the statistical analysis did not show differences in the number of spikes between the two cultivars. Only in Cigliano had the "Illico" cultivar a higher number of spikes m^{-2} than the hybrid cultivar, in 2013–2014 on WEEDED plots and in 2015–2016 on NOT-WEEDED plots. In both sites, the highest spike density values were recorded during the 2015–2016 growing season, while the lowest values were recorded in 2012–2013 at Cigliano site (Table 9).

Table 9. Spikes density recorded at the two locations in the studied period.

Location	Year	Spikes Density (Spikes m^{-2})			
		WEEDED		NOT-WEEDED	
		"Hystar"	"Illico"	"Hystar"	"Illico"
	2012–2013	353.12	407.81	343.75	416.87
Cigliano	2013–2014	463.90 *	591.85 * ψ	508.66	477.14 ψ
	2015–2016	686.61	723.21	601.82 *	723.21 *
	2013–2014	467.16	448.57	485.45	565.25
Grugliasco	2014–2015	428.26	448.57	375.72	448.21
	2015–2016	500.41	500.00	503.74	514.29

Notes: * Statistical differences between cultivars within the same treatment (WEEDED/NOT WEEDED); ψ Statistical differences between treatments within each cultivar (WEEDED vs NOT- WEEDED).

3.2.4. Crop Vigour

The crop vigour was determined taking NDVI measurements on the fields at different moments of the crop cycle (Tables 10 and 11). Overall, only limited differences were observed between the cultivars over the years, regardless of the presence of weeds. These differences, when statistically relevant, occurred in particular in the period ranging from the end of the winter season to the beginning of spring time (March and April). In Cigliano, in the first two seasons the differences are likely attributable mainly to the presence of weeds within the plots, while in 2015/2016 season they could be also associated with the higher plant density measured on "Illico" plots (Table 9). During the 2013/2014 growing season the NDVI values observed in NOT-WEEDED plots resulted in being significantly higher than those observed in WEEDED plots, in particular on plots cultivated with the hybrid cultivar. On these plots a weed coverage of about 50% was measured at the time of second assessment.

At the Grugliasco site, observing the data presented in Table 11, it is possible to note differences in NDVI values measured in March over the years both in WEEDED and NOT-WEEDED plots. In the first season (2013/2014) on March 12 NDVI values ranged from 0.66 to 0.77. At a similar period of time for the two following seasons, the NDVI values ranged instead from 0.23 to 0.43. The reasons forf these relevant differences are attributable primarily to the late sowing period and secondarily to low rainfall. Wheat was sowed on October 22/21 in 2013/2014 and 2015/2016, while only on November 2, in 2014/2015. The lowest NDVI values recorded in the last two seasons reflected an initial scarce growth of both cultivars, due, in 2014/2015, to the retarded sowing, while in 2015/2016 because of the scarcity of rainfall over the first three months after sowing. In 2014/2015, on NOT- WEEDED plots, at March and April assessments, the "Illico" cultivar showed significantly higher NDVI values compared to those observed in "Hystar". As the two cultivars had a similar spike density, this difference was likely due to the higher weed density on "Illico" plots (> 350 plants m^{-2}) compared to "Hystar" plots (172 plants m^{-2}) (Table 4). In the following season, at the first NDVI assessments carried out in February and March, on WEEDED plots the "Illico" cultivar showed higher NDVI values than those observed in the "Hystar" cultivar.

Table 10. Normalized Difference Vegetation Index (NDVI) values recorded at Cigliano in the period 2012–2013, 2013–2014 and 2015–2016.

		NDVI			
		WEEDED		NOT-WEEDED	
	Date	"Hystar"	"Illico"	"Hystar"	"Illico"
2012/2013	Apr 2	0.63 *	0.73 *	0.68	0.73
	Apr 15	0.71	0.78	0.77	0.79
	May 6	0.82	0.87	0.82	0.85
	May 20	0.81	0.81	0.83	0.83
	May 28	0.78	0.81	0.81	0.82
2013/2014	Mar 7	0.66	0.60	0.67	0.72
	Mar 20	0.68 *	0.72 *	0.74 *	0.86 *
	Mar 31	0.69 *ψ	0.75 *ψ	0.81 ψ	0.83 ψ
	Apr 8	0.69 ψ	0.77 ψ	0.82 ψ	0.83 ψ
	Apr 16	0.67 ψ	0.75	0.80 ψ	0.78
	Apr 23	0.72 ψ	0.79	0.82 ψ	0.81
	May 5	0.74	0.78	0.80	0.77
	May 13	0.68	0.66	0.68	0.68
2015/2016	Feb 23	0.62	0.57	0.63	0.58
	Mar 4	0.66 *	0.58 *	0.68	0.60
	Mar 15	0.73	0.66	0.73	0.67
	Mar 24	0.73	0.71	0.75	0.72
	Apr 6	0.73	0.73	0.75	0.75
	Apr 15	0.81	0.80	0.81	0.83
	May 3	0.87	0.88	0.86	0.86
	May 13	0.85	0.86	0.86	0.86

Notes: * Statistical differences between cultivars within the same treatment (WEEDED/NOT- WEEDED); ψ Statistical differences between treatments within each cultivar (WEEDED vs NOT- WEEDED).

Table 11. NDVI values recorded at Grugliasco in the period 2013–2014, 2014–2015 and 2015–2016.

		NDVI			
		WEEDED		NOT-WEEDED	
	Date	"Hystar"	"Illico"	"Hystar"	"Illico"
2013/2014	Mar 12	0.71	0.77	0.66	0.76
	Mar 24	0.81	0.79	0.79	0.86
	Apr 4	0.76 ψ	0.82 ψ	0.88 ψ	0.90 ψ
	Apr 11	0.70 *ψ	0.77 *	0.79 ψ	0.82
	Apr 22	0.77 ψ	0.79	0.84 ψ	0.83
	May 8	0.75	0.77	0.78	0.78
2014/2015	Mar 10	0.25 *	0.37 *	0.23 *	0.43 *
	Mar 23	0.44	0.58	0.41 *	0.61 *
	Apr 1	0.52	0.67	0.48 *	0.71 *
	Apr 14	0.59	0.69	0.62 *	0.75 *
	Apr 24	0.75	0.76	0.77	0.81
	May 5	0.78	0.76	0.79	0.77
	May 14	0.74	0.74	0.72	0.73
2015/2016	Feb 26	0.26 *	0.38 *	0.28 *	0.33 *
	Mar 8	0.33 *	0.50 *	0.36	0.39
	Mar 22	0.47 *	0.68 *	0.55	0.56
	Apr 5	0.69	0.75	0.76	0.72
	Apr 14	0.76	0.76	0.81	0.74
	May 5	0.83	0.81	0.86	0.84
	May 19	0.78	0.80	0.79	0.81

Notes: * Statistical differences between cultivars within the same treatment (WEEDED/NOT- WEEDED); ψ Statistical differences between treatments within each cultivar (WEEDED vs NOT-WEEDED).

4. Discussion

Wheat hybrid cultivars have become of great interest in the last decade. The global food cereal demand for both human and animal consumption is constantly increasing as world population rises and it will reach 3 billion tonnes in 2050 [21,26]. Unlike other cereals, such as rice and maize, where the spread of hybrids is already affirmed, research activities and large-scale adoption in wheat are still relatively limited [20]. The main advantage attributable to hybrid cultivars is their high productivity compared to the conventional cultivars despite a considerable reduced seeding rate. A lower seeding rate, which in our experiment was 33% of the conventional variety, leaves more soil free from crop coverage, particularly during the initial part of the season. This could represent an opportunity for weeds, which have more space free for growing. Considering that in Northern Italy most of the wheat farmers apply post-emergence herbicides, from October to March, weeds are free to develop. In hybrid cultivars we might expect a more evident development of weeds during this period, with potential yield losses, competition, and greater weed seed burden for the following seasons.

Previous studies conducted on different wheat genotypes reported that with doubling the seeding rate, yield losses were significantly reduced due to the suppression of weeds [27].

In this study, weed infestation at the two sites was mainly characterized by the presence of weeds with early growth pattern, such as *Stellaria media*, *Veronica persica*, *Matricharia chamomilla*, *Viola arvensis*, and *Ranunculus repens*. These species have a lower competitive ability than other more troublesome wheat weed species (eg. *Galium aparine* L., *Avena sterilis* L., *Lolium multiflorum* Lamark, *Papaver rhoeas* (L.) [28]. Overall, weed infestation at Grugliasco was much higher than that observed at Cigliano and some weeds were not typical cereal weeds (eg. *Ranunculus repens*, *Rumex* spp.). This difference is plausibly related to the agronomic history of the fields while over the last few years, these fields, until 2012/2013, were covered by a permanent grassland. It is well known how the agricultural management system may affect the weed seed bank density as well as its composition [29]. If inserted within a rotation, meadows have generally a cleansing effect on the seed bank [30]; however, in annual crops that follow a meadow, some weeds that are not typical of the weed communities infesting these crops may spread and become abundant in the first seasons. Grass in the rotation can potentially increase some weed species such as *Poa* spp. and *Matricaria* spp. [31]. Similarly, the repeated application of herbicides for weed control in conventional systems may lead to a shift of weed composition, in terms of seed numbers and weed species [32]. When herbicides are the major agricultural tool, a reduction in species composition is also expected [29].

Weed crop competition generally causes a significant reduction in crop yield [33,34]. At both sites, the limited weed infestation did not lead to statistically significant yield reduction in either cultivar. The reduced seeding rate adopted in the hybrid cultivar left more bared soil, however this free space was not occupied by weeds as winter cereals show a certain ability to compete with weeds, but the magnitude of this ability depends on the cultivar [12,13,34]. The new modern cultivars are more productive, shorter, and with a higher harvest index than the older ones, but their yield potential is only expressed when effective chemical weed control programs are adopted [27]. A high crop density can generally hamper the growth of some weed species with the exception of those able to overhang the crop canopy [28]. Weiner et al., [35] suggested that a more crowded and uniform distribution of the crop may represent an effective strategy of weed control. Moreover, competition may affect wheat yield especially when it occurs late in the cycle [33,35]. However, if weeds are not highly competitive or have an early period of growth, they have scarce effect on crop yield [28]. Our results showed that despite the initial reduced seeding rate, at the early dough stage, no differences in spike density were observed between hybrid and conventional cultivar. In our experimental conditions, the hybrid cultivar fully recovered the initial gap exploiting its high tillering ability. Overall, the two cultivars had generally not dissimilar yields within the same year and only showed variations across the years due to the differences in the meteorological conditions. Even in the absence of chemical control, the wheat did not face severe weed competition, regardless of the cultivar.

Over the years, at both sites, "Illico" cultivar showed the highest test weight values. Occasionally, the test weight was significantly higher in weeded plots than plots not weeded. In these cases, the highest weed density or a more relevant weed cover was observed. Considering the observed lack of differences between varieties in terms of yield, the impact of weeds appears to be more qualitative than quantitative. The differences observed in this yield parameter are related to the specific characteristics of each cultivar and may be influenced by grain moisture [36–39]. The presence of weeds is generally associated with an increase in the humidity rate of the grain [40]. This is because at the time of harvest the crop has already ended its cycle while weeds may be still in active photosynthetic activity [40]. Without the presence of weed infestation, "Illico" had the driest grain compared to the hybrid cultivar. On the contrary, on infested plots the presence of weeds probably increased the humidity rate of the conventional cultivar, bringing it close to the moisture values detected on the hybrid variety, without statistical differences. The presence of weeds may affect the moisture level of the crop microclimate favouring the development of fungal disease, even considering that many weeds are recognized as reservoirs or inoculum of many fungal species [41]. Even a more humid grain at harvest may boost the development of mycotoxigenic fungi [42], making a good weed control, essential.

NDVI is one of the most used spectral reflectance indices. It is commonly used to estimate biomass, LAI (Leaf Area Index), photosynthesis and yield in many cereals, including wheat [43,44]. In our study, NDVI differences between cultivars were generally encountered at the assessments carried out during spring time, from February to April. In these assessments, "Hystar" always had the lower values compared to the conventional cultivar. Differences of NDVI values within each cultivar attributable to the presence (or not) of weeds were observed in both sites only during the 2013–2014 growing season. The NDVI measurements pointed out that even in the case of delayed sowing or a dry period in the first part of the growing seasons, the hybrid cultivar was able to recover the initial density gap when the climatic conditions returned to be favourable to the growth. We may also consider that dryness condition at the early growth stages may give cereals an advantage against some weeds [45], while germination of weed seeds is reduced in the case of low soil moisture [46]. When weed flora is composed of early emerging and less competitive weeds, we might not expect yield contractions even with a high infestation density. The pedoclimatic conditions are also an important factor to be taken into account. In our experimental conditions, despite the initial reduced sowing rate, hybrids were able to achieve the same crop density as the conventional cultivar even in the case of delayed sowing time or the driest meteorological conditions. From the agronomic point of view, in order to minimize the initial risk of less homogeneous soil cover during the first stages of growth of hybrid wheat, it is important, particularly in the cooler environments, to anticipate the sowing time and management of nitrogen fertilization at sowing or at vegetative restart, in order to enhance a proper and quick crop tillering.

In conclusion, on fields characterized by a reduced weed pressure and in the case of weed infestation mostly represented by early emerging weeds, hybrids cultivar may not be significantly affected by yield losses. As the magnitude of weed infestation and its competitiveness are affected by crop rotation history and weed flora composition, the initial hypothesis, postulating that a hybrid variety would be more affected by weed infestation than a comparable conventional variety, was not supported.

Author Contributions: Conceptualization, M.M. and M.B.; data curation, M.M. and S.F.; formal analysis, M.M.; funding acquisition, F.V. and M.B., investigation, M.M. and S.F.; methodology, M.M. and F.V.; supervision, F.V. and M.B.; writing—original draft preparation, M.M.; writing—review and editing, M.M, F.V., M.B. and S.F. All authors have read and agreed to the published version of the manuscript.

Funding: This research received no external funding.

Acknowledgments: The authors thank all the lab. and field technicians who made a valuable contribution during the study. A further acknowledgment needs to be addressed to all farmers who hosted the experimental studies on their fields and collaborated strictly with the present research team through all the study.

Conflicts of Interest: The authors declare no conflict of interest.

References

1. Gianessi, L.P. The increasing importance of herbicides in worldwide crop production. *Pest Manag. Sci.* **2013**, *69*, 1099–1105. [CrossRef] [PubMed]
2. Kraehmer, H.; Laber, B.; Rosinger, C.; Schulz, A. Herbicides as Weed Control Agents: State of the Art: I. Weed Control Research and Safener Technology: The Path to Modern Agriculture. *Plant Physiol.* **2014**, *166*, 1119. [CrossRef] [PubMed]
3. MacDonald, J.F. Herbicide Resistance in Plants. In *Biotechnology and Sustainable Agriculture—Policy Alternatives*; NABC: Ithaca, NY, USA, 1989.
4. Heap, I. The International Survey of Herbicide Resistant Weeds. In Proceedings of the Brighton Crop Protection Conference, Brighton, UK, 19–20 November 2019; BCPC: Brighton, UK, 2019.
5. Heap, I. Global perspective of herbicide-resistant weeds. *Pest Manag. Sci.* **2014**, *70*, 1306–1315. [CrossRef] [PubMed]
6. Upadhyaya, M.K.; Blackshaw, R.E. *Non-Chemical Weed Management: Principles, Concepts and Technology*; CABI: Wallingford, Oxfordshire, UK, 2007; ISBN 1-84593-290-0.
7. Van der Meulen, A.; Chauhan, B.S. A review of weed management in wheat using crop competition. *Crop Prot.* **2017**, *95*, 38–44. [CrossRef]
8. Jannink, J.-L.; Orf, J.H.; Jordan, N.R.; Shaw, R.G. Index Selection for Weed Suppressive Ability in Soybean Contribution No. 00-13-0146 of the Minnesota Agric. Exp. Stn., Univ. of Minnesota, St. Paul, MN. 1 Mention of any product is for scientific purposes only and does not imply endorsement by the Minnesota Agricultural Experiment Station. *Crop Sci.* **2000**, *40*, 1087–1094.
9. Zhao, D.L. Weed Competitiveness and Yielding Ability of Aerobic Rice Genotypes. Ph.D. Thesis, Wageningen University, Wageningen, The Netherlands, 2006.
10. Jordan, N. Prospects for weed control through crop interference. *Ecol. Appl.* **1993**, *3*, 84–91. [CrossRef]
11. Ni, H.; Moody, K.; Robles, R.; Paller, E.; Lales, J. Oryza sativa plant traits conferring competitive ability against weeds. *Weed Sci.* **2000**, *48*, 200–204. [CrossRef]
12. Mason, H.E.; Spaner, D. Competitive ability of wheat in conventional and organic management systems: A review of the literature. *Can. J. Plant Sci.* **2006**, *86*, 333–343. [CrossRef]
13. Mason, H.E.; Navabi, A.; Frick, B.L.; O'Donovan, J.T.; Spaner, D.M. The Weed-Competitive Ability of Canada Western Red Spring Wheat Cultivars Grown under Organic Management. *Crop Sci.* **2007**, *47*, 1167–1176. [CrossRef]
14. Sardana, V.; Mahajan, G.; Jabran, K.; Chauhan, B.S. Role of competition in managing weeds: An introduction to the special issue. *Crop Prot.* **2017**, *95*, 1–7. [CrossRef]
15. Davies, D.; Welsh, J. *Weed Control in Organic Cereals and Pulses*; Chalcombe Publications: Bedfordshire, UK, 2002; pp. 77–114.
16. Holt, J.S. Plant responses to light: A potential tool for weed management. *Weed Sci.* **1995**, *43*, 474–482. [CrossRef]
17. Singh, B.; Dhaka, A.; Pannu, R.; Kumar, S. Integrated weed management-a strategy for sustainable wheat production-a review. *Agric. Rev.* **2013**, *34*, 243–255. [CrossRef]
18. Borger, C.P.D.; Hashem, A.; Pathan, S. Manipulating crop orientation to suppress weeds and increase crop yield. *Weed Sci.* **2010**, *58*, 174–178. [CrossRef]
19. Alexandratos, N.; Bruinsma, J. *World Agriculture towards 2030/2050: The 2012 Revision 2012*; FAO: Roma, Italy, 2012.
20. Whitford, R.; Fleury, D.; Reif, J.C.; Garcia, M.; Okada, T.; Korzun, V.; Langridge, P. Hybrid breeding in wheat: Technologies to improve hybrid wheat seed production. *J. Exp. Bot.* **2013**, *64*, 5411–5428. [CrossRef]
21. FAO. *The Future of Food and Agriculture Trends and Challenges*; FAO: Roma, Italy, 2017.
22. Informatore Agrario. *Frumento Ibrido: Cos'è e Quali Benefici Apporta*; Informatore Agrario-Supplemento, Edizioni l'informatore agrario: Verona, Italy, 2015; pp. 4–5.
23. Tartarini, E. *Superfici a Frumento Ibrido in Europa ed in Italia (Total Area Cultivated with Hybrids Wheat in Europe and Italy)*; RV Venturoli srl: Pianoro (BO), Italy, 2019.
24. Govaerts, B.; Verhulst, N. *The Normalized Difference Vegetation Index (NDVI) Greenseeker (TM) Handheld Sensor: Toward the Integrated Evaluation of Crop Management Part A: Concepts and Case Studies*; International Maize and Wheat Improvement Center: Mexico City, Mexico, 2010.
25. Lancashire, P.D.; Bleiholder, H.; Boom, T.V.D.; Langelüddeke, P.; Stauss, R.; Weber, E.; Witzenberger, A. A uniform decimal code for growth stages of crops and weeds. *Ann. Appl. Biol.* **1991**, *119*, 561–601. [CrossRef]
26. FAO. *Global Agriculture towards 2050*; FAO: Roma, Italy, 2009.

Agronomy **2020**, *10*, 526

27. Lemerle, D.; Verbeek, B.; Cousens, R.; Coombes, N. The potential for selecting wheat varieties strongly competitive against weeds. *Weed Res.* **1996**, *36*, 505–513. [CrossRef]
28. Wilson, B.; Wright, K.J. Predicting the growth and competitive effects of annual weeds in wheat. *Weed Res.* **1990**, *30*, 201–211. [CrossRef]
29. Menalled, F.D.; Gross, K.L.; Hammond, M. Weed aboveground and seedbank community responses to agricultural management systems. *Ecol. Appl.* **2001**, *11*, 1586–1601. [CrossRef]
30. Tomasoni, C.; Borrelli, L.; Pecetti, L. Influence of fodder crop rotations on the potential weed flora in the irrigated lowlands of Lombardy, Italy. *Eur. J. Agron.* **2003**, *19*, 439–451. [CrossRef]
31. Davies, D. Changes in weed populations in the conversion of two arable farms to organic farming. In Proceedings of the Brighton Crop Protection Conference: Weeds, Brighton, UK, 17–20 November 1997; pp. 973–978.
32. Ball, D.A. Weed Seedbank Response to Tillage, Herbicides, and Crop Rotation Sequence. *Weed Sci.* **1992**, *40*, 654–659. [CrossRef]
33. Das, T.; Yaduraju, N. Effect of weed competition on growth, nutrient uptake and yield of wheat as affected by irrigation and fertilizers. *J. Agric. Sci.* **1999**, *133*, 45–51. [CrossRef]
34. Siddiqui, I.; Bajwa, R.; Huma, Z.-E.; Javaid, A. Effect of six problematic weeds on growth and yield of wheat. *Pak. J. Bot.* **2010**, *4*, 2461–2471.
35. Weiner, J.; Griepentrog, H.-W.; Kristensen, L. Suppression of weeds by spring wheat *Triticum aestivum* increases with crop density and spatial uniformity. *J. Appl. Ecol.* **2001**, *38*, 784–790. [CrossRef]
36. Pushman, F.M. The effects of alteration of grain moisture content by wetting or drying on the test weight of four winter wheats. *J. Agric. Sci.* **1975**, *84*, 187–190. [CrossRef]
37. Tkachuk, R.; Kuzina, F. Wheat: Relations between some physical and chemical properties. *Can. J. Plant Sci.* **1979**, *59*, 15–20. [CrossRef]
38. Troccoli, A.; Borrelli, G.M.; De Vita, P.; Fares, C.; Di Fonzo, N. Mini Review: Durum Wheat Quality: A Multidisciplinary Concept. *J. Cereal Sci.* **2000**, *32*, 99–113. [CrossRef]
39. Fogliatto, S.; Milan, M.; De Palo, F.; Ferrero, A.; Vidotto, F. Effectiveness of mechanical weed control on Italian flint varieties of maize. *Renew. Agric. Food Syst.* **2018**, *34*, 447–459. [CrossRef]
40. Burnside, O.; Wicks, G. The effect of weed removal treatments on sorghum growth. *Weeds* **1967**, *15*, 204–207. [CrossRef]
41. Dastjerdi, R. High Fumonisin Content in Maize: Search for Source of Infection and Biological Function. Ph.D. Thesis, Niedersächsische Staats-und Universitätsbibliothek Göttingen, Göttingen, Germany, 2014.
42. Ono, E.; Sasaki, E.; Hashimoto, E.; Hara, L.; Correa, B.; Itano, E.; Sugiura, T.; Ueno, Y.; Hirooka, E. Post-harvest storage of corn: Effect of beginning moisture content on mycoflora and fumonisin contamination. *Food Addit. Contam.* **2002**, *19*, 1081–1090. [CrossRef]
43. Aparicio, N.; Villegas, D.; Casadesus, J.; Araus, J.L.; Royo, C. Spectral Vegetation Indices as Nondestructive Tools for Determining Durum Wheat Yield. *Agron. J.* **2000**, *92*, 83–91. [CrossRef]
44. Royo, C.; Aparicio, N.; Villegas, D.; Casadesus, J.; Monneveux, P.; Araus, J.L. Usefulness of spectral reflectance indices as durum wheat yield predictors under contrasting Mediterranean conditions. *Int. J. Remote Sens.* **2003**, *24*, 4403–4419. [CrossRef]
45. Pavlychenko, T.K.; Harrington, J.B. Competitive efficiency of weeds and cereal crops. *Can. J. Res.* **1934**, *10*, 77–94. [CrossRef]
46. Locke, M.A.; Reddy, K.N.; Zablotowicz, R.M. Weed management in conservation crop production systems. *Weed Biol. Manag.* **2002**, *2*, 123–132. [CrossRef]

 *agronomy*

Article

Optimizing Nitrogen Fertilization to Improve Qualitative Performances and Physiological and Yield Responses of Potato (*Solanum tuberosum* L.)

Sara Lombardo, Gaetano Pandino *  **and Giovanni Mauromicale**

Dipartimento di Agricoltura, Alimentazione e Ambiente (Di3A), University of Catania, via Valdisavoia 5, 95123 Catania, Italy; sara.lombardo@unict.it (S.L.); g.mauromicale@unict.it (G.M.)
* Correspondence: g.pandino@unict.it; Tel.: +39-095-478-3449

Received: 18 February 2020; Accepted: 29 February 2020; Published: 4 March 2020

Abstract: Potato is often produced by adopting high nitrogen (N) external inputs to maximize its yield, although the possible agronomic and qualitative benefits of a N over-fertilization to the crop are scarcely demonstrated. Therefore, our aim was to determine, over two years, the effect of three N fertilization rates (0, 140 and 280 kg ha^{-1}, referred to as N0, N140 and N280) simultaneously on the crop physiology, yield components, N use efficiency and tuber chemical composition of cv. Bellini. Throughout the field monitoring, our data highlighted that N140 provided an improvement of the crop physiology, as expressed in terms of leaf photosynthesis rate and Soil Plant Analysis Development (SPAD) readings, than the other N fertilization rates. In addition, regardless of year and as compared to N0 and N280, the supply of 140 kg N ha^{-1} also ensured the highest yield and an intermediate value of the nitrogen use efficiency (59.1 t ha^{-1} and 37.1 kg tuber dry weight kg N^{-1}, respectively), together with nutritionally relevant tuber qualitative traits, i.e. high levels of dry matter, starch (by an enzymatic/spectrophotometric method), total polyphenols (by Folin-Ciocalteu assay) and ascorbic acid [by high-performance liquid chromatography (HPLC) analysis], and a low nitrate amount (by an ion-selective electrode method) (16.6%, 634-3.31-0.61 and 0.93 g kg^{-1} of dry matter, respectively). Therefore, although a certain interaction between N fertilization rate and year was observed, our findings demonstrated that a conventional N fertilization rate (280 kg ha^{-1}) is unnecessary from both agronomic and qualitative standpoints. This is of considerable importance in the perspective to both limit environmental pollution and improve growers' profits by limiting N external inputs to potato crops.

Keywords: potato; nitrogen fertilization rate; photosynthesis rate; SPAD readings; tuber yield; nitrogen efficiency indices; tuber nutritional composition

1. Introduction

The efforts of modern agriculture are primarily focused on reducing the environmental impact of crop-management practices, while ensuring both high and stable yields and high produce quality. In this way, the balance of nitrogen (N) supply by fertilization is fundamental for establishing sustainable farming systems, given its role as both an environmental pollutant [1] and an essential plant macronutrient to promote crop quantitative and qualitative performances [2]. Sufficient N availability has a positive effect on plant growth and development, being involved in protein and chloroplast structure; while an excessive N supply leads to an over-emphasis on vegetative growth and a detrimental effect of root or fruit development [3]. Therefore, the efficient use of N is fundamental with respect to potential impacts on both environmental safety and crop performance.

Since Mediterranean soils are often characterized by low organic matter and thus low N reserves [4], growers often adopt irrational N fertilization programs in order to maximize the productivity of several crops, including potato (*Solanum tuberosum* L.) [5]. This is the most important non-grain food in the world, resulting fourth in terms of production quantity [6]. In the Mediterranean Basin, potato cultivation occupies an area of about 1 Mha producing ~31 Mt of tubers [6]. Compared with Northern European countries, tuber yields in the Mediterranean area are low. This is because the environmental (most importantly water availability and temperature) and agro-economic conditions are less favourable than those in the Northern European countries where most of the cultivars have been developed [7]. In southern Italy (Sicily, Campania and Apulia), as in other Mediterranean coastal areas, such as North African countries, Cyprus and Turkey, the potato crop is not grown in the usual main cycle (spring–summer), owing to the high temperatures and considerable demand for irrigation water, but is mostly cropped in a winter–spring cycle (planting from November to January and harvesting from March to early June) with the aim of obtaining an early product [8–10]. This is highly appreciated for its specific qualitative traits [11,12] and so profitably exported to northern European countries for fresh consumption [13]. In addition, early potato tubers are also increasingly serving as feedstock for industrial products [14,15]. This makes early potato production even more attractive in the Mediterranean Basin. However, the potential yield is never fully obtained in natural productive systems since biotic and abiotic factors may negatively affect plant growth and tuber development [16]. This growing season cycle is often characterized by a relatively low temperature, a short photoperiod and limited solar radiation, which are conditions with an appreciable effect on plant growth, substantially modifying the morphological and phenological characteristics of the plants (for example, most potato cultivars do not flower) compared to those cultivated in the spring–summer cycle [17]. Apart from the cultivar choice, crop protection and irrigation, an important agronomic measure for improving early potato production is represented by an adequate nutrient management [16]. Particularly, due to the high nutritional requirements during vegetative growth and tuber bulking, as well as to meet the quality standards demanded by both the fresh vegetable market and processing industry, early potato growers typically rely heavily on the use of inorganic fertilizers to maximize their incomes. Potato is known to have a relatively low N uptake efficiency ranging between 50% and 60% [18], due to its shallow root system which is less efficient in taking up N than other crops like wheat, maize or sugar beet [19]. Therefore, it needs adequate levels of N for a fast plant cycle and plant growth rate [8], in order to promote both earliness and high-profitable yields [13]. Indeed, N has positive effect on both the number of emerging leaves and the rate of leaf expansion and, hence, on the canopy development of the plant [20] and on the photosynthesis efficiency by increasing the intercepted radiation [21,22]. This has a decisive impact on dry matter partitioning to the tubers, tuber bulking and, of course, on tuber yield [23]. By contrast, N over-fertilization may promote an excessive stolon and leaf growth at the expense of tuber development/maturity and quality [24]. Fertilizers are generally used inefficiently by the crop, also due to large N losses through seepage or percolation, particularly when conventional irrigation methods, e.g. furrows or sprinklers, are used [25]. Hence, in the latest years the improvement of N use efficiency for the potato crop is a priority for researchers [19,26,27]. Several N fertilization rates have been suggested as optimal for potato production; in some European countries and the USA the recommended N fertilization rates vary from 70 to 330 kg ha^{-1}, and the most economically efficient rates from 147 to 201 kg ha^{-1} [24]. Nevertheless, literature still lacks on comprehensive studies including several aspects, from physiology to yield and quality responses of potato crop in relation to N fertilization. In addition, despite some works dealing with main potato crop response to N fertilization [10,16,20,21,24,28], no attempts have been focused on defining the effects of different N fertilization rates on both the crop physiology, yield and tuber chemical composition of early potato. Indeed, the environmental conditions associated with early potato production substantially modify the morphology and phenology of the crop, and thus the tubers are essentially immature and so differ qualitatively from those produced in the main crop cycle [11]. As a result, little of the literature describing the characteristics of main crop potatoes can be used to make inferences regarding early potato cultivation. Taking into account all these considerations, in the present research we have investigated whether it is possible to reduce the N fertilization rate, while keeping yield reduction to a minimum and having positive effects on certain tuber qualitative traits. In this respect, this study represents the first comprehensive approach to detect simultaneously agronomic and qualitative performances of this crop under different N fertilization rates. Since the annual weather variations have also a considerable influence over these traits, the experiments were replicated over two years in a major potato production area in the Mediterranean Basin.

2. Materials and Methods

2.1. Field Experimental Design, Plant Material and Management Practices

The experiment spread over two years (2014 and 2015) at a commercial farm located on the coastal plain of Siracusa (37°01′ N, 15°12′ E, 30 m above sea level). The soil, moderately deep, is classified as calcixerollic xerochrepts type [29], with pH 7.7 and a soil composition of 48% sand (2–0.02 mm), 18% silt (0.02–0.002 mm), 34% clay (<0.002 mm), 6% limestone, 1.8% organic matter, 0.2% total nitrogen, 28 mg kg^{-1} of available P_2O_5 and 180 mg kg^{-1} of exchangeable K_2O. The soil characteristics may be considered strongly representative of the potato cultivation area in Sicily [30,31]. A layer, 0.25 m thick (from −0.05 to −0.30 m), where about 90% of active potato roots were located, was considered for the soil analysis. Soil minerals analyses were obtained according to procedures approved by the Italian Society of Soil Science [32], whereas the remaining analyses were carried out using widely employed and adopted methods in Italy [33].

Disease-free, non-pre-sprouted "seed" tubers of cv. Bellini, from a single seed lot, were manually planted on 18 January 2014 and 25 January 2015. This cultivar was recently introduced for conventional production of early potato in the Mediterranean Basin, where it has shown a good adaptation to the pedoclimatic conditions. It has yellow skin and pulp, and is a B cooking type (i.e., multi-purpose cooking) according to the EAPR (European Association for Potato Research) cooking-type scale.

In this experiment three nitrogen fertilization rates were compared: 0 (as control), 140 and 280 kg N ha^{-1}, hereafter referred to as N0, N140 and N280, respectively. In particular, N280 represents the conventional N fertilization rate commonly adopted by Sicilian producers for enhancing yields; while N140 was formulated on the basis of the N uptake by potato crop with target yields of 40–50 t ha^{-1} [34], the available soil N during the growing season (equal to 70 kg ha^{-1}; see Section 2.4) and N fertilization efficiency (equal to 90%, due to the modality of N fertilization).

A randomized complete-block design with four replicates was adopted. Each plot size was 4.2 × 4.2 m, with 84 plants and consisted of six rows. Whole tubers were planted at 0.3 m intervals in rows 0.70 m apart, corresponding to a planting density of 4.76 plants m^{-2}. The two external rows and two plants on each row-end were used as border to minimize contamination from adjacent nitrogen treatments. The two middle rows per plot were harvested to assess the yield. N was soil-applied incorporated by mineral source (ammonium nitrate, at 26% of N) and the total amount was split in 2 applications (10 and 40 d after transplant). Besides the different nitrogen fertilization rates, prior to planting, all plots received the same base fertilization consisting of 80 kg P_2O_5 ha^{-1} and 140 kg K_2O ha^{-1} as mineral superphosphate (19% of P_2O_5) and potassium sulphate (50% of K_2O), respectively. The experimental field had been cultivated in a potato-lettuce(fennel)-carrot rotation for almost 20 years, as commonly used in the cultivation area. Obviously, in both years, the previous crop was carrot (fertilized with 120 kg N ha^{-1}), and the area used for our experiments was different to avoid cumulative effects of nitrogen fertilizer treatments over years. Drip irrigation was provided once the accumulated daily evaporation rate (derived from measurements of an unscreened class A-Pan evaporimeter) had reached about 30 mm. Over the crop cycle 220 (2014) and 160 mm (2015) irrigation water were provided. Five (2014) and four (2015) irrigation applications were performed. Weed and pest control followed standard commercial practice.

2.2. Crop Physiology

The physiological measurements were made from the youngest fully expanded leaf (usually the 3rd or 4th leaf from the apex). They were determined at 80, 90, 111 and 119 days after planting (DAP) in 2014 and at 89, 107, 116 and 123 DAP in 2015. At each time point, measurements per each N fertilization rate were taken in duplicate on the same leaves of five plants per plot, previously marked for the purpose. Leaf Soil Plant Analysis Development (SPAD) readings were performed using a portable absorbance-based chlorophyll meter (SPAD-502 model, Konica Minolta, Sakai, Osaka, Japan). The measurements were conducted between 10:00 and 12:00 h (local solar time). The SPAD readings provide an indication of crop greenness and N requirements, while having a strong and non-linear correlation with chlorophyll foliar content in potato [35]. Photosynthesis rate was measured by a LI-6200 closed gas-exchange system (LI-Cor Inc., Lincoln, NE, USA) using a 250 cm^3 chamber in the closed circuit mode. Instantaneous gas-exchange measurements were made in the morning (at 10:00 and 12:00), closely matching the respective growth chamber CO_2 conditions, under clear sunny meteorological conditions. Days on which photosynthesis rate was measured were typically clear sunny days characterized by a Photosynthetically Active Radiation (PAR) ≤1800 mmol photons m^{-2} s^{-1}. Air temperatures varied only slightly during each measuring hour, but ranged between 19 and 28 °C during the period of measurements.

2.3. Crop Yield and Its Components

In both years, for the determination of yield and its components, tubers (from each plot and replicate) were harvested manually when about 70% of haulms were fully desiccated (i.e., at 125 and 130 DAP in 2014 and 2015, respectively), and the number and weight of both marketable and unmarketable tubers per plant were determined. Tubers which were greened, misshapen or displayed pathological damage were classed as unmarketable, as well as those with weight lower than 20 g. This allowed the calculation of the number of tubers per plant (NTP), mean tuber weight (MTW) and marketable yield (MY). The yield of unmarketable tubers was very low (below 1.0%) and hence excluded from the data.

2.4. Nitrogen Crop Uptake and Nitrogen Efficiency Indices

For the determination of crop N uptake (CNU), only the N in tubers was considered under the assumption that contents in roots were negligible and contents in the aboveground shoot mass had been mostly translocated before dieback of the haulm [36]. Nitrogen concentration was determined by the Kjeldahl method, using fresh tubers collected at harvest, which were oven-dried and finely ground through a mill (IKA, Labortechnick, Staufen, Germany) with a 1.0 mm sieve.

The efficiency of N was calculated in terms of NUE (Nitrogen Use Efficiency), NUtE (Nitrogen Utilization -Efficiency) and NUpE (Nitrogen Uptake Efficiency). Firstly, available soil nitrogen was calculated as the sum of mineral N (NO_3-N and NH^4-N) in the 0–0.40 m soil layer at sowing, plus the N released through mineralization during the growing season, and the N supplied through fertilization [37]. The initial mineral nitrogen content of the soil profile (0–0.40 m) was set at 1.2% of total N, determined by the Kjeldahl method. N soil mineralization per month was calculated according to Gariglio et al. [38], from total N content corrected by an N mineralization factor as a function of soil temperature. On the basis of this procedure and considering the losses for volatilization, leaching and microbial mineralization [39], the quantity of available mineral nitrogen in the soil (N_A) for the crop cycle was equal to about 70 kg ha^{-1} in both years. This N availability was similar to that recently reported by Ierna and Mauromicale [18], in field experiments carried out in the same cultivation area.

Therefore, the following indices were calculated:

NUE [expressed as kg tuber dry matter content (DM) kg N] = marketable dry tuber yield/NA

NUpE (expressed as kg N kgN^{-1}) = CNU/NA

NUtE (expressed as kg tuber DM kg N^{-1}) = marketable dry tuber yield/CNU

Fertilizer use efficiency (FUE) was also calculated by using the following formula:

FUE (expressed as kg kg^{-1}) = [yield of fertilized plot (kg) − yield of unfertilized plot (kg)]/N fertilizer rate applied (kg)

2.5. Tuber Chemical Composition

A total of 24 samples (2 years × 3 N fertilization rates × 4 replicates) were analyzed per each chemical determination. Each sample consisted of 15 marketable tubers, which were washed with tap water, dried with tissue paper, diced and immediately blended using a domestic food processor at 0 °C (Kenwood Multipro, Milan, Italy). Finally, per each replicate an amount (500 g) of the resulting slurry was freeze-dried (Christ freeze drier, Osterode am Harz, Germany) and stored at −20 °C until analysis of ascorbic acid, total polyphenols, antioxidant activity, starch and simple sugars, while the remaining portion was oven-dried at 65 °C (Binder, Milan, Italy), until a constant weight was reached, in order to determine the DM. Then, the dehydrated material was used for the determination of total protein and nitrate.

Kits for the enzymatic and spectrophotometric determination of total starch and simple sugars were obtained from Megazyme International Ireland Ldt. (Bray, Co. Wicklow, Ireland), as well as sugars standards. 5-O-caffeoylquinic acid (used as standard for the determination of total polyphenols content) was obtained from Extrasynthese (Lyon, France). All the other reagents and solvents adopted were purchased from Sigma-Aldrich (Milan, Italy) and were of analytical or high-performance liquid chromatography (HPLC) grade. Bidistilled water was used throughout this research.

Total protein content was determined according to Snyder and Desborough [40], reading the absorbance at 595 nm with a Shimadzu 1601 ultraviolet (UV)-visible spectrometer (Shimadzu Corp., Tokyo, Japan).

For analysis of simple sugars (glucose, fructose and sucrose), sample preparation was carried out following the protocol described in Megazyme Assay Kit K-SUFRG [41], while total starch determination was performed according to the protocol described in Megazyme Total Starch Assay Kit AA/AMG [41].

The ascorbic acid determination was carried out by HPLC as described by Lombardo et al. [42].

Total polyphenols content was quantified spectrophotometrically as reported by Lombardo et al. [12]; using the same extracts, the antioxidant activity was evaluated and expressed as percentage inhibition of DPPH (2,2-diphenyl-1picrylhydrazyl) [43], obtained by the following equation: [AC$_0$−AS$_{30}$/AC$_0$] × 100, where AC$_0$ is the absorbance of a blank control at the beginning of the assay and AS$_{30}$ the sample absorbance after 30 min.

Nitrate content determination was performed using an ion selective electrode method [44].

Excepted for DM and antioxidant activity, reported as %, all the other chemical traits were expressed as g kg^{-1} of DM.

2.6. Statistical Analysis

Given the normality of distributions (Shapiro and Wilks test) [45] and the homogeneity of variances (Levene's test) [46], the data were generally subjected to a two-way analysis of variance (ANOVA), based on a factorial combination of three N fertilization rates × two years, By contrast, for the data of photosynthesis rate and SPAD readings ANOVA was performed separately per year (since the considered measurement times did not intercept the same phenological phases in both years) and was based on a factorial combination of three N fertilization rates × four measurement times. Means were separated by a least significant difference (LSD) test, when the *F*-test was significant. For DM and antioxidant activity, the % values were subjected to Bliss transformation prior to analysis and then to ANOVA; however, untransformed data (thus expressed as %) for these traits were reported and discussed. All calculations and analyses were performed using CoStat® version 6.003 (CoHort Software, Monterey, CA, USA).

2.7. Weather Conditions

Meteorological data were monitored during each growing season (from January to May) in the years of study by a meteorological station (Mod. Multirecorder 2.40; EGT, Florence, Italy) sited within 250 m of the experimental field. The total rainfall during 2014 (January–May) was below average (68 mm versus a long-term mean of 179 mm), while the mean minimum temperature was significantly higher as compared to the long-term period (12.5 versus 9.8 °C) (Table 1). Total rain recorded (160 mm) in 2015 did not substantially differed from the long-term climate, experiencing about 92 mm more rain than in the first year. In both 2014 and 2015, the mean daily minimum temperature was higher than long-term average. By contrast, mean daily maximum temperature was slightly below that recorded in the long-term period (17.3 versus 18.8 °C) in 2015 (Table 1).

Table 1. Rainfall, mean minima and maxima temperatures during the 'early' potato growing season in the years of study as compared to the long-term period (1977–2006).

Month	2014			2015			Long-Term Period		
	Rainfall (mm)	Min. Air Temp. (°C)	Max. Air Temp. (°C)	Rainfall (mm)	Min. Air Temp. (°C)	Max. Air Temp. (°C)	Rainfall (mm)	Min. Air Temp. (°C)	Max. Air Temp. (°C)
January	21	11.1	16.5	46	8.8	15.2	65	7.1	15.4
February	23	11.0	16.9	73	8.5	13.7	38	7.6	16.2
March	15	10.7	16.6	31	9.5	16.1	25	8.8	17.7
April	8	13.2	19.7	0	13.4	18.7	31	10.9	20.2
May	1	16.4	22.0	10	15.3	22.8	20	14.4	24.3
Total/mean	68	12.5	18.3	160	11.4	17.3	179	9.8	18.8

3. Results and Discussion

3.1. Physiological Traits

According to ANOVA results (Table 2) and regardless of the specific yearly trend, at each measurement time leaves from N0 plots displayed a lower photosynthetic rate than those from the N fertilized plots (i.e., N140 and N280) (Figure 1). This can be related to the decisive impact of N fertilization on the canopy development of the plant [20], as here expressed by the influence of N fertilization rate on the level of aboveground dry biomass (Table 3). Indeed, this was higher in the N fertilized plots than in N0 ones (Table 4). The differences in terms of photosynthetic rate between N140 and N280 plots were not significant across all the measurement times (Figure 1). It is likely that this depended on the increased stomatal resistance in the N280 plants, resulting from their higher evapotraspirative demand due to their higher aboveground dry biomass. A similar trend was highlighted for SPAD values, a non-destructive and instant indicator of potato plant N status [22]. Indeed, at each measurement time, in both 2014 and 2015 N0 plants had the lowest SPAD values than those fertilized. Nevertheless throughout the field monitoring, no statistical differences were observed between N140 and N280 in terms of SPAD readings (Figure 1). This indicates that N140 may represent a threshold over that early potato does not benefit from higher N fertilization rate (N280) in terms of SPAD readings improvement.

Table 2. Summary of statistical significance from analysis of variance for photosynthesis rate and Soil Plant Analysis Development (SPAD) readings.

		Source of Variation		
Variable	**Year**	**Nitrogen Fertilization Rate (N)**	**Measurement Time (M)**	**(N) × (M)**
Photosynthesis Rate	2014	***	***	*
		***	***	*
SPAD Readings	2015	***	***	***
		***	***	**

*, ** and *** indicate significant at $p < 0.05$, 0.01 and 0.001, respectively. For both years, degrees of freedom were equal to 2–3 and 6 for (N), (M) and (N) × (M) interaction, respectively.

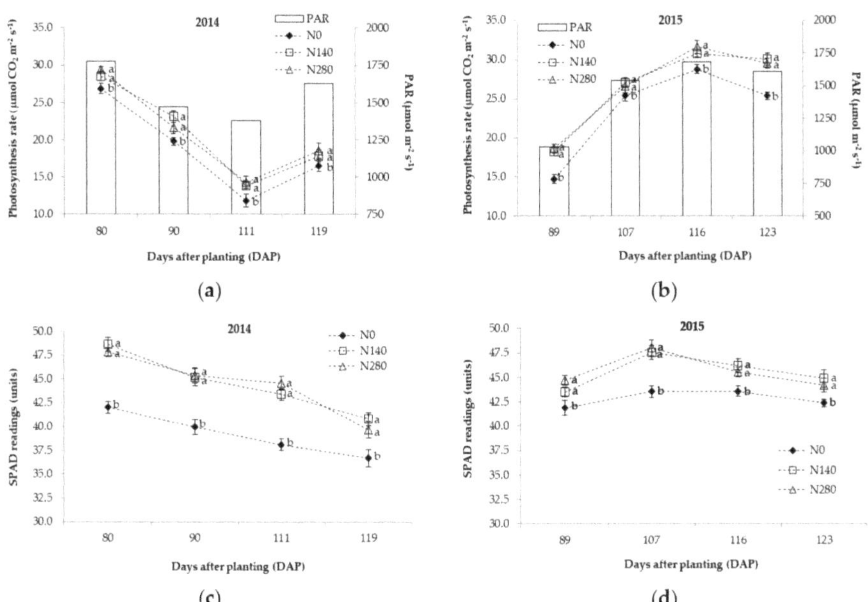

Figure 1. Photosynthesis rate and SPAD readings of early potato leaves as affected by 'N fertilization rate x measurement data' interaction. Different letters within each measurement time (expressed as day after planting, DAP) indicate statistically significant differences among the N fertilization rates (least significant difference (LSD) test, $p < 0.05$).

Table 3. Summary of statistical significance from analysis of variance for the aboveground dry biomass, yield components, N efficiency indices and tuber chemical characteristics.

Variable [a]	Source of Variation		
	Nitrogen fertilization Rate (N)	Year (Y)	(N) × (Y)
ADB	*	**	NS
MY	***	**	*
MTW	***	***	***
NTP	**	NS	NS
CNU	***	***	**
NUE	***	**	*
NUpE	***	***	***
NUtE	NS	***	**
FUE	***	**	***
DM	**	NS	*
Starch	***	***	*
Sucrose	***	**	NS
Glucose	**	NS	NS
Fructose	**	NS	NS
Total protein	***	NS	NS
Total polyphenols	***	***	*
Nitrate	***	***	*
Ascorbic acid	***	**	***
Antioxidant activity	***	***	**

*, ** and *** indicate significant at $p < 0.05$, 0.01 and 0.001, respectively; NS = not significant. Degrees of freedom were equal to: 1 for (Y); 2 for both (N) and (N) × (Y) interaction. [a] ADB: Aboveground Dry Biomass; MY: Marketable Yield; MTW: Mean Tuber Weight; NTP: Number of Tubers Plant^{-1}; CNU: Crop Nitrogen Uptake; NUE: Nitrogen Use Efficiency; NUtE: Nitrogen Utilization Efficiency; NUpE: Nitrogen Uptake Efficiency; FUE: Fertilizer Use Efficiency; DM: Dry Matter.

Table 4. Aboveground dry biomass, yield components, N efficiency indices and tuber chemical characteristics of "early" potato as affected by the main effects. Different letters between years or among N fertilization rates within the same row show significant differences (LSD test, $p < 0.05$).

Variable	Year		N Fertilization Rate		
	2014	2015	N0	N140	N280
ADB (t ha^{-1} DM)	1.02 ± 0.02 a	0.85 ± 0.04 b	0.78 ± 0.02 c	0.99 ± 0.03 b	1.10 ± 0.03 a
MY (t ha^{-1}) [a]	55.5 ± 1.2 a	48.3 ± 2.0 b	45.8 ± 1.0 c	59.1 ± 1.3 a	50.8 ± 2.0 b
MTW (g)	127 ± 4 a	111 ± 6 b	108 ± 4 b	137 ± 4 a	112 ± 5 b
NTP (no. plant^{-1})	9.7 ± 0.3 a	9.6 ± 0.2 a	9.4 ± 0.3 b	9.6 ± 0.2 b	10.1 ± 0.4 a
CNU (kg ha^{-1})	145 ± 5 a	128 ± 3 b	113 ± 4 b	147 ± 4 a	148 ± 5 a
NUE (kg tuber DW kg N^{-1})	66.0 ± 1.0 a	56.6 ± 1.3 b	113.3 ± 0.9 a	46.8 ± 0.4 b	23.8 ± 1.0 c
NUpE (kg N kg N^{-1})	0.98 ± 0.08 a	0.85 ± 0.036 b	1.61 ± 0.09 a	0.70 ± 0.08 b	0.42 ± 0.04 c
NUtE (kg tuber DW kg N^{-1})	64.4 ± 0.7 a	64.3 ± 0.6 a	70.3 ± 0.8 a	66.6 ± 0.6 a	56.2 ± 0.7 b
FUE (kg kg^{-1})	67.7 ± 1.3 a	45.0 ± 1.0 b	-	95.0 ± 1.6 a	17.7 ± 1.0 b
Dry matter (DM) (%)	16.8 ± 1.0 a	16.8 ± 0.9 a	17.2 ± 1.0 a	16.6 ± 0.5 b	16.4 ± 0.6 b
Starch (g kg^{-1} DM)	632 ± 8 a	604 ± 7 b	640 ± 8 a	634 ± 6 a	580 ± 6 b
Sucrose (g kg^{-1} DM)	11.7 ± 0.5 a	11.9 ± 0.6 a	13.1 ± 0.6 a	11.4 ± 0.4 b	11.0 ± 0.3 b
Glucose (g kg^{-1} DM)	6.1 ± 0.1 a	5.9 ± 0.3 a	6.7 ± 0.4 a	5.8 ± 0.2 b	5.6 ± 0.3 b
Fructose (g kg^{-1} DM)	2.0 ± 0.2 a	2.1 ± 0.3 a	2.4 ± 0.4 a	1.8 ± 0.1 b	1.9 ± 0.1 b
Total protein(g kg^{-1} DM)	89 ± 4 a	90 ± 3 a	82 ± 4 b	86 ± 4 b	100 ± 5 a
Total polyphenols (g kg^{-1} DM)	3.17 ± 0.07 b	3.83 ± 0.10 a	3.86 ± 0.09 a	3.31 ± 0.08 b	3.33 ± 0.07 b
Nitrate (g kg^{-1} DM)	1.03 ± 0.7 a	0.88 ± 0.09 b	0.86 ± 0.09 c	0.93 ± 0.06 b	1.08 ± 0.04 a
Ascorbic acid (g kg^{-1} DM)	0.60 ± 0.05 a	0.63 ± 0.05 a	0.71 ± 0.04 a	0.61 ± 0.07 b	0.52 ± 0.08 c
Antioxidant activity (%$_{inhibition}$ DPPH)	55.6 ± 1.2 b	61.1 ± 1.2 a	62.3 ± 1.2 a	58.3 ± 1.0 b	54.4 ± 1.3 c

Data are mean ± standard deviation, n = 12 and 8 for year and N fertilization rate, respectively. [a] See Table 3 for the list of acronyms.

For both the physiological traits under study, it was evident a different annual trend (Figure 1). In 2014 the photosynthesis rate significantly declined from 80 to 119 DAP by an extent between 37.0

(N280) and 38.5% (N0). By contrast in 2015, with increasing plant age, photosynthesis rate exhibited a bell-shaped curve (Figure 1), increasing up to a complete canopy developing (116 DAP) and declining thereafter (123 DAP). Reason for the different trend of photosynthesis rate in the two years primarily is that the considered measurement times did not intercept the same crop phenological phases in both years. Indeed, due to the highest mean temperatures during the initial months (late January and February) in 2014, potato plants emerged 10 days early than in 2015 (data not shown), and therefore potato leaves had a presumable fully photosynthetic capacity earlier in 2014. This may explain why potato plants recorded a high photosynthesis rate early (at 80 DAP) in 2014, while similar values (equal to about 28 μmol CO_2 m^{-2} s^{-1}) were reached later in 2015 (between 107 and 116 DAP). The different annual trend of photosynthesis rate may be also related to the PAR values registered at each measurement time, since a strongly and positive correlation between these parameters was highlighted in both 2014 and 2015 (respectively, $r = 0.94$ and 0.99, $p < 0.01$).

The SPAD readings showed a similar trend to that of the photosynthesis rate recorded in the same year (Figure 1). According to literature data [22,47], in 2014 the SPAD values decreased linearly and significantly with plant age, with a more marked reduction for the fertilized plots than unfertilized ones (17% vs. 13%) from 80 to 119 DAP (Figure 1). This could be related to N remobilization from the oldest to the youngest leaves [47]. By contrast, this decreasing tendency of SPAD values across the field monitoring was less evident in 2015 (Figure 1). Indeed, SPAD readings time-course is also largely dependent on many external factors, among which weather conditions and light intensity are important [48]. The effect of meteorological conditions experienced in 2014, characterized by higher mean temperatures throughout the monitoring period, caused higher SPAD values at 80 DAP (46.2 SPAD units, on the average of N fertilization rates) than those reported by Mauromicale et al. [22]. Indeed, air temperature together with solar radiation is the main external factor affecting physiological processes, among which foliar chlorophyll concentration. As a consequence, in 2015 early potato plants reached a similar SPAD values only at 107 DAP.

Finally, based on our results and regardless of the yearly trend, a conventional N fertilization rate (N280), commonly adopted in the Mediterranean Basin, was not associated to an improvement of both photosynthesis rate and SPAD value in the early potato crop.

3.2. Aboveground Dry Biomass and Yield Components

Aboveground dry biomass and yield components were all influenced by the N fertilization rate, while 'N fertilization rate × year' interaction was significant only for MY and MTW (Table 3).

Regardless of year, as highlighted in a work carried out by Fontes et al. [24], N0 plots reported the lowest MYs (45.8 t ha^{-1}) than the N fertilized ones (55.0 t ha^{-1}, on average) (Table 4).

On the basis of our results on MY, N140 represented the best treatment. However, the reasonable values reached under N0 (abundantly above the Italian mean yield equal to 27.7 t ha^{-1}, according to data provided by FAO) [6] may be ascribed to the residual soil fertility (about 70 kg ha^{-1} in both years) of the studied cultivation area, where every year conspicuous amounts of N fertilizers were applied to vegetable crops. On the contrary, we demonstrated as an over-fertilization to the early potato is not necessary, since N280 either decreased (in 2014) or maintained stable (in 2015) the MY as compared to N140 (Figure 2). In researches carried out in similar Mediterranean environments [13,49], potato yields increased with increasing nitrogen rate up to 120 kg ha^{-1}, but did not change further with higher N fertilization rates. In particular, the high MY (66.3 t ha^{-1}) achieved by N140 plots in 2014 is attributable to the highest MTW reported (Figure 2). Indeed, several researchers [35,50] have highlighted as sufficient N fertilizer amounts reduced the small size tuber fraction. In addition, N140 showed an intermediate canopy development (here expressed by the level of ADB) between N0 and N280, which was able to ensure acceptable values of photosynthesis rate and therefore high MYs (Table 4). By contrast, the higher ADB observed for N280 plots may explain the lowest MY values recorded (Table 4), since tuberization can even be suppressed or delayed by high N supply in favour of higher shoot growth [51]. This may have reduced the carbohydrate translocation in the tubers, as demonstrated by the lower MTW under N280 in both years.

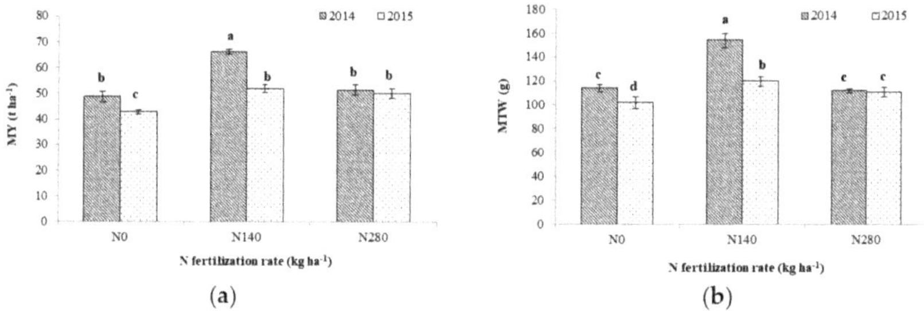

Figure 2. Marketable yield (MY) and mean tuber weight (MTW) of early potato as affected by 'N fertilization rate x year' interaction. Data are mean ± standard deviation of four individual plots ($n = 4$). Different letters within each parameter indicate statistically significant differences (LSD test, $p < 0.05$).

Regardless of N fertilization rate, the MY reduction from 2014 to 2015 is attributable to the decrease of the MTW (Figure 2). Indeed, the correlation analysis between MY and MTW showed a very strong correlation ($R^2 = 0.98$ ***), while that between MY and NTP was not significant ($R^2 = 0.06$). This observed year-to-year variation in MY could be attributable to the meteorological conditions experienced throughout the field trials (Table 1). Basically, as compared to the long-term period, the more favourable mean air temperatures experienced in 2014 may have improved crop productivity by sustaining the increase in MTW (Figure 2); by contrast, the NTP was unaffected by seasonal conditions. Indeed, according to De la Morena et al. [52], N has the greatest impact on the average tuber weight rather than the other components of potato yields.

3.3. Crop Nitrogen Uptake and N Efficiency Indices

It is particularly crucial to estimate both crop N uptake and N efficiency indices for potato crop, considering that it has a shallow and slow-growing root system with possible repercussions for nutrients uptake. On the whole, our results on CNU were comparable with those reported by Darwish et al. [48] obtained also in the Mediterranean environment. Here, as expected, CNU was higher in the N fertilized plots (148 kg ha⁻¹, on average) than in N0 (113 kg ha⁻¹) (Table 4), but the magnitude of the effect of N fertilization rate was year-dependent. Indeed, in 2014 CNU was highest under N140, while in 2015 under N280 (Table 5). The highest CNU reported for the N140 plots may also explain the highest MYs observed (Figure 2). In general, CNU decreased from 2014 to 2015 (Table 5). Indeed, N uptake is favored when irrigation or soil humidity is close to 100% of maximum evapotranspiration [53], but high rainfall (as experienced in 2015) may lead to significantly higher N leaching losses [54].

Accordingly, the three N efficiency indices studied (NUE, NUpE, NUtE) tended to decline with increasing N fertilization rates (Table 4), but the extent of the impairment was year-dependent (Table 5) as also reported by Xu et al. [55] and Meise et al. [35]. The low NUE values of potato, as compared to other crops, can be related to its shallow rooting system that leads to a restricted uptake and, thus, use of N [56]. In addition, the water regime can significantly affect the NUE [16]. This may explain the general decrease of NUE from 2014 to 2015, since the higher rainfall in 2015 may have promoted nitrate leaching out of the rooting zone. NUE depends on two processes: the uptake efficiency (NUpE, the ability to remove N from the soil) and the utilization efficiency (NUtE, the ability to use the absorbed N to produce yield) [36]. Indeed, our results on NUE were strongly correlated with those of NUpE ($r^2 = 0.99$ ***) and NUtE ($r^2 = 0.89$ ***). Therefore, the lowest NUE values under N280 may be associated with the reduced N uptake ability and the lowered transport/redistribution of nutrients, resulting in lower yields than N140. Also Tyler et al. [57] and Zvomuya et al. [58] revealed decreasing values of both NUpE and NUtE with growing N fertilization amounts.

Table 5. CNU, NUE, NUpE, NUtE and FUE as affected by 'N fertilization rate × year' interaction. Different letters within each column indicate significant statistical differences (LSD test, $p < 0.05$).

Year	N Fertilization Rate	CNU (kg ha^{-1})	NUE (kg Tuber DM kg N^{-1})	NUpE (kg N kg N^{-1})	NUtE (kg Tuber DM kg N^{-1})	FUE (kg kg^{-1})
2014	N0	120.5 ± 2.5 d	122.1 ± 1.3 a	1.72 ± 0.06 a	70.9 ± 2.3 a	-
	N140	162.6 ± 3.0 a	52.0 ± 1.8 c	0.77 ± 0.08 c	67.1 ± 2.0 b	125.7 ± 2.0 a
	N280	150.9 ± 4.0 b	23.8 ± 1.0 e	0.43 ± 0.05 e	55.2 ± 1.9 d	9.6 ± 0.3 d
2015	N0	104.8 ± 2.9 e	104.5 ± 1.8 b	1.5084 ± 0.05 b	69.7 ± 1.6 a	-
	N140	132.2 ± 3.0 c	41.6 ± 1.5 d	0.63 ± 0.05 d	66.0 ± 2.5 b	64.3 ± 1.2 b
	N280	145.7 ± 2.2 bc	23.8 ± 1.2 e	0.42 ± 0.05 e	57.1 ± 2.5 c	25.7 ± 1.8 c

Data are mean ± standard deviation of 4 individual plots ($n = 4$).

From an agronomic fertilizer use efficiency standpoint, N140 was most efficient than N280 in both years (125.7 vs. 9.6 kg kg^{-1} in 2014 and 64.3 vs. 25.7 kg kg^{-1} in 2015). Since agronomic fertilizer FUE is defined as the increase in yield of the harvested portion of the crop per unit of fertilizer applied, our results suggest that applying N fertilizer in high amounts (N280) might have resulted in more N losses. In 2014 the wider difference, in terms of FUE, between N140 and N280 is directly correlated to the trend of yield and its components (particularly, MTW) as compared to 2015.

3.4. Tuber Chemical Traits

The results about the effects of N fertilization rate on the chemical parameters of early potato tubers were reported in Tables 4 and 6.

Table 6. Chemical composition of early potato tubers as affected by the interaction between year and nitrogen fertilization rate. Different letters within each column indicate statistically significant differences (LSD test, $p < 0.05$).

Year	N Fertilization Rate	Dry Matter (DM) (%)	Starch (g kg^{-1} DM)	Nitrate (g kg^{-1} DM)	Ascorbic Acid (g kg^{-1} DM)	Total Polyphenols (g kg^{-1} DM)	Antioxidant Activity (%$_{inhibition\ DPPH}$)
2014	N0	17.5 ± 1.0 a	647 ± 4 ab	0.94 ± 0.06 c	0.67 ± 0.02 b	3.42 ± 0.11 b	58.2 ± 1.2 bc
	N140	16.5 ± 1.5 cd	657 ± 8 a	1.01 ± 0.03 b	0.58 ± 0.04 d	3.04 ± 0.05 c	55.9 ± 0.4 c
	N280	16.3 ± 0.9 d	592 ± 9 cd	1.15 ± 0.09 a	0.54 ± 0.06 e	3.05 ± 0.08 c	52.5 ± 0.8 d
2015	N0	17.0 ± 0.5 b	632 ± 5 b	0.78 ± 0.06 e	0.75 ± 0.04 a	4.31 ± 0.20 a	66.4 ± 0.7 a
	N140	16.8 ± 0.7 bc	612 ± 11 c	0.85 ± 0.04 d	0.64 ± 0.03 c	3.58 ± 0.13 b	60.7 ± 0.6 b
	N280	16.6 ± 0.1 c	569 ± 10 d	1.02 ± 0.04 b	0.49 ± 0.05 f	3.61 ± 0.07 b	56.3 ± 0.4 c

Data are mean ± standard deviation of four individual plots ($n = 4$).

Overall, the tuber chemical traits, except for the total protein and soluble sugars content, was significantly affected by 'N fertilization rate x year' interaction. Firstly, N fertilization rate influenced the tuber DM level, one of the most important traits of potato tubers in the context of domestic cooking and industrial processing quality [59]. In both years the DM for N280 did not differ significantly from that shown for N140 (Table 6). Indeed, nitrogen is essential for potato canopy growth, but its over-supply could delay maturity and thus may reduce DM and starch levels [16]. In addition, a higher N fertilization rate results in tubers with immature skin prone to bruising and susceptibility to shatter bruise [60]. The range in DM (<20%) here observed was consistent with previous results reported in literature [41], making our samples more suitable for processing into boiled and frozen products according to the classification given by Cacace et al. [61]. A variation in the DM values over the two years was also observed and it can be ascribed to the meteorological conditions (temperature and rainfall). In particular, the cooler temperatures experienced in 2015 may be responsible for a longer time necessary for the interception of global solar radiation flux density and conversion of intercepted radiation into DM [62].

Sugars make a major contribution to the overall tuber DM, mostly deposited in the form of starch. This was found markedly affected by the 'N fertilization rate x year' interaction (Table 6). In both years, it was decreased by the conventional fertilization rate (N280) (592 and 569 g kg^{-1} of DM in 2014 and 2015, respectively), confirming the trend here reported for the tuber DM level. By contrast, tubers grown under N140 had comparable starch levels compared to those unfertilized (N0). A previous study reported an inverse relationship between nitrogen fertilization rate and levels of DM and starch [63]. This is in agreement with the "carbon/nitrogen balance" theory [64], which proposes that when N availability limits plant growth the metabolism shifts towards carbon rich compounds, such as starch. Potato tubers also contain considerable amount of soluble sugars, mainly sucrose, glucose and fructose, which have impact on their processing [59]. In particular, high amounts of sucrose (a non-reducing sugar), glucose and fructose (reducing sugars) in potato tubers are undesired for processing at high temperatures because reducing sugars are precursors of the Maillard reaction and sucrose is the main source of reducing sugars during its enzyme-catalyzed hydrolysis [65]. In this study, sucrose was the most and fructose the least abundant (Table 4). Their level, similar to the data reported by Lombardo et al. [41], was significantly influenced only by the N fertilization rate. This result is consistent with the above-cited "carbon/nitrogen balance" theory [64], since potato tubers grown under limited N availability (N0 in the present study) accumulated more soluble sugars than those grown under N140 and N280 (Table 4).

Although potato tubers are commonly regarded as a source of sugars, they also contain a good level of protein that can vary due to several pre-harvest factors [66]. In this study, this qualitative parameter was strongly influenced by the N fertilization rate and, as expected, it was significantly higher in the tubers grown under N280 (100 g kg^{-1} DM) than in those grown under N140 and N0 (86 and 82 g kg^{-1} DM, respectively; Table 4). Indeed, according to Wang et al. [67], a low fertilization supply decreased tuber nitrogenous compounds due to the dilution effect caused by higher DM accumulation.

The conventional N fertilization rate (N280), commonly adopted in the Mediterranean Basin for potato cultivation, also markedly enhanced the nitrate level in the early potato tubers (Tables 4 and 6). The higher nitrate level experienced in 2014 are in agreement with Sadej and Namiotko [68]. Indeed, both the higher temperatures and lower total rainfall in 2014 may have favoured the uptake of nitrates by plants. Although nitrate content is less crucial for tuber quality than DM level, it may impair food safety due to possible hazard to human health driven by the increasing volume of potato tubers consumption [69]. Therefore, also due to the lack of threshold limits in European Union (EU) legislation, some countries have already introduced inland regulations limiting nitrate content in commercialized potato tubers. As an instance, in Germany (the most important market for the exported Mediterranean early potato) only tubers with less than 200 mg kg^{-1} of fresh weight are accepted. Here, no value exceeded such threshold limit since the maximum level of tuber nitrate content was 1.15 g kg^{-1} DM under N280, with a DM of 16.3%, which corresponds to 189 mg kg^{-1} of fresh weight. However, the optimized fertilization rate (140 kg N ha^{-1}) is preferable with a perspective of a lower nitrate level in the tubers (0.93 g kg^{-1} DM, on average of years) combined with an acceptable productive yield and a minor environmental pollution.

Potato tubers are, due to their consumption rate, one of the major sources of antioxidant compounds in the human diet. The early potato tubers contain high levels of ascorbic acid and polyphenols [11,28]. The ascorbic acid is an inhibitor of enzymatic browning, therefore its presence helps to reduce some post-harvest qualitative losses [70,71]; while the polyphenols are associated with a range of health-promoting properties [72]. Here, ascorbic acid and total polyphenol amounts were both influenced by 'N fertilization rate × year' interaction (Table 6). In particular, it is apparent from our results that in both years the conventional N fertilization rate (280 kg N ha^{-1}) had a negative effect on the level of ascorbic acid (Table 6). A major amount of the soil N available to the crop may likely stimulate leaf growth, and thereby enhance the photosynthetic rate and the production of the sugars needed for ascorbic acid synthesis [11], but at the same time the increased plant foliage to high N fertilization rates may reduce the light intensity and accumulation of ascorbic acid in plant shaded parts [73]. Hence, the choice of the N fertilization rate may have a significant effect on the balance between these two phenomena, i.e., leaf growth and shading, and thereby on the ascorbic acid synthesis and accumulation in potato tubers. In this regard, N140 seems to be the right compromise with the aim of obtaining tubers with high health-promoting properties, since the conventional N fertilization rate (N280) increased the concentration of nitrates and simultaneously decreased that of ascorbic acid (Table 6). The differences among the studied N fertilization rates on the ascorbic acid accumulation was particularly notable in 2015 (Table 6), probably due to the higher rainfall level and lowest minima air temperatures experienced (Table 1).

The content of total polyphenols was also evaluated in the present study (Tables 4 and 6). Consistently with the results obtained for the ascorbic acid, in both 2014 and 2015 N0 tubers showed the highest total polyphenol amounts compared to those of N140 and N280. Our results are in agreement with Lachman et al. [74], which report an increase in the total polyphenols content under low N supply or deficiency due to the increased activity of phenylalanine ammonia lyase (PAL), the key-enzyme for their biosynthesis. With respect to human health, increasing the total polyphenol content through the proper choice of N fertilization rate could be of importance in diets which are dominated by potatoes. Also for the total polyphenols content, the differences among the tested N fertilization rates were notably evident in 2015 as a response to the meteorological conditions during the field trials. In particular, this may be due to the higher level of precipitation experienced in 2015 (Table 1), as total polyphenol content is known to be enhanced by high humidity [75]. Finally, the N fertilization rate also affected the antioxidant activity of early potato tubers (Tables 4 and 6). This parameter tended to be higher in N0 tubers grown than in N140 and N280 ones, confirming the trend observed for both ascorbic acid and total polyphenol levels. However, it is noteworthy that N140 tubers displayed a higher antioxidant activity than N280 ones (58.3% vs. 54.4%, on average of years). In addition, all the N fertilization rates under study harboured greater antioxidant activity in 2015, as a response to the abiotic stresses induced by meteorological conditions. This was also corroborated by the larger difference between N0 and N280, as highlighted for the total polyphenols content (Tables 4 and 6).

4. Conclusions

Recently, agronomic research has increasingly been directed at finding management practices that maximize crop production and enhance product quality, while minimizing the environmental impact. This is particularly true for the potato, a crop that requires significant external inputs during both vegetative growth and tuber bulking, to meet the yield and qualitative standard levels demanded by either the fresh market or processing industry. Hence, potato growers often tend to use huge amounts of inorganic (especially nitrogenous) fertilizers to maximize their incomes. In this sense, for the first time, this study provided comprehensive data on both agronomic, N use efficiency and tuber qualitative traits of the early crop potato under different N fertilization rates with the aim to investigate whether it is possible to reduce the N fertilization rate without implications on the aforementioned characteristics. In particular, we highlighted that an optimal N fertilization rate (140 kg ha^{-1}, based on soil nitrogen balance, crop rotation and potato requirements) may ensure a high yield and a limited reduction of N use efficiency, combined with important nutritional traits of the tubers, e.g., a high level of dry matter, starch, total polyphenols and ascorbic acid, and a low nitrate amount, as compared to the unfertilized and over-fertilized plots. These results may have positive repercussions for potato cultivation, allowing farmers to increase their incomes through better tuber quality and lower production costs. In addition, our findings are relevant in the perspective to limit environmental pollution by reducing the N fertilization rate to the early potato crop, since growers often adopt N over-fertilization (280 kg ha^{-1} or more) without a scientifically supported basis. Taking into account that the experimental field-trials were carried in a typical potato cultivation area in Sicily (soil characteristics, climate, crop rotation and management), which is also representative of the potato cultivation in the Mediterranean basin, we reasonably considered the fertilization rate of 140 kg N ha^{-1} as a recommendable target dose in similar soils, with a N availability equal to ~70 kg ha^{-1}. Future research studies are, however, necessary to assess the behavior of other cultivars, as well as to deepen insights into the possible interaction of N fertilization with other agronomic practices (e.g., irrigation) in terms of yield, N use efficiency and tuber quality performances of early potato crop.

Author Contributions: All the authors equally contributed to the research plan and to write the manuscript. In particular, G.M. was responsible for funding acquisition; S.L. and G.P. were responsible for laboratory analyses of samples and validation of analytical methods. All authors have read and agreed to the published version of the manuscript.

Funding: This research was funded by "Utilizzo integrato di approcci tecnologici innovativi per migliorare la shelf-life e preservare le proprietà nutrizionali di prodotti agroalimentari (SHELF-LIFE)", project cod. PON02_00451_3361909.

Acknowledgments: The authors particularly thank Angelo Litrico (Catania University) for his excellent technical assistance in conducting the field trial.

Conflicts of Interest: The authors declare no conflict of interest.

References

1. Savci, S. An agricultural pollutant: Chemical fertilizer. *Int. J. Environ. Sci. Dev.* **2012**, *3*, 77–80. [CrossRef]
2. Stefanelli, D.; Goodwin, I.; Jones, R. Minimal nitrogen and water use in horticulture: Effects on quality and content of selected nutrients. *Food Res. Int.* **2010**, *43*, 1833–1843. [CrossRef]
3. Mengel, K.; Kirkby, E.A.; Kosegarten, H.; Appel, T. *Principles of Plant Nutrition*; Kluwer Academic Publishers: Dordrecht, The Netherlands, 2001.
4. Ryan, J.; Masri, S.; Ceccarelli, S.; Grando, S.; Ibrikci, H. Differential responses of barley landraces and improved barley cultivars to nitrogen-phosphorus fertilizer. *J. Plant Nutr.* **2008**, *31*, 381–393. [CrossRef]
5. Karyotis, T.; Güçdemir, İ.; Akgül, S.; Panagopoulos, A.; Karyoti, K.; Demir, S.; Kasacı, A. Nitrogen fertilization plans for the main crops of Turkey to mitigate nitrates pollution. *Eurasian J. Soil Sci.* **2014**, *3*, 13–24. [CrossRef]
6. FAO Statistical Database. Available online: http://www.faostat.org/ (accessed on 5 July 2019).
7. Frusciante, L.; Barone, A.; Carputo, D.; Ranalli, P. Breeding and physiological aspects of potato cultivation in the Mediterranean region. *Potato Res.* **1999**, *42*, 265–277. [CrossRef]
8. Ierna, A.; Pandino, G.; Lombardo, S.; Mauromicale, G. Tuber yield, water and fertilizer productivity in early potato as affected by a combination of irrigation and fertilization. *Agric. Water Manag.* **2011**, *101*, 35–41. [CrossRef]
9. Cantore, V.; Wassar, F.; Yamac, S.S.; Sellami, M.H.; Albrizio, R.; Stellacci, A.M.; Todorovic, M. Yield and water use efficiency of early potato grown under different irrigation regimes. *Int. J. Plant Prod.* **2014**, *8*, 409–428.
10. Meise, P.; Seddig, S.; Uptmoor, R.; Ordon, F.; Schum, A. Impact of nitrogen supply on leaf water relations and physiological traits in a set of potato (*Solanum tuberosum* L.) cultivars under drought stress. *J. Agron. Crop. Sci.* **2018**, *204*, 59–374. [CrossRef]
11. Lombardo, S.; Lo Monaco, A.; Pandino, G.; Parisi, B.; Mauromicale, G. The phenology: Yield and tuber composition of 'early' crop potatoes: A comparison between organic and conventional cultivation systems. *Renew. Agric. Food Syst.* **2013**, *28*, 50–58. [CrossRef]
12. Lombardo, S.; Pandino, G.; Mauromicale, G. The influence of growing environment on the antioxidant and mineral content of "early" crop potato. *J. Food Compos. Anal.* **2013**, *32*, 28–35. [CrossRef]
13. Ierna, A.; Mauromicale, G. Potato growth, yield and water productivity response to different irrigation and fertilization regimes. *Agric. Water Manag.* **2018**, *201*, 21–26. [CrossRef]
14. Izmirlioglu, G.; Demirci, A. Enhanced bio-ethanol production from industrial potato waste by statistical medium optimization. *Int. J. Mol. Sci.* **2015**, *16*, 24490–24505. [CrossRef]
15. Jagatee, S.; Behera, S.; Dash, P.K.; Sahoo, S.; Mohanty, R.C. Bioprospecting starchy feedstocks for bioethanol production: A future perspective. *JMRR* **2015**, *3*, 24–42.
16. Koch, M.; Naumann, M.; Pawelzik, E.; Gransee, A.; Thiel, H. The importance of nutrient management for potato production Part I: Plant nutrition and yield. *Potato Res.* **2019**. [CrossRef]
17. Mauromicale, G.; Signorelli, P.; Ierna, A.; Foti, S. Effects of intraspecific competition on yield of early potato grown in Mediterranean environment. *Am. J. Potato Res.* **2003**, *80*, 281–288. [CrossRef]
18. Ierna, A.; Mauromicale, G. Sustainable and profitable nitrogen fertilization management of potato. *Agronomy* **2019**, *10*, 582. [CrossRef]

19. Milroy, S.P.; Wang, P.; Sodras, V.O. Defining upper limits of nitrogen uptake and nitrogen use efficiency of potato in response to crop N supply. *Field Crop. Res.* **2019**, *239*, 38–46. [CrossRef]
20. Ospina, C.A.; Lammerts van Bueren, E.T.; Allefs, J.J.H.M.; Engel, B.; van der Putten, P.E.L.; van der Linden, C.G.; Struik, P.C. Diversity of crop development traits and nitrogen use efficiency among potato cultivars grown under contrasting nitrogen regimes. *Euphytica* **2014**, *199*, 13–29. [CrossRef]
21. Vos, J.; van der Putten, P.E.L. Effect of nitrogen supply on leaf growth, leaf nitrogen economy and photosynthetic capacity in potato. *Field Crop. Res.* **1998**, *59*, 63–72. [CrossRef]
22. Mauromicale, G.; Ierna, A.; Marchese, M. Chlorophyll fluorescence and chlorophyll content in field-grown potato as affected by nitrogen supply, genotype, and plant age. *Photosynthetica* **2006**, *44*, 76–82. [CrossRef]
23. Ahmed, A.; Zaki, M.; Shafeek, M.; Helmy, Y.; El-Baky, M.A. Integrated use of farmyard manure and inorganic nitrogen fertilizer on growth, yield and quality of potato (*Solanum tuberosum* L.). *Int. J. Curr. Microbiol. App. Sci.* **2015**, *4*, 325–349.
24. Fontes, P.C.R.; Braun, H.; Busato, C.; Cecon, P.R. Economic optimum nitrogen fertilization rates and nitrogen fertilization rate effects on tuber characteristics of potato cultivars. *Potato Res.* **2010**, *53*, 167–179. [CrossRef]
25. Foti, S.; Mauromicale, G.; Ierna, A. Influence of irrigation regimes on growth and yield of potato cv. Spunta. *Potato Res.* **1995**, *38*, 307–318. [CrossRef]
26. Getahun, B.B. Potato breeding for nitrogen-use efficiency: Constraints, achievements, and future prospects. *J. Crop. Sci. Biotechnol.* **2018**, *21*, 269–281. [CrossRef]
27. Tiwari, J.K.; Plett, D.; Garnett, T.; Chakrabarti, S.K.; Singh, R.K. Integrated genomics, physiology and breeding approaches for improving nitrogen use efficiency in potato: Translating knowledge from other crops. *Funct. Plant Biol.* **2018**, *45*, 587–605. [CrossRef]
28. Bélanger, G.; Walsh, J.R.; Richards, J.E.; Milburn, P.H.; Ziadi, N. Yield response of two potato cultivars to supplemental irrigation and N fertilization in New Brunswick. *Am. J. Potato Res.* **2000**, *77*, 11–21. [CrossRef]
29. Soil Survey Staff. *Soil Taxonomy: A Basic System of Soil Classification for Making and Interpreting Soil Surveys*, 2nd ed.; U.S. Gov. Print. Office: Washington, DC, USA, 1999.
30. Fierotti, G.; Dazzi, C.; Raimondi, S. *Carta dei suoli della Sicilia 1:250.000*; Regione Siciliana and Università degli Studi di Palermo: Palermo, Italy, 1988.
31. Foti, S. Early potatoes in Italy with particular reference to Sicily. *Potato Res.* **1999**, *42*, 229–240. [CrossRef]
32. Italian Society of Soil Science. *Metodi Normalizzati di Analisi del Suolo*; Edagricole: Bologna, Italy, 1985.
33. UNICHIM. *Analisi dei Terreni Agrari*; Unichim: Milano, Italy, 1985.
34. Mauromicale, G.; Ierna, A. Patata primaticcia. In *Fisionomia e profili di qualità dell'orticoltura meridional*; Bianco, V.V., La Malfa, G., Tudisca, G., Eds.; Arti Grafiche Siciliane: Palermo, Italy, 1999; pp. 275–296.
35. Uddling, J.; Gelang-Alfredsson, J.; Piikki, K.; Pleijel, H. Evaluating the relationship between leaf chlorophyll concentration and SPAD-502 chlorophyll meter readings. *Photosynth. Res.* **2007**, *91*, 37–46. [CrossRef]
36. Meise, P.; Seddig, S.; Uptmoor, R.; Ordon, F.; Schum, A. Assessment of yield and yield components of starch potato cultivars (*Solanum tuberosum* L.) under nitrogen deficiency and drought stress conditions. *Potato Res.* **2019**, *62*, 193–220. [CrossRef]
37. Moll, R.H.; Kamprath, E.J.; Jackson, W.A. Analysis and interpretation of factors which contribute to efficiency of nitrogen utilization. *Agron. J.* **1982**, *74*, 562–564. [CrossRef]
38. Gariglio, N.F.; Pilatti, R.A.; Baldi, B.L. Using nitrogen balance to calculate fertilization in strawberries. *HortTechnology* **2000**, *10*, 147–150. [CrossRef]
39. Ceccon, P.; Fagnano, M.; Grignani, C.; Monti, M.; Orlandini, S. *Agronomia*; Edises: Napoli, Italy, 2017.
40. Snyder, J.C.; Desborough, S.L. Rapid estimation of potato tuber total protein content with coomassie brilliant blue g-250. *Appl. Genet.* **1978**, *52*, 135–139. [CrossRef] [PubMed]
41. Lombardo, S.; Pandino, G.; Mauromicale, G. The effect on tuber quality of an organic versus a conventional cultivation system in the early crop potato. *J. Food Comp. Anal.* **2017**, *62*, 189–196. [CrossRef]
42. Lombardo, S.; Pandino, G.; Mauromicale, G. The nutraceutical response of two globe artichoke cultivars to contrasting NPK fertilizer regimes. *Food Res. Int.* **2015**, *76*, 852–859. [CrossRef]
43. Brand-Williams, W.; Cuvelier, M.E.; Berset, C. Use of a free radical method to evaluate antioxidant activity. *Lebensm. Wiss. Technol.* **1995**, *22*, 25–30. [CrossRef]
44. Wilhelm, W.W.; Arnold, S.L.; Schepers, J.S. Using a nitrate specific ion electrode to determine stalk nitrate-nitrogen concentration. *Agron. J.* **2000**, *92*, 186–189.

45. Shapiro, S.S.; Wilk, M.B. Analysis of variance test for normality (complete samples). *Biometrika* **1965**, *52*, 591–611. [CrossRef]

46. Levene, H. Robust tests for equality of variances. In *Contributions to Probability and Statistics*; Olkin, I., Ghurye, S.G., Hoeffding, W., Madow, W.G., Mann, H.B., Eds.; Stanford University Press: Stanford, CA, USA, 1960; pp. 278–292.

47. Busato, C.; Fontes, P.C.R.; Braun, H.; Cecon, P.R. Seasonal variation and threshold values for chlorophyll meter readings on leaves of potato cultivars. *J. Plant Nutr.* **2010**, *33*, 2148–2156. [CrossRef]

48. Gianquinto, G.; Goffart, J.P.; Olivier, M.; Guarda, G.; Colauzzi, M.; Dalla Costa, L.; Delle Vedove, G.; Vos, J.; MacKerron, D.K.L. The use of hand-held chlorophyll meters as a tool to assess the nitrogen status and to guide nitrogen fertilization of potato crop. *Potato Res.* **2004**, *47*, 35–80. [CrossRef]

49. Darwish, T.M.; Atallah, T.W.; Hajhasan, S.; Haidar, A. Nitrogen and water use efficiency of fertigated processing potato. *Agric. Water Manag.* **2006**, *85*, 95–104. [CrossRef]

50. Badr, M.A.; El-Tohamy, W.A.; Zaghloul, A.M. Yield and water use efficiency of potato grown under different irrigation and nitrogen levels in an arid region. *Agric. Water Manag.* **2012**, *110*, 9–15. [CrossRef]

51. Zebarth, B.J.; Rosen, C.J. Research perspective on nitrogen bmp development for potato. *Am. J. Potato Res.* **2007**, *84*, 3–18. [CrossRef]

52. De la Morena, I.; Guillén, A.; del Moral, L.F.G. Yield development in potatoes as influenced by cultivar and the timing and level of nitrogen fertilization. *Am. Potato J.* **1994**, *71*, 165–173. [CrossRef]

53. Dalla Costa, L.; Delle Vedove, G.; Gianquinto, G.; Giovanardi, R.; Peressotti, A. Yield, water use efficiency and nitrogen uptake in potato: Influence of drought stress. *Potato Res.* **1997**, *40*, 19–34. [CrossRef]

54. Neumann, A.; Torstensson, G.; Aronsson, H. Nitrogen and phosphorus leaching losses from potatoes with different harvest times and following crops. *Field Crop. Res.* **2012**, *133*, 130–138. [CrossRef]

55. Xu, G.; Fan, X.; Miller, A.J. Plant nitrogen assimilation and use efficiency. *Annu. Rev. Plant Biol.* **2012**, *63*, 153–182. [CrossRef]

56. Iwama, K. Physiology of the potato: New insights into root system and repercussions for crop management. *Potato Res.* **2008**, *51*, 333–353. [CrossRef]

57. Tyler, K.B.; Broadbent, F.E.; Bishop, J.C. Efficiency of nitrogen uptake by potatoes. *Am. Potato J.* **1983**, *60*, 261–269. [CrossRef]

58. Zvomuya, F.; Rosen, C.J.; Creighton Miller, J. Response of Russet Norkotah clonal selections to nitrogen fertilization. *Am. J. Potato Res.* **2002**, *79*, 231–239. [CrossRef]

59. Solaiman, A.H.M.; Nishizawa, T.; Roy, T.S.; Rahman, M.; Chakraborty, R.; Choudhury, J.; Choudhury, J.; Sarkar, M.D.; Hasanuzzama, M. Yield, dry matter: Specific gravity and color of three Bangladesh local potato cultivars as influenced by stage of maturity. *J. Plant Sci.* **2015**, *10*, 108–115. [CrossRef]

60. Dean, B.B.; Thomton, R.E. *The Specific Gravity of Potatoes*; Washington State Univ. Cooperative Extension Bulletin #1541: Pullman, WA, USA, 1992.

61. Cacace, J.E.; Huarte, M.A.; Monti, M.C. Evaluation of potato cooking quality in Argentina. *Am. Potato J.* **1994**, *71*, 145–153. [CrossRef]

62. Pereira, A.B.; Villa Nova, N.A.; Ramos, V.J.; Pereira, A.R. Potato potential yield based on climatic elements and cultivar characteristics. *Bragantia* **2008**, *67*, 327–334. [CrossRef]

63. Locascio, S.J.; Wiltbank, W.J.; Gull, D.D.; Maynard, D.N. Fruit and vegetable quality as affected by nitrogen nutrition. In *Nitrogen in Crop Production*; Hauck, R.D., Ed.; American Society of Agronomy: Madison, WI, USA, 1984; pp. 617–641.

64. Bryant, J.P.; Chapin, F.S.; Klein, D.R. Carbon/nutrient balance of boreal plants in relation to vertebrate herbivory. *Oikos* **1983**, *40*, 357–368. [CrossRef]

65. van Eck, H.A. Genetics of morphological and tuber traits. In *Potato Biology and Biotechnology: Advances and Perspectives*; Vreugdenhil, D., Bradshaw, J., Gebhardt, C., Govers, F., MacKerron, D.K.L., Taylor, M., Ross, H., Eds.; Elsevier: Oxford, UK, 2007; pp. 91–111.

66. Bártová, V.; Bárta, J.; Diviš, J.; Švajner, J.; Peterka, J. Crude protein content in tubers of starch processing potato cultivars in dependence on different agro-ecological conditions. *J. Cent. Eur. Agric.* **2009**, *10*, 57–66.

67. Wang, Z.; Li, S.; Malhi, S. Effects of fertilization and other agronomic measures on nutritional quality of crops. *J. Sci. Food Agric.* **2008**, *88*, 7–23. [CrossRef]

68. Sadej, W.; Namiotko, A. Nitrates content in potato tubers cultivated under various fertilization systems. *Ecol. Chem. Eng. A* **2011**, *18*, 1123–1130.

69. Santamaria, P. Nitrate in vegetable: Toxicity, content, intake and EC regulation. *J. Sci. Food Agric.* **2006**, *86*, 10–17. [CrossRef]

70. Licciardello, F.; Lombardo, S.; Rizzo, V.; Pitino, I.; Pandino, G.; Strano, M.G.; Muratore, G.; Restuccia, C.; Mauromicale, G. Integrated agronomical and technological approach for the quality maintenance of ready-to-fry potato sticks during refrigerated storage. *Postharvest Biol. Technol.* **2018**, *136*, 23–30. [CrossRef]

71. Rizzo, V.; Amoroso, L.; Licciardello, F.; Mazzaglia, A.; Muratore, G.; Restuccia, C.; Lombardo, S.; Pandino, G.; Strano, M.G.; Mauromicale, G. The effect of sous vide packaging with rosemary essential oil on storage quality of fresh-cut potato. *LWT Food Sci. Technol.* **2018**, *94*, 111–118. [CrossRef]

72. Brown, C.R. Antioxidants in potato. *Am. J. Potato Res.* **2005**, *82*, 163–172. [CrossRef]

73. Lee, S.K.; Kader, A. Preharvest and postharvest factors influencing vitamin C content of horticultural crops. *Postharvest Biol. Technol.* **2000**, *20*, 207–220. [CrossRef]

74. Lachman, J.; Hamouz, K.; Čepl, J.; Pivec, V.; Šulc, M.; Dvořák, P. The effect of selected factors on polyphenol content and antioxidant activity in potato tubers. *Chem. Listy* **2006**, *100*, 522–527.

75. Hamouz, K.; Čepl, J.; Dvořák, P. Influence of environmental conditions on the quality of potato tubers. *Hortic. Sci.* **2005**, *32*, 89–95. [CrossRef]

 *agronomy*

Review

Integrated Weed Management in Herbaceous Field Crops

Aurelio Scavo * and **Giovanni Mauromicale**

Department of Agriculture, Food and Environment (Di3A), University of Catania, 95123 Catania, Italy;
g.mauromicale@unict.it
* Correspondence: aurelio.scavo@unict.it; Tel.: +390954783415

Received: 10 February 2020; Accepted: 24 March 2020; Published: 27 March 2020

Abstract: Current awareness about the environmental impact of intensive agriculture, mainly pesticides and herbicides, has driven the research community and the government institutions to program and develop new eco-friendly agronomic practices for pest control. In this scenario, integrated pest management and integrated weed management (IWM) have become mandatory. Weeds are commonly recognized as the most important biotic factor affecting crop production, especially in organic farming and low-input agriculture. In herbaceous field crops, comprising a wide diversity of plant species playing a significant economic importance, a compendium of the specific IWM systems is missing, that, on the contrary, have been developed for single species. The main goal of this review is to fill such gap by discussing the general principles and basic aspects of IWM to develop the most appropriate strategy for herbaceous field crops. In particular, a 4-step approach is proposed: (i) prevention, based on the management of the soil seedbank and the improvement of the crop competitiveness against weeds, (ii) weed mapping, aiming at knowing the biological and ecological characteristics of weeds present in the field, (iii) the decision-making process on the basis of the critical period of weed control and weed thresholds and iv) direct control (mechanical, physical, biological and chemical). Moreover, the last paragraph discusses and suggests possible integrations of allelopathic mechanisms in IWM systems.

Keywords: sustainable agriculture; integrated weed management; yield losses; preventive weed control; mechanical weed control; physical weed control; biological weed control; herbicides; allelopathy

1. Introduction

Herbaceous field crops include several hundred plants species diffused worldwide, of which about 100–200 play a significant economic importance, especially in developing countries. Among them, only 15–20 species play a key role for the global economy, with about 1600 million ha of harvested area. Herbaceous field crops can be classified based on taxonomy, life span cycle, climate, season, human uses and plant part used (Figure 1). It is now well recognized that weeds are the most important biotic factor affecting their growth and yield [1]. On average, Oerke [2] calculated a potential loss of 34% of crop production caused by weed pressure, followed by −18% from animal pests and −16% from pathogens. Furthermore, he estimated, as follows, the potential losses of six major herbaceous field crops: wheat −23%, rice −37%, maize −40%, potato −30%, soybean −37% and cotton −36%. The annual global economic loss caused by weeds was estimated by Appleby et al. [3] at more than 100 billion US dollars, while Kraehmer and Baur [4] assessed their control global cost as running into the $ billions. For this reason, and considering also that weeds are a dynamic threat, weed control has always been placed in the center of the agricultural activity by farmers since ancient times. Nowadays, weed management in cropping systems branches out into two different directions corresponding to distinct approaches [5]:

in one scenario, the widespread use of synthetic herbicides, while in the other, weed suppression is largely based on mechanical, physical and ecological methods. The former direction has been the most adopted by developed countries after World War II with the aim of increasing yields. This approach, however, has caused considerable negative effects on environmental, human and animal health. Moreover, the improper utilization of herbicides in agroecosystems was accompanied by a dramatic increase of herbicide-resistant weeds, including those with multiple herbicide resistances, and effects on non-target organisms, as well as the development of a substitution weed flora and weed population shifts that contribute to make herbicide-dependent cropping systems more vulnerable [6,7]. These concerns have led, since the 1980s, to a growing public awareness of the adverse environmental effects of pesticides, including herbicides, typical of the conventional agriculture devoted to yield maximization [8]. In this context, the second scenario started to acquire more importance, driven by public opinion, agricultural policies and the scientific community. The aim of agriculture at present is to obtain a crop production programmed in quantity, quality and time while preserving the environment. In order to reduce the adoption of pesticides in favor of sustainable and eco-friendly agronomic practices for pest control, in 1991, the United Nations Conference on Environment and Development elected the Integrated Pest Management (IPM) as the preferred strategy for sustainable agriculture [9]. In 2009, the IPM, including the Integrated Weed Management (IWM), became mandatory in the European Union after the Directive 2009/128/EC on sustainable use of pesticides [10]. IWM play a cardinal role for the weed management of advanced cropping systems of developed countries, especially in the European Union, while on the contrary, it is still little adopted in developing countries. The increasing worldwide interest in IWM by the scientific community is demonstrated by Figure 2, which reports the number of journal papers using the keywords "integrated", "weed" and "management" on the Scopus® database. In this graph, it is possible to observe an exponential growth, still ongoing, since 1965, which corresponds to the period of the policies of Agenda 21, especially in the United States. The increased interest of researchers is probably also linked to the development and growth of organic farming, low-input and conservative agriculture, in which weed management is essentially based on IWM practices. Specific IWM systems have been developed for selected herbaceous field crops such as soybean [11], wheat [12], maize [13], rice [14], cotton [15], several horticultural species [16,17], etc. However, a compendium of these IWM systems lacks in literature and it could be important to help farmers in developing the most suitable IWM strategy applicable to such crops.

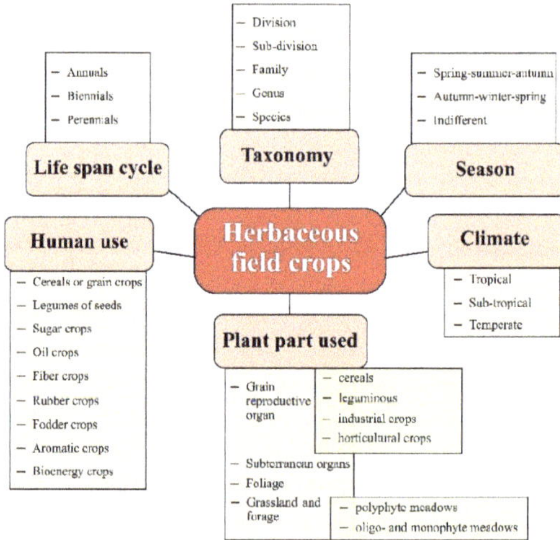

Figure 1. Criteria for classification of herbaceous field crops.

Number of journal papers

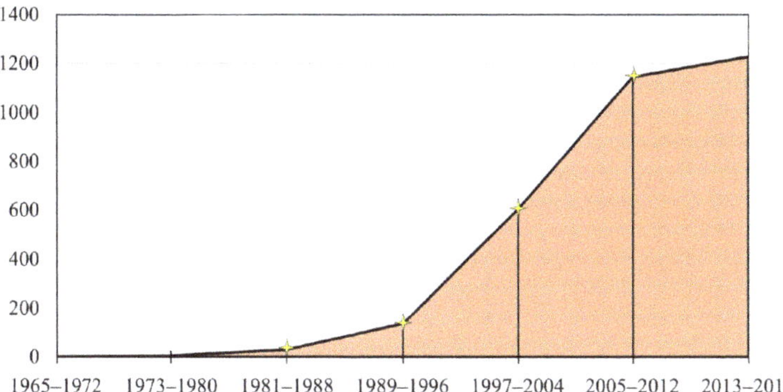

Figure 2. Number of journal papers among the past 54 years accessed on Scopus® using the search terms "integrated", "weed" and "management", arranged every 8 years (7 in the last period).

This review focuses on the general principles and basic aspects of IWM under a holistic approach to develop the most appropriate IWM strategy for herbaceous field crops. After an overview of preventive control methods focused on the management of the soil seedbank and the improvement of the crop competitiveness against weeds, a synthesis of the decision-making process is provided through the development of weed thresholds. In this regard, particular attention has been given to field weed mapping and the critical period of weed control (CPWC). Then, the direct control methods (mechanical, physical, biological and chemical) are presented separately for simplicity and to make the reading easier, but many examples of possible combinations are suggested. Finally, a description of the latest updates of allelopathy for weed control and its possible integration to an IWM strategy for herbaceous field crops is reported, with a view of sustainability.

2. Weeds in Agroecosystems

Weeds are generally referred to as *strictu sensu*, closely linked to agricultural activities. However, the concept of weed is relative and not absolute. Many definitions of weed, in fact, have been proposed by the scientific community under different points of view: agronomic, biological, ecological, etc. Nowadays, the definitions commonly adopted are those provided by the European Weed Research Society in 1986 ("any plant or vegetation, excluding fungi, interfering with the objectives or requirements of people") and by the Weed Science Society of America in 1989 ("a plant growing where is it not desired"). In this review, we consider weed as only the autotrophic higher plants, except for some heterotrophic parasitic plants such as *Cuscuta* spp., *Orobanche* spp., etc. Given the high biodiversity of weeds, Baker [18] produced a series of characteristics that might be expected in "the ideal weed". Among them, those to be taken more into account for weed management are:

The ability to germinate under adverse environmental conditions.
The ability to produce copious and diversified propagation organs, as well as the presence of mechanisms allowing to launch them at a distance and maintain long-viable seeds.
The high production of seeds (e.g., more than 190,000 seeds plant^{-1} for *Amaranthus retroflexus* L. and *Portulaca oleracea* L.) and discontinuous germination.
The rapid growth from the vegetative phase to flowering.
The highly competitive capacity and allelopathic activity.

These aspects are of key importance for a better setup and performance of an IWM strategy.

Harmful and Beneficial Effects of Weeds in Agroecosystems

The presence of weeds is often associated with a series of harmful aspects both in agro- and eco-systems, of which, the most important and widespread one is the reduction of crop yield. An exhaustive list of yield losses with relative costs was provided by Zimdahl [19]. Crop yield losses are caused by phenomena of weed competition, allelopathy and parasitism. Since in nature competition and allelopathy interact with high synergism, it should be noted that in the former, a vital resource for life (e.g., water, light, nutrient, space, etc.) is reduced or removed by another plant sharing the same habitat, while the latter implies the release of chemical substances with positive/stimulating or negative/inhibiting effects into the environment [20]. Qasem and Foy [21] identified and reported over 240 weeds with allelopathic properties on crops. Given the difficulty in distinguishing and separately describing allelopathic effects from those of competition, Muller [22] proposed the term "interference" to indicate the total adverse effect, allelopathy + competition, of one plant on another [19]. The level of crop–weed interference is determined by many factors acting additively, antagonistically or synergistically, and is closely linked to the genotype of both weed and crop (e.g., relative growth rates, development of the root system, time of emergence, seed size, seedling vigor, etc.) as well as to agronomic and environmental variables. In agricultural cropping systems, a complete crop failure (100% yield loss) occurs in the absence of weed control. Under a reductive approach, as plant density increases, crop yield gradually decreases. In order to better understand the effects of weed presence on crop production, since the 1980s, a series of bioeconomic and predictive yield models have been designed with the aim of developing economic weed thresholds as a basis for weed management decisions. Some of the most important empirical and ecophysiological models of crop–weed interference are reported in Table 1.

Other damages caused by weeds are related to the qualitative depletion of agricultural products in terms of food contamination or by acting directly on the dietary quality of the product. Moreover, weeds can harbor insect pests and other crop pathogens [21], increase production and processing costs (e.g., interference with agricultural operations such as mechanical tillage), decrease land value (especially perennial and parasitic weeds) and reduce crop choice, interfere with water management (e.g., increased evapotranspirative water losses, reduced water flow in irrigation ditches, etc.) and human aims in recreative areas and cause different kinds of allergic reactions in humans (several Poaceae species, *Parietaria officinalis* L., etc.) [19].

However, particularly when occurring at low densities, the presence of weeds also provides a series of agronomic and ecological (i.e., increasing of biodiversity) benefits. Weeds with a deep and extensive root system can reduce soil erosion and mineral nutrient leaching, conserve soil moisture and improve soil structure. The reduction of soil erosion is due on one side to the decrease of pouring rain action, and on the other side, to the fibrous and branched root system of monocotyledonous weeds such as *Digitaria* spp., *Cynodon* spp., *Agropyron* spp., *Echinochloa crus-galli* (L.) P. Beauv., etc. Such weeds, thanks to root branching and deepening, may help to increase water infiltration into the soil and improve the water holding capacity and soil structure. Regarding the latter aspect, it can be explained not only in physical terms, but also through the root exudation process which promotes the formation of aggregates thanks to the adsorption of rhizodeposits (e.g., ions such as Ca^{2+}, Fe^{2+}, Al^{3+}, K^+, mucillages and several organic acids) with colloids, and the stimulation of microorganisms [23]. In addition, the joint action of root exudates and weed living and dead mulch contribute to enhance the soil organic matter content. Nevertheless, in some cases, a moderate presence of weeds is reported to increase the soil nitrogen level by reducing nitrates losses via leaching and by the N_2 fixation of Fabaceae species with rhizosphere bacteria. Kapoor and Ramakrishnan [24], for example, found a significant increase of wheat dry weight yield when grown in association with *Medicago polyceratia* (L.) Trautv. For these reasons, in advanced cropping systems, weeds are seen as an integral part of the agroecosystem and thus, they should not be conceived as entities to be eliminated, but entities with many agroecological roles that must be managed. According to the "ecological restoration" concept of weed management proposed by Jordan and Vatovec [25], weeds should be accepted as a normal and manageable part of the agroecosystem and weed management should aim to reduce harmful effects and increase benefits resulting from this flora.

Table 1. List of major empirical and ecophysiological models estimating crop yield loss (Y) to weed density.

Model	Data	Type of Function	Reference
(A) Empirical models			
$Y = \frac{iD}{1+iD}$	D = weed density i = yield loss per weed m^{-2} as D$\to 0$	Rectangular hyperbola with one parameter	[26]
$Y = \frac{iD}{1+\frac{iD}{A}}$	A = maximum yield loss as D$\to \infty$	Rectangular hyperbola with two parameters	[27]
$Y = b_0 + b_1 X_1 + b_2 \sqrt{X_2}$	b_0 = Y intercept b_1 = regression coefficient for X_1 X_1 = time interval between weed and crop emergence b_2 = regression coefficient for density X_2 = weed density (plants m^{-2})	Linear through multiple-regression model	[28]
$Y = \frac{iD}{e^{CT}+\frac{iD}{A}}$	T = time interval between weed and crop emergence C = nonlinear regression coefficient	Rectangular hyperbola with three parameters and sigmoidal relationship between C and T	[29]
$Y = \frac{jD_c}{1+\frac{jD_c}{Y_{max}}} \times \left(1 - \frac{iD_w}{1+\frac{iD_w}{a}}\right)$	D_c = crop density D_w = weed density Y_{max} = maximum crop yield	Rectangular hyperbola consisting in two linked hyperbolic equations	[30]
(B) Ecophysiological models			
$Y = \frac{qL_w}{1+(q-1)L_w}$	L_w = relative leaf area of the weed q = relative damage coefficient of the weed on the crop	Rectangular hyperbola with one parameter	[31]
$Y = \frac{qL_w}{1+\left(\frac{q}{m}-1\right)L_w}$	m = maximum yield loss caused by weeds	Rectangular hyperbola with two parameters	[32]

3. Development of an IWM Strategy

Within this context grows and develops the concept of IWM, a systematic weed management approach combining monitoring, prevention and control and not based on the complete eradication of weeds, but rather on their control below thresholds that are agronomically, environmentally and economically acceptable. Numerous definitions of IWM have been provided in the last decades, with agronomic, economic and/or ecological goals incorporated [33]. It can be simply defined as a component of IPM consisting in the combination of preventive practices and different control methods (mechanical, physical, biological and chemical) under a medium–long-term strategy [8]. The basic principle is that none of these individual methods on their own, except for chemical ones, are able to provide an adequate control of weed flora. On the contrary, they should be implemented and integrated in a multi-dimensional regime. The integration of indirect and direct control methods depends on the weed species, climatic conditions (e.g., solar radiation, temperature, rainfall regime and wind intensity), soil exposure and texture, irrigation method used, form of plant farming, socio-economic constraints and farmer's expectations [34]. Therefore, an IWM program is not absolute, but it needs to be adjusted according to the context-specific requirements and from year to year [35]. Several IWM systems have been combined, as suggested by Harker and O'Donovan [33]: many of these systems involve chemical–physical and chemical–cultural methods, while very few combine all weed management methods; indeed, the so-called integrated herbicide management, a "rationale" chemical weed control, is still the most adopted in advanced agroecosystems, despite the fact that it is not an IWM program *strictu sensu* [36]. Contrary to conventional weed control, in the IWM, the adoption of synthetic herbicides is strongly reduced in favor of a mixture of control methods that minimize the environmental impact. In general, an IWM system for herbaceous field crops should consider four main steps: (i) prevention, (ii) weed scouting and mapping, (iii) the decision-making process and (iv) the direct control (Figure 3). The first three steps involve the so-called proactive strategies, while the direct control is a set of reactive measures. The proactive strategies are based on the creation of an ecological environment unfavorable to the introduction, growth, spread and competition of weeds through various weed-suppressive agronomic practices, with the aim of making enough reactive measures have a lower impact on the environment [37]. The reduction of the soil seedbank and the increase of crop competitive ability are the main goals of the proactive strategies. Thereafter, the knowledge of the biological characteristics and ecological behaviors of weeds by means of field scouting and mapping in order to make a weed control decision based on weed patches and thresholds is essential [33,35,37]. Finally, the reactive measures coincide essentially with the direct control, which is mainly represented by mechanical, physical, biological and chemical methods.

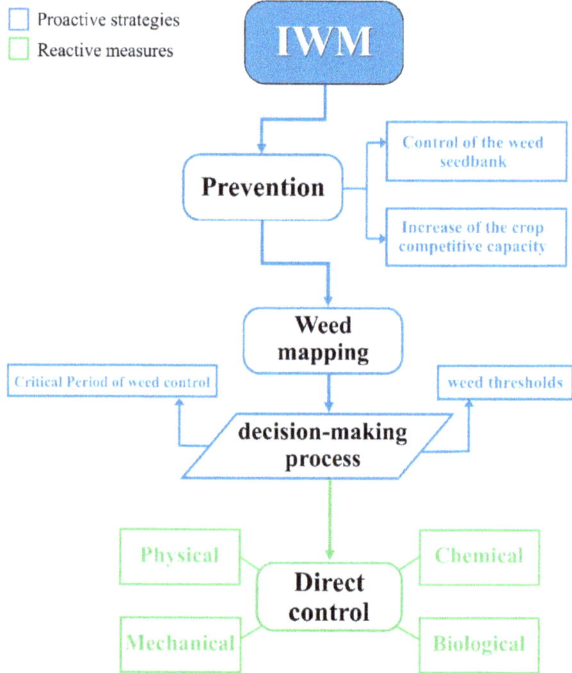

Figure 3. Proactive and reactive tactics of an Integrated Weed Management (IWM) strategy. At the base there is prevention, which should be combined with direct control (integration both intra-preventive methods and inter-preventive/direct ones) after an appropriate decision-making process closely linked to the specific weed flora.

4. Preventive Methods

Preventive methods, often referred to as cultural methods, include those strategies or agronomic choices aimed at preventing weed germination, emergence, growth, diffusion and dispersal [38]. These goals could be reached by reducing the soil weed seedbank and increasing the crop competitive capacity (Table 2).

Table 2. Main effect and description of preventive methods involved in the integrated weed management of herbaceous field crops.

Action	Main Effect	Description
(A) Control of the soil weed seedbank		
Crop rotation	Reduction in weed emergence and germination	The diversification of the crop sequence prevents weeds from adapting and establishing, thus disrupting the establishment of a specialized flora in favour of a multifaceted weed community composed by many species each present at low density.
Stale seedbed	Reduction in weed emergence	An earlier seedbed preparation combined with a light irrigation or rainfall and followed by a mechanical, physical or chemical weed control, limits weed emergence in early stages of the crop growing period.
Soil solarization	Reduction in weed germination	Solarization allows reaching 50–55 °C at 5 cm soil depth and more than 40 °C in the surface layers, thus preventing seed germination by thermal killing of germinating seeds or inducing seed dormancy.
Good agronomic practices	Reduction in seedbank input	Adoption of certified seeds with high pureness rate, cleaning equipment and mechanical tools before moving from field to field, avoid transportation of soil from weed-infested areas, use well-composted manure, filtering irrigation water, field sanification (including uncultivated areas) before weed reproduction.
Ploughing	Increase in seedbank output	Ploughing, by influencing the vertical distribution of the seedbank, on one side decreases the germination of buried weed seeds and, on the other side, increases predation and physiological death of weed seeds on the soil surface.
Cover cropping, mulching, intercropping and green manuring	Reduction in weed emergence	Living mulches between rows and buried or shallow dead mulches prevent weed germination physically and chemically through allelopathy.
(B) Increase of the crop competitive capacity		
Choice of weed-competitive cultivars	Increase in speed soil cover rates in early stages	Choice of cultivars with high root development, early vigour, faster seedling emergence, high growth rates, wide leaf area and allelopathic ability.
Crop density	Reduction in weed emergence and biomass	The increase in crop density and the reduction of row spacing influence the weed-crop competition in favour of the crop.
Spatial patterns and plant arrangement	Improvement in crop competitive ability for the whole cycle	Narrow-row spacing, bidirectional sowing, twin-row system, etc., contribute in smothering weeds.
Crop planting/sowing date	Improvement in crop competitive ability in early stages	A planting/sowing date in correspondence of the most suitable meteorological conditions allows the crop germinating/emerging before weeds and, thus, competing better for nutrients, water, light and space.
Crop transplant	Improvement in crop competitive ability in early stages	Transplanted crops have a shorter critical period and an easier mechanical or chemical control than sown crops.

4.1. Control of the Soil Weed Seedbank

The soil seedbank is the reserve of all viable (dormant as well as ready to germinate) weed seeds stored in the soil and, in agroecosystems, represents the primary source of new infestations because the real weed flora derives almost exclusively from the potential weed population communities [39]. For this reason, getting its control under an acceptable level (<20 million weed seeds ha^{-1}) is of key importance for the weed populations occurring in a field and for the subsequent weed management. In addition to the size, farmers should also consider the composition, the vertical distribution and the dynamic of the seedbank. The main objective is to decrease weed seeds' input, increase the output and reduce the level of residual seed emergence [40].

Every IWM system is based on the prevention of weeds' adaptation. It is well known that monoculture and the repeated succession for years of the same weed control practices lead to the development of a specialized flora more and more resistant from season to season to such practices. Therefore, farmers should pursue the maximum possible diversification of the cropping system to disrupt the establishment of a specialized flora in favor of a multifaceted weed community composed of many species, each present at low density [6]. Diversification of the crop sequence, i.e., crop rotation, allows for rotating herbicide choices, varying kinds of tillage, fertilization, seeding rate and row spacing [15,41]. Moreover, since the weeds' life cycle is closely correlated to that of the crop (e.g., perennial weeds are more common in perennial crops while annual weeds are mostly found in annual crops), crop rotation prevents weeds from adapting and establishing [35]. The effects of crop rotation can also be observed in terms of reduction of the seedbank size. This effect, of course, increases when combined with tillage, as demonstrated by Cardina et al. [42] and Dorado et al. [43]. Numerous crop rotation systems have been suggested for herbaceous field crops, generally based on the cereal-leguminous or nutrient-depleting and nutrient-building, or even high–low competitive crops' alternance.

Other valuable preventive methods commonly reported for the reduction of the weed seedbank are the soil solarization and the stale seedbed. Despite the fact that soil solarization is often considered a direct and physical weed control method, we prefer to include it among preventive methods, considering that its phytotoxic effect is exerted on the soil seedbank. Such a technique entails covering ploughed, levelled and wet soil with transparent polyethylene film during the hot season of the year, for at least four weeks, in order to capture the solar radiation and warm the soil [44]. The solarization allows for reaching more than 40 °C in the surface layers of the soil, and even 50–55 °C at 5 cm [44], which is lethal to many soil-borne pests (mainly fungi and nematodes) and weed seeds by preventing their germination. The application of soil solarization is normally restricted in greenhouse conditions [45], but it is reported to be one of the most effective methods of parasitic plants control, especially from the *Orobanche* and *Phelipanche* genus, in field crops [46,47]. The phytotoxic process involved in soil solarization is due to the thermal breaking of seed dormancy followed by thermal killing, the direct thermal killing of germinating seeds or even the indirect effects via microbial attack of seeds weakened by sub-lethal temperature [48]. Annual weeds are the most sensitive to solarization, while perennials reproduced vegetatively (by rhizomes, tubers, etc.) are generally tolerant, probably due to the limited penetration of heat in soil beyond a 10 cm depth and to their ability in rapidly regenerating from partially damaged underground organs [49]. The economic and agronomic suitability of soil solarization is explicated in climatic zones such as the Mediterranean, the tropical and sub-tropical regions where, during summer months, air temperature goes up to 40 °C and there is little cropping activity, especially if integrated with the control of soil-borne pathogens crops [47]. The stale seedbed, which is one of the most common techniques practiced for a wide number of herbaceous field crops, consists in the earlier seedbed preparation (at least 2–3 weeks before crop emergence depending on the plant species) combined with a light irrigation or rainfall to allow weed emergence and is then killed mechanically through shallow tillage, physically by flaming or chemically by means of nonselective herbicides [50]. This technique is effective mainly on the weed species characterized by initial low dormancy, requiring light to germinate and present on the soil surface, such as *Amaranthus* spp., *P. oleracea*, *Sorghum halepense*

(L.) Pers., *Digitaria* spp., *Capsella bursa-pastoris* (L.) Medik, etc. The stale seedbed on one side reduces the weed seedbank, while on the other, limits weed emergence. Furthermore, it is a preventive method that assures a competitive advantage to the crop by reducing weed pressure at the beginning of the growing period when weed damages are the highest.

In order to avoid or reduce the introduction of new weed seeds in the soil seedbank, several agronomic choices are commonly suggested: adopting seeds with a high pureness rate, cleaning equipment and mechanical tools before moving from field to field, avoiding transportation of soil from weed-infested areas, using well-composted manure when adopted, adopting localized irrigation and fertilization and filtering irrigation water [16,35]. A valid tool is provided by field sanification before weed reproduction and spread throughout the whole farm area, including uncultivated areas (field banks, paths, water channels, etc.). Another strategy for the control of the soil seedbank is maximizing outputs which are represented by seed germination, physiological death, predation and biological death caused by various pathogens. This objective is generally pursued, influencing the vertical distribution of the seedbank and leaving as many weed seeds as possible on the soil surface. In this regard, tillage plays a strategic role and will be discussed in the "mechanical control" section.

Cover cropping, mulching, intercropping and green manuring are efficient tactics in reducing weed emergence. Even though indicated separately as independent techniques, indeed they are different facets belonging to cover cropping, namely the mono- or inter-cropping of herbaceous plants either for a part or an entire year with the aim of enhancing yields [51,52]. Cover cropping is generally used in conservative agricultural systems or organic farming, where the presence of cover crops is often negatively correlated to weed biomass. Berti et al. [53], for example, reported that the integration of cover cropping and zero tillage produces a more efficient weed control than the single techniques thanks to the joint action of plant residues and allelochemicals released into the soil, which together inhibits weed seed germination and emergence. The use of cover crops is also suggested in conventional agriculture for herbaceous field crops due to the significant positive effects in enhancing soil fertility and reducing soil erosion, in addition to weed suppression. Cover crops can act as living mulches if intercropped with the cash crop, as well as dead mulches by living plant resides on place or green manures by ploughing down the resides [54]. In all cases, they prevent weed emergence both physically and chemically [55]: the former by increasing the competition with weeds for space, water, light and nutrients, while the latter through the release of phytotoxic compounds able to inhibit seed germination, weed emergence, establishment and early growth. The herbicidal potential of cover crops is closely dependent on cover crop genotype and management (e.g., sowing date, date of incorporation, agricultural practices), weed community composition, environmental and pedological conditions, amount of the plant residues and rate of decomposition [23,56]. Several practical applications of cover cropping for field herbaceous field crops have been suggested: rye, wheat, sorghum, oat, hairy vetch, subterranean clover and alfalfa cover crops are indicated by numerous authors, in different agricultural systems, to exert significant effects on weed control in cotton, maize, soybean and tomato [57–59].

4.2. Increase of the Crop Competitive Capacity

The second strategy to reduce the germination, emergence and diffusion of weeds is the increase of the crop competitive capacity. It is important to underline that such a set of agronomic choices/strategies by itself does not provide a satisfactory level of weed control, but it is effective only if the other preventive methods have been well carried out. This phase is focused on the interference relationships between crop and weeds. The main goal is to have a crop be able to cover the soil as fast as possible, which depends essentially on four factors: (1) genetic traits of the crop, (2) ideal spatial arrangement of plants, (3) optimal crop density and (4) fast seedling emergence. Such goals, therefore, can be realized through the varietal selection and the choice of the crop sowing date, density and spatial patterns

(Table 2). The review by Sardana et al. [60] and the whole correlated Special Issue is suggested for further reading.

In addition to the yields, qualitative characteristics of products and resistance to pathogens, crop varieties should also be chosen in relation to the morpho-physiological traits (e.g., root development, early vigor, faster seedling emergence, high growth rates, wide leaf area and allelopathic ability), conferring the conditions to better compete with weeds, although such traits are often closely affected by environmental conditions [41]. In conventional agriculture, the use of highly competitive cultivars helps in reducing herbicide adoption and labor costs but it is clear that this approach is increasingly important in organic and low-input agricultural systems. Many herbaceous field crops have been addressed by breeders for their weed competitiveness. The choice of weed-suppressive genotypes is widely reported for wheat [61], rice [62], maize [63], soybean [64], cotton [65], barley [66], etc.

The effects of competitive genotypes on weed control become more significant if integrated with agronomic manipulations such as crop density, sowing date and spatial patterns. Generally, an increased crop density and reduced row spacing help in reducing weed emergence and biomass, especially in the early phases of the biological cycle, by influencing weed-crop competition in favor of the crop. However, a crop density too high hinders the use of cultivators and other mechanical weeding operations and could lead to intraspecific competition phenomena and lower yields. The relationship between crop density and crop yield can be either asymptotic or parabolic [67]. The optimal density for weed suppression is unknown for most crops, but the mathematical models described in Table 1 and weed thresholds addressed in the next paragraph may help in the decision process. Plant-to-plant spacing is another factor influencing weed suppression, with particular reference to the starting time and the duration of the critical period. Benefits deriving from narrow-row spacing are rapid canopy closure, suppression of late-emerging weeds or weeds not killed by a postemergence herbicide application, and short CPWC [13,15]. Change in plant arrangement (e.g., bidirectional sowing, twin-row system, etc.) contributes in smothering weeds [68].

Weed emergence and composition is significantly influenced by the crop's planting date. In general, it would be appropriate to choose the crop's planting/sowing date allowing suitable meteorological conditions (temperature, soil water and oxygen content, light) for a fast germination and emergence. Indeed, a rapid germination and emergence provides a competitive advantage to the crop because it will be able to accumulate nutrients, water, light and space earlier than weeds. Furthermore, weeds emerging before the crop tend to produce more seeds, have higher shoot weights and cause greater yields than weeds emerging after the crop [28,69]. In certain situations, the relative time of emergence of weeds contributes to yield losses more than plant density. O'Donovan et al. [28], for example, found that for every day wild oat emerged before wheat and barley, crop yield loss increased by about 3%. Bosnic and Swanton [69] reported that at similar densities, corn's yield losses ranged from 22% to 36% when barnyard grass emerged before the crop, while they decreased to ~6% when it emerged after.

An additional tool for increasing the crop competitive capacity is the use of transplanted crops, primarily due to their shorter critical period and easier mechanical or chemical control than sown crops [41]. Transplant is commonly adopted for horticultural species (usually Solanaceae, Cucurbitaceae and Asteraceae families), which are generally poor competitors to weeds, and rice among field crops, mainly in Asia. However, crop transplant is generally limited for herbaceous field crops under an IWM system due to the high costs of transplanted crops and the need to have an adequate inter-row spacing [70].

5. The Decision-Making Process: from Weed Mapping to Weed Thresholds

After prevention, a rational IWM system must predict the knowledge of the biological and ecological characteristics of weeds to guide the decision-making process and increase the efficiency of direct control methods. Information concerning weed abundance and community composition indicate whether preventive tactics are working over the medium–long period, whether adjustments in control tactics need to be carried out and whether there are new weed species to control before diffusion and widespread [71]. To track these parameters, several field mapping and scouting methods can be used, based on time and money available and level of precision needed. In general, in order to achieve the maximum possible representativeness of the survey, the size of the survey area in which the sampling is carried out should never be lower than the minimum area. Among the different definitions provided, Müeller-Dombois and Ellenberg [72] suggested that the minimum area is the smallest area in which the species composition of a plant community is adequately represented. However, despite the numerous botanical studies on species–area relationships, only a few experiments have been directly aimed at the minimum area assessment in agroecosystems, mainly for arid regions of the Mediterranean [73]. Practically, the entire field area should be walked in a zigzag or "W" pattern, imaginatively divided in regular quadrats and weed samples collected in a 1 m^2 plot for each quadrat. A completely randomized block or a nested-plot survey design can be adopted [72]. Nowadays, computer-drawn maps recognized by satellite-assisted systems, sensor-driven automated weed detection with earth-bound or multispectral cameras are available and recommended, especially for large fields [74]. Useful information on seed persistence in the soil, temporal patterns of weed seed rain and weed emergence should be assessed from the soil seedbank analysis.

Once major weeds have been identified and their ecological aspects (kind of reproduction and propagule dispersion, temporal pattern of emergence, duration of the biological cycle, etc.) determined, it is necessary to establish the need for and timing of weed control. Weed thresholds provide information on the need for weed control. In weed science, weed threshold is a point at which weed density causes important crop losses [11]. Among the different weed thresholds suggested by scientists, the economic damage threshold is considered the most suitable in an IWM system. It is the weed density at which the costs of weed control are equal to or lower than the increase in crop value from control [75]. In other words, it refers to the weed densities at which they cause considerable yield losses and hence the weed control becomes economical [15]. Practically, the economic damage threshold presents two main concerns: it measures only a single year of weed effects based on a single weed species, resulting in a difficulty in distinguishing the competitive effect of one weed on another [75]. Moreover, because of the dynamicity of weed emergence during a crop season, the economic damage threshold is useless, if taken alone, for the determination of "when" to intervene. For a deeper revision of weed thresholds, the review by Swanton et al. [76] is recommended.

The timing of weed control, however, can be obtained by identifying the CPWC, defined as a period in the crop growth cycle during which weeds must be controlled to prevent crop yield losses. It is expressed as the days after crop emergence: weeds that germinate before and after this period do not cause significant yield reductions and may not be controlled. Therefore, CPWC is a helpful tool in IWM, after preventive measures, and is associated with postemergence weed control to avoid unnecessary herbicide applications [15]. Functionally, CPWC represents the time interval between two measured crop–weed interference components: the critical timing of weed removal (CTWR) and the critical weed-free period (CWFP) [77]. CTWR, which is based on the so-called weedy curve (descending line), is the maximum amount of time in which early season weed competition can be tolerated by the crop before an acceptable yield loss of 5% and indicates the beginning of the CPWC. CWFP, determined from the weed-free curve (ascending line), is the minimum weed-free period required from the time of planting to prevent more than 5% yield loss and determines the end of the CPWC. Table 3 reports the CPWC for some of the most important herbaceous field crops; however, the CPWC can be variable, also depending on crop variety, major weed species and their initial densities, agronomic characteristics of the crop (e.g., density, spatial arrangement, row spacing, etc.), preventive methods applied before sowing or transplant and climatic conditions [15,78].

6. Direct Methods

The direct methods include mechanical, physical, biological and chemical weed control aimed at managing the emerged weed flora. The direct control is the last step of an IWM strategy, and for this reason, its efficiency increases if commensurate with weed mapping and if preventive methods have been well carried out. Table 4 reports some examples of applied combinations of direct methods.

6.1. Mechanical Control

Mechanical methods for weed control can be classified in relation to the execution period (autumn, winter, springer and summer), the soil depth (shallow when <25 cm, medium if ranging from 25 to 40 cm and deep when >40 cm), the mode of action towards crop row (inter- or intra-row tools) and the presence/absence of the crop. Thanks to the boom in organic farming which occurred over the last years, both the agricultural machinery companies and the scientific community reached important technological advances in mechanical tools such as torsion weeders, finger weeders, brush weeders, weed blower and flex-time harrow for the intra-row weed control [79]. Furthermore, a series of robotic solutions (e.g., electronic sensors, cameras, satellite imagery, Global Positioning System based guidance systems, etc.) have been developed for the equipment of weed control machines, especially for an automated management in field conditions, with the aim of increasing productivity and minimizing labor cost [80]. Despite these important advancements, mechanical methods still have some limitations: high initial price and management costs for labor and carburant, poor effectiveness on intra-row weeds and high dependence on pedoclimatic conditions (mainly soil texture and moisture), weed species and growth stage.

Table 3. Critical period of weed control (CPWC) of some herbaceous field crops.

Common Name	Binomial Name	CPWC	Reference
canola	*Brassica napus* L.	17–38 DAE	[81]
carrot	*Daucus carota* L.	up to 930 GDD when seeded in late April 414 to 444 GDD when seeded in mid to late May	[82]
chickpea	*Cicer arietinum* L.	from 17–24 to 48–49 DAE	[83]
corn	*Zea mays* L.	from the 3rd to 10th leaf stage	[78]
cotton	*Gossypium hirsutum* L.	from 100–159 to 1006–1174 GDD	[84]
leek	*Allium porrum* L.	7–85 DAE	[85]
lentil	*Lens culinaris* Medik.	447–825 GDD	[86]
penaut	*Arachis hypogaea* L.	3–8 weeks after planting	[87]
potato	*Solanum tuberosum* L.	from 19–24 to 43–51 DAE	[88]
red pepper	*Capsicum annuum* L.	0–1087 GDD (from germination to harvest)	[89]
rice	*Oryza sativa* L.	30–70 days after transplant	[14]
soybean	*Glycine max* (L.) Merr.	up to 30 DAE	[90]
sunflower	*Helianthus annuus* L.	14–26 DAE without preherbicide treatment 25–37 DAE with preherbicide treatment	[91]
tomato	*Solanum lycopersicum* L.	28–35 days after planting	[92]
white bean	*Phaseolus vulgaris* L.	from the second-trifoliolate and first-flower stages of growth	[93]
winter wheat	*Triticum aestivum* L.	506–1023 GDD	[94]

Note: DAE: day after emergence; GDD: growing degree days, calculated as $((T_{max} + T_{min})/2 - T_b)$.

In organic farming, low-input and conservative agriculture systems, tillage is the major way to control weeds, but it is also widely adopted in conventional agriculture for its many positive effects: seedbed preparation, control of soil erosion and evapotranspirative water losses, improvement of soil structure, aeration and water infiltration, deepening of roots, burial of plant residues and fertilizers, etc. In herbaceous field crops, generally, the soil is first plowed up to 30–40 cm to cut and/or invert the soil and bury plant residues; then, the soil upper layer is shallow-tilled repeatedly by harrowing, rototiller, etc., to clean the field before sowing or planting [15]. Normally, weed mechanical control is also carried out in postemergence between or inside rows. When applied in pre-emergence, the main goal of tillage is to control the soil weed seedbank and to give the crop a better start to compete against weeds during the first stages. The herbicidal activity of tillage is exerted by affecting the vertical distribution of the seedbank: on one side, the germination of weed seeds buried into the soil decreases significantly due to changes in microclimatic patterns (temperature, aeration, light), while on the other side, predation and physiological death of weed seeds and vegetative propagules on the soil surface increases [40,95]. Information on the differences between tillage systems (zero, minimum and conventional) on weed density and diversity indices are contrasting, probably due to the differences in climatic conditions, soil characteristics and agronomic practices of the areas where the experiments were conducted. Reduced or zero tillage are often associated to an increased seedbank size and species composition in the surface soil layer [96]. Weed density and species richness also increased when converting from conventional to zero tillage. Nevertheless, biennial and perennial weeds are reported to be dominant under conservation tillage, such as zero tillage, due to the non-disruption of their root systems, while annual weeds are likely to increase under conventional tillage because they are able to germinate from various depths [97]. In a 35-year field experiment of crop rotation and tillage systems, Cardina et al. [42] found the highest seedbank size in zero tillage, with a decline as tillage intensity increased. To the contrary, Mas and Verdú [98] indicated the zero tillage as the best systems of weed management because they prevent the domination of the weed flora by only a few species.

Mechanical control plays a key role in the IWM because it almost always becomes part of the combination of different methods. For example, the integration of zero tillage and cover cropping, thanks to an increased amount of weed seeds and plant residues on the soil surface, combined with the release of allelochemicals into the soil, is reported to improve the weed control effectiveness [53]. In addition to cover cropping, tillage is often combined with the stale seedbed in pre-emergence after the preparation of the seedbed, with the tactics of crop competitiveness increasing (especially crop density and spatial arrangement) or with other direct methods, as discussed below and in Table 4.

6.2. Physical Control

Since mulching and solarization were included in the preventive methods, because their herbicidal activity is related to the control of the soil seedbank, the direct physical methods discussed here refer to the thermal control. Based on their mode of action, thermal methods can be classified as direct heating methods (flaming, hot water, hot hair, steaming, infrared weeders), indirect heating methods (electrocution, microwaves, ultraviolet light, laser radiation) and freezing by liquid nitrogen or carbon dioxide snow [99]. Among them, indirect heating methods, and mainly microwaves, laser radiation and ultraviolet light, are still at an early experimental stage. All these methods are characterized by a high initial cost of the machine, high treatment frequency, high costs for fuels and requirement of specialized labor. By contrast, they can be used when the soil is too moist for mechanical weeding, can be applied without soil disturbance and are effective against those weeds that have developed resistance to herbicides. Freezing has been used primarily in laboratory experiments [100], but in the current state-of-the-art, its adoption in field conditions remains not applicable and sustainable. Flaming is the most commonly applied thermal method and thus, deserves particular attention.

Flame weeding is a direct thermal method commonly used in organic farming which relies on propane gas burners or, recently, renewable alternatives such as hydrogen [101], to generate combustion temperatures up to 1900 °C. Once the foliar contact with the target plant occurs, the temperature of the exposed plant tissues raises rapidly up to ~50 °C inside plant cells, causing a denaturation and aggregation (i.e., coagulation) of membrane proteins [101]. The disruption of cell membranes results in a loss of cell function, thus causing intracellular water expansion, dehydration of the affected tissue and finally desiccation [102]. As a result of this, flamed weeds can die normally within 2 to 3 days or their competitive ability against the crop could be severely reduced. Flaming should not be confused with burning, since plant tissues do not ignite but heat rapidly up to the point of rupturing cell membranes [102]. The effectiveness of flaming is closely influenced by weed species and seedling size (generally, dicot species are more sensible than monocot ones), weed growth stage (seedlings at the early growth stages such as the fourth-fifth leaves are more susceptible) and regrowth potential, as well as techniques of flaming (e.g., temperature, exposure time, energy input, etc.) [101,103]. A wide number of annul weeds are significantly controlled by flaming in maize, cotton, soybean, sorghum and various horticultural species fields, including redroot pigweed (*A. retroflexus*), barnyard grass (*E. crus-galli*), common lambsquarters (*Chenopodium album* L.), velvetleaf (*Abutilon theophrasti* Medik.), shepherd's purse (*C. bursa-pastoris*), yellow foxtail (*Setaria glauca* (L.) Beauv.), field bindweed (*Convolvulus arvensis* L.), venice mellow (*Hibiscus trionum* L.), kochia (*Kochia scoparia* (L.) Schrad.), etc. The thermal control of these weeds can be done prior to sowing, in pre-emergence or in postemergence [103]. In the first two cases, typical of fast-growing crops, flame weeding is commonly integrated with the stale seedbed, which allows a significant decrease of the first flush of weeds [38]. This is a sort of temporal selectivity. When applied after crop emergence, typical of slow-growing crops where later flushes of weeds can cause serious competition problems, flaming can be done directed or shielded. Directed flaming is suggested for heat-resistant crops (e.g., cotton, corn, sugarcane, etc.) and provides an intra-row weed control, while inter-row weeds can be effectively managed by conventional mechanical methods [50]. Angling from 22.5° to 45° to the horizontal, shielding or parallel burner systems are used for heat-sensitive crops to control the weeds between the rows [38,103]. Several attempts to estimate the demand in propane doses ha^{-1} or the costs of flaming operation ha^{-1} have been proposed [11]; undoubtedly, flame weeding is less expensive than organic herbicides and reduces the need for hand weeding, mainly in low-input agriculture. Flaming is commonly combined with the stale seedbed in pre-emergence or with mechanical methods such as hoeing or cultivators in postemergence (Table 4). Several researches reported interesting results on the combination of preventive and direct methods [50]. Suggested and common integrations involving physical control are crop rotation/cover cropping/torsion or finger weeders combined with flaming or else stale seedbed/flaming/crop density and fertilizers' placement/interrow hoeing/herbicides at low rates.

6.3. Biological Control

According to the European Weed Research Society, "biological weed control is the deliberate use of endemic or introduced organisms (primarily phytophagous arthropods, nematodes and plant pathogens) for the regulation of target weed populations". The Weed Science Society of America defined the biological control of weeds as "the use of an agent, a complex of agents, or biological processes to bring about weed suppression", specifying that all forms of macrobial and microbial organisms are considered as biological control agents. Cordeau et al. [104] grouped biocontrol agents in macro-organisms (e.g., predators, parasitoid insects and nematodes), microorganisms (e.g., bacteria, fungi and viruses), chemical mediators (e.g., pheromones) and natural substances (originated from plant or animal). In this review, the latter category will be discussed as an "allelopathic" tool in the last paragraph.

The biological control has gained a particular and worldwide attention since the 1980s from researchers, industrial companies and stakeholders, parallel to the growth of organic farming under a sustainable agriculture perspective. Using the information reported in the fifth edition of "Biological control of weeds: a world catalogue of agents and their target weeds", Schwarzländer et al. [105] stated that (i) the five countries/regions most active in biocontrol research and releases are Australia, North America, South Africa, Hawaii and New Zealand (in decreasing order), that (ii) three insect orders (Coleoptera, Lepidoptera and Diptera) comprised about 80% of all biocontrol agent species released and that (iii) 66% of the weeds targeted for biological control experienced some level of control. Exhaustive reviews and lists of practical applications are reported by Charudattan [106], Müller-Schärer and Collins [5] and Sheppard et al. [107]. Despite the increasing interest in biological control tools, the market share of bioherbicides (i.e., products of natural origin for weed control) represents less than 10% among all kinds of biopesticides (biofungicides, biobactericides, bioinsecticides and bionematicides) [106]. Most bioherbicides actually available as commercial formulates are mycoherbicides such as DeVine®, Collego®, Smoulder®, Chontrol®, etc. In addition to their public acceptance and environmentally friendly behavior, bioherbicides offer new modes of actions and molecular target sites compared to synthetic herbicides [108]. However, the low number of commercial formulates is explained by their shorter half-life and lower reliability of field efficiency than chemicals, as well as by the need to be formulated with co-formulants and encapsulated, processes which require a great effort in terms of coordination between public and private groups, costs and time [109,110]. Indeed, among all the bioherbicide projects underway, only 8% were successful, with 91.5% of them remaining not applicable [106].

Table 4. Examples of applied combinations of direct methods for integrated weed management systems.

Methods Involved	Type of Integration	Description	Reference
Mechanical–Physical	Hoeing–Brush weeding	A combined hoeing close to the row plus vertical brush weeding increases weed control efficiency.	[111]
Physical–Mechanical	Banded flaming–Cultivator	A banded flaming intra-row followed by aggressive mechanical cultivation inter-row provides over 90% of weed control in organic maize.	[112]
Mechanical–Biological	Reduced tillage–Bioherbicides	In zero- or minimum-tillage systems, weed seeds concentrate in the upper soil layer, thus allowing the surface application of bioherbicides with seed-targeting agents.	[113]
Biological–Chemical	Bioherbicide–Herbicide	Combining the pre-emergence inoculation with the fungal pathogen *Pyrenophora semeniperda* and post-emergence imazapic application limits the spread of cheatgrass.	[114]
Chemical–Mechanical	Herbicides–Hoeing	The integration of herbicides intra-row and hoeing inter-row allows halving herbicide's amount in maize, sunflower and soybean, with no loss in weed control and crop yield.	[115]
Chemical–Mechanical	Herbicides–Ploughing	The integration of pre-sowing and pre-emergence herbicides with post-emergence inter-row cultivation increases yields and reduces total weed density in a cotton-sugar beet rotation.	[116]

In an IWM system, where the final goal is not the complete eradication of weeds but their control below acceptable thresholds, biological methods need to be integrated with other weed management tactics to produce acceptable levels of control. The use of an inoculative, inundative or conservative approach is closely related to the site-specific conditions: biology and population dynamics of the weed flora (field mapping plays a key role in this respect), crop species and variety, agronomic practices and weed management techniques adopted [5]. Several examples of systemic combinations of bioherbicides with synthetic herbicides and other weed control methods have been provided [5,104,106]. Müller-Schärer and Collins [5] distinguished a horizontal integration, aimed at controlling different weed species in one crop, and a vertical integration against a single weed species. Since harmful effects of weeds in agroecosystems are often caused by the presence of a multifaceted weed flora, the horizontal approach involving the joint application of synthetic herbicides at low rates with pathogens and bioherbicides, or the combination of bioherbicides with mechanical methods, is the most common practical application of biological control under an IWM strategy in open fields.

6.4. Chemical Control

Chemical control is based on the use of herbicides, i.e., chemical substances (organic or inorganic) used to kill or suppress the growth of plants (Weed Science Society of America). In intensive cropping systems, herbicides are the backbone of weed management because they are the most effective weed control tool, allow flexibility in weed management, significantly increase crop production and require less costs and human efforts [95]. A wide number of herbicides have been produced and are currently under development for herbaceous field crops. Herbicides can be classified according to chemical family, time of application (preplant, pre-emergence and postemergence), mechanism of action, formulation, site of uptake and selectivity [19]. The choice of herbicide is based on crop genotype, weed spectrum and specific pedo-climatic conditions. Continuous and frequent application of the same herbicide in the same crop at the same area induced resistance in many weeds. Herbicide resistance (HR) is defined as the survival of a segment of the population of a weed species following an herbicide dose lethal to the normal population [117] due to genetic mutations or adaptive mechanisms. Resistance develops when these mutations increase over time after each herbicide application until they become predominant. Nowadays, there are globally 510 unique cases (species × site of action) of HR weeds from 262 species (152 dicots and 110 monocots) to 23 of the 26 known herbicide sites of action [118]. Among the biological mechanisms involved in HR (e.g., overexpression of wild-type herbicide-target-site proteins, deactivation or reduced activation of herbicide molecules, altered herbicide absorption, translocation or sequestration), the enhanced metabolism by alteration of target sites is the most common mechanism [119]. Therefore, HR is closely linked to their mode or site of action and weeds evolve more resistance to some herbicides site of actions than others. In relation to the site/mechanism of action, herbicides are classified into seven groups [120]: light-dependent herbicides (inhibitors of photosynthesis, inhibitors of pigment production, cell membrane disruptors and inhibitors), fatty acid biosynthesis inhibitors, cell growth inhibition, auxin-like action-growth regulators, amino acid biosynthesis inhibitors, inhibitors of respiration and unknown mechanism of action.

The concept of resistance should not be confused with that of tolerance, defined by Penner [117] as "survival of the normal population of a plant species following a herbicide dosage lethal to other species", and by LeBaron and Gressel [121] as "the natural and normal variability of response to herbicides that exists within a species and can easily and quickly evolve". In the last years, many conventionally bred (CHT) and genetically modified herbicide-tolerant (GMHT) crops have been commercially grown thanks to their low cost, simplified, more flexible and selective weed management, their good compatibility with reduced-tillage systems and possibility to control congeneric weeds to the crop [7]. Some examples of GMHT herbaceous field crops are cotton, oilseed rape, rice, maize, sugarbeet, canola, alfalfa and soybean. However, the use of CHT and GMHT crops accelerated the selection of HR weeds, which in fact increased dramatically in the last decade [118]. In addition, the continuative adoption of the same herbicide and the use of HR and GMHT crops has led to a greater selection pressure and to shifts in the weed species community, especially in major herbaceous field crops [14,95]. In order to avoid such problems, it is of key importance to not only integrate chemical control with other methods within an IWM strategy, but also apply herbicides after overcoming the economic damage threshold, as well as use the correct rates, rotations, mixtures and sequences. Use of reduced rates is generally reported to offer good effectiveness in weed control without yield losses; however, factors such as climatic conditions (temperature, solar radiation, air and soil moisture), droplet size, spray volume, herbicide formulation, etc., may affect results because a full rate applied at sub-optimal conditions may be less effective than a low rate at optimal conditions [122]. Granule formulations or microencapsulation of herbicides, for example, provides a better weed control than liquid formulations in no-till or reduced cropping systems, probably due to their higher movements through soil layers [95]. Nevertheless, the weed flora composition should also be taken into account, since a lower rate of one herbicide may be more effective than a full rate of another herbicide [7]. In model-based approaches, several mathematical models have been suggested to calculate the dose of herbicide required to limit crop yield loss to less than a given level, generally by using symmetrical sigmoidal curves [123]. Rotation of herbicides with different modes/sites of action and herbicide mixtures are widely recommended to prevent HR [122].

Major chemical control integrations are those with the stale seedbed [50], mechanical methods and cover cropping [124] (Table 4). Several inter-row tillage operations, such as ploughing or hoeing, can be combined with pre-sowing/pre-emergence or postemergence herbicides with the aim of reducing rates without decreasing weed control efficiency and crop yield [115,116]. Concerning cover cropping, amounts too high of cover crop residues on one hand can reduce the efficiency of herbicides by intercepting from 15% to 80% of the applied rate or by enhancing the soil microbial activity, while on the other hand, can increase the herbicidal effect on surface-germinating seeds thanks to the herbicide adsorption by residues near the germinating seeds [95]. A few attempts of chemical–biological integration have been carried out, like Ehlert et al. [114], but the modest results on one side and the high costs of bioherbicides on the other side, have made this combination poorly adaptable and little diffused in field conditions.

7. Allelopathic Mechanisms for Weed Control

Given the keen interest in eco-friendly practices for weed control, the use of allelopathy is gaining in popularity. Secondary metabolites released by plants into the environment are named allelochemicals. They are defense compounds belonging to a wide range a chemical classes, mainly phenolic compounds and terpenoids [20]. Comprehensive lists of plant allelochemicals can be found in Macías et al. [125] and Scavo et al. [20]. The synthesis of these compounds in the donor plant and their effect on the target plant, is closely influenced by several abiotic (e.g., solar radiation and light quality, temperature, soil moisture, mineral availability, soil characteristics, etc.) and biotic (e.g., plant genotype, organ and density, diseases and pathogens attacks) factors [20]. Moreover, plants under stress conditions generally increase the production of allelochemicals and, at the same time, become more sensitive to such compounds. Allelochemicals occur in any plant organ (leaves, stems, roots, rhizomes, seeds, flowers, fruits, pollen) and can be released through volatilization from living parts of the plant, leaching from plant foliage, decomposition of plant material and root exudation [20]. Modes of action can be either direct or indirect and refer to the alteration of cell division, elongation and structure, membrane stability and permeability, activity of various enzymes, plant respiration and photosynthesis, protein synthesis and nucleic acid metabolism, etc., that as the final result means inhibition of seed germination and low seedling growth [20].

Many herbaceous field crops show allelopathic traits [126]. Most of them belong to the Poaceae family, such as wheat, rice, maize, barley, sorghum, oat, rye and pearl millet. However, other important herbaceous crops including sunflower, tobacco, sweet potato, alfalfa, subterranean clover, coffee and several legume species, also possess allelopathic properties. The allelopathic mechanisms can be managed and used in agroecosystems for weed management through (1) the inclusion of allelopathic crops in crop rotations, (2) the use of their residues for cover cropping and (3) the selection of the most active allelochemicals and their use as bioherbicides (Table 5). Their efficacy, of course, is clearly weak if done alone, becoming more effective when combined within an IWM strategy.

The above-mentioned effects of crop rotation can be further exacerbated by including an allelopathic crop within a crop rotation in order to overcome the autotoxicity and decrease the pressure of plant pests [127]. In particular, allelochemicals exuded into the rhizosphere exert, directly and/or indirectly (by microbial interactions), inhibitory effects on seed germination and weed density [23]. For this reason, several crop sequences such as soybean–wheat–maize [128], sugar beet–cotton [129], sunflower–wheat [130], etc., are suggested. In a recent study, Scavo et al. [39] demonstrated that *Cynara cardunculus* L. cropping for three consecutive years significantly reduced the number of seeds in the soil seed bank, while showing a positive effect on some bacteria involved in the soil N-cycle.

Table 5. Practical applications of allelopathy for sustainable weed management.

Technique	Allelopathic Source	Target Weeds	Description	Reference
Crop rotation	*Glycine max* (L.) Merr., *Triticum aestivum* L.	*Setaria faberi* Herrm.	Corn following wheat in a soybean–wheat–corn rotation significantly reduced giant foxtail population.	[128]
Intercropping	*Vigna mungo* (L.) Hepper	*Echinochloa colona* (L.) Link, *Digitaria sanguinalis* (L.) Scop, *Setaria glauca* (L.) Beauv.	Intercropping black gram in a rice field was very effective in suppressing weeds and increasing crop yields.	[131]
Mulching	*Sorghum bicolor* (L.) Moench	*Cyperus rotundus* L., *Trianthema portulacastrum* L., *Cynodon dactylon* (L.) Pers., *Convolvulus arvensis* L., *Dactyloctenium aegyptium* (L.) Willd., *Portulaca oleracea* L.	Surface-applied sorghum mulch at sowing in maize reduced weed density and dry weight.	[132]
Green manure	*Brassica nigra* L.	*Avena fatua* L.	Soil incorporation of both roots and shoots of black mustard significantly decreased wild oat emergence, height and dry weight per plant.	[133]
Bioherbicide	*Juglans nigra* L.	*Conyza canadensis* (L.) Cronquist, *C. bonariensis*, *P. oleracea*, *Ipomoea purpurea* (L.) Roth	The black walnut extract-based commercial product (NatureCur®) decreased the germination and seedling growth of target weeds.	[134]
Water extract + Herbicide	*S. bicolor*, *Helianthus annuus* L., *Brassica campestris* L.	*T. portulacastrum*, *C. rotundus*, *Chenopodium album* L., *Cronopus didymus* L.	The combined application of a mixed water extract from sorghum, sunflower and mustard with pendimethalin allows for reducing herbicide rate.	[135]

The use of allelopathic cover crops, such as subterranean clover, alfalfa, oat, rye, sorghum, chickpea, summer squash, etc., is an effective weed management strategy in low-input agricultural systems and mainly in organic farming [55]. The scientific literature is full of research concerning the allelopathic intercropping, as well as the adoption of surface-applied or soil-incorporated mulching from allelopathic species [127,136]. In the case of mulching, several authors suggest the combined application of various allelopathic materials to increase the efficiency in weed management, due to the synergistic effect of diverse allelochemicals. The soil surface-placed allelopathic mulching can be integrated with no-tillage or reduced tillage [54]. Other implications and technical suggestions of allelopathic cover cropping are available in Kruidhof et al. [137].

The selection of active allelochemicals and their potential use as bioherbicides is one of the most popular sectors in the field of allelopathy among the last years [125]. Advantages and disadvantages derived from bioherbicides are reviewed by Dayan et al. [109]. Some of the most active allelochemicals are phenolics (e.g., vanillic acid, *p*-hydroxybenzoic acid), flavonoids (e.g., kaempferol, quercetin, naringenin), cinnamic acid derivatives (e.g., chlorogenic acid, ferulic acid, caffeic acid, sinapic acid, *p*-coumaric acid), coumarins (e.g., umbelliferone, esculetin, scopoletin) and sesquiterpene lactones (e.g., artemisinin, centaurepensin, cynaropicrin) [20]. Juglone, a naphthoquinone widely abundant in the Juglandaceae family (notably *Juglans nigra* L. and *J. regia*) and ailanthone, a quassinoid exudated by ailanthus (*Ailanthus altissima* (Mill.) Swingle), are two well-known allelochemicals subjected to intense research activity. Several black walnut and ailanthus extract-based products were found to show a good potential as pre- and post-emergence bioherbicides, although are not yet registered. Most allelochemicals are water-soluble and, for this reason, they are commonly used as water extracts, which is also the easiest and the cheapest way to extract these compounds. Despite the high interest in this field, only very few plant-based bioherbicides are available for commercial use. The steps of producing a commercially formulated bioherbicide can be summarized as follows: (i) identification of an allelopathic behavior in a determined plant, (ii) identification of most active allelochemicals involved, (iii) extraction, purification and selection of these compounds, (iv) screening of the in vitro and in vivo allelopathic activity of crude extracts and pure compounds, both in pre- and post-emergence, (v) identification of the most allelopathic genotypes within the plant species, (vi) selection of the best harvest time of plant material in relation to abiotic and biotic factors and (vii) industrial processing in obtaining a commercially formulated bioherbicide. For example, the herbaceous field crop *C. cardunculus* was recently studied for the biological control of weeds, following a step-by-step approach. The allelopathic effects of the three *C. cardunculus* botanical varieties (globe artichoke, wild and cultivated cardoon) leaf aqueous extracts, at first, were evaluated on seed germination and seedling growth of some cosmopolitan weeds [138,139]. In a second phase, the set-up of the most efficient extraction method of its allelochemicals in terms of costs, yields and inhibitory activity was realized, selecting dried leaves as the best plant material and ethanol and ethyl acetate as the best solvents [140]. Moreover, new *C. cardunculus* allelochemicals (cynaratriol, deacylcynaropicrin, 11,13-dihydro-deacylcynaropicrin and pinoresinol) were purified [141]. Then, after the development of a new ultra-high-performance liquid chromatography-tandem mass spectrometry analysis method, the influence of genotype and harvest time was studied on the phytotoxicity, amount and composition of its allelochemicals [142].

8. Conclusions

Weeds are the main biotic drawback to crop yield in agroecosystems. Nowadays, following the request for setting up eco-friendly weed control practices which are agronomically and economically sustainable, the IWM system has become a consolidated approach, especially in organic agriculture and, more generally, in low-input agricultural systems. In herbaceous field crops cultivated conventionally, effective weed management without herbicide use cannot be conceivable and, for this reason, there is a need to integrate different tactics (e.g., stale seedbed/weed thresholds/combined directs methods, soil solarization/CPWC/herbicides, etc.) under a holistic approach in order to reduce the adoption of chemical tools. Furthermore, IWM must remain flexible to adapt to changing environmental and socio-economic factors and to readjust after a period of time. Integrating control methods very diverse from each other is certainly very difficult and requires support by research, especially for the development of long-term experiments, policies and incentives.

Author Contributions: Conceptualization, A.S. and G.M.; resources, data curation, writing and original draft preparation, illustrations and tables preparation, A.S.; review, editing and supervision, G.M. All authors have read and agreed to the published version of the manuscript.

Funding: This research received no external funding.

Conflicts of Interest: The authors declare no conflict of interest.

References

1. Oerke, E.-C.; Dehne, H.-W. Safeguarding production—Losses in major crops and the role of crop protection. *Crop Prot.* **2004**, *23*, 275–285. [CrossRef]
2. Oerke, E.-C. Crop losses to pests. *J. Agric. Sci.* **2006**, *144*, 31–43. [CrossRef]
3. Appleby, A.P.; Muller, F.; Carpy, S. Weed control. In *Agrochemicals*; Muller, F., Ed.; Wiley: New York, NY, USA, 2000; pp. 687–707.
4. Kraehmer, H.; Baur, P. *Weed Anatomy*; Wiley-Blackwell: Chichester, UK, 2013.
5. Müller-Schärer, H.; Collins, A.R. Integrated Weed Management. In *Encyclopedia of Environmental Management*; Jorgensen, S.E., Ed.; Taylor and Francis: New York, NY, USA, 2012. [CrossRef]
6. Buhler, D.D.; Liebman, M.; Obrycki, J.J. Theoretical and practical challenges to an IPM approach to weed management. *Weed Sci.* **2000**, *48*, 274–280. [CrossRef]
7. Lamichhane, J.R.; Devos, Y.; Beckie, H.J.; Owen, M.D.K.; Tillie, P.; Messéan, A.; Kudsk, P. Integrated weed management systems with herbicide-tolerant crops in the European Union: Lessons learnt from home and abroad. *Crit. Rev. Biotechnol.* **2017**, *37*, 459–475. [CrossRef] [PubMed]
8. Swanton, C.J.; Weise, S.F. Integrated weed management: The rationale and approach. *Weed Technol.* **1991**, *5*, 657–663. [CrossRef]
9. Agenda 21. Programme of Action for Sustainable Development. In Proceedings of the United Nations Conference on Environment & Development, Rio de Janerio, Brazil, 3–14 June 1992; Available online: https://sustainabledevelopment.un.org/content/documents/Agenda21.pdf (accessed on 29 December 2019).
10. OJEU. Directive 2009/128/EC of the European Parliament and of the Council of 21 October 2009 establishing a fr amework for Community action to achieve the sustainable use of pesticides. *Off. J. Eur. Union* **2009**, *309*, 71–86.
11. Knezevic, S.Z. Integrated weed management in soybean. In *Recent Advances in Weed Management*; Chauhan, B.S., Mahajan, G., Eds.; Springer: New York, NY, USA, 2014; pp. 223–237. [CrossRef]
12. Marwat, K.B.; Khan, M.A.; Hashim, S.; Nawab, K.; Khattak, A.M. Integrated weed management in wheat. *Pak. J. Bot.* **2011**, *43*, 625–633.
13. Norsworthy, J.K.; Frederick, J.R. Integrated weed management strategies for maize (*Zea mays*) production on the southeastern coastal plains of North America. *Crop Prot.* **2005**, *24*, 119–126. [CrossRef]
14. Mahajan, G.; Chauhan, B.S.; Kumar, V. Integrated weed management in rice. In *Recent Advances in Weed Management*; Chauhan, B.S., Mahajan, G., Eds.; Springer: New York, NY, USA, 2014; pp. 125–153. [CrossRef]
15. Doğan, M.N.; Jabran, K.; Unay, A. Integrated weed management in cotton. In *Recent Advances in Weed Management*; Chauhan, B.S., Mahajan, G., Eds.; Springer: New York, NY, USA, 2014; pp. 197–222. [CrossRef]
16. Pannacci, E.; Lattanzi, B.; Tei, F. Non-chemical weed management strategies in minor crops: A review. *Crop Prot.* **2017**, *96*, 44–58. [CrossRef]
17. Robinson, D.E. Integrated Weed Management in Horticultural Crops. In *Recent Advances in Weed Management*; Chauhan, B.S., Mahajan, G., Eds.; Springer: New York, NY, USA, 2014; pp. 239–254. [CrossRef]
18. Baker, H.G. Characteristics and modes of origin of weeds. In *The Genetics of Colonizing Species*; Baker, H.G., Stebbins, G.L., Eds.; Academic Press: New York, NY, USA, 1965; pp. 147–172.
19. Zimdahl, R.L. *Fundamentals of Weed Science*, 5th ed.; Academic Press: San Diego, CA, USA, 2018.
20. Scavo, A.; Restuccia, A.; Mauromicale, G. Allelopathy: General principles and basic aspects for agroecosystem control. In *Sustainable Agriculture Reviews*; Gaba, S., Smith, B., Lichtfouse, E., Eds.; Springer: Cham, Switzerland, 2018; Volume 28, pp. 47–101. [CrossRef]

21. Qasem, J.R.; Foy, C.L. Weed allelopathy, its ecological impacts and future prospects: A review. *J. Crop. Prod.* **2001**, *4*, 43–119. [CrossRef]
22. Muller, C.H. Allelopathy as a factor in ecological process. *Vegetatio* **1969**, *18*, 348–357. [CrossRef]
23. Scavo, A.; Abbate, C.; Mauromicale, G. Plant allelochemicals: Agronomic, nutritional and ecological relevance in the soil system. *Plant Soil* **2019**, *442*, 23–48. [CrossRef]
24. Kapoor, P.; Ramakrishnan, P.S. Studies on crop-legume behaviour in pure and mixed stands. *Agro-Ecosyst.* **1975**, *2*, 61–74. [CrossRef]
25. Jordan, N.; Vatovec, C. Agroecological benefits from weeds. In *Weed Biology and Management*; Inderjit, Ed.; Springer: Dordrecht, The Netherlands, 2004; pp. 137–158. [CrossRef]
26. Spitters, C.J.T.; Kropff, M.J.; de Groot, W. Competition between maize and *Echinochloa crus-galli* analysed by a hyperbolic regression model. *Ann. Appl. Biol.* **1989**, *115*, 541–551. [CrossRef]
27. Cousens, R. A simple model relating yield loss to weed density. *Ann. Appl. Biol.* **1985**, *107*, 239–252. [CrossRef]
28. O'Donovan, J.T.; Remy, E.A.S.; O'Sullivan, P.A.; Dew, D.A.; Sharma, A.K. Influence of the relative time of emergence of wild oat (*Avena fatua*) on yield loss of barley (*Hordeum vulgare*) and wheat (*Triticum aestivum*). *Weed Sci.* **1985**, *33*, 498–503. [CrossRef]
29. Cousens, R.; Brain, P.; O'Donovan, J.T.; O'Sullivan, A. The use of biologically realistic equations to describe the effects of weed density and relative time of emergence on crop yield. *Weed Sci.* **1987**, *35*, 720–725. [CrossRef]
30. Jasieniuk, M.; Maxwell, B.D.; Anderson, R.L.; Evans, J.O.; Lyon, D.J.; Miller, S.D.; Don, W.; Morishita, D.W.; Ogg, A.G.; Seefeldt, S.S.; et al. Evaluation of models predicting winter wheat yield as a function of winter wheat and jointed goatgrass densities. *Weed Sci.* **2001**, *49*, 48–60. [CrossRef]
31. Kropff, M.J.; Spitters, C.J.T. A simple model of crop loss by weed competition from early observations on relative leaf area of the weeds. *Weed Res.* **1991**, *31*, 97–105. [CrossRef]
32. Kropff, M.J.; Lotz, L.A.P. Systems approaches to quantify crop-weed interactions and their application in weed management. *Agric. Syst.* **1992**, *40*, 265–282. [CrossRef]
33. Harker, K.N.; O'Donovan, J.T. Recent weed control, weed management, and integrated weed management. *Weed Technol.* **2003**, *27*, 1–11. [CrossRef]
34. Gunsolus, J.L.; Buhler, D.D. A risk management perspective on integrated weed management. *J. Crop. Prod.* **1999**, *2*, 167–187. [CrossRef]
35. Knezevic, S.Z.; Jhala, A.; Datta, A. Integrated weed management. In *Encyclopedia of Applied Plant Sciences*, 2nd ed.; Thomas, B., Murray, B.G., Murphy, D.J., Eds.; Academic Press: Waltham, MA, USA, 2017; Volume 3, pp. 459–462.
36. Harker, K.N.; O'Donovan, J.T.; Blackshaw, R.E.; Beckie, H.J.; Mallory-Smith, C.; Maxwell, B.D. Our view. *Weed Sci.* **2012**, *60*, 143–144. [CrossRef]
37. Zanin, G.; Ferrero, A.; Sattin, M. La gestione integrata delle malerbe: Un approccio sostenibile per il contenimento delle perdite di produzione e la salvaguardia dell'ambiente. *Ital. J. Agron.* **2011**, *6*, 31–38. [CrossRef]
38. Bond, W.; Grundy, A.C. Non-chemical weed management in organic farming systems. *Weed Res.* **2001**, *41*, 383–405. [CrossRef]
39. Scavo, A.; Restuccia, A.; Abbate, C.; Mauromicale, G. Seeming field allelopathic activity of *Cynara cardunculus* L. reduces the soil weed seed bank. *Agron. Sustain. Dev.* **2019**, *39*, 41. [CrossRef]
40. Gallandt, E.R. How can we target the weed seedbank? *Weed Sci.* **2006**, *54*, 588–596. [CrossRef]
41. Bàrberi, P. Weed management in organic agriculture: Are we addressing the right issues? *Weed Res.* **2002**, *42*, 177–193. [CrossRef]
42. Cardina, J.; Herms, C.P.; Doohan, D.J. Crop rotation and tillage systems effects on weed seedbanks. *Weed Sci.* **2002**, *50*, 448–460. [CrossRef]
43. Dorado, J.; Del Monte, J.P.; López-Fando, C. Weed seed bank response to crop rotation and tillage in semiarid agro-ecosystems. *Weed Sci.* **1999**, *47*, 67–73. [CrossRef]
44. Mauromicale, G.; Lo Monaco, A.; Longo, A.M.G.; Restuccia, A. Soil solarization, a nonchemical method to control branched broomrape (*Orobanche ramosa*) and improve the yield of greenhouse tomato. *Weed Sci.* **2005**, *53*, 877–883. [CrossRef]

45. Lombardo, S.; Longo, A.M.G.; Lo Monaco, A.; Mauromicale, G. The effect of soil solarization and fumigation on pests and yields in greenhouse tomatoes. *Crop Prot.* **2012**, *37*, 59–64. [CrossRef]

46. Mauro, R.; Lo Monaco, A.; Lombardo, S.; Restuccia, A.; Mauromicale, G. Eradication of Orobanche/Phelipanche spp. seedbank by soil solarization and organic supplementation. *Sci. Hortic.* **2015**, *193*, 62–68. [CrossRef]

47. Mauromicale, G.; Restuccia, G.; Marchese, M. Soil solarization, a non-chemical technique for controlling *Orobanche crenata* and improving yield of faba bean. *Agronomie* **2001**, *21*, 757–765. [CrossRef]

48. Rubin, B.; Benjamin, A. Solar heating of the soils: Involvement of environmental factors in the weed control process. *Weed Sci.* **1984**, *32*, 138–142. [CrossRef]

49. Horowitz, M.; Regve, Y.; Herzlinger, G. Solarization for weed control. *Weed Sci.* **1983**, *31*, 170–179. [CrossRef]

50. Melander, B.; Rasmussen, I.A.; Bàrberi, P. Integrating physical and cultural methods of weed control—Examples from European research. *Weed Sci.* **2005**, *53*, 369–381. [CrossRef]

51. Mauromicale, G.; Occhipinti, A.; Mauro, R.P. Selection of shade-adapted subterranean clover species for cover cropping in orchards. *Agron. Sustain. Dev.* **2010**, *30*, 473–480. [CrossRef]

52. Mauro, R.P.; Occhipinti, A.; Longo, A.M.G.; Mauromicale, G. Effects of shading on chlorophyll content, chlorophyll fluorescence and photosynthesis of subterranean clover. *J. Agron. Crop. Sci.* **2011**, *197*, 57–66. [CrossRef]

53. Berti, A.; Onofri, A.; Zanin, G.; Sattin, M. Sistema integrato di gestione della lotta alle malerbe (IWM). In *Malerbologia*; Catizone, P., Zanin, G., Eds.; Pàtron Editore: Bologna, Italy, 2001; pp. 659–710.

54. Teasdale, J.R.; Mohler, C.L. The quantitative relationship between weed emergence and the physical properties of mulches. *Weed Sci.* **2000**, *48*, 385–392. [CrossRef]

55. Lemessa, F.; Wakjira, M. Mechanisms of ecological weed management by cover cropping: A review. *J. Biol. Sci.* **2014**, *14*, 452–459. [CrossRef]

56. Bàrberi, P.; Mazzoncini, M. Changes in weed community composition as influenced by cover crop and management system in continuous corn. *Weed Sci.* **2001**, *49*, 491–499. [CrossRef]

57. Abdin, O.A.; Zhou, X.M.; Cloutier, D.; Coulman, D.C.; Faris, M.A.; Smith, D.L. Cover crops and interrow tillage for weed control in short season maize (*Zea mays*). *Eur. J. Agron.* **2000**, *12*, 93–102. [CrossRef]

58. Bernstein, E.R.; Stoltenberg, D.E.; Posner, J.L.; Hedtcke, J.L. Weed community dynamics and suppression in tilled and no-tillage transitional organic winter rye-soybean systems. *Weed Sci.* **2014**, *62*, 125–137. [CrossRef]

59. Norsworthy, J.K.; McClelland, M.; Griffith, G.; Bangarwa, S.K.; Still, J. Evaluation of cereal and Brassicaceae cover crops in conservation-tillage, enhanced, glyphosate-resistant cotton. *Weed Technol.* **2011**, *25*, 6–13. [CrossRef]

60. Sardana, V.; Mahajan, G.; Jabran, K.; Chauhan, B.S. Role of competition in managing weeds: An introduction to the special issue. *Crop Prot.* **2017**, *95*, 1–7. [CrossRef]

61. Korres, N.E.; Froud-Williams, R.J. Effects of winter wheat cultivars and seed rate on the biological characteristics of naturally occurring weed flora. *Weed Res.* **2002**, *42*, 417–428. [CrossRef]

62. Caton, B.P.; Cope, A.E.; Mortimer, M. Growth traits of diverse rice cultivars under severe competition: Implications for screening for competitiveness. *Field Crops Res.* **2003**, *83*, 157–172. [CrossRef]

63. Mhlanga, B.; Chauhan, B.S.; Thierfelder, C. Weed management in maize using crop competition: A review. *Crop Prot.* **2016**, *88*, 28–36. [CrossRef]

64. Jannink, J.L.; Orf, J.H.; Jordan, N.R.; Shaw, R.G. Index selection for weed suppressive ability in soybean. *Crop Sci.* **2000**, *40*, 1087–1094. [CrossRef]

65. Molin, W.T.; Boykin, D.; Hugie, J.A.; Ratnayaka, H.H.; Sterling, T.M. Spurred anoda (*Anoda cristata*) interference in wide row and ultra narrow row cotton. *Weed Sci.* **2006**, *54*, 651–657. [CrossRef]

66. Watson, P.R.; Derksen, D.A.; Van Acker, R.C. The ability of 29 barley cultivars to compete and withstand competition. *Weed Sci.* **2006**, *54*, 783–792. [CrossRef]

67. Willey, R.W.; Heath, S.B. The quantitative relationship between plant population density and crop yield. *Adv. Agron.* **1969**, *21*, 281–321. [CrossRef]

68. Locke, M.A.; Reddy, K.N.; Zablotowicz, R.M. Weed management in conservation crop production systems. *Weed Biol. Manag.* **2002**, *2*, 123–132. [CrossRef]

69. Bosnic, A.C.; Swanton, C.J. Influence of barnyardgrass (*Echinochloa crusgalli*) time of emergence and density on corn (*Zea mays*). *Weed Sci.* **1997**, *45*, 276–282. [CrossRef]

70. Bastiaans, L.; Paolini, R.; Baumann, D.T. Focus on ecological weed management: What is hindering adoption? *Weed Res.* **2008**, *48*, 481–491. [CrossRef]

71. Liebman, M.; Bastiaans, L.; Baumann, D.T. Weed management in low-external-input and organic farming systems. In *Weed Biology and Management*; Inderjit, Ed.; Springer: Dordrecht, The Netherlands, 2004; pp. 285–315. [CrossRef]

72. Müeller-Dombois, D.; Ellenberg, H. *Aims and Methods of Vegetation Ecology*; Wiley: New York, NY, USA, 1974.

73. Cristaudo, A.; Restuccia, A.; Onofri, A.; Lo Giudice, V.; Gresta, F. Species–area relationships and minimum area in citrus grove weed communities. *Plant Biosyst.* **2015**, *149*, 337–345. [CrossRef]

74. Krähmer, H.; Andreasen, C.; Economou-Antonaka, G.; Holec, J.; Kalivas, D.; Kolářová, M.; Novák, R.; Panozzo, S.; Pinke, G.; Salonen, J.; et al. Weed surveys and weed mapping in Europe: State of the art and future tasks. *Crop Prot.* **2020**, *129*, 105010. [CrossRef]

75. Cousens, R. Theory and reality of weed control thresholds. *Plant Prot. Q.* **1987**, *2*, 13–20.

76. Swanton, C.J.; Weaver, S.; Cowan, P.; Van Acker, R.; Deen, W.; Shreshta, A. Weed thresholds: Theory and applicability. *J. Crop. Prod.* **1999**, *2*, 9–29. [CrossRef]

77. Knezevic, S.Z.; Datta, A. The critical period for weed control: Revisiting data analysis. *Weed Sci.* **2015**, *63*, 188–202. [CrossRef]

78. Hall, M.R.; Swanton, C.J.; Anderson, G.W. The critical period of weed control in grain corn (*Zea mays*). *Weed Sci.* **1992**, *40*, 441–447. [CrossRef]

79. Van der Weide, R.Y.; Bleeker, P.O.; Achten, V.; Lotz, L.A.P.; Fogelberg, F.; Melander, B. Innovation in mechanical weed control in crop rows. *Weed Res.* **2008**, *48*, 215–224. [CrossRef]

80. Slaughter, D.C.; Giles, D.K.; Downey, D. Autonomous robotic weed control systems: A review. *Comp. Electron. Agric.* **2008**, *61*, 63–78. [CrossRef]

81. Martin, S.G.; Van Acker, R.C.; Friesen, L.F. Critical period of weed control in spring canola. *Weed Sci.* **2001**, *49*, 326–333. [CrossRef]

82. Swanton, C.J.; O'Sullivan, J.; Robinson, D.E. The critical weed-free period in carrot. *Weed Sci.* **2010**, *58*, 229–233. [CrossRef]

83. Mohammadi, G.; Javanshir, A.; Khooie, F.R.; Mohammadi, S.A.; Salmasi, S.Z. Critical period of weed interference in chickpea. *Weed Res.* **2005**, *45*, 57–63. [CrossRef]

84. Bukun, B. Critical periods for weed control in cotton in Turkey. *Weed Res.* **2004**, *44*, 404–412. [CrossRef]

85. Tursun, N.; Bukun, B.; Karacan, S.C.; Ngouajio, M.; Mennan, H. Critical period for weed control in leek (*Allium porrum* L.). *Hortscience* **2007**, *42*, 106–109. [CrossRef]

86. Smitchger, J.A.; Burke, I.C.; Yenish, J.P. The critical period of weed control in lentil (*Lens culinaris*) in the Pacific Northwest. *Weed Sci.* **2012**, *60*, 81–85. [CrossRef]

87. Everman, W.J.; Clewis, S.B.; Thomas, W.E.; Burke, I.C.; Wilcut, J.W. Critical period of weed interference in peanut. *Weed Technol.* **2008**, *22*, 63–67. [CrossRef]

88. Ahmadvand, G.; Mondani, F.; Golzardi, F. Effect of crop plant density on critical period of weed competition in potato. *Sci. Hortic.* **2009**, *121*, 249–254. [CrossRef]

89. Tursun, N.; Akinci, I.E.; Uludag, A.; Pamukoglu, Z.; Gozcu, D. Critical period for weed control in direct seeded red pepper (*Capsicum annuum* L.). *Weed Biol. Manag.* **2012**, *12*, 109–115. [CrossRef]

90. Van Acker, C.R.; Swanton, C.J.; Weise, S.F. The critical period of weed control in soybean [*Glycine max.* (L) Merr.]. *Weed Sci.* **1993**, *41*, 194–200. [CrossRef]

91. Knezevic, S.Z.; Elezovic, I.; Datta, A.; Vrbnicanin, S.; Glamoclija, D.; Simic, M.; Malidza, G. Delay in the critical time for weed removal in imidazolinone-resistant sunflower (*Helianthus annuus*) caused by application of pre-emergence herbicide. *Int. J. Pest Manag.* **2013**, *59*, 229–235. [CrossRef]

92. Weaver, S.E.; Tan, C.S. Critical period of weed interference in transplanted tomatoes (*Lycopersicum esculentum*): Growth analysis. *Weed Sci.* **1983**, *31*, 476–481. [CrossRef]

93. Woolley, B.L.; Michaels, T.E.; Hall, M.R.; Swanton, C.J. The critical period of weed control in white bean (*Phaseolus vulgaris*). *Weed Sci.* **1993**, *41*, 180–184. [CrossRef]

94. Welsh, J.P.; Bulson, H.A.J.; Stopes, C.E.; Froud-Williams, R.J.; Murdoch, A.J. The critical weed-free period in organically-grown winter wheat. *Ann. Appl. Biol.* **1999**, *134*, 315–320. [CrossRef]

95. Eslami, S.V. Weed management in conservation agriculture systems. In *Recent Advances in Weed Management*; Chauhan, B.S., Mahajan, G., Eds.; Springer: New York, NY, USA, 2014; pp. 87–124. [CrossRef]

96. Feldman, S.R.; Alzugaray, C.; Torres, P.S.; Lewis, P. The effect of different tillage systems on the composition of the seedbank. *Weed Res.* **1997**, *37*, 71–76. [CrossRef]

97. Manley, B.S.; Wilson, H.P.; Hines, T.E. Management programs and crop rotations influence populations of annual grass weeds and yellow nutsedge. *Weed Sci.* **2002**, *50*, 112–119. [CrossRef]

98. Mas, M.T.; Verdú, A.M.C. Tillage system effects on weed communities in a 4-year crop rotation under Mediterranean dryland conditions. *Soil Tillage Res.* **2003**, *74*, 15–24. [CrossRef]

99. Rask, A.M.; Kristoffersen, P. A review of non-chemical weed control on hard surface. *Weed Res.* **2007**, *47*, 370–380. [CrossRef]

100. Jitsuyama, Y.; Ichikawa, S. Possible weed establishment control by applying cryogens to fields before snowfalls. *Weed Technol.* **2011**, *25*, 454–458. [CrossRef]

101. Ascard, J. Effects of flame weeding on weed species at different developmental stages. *Weed Res.* **1995**, *35*, 397–411. [CrossRef]

102. Laguë, C.; Gill, J.; Péloquin, G. Thermal control in plant protection. In *Physical Control Methods in Plant Protection*; Vincent, C., Panneton, B., Fleurat-Lessard, F., Eds.; Springer: Berlin/Heidelberg, Germany, 2001; pp. 35–46. [CrossRef]

103. Cisneros, J.J.; Zandstra, B.H. Flame weeding effects on several weed species. *Weed Technol.* **2008**, *22*, 290–295. [CrossRef]

104. Cordeau, S.; Triolet, M.; Wayman, S.; Steinberg, C.; Guillemin, J.P. Bioherbicides: Dead in the water? A review of the existing products for integrated weed management. *Crop Prot.* **2016**, *87*, 44–49. [CrossRef]

105. Schwarzländer, M.; Hinz, H.L.; Winston, R.L.; Day, M.D. Biological control of weeds: An analysis of introductions, rates of establishment and estimates of success, worldwide. *BioControl* **2018**, *63*, 319–331. [CrossRef]

106. Charudattan, R. Biological control of weeds by means of plant pathogens: Significance for integrated weed management in modern agro-ecology. *BioControl* **2001**, *46*, 229–260. [CrossRef]

107. Sheppard, A.W.; Shaw, R.H.; Sforza, R. Top 20 environmental weeds for classical biological control in Europe: A review of opportunities, regulations and other barriers to adoption. *Weed Res.* **2006**, *46*, 93–117. [CrossRef]

108. Duke, S.O.; Dayan, F.E.; Romagni, J.G.; Rimando, A.M. Natural products as sources of herbicides: Current status and future trends. *Weed Res.* **2000**, *40*, 99–111. [CrossRef]

109. Dayan, F.E.; Owens, D.K.; Duke, S.O. Rationale for a natural products approach to herbicide discovery. *Pest Manag. Sci.* **2012**, *68*, 519–528. [CrossRef]

110. Mejías, F.J.R.; Trasobares, S.; López-Haro, M.; Varela, R.M.; Molinillo, J.M.G.; Calvino, J.J.; Macías, F.A. In situ eco encapsulation of bioactive agrochemicals within fully organic nanotubes. *ACS Appl. Mater. Interfaces* **2019**, *11*, 41925–41934. [CrossRef]

111. Melander, B.; Rasmussen, G. Effects of cultural methods and physical weed control on intrarow weed numbers, manual weeding and marketable yield in direct-sown leek and bulb onion. *Weed Res.* **2001**, *41*, 491–508. [CrossRef]

112. Stepanovic, S.; Datta, A.; Neilson, B.; Bruening, C.; Shapiro, C.A.; Gogos, G.; Knezevic, S.Z. Effectiveness of flame weeding and cultivation for weed control in organic maize. *Biol. Agric. Hortic.* **2016**, *32*, 47–62. [CrossRef]

113. Wagner, M.; Mitschunas, N. Fungal effects on seed bank persistence and potential applications in weed biocontrol: A review. *Basic Appl. Ecol.* **2008**, *9*, 191–203. [CrossRef]

114. Ehlert, K.A.; Mangold, J.M.; Engel, R.E. Integrating the herbicide imazapic and the fungal pathogen *Pyrenophora semeniperda* to control *Bromus tectorum*. *Weed Res.* **2014**, *54*, 418–424. [CrossRef]

115. Pannacci, E.; Tei, F. Effects of mechanical and chemical methods on weed control, weed seed rain and crop yield in maize, sunflower and soybean. *Crop Prot.* **2014**, *64*, 51–59. [CrossRef]

116. Vasileiadis, V.P.; Froud-Williams, R.J.; Eleftherohorinos, I.G. Tillage and herbicide treatments with inter-row cultivation influence weed densities and yield of three industrial crops. *Weed Biol. Manag.* **2012**, *12*, 84–90. [CrossRef]

117. Penner, D. Herbicide action and metabolism. In *Turf Weeds and Their Control*; ASA, CSSA, Eds.; American Society of Agronomy and Crop Science Society of America: Madison, WI, USA, 1994; pp. 37–70. [CrossRef]

118. Heap, I. The International Survey of Herbicide Resistant Weeds. Available online: www.weedscience.org (accessed on 29 December 2019).

119. Christoffers, M.J.; Nandula, V.K.; Mengistu, L.W.; Messersmith, C.G. Altered herbicide target sites: Implications for herbicide-resistant weed management. In *Weed Biology and Management*; Inderjit, Ed.; Springer: Dordrecht, The Netherlands, 2004; pp. 199–210. [CrossRef]

120. Devine, M.; Duke, S.O.; Fedtke, C. *Physiology of Herbicide Action*; PTR: Prentice Hall, NJ, USA, 1993.

121. LeBaron, H.; Gressel, J. *Herbicide Resistance in Plants*; John Wiley and Sons: New York, NY, USA, 1982.

122. Holm, F.A.; Kirkland, K.J.; Stevenson, F.C. Defining optimum herbicide rates and timing for wild oat (*Avena fatua*) control in spring wheat (*Triticum aestivum*). *Weed Technol.* **2000**, *14*, 167–175. [CrossRef]

123. Kim, D.S.; Brain, P.; Marshall, E.J.P.; Caseley, J.C. Modelling herbicide dose and weed density effects on crop: Weed competition. *Weed Res.* **2002**, *42*, 1–13. [CrossRef]

124. Chauhan, B.S.; Abugho, S.B. Interaction of rice residue and PRE herbicides on emergence and biomass of four weed species. *Weed Technol.* **2012**, *26*, 627–632. [CrossRef]

125. Macías, F.A.; Molinillo, J.M.G.; Varela, R.M.; Galindo, J.C.G. Allelopathy—A natural alternative for weed control. *Pest Manag. Sci.* **2007**, *63*, 327–348. [CrossRef]

126. Tesio, F.; Ferrero, A. Allelopathy, a chance for sustainable weed management. *Int. J. Sust. Dev. World* **2010**, *17*, 377–389. [CrossRef]

127. Farooq, M.; Jabran, K.; Cheema, Z.A.; Wahid, A.; Siddique, K.H. The role of allelopathy in agricultural pest management. *Pest Manag. Sci.* **2011**, *67*, 493–506. [CrossRef]

128. Schreiber, M.M. Influence of tillage, crop rotation, and weed management on giant foxtail (*Setaria faberi*) population dynamics and corn yield. *Weed Sci.* **1992**, *40*, 645–653. [CrossRef]

129. Kalburtji, K.L.; Gagianas, A. Effects of sugar beet as a preceding crop on cotton. *J. Agron. Crop Sci.* **1997**, *178*, 59–63. [CrossRef]

130. Alsaadawi, I.S.; Sarbout, A.K.; Al-Shamma, L.M. Differential allelopathic potential of sunflower (*Helianthus annuus* L.) genotypes on weeds and wheat (*Triticum aestivum* L.) crop. *Arch. Agron. Soil Sci.* **2012**, *58*, 1139–1148. [CrossRef]

131. Midya, A.; Bhattacharjee, K.; Ghose, S.S.; Banik, P. Deferred seeding of blackgram (*Phaseolus mungo* L.) in rice (*Oryza sativa* L.) field on yield advantages and smothering of weeds. *J. Agron. Crop Sci.* **2005**, *191*, 195–201. [CrossRef]

132. Cheema, Z.A.; Khaliq, A.; Saeed, S. Weed control in maize (*Zea mays* L.) through sorghum allelopathy. *J. Sustain. Agric.* **2004**, *23*, 73–86. [CrossRef]

133. Turk, M.A.; Tawaha, A.M. Allelopathic effect of black mustard (*Brassica nigra* L.) on germination and growth of wild oat (*Avena fatua* L.). *Crop Prot.* **2003**, *22*, 673–677. [CrossRef]

134. Shrestha, A. Potential of a black walnut (*Juglans nigra*) extract product (NatureCur®) as a pre- and post-emergence bioherbicide. *J. Sustain. Agric.* **2009**, *33*, 810–822. [CrossRef]

135. Jabran, K.; Cheema, Z.A.; Farooq, M.; Hussain, M. Lower doses of pendimethalin mixed with allelopathic crop water extracts for weed management in canola (*Brassica napus*). *Int. J. Agric. Biol.* **2010**, *12*, 335–340.

136. Jabran, K.; Mahajan, G.; Sardana, V.; Chauhan, B.S. Allelopathy for weed control in agricultural systems. *Crop Prot.* **2015**, *72*, 57–65. [CrossRef]

137. Kruidhof, H.M.; Bastiaans, L.; Kropff, M.J. Cover crop residue management for optimizing weed control. *Plant Soil* **2009**, *318*, 169–184. [CrossRef]

138. Scavo, A.; Restuccia, A.; Pandino, G.; Onofri, A.; Mauromicale, G. Allelopathic effects of *Cynara cardunculus* L. leaf aqueous extracts on seed germination of some Mediterranean weed species. *Ital. J. Agron.* **2018**, *13*, 119–125. [CrossRef]

139. Scavo, A.; Pandino, G.; Restuccia, A.; Lombardo, S.; Pesce, G.R.; Mauromicale, G. Allelopathic potential of leaf aqueous extracts from *Cynara cardunculus* L. on the seedling growth of two cosmopolitan weed species. *Ital. J. Agron.* **2019**, *14*, 78–83. [CrossRef]

140. Scavo, A.; Pandino, G.; Restuccia, A.; Mauromicale, G. Leaf extracts of cultivated cardoon as potential bioherbicide. *Sci. Hort.* **2020**, *261*, 109024. [CrossRef]

141. Scavo, A.; Rial, C.; Molinillo, J.M.G.; Varela, R.M.; Mauromicale, G.; Macias, F.A. The extraction procedure improves the allelopathic activity of cardoon (*Cynara cardunculus* var. *altilis) leaf allelochemicals. Ind. Crops Prod.* **2019**, *128*, 479–487. [CrossRef]
142. Scavo, A.; Rial, C.; Varela, R.M.; Molinillo, J.M.G.; Mauromicale, G.; Macias, F.A. 2019 Influence of genotype and harvest time on the *Cynara cardunculus* L. sesquiterpene lactone profile. *J. Agric. Food Chem.* **2019**, *67*, 6487–6496. [CrossRef]

Article

Grazing and Cutting under Different Nitrogen Rates, Application Methods and Planting Density Strongly Influence Qualitative Traits and Yield of Canola Crop

Sajjad Zaheer [1,2], Muhammad Arif [2], Kashif Akhtar [3], Ahmad Khan [2], Aziz Khan [4], Shahida Bibi [5], Mehtab Muhammad Aslam [6], Salman Ali [2], Fazal Munsif [2], Fazal Jalal [7], Noor Ul Ain [8], Fazal Said [7], Muhammad Ali Khan [6], Muhammad Jahangir [9] and Fan Wei [1,*]

[1] Guangxi Key laboratory of medicinal resources protection and genetic improvement, Guangxi botanical garden of medicinal plants, Nanning 530005, China; Sajjadzaheer15@gmail.com
[2] Department of Agronomy, The University of Agriculture Peshawar-Pakistan, Peshawar 25000, Pakistan; marifkhan75@aup.edu.pk (M.A.); ahmad0936@yahoo.com (A.K.); salmankhan@aup.pk (S.A.); munsiffazal@yahoo.com (F.M.)
[3] Institute of Nuclear Agriculture Sciences, College of Agriculture and Biotechnology, Zhejiang University, Hangzhou 310058, China; kashif@zju.edu.cn
[4] College of Agriculture, Guangxi University, Nanning 530005, China; azizkhanturlandi@gmail.com
[5] Department of Weed Science, The University of Agriculture Peshawar-Pakistan, Peshawar 25000, Pakistan; shahida@aup.edu.pk
[6] Center for Plant Water-Use and Nutrition Regulation, College of Life Sciences, Joint International Research Laboratory of Water and Nutrient in Cops, Fujian Agriculture and Forestry University, Fuzhou 350002, Fujian, China; mehtabmuhammadaslam@yahoo.com (M.M.A.); malikhan@awkum.edu.pk (M.A.K.)
[7] Department of Agriculture, Abdul Wali khan University Mardan, Mardan 23200, Pakistan; fazaljalal@awkum.edu.pk (F.J.); fazal@awkum.edu.pk (F.S.)
[8] FAFU and UIUC-SIB joint center for genomics and biotechnology, Fujian Agriculture and Forestry University, Fuzhou 350002, Fujian, China; noorulainali001@gmail.com
[9] Department of Horticulture, The University of Agriculture Peshawar-Pakistan, Peshawar 25000, Pakistan; jahangir.hayat44@gmail.com
* Correspondence: 15277125531@163.com; Tel.: +86-152-7712-5531

Received: 28 January 2020; Accepted: 11 March 2020; Published: 16 March 2020

Abstract: Canola crop has the potential for both seeds and grazing. Optimal planting density, time of nitrogen (N) fertilizer application and rates are the major aspects for successful qualitative traits and canola yield formation. In this content, optimization of planting density, N levels and its time of application in dual purpose canola are needed. This study was carried out in RCB design with split pot arrangement having three repeats during winter 2012–2013 and 2013–2014. The study evaluated N levels (120 and 80 kg N ha^{-1}), cutting treatment, N application timings and planting density (20 and 40 plants m^{-2}) effects on qualitative traits and yield of canola. No-cut treatment had 7.02%, 2.46%, and 4.26% higher, glucosinolates, oil, and protcin content with 31.3% and 30.5% higher biological and grain yield respectively, compared with grazed canola. Compared with no-cut canola, grazed canola resulted in 7.74% of higher erucic acid. Further, application of N at 120 kg N ha^{-1} had 8.81%, 5.52%, and 6.06% higher glucosinolates, percent protein, and seed yield, respectively than 80 kg N ha^{-1}. In-addition, the application of N into two splits was most beneficial than the rest application timings. Cutting had 15% reduction in grain yield of canola and fetched additional income of 143.6 USD compared with no-cut. Grazing resulted in a 23% reduction in grain yield while had additional income of 117.7 USD from fodder yield. Conclusively, the application of N in two splits at 120 kg N ha^{-1} combined with 20 plants m^{-2} is a promising strategy to achieve good qualitative attributes and canola yield under dual purpose system.

Keywords: dual purpose canola; nitrogen fertilizer; planting density; oil content; grazing

---

1. Introduction

The domestic production can only meet 29% of the total edible oil requirements of Pakistan and the remaining 71% was mad through imports [1]. Like other developing countries, Pakistan is also deficit in edible oil production. According to a study conducted the consumer's demand has steadily increased from 0.3 million tons to 2.764 tons during the last two and half decades. The average yield of canola in Pakistan is 839 kg ha^{-1} [2], which is very low compared with other agriculturally advanced countries. The European countries have a yield level of 3500 kg ha^{-1}; Canada 3200 kg ha^{-1}; and Australia 2000 kg ha^{-}1 for canola crop [3]. To cut down these gaps concrete efforts are needed to increase its local production. Canola is an improved form of conventional rape seed variety developed through genetic engineering having erucic acid less than 2% and 30 μmolg^{-1} glucosinolates, which are considered the safe limits for health [4]. As compared to other oil crops, it contains less amount of cholesterol [3]. Seed oil concentration was inversely proportional to seed protein concentration in mustard and canola genotypes. Increase in seed yield increased the oil concentration, but decreased the protein concentration [5].

Optimum amount of nutrients supply at proper time to any crop is important [6]. Canola crop requires a higher amount of nutrients, and available nitrogen (N) compared with cereals [7]. Split application of N fertilizer has become more popular in terms of high nitrogen use efficiency. An appropriate rate and timing of N fertilizer application is one of the most important aspects of successful canola production [3]. Canola yield is strongly correlated with biotic and abiotic factors; one of the factors that availability of nutrients especially N is the key driver for improving root growth, leaf photosynthetic rate, biomass production, and yield [8]. N fertilizer boosts yield improving thousand seed weights, seeds pod^{-1}, and pod number plant^{-1} [9]. Dual purpose cropping is the use of crops for fodder purpose at vegetative stage grazed by animals. The regrowth of plants after cutting or grazing strongly relies on the regenerative ability of the species. After being in stress condition, soon after grazing crops N fertilization is needed for growth improvement, thus N fertilizer selection is a good choice for growth improvement. Increasing N level from normal or recommended dozes boosts the overall plant health and seed production [9]. Considering DP canola, N fertilizer might be increased for better re-growth or regeneration of canola. Moreover, N is the most volatile and due to high losses, N efficiency becomes lesser and thereby affects plant normal functioning [10]. Winter or long-season spring canola with proper N rates can be sown to produce high-quality forage for grazing or fodder for cut and carry and recover from grazing to produce a high grain yield (4 t/ha) with good oil content (47%) [11]. Canola crop has the potential to produce grains and to graze. Although, canola crop is good for forage. Canola grazing is only one of the assortments of choice to the farmers to perk up farm economics and productivity.

However, oil quality is also important, and contains less amount of cholesterol, which is good for human health. Studies regarding N rate, timing, and planting density under dual purpose canola production are lacking. The objectives of this study were to explore canola qualitative and yield attributes to different N rate, application, and planting density under dual purpose canola technology. This study also explores the quantitative relationship among N level, application time and planting density for dual purpose use of canola. Furthermore, the combination of all above discussed factors will make an understanding of the effect of grazing on the yield and quality of canola crop and even development of commercial grazing practices. The tested hypothesis was that N rate, time of application, and planting density would improve canola qualitative traits and yield under dual purpose technology.

2. Materials and Methods

2.1. Experimental Site

The present study was conducted at New Developmental Farm of Agricultural University Peshawar during Rabi 2012–2013. The research farm is located about 300 m above the sea level, while the site has 34° N latitude and 72° E longitude. The soil of the site was clay loamy having pH values ranges between 7.0–7.5. Temperature (°C) and rainfall (mm) during the crop growing season have been shown in Figure 1. Weather data were collected from the meteorological station located near the experimental site.

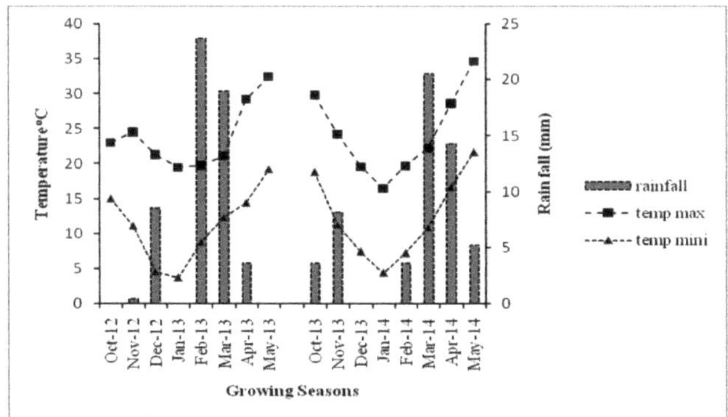

Figure 1. Annual rainfall, maximum and minimum temperature of the experimental site for 2012–2013 and 2013–2014.

2.2. Treatments and Methods

The experiment contained two levels (80 and 120 kg ha^{-1}) of nitrogen levels (NL), cuttings (C; (cut, no cut and grazing), application timing (NT; (a) full application at sowing, (b) half dose of N both at sowing and start of rosette stage, (c) one third dose of N each at sowing, start of rosette stage and soon after cut at late rosette stage (60 days after sowing) and plant density (PD: 20 and 40 plants m^{-2}) (Table 1). In 2012–2014, the experiment was conducted in RCB design with split plot arrangement having three replications. Cutting and N levels (urea as a source of nitrogen was applied as top dressed) were assigned to the main plot while N application timings and planting densities were allotted to sub-plot. Cutting and grazing were done 60 days after sowing. Cutting was done manually by cutting the crop at about 10 cm above the ground. However, grazing was done through sheep for predetermined time.

Table 1. The detailed presentation of experimental treatments.

Main Plot Factors	Treatment Levels	Sub Plot Factors	Treatment Levels
Cuttings (C)	No-cut (C1)	Nitrogen application timings (NT)	(1) full application at sowing (NT$_1$)
	Cut (C2)		(2) half dose of N both at sowing and start of rosette stage (NT$_2$)
	Grazing (C3)		(3) one third dose of N each at sowing, start of rosette stage and soon after cut at late rosette stage (60 days after sowing) (NT$_3$)
Nitrogen Levels (NL)	80 kg ha^{-1} (NL$_1$)	Planting density (PD)	20 plants m^{-2} (PD$_1$)
	120 kg ha^{-1} (NL$_2$)		40 plants m^{-2} (PD$_2$)

2.3. Field Preparation and Cultural Practices

Cultivar Abasin-95 was sown with a uniform seed rate of 8 kg ha^{-1}. Row to row distance of 50 cm was maintained with a subplot size of 10.5 m^2, having 7 rows, and 3 m long. Before sowing, a fine seedbed was prepared by ploughing the field with cultivator followed by rotavator. A basal dose of phosphorus at 60 kg ha^{-1} was applied in the form of single super phosphate. Nitrogenous fertilizer was applied in the form of urea. Weeds were controlled manually by hoeing, when the crop reached 6–8 cm height. The field was harvested on 10 April each year. All cultural practices were carried out uniformly in all plots.

2.4. Grazing Management

Sheep stock was arranged for grazing canola from nearby village. The sheep were allowed to graze a normal canola field about five days before the treatment grazing to acclimatize them to canola/brassica consumption. The sheep were controlled with the help of fences from going to other treatment plots. Animals for grazing were allowed in noon time because of much frost in morning time during grazing period in the month of December in both years.

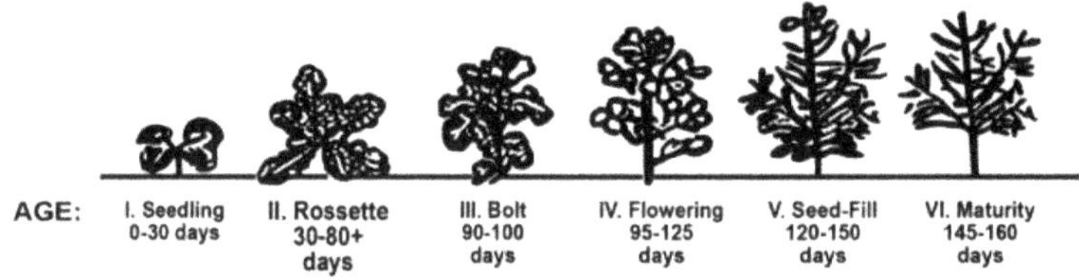

Canola growth stages.

2.5. Observations

2.5.1. Quality Attributes of Canola

Protein, percent oil content, erucic acid, and glucosinolates were determined by collecting randomly seed samples in each plot and were analyzed by Full Option Science System (FOSS) Routine Near Measurement System (35RP-3752F) TR-3657-C Model 6500, at oilseed laboratory, Nuclear Institute for Food and Agriculture, Peshawar (NIFA). Near infrared reflectance (NIR) spectroscopy is a quick and whole seed analyzing method, which does not require any sample preparation or chemicals [12].

2.5.2. Canola Yield

Biomass yield was determined by harvesting of four central rows, dried and weighted. While in order to determine grain yield, bundles from the same central four rows were threshed, seeds were weighed and the data were converted to kg ha^{-1} [13].

2.5.3. Statistical Analysis

The data were statistically analyzed over years using ANOVA techniques appropriate for RCB design with split plot arrangement using SPSS software (SPSS Inc., Chicago, IL, USA). Means were compared using LSD test at 0.05 level of probability, when the *F*-values were significant [14].

3. Results

3.1. Treatments Interactions

The significance ANOVA for main factors and interaction is presented in Table 2. The C × NL significantly affected on Glucosinolates, while the rest of treatments interaction were not significant. Oil content were significantly affected by the C × NL, C × NT, and NT×PD treatments. Protein content was not significant throughout the treatments' interactions. Eurcic acid showed significant differences on C × PD treatment, while biological yield was significant on C × NL, and NT × NL, and grain yield was on C × NT, and NT × PD treatments.

Table 2. Mean square table for crop yield, qualitative attributes, glucosinulates, and erucic acid for the years 2012–14.

SOV	Grain Yield	Biological Yield	Oil Content	Protein Content	Glucosinulates Content	Eurcic Acid
Year	134 ns	7.397 *	1598.49 *	10,560.2	799.26 *	346.56 *
Cuttings (C)	1,482,015 *	1.383 *	18.93 *	27.7 ns	330.24 *	115.59 *
N-levels (NL)	204,857 *	5,140,844 ns	243.63 *	135.9	1547.22 *	22.556 ns
Planting densities (PD)	45,182 ns	4,630,281 *	18.84 *	1.2 ns	1.23 ns	6.476 ns
Nitrogen timings (NT)	30,938 ns	6,378,573 *	23.12 *	1.9 ns	6.18 ns	14.529 ns
C × NL	ns	*	*	ns	*	ns
C × NT	*	ns	*	ns	ns	*
C × PD	ns	ns	ns	ns	ns	ns
NL × PD	ns	ns	ns	ns	ns	ns
NT × NL	ns	*	ns	ns	ns	ns
NT × PD	*	ns	*	ns	ns	*
C × NT × NL	ns	ns	ns	ns	ns	ns
C × NT × PD	ns	ns	ns	ns	ns	ns
NT × NL × PD	ns	ns	ns	ns	ns	ns
C × NL × NT × PD	ns	ns	ns	ns	ns	ns

Note: ns = non-significant, * = Significant at 5% level of probability.

3.2. Crop Yield

The significance ANOVA for main factors is presented in Table 2. Canola biomass yield was significantly affected by C, PD, and year, whereas NL and NT did not affect biomass yield (Table 3). C_3 and C_2 decreased biological yield by 11.74% and 31.2% compared with that of C_1. Plants grown at PD_2 had 5.3% higher biological yield than PD_1. The C, NL, and NT significantly influenced grain yield of canola, whereas the effect of planting density and year remained unaffected on grain yield (Table 4). The interactions among C × NT and NT × PD were significant. Higher grain yield was produced in C_1 plots, followed by C_2 plots, whereas lower grain yield resulted in C_3 plots. Grain yield was higher when N was applied at NT_2 as compared to NT_3.

Table 3. Biomass yield (kg ha^{-1}) of canola under cutting treatments, nitrogen levels and application timings under different planting densities.

Variables	2012–2013	2013–2014	Two Years Average
Cutting treatments (C)			
C_1	9695 a	9353 a	9524a
C_2	7272 b	9774 a	8523 b
C_3	6080 c	8432 b	7256 c
LSD $_{(0.05)}$	598	631	407
Nitrogen levels (NL)			
NL_1	8099	8794 b	8446
NL_2	7932	9578 a	8755
Significance level	ns	*	ns
N application timings (NT)			
NT_1	7974 ab	9023	8496
NT_2	7700 b	9428	8564
NT_3	8372 a	9107	8739
LSD $_{(0.05)}$	500	506	352
Planting density (PD)			
PD_1	7785 b	8584 b	8454 b
PD_2	8264 a	9787 a	8747 a
Significance level	*	*	*
Year (*)	8016 b	9186 a	

Note: Cuttings (C): no-cut (C_1), cutting (C_2), grazing (C_3); Nitrogen levels (NL): 80 kg ha^{-1}(NL_1), 120 kg ha^{-1} (NL_2); Nitrogen timings: Full application at sowing (NT_1), half dose of N both at sowing and start of rosette stage (NT_2), one third dose of N each at sowing, start of rosette stage and soon after cut at late rosette stage (60 days after sowing (NT_3); Planting density (PD): 20 plants m^{-2} (PD_1), 40 plants m^{-2} (PD_2). Means of the same category followed by different letters are significantly different at 5 % level of probability using LSD $_{(0.05)}$ test. ns = non-significant, * = Significant at 5% level of probability.

Table 4. Grain yield (kg ha^{-1}) of canola under cutting treatments, nitrogen levels and application timings under different planting densities.

Variables	2012–2013	2013–2014	Two Years Average
Cutting treatments (C)			
C_1	1291 a	1128 a	1210 a
C_2	983 b	1068 b	1025 b
C_3	886 c	968 c	927 c
LSD $_{(0.05)}$	72.0	54.1	42.2
Nitrogen levels (NL)			
NL_1	1022 b	1024 b	1023 b
NL_2	1084 a	1086 a	1085 a
Significance level	*	*	*
N application timings (NT)			
NT_1	1021 b	1054 a	1045 ab
NT_2	1029 b	1058 a	1078 a
NT_3	1109 a	1052 a	1039 b
LSD $_{(0.05)}$	42.5	64.6	38.3
Planting density (PD)			
PD_1	1066	1057	1039
PD_2	1044	1053	1068
Significance level	ns	ns	ns
Year (ns)	1053	1055	

Note: Cuttings (C): no-cut (C_1), cutting (C_2), grazing (C_3); Nitrogen levels (NL): 80 kg ha^{-1}(NL_1), 120 kg ha^{-1} (NL_2); Nitrogen timings: Full application at sowing (NT_1), half dose of N both at sowing and start of rosette stage (NT_2), one third dose of N each at sowing, start of rosette stage and soon after cut at late rosette stage (60 days after sowing (NT_3); Planting density (PD): 20 plants m^{-2} (PD_1), 40 plants m^{-2} (PD_2). Means of the same category followed by different letters are significantly different at 5 % level of probability using LSD $_{(0.05)}$ test. ns = non-significant, * = Significant at 5% level of probability.

3.3. Quality Parameters of Canola

3.3.1. Oil Content (%) and Protein Content (%)

Cuttings, PD, NL, NT, and years significantly affected oil content of canola crop (Table 5). A 1.4% higher oil content was resulted in C_1 followed by C_2. The C_3 plots substantially reduced oil content. Application of NL1 had 5.07% of higher oil content compared with NL_2. Oil content was higher NT_2 followed by NT_3 and NT_1. Between planting densities, plants under PD_2 had higher oil content compared with PD_1. The C × NL interaction showed an obvious reduction in oil content for C_2 and C_3 plots under NL_2, whereas the oil content of all cutting treatments remained unchanged under NL_1 application (Figure 2A). Interaction between C × NT showed that canola oil content decreased in C_3 plots at NT_2 than C_1 and C_2. On other hand, oil content of all cutting treatments remained unchanged under three equal splits application of N (Figure 2B). Interaction between NT and PD indicated that oil content was reduced for PD_1 with sole N at sowing than two or three splits and planting densities (Figure 2C). Cuttings, NL and years had remarkedly effects on grain protein content of canola seeds while the effects of NT and PD were insignificant (Table 5). Higher protein content was noted for cut plots which were statistically at par with C_3 plots, while C_1 plots had lower crude protein content in canola seed. The application of NL_2 had 5.5% higher crude protein content compared with that of NL_1 (Table 6).

Table 5. Canola oil content (%) in response to cutting treatments, nitrogen levels and application timings under varying planting densities.

Variables	2012–2013	2013–2014	Two Years Average
Cutting treatments (C)			
C_1	40.78 a	45.95 a	43.37 a
C_2	40.03 b	45.51 ab	42.77 b
C_3	39.51 b	45.18 b	42.35 c
LSD $_{(0.05)}$	0.67	0.59	0.42
Nitrogen levels (NL)			
NL_1	41.13 a	46.66 a	43.89 a
NL_2	39.09 b	44.44 b	41.77 b
Significance level	*	*	*
N application timings (NT)			
NT_1	39.81 b	45.33 ab	42.26 b
NT_2	39.77 b	45.09 b	43.39 a
NT_3	40.74 a	46.22 a	42.83 ab
LSD $_{(0.05)}$	0.75	0.90	0.58
Planting density (PD)			
PD_1	40.00	45.35	42.53 b
PD_2	40.21	45.76	43.12 a
Significance level	*	ns	*
Year (*)	40.11 b	45.55 a	

Note: Cuttings (C): no-cut (C_1), cutting (C_2), grazing (C_3); Nitrogen levels (NL): 80 kg ha^{-1}(NL_1), 120 kg ha^{-1} (NL_2); Nitrogen timings: Full application at sowing (NT_1), half dose of N both at sowing and start of rosette stage (NT_2), one third dose of N each at sowing, start of rosette stage and soon after cut at late rosette stage (60 days after sowing (NT_3); Planting density (PD): 20 plants m^{-2} (PD_1), 40 plants m^{-2} (PD_2). Means of the same category followed by different letters are significantly different at 5 % level of probability using LSD $_{(0.05)}$ test. ns = non-significant, * = Significant at 5% level of probability.

Table 6. Canola protein content (%) in response to cutting treatments, nitrogen levels and application timings under varying planting densities.

Variables	2012–2013	2013–2014	Two Years Average
Cutting treatments (C)			
C_1	36.26	21.56 b	28.91 b
C_2	37.25	23.00 a	30.14 a
C_3	36.19	23.21 a	29.70 ab
LSD $_{(0.05)}$	2.01	1.07	1.06
Nitrogen levels (NL)			
NL_1	35.90	21.69 b	28.79 b
NL_2	37.26	23.50 a	30.38 a
Significance level	ns	*	*
N application timings (NT)			
NT_1	35.90 b	22.93	29.44
NT_2	36.45 ab	22.57	29.76
NT_3	37.37 a	22.58	29.55
LSD $_{(0.05)}$	1.42	0.83	0.81
Planting density (PD)			
PD_1	36.24	23.03 a	29.5
PD_2	36.92	22.16 b	29.66
Significance level	*	*	ns
Year (*)	22.59 b	36.58 a	

Note: Cuttings (C): no-cut (C_1), cutting (C_2), grazing (C_3); Nitrogen levels (NL): 80 kg ha^{-1}(NL_1), 120 kg ha^{-1} (NL_2); Nitrogen timings: Full application at sowing (NT_1), half dose of N both at sowing and start of rosette stage (NT_2), one third dose of N each at sowing, start of rosette stage and soon after cut at late rosette stage (60 days after sowing (NT_3); Planting density (PD): 20 plants m^{-2} (PD_1), 40 plants m^{-2} (PD_2). Means of the same category followed by different letters are significantly different at 5 % level of probability using LSD $_{(0.05)}$ test. ns = non-significant, * = Significant at 5% level of probability.

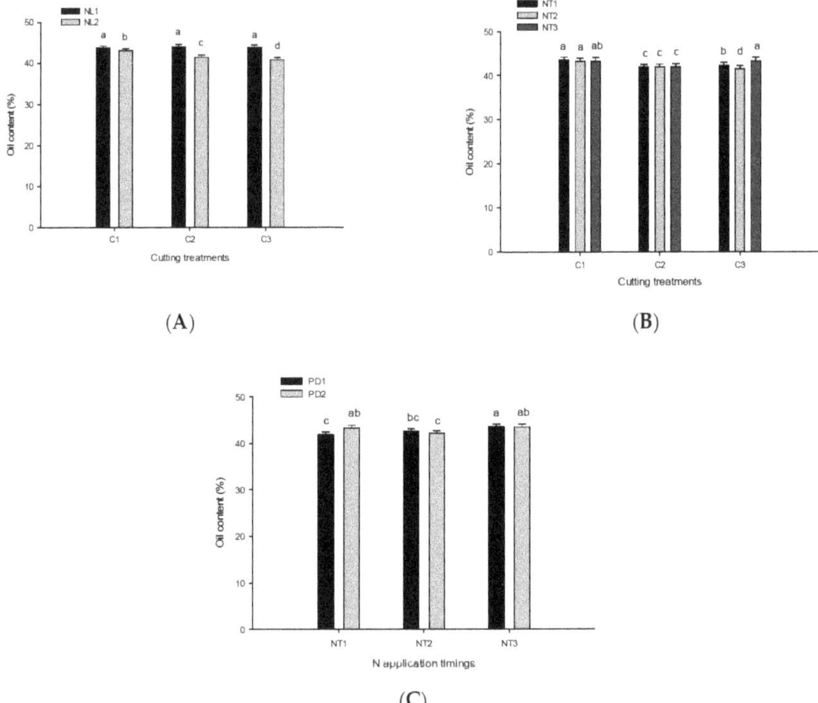

Figure 2. The interactive effects of nitrogen levels and cutting treatments (**A**), nitrogen application timings and cutting treatments (**B**), and nitrogen application timings and planting densities (**C**) for oil content. Vertical bars represent Standard Error. Note: Cuttings (C): no-cut (C_1), cutting (C_2), grazing (C_3); Nitrogen levels (NL): 80 kg ha^{-1} (NL_1), 120 kg ha^{-1} (NL_2); Nitrogen timings: Full application at sowing (NT_1), half dose of N both at sowing and start of rosette stage (NT_2), one third dose of N each at sowing, start of rosette stage and soon after cut at late rosette stage (60 days after sowing (NT_3); Planting density (PD): 20 plants m^{-2} (PD_1), 40 plants m^{-2} (PD_2).

3.3.2. Erucic acid Content (%)

Canola erucic acid content was significantly influenced by C and NT, while the effect of year, NL and PD was non-significant (Table 7). Erucic acid content was 7.7% higher in C_3 than other cutting treatments. Similarly, NT_1 had higher erucic acid content than NT_2 or NT_3. In addition, in the first year the content was 8.07% higher than 2nd year of the study. However, significant C × PD interaction revealed that erucic acid content increased with imposition of C_2 and C_3 than C_1 under PD_1. No or least variation in erucic acid content in all cutting treatments was noted under PD_2 treatment (Figure 3).

Table 7. Erucic acid (%) canola as affected by cutting treatments, nitrogen levels and application timings under varying planting densities.

Variables	2012–2013	2013–2014	Two Years Average
Cutting treatments (C)			
C_1	32.3 b	31.2 b	31.0 b
C_2	34.1 ab	33.1 ab	32.9 a
C_3	34.7 a	33.6 a	33.4 a
LSD $_{(0.05)}$	1.94	1.91	1.26
Nitrogen levels (NL)			
NL_1	33.4	32.3	32.1
NL_2	34.1	32.9	32.8
Significance level	ns	ns	ns
N application timings (NT)			
NT_1	33.9	32.8	32.9 a
NT_2	33.9	32.8	32.0 b
NT_3	33.3	32.1	32.4 ab
LSD $_{(0.05)}$	ns	ns	0.805
Planting density (PD)			
PD_1	33.7	32.6	32.3
PD_2	33.7	32.5	32.6
Significance level	ns	ns	ns
Year (*)	33.72 a	31.2 b	

Note: Cuttings (C): no-cut (C_1), cutting (C_2), grazing (C_3); Nitrogen levels (NL) 80 kg ha^{-1}(NL_1), 120 kg ha^{-1} (NL_2); Nitrogen timings: Full application at sowing (NT_1), half dose of N both at sowing and start of rosette stage (NT_2), one third dose of N each at sowing, start of rosette stage and soon after cut at late rosette stage (60 days after sowing (NT_3); Planting density (PD): 20 plants m^{-2} (PD_1), 40 plants m^{-2} (PD_2). Means of the same category followed by different letters are significantly different at 5% level of probability using LSD $_{(0.05)}$ test. ns = non-significant, * = Significant at 5% level of probability.

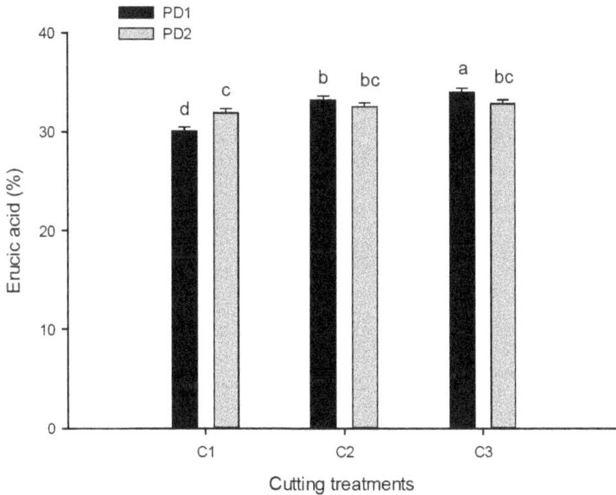

Figure 3. The interactive effect of planting density and cutting treatments of Erucic acid (%) for canola seed. Vertical bars represent Standard Error. Note: Cuttings (C): no-cut (C_1), cutting (C_2), grazing (C_3); Planting density (PD): 20 plants m^{-2} (PD_1), 40 plants m^{-2} (PD_2).

3.3.3. Glucosinolates Concentration

Canola glucosinolates concentration was substantially impacted by C, NL and year, while the effect of NT and, PD and year was not significant (Table 8). Glucosinolates content was 4% higher in C_1 followed by C_2, while C_3 resulted in lower content of glucosinolates. Similarly, glucosinolates concentration increased with the increased of N level from 80 to 120 kg ha^{-1}. Glucosinolates content was 6% higher in second year compared with compared with first year. The C x NL interaction indicated that increasing N from 80 upto 120 kg ha^{-1} increased the glucosinolates contents in all cutting treatments. However, the glucosinolates were markedly lower in C_3 and C_2 than C_1 under NL_1 (Figure 4).

Table 8. Canola glucosinolates content (μ mol g^{-1}) as affected by cutting treatments, nitrogen levels and application timings under varying planting densities.

Variables.	2012–2013	2013–2014	Two Years Average
Cutting treatments (C)			
C_1	59.97 b	66.18 a	64.48 a
C_2	62.78 a	63.67 b	61.82 b
C_3	58.03 c	62.46 b	60.25 c
LSD $_{(0.05)}$	1.88	1.91	1.26
Nitrogen levels (NL)			
NL_1	57.59 b	61.42 a	59.61 b
NL_2	62.93 a	66.79 b	64.86 a
Significance level	*	*	*
N application timings (NT)			
NT_1	60.07	63.74	62.45
NT_2	60.27	64.26	62.23
NT_3	60.45	64.32	61.87
LSD $_{(0.05)}$	1.07	1.51	0.99
Planting density (PD)			
PD_1	60.76	64.29	62.11
PD_2	59.76	63.93	62.26
Significance level	ns	ns	ns
Year (*)	60.26 b	64.11 a	

Note: Cuttings (C): no-cut (C_1), cutting (C_2), grazing (C_3); Nitrogen levels (NL) 80 kg ha^{-1}(NL_1), 120 kg ha^{-1} (NL_2); Nitrogen timings: Full application at sowing (NT_1), half dose of N both at sowing and start of rosette stage (NT_2), one third dose of N each at sowing, start of rosette stage and soon after cut at late rosette stage (60 days after sowing (NT_3); Planting density (PD): 20 plants m^{-2} (PD_1), 40 plants m^{-2} (PD_2). Means of the same category followed by different letters are significantly different at 5 % level of probability using LSD $_{(0.05)}$ test. ns = non-significant, * = Significant at 5% level of probability.

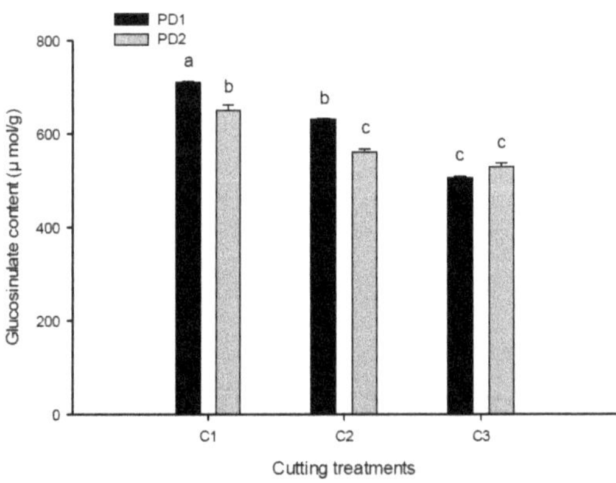

Figure 4. Interaction between cutting treatments and nitrogen levels for glucosinulates content. Vertical bars represent Standard Error. Note: Cuttings (C): no-cut (C_1), cutting (C_2), grazing (C_3); Planting density (PD): 20 plants m^{-2} (PD_1), 40 plants m^{-2} (PD_2).

3.4. Economic Analysis

Economic analysis showed that C_2 produced higher net income (USD 1202.1) compared to that of C_1 (USD 1057.2) (Table 9). However, in C_1 plots net income was higher compared to C_3 (USD 121). C_2 and C_3 plots reduced grain yield by 15% and 23%, respectively, than C_1 plots. Likewise, value cost ratio (VCR) of C_2 was higher compared with C_3 and C_1 plots. Furthermore, higher VCR (4.16) was found in C_2 where N at the rate of 80 kg ha^{-1} in NT_2 with high density (40 plants m^{-2}) was followed by C_2 and received at NL_1 with PD_2. The lower VCR (1.58) in C_3 received at NL_2 in NT_3 in PD_1 (Table 10). The net income from fodder of canola in C_2 plots were recorded on the basis of area as per usual practice at the same farm.

Table 9. Economic analysis of dual-purpose canola.

Yield, Value or Cost	C_1	C_2	C_3
Forage yield (kg ha^{-1})	0.0	13.9	14.1
Grain yield (kg ha^{-1})	12.4	10.5	9.5
Straw yield (kg ha^{-1})	92.1	74.3	63.8
Forage value (USD)	0.0	69.5	70.3
Grain value (USD)	884.5	1067.1	863.4
Straw value (USD)	552.4	446.1	383.1
Gross income (USD)	1436.9	1582.8	1316.9
Net income over control (USD)	1057.2	1202.1	936.2
Value cost ratio (VCR %)	2.78	3.1	2.4

Note: Cuttings (C): no-cut (C_1), cutting (C_2), grazing (C_3).

Table 10. The interactive effects of different factors over economic benefits (VCR values).

Plant density m^{-2}	Cutting	N rate	NT_1	NT_2	NT_3
PD_1	C_1	NL_1	2.95	2.57	2.82
	C_1	NL_2	2.59	2.21	2.71
	C_2	NL_1	3.00	3.35	3.00
	C_2	NL_2	2.71	3.36	3.17
	C_3	NL_1	3.09	3.35	2.92
	C_3	NL_2	2.85	2.22	1.58
PD_2	C_1	NL_1	2.60	3.30	3.10
	C_1	NL_2	3.02	2.69	2.80
	C_2	NL_1	3.52	4.16	2.57
	C_2	NL_2	3.22	3.06	2.83
	C_3	NL_1	2.18	2.63	2.07
	C_3	NL_2	2.59	2.23	1.96

Note: Cuttings (C): no-cut (C_1), cutting (C_2), grazing (C_3); Nitrogen levels (NL) 80 kg ha^{-1}(NL_1), 120 kg ha^{-1} (NL_2); Nitrogen timings: Full application at sowing (NT_1), half dose of N both at sowing and start of rosette stage (NT_2), one third dose of N each at sowing, start of rosette stage and soon after cut at late rosette stage (60 days after sowing (NT_3); Planting density (PD): 20 plants m^{-2} (PD_1), 40 plants m^{-2} (PD_2).

4. Discussion

4.1. Crop Yield

4.1.1. Effects of Grazing and Cut on Canola Yield

Biological yield is the function of increase rate and growth duration both of which indicate the possibility for improved yield. In this study, biological yield was higher in no-cut plots followed by cut and grazed plots. Biological yield recovered rapidly in cut and grazed plots but the removal of branches in initial grazing and cuttings had led to the differences among the means. Delay in flowering may affect biological yield, when grazing removed the main auxiliary buds from the stems [15]. In our study biological yield was higher in second year (2013–2014) compared with first year (2012–2013), might be due to that in the second year, canola was sown for 15 days earlier than first year. Earlier sown plans of dual-purpose cropping for better results [16]. Reduction in biological yield was associated with the removal of branches in flowering, and not affected on seed yield [17]. Gross marginal value of DP canola is greater than grains only. In most cases grazing during early growth stages does not show significant results in the reduction of seed yield [18,19], while grazing after vegetative stage caused reduction up to 25% or even more [15,16]. Cutting treatments caused significant reduction in grain yield, might be due to less re-growth ability of plants in cut and grazed plots, and they were unable to regenerate quickly and reach to the growth of plants of no cut plots. The possible reason for this substantial decrease in yield of grazed and cut plots might be the removal of main stem either manually or by sheep grazing.

4.1.2. Effects of N fertilizer, Application Timings on Canola Yield

Canola crop responded well to N application timings. In these experiments, 3% and 1% increase in biological yield of canola was recorded in plots where N was applied in two or three splits as compared with sole application, respectively. The increase in yield with split application of N at seedling, rosette stage and early flowering [20]. Similarly, application of N at the rate of 100 kg ha^{-1} in split form (half each at sowing and soon after grazing) increased the biological yield up to 3.95 t ha^{-1} [21], and nitrogen in split form resulted better than sole application [22].

Canola crop requires high amount of N fertilizer compared to cereals to produce high yields [23]. In our study, seed yield increased by 6 % when N rate mounted from 80 to 120 kg ha^{-1}. Higher yields of canola achieved in the current study also highlights the high levels of N fertilizer which must be applied to achieve enhanced seed yield. In general, for canola crop 80 kg N ha^{-1} may be applied in the growing season for each 1 t/ha predictable yield [21]. Several studies observed no significant results with further (200 kg ha^{-1}) increase in N levels [3,24]. While, in our study, the improvement in grain yield can be attributed to N fertilizer at 120 kg ha^{-1} in two splits half at sowing while remaining at rosette stage. The split application of N fertilizer provides flexibility in their fertilizer program, and attracts farmers. Further, higher grain yield was noted in plots where N was applied in two splits as compared with single fertilization of N. The splits application of N fertilizer benefited crop growth and ensure availability of nutrients at two splits, one at sowing and second at rosette stage which may result in higher grain yield of faba bean [25,26].

4.1.3. Effects of Planting Densities on Canola Yield

Optimum planting is important to attain high yield and is a best option for reducing lodging among the plants. Biological yield increased by 3.3% with planting density of 40 plants m^{-2} compared to 20 plants m^{-2}. This difference may be mainly due to increase in plants per unit area. Dahmardeh et al. compared three planting densities (12.5, 16.7, and 20 plants m^{-2}) and found that biological yield of canola was highest for 20 plant m^{-2} compared to other planting density [27]. The inter-competition for nutrients among the plants might be a reason for the lower biological yield. Planting density is an important factor which determines the yield and which is individually affected by the climatic conditions and production system of an area as well [28]. In our study, planting density had non-significant effects on grain yield which indicated that 20 and 40 plants m^{-2} gave same results for grain yield of canola. However, higher yield in least densities indicated the proper utilization and maximum facilitation of nutrients [3].

4.2. Qualitative Traits of Canola

4.2.1. Effects of Grazing and Cut on Quality Traits of Canola

The improvement in quality of seed is the primary objective of breeding oil seed crops to fulfill upcoming edible oil requirements [29]. Oil content is mainly related with genetics for most of the species and varieties but the role of environment cannot be ignored. Cutting and grazing declined oil content of canola. We do not agree with the findings of Kirkegaard et al. that allowing sheep for grazing before bud elongation had no impact on oil content of canola seed [16]. Protein content in cut plots was higher than no cut and grazed plots while, glucosinolates was maximum in no cut plots compared to grazed and cut plots. However, year had significant effect on oil content of canola. Almost 113.6% higher oil content was recorded in second year compared to that of first year. Likewise, in second year protein content was increased up to 161.9% as compared to first year.

4.2.2. Effects of N Fertilizer, Application Timings on Quality Traits of Canola

The N fertilizer had a negative correlation with oil content of the seeds [30,31]. Increasing N level from 80 to 120 kg ha^{-1} decreased the oil content of canola seed. Likewise, higher oil content was measured in plots where N was applied in splits compared with sole N application. Further, seed oil content of canola reduced significantly with increase in N levels from 0 to 200 kg ha^{-1} [32,33], and the highest oil content (43.08%) in plots with low N rates (50 kg ha^{-1}) while lowest oil content (38.64%) was recorded with high levels of N fertilizer (200 kg ha^{-1}) [34]. However, the reduction in oil content due to increase in N levels [31,34]. For example, the accessibility of sugar for oil synthesis becomes less with increase in N rates, that the application of high rates of nitrogen fertilizer increased the amount of N containing protein; so this protein development goes through a competition for photosynthesis, as a consequence of less amount of the later is obtainable for fats production [32,35,36]. This inverse relationship between oil and protein content with increase in N levels may also be the possible reason [37]. However, our data did not agree with the findings of Brennan et al. (2000) who concluded that oil content is not going to decrease with increased in N rates [38]. It is also noticeable that protein content of canola seed improved with rising levels of nitrogen.

Nitrogen is the integral part of protein structures and involved in many other plant metabolic processes. Thus, increasing N levels increased the protein content of canola seeds. N at 120 kg N ha^{-1} application had higher protein content over 80 kg ha^{-1}. The higher protein value is the evidence of negative correlation between oil content and protein content. Both these are inversely proportional to each other [6]. Increased N supply helps in increasing protein synthesis without compromising oil content reduction [18]. Similarly, split application resulted in higher protein contents compared with sole N application. The protein content of canola increased from 22.7% to 23.7% with the increasing N rates from 80 to 160 kg ha^{-1} [39]. Glucosinolates contents were significantly affected by N levels but N application timing had a non-significant effect on glucosinolates contents. Glucosinolates contents increased from 59 to 64 µmol g^{-1} with increasing N from 80 to 120 kg ha^{-1}. These data indicated that increasing N levels significantly increased glucosinolates contents. The increase in glucosinolates contents due to N fertilization was also reported by [35,40–42]. Glucosinolates structure contains N therefore high N concentration may be influenced by the addition of N fertilizers [43].

4.2.3. Effects of Planting Densities on Quality Traits of Canola

Competition among the plants due to high planting density can result in poor quality attributes, impairs plant growth, reduced biomass formation, and consequently yield loss due to low nutrients uptake, and disruption in leaf structural and functional characteristics [44–46]. Increasing density decreases oil content may be due to inter plant competition for nutrients. Plants grown at 40 plants m^{-2} had higher oil content compared to 20 plants m^{-2}. In contrary, no significant variation in oil content with increase in planting densities from 45 to 80 plants m^{-2} [47].

4.3. Interaction Effect of Factors

The proper nutrients at proper time to any crop is important [6], and canola crop requires a higher amount of nutrients, and available nitrogen (N) compared with that of cereals [7]. Split application of N fertilizer has become more popular in terms of high nitrogen use efficiency. Therefore, an appropriate rate and timing of N fertilizer application is one of the most important aspects of successful canola production [3]. These findings support our results that the interactive effects of NL2 with NT2 at PD1 were more qualitative and productive for canola under dual purpose.

4.4. Economics Benefits

The effectiveness of dual-purpose canola can be predictable by the economic analysis of no-cut, cut and grazing systems. Cut plots produced higher net income as compared to no-cut and grazed plots. The higher net income and thus higher value cost ratio (VCR) value of cut plots was due to high fodder and grains as compared to no-cut system. However, grazed plots reduced VCR value by 10% and net income by 69% as compared to no-cut plots. Our results are against with the findings that higher net income ($240 to $500) for grazed plots as compared to no-cut or grains only system [47].

5. Conclusions

The study revealed that integration of DP cropping would increase farm productivity, profitability and flexibility of the farm operations. It is an innovation which captures more food by increasing crop and livestock production on the same farm. C_2 caused a 15% reduction in grain yield of canola; however, it fetched additional income of USD 143.6 compared to C_1. In case of C_3, 23% reduction was resulted in grain yield of canola with income of USD 117.7 from fodder yield of the same canola. Treatment NL_2 produced higher seed yields and improved quality traits of canola compared with NL_1. Further, NL_2 increased grain yield and qualitative parameters of canola. Crops under PD_2 produced more biological yield compared to PD_1. However, seed yield was higher at PD_1. Dual purpose cropping is a classical way which can contribute to continued development of sustainable agriculture systems. Currently, Pakistan is facing a serious shortage of edible oils and food insecurity threats. Therefore, the use of application of N in two splits at 120 kg N ha^{-1} coupled with 20 plants m^{-2} is a good option to achieve better qualitative attributes and high yield of canola under dual purpose system.

Author Contributions: S.J., performed experiment, K.A., F.S., and M.J., help in data analysis and making figures; M.A., supervision; A.K. (Ahmad Khan), and A.Z., S.A., A.K. (Aziz Khan), Methodology and software; S.B., and F.M., worked as investigators; S.J., write final full length draft; M.M.A., F.J., N.A., and M.A.K., revised and edit English language; F.W., revised article and provide funding; S.Z., Data curation; Writing—Originial draft; N.U.A., Writing—Review & editing. All authors have read and agreed to the published version of the manuscript.

Funding: This review is financially supported by the Guangxi Innovation-Driven Development Project (GuiKe AA18242040, Guangxi Natural Science Foundation (2018GXNSFBA294016).

Conflicts of Interest: The authors declare no conflict of interest.

References

1. Barlog, P.; Grzebisz, W. Effect of timing and nitrogen fertilizer application on winter oilseed rape (*Brassica napus* L.). I. Growth Dynamics and Seed Yield. *J. Agron. Crop. Sci.* **2004**, *190*, 305–310. [CrossRef]

2. GOP. *Economic Survey Report, 2003–2004*; Govt. of Pakistan, Finance Division: Islamabad, Pakistan, 2007; pp. 3–4.

3. Cheema, M.A.; Malik, M.A.; Hussain, A.; Shah, S.H.; Basra, S.M.A. Effects of time and rate of nitrogen and phosphorus application on growth and seed and oil yields of canola. *J. Agron. Crop. Sci.* **2001**, *186*, 103–110. [CrossRef]

4. Grombacher, A.; Nelson, L. *Canola Production. A University of Nebraska Neb Guide Publication, G92-1076-A*; Institute of Agriculture and Natural Resources, University of Nebraska-Lincoln: Lincoln, NE, USA, 1992.

5. Gunasekera, C.P.; Martin, L.D.; Siddique, K.H.M.; Walton, G.H. Genotype by environment interactions of indian mustard (*Brassica juncea* L.) and canola (*Brassica napus* L.) in mediterranean-type environments: Ii. Oil and protein concentrations in seed. *Eur. J. Agron.* **2006**, *25*, 13–21. [CrossRef]

6. Hao, X.C.; Chang, C.; Travis, G.J. Short communication: Effect of long-term cattle manure application on relations between nitrogen and oil content in canola seed. *J. Plant Nutr. Soil Sci.* **2004**, *167*, 214–215. [CrossRef]

7. Hocking, P.J.; Strapper, M. Effect of sowing time and nitrogen fertilizer on canola and wheat, and nitrogen fertilizer on Indian mustard. II. Nitrogen concentrations, N accumulation, and N use efficiency. *Aust. J. Agric. Res.* **2001**, *52*, 635–644. [CrossRef]

8. Faraji, A.; Sadeghi, S.; Asadi, M. Evaluation of nitrogen and irrigationeffects on yield and yield components of canola varieties in Gonbad. *J. Agric. Res. Nat. Resour.* **2005**, *12*, 63–72.

9. Wright, G.C.; Smith, C.J.; Woodroffe, M.R. The effect of irrigation and nitrogen fertilizer on rapeseed (*Brassica napus*) production in South- Eastern Australia, I. Growth and seed yield. *Irrig. Sci.* **1988**, *9*, 1–13. [CrossRef]

10. Sommer, S.G.; Schjoerring, J.K.; Denmead, O.T. Ammonia emission from mineral fertilizers and fertilized crops. *Adv. Agron.* **2004**, *82*, 557–622.

11. Kirkegaard, J.A.; Sprague, S.J.; Hamblin, P.J.; Graham, J.M.; Lilley, J.M. Refining crop and livestock management for dual-purpose spring canola (*Brassicanapus*). *Crop Pasture Sci.* **2006**, *63*, 429–443. [CrossRef]

12. Daun, J.K.; Clear, K.M.; Williams, P. Comparison of three whole seed near infrared analyzers for measuring quality components of canola seed. *J. Am. Oil Chem. Soc.* **1994**, *71*, 1063–1068. [CrossRef]

13. Canola Council of Canada. *The Grower's Manual (Online)*; Canola Council of Canada: Winnipeg, MB, Canada, 2005.

14. Jan, M.; Shah, P.; Hollington, P.; Khan, M.; Sohail, Q. Agriculture research: Design and analysis, a monograph. *NWFP Agric. Univ. Pesh. Pak.* **2009**, *1*, 232–250.

15. Kirkegaard, J.; Sprague, S.J.; Marcroft, S.; Potter, T.D.; Graham, J.; Virgona, J. Identifying canola varieties for dual-purpose use. In *Global Issues—Paddock Action, Proceedings of the 14th Australian Agronomy Conference, Adelaide South, Australia, 21–25 September 2008*; Australian Society of Agronomy: Gosford, Australia, 2008.

16. Khan, A.H.; Khalil, I.A.; Shah, H. Nutritional yield and oil quality of canola cultivars grown in NWFP. *Sarhad J. Agric.* **2004**, *20*, 287–290.

17. Susko, D.J.; Superfisky, B. A comparison of artificial defoliation techniques using canola (*Brassica napus*). *Plant Ecol.* **2009**, *202*, 169–175. [CrossRef]

18. McCormick, J.I.; Virgona, J.M.; Kirkegaard, J.A. Growth and yield of dual-purpose canola (*Brassica napus*) under drier inland seasonal conditions of south-eastern Australia. *Crop Pasture Sci.* **2012**, *63*, 635–646. [CrossRef]

19. Salehian, H.; Rafiey, M.; Fathiand, G.; Siadat, S.A. Effect of plant density on growth and seed yield of colza varieties under Andimeshk conditions. In Proceedings of the 7th Iranian Crop Sciences Congress, Karaj, Iran, 15 March 2007.

20. Sprague, S.J.; Kirkegaard, J.A.; Graham, J.M.; Dove, H.; Kelman, W.M. Crop and livestock production for dual-purpose winter canola (*Brassica napus*) in the high-rainfall zone of south-eastern Australia. *Field Crop. Res.* **2014**, *156*, 30–39. [CrossRef]

21. Al-Jaloud, A.A.; Hussain, G.; Karimulla, S.; Alhamdi, A.H. Effect of irrigation and nitrogen on yield and yield components of two rapeseed cultivars. *Agric. Water Manag.* **1996**, *30*, 57–68. [CrossRef]

22. Balint, T.; Rengel, Z.; Allen, D. Australia canola germplasm differs in nitrogen and sulphur defficiency. *Aust. J. Agric. Res.* **2008**, *59*, 1164–1174. [CrossRef]

23. Buttar, G.S.; Thind, H.S.; Aujla, M.S. Methods of planting and irrigation at various levels of nitrogen affect the seed yield and water use efficiency in transplanted oil seed rape. *Agric. Water Manag.* **2005**, *85*, 253–260. [CrossRef]

24. Cheema, M.A. Production Efficiency of Canola (*Brassica napus* L.) cv. Shiralee under Different Agro-Management Practices. Ph.D. Thesis, Department of Agronomy, University of Agriculture, Faisalabad, Pakistan, 1999.

25. Zaman, M. Effect of Rate and Time of Nitrogen Application on Growth, Seed Yield and Oil Contents of Canola (*Brassica napus* L.). Bachelor's Thesis, Department of Agronomy, University of Agriculture, Faisalabad, Pakistan, 2003.

26. Dahmardeh, M.; Mahmood, R.; Valizadeh, J. Effect of plant density and cultivars on growth, yield and yield components of faba bean (*Viciafaba* L.). *Afr. J. Biotechnol.* **2010**, *9*, 8643–8647.

27. Shengwu, H.; Ovesna, J.; Kueera, L.; Kueera, V.; Vyvadilova, M. Evaluation of genetic diversity of Brassica napusgermplasm from China and Europe assessed by RAPD markers. *Plant Soil Environ.* **2003**, *49*, 106–113.

28. Kirkegaard, J.A.; Sprague, S.J.; Lilley, J.M.; McCormick, J.I.; Virgona, J.M.; Morrison, M.J. Physiological response of spring canola (*Brassica napus*) to defoliation in diverse environments. *Field Crop. Res.* **2012**, *125*, 61–68. [CrossRef]

29. Söchtling, H.P.; Verret, J.A. Effects of different cultivation systems (soil management and nitrogen fertilization) on the epidemics of fungal diseases in oilseed rape (*Brassica napus* L. var. *napus*). *J. Plant Dis. Prot.* **2004**, *111*, 1–29. [CrossRef]

30. Rathke, G.W.; Christen, O.; Diepenbrock, W. Effects of nitrogen source and rate on productivity and quality of winter oilseed rape (*Brassica napus* L.) grown in different crop rotations. *Field Crop. Res.* **2005**, *94*, 103–113. [CrossRef]

31. Akhtar, K.; Wang, W.; Khan, A.; Ren, G.; Afridi, M.Z.; Feng, Y.; Yang, G. Wheat straw mulching offset soil moisture deficient for improving physiological and growth performance of summer sown soybean. *Agric. Water Manag.* **2019**, *211*, 16–25. [CrossRef]

32. Saleem, M.; Cheema, A.M.; Malik, A.M. Agro-EconomicAssessment of Canola Planted under Different Levels of Nitrogenand Row Spacing. *Int. J. Agric. Biol.* **2001**, *3*, 27–30.

33. Asare, E.; Scarisbrick, D.H. Rate of nitrogen and sulphur fertilizers on yield, yield components and seed quality of oilseed rape (*Brassica napus* L.). *Field Crop. Res.* **1995**, *44*, 41–46. [CrossRef]

34. Omirou, M.D.; Papadopoulou, K.K.; Papastylianou, I.; Constantinou, M.; Karpouzas, D.G.; Asimakopoulos, I.; Ehaliotis, C. Impact on nitrogen and sulfur fertilization on the composition of glucosinolates in relation to sulfur assimilation in different plant organs of broccoli. *J. Agric. Food Chem.* **2009**, *57*, 9408–9417. [CrossRef]

35. Holmes, M.R.J. Nitrogen. In *Nutrition of the Oilseed Rape Crop*; Applied Science Publishers: Barking, UK, 1980; pp. 21–67.

36. Kutcher, H.R.; Malhi, S.S.; Gill, K.S. Topography and management of nitrogen and fungicide affects diseases and productivity of canola. *Agron. J.* **2005**, *97*, 533–541. [CrossRef]

37. Brennan, R.F.; Mason, M.G.; Walton, G.H. Effect of nitrogen fertilizer on the concentrations of oil and protein in canola (*Brassica napus*) seed. *J. Plant Nutr.* **2000**, *23*, 339–348. [CrossRef]

38. Ahmad, G.; Jan, A.; Arif, M.; Jan, M.T.; Khattak, R.A. Influence of nitrogen and sulfur fertilization on quality of canola (*Brassica napus* L.) under rain fed conditions. *J. Zhejiang Univ. Sci. B* **2007**, *8*, 731–737. [CrossRef]

39. Narits, L. Effect of nitrogen rate and application time to yield and quality of winter oilseed rape (*Brassica napus* L. *var. oleiferasubvar. biennis*). *Agron. Res.* **2010**, *8*, 671–686.

40. Chen, X.J.; Zhu, Z.J.; Ni, X.L.; Qian, Q.Q. Effect of Nitrogen and Sulfur Supply on Glucosinolates in *Brassica campestris* ssp. *chiensis*. *Agric. Sci. China* **2006**, *5*, 603–608. [CrossRef]

41. Zhao, F.; Evans, E.J.; Bilsborrow, P.E.; Syers, J.K. Influence of nitrogen and sulphur on the glucosinolates profile of rapeseed (*Brassica napus* L.). *J. Sci. Food Agric.* **2006**, *64*, 295–304. [CrossRef]

42. Falk, K.L.; Tokuhisa, J.G.; Gershenzon, J. The effect of sulfur nutrition on plant glucosinolates content: Physiology and molecular mechanisms. *Plant Biol.* **2007**, *9*, 573–581. [CrossRef]

43. Saeedi, F.H.; Bagheri, H.; Rasoli, S.A.; Ghasemi, J. The effect of planting density on morphological traits of spring canola cultivars in Chaloos region of Iran. *Int. J. Agron. Plant Prod.* **2013**, *4*, 911–914.

44. Khan, A.; Zheng, J.; Tan, D.K.Y.; Khan, A.; Akhtar, K.; Kong, X.; Fahad, S. Changes in Leaf Structural and Functional Characteristics when Changing Planting Density at Different Growth Stages Alters Cotton Lint Yield under a New Planting Model. *Agronomy* **2019**, *9*, 859. [CrossRef]

45. Khan, A.; Kong, X.; Najeeb, U.; Zheng, J.; Tan, D.K.Y.; Akhtar, K.; Zhou, R. Planting density induced changes in cotton biomass yield, fiber quality, and phosphorus distribution under beta growth model. *Agronomy* **2019**, *9*, 500. [CrossRef]

46. Khan, A.; Wang, L.; Ali, S.; Tung, S.A.; Hafeez, A.; Yang, G. Optimal planting density and sowing date can improve cotton yield by maintaining reproductive organ biomass and enhancing potassium uptake. *Field Crop. Res.* **2017**, *214*, 164–174. [CrossRef]

47. Gan, Y.; Malhi, S.S.; Brandt, S.; Katepa-Mupondwa, F.; Stevenson, C. Nitrogen use efficiency and nitrogen uptake of juncea Canola under diverse environments. *Agron. J.* **2007**, *100*, 285–295. [CrossRef]

 *agronomy*

Article

Use of Plant Water Extracts for Weed Control in Durum Wheat (*Triticum turgidum* L. Subsp. *durum* Desf.)

Alessandra Carrubba [1,*] **, Andrea Labruzzo** [1]**, Andrea Comparato** [1]**, Serena Muccilli** [2] **and Alfio Spina** [3]

[1]   DSAAF—Department of Agriculture, Food and Forest Sciences, University of Palermo, 90128 Palermo (PA), Italy; andrea.labruzzo@unipa.it (A.L.); andreacomparato88@gmail.com (A.C.)

[2]   CREA—Council for Agricultural Research and Agricultural Economy Analysis – Research Center for Olive, Citrus and Tree Fruits, 95024 Acireale (CT), Italy; serenamuccilli@hotmail.com

[3]   CREA—Council for Agricultural Research and Agricultural Economy Analysis – Research Center for Cereal and Industrial Crops, 95024 Acireale (CT), Italy; alfio.spina@crea.gov.it

*   Correspondence: alessandra.carrubba@unipa.it; Tel.: +39-091-23862208

Received: 1 February 2020; Accepted: 2 March 2020; Published: 6 March 2020

Abstract: The use of plant water extracts to control weeds is gaining attention in environmentally-friendly agriculture, but the study of the effect that such extracts may exert on the yield of durum wheat is still unexplored. In 2014 and 2016, the herbicidal potential of several plant water extracts was field tested on durum wheat (cv Valbelice). In 2014, extracts obtained from *Artemisia arborescens*, *Rhus coriaria*, *Lantana camara*, *Thymus vulgaris*, and *Euphorbia characias* were used, whereas in 2016 only *A. arborescens* and *R. coriaria* were tested as "donor" plants. In both years, weed incidence was evaluated, together with the major yield parameters of wheat. None of the treatments (including chemicals) could eradicate weeds from the field. In 2014, dicots were in general prevailing in plots treated with extracts of *E. characias*, while monocots prevailed after treatments with *L. camara* and *R. coriaria*. In 2016, lower weed biomass and diversity level were found, and only *Avena* and *Phalaris* were detected at harvest time. Treatment with plant water extracts affected grain yields, but it seems likely that those effects are not due to the diverse incidence of weeds in treated and untreated plots, rather to some direct action exerted by allelopathic substances.

Keywords: cereal crops; plant water extracts; bioherbicides; weed management; allelopathy

1. Introduction

The use of environmentally-friendly methods for weed control is gaining attention in agricultural practice. By one side, the widespread use of synthetically-derived herbicides has caused a number of adverse effects, such as the high persistence of herbicides in the environment and in the food chain, and the development of highly resistant weed populations. Second, there are some special cropping systems, such as those addressed to organic production, where the use of synthetic chemicals is banned.

In this general frame, an increased number of farmers are seeking alternative technical choices for weed management [1]. Many methods have been suggested in time, with contrasting results according to the chosen method, the weed population, and the expected results [2–4]. Among these new techniques, allelopathy plays an increasingly important role [5,6].

The term "allelopathy" indicates a complex of effects exerted by one plant species (the donor) to another one (the acceptor), through the release into the environment of a number of chemical substances, termed allelochemicals. This transmission can occur in several ways: The substances may be released directly and continuously by the living plants in the form of volatile compounds emitted into the atmosphere, as root exudates into the soil, or as chemicals formed by the microbial degradation of plant residuals [7].

Generally speaking, the allelopathic phenomena may act both on seed germination and on the whole development cycle of the plant, with deterrent effects, e.g., on photosynthesis. The secondary metabolites with allelopathic action belong to many chemical families, the most important being phenols, flavonoids, terpenoids, glucosynolates, benzoxaquinones, and cyanogenic compounds [8].

The possibility to use allelochemicals for environmentally-friendly weed control is not new, being suggested by many authors since the early 2000s [9,10]. Many different methods of use are possible, differentiated by both timing and way of application. In addition, the presowing (or pretransplant) distribution of the "donor" material followed by burying [11], those compounds have been suggested as post-emergence treatments, as usual herbicides.

Studies about this topic may be addressed to two directions: To verify the effects of the plant extracts against the germination of weed seeds, or to test such effects against adult weed plants. Research in the first direction may have a great practical importance, since this action towards soil seed bank could limit the wave of weeds emergence, at least until the crop has developed a good competitiveness. Yet, not many experiments have investigated the application in field conditions of the results from in vitro experiments. Many factors may play a role in modifying these results, such as the interaction with soil organic matter, and micro- and meso-organisms [12]. Although allelopathy is claimed to have a strategic ecological role in natural conditions [13], the possibility to use the allelopathic compounds as a resource for in field weed management strategy is still fairly unexplored.

In biological essays, the most proper operating way to assess allelopathic activity should be the isolation of the one (or few) active molecule(s) which are directly responsible for the given biological activity, and some author [14] advices that only such isolated chemical compound(s) should be termed "allelochemical". Anyway, this field of study is huge, and the starting point is undoubtedly the individuation of effective crop/donor plant combinations. Hence, many researchers all over the world have started to study the efficacy of crude plant extracts or plant parts against many common pests.

In cereal crops, the need for suitable tools for weed management has an outstanding relevance. Indeed, competition with weeds not only is one of the biggest causes of yield losses, but also an issue for quality achievement [15]. Among cereal crops, durum wheat (*Triticum turgidum* L subsp. *durum* Desf.) is used all over the word mainly to produce pasta [16]. This crop reaches an outstanding importance in all Mediterranean countries, where it is cultivated for making high-quality products including, besides pasta, also bulghur, couscous, and bread. Most of the market value of these items relies upon the quality level reached by the harvested grain [17,18], and weed control bears a deep importance in ensuring a high-quality level product [19]. Additionally, the present large demand of organic cereals, associated with the establishment of a public compensation payment system, create a favorable context to promote organic arable farming systems. These systems will face technical problems such as weed control [20], which affect economic viability and may greatly influence cereals quality.

Weed control by means of allelopathic substances was evaluated in soft wheat (*Triticum aestivum* L.), using water extracts obtained from a number of plants including *Sorghum bicolor* (L.) Moench, *Helianthus annuus* L., *Parthenium hysterophorus* L., *Oryza sativa* L., etc., alone, in mixture, or coupled with different chemical herbicides [21–24]. The results from these experiments not only varied according to weed species and conditions (time and rate) of supply, but also according to their various possible combinations.

As far as we know, the effects that crude plant extracts may exert on the yield of durum wheat are still unexplored. This work was carried out in 2014 and 2016, with the goal of evaluating the activity of water extracts obtained by different plant (donor) species towards the weed population of durum wheat (cv Valbelice—acceptor), and to further evaluate the effect exerted by such extracts on crop growth and yield.

2. Materials and Methods

2.1. Preparation and Use of Plant Water Extracts

The donor ("allelopathic") species used for the preparation of water extracts were chosen based on their assessed biological activity and the availability of plant biomass. Five donor plants (*A. arborescens, R. coriaria, L. camara, T. vulgaris,* and *E. characias*) were tested in the first year, whereas in 2016 the field trial was restricted to *A. arborescens* and *R. coriaria*, that had previously proved interesting inhibitory activity of seed germination, coupled with significant effects on some qualitative parameters of durum wheat [25,26].

Artemisia arborescens (Vaill.) L. is a shrub from *Asteraceae*, widespread, and spontaneous in the Mediterranean environments. The essential oil extracted from this species has been the subject of many studies that have assessed its strong biocidal activity towards many micro-organisms and weeds [27,28]. Less information is available about *A. arborescens* water extracts, although their biological activity is demonstrated. Some preliminary studies have assessed their ability in inhibiting in vitro germination of some weeds [29]. Two lignans, ashantin and sesamin, were detected after a phytotoxicity bioassay-guided isolation of *A. arborescens* extracts [30]. Extracts from leaves of *A. arborescens* at very low concentration were found to be responsible for an enhancement of growth in some ornamental plants and, as such, are patented in the US (US 5434122A).

Rhus coriaria L. is a small shrub from *Anacardiaceae*, and is widely distributed inside wild Sicilian flora. The plant may reach 3 m in height and is considered a noxious weed. Its fruits are red-brownish drupas, that are toxic when consumed fresh, but after drying are largely used in Middle-Eastern cooking to season soups and vegetables. In Sicily, the plant was formerly cultivated in specialized areas named "sommaccheti" (from its local name "Sommacco"), to use in tannery its tannin-rich bark and leaves. This practice is now obsolete, due to the substitution of natural tannins with the analogous synthetically-derived items [31]. In regards to *R. coriaria* chemical composition, available information is mainly focused on fruits (the most commonly used part of the plant), rich in volatile aromatic compounds endowed with strong antioxidant ant antimicrobial activities [31–33]. Less research is available about other parts of the *R. coriaria* plant, although gallotannins and flavonoid derivatives were the most representative compounds in aqueous extracts of sumac leaves [34,35].

Lantana camara L. is a perennial from *Verbenaceae*, native to Central and Southern America and further introduced all over the world especially for ornamental purposes. The plant is strongly invasive, and is considered a noxious weed in many tropical and sub-tropical areas throughout the world [36]. Its essential oil, mainly extracted from the leaves, has shown a strong insecticidal activity, but the allelopathic effect of its leaves proved significant, as well [37]. The leaves contain phenolic compounds, mainly phenolic acids and flavonoids [38,39].

Thymus vulgaris L., belonging to the family *Labiatae*, is a small shrub with woody branches, and is spontaneous in sunny environments of Mediterranean areas. Its essential oil, obtained mainly from the flowering tops, is widely used for its antibacterial, antifungal, and antiviral properties [40,41]. In addition to essential oil, other components have been individuated in thyme, such as phenolic compounds (mostly phenolic acids and flavonoids), that are probably the active agents of many biological activities ascribed to *T. vulgaris* aqueous extracts [42].

Euphorbia characias L. is an evergreen shrub typical of the Mediterranean maquis, sometimes growing higher than 1 m. The plant grows well in dry areas and may tolerate rather long drought periods. All plant parts are toxic, above all because of its whitish latex, which is irritant by contact, but just for this reason in folk medicine it is used to treat warts. In some areas of the Mediterranean, the latex was used for illegal fishing, especially in Sicily where it was used to catch eels in sweet water pools (locally termed "nache") [43]. The toxicity of the latex from *Euphorbiaceae* is well known [44]; in *E. characias*, two toxic lectins [45] and many bioactive compounds such as polycyclic diterpenoids, bicyclic diterpenes, tocopherols, and sterols have been isolated [46].

The extracts that were used for the trials were prepared in the labs of the Department of Agricultural, Food and Forest Sciences of the University of Palermo, using plant material (including both leaves and inflorescences) picked from wild (*A. arborescens, E. characias, R. coriaria*) or cultivated plants (*T. vulgaris, L. camara*) growing near Ciminna (Palermo, Sicily) and Sparacia (Cammarata, Agrigento, Sicily). All plant material was first air dried at room temperature for at least five days.

To obtain water extracts, 1 kg of each dried product was soaked in 10 L of distilled water (weight/volume ratio: 1/10), and put in constant stirring with a speed rotation of 70 rounds/min for at least 10 h. At the end of the extraction process, the mass was filtered through filter paper (Whatman n. 4), and the obtained extracts were refrigerated at 4 °C until used. The dry matter concentration of each extract (Table 1) was measured after desiccation in the stove for 24 h at 105 °C.

Table 1. Concentration of the used extracts (% w/v) ([a]).

Plant species	Concentration (% w/v)
Rhus coriaria	8.75
Artemisia arborescens	18.82
Euphorbia characias	2.27
Lantana camara	6.14
Thymus vulgaris	22.33

([a]) weight/volume percentage

2.2. Field Management

The field trials were carried out in the experimental farm "Sparacia" (Cammarata, AG, Sicily; 37° 38′ N–13° 46′ E; 415 m a.s.l.), of the Department of Agricultural, Food and Forest Sciences of the University of Palermo. The chosen durum wheat variety was the cv Valbelice (0111 × BC5), obtained in 1992 by the same Department. In both years, the preceding crop was Berseem clover (*Trifolium alexandrinum* L.). Durum wheat was cultivated accordingly to the cropping techniques ordinarily applied in the cereal areas of the site. Hence, soil was prepared by means of a summer work (25–30 cm deep), followed by two shallow harrowings. Sowing was made mechanically on 19 December, 2013 and 22 December, 2015, spreading at a soil depth of about 5 cm, and on rows 30 cm apart, an amount of seed aimed to obtain a seeding density of 350 viable seeds per m^2 (about 200 kg ha^{-1}). At sowing time, 1.5 t ha^{-1} of diammonium phosphate (18/46) were distributed. Next, after the crop had reached the stage of full tillering, 1.1 t ha^{-1} of urea (46) were additionally spread. Eight treatments were tested in the first year (five plant water extracts; one chemical herbicide; two controls), and five treatments (two plant water extracts; one chemical herbicide; two controls) were tested in 2016. The experimental plots were arranged in the field according to a randomized block design with three repetitions; each treatment was applied on nine durum wheat rows 1.70 m in length (size of plots 2.67 × 1.70 m = 4.54 m^2). In order to avoid interference phenomena between the treated plots, an essay area (1.20 × 1.67 m = 2.00 m^2) was delimited within each plot, and all surveys on both weeds and durum wheat were taken therein. Treatments with plant extracts were applied twice, distributing in crop post-emergence, 4 L m^{-2} of each previously prepared extract. In the first year, the first treatment was applied on 13 January, 2014 (i.e., after 25 DAS—days after sowing), when wheat was at the stage of 2–3 true leaves unfolded (Zadoks' scale: Z13) [47], whereas the second was applied on 13 March, 2014 (88 DAS), when the crop was entering the full stem elongation stage (Zadoks' scale: Z31). In the second year, the same crop development stages were detected on 12 February, 2016 and 4 April, 2016, and in those dates the planned treatments were consequently applied.

In the appositely planned plots, chemical weeding was performed only once, contemporarily to the first distribution (in different plots) of water extracts. The chemical herbicide was a mixture of mesosulfuron-methyl 3% + iodosulfuron-methyl-sodium 0.6% + mefenpir-diethyl 9% (ATLANTIS®), distributed by Bayer®for post-emergence weeding against all graminaceous weeds and some important dicots. In compliance with label recommendation, chemical herbicide was distributed at a supply rate of 1.2 L ha^{-1} (formulated product).

For comparison, a group of plots (further named "untreated") were left undisturbed, i.e., without any weeding operation. Furthermore, in order to verify if the additional amount of water contained in the extracts could have a stimulating effect on plant growth (both on durum wheat and weeds), and to allow separating this effect, if any, an additional control plot was set up, where 4 L m^{-2} of water were spread twice, contemporarily to the distribution of extracts in the other plots.

In either years, durum wheat was harvested in the second half of June. At harvest time, each essay area was harvested separately, and the total obtained biomass (durum wheat and weeds) was sorted by botanical species and weighed. Samples of both durum wheat and weeds were dried in the stove (24 h at 105 °C, until constant weight) to determine their moisture content, in order to convert all biomass measurements in dry matter. In wheat, the number of spikes per unit area (m^2) was measured. Thereafter, wheat biomass was partitioned between grain and straw (g m^{-2}) and the Harvest Index value (%) for each treatment, given by the percent ratio between grain and total biomass (including grain) yield, was calculated. On a representative sample of 30 spikes per plot (including controls), the number of spikelets per spike and number and weight of seeds per spike were counted. On a representative sample of kernels per each plot, thousand kernel weight (TKW; g) was measured, and the moisture level of kernels was assessed after drying in the stove (24 h at 105 °C, until constant weight). For consistency, all weight data were reported as dry matter.

To evaluate the success of treatments against weed population, the weed suppression ability for each treatment (S$_t$) was calculated, applying the formula (modified from [48]):

$$S_t = 100(W_u - W_t)/W_u \tag{1}$$

where W_u is the weight of weed biomass (g m^{-2} of dm) found at harvest time in unrestricted conditions of weed growth (untreated plots), and W_t is the weight of weed biomass measured in every treated plot.

From the first treatment and throughout all crop cycle until harvest, the growth and phytosanitary conditions of wheat were checked by means of periodical field surveys. The presence of weeds and their botanical composition were checked, as well. With this purpose, in each survey, two 50-cm long row segments were randomly chosen in each plot. All plants growing in these lengths (including both wheat and weeds) were counted, and weeds were botanically determined. The number of retrieved species, throughout all crop cycle and in each cropping condition (treated and untreated plots) was used as "richness" index [49].

The degree of diversity inside each plot was evaluated on each survey date through the Shannon–Weiner index H' [49]:

$$H' = \sum_{p=1}^{s} [(p_i) \times \ln(p_i)] \tag{2}$$

where s is the number of retrieved species, p_i is the frequency of the individuals of the species i ($p_i = n_i/N$, being n_i the number of individuals of the species I, and N the total number of individuals of all species).

2.3. Experimental Site and Climatic Details

The trial environment is typical of the inner hilly areas of Sicily (meso-thermo-Mediterranean climate), with a long and dry summer period and a colder winter, with few snow days and irregular rainfall. In the trial area, the average year rainfall is about 500–650 mm, mostly occurring in the autumn-winter period, whereas the spring rainfall amount is about 20% of year rainfall. Summer months are mostly dry, with no more than 10% rainy days, that are mostly torrential. In the first year (Figure 1a), the total rainfall amount reached 390 mm, distributed throughout the whole winter period, mostly between the end of January and the first ten days of February, and in the first ten days of March. As usual in the trial site, the temperatures were fairly high, with minimum values around 2 °C in December, January, and February, and maximum values spanning between 12 and 32 °C at the end of crop cycle. In 2016 (Figure 1b) the rainfall amount was lower (209 mm from December to June) and lower temperatures were recorded in winter (throughout the second half of January to the first half of February) and in early spring (March).

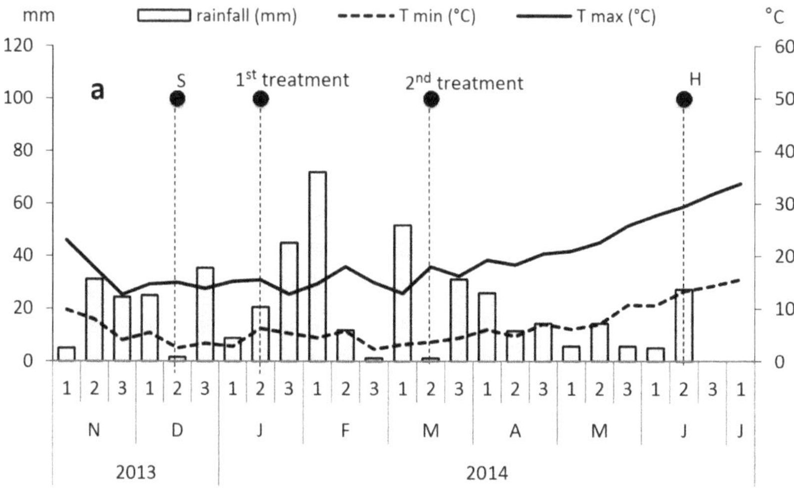

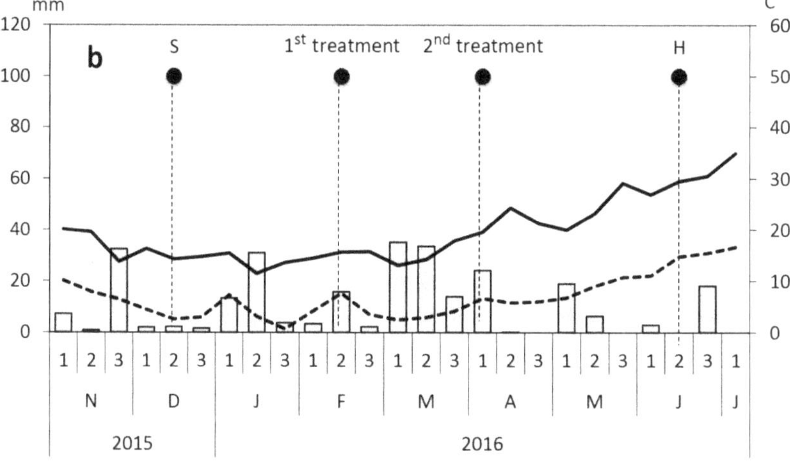

Figure 1. Cumulated rainfall (mm; bars) and mean maximum (°C; solid lines) and minimum (°C; dotted lines) temperatures recorded in 2013–2014 (a) and 2015–2016 (b) at Sparacia (Cammarata, AG, Italy). S: Sowing time; H: Harvest time; first treatment, second treatment: Dates of the two treatments with plant extracts. Data are reported as ten day values from November (N) to July (J) in both trial years.

2.4. Statistical Data Management

Field data were managed according to the chosen experimental layout (RCB with three repetitions), using the GLM procedure of the statistical package Minitab v. 17.1.0 (Minitab Inc., State College, PA, USA, 2013). All yield and biomass variables (height of plants, yield, and yield components) measured on wheat were considered as dependent variables, whereas year (Y) and treatments (T) were set as independent variables. All data were submitted to a preliminary individual ANOVA on a per-year basis; the comparison between years was performed only on the treatments in common (*A. arborescens*, *R. coriaria*, chemical, water, and untreated). In all cases, when the F-test indicated statistical significance at the $p \leq 0.05$ level, Tukey's HSD test was used to evidence the differences among mean values [50]. Shannon Wiener's index was calculated by means of the PAST software [51].

3. Results

3.1. Effects of Treatments on Durum Wheat Growth and Yield

In the first trial year, the crop was favored by the satisfactory rainfall amounts, and at harvest time, plants (except for those treated with *A. arborescens* extracts) were higher than 130 cm (Figure 2). Contrastingly, in 2016, the crop was somehow constrained by the weather conditions, and plants always were shorter than 120 cm (Figure 2; Table 2).

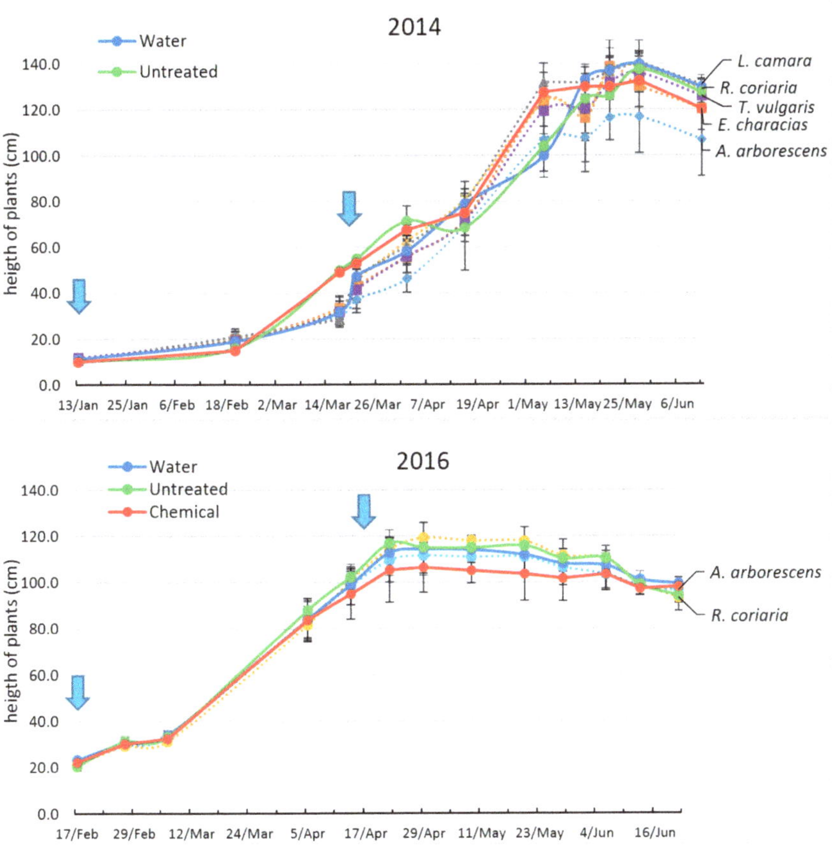

Figure 2. Height of wheat plants measured throughout the trial periods in 2014 and 2016. For each treatment and survey date, each point is the average of five measurements ± standard deviation. Solid lines refer to the controls (water, blue; untreated, green; chemical, red); dotted lines refer to the treatments with water extracts (labels at the last point of the line). In each panel, the arrows indicate the date of the two treatments.

Table 2. Mean effects ± standard deviations of treatment with plant extracts on biometrical and yield traits in durum wheat (cv Valbelice), in comparison with two controls and one chemical weeding, and calculated F values for the treatments. Within each group (2014, 2016, Year, and Treatment), values followed by the same letter are statistically not different ($p \leq 0.05$, Tukey's HSD test).

Variability Source	Plants height at Harvest Time (cm)		Plant Population at Harvest Time (n. plants m^{-2})	Grain yield (g m^{-2})		Spikes (n m^{-2})	Spikelets (n plant^{-1})		Tillers (n plant^{-1})		TKW (g)		HI (% dm)	
				2014										
Water	139.8 ± 0.3	A	81.7 ± 11.7	441.6 ± 56.0	A	293.0 ± 35.0	20.4 ± 0.2	AB	3.58 ± 0.14	A	46.6 ± 0.2	A	51.2 ± 2.1	
Untreated	137.5 ± 5.5	A	110.0 ± 23.3	392.0 ± 59.4	AB	261.5 ± 28.5	19.2 ± 0.5	AC	2.45 ± 0.68	B	43.9 ± 2.0	A	53.2 ± 0.5	
Chemical	132.5 ± 2.5	AB	105.0 ± 5.0	406.5 ± 8.3	AB	284.5 ± 13.5	18.4 ± 0.8	AC	2.71 ± 0.00	AB	42.8 ± 0.6	A	42.2 ± 1.3	
A. arborescens	116.8 ± 14.3	B	98.3 ± 11.7	256.4 ± 62.0	B	268.0 ± 3.0	17.0 ± 1.8	C	2.73 ± 0.15	AB	36.1 ± 2.0	B	41.5 ± 9.4	
R. coriaria	138.8 ± 8.3	A	93.3 ± 6.7	430.3 ± 57.6	AB	312.0 ± 46.0	20.8 ± 0.7	A	3.38 ± 0.74	AB	44.3 ± 3.2	A	43.5 ± 3.7	
E. characias	130.3 ± 7.8	AB	86.7 ± 6.7	383.1 ± 112.3	AB	313.5 ± 14.5	18.2 ± 1.5	AC	3.64 ± 0.25	A	43.5 ± 0.0	A	44.8 ± 0.6	
T. vulgaris	135.8 ± 2.8	AB	100.0 ± 0.0	365.8 ± 61.9	AB	304.5 ± 5.5	17.8 ± 0.5	BC	3.05 ± 0.06	AB	44.7 ± 1.7	A	50.8 ± 11.7	
L. camara	140.8 ± 3.8	A	108.3 ± 1.7	469.8 ± 8.6	A	317.0 ± 5.0	19.5 ± 0.1	AC	2.93 ± 0.09	AB	46.4 ± 2.5	A	46.5 ± 3.8	
2014 F value $_{(7, 16)}$ (a)	3.79 *		2.65 n.s.	3.30 *		2.36 n.s.	5.64 **		2.94 *		9.43 ***		1.87 n.s.	
				2016										
Water	104.6 ± 10.9		64.6 ± 5.5	145.8 ± 105.0		254.0 ± 103.0	16.5 ± 0.8		3.92 ± 1.57		31.8 ± 4.0		25.0 ± 12.5	
Untreated	95.1 ± 8.5		75.7 ± 16.7	167.1 ± 29.1		246.8 ± 39.3	16.3 ± 1.1		3.35 ± 0.84		37.3 ± 1.8		33.6 ± 5.0	
Chemical	100.1 ± 8.1		66.7 ± 5.5	116.5 ± 89.9		213.7 ± 106.9	16.4 ± 1.4		3.27 ± 1.56		32.4 ± 4.9		23.8 ± 17.0	
A. arborescens	95.7 ± 4.8		62.5 ± 14.6	110.6 ± 93.6		234.8 ± 93.0	16.0 ± 0.6		3.53 ± 1.27		33.1 ± 9.7		23.2 ± 16.8	
R. coriaria	103.2 ± 5.3		68.8 ± 6.3	192.6 ± 58.7		275.5 ± 43.6	16.4 ± 1.3		4.09 ± 0.44		37.1 ± 2.9		33.0 ± 2.6	
2016 F value $_{(4, 10)}$ (b)	<1 n.s.		<1 n.s.	<1 n.s.		<1 n.s.	<1 n.s.		<1 n.s.		<1 n.s.		<1 n.s.	
				Year (Y) (c)										
Mean 2014 (c)	133.1 ± 11.0	A	97.7 ± 15.2 A	385.4 ± 82.2	A	283.8 ± 31.1	19.2 ± 1.6	A	2.99 ± 0.61		42.7 ± 4.0 A		46.3 ± 6.4	A
Mean 2016	99.7 ± 7.7	B	67.6 ± 10.3 B	146.5 ± 74.9	B	245.0 ± 73.0	16.3 ± 0.9	B	3.64 ± 1.08		34.3 ± 5.2 B		27.7 ± 11.5	B
Y F value $_{(1,20)}$ (c)	135.25 ***		45.76 ***	92.88 ***		2.94 n.s.	56.91 ***		3.69 n.s.		32.29 ***		29.86 ***	
				Treatment (T) (c)										
Water	122.2 ± 20.5	A	73.1 ± 12.4	293.7 ± 178.6	AB	273.5 ± 72.0	18.4 ± 2.2	A	3.76 ± 1.01		39.2 ± 8.5		38.10 ± 16.44	
Untreated	116.3 ± 24.1	AB	92.8 ± 26.1	279.5 ± 130.1	AB	254.2 ± 31.8	17.8 ± 1.7	AB	2.92 ± 0.87		40.6 ± 4.0		43.40 ± 11.24	
Chemical	116.3 ± 18.5	AB	85.8 ± 21.5	261.5 ± 168.8	AB	249.1 ± 78.4	17.4 ± 1.5	AB	2.94 ± 0.94		37.6 ± 6.5		32.98 ± 14.76	
A. arborescens	106.2 ± 14.9	B	80.4 ± 22.9	183.5 ± 106.9	B	251.4 ± 61.6	16.5 ± 1.3	B	3.25 ± 1.05		34.6 ± 6.5		32.31 ± 15.73	
R. coriaria	121.0 ± 20.4	A	81.0 ± 14.7	311.4 ± 140.2	A	293.8 ± 44.8	18.6 ± 2.6	A	3.69 ± 0.64		40.7 ± 4.8		38.26 ± 6.41	

Table 2. *Cont.*

Variability Source	Plants height at Harvest Time (cm)	Plant Population at Harvest Time (n. plants m^{-2})	Grain yield (g m^{-2})	Spikes (n m^{-2})	Spikelets (n plant^{-1})	Tillers (n plant^{-1})	TKW (g)	HI (% dm)
T F value $_{(4, 20)}$ (c)	3.83 *	2.16 n.s.	3.20 *	<1 n.s.	4.04 *	1.10 n.s.	2.39 n.s.	1.42 n.s.
				Year (Y) × Treatment (T) (c)				
YxT F value $_{(4,20)}$ (c)	1.48 n.s.	<1 n.s.	<1 n.s.	<1 n.s.	2.60 n.s.	<1 n.s.	1.80 n.s.	<1 n.s.

(a): Results of univariate ANOVA for 2014 data (DF: 7, 16)
(b): Results of univariate ANOVA for 2016 data (DF: 4, 10)
(c): Means and ANOVA are referred only to treatments in common to both years
Significance of F values: *: P ≤ 0.05; **: P ≤ 0.01; ***: P ≤ 0.001; n.s.: not significant

In both years no apparent difference in plant height could be noted among treatments, except for some advantage of untreated and chemically treated plots in the earlier growth stages, that was, however, balanced as growth season was going on. At harvest time in both years, the lowest height values were found on wheat treated with extract of *A. arborescens*, although in 2016 this value was not statistically different from the others (Table 2).

Among the treatments tested in either years, none of the examined biometrical and yield traits (Table 2) showed at ANOVA a significant Year × Treatment interaction. That means, differences among treatments were not significantly affected by the experimental year, and the "year" effect was the same in all tested treatments.

The main effect of both years and treatments were otherwise significant in many cases. Grain yield, number of spikes per plant, number of spikelets per spike, TKW, and HI were all higher in 2014 than in 2016.

Grain yield (Table 2) in the first year was more than twice as in 2016. Concerning the main effect of treatments, those with plant extracts reached the highest and the lowest yield value, respectively for *A. arborescens* (183.5 g m^{-2}) and *R. coriaria* (311.4 g m^{-2}). In the first year, however, the highest yield value (470 g m^{-2} dm) was found in the *L. camara* treatment (tested only in 2014), and the lowest (256.4 g m^{-2} dm) in the *A. arborescens* treatment. The number of spikes per area unit counted at harvest time averaged values between 238 and 287 in 2014 and 214 to 275 in 2016, without showing any significant difference among years or treatments. The number of spikelets per spike exhibited the highest mean value in the plants previously treated with extracts of *R. coriaria*. The number of tillers per plant was higher in the water control and in the treatment with *R. coriaria* extracts; high values were retrieved also in the treatments with *E. characias* and *T. vulgaris*, which however were excluded from pooled ANOVA, being tested only in 2014. The highest mean value of TKW was recorded in the first year (42.7 g), when significant differences showed up between the treatment with *A. arborescens* (36.1 g) and all the other treatments.

The HI (%) showed significant differences only between years, being almost unaffected by treatments. On average, HI ranged between 25% (water control in 2016) and 53.2% (untreated plots in 2014). Values of HI > 50%, demonstrating that more than half of the produced biomass was represented by grain, were obtained in three cases only, all of them in 2014: Water control (51.2%), untreated (53.2%), and *T. vulgaris* (50.8%).

3.2. Effect of Treatments on Weed Population

The values of total dry biomass (wheat + weeds) recorded at harvest time in treated and untreated plots (Figure 3) were submitted to ANOVA both as cumulated values and sorted between components, i.e., accounting for wheat biomass and weeds biomass, separately (ANOVA results not shown). The factor "year" resulted significant in all analyses, whereas treatments and Y × T interaction were highly significant only on dry matter values of measured weed biomass. Hence, all measured biomass values (wheat, weeds, and the sum of both) were, on average, significantly higher in 2014 than in 2016, but the effect of treatments was significant only on weeds biomass, and such effect was variable according to the year. In both years, although there was no noticeable presence of weeds in the chemically treated plot, neither wheat biomass nor wheat grain yield were significantly higher after chemical weeding. In 2014, the highest weed biomass was retrieved in the untreated plots (255 g m^{-2} dm, sharing 25.7% of total biomass) and in the control plots with water (245 g m^{-2}, i.e., 22% of total mean biomass). In 2016, weed incidence in the control plots was comparatively lower (in the water controls 43.4 g m^{-2}, i.e., 7.8% of total biomass, and in the untreated plots 27.7 g m^{-2}, i.e., 5.2% of total biomass). In 2016, the highest weed biomass was, however, measured in the *A. arborescens* treatments (47.8 g m^{-2}, i.e., 10.2% of total biomass). Except for chemical and controls, the trend of weeds incidence on total biomass in the first year was *R. coriaria* (11.3%) < *E. characias* (15.4%) < *L. camara* (17.4%) < *T. vulgaris* (18.4%) < *A. arborescens* (20.5%). In the second year, when only two water extract treatments were tested, the trend was confirmed as *R. coriaria* (3.9%) < *A. arborescens* (10.2%) (Figure 3).

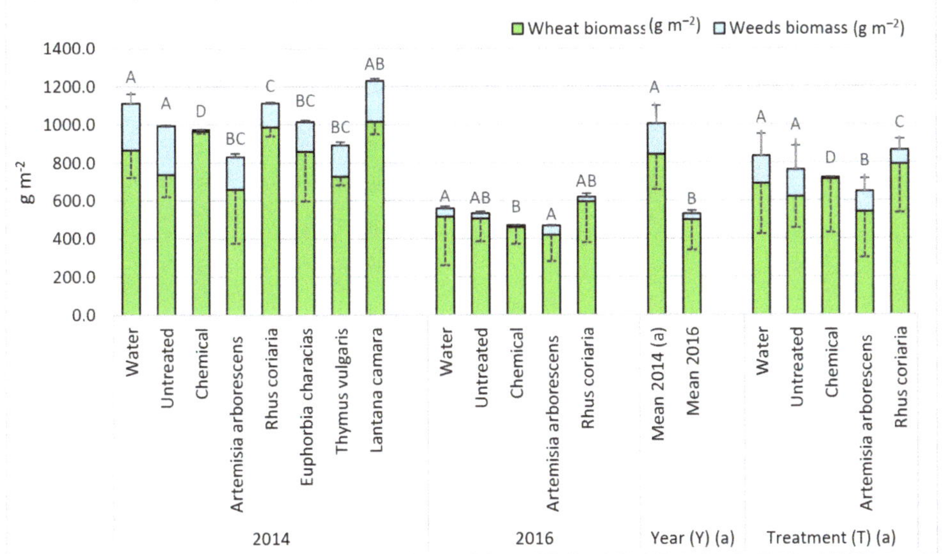

Figure 3. Total biomass (g m⁻² dm) of wheat and weeds measured at harvest time in treated and untreated plots. Mean values marked with "a" refer only to the treatments common to both years. Error bars indicate the standard deviation of each mean, respectively for wheat (downward dotted lines) and weeds (upward entire lines). In each group (2014, 2016, Year, and Treatment), values of weeds biomass accompanied by the same letter are not different at $p \leq 0.05$ (HSD Tukey's test).

The weed suppression index ($S_t\%$) calculated on data obtained at harvest time (Figure 4) illustrates the overall effect exerted by each treatment compared to the untreated plots. The highest suppression ability was found on the chemically treated plots, that constrained weed incidence of 96.5% in the first year and 65.2% in 2016. In 2014, all treatments with water extracts gained statistically non different values of the S_t index, ranging from 50.8% (*R. coriaria*) to 16.0% (*L. camara*). In 2016, the trend was markedly different, and the most effective treatment (*R. coriaria*) suppressed weeds of 13.4% only, whereas *A. arborescens* appeared to exert a stimulating effect on weed presence, even higher than the effect exerted by water alone.

For a deeper insight of the mechanism underlying the comparison and persistence of weeds, the data retrieved in both years throughout the crop cycle were taken into consideration. Figure 5 shows the time pattern of appearing and duration inside the single plots of the retrieved weed species, as sum of the three repetitions, irrespective of their weight incidence. The botanical composition of the weeds detected at harvest time showed a differentiation among years. In the first year, in all treated plots it was possible to observe how the appearance of wild oat was definitively delayed with respect to the controls. In 2016, this outcome was confirmed for *A. arborescens*, whereas in plots treated with *R. coriaria* extracts, the appearance of *Avena fatua* was almost simultaneous to that recorded on the water control. In 2014, when the weed biomass at harvest time was much higher than in 2016, in rather all plot, irrespective if treated or not, the appearance of weeds was delayed. Contrastingly, in 2016 weed appearance was earlier, but most of weed species disappeared throughout wheat cycle, and weed biomass at harvest time was almost totally composed by *A. fatua*.

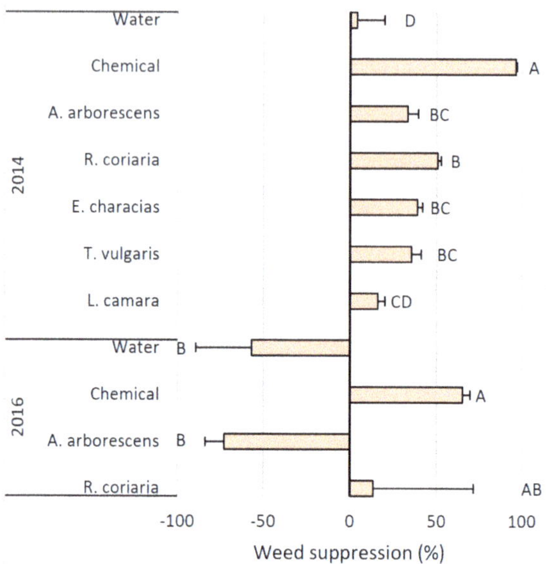

Figure 4. Weed suppression indices (S_t %) calculated in 2014 and 2016 for all treatments, including chemical and water control. Error bars indicate the standard deviation of each mean. In each year, values accompanied by the same letter are not different at $p \leq 0.05$ (HSD Tukey's test).

This outcome is also evidenced in the graphs in Figure 6, where the detected trend of the number of weed species per area unit (species richness) is reported, throughout all survey dates, and Figure 7, which illustrates the trend over time of the calculated Shannon's index in all plots. In 2014, species richness was initially low, and then shifted to higher values until harvest. Appreciable variations were found among treatments and, noticeably, the chemical treatment showed constantly the lowest values. Monocots and dicots where found in rather the same proportions in both controls and in the plots treated with *A. arborescens* and *T. vulgaris*, whereas a sharp prevalence of dicots was found on *E. characias*, and monocots were definitively prevailing in *L. camara* and *R. coriaria*. Among monocots, *Avena fatua* and *Phalaris paradoxa* showed the highest incidence, sharing from 12.7 to 37.4% of total dm weed biomass. Among dicots, wild dill (*Ridolfia segetum*) was certainly the most relevant, found in all plots with highly sized plants, where it represented 23% to 30% of total weed biomass. A significant presence (36%) of *Polygonum aviculare* was found in the plots treated with *E. characias* extracts. In 2016, the opposite trend was evidenced, and weed species number decreased over time. A more simplified weed flora was assessed, and only *Phalaris* and *Avena* were retrieved at harvest time.

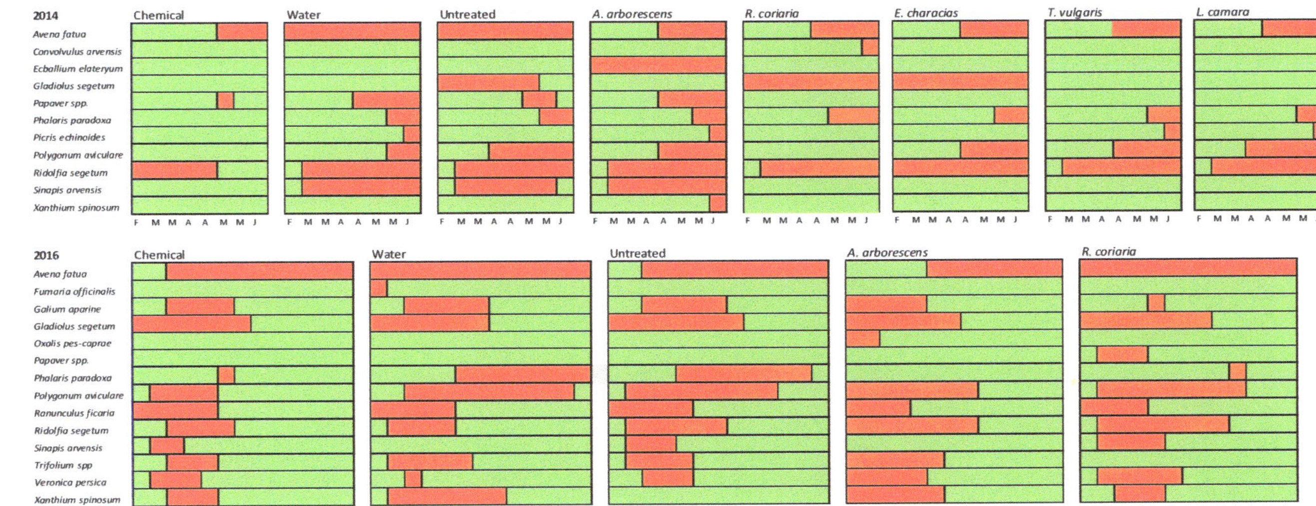

Figure 5. Time pattern of field emergence of weeds in durum wheat treated with five (2014) and two (2016) plant water extracts, compared with an untreated control, a chemical herbicide, and a control with only water. For each weed species, red areas mark the observed presence in the field from February (F) to June (J).

Since the Shannon index not only takes into account the number of species, but also the total number of individuals, it may be considered as a representation of the degree of botanical diversity inside each plot. In 2014, the diversity index was rather constantly higher in the untreated plots, and constantly lower in the chemically treated ones. All treatments took intermediate values between these two extreme series; a slight advantage of the *R. coriaria* treatments over the other treatments was detectable, but it must be noticed that in the last part of wheat cycle all water extract treatments exhibited high, and similar, values. In 2016, the diversity index showed a decreasing trend from March onward, homogeneous among all treatments (including chemical) and the controls.

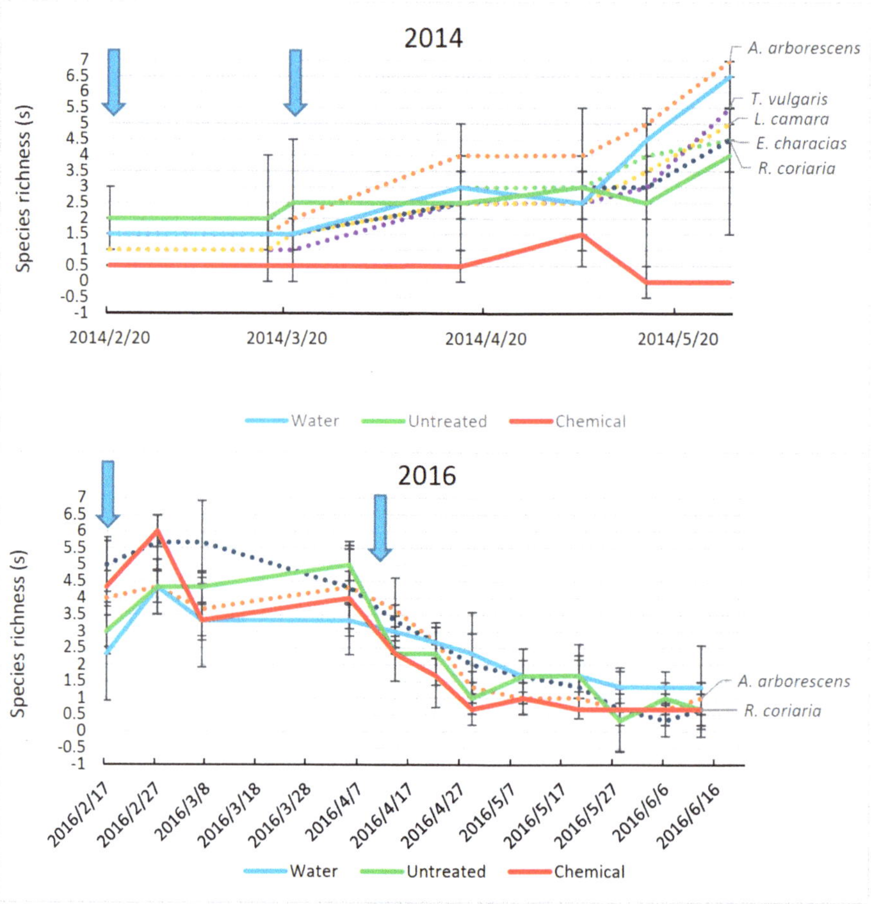

Figure 6. Species richness (n of detected weed species m^{-2}) in durum wheat treated with five (2014) and two (2016) plant water extracts, compared with an untreated control (green line), a chemical herbicide (red line), and a control with only water (blue line). Each value is the mean of three repetitions ± standard deviation. Arrows indicate the date of treatments.

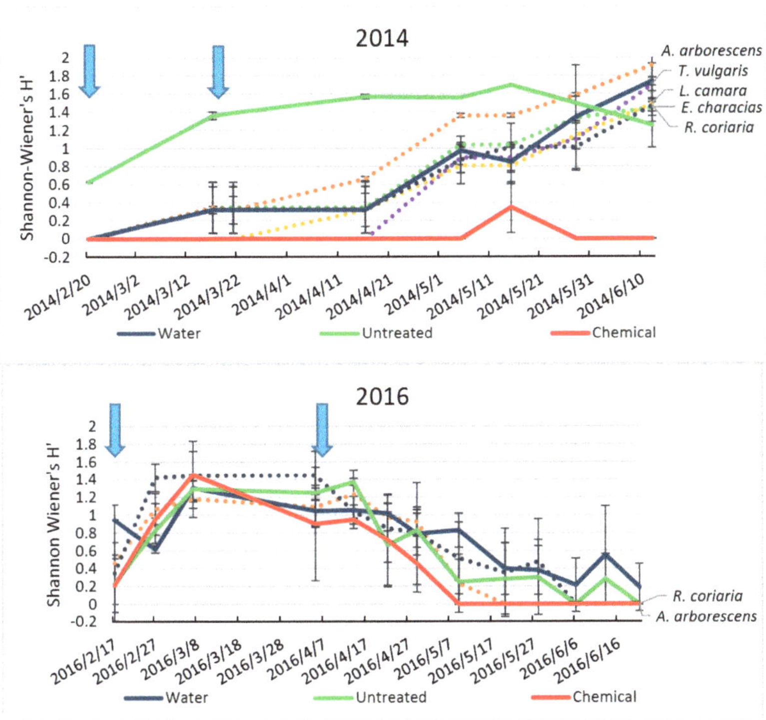

Figure 7. Species diversity (Shannon-Wiener index, *H′*) in 2014 and 2016 in durum wheat treated with five (2014) and two (2016) plant water extracts, compared with an untreated control (green line), a chemical herbicide (red line) and a control with only water (blue line). Each value is the mean of three repetitions ± standard deviation. Arrows indicate the date of treatments.

4. Discussion

This work was aimed to evaluate, in field conditions, the effects on durum wheat of several water plant extracts, applied for weed control. With this purpose, not only the bare conditions of the presence/absence of weeds were accounted for, but also the possible interactions between the supplied extracts and the major growth and yield parameters of the crop.

The effect of treatments on weed population was variable between years. In 2014, dicots were in general prevailing in plots treated with extracts of *E. characias*, while monocots prevailed after treatments with *L. camara* and *R. coriaria*. In 2016, when a generally lower weed biomass was present, also a lower diversity level was found, and only the most competitive weed species (*Avena fatua* and *Phalaris paradoxa*) were detected at harvest time. The marked variability expressed by the *A. arborescens* extract on weeds, as revealed by the opposite directions shown by the calculated suppression index in the two years, may be possibly explained by a toxic effect exerted by this extract against wheat in both years and especially in 2016, when this treatment probably induced a less dense wheat canopy (fewer and shorter plants), which allowed weeds to grow and develop even more than in the untreated control.

In general, none of the tested treatments (including chemicals) was able to eradicate weeds from the field, and weeds were retrieved at harvest time in all plots. Hence, although chemically-treated plots showed in both years the highest suppression ability, some lately-sprouting weeds were found also therein. However, the fact that in both years grain yield was not significantly different between chemically treated plots and untreated ones, demonstrates that, in the chosen wheat genotype, weed control using chemical herbicide does not necessarily result in a significant increase in grain yield.

Total weed biomass did not appear to be a determinant factor in assessing wheat yields, showing on average—opposite to what was expected—the highest values in the most productive year and treatments. Both measurements (grain yield and weed biomass) were, however, significantly different according to the tested treatment. On average, *R. coriaria* always exerted a positive effect on wheat yields, and *A. arborescens* always a negative effect. A possible explanation could be that the retrieved yield differences are a consequence of the distribution of plant extract itself, rather than an effect exerted on weed biomass. An effect of *R. coriaria* extract on several quality parameters of durum wheat has been already assessed by previous experiments [26]. Further research is needed to explore these aspects.

Noticeable differences resulted in the date of appearance of major weeds, whose flush of emergence was generally earlier in 2016 than in 2014. In all treated plots in the first year, the appearance of wild oat (*Avena fatua*) was delayed with respect to the controls, but this trend in 2016 was confirmed only on plots treated with extracts of *A. arborescens*. Since wild oat and *Phalaris* spp. are among the most noxious weeds in wheat, if confirmed by further experiences, this outcome would have a great practical relevance. The delay of weed emergence is claimed to be a major factor in improving yield levels, since a longer time is at the crop's disposal to enhance its competitivity [52].

The competitive ability of the selected durum wheat genotype (cv Valbelice) resulted in a higher yield capacity even in the presence of a significant weed biomass. To explore this aspect, plant traits correlated to crop competitiveness, i.e., plant population, plant height, and tillering [53,54] were taken into consideration. All of them expressed large differences in consequence of the different climatic pattern of the two years (Y factor always highly significant). As such, climatic conditions acted giving wheat a higher competitive ability in the first year (height values always higher; plant population higher). An advantage of taller plants was evident in both years and in all circumstances, since a general trend of higher productivity with higher plants was rather always recognizable. Similarly, the yield disadvantage of shorter plant size, as retrieved in plots treated with *A. arborescens* extracts, was evident as well.

Both tiller number per plant and number of spikes per area unit resulted to be mostly density-dependent, and did not seem associated with reduced weeds.

5. Conclusions

Although certainly preliminary, this work represents a step forward in the study of weed management through allelochemicals. Although the herbicidal effectiveness of the studied extracts under the given experimental conditions was rather limited, water plant extracts confirmed exerting different—and not always predictable—effects on crop yield and development. By one side, it must be stressed that the goal of weeding is no longer the complete eradication of weeds, rather the containment of weeds population beyond an "acceptability" threshold [55–59]. By another side, the occurrence of significant effects of these extracts on crop open the way to a huge field of investigations involving agronomical, physiological, and biochemical issues. Further studies are necessary, using a broader range of crops and allelochemicals, and pointing out in detail doses and methods of application of the supplied compounds.

Author Contributions: Conceptualization, A.C. (Alessandra Carrubba) and A.S.; data curation, A.C. (Alessandra Carrubba), A.L., A.C. (Andrea Comparato), S.M., and A.S.; formal analysis, A.C. (Alessandra Carrubba); investigation, A.C. (Alessandra Carrubba) and A.S.; methodology, A.C. (Alessandra Carrubba) and A.S.; resources, A.C. (Alessandra Carrubba); software, A.C. (Alessandra Carrubba), A.L., S.M., and A.S.; supervision, A.C. (Alessandra Carrubba); validation, A.C. (Alessandra Carrubba), A.L., A.C. (Andrea Comparato), S.M., and A.S.; visualization, A.C. (Alessandra Carrubba) and A.S.; writing—original draft, A.C. (Alessandra Carrubba); writing—review and editing, A.C. (Alessandra Carrubba) and A.S. All authors have read and agreed to the published version of the manuscript.

Funding: This research received no external funding.

Conflicts of Interest: The authors declare no conflict of interest.

References

1. Öz atalbaş, O. Current Status of Advisory and Extension Services for Organic Agriculture in Europe and Turkey. In *Organic Agriculture Towards Sustainability*; Pilipavicius, V., Ed.; InTech Open publ.: London, UK, 2014; pp. 67–87. [CrossRef]
2. Barberi, P. Weed management in organic agriculture: Are we addressing the right issues? *Weed Res.* **2002**, *42*, 177–193. [CrossRef]
3. Bond, W.; Turner, R.J.; Grundy, A.C. *A Review of Non-Chemical Weed Management*; HDRA, The Organic Organisation, Ryton Organic Gardens: Coventry, UK, 2003.
4. Carrubba, A.; Militello, M. Nonchemical weeding of medicinal and aromatic plants. *Agron. Sustain. Dev.* **2013**, *33*, 551–561. [CrossRef]
5. Vyvyan, J.R. Allelochemicals as leads for new herbicides and agrochemicals. *Tetrahedron* **2002**, *58*, 1631–1646. [CrossRef]
6. Bhowmik, P.C. Challenges and opportunities in implementing allelopathy for natural weed management. *Crop Prot.* **2003**, *22*, 661–671. [CrossRef]
7. Mallik, A.U. Introduction: Allelopathy research and application in sustainable agriculture and forestry. In *Allelopathy in Sustainable Agriculture and Forestry*; Zeng, R.S., Mallik, A.U., Luo, S.M., Eds.; Springer: New York, NY, USA, 2008; pp. 1–10.
8. Soltys, D.; Krasuska, U.; Bogatek, R.; Gniazdowska, A. Allelochemicals as Bioherbicides—Present and Perspectives. In *Herbicides—Current Research and Case Studies in Use*; Price, A.J., Kelton, J.A., Eds.; IntechOpen: London, UK, 2013; pp. 517–542. [CrossRef]
9. Dudai, N.; Poljakoff-Mayber, A.; Mayer, A.M.; Putievsky, E.; Lerner, H.R. Essential oils as allelochemicals and their potential use as bioherbicides. *J. Chem. Ecol.* **1999**, *25*, 1079–1089. [CrossRef]
10. Duke, S.O.; Rimando, A.M.; Baerson, S.R.; Scheffler, B.E.; Ota, E.; Belz, R.G. Strategies for the use of natural products for weed management. *J. Pestic. Sci.* **2002**, *27*, 298–306. [CrossRef]
11. De Mastro, G.; Fracchiolla, M.; Verdini, L.; Montemurro, P. Oregano and its potential use as bioherbicide. *Acta Hortic.* **2006**, *723*, 335–345. [CrossRef]
12. Khan, K. Factors affecting phytotoxic activity of allelochemicals in soil. *Weed Biol. Manag.* **2004**, *4*, 1–7.
13. Khan, S.; Ali, K.W.; Shinwari, M.I.; Khan, R.A.; Rana, T. Environmental, ecological and evolutionary effects of weeds allelopathy. *Int. J. Botany Stud.* **2019**, *4*, 77–84.
14. Duke, S.O. Proving Allelopathy in Crop–Weed Interactions. *Weed Sci.* **2015**, *63*, 121–132. [CrossRef]
15. Jabran, K.; Mahmood, K.; Melander, B.; Bajwa, A.A.; Kudsk, P. Weed Dynamics and Management in Wheat. *Adv. Agron.* **2017**, *145*, 97–166. [CrossRef]
16. Palumbo, M.; Spina, A.; Boggini, G. Bread-making quality of Italian durum wheat (Triticum durum Desf.) cultivars. *Ital. J. Food Sci.* **2002**, *2*, 123–133.
17. Rharrabti, Y.; Villegas, D.; Royo, C.; Martos-Núñez, V.; García del Moral, L.F. Durum wheat quality in Mediterranean environments. II. Influence of climatic variables and relationships between quality parameters. *Field Crop Res.* **2003**, *80*, 133–140. [CrossRef]
18. Giannone, V.; Giarnetti, M.; Spina, A.; Todaro, A.; Pecorino, B.; Summo, C.; Caponio, F.; Paradiso, V.M.; Pasqualone, A. Physico-chemical properties and sensory profile of durum wheat Dittaino PDO (Protected Designation of Origin) bread and quality of re-milled semolina used for its production. *Food Chem.* **2018**, *241*, 242–249. [CrossRef]
19. Bulut, S.; Öztürk, A.; Karaoğlu, M.M.; Yildiz, N. Effects of organic manures and non-chemical weed control on wheat. II. Grain quality. *Turk. J. Agric. For.* **2013**, *37*, 271–280. [CrossRef]
20. Alföldi, T.; Fliessbach, A.; Geier, U.; Kilcher, L.; Niggli, U.; Pfiffner, L.; Stolze, M.; Willer, H. Organic agriculture and the environment. In *Organic Agriculture, Environment and Food Security*; El-Hage Scialabba, N., Hattam, C., Eds.; Environment and Natural Resources Series 4; FAO: Rome, Italy, 2002; Chapter 2.
21. Cheema, Z.A.; Iqbal, M.; Ahmad, R. Response of Wheat Varieties and some Rabi Weeds to Allelopathic Effects of Sorghum Water Extract. *Int. J. Agric. Biol.* **2002**, *1*, 52–55.
22. Khan, R.; Khan, M.A. Weed Control Efficiency of Bioherbicides and Their Impact on Grain Yield of Wheat (Triticum aestivum L.). *Eur. J. Appl. Sci.* **2012**, *4*, 216–219. [CrossRef]

23. Khan, M.A.; Afridi, R.A.; Hashim, S.; Khattak, A.M.; Ahmad, Z.; Wahid, F.; Chauhan, B.S. Integrated effect of allelochemicals and herbicides on weed suppression and soil microbial activity in wheat (Triticum aestivum L.). *Crop Prot.* **2016**, *90*, 34–39. [CrossRef]

24. Naeem, M.; Cheema, Z.A.; Ihsan, M.Z.; Hussain, Y.; Mazari, A.; Abbas, H.T. Allelopathic effects of different plant water extracts on yield and weeds of wheat. *Planta Daninha* **2018**, *36*, e018177840. [CrossRef]

25. Labruzzo, A.; Carrubba, A.; Di Marco, G.; Ebadi, M.T. Herbicidal potential of aqueous extracts from Melia azedarach L., Artemisia arborescens L., Rhus coriaria L. and Lantana camara L. *Allelopath. J.* **2017**, *41*, 81–92. [CrossRef]

26. Carrubba, A.; Comparato, A.; Labruzzo, A.; Muccilli, S.; Giannone, V.; Spina, A. Quality characteristics of wholemeal flour and bread from durum wheat (Triticum turgidum L subsp. durum Desf.) after field treatment with plant water extracts. *J. Food Sci.* **2016**, *81*, C2158–C2166. [CrossRef] [PubMed]

27. Militello, M.; Settanni, L.; Aleo, A.; Mammina, C.; Moschetti, G.; Giammanco, G.M.; Blazquez, M.A.; Carrubba, A. Chemical composition and antibacterial potential of Artemisia arborescens L. essential oil. *Curr. Microbiol.* **2011**, *62*, 1274–1281. [CrossRef] [PubMed]

28. Militello, M. Studi Fitochimici e Agronomici su Artemisia Arborescens L. (Asteraceae) Della flora Spontanea Siciliana e Attività Biocida Degli oli Essenziali. Ph.D. Thesis, Università di Palermo, Palermo, Italy, 2012. (In Italian with English Abstract).

29. Militello, M.; Carrubba, A. Biological activity of extracts from Artemisia arborescens (Vaill.) L.: An overview about insecticidal, antimicrobial, antifungal and herbicidal properties. In *Natural Products: Research Reviews*; Gupta, V.K., Ed.; Daya Publishing House—Astral International Pvt. Ltd.: New Delhi, Indian, 2016; Volume 3, pp. 361–387.

30. Labruzzo, A.; Cantrell, C.L.; Carrubba, A.; Ali, A.; Wedge, D.E.; Duke, S.O. Phytotoxic lignans from Artemisia arborescens. *Nat. Prod. Commun.* **2018**, *3*, 237–240. [CrossRef]

31. Rayne, S.; Mazza, G. Biological activities of extracts from Sumac (Rhus spp.): A review. *Plant Foods Hum. Nutr.* **2007**, *62*, 165–175. [CrossRef] [PubMed]

32. Giovanelli, S.; Giusti, G.; Cioni, P.L.; Minissale, P.; Ciccarelli, D.; Pistelli, L. Aroma profile and essential oil composition of Rhus coriaria fruits from four Sicilian sites of collection. *Ind. Crop Prod.* **2017**, *97*, 166–174. [CrossRef]

33. Özcan, M. Antioxidant Activities of Rosemary, Sage, and Sumac Extracts and Their Combinations on Stability of Natural Peanut Oil. *J. Med. Food* **2003**, *3*, 267–270. [CrossRef] [PubMed]

34. Regazzoni, L.; Arlandini, E.; Garzon, D.; Santagati, N.A.; Beretta, G.; Maffei Facino, R. A rapid profiling of gallotannins and flavonoids of the aqueous extract of Rhus coriaria L. by flow injection analysis with high-resolution mass spectrometry assisted with database searching. *J. Pharmaceut. Biomed.* **2013**, *72*, 202–207. [CrossRef]

35. Romeo, F.V.; Ballistreri, G.; Fabroni, S.; Pangallo, S.; Li Destri Nicosia, M.G.; Schena, L.; Rapisarda, P. Chemical characterization of different Sumac and Pomegranate extracts effective against Botrytis cinerea Rots. *Molecules* **2015**, *20*, 11941–11958. [CrossRef]

36. Sharma, G.P.; Raghubanshi, A.S.; Singh, J.S. Lantana invasion: An overview. *Weed Biol. Manag.* **2005**, *5*, 157–165. [CrossRef]

37. Talukdar, D. Allelopathic effects of Lantana camara L. on Lathyrus sativus L.: Oxidative imbalance and cytogenetic consequences. *Allelopath. J.* **2013**, *31*, 71–90.

38. Kumar, S.; Sandhir, R.; Ojha, S. Evaluation of antioxidant activity and total phenol in different varieties of Lantana camara leaves. *BMC Res. Notes* **2014**, *7*, 560. [CrossRef] [PubMed]

39. Costa, J.G.M.; Rodrigues, F.F.G.; Sousa, E.O.; Junior, D.M.S.; Campos, A.R.; Coutinho, H.D.M.; de Lima, S.G. Composition and larvicidal activity of the essential oils of Lantana camara and Lantana montevidensis. *Chem. Nat. Compd.* **2010**, *46*, 313–315. [CrossRef]

40. El Abed, N.; Kaabi, B.; Smaali, M.I.; Chabbouh, M.; Habibi, K.; Mejri, M.; Marzouki, M.N.; Ahmed, S.B.H. Chemical composition, antioxidant and antimicrobial activities of Thymus capitata essential oil with its preservative effect against Listeria monocytogenes inoculated in minced beef meat. *Evid.-Based Compl. Altern.* **2014**, 152487. [CrossRef]

41. De Martino, L.; Bruno, M.; Formisano, C.; De Feo, V.; Napolitano, F.; Rosselli, S.; Senatore, F. Chemical composition and antimicrobial activity of the essential oils from two species of Thymus growing wild in Southern Italy. *Molecules* **2009**, *14*, 4614–4624. [CrossRef]

42. Lorenzo, J.M.; Khaneghah, A.M.; Gavahian, M.; Marszałek, K.; Eş, I.; Munekata, P.E.S.; Ferreira, I.C.F.R.; Barba, F.J. Understanding the potential benefits of thyme and its derived products for food industry and consumer health: From extraction of value-added compounds to the evaluation of bioaccessibility, bioavailability, anti-inflammatory, and antimicrobial activities. *Crit. Rev. Food Sci.* **2018**, *17*. [CrossRef]

43. Savo, V.; La Rocca, A.; Caneva, G.; Rapallo, F.; Cornara, L. Plants used in artisanal fisheries on the Western Mediterranean coasts of Italy. *J. Ethnobiol. Ethnomed.* **2013**, *9*, 9. [CrossRef]

44. de Souza, L.S.; Puziol, L.C.; Tosta, C.L.; Bittencourt, M.L.F.; Ardisson, J.S.; Kitagawa, R.R.; Filgueiras, P.R.; Kuster, R.M. Analytical methods to access the chemical composition of an Euphorbia tirucalli anticancer latex from traditional Brazilian medicine. *J. Ethnopharmacol.* **2019**, *237*, 255–265. [CrossRef]

45. Barbieri, L.; Falasca, A.; Franceschi, C.; Licastro, F.; Rossi, C.A.; Stirpe, F. Purification and properties of two lectins from the latex of the euphorbiaceous plants Hura crepitans L. (sand-box tree) and Euphorbia characias L. (Mediterranean spurge). *Biochem. J.* **1983**, *215*, 433–439. [CrossRef]

46. Pintus, F.; Spanò, D.; Mascia, C.; Macone, A.; Floris, G.; Medda, R. Acetylcholinesterase inhibitory and antioxidant properties of Euphorbia characias latex. *Rec. Nat. Prod.* **2013**, *2*, 147–151.

47. Zadoks, J.C.; Chang, T.T.; Konzak, C.F. A decimal code for the growth stages of cereals. *Weed Res.* **1974**, *14*, 415–421. [CrossRef]

48. Hoad, S.; Topp, C.; Davies, K. Selection of cereals for weed suppression in organic agriculture: A method based on cultivar sensitivity to weed growth. *Euphytica* **2008**, *163*, 355–366. [CrossRef]

49. Légère, A.; Stevenson, F.C.; Benoit, D.L. Diversity and assembly of weed communities: Contrasting responses across cropping systems. *Weed Res.* **2005**, *45*, 303–315. [CrossRef]

50. Gomez, K.A.; Gomez, A.A. *Statistical Procedures for Agricultural Research*; John Wiley & Sons Inc.: New York, NY, USA, 1984.

51. Hammer, Ø. PAST PAleontological Statistics, Version 3.25, Reference Manual 1999–2019. Available online: https://folk.uio.no/ohammer/past/past3manual.pdf (accessed on 10 January 2020).

52. Fahad, S.; Hussain, S.; Chauhan, B.S.; Saud, S.; Wu, C.; Hassan, S.; Tanveer, M.; Jan, A.; Huang, J. Weed growth and crop yield loss in wheat as influenced by row spacing and weed emergence times. *Crop Prot.* **2015**, *71*, 101–108. [CrossRef]

53. Andrew, I.K.S.; Storkey, J.; Sparkes, D.L. A review of the potential for competitive cereal cultivars as a tool in integrated weed management. *Weed Res.* **2015**, *55*, 239–248. [CrossRef]

54. Lemerle, D.; Verbeek, B.; Cousens, R.D.; Coombes, N.E. The potential for selecting wheat varieties strongly competitive against weeds. *Weed Res.* **1996**, *36*, 505–513. [CrossRef]

55. Hurle, K. Concepts in weed control—How does biocontrol fit in? *Pest Manag. Rev.* **1997**, *2*, 87–89. [CrossRef]

56. Beringer, J. Farming for biodiversity—The nature of intervention. *Outlook Agric.* **2001**, *30*, 7–9. [CrossRef]

57. García-Martín, A.; López-Bellido, R.J.; Coleto, J.M. Fertilisation and weed control effects on yield and weeds in durum wheat grown under rain-fed conditions in a Mediterranean climate. *Weed Res.* **2007**, *47*, 140–148. [CrossRef]

58. Sims, B.; Corsi, S.; Gbehounou, G.; Kienzle, J.; Taguchi, M.; Friedrich, T. Sustainable Weed Management for Conservation Agriculture: Options for Smallholder Farmers. *Agriculture* **2018**, *8*, 118. [CrossRef]

59. Bajwa, A.A.; Khan, M.J.; Bhowmik, P.C.; Walsh, M.; Chauhan, B.S. Sustainable Weed Management. In *Innovations in Sustainable Agriculture*; Farooq, M., Pisante, M., Eds.; Springer Nature: Basel, Switzerland, 2019; pp. 249–286. [CrossRef]

 *agronomy*

Article

Comparative Yield, Fiber Quality and Dry Matter Production of Cotton Planted at Various Densities under Equidistant Row Arrangement

Nangial Khan [1], Fangfang Xing [1], Lu Feng [1,2], Zhanbiao Wang [1,2], Minghua Xin [1], Shiwu Xiong [1], Guoping Wang [1], Huanxuan Chen [1,2], Wenli Du [1] and Yabing Li [1,2,*]

[1] State Key Laboratory of Cotton Biology, Institute of Cotton Research of Chinese Academy of Agricultural Sciences, Anyang 455000, China; nangialkhan@hotmail.com (N.K.); xingyue530@163.com (F.X.); fenglucri@126.com (L.F.); wang_zhanbiao@126.com (Z.W.); xinminghua985@126.com (M.X.); 18703673670@163.com (S.X.); Zmswgp@126.com (G.W.); chenhuanxuan123@163.com (H.C.); w1233216917@163.com (W.D.)

[2] State Key Laboratory of Cotton Biology, Zhengzhou Research Base, Zhengzhou University, Zhengzhou 450001, China

* Correspondence: criliyabing1@163.com; Tel.: +86-0372-256-2293

Received: 20 December 2019; Accepted: 3 February 2020; Published: 5 February 2020

Abstract: The number of cotton plants grown per unit area has recently gained attention due to technology expense, high input, and seed cost. Yield consistency across a series of plant populations is an attractive cost-saving option. Field experiments were conducted to compare biomass accumulation, fiber quality, leaf area index, yield and yield components of cotton planted at various densities (D1, 1.5; D2, 3.3; D3, 5.1; D4, 6.9; D5, 8.7; and D6, 10.5 plants m^{-2}). High planting density (D5) produced 21% and 28% more lint yield as compared to low planting density (D1) during both years, respectively. The highest seed cotton yield (4662 kg/ha) and lint yield (1763 kg/ha) were produced by high plant density (D5) while the further increase in the plant population (D6) decreased the yield. The increase in yield of D5 was due to more biomass accumulation in reproductive organs as compared to other treatments. The highest average (19.2 V_A gm m^{-2} d^{-1}) and maximum (21.8 V_M gm m^{-2} d^{-1}) rates of biomass were accumulated in reproductive structures. High boll load per leaf area and leaf area index were observed in high planting density as compared to low, while high dry matter partitioning was recorded in the lowest planting density as compared to other treatments. Plants with low density had 5% greater fiber length as compared to the highest plant density, while the fiber strength and micronaire value were 10% and 15% greater than the lowest plant density. Conclusively, plant density of 8.7 plants m^{-2} is a promising option for enhanced yield, biomass, and uniform fiber quality of cotton.

Keywords: cotton; plant density; biomass accumulation; yield; fiber quality

1. Introduction

Cotton is an important cash crop grown worldwide as a major source of fiber [1]. Cotton is perennial but commercially grown as an annual crop and has indeterminate growth. China is the largest cotton-producing country in the world by contributing about 30% of the world's cotton production [2]. Henan Province is one of the major cotton growing provinces of China, with more than 400 thousand hectares of land [3]. Plant density determination is one of the most important practices for increasing yield of cotton [4]. Plant density is the key factor for optimizing structures and increasing the photosynthetic capacity of the cotton canopy. High planting density has become common in cotton production systems. It has been reported that both too high and too low plant density reduces cotton yield by affecting light penetration and moisture availability, further influencing plant height, architecture, boll behavior, and crop maturity. An optimum plant density not only improves the yield and fiber quality of cotton but also reduces input costs by minimizing seed rate and fertilizer application without decreasing yield [5]. Low plant density produced a higher number of heavy bolls per plant, while both the number and weight of bolls reduced with increasing plant density [6,7]. Currently, suggested and practiced plant densities in China are 5.3×10^4 to 7.5×10^4 plants ha^{-1} in the Yellow River Valley [8], 3.0×10^4 plants ha^{-1} in the Yangtze River Valley [9], and 22.7×10^4 plants ha^{-1} in the Northwest region. The difference between the plant densities among various locations is due to difference in climatic conditions which affect the yield and fiber quality of cotton.

Biomass accumulation in the cotton plant during the early growth period is an important factor for final yield determination. More biomass accumulation in early stages helps in better establishment of a crop while accumulation at late growth stages increases assimilation to the reproductive organs, resulting in a higher yield and quality of cotton [10]. Cotton plants accumulate more biomass in vegetative organs due to its indeterminate nature. More assimilate accumulation to vegetative and reproductive organs increases the shedding of fruit and leaves [11,12]. At maturity, the aboveground biomass becomes lower than the total due to the shedding of leaves and fruits [13]. Previous studies have confirmed that optimum plant density is the critical factor for establishing optimal canopy structure and leaf area index (LAI). Optimal LAI determines light penetration in the canopy [14–16]. Several researchers have examined the relationship between the plant density, LAI, and cotton production [17–19] and found that an increase in plant density results in higher LAI, while too-high LAI caused shading and reduced the yield [20,21]. Both LAI and yield increases slowly with an increase in plant density [22]. Fiber quality indicators including fiber strength, fitness, length, uniformity index, and fineness are negatively affected by environmental and genetic factors as well as poor management practices at flowering and boll formation stages [23,24]. Similarly, fiber quality is affected by plant density, irrigation, fertilization, and weather changes [23,25]. This study is conducted with the aim to assess the response of cotton yield and fiber quality, biomass accumulation, and partitioning of various plant densities to identify technological alternatives to make efficient use of land and increase yield and profitability of cotton.

2. Materials and Methods

2.1. Experimental Site

The study was conducted in 2016 and 2017 at the experimental station of the Institute of Cotton Research of Chinese Academy of Agricultural Sciences in Anyang, Henan, China (36°06′ N, 114°21′ E). The soil was medium loam in texture with a total N of 0.65 g kg^{-1}, P of 0.01 g kg^{-1} and K of 0.11 g kg^{-1}. The monthly average temperature and relative humidity data of both years of cotton growing seasons are presented in Figure 1. The average temperature during the cotton growing season was 22 °C and 23 °C in 2016 and 2017, respectively. Annual rainfall was 713 mm in 2016 and 585 mm in 2017. Annual sunshine hours were 1737 h in 2016 and 1838 h in 2017. The average air temperature at the seedling and reproductive stages was cooler as compared to other growth stages. The overall cotton growing season in 2016 was cooler with more rainfall as compared to 2017.

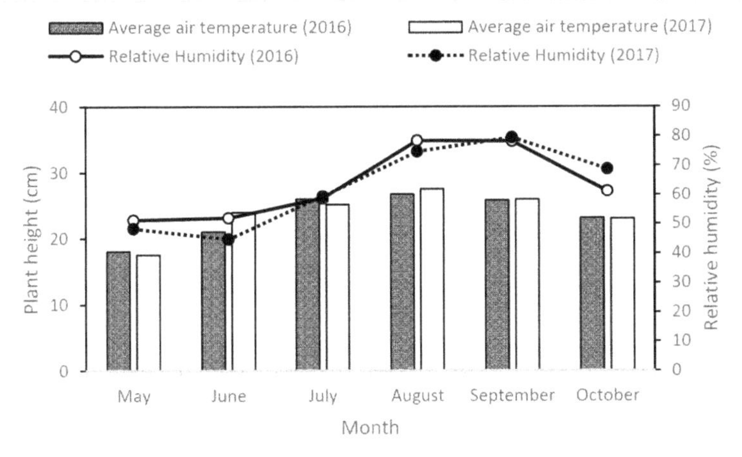

Figure 1. Monthly average air temperature and relative humidity in the 2016 and 2017 growing seasons.

2.2. Experimental Design

The experiment was conducted in a randomized complete block design (RCBD). Six plant densities (D1, 1.5; D2, 3.3; D3, 5.1; D4, 6.9; D5, 8.7; and D6, 10.5 plants m^{-2}) were plotted randomly in three replications on clay loam soil. Each experimental plot area was 64 m^2 with 8 m in length and width. Each plot consisted of 10 rows, with a row spacing of 0.8 m, which was constant for all plant densities. Seeds of cotton mid maturity cultivar SCRC28 were sown by hand on flat beds with plastic mulching to conserve soil moisture from evaporation. Plastic mulch was removed after one month of full emergence. Seedlings were thinned to the required plant densities after three weeks of emergence. During both years, the land was prepared by ploughing, and irrigated in early spring before sowing. Sowing was done during the growing season on 22 April in 2016 and 2017.

A basal dose of 225 kg N ha^{-1}, 150 kg P$_2$O$_5$ ha^{-1}, and 225 kg K$_2$O ha^{-1} was applied to the field before sowing. Irrigation was applied by flooding during the flowering stage at a total volume of approximately 45 m^3. Crop management practices such as weeding, hoeing, pesticides, and irrigation were performed in a timely manner to enhance crop growth.

2.3. Data Collection

Data were recorded on cotton leaf area index, biomass accumulation at critical stages of crop growth, fiber quality, yield and yield components (boll m^{-2} and boll weight) during 2016 and 2017 at different days after emergence.

2.3.1. Yield and Yield Components

Seed cotton yield (kg/ha) and lint yield (kg/ha) were recorded by hand-harvesting three times from each treatment. The boll moisture was reduced to less than 11% by air-drying and seed cotton of 100 bolls at first harvest were sampled for boll weight. Weight of single boll was calculated by dividing total seed cotton yield of 100 bolls by the total number of bolls. Lint percentage was calculated from lint yield of 100 bolls divided by seed cotton weight of 100 bolls.

2.3.2. Biomass Accumulation and Partitioning

The dry weight of cotton plants was recorded seven times during the growing season with an interval of 15 days at 42 days after emergence (DAE), 57 DAE, 72 DAE, 87 DAE, 102 DAE, 117 DAE, and 132 DAE. Three plants from each plot of three replications were uprooted randomly and dissected into the underground part (roots), leaves, stem, and reproductive structures. Samples were quickly placed for 30 min in an electric fan-assisted oven at 105 °C in order to stop metabolism. Samples were dried at 80 °C for 48 h to attain a constant weight. Dry matter partitioning was calculated by the ratio of the dry weight of reproductive organs (DWRO) (squares, flowers, green, and open bolls) to plant total biomass while boll load was calculated by dividing DWRO by leaf area. A logistic regression equation was used to describe biomass accumulation [26].

$$Y = \frac{A}{1 + be^{-kt}} \tag{1}$$

In Equation (1) Y (kg) is the biomass, A (kg) the maximum biomass, t (d) is the number of days after emergence (DAE) while a and b are constants.

From Formula (1), the following equations were calculated:

$$t_o = \frac{\ln b}{k} (t_o = t) \tag{2}$$

$$t_1 = \frac{\ln b - \ln\left(2 + \sqrt{3}\right)}{k} \tag{3}$$

$$t_2 = \frac{\ln b + \ln\left(2 + \sqrt{3}\right)}{k} \tag{4}$$

$$V_M = \frac{Ak}{4} \tag{5}$$

$$\Delta t = t_2 - t_1 \tag{6}$$

$$V_A = \frac{Y_2 - Y_1}{\Delta t} \tag{7}$$

In the above equations, V_M (kg ha^{-1} d^{-1}) is the highest rate of biomass accumulation, and t (d) is the maximum biomass fast accumulation period. Y_1 and Y_2 are the biomass at t_1 and t_2. V_A indicates the average biomass accumulation from t_1 to t_2 and Δt (d) is the total period of average biomass accumulation.

2.3.3. Leaf Area Index

LAI of cotton plants were calculated by taking photos of leaves through a scanning machine (Phantom p800xl, MiCROTEK, Shanghai, China) and leaf area was calculated by using Image-Pro Plus 7.0 (Media Cybernetics, Rockville, MD, USA). The LAI was determined by dividing the total plant leaf area per unit ground area.

2.3.4. Fiber Quality

Fiber quality, including fiber length (mm), fiber uniformity, fiber strength (cN tex^{-1}), and fiber micronaire, were assessed by the Supervision, Inspection and Test Center of Cotton Quality, Ministry of Agriculture, in Anyang, Henan province of China using a high volume instrument (HVI-900) (Changing Technologies, Mainland, China) according to the internationally accepted ICC standard.

2.3.5. Statistical Analysis

Microsoft Excel 365 (Microsoft, Bothell, WA, USA) as used for the processing of data. SPSS 19.0 (SPSS Inc. Chicago, IL, USA) and Origin 2016 (OriginLab Corporation, Northampton, MA, USA) were used for the analysis of data. Figures were plotted by using Origin 2016. Duncan's multiple range test at 5% probability level was used to test differences among mean values.

3. Results

3.1. Yield and Yield Components

Yield and yield components of cotton varied with plant density. Seed cotton yield and lint yield along with yield components were significantly affected by plant density except boll weight and lint percentage in both years (Tables 1 and 2). During both years, D5 plant density (PD) produced the highest seed cotton and lint yield as compared to other plant densities. Highest seed cotton yield of 4662 kg ha^{-1} and highest lint yield 1763 kg ha^{-1} was produced by D5, which was followed by D4, D6, D3, D2, and D1. The highest lint percentage (43.5%) was recorded at D1, followed by D2, D3, D4, D5, and D6. The boll density per unit ground area generally increased with increasing plant density but the boll density of individual plants decreased with increasing plant density. More number of bolls m^{-2} (105.4) was produced by D6 in 2016, while in 2017 more bolls m^{-2} (75.7) was produced by D5. During both years, bigger bolls were produced by D1 as compared to other treatments.

Table 1. Comparison of boll m^{-2} and boll weight (g) at various plant densities in 2016 and 2017 in the cotton growing season.

Treatment	Boll (m^2)	Boll Weight (g)
Year 2016		
Plant Density (PD)		
D1	64.3f	6.2a
D2	72.7e	5.8a
D3	82.4d	5.7a
D4	90.1c	5.7a
D5	99b	5.7a
D6	104.4a	5.6a
Year 2017		
Plant Density (PD)		
D1	46.5e	6a
D2	51.3d	6a
D3	59.5c	5.9a
D4	64.4b	5.9a
D5	75.7a	5.7ab
D6	66.4b	5.6b
ANOVA		
Y	0.1509	0.3616
D	0.0061	0.5045
Y × D	0.0001	<0.0001

Means followed by the same letters within the same category are statistically similar according to Duncan's multiple range test at $p < 0.05$.

Table 2. Comparison of seed cotton and lint yield of various plant densities in 2016 and 2017 in the cotton growing season.

Treatment	Seed Cotton Yield (kg ha^{-1})	Lint Yield (kg ha^{-1})	Lint Percentage (%)
D1	3258e	1389e	42.8a
D2	3598d	1490d	41.6ab
D3	3989c	1574c	39.5bc
D4	4304b	1669b	38.8c
D5	4662a	1763a	37.9c
D6	4259b	1609bc	37.8c
ANOVA			
Y	0.0008	<0.0001	0.8173
D	0.0005	0.0002	0.0700
Y × D	0.1827	0.7746	0.2969

Means followed by the same letters within the same category are statistically similar according to Duncan's multiple range test at $p < 0.05$.

3.2. Biomass Accumulation

Cotton plant biomass accumulation (CPB) was significantly affected by plant density and followed a normal logistic model by DAE (Figure 2). CPB increased as plant density increased and differences were found between the different densities. The D6 plant density had more CPB accumulation as compared to D1, D2, D3, D4, and D5 during both years. Vegetative organ biomass (VOB) during 2016 and 2017 was positively affected by plant density (Tables 3 and 4). The VOB increased linearly with the increase in plant density. The highest PD, D6, produced more VOB as compared to other plant densities while individual plant VOB decreased as density increased due to resource competition among plants. Reproductive growth of cotton started from the appearance of the first square. Less biomass accumulated to reproductive organs of cotton which increases linearly with further growth. Treatment D5 produced more ROB in 2016 and 2017, followed by D6, D4, D3, D2, and D1.

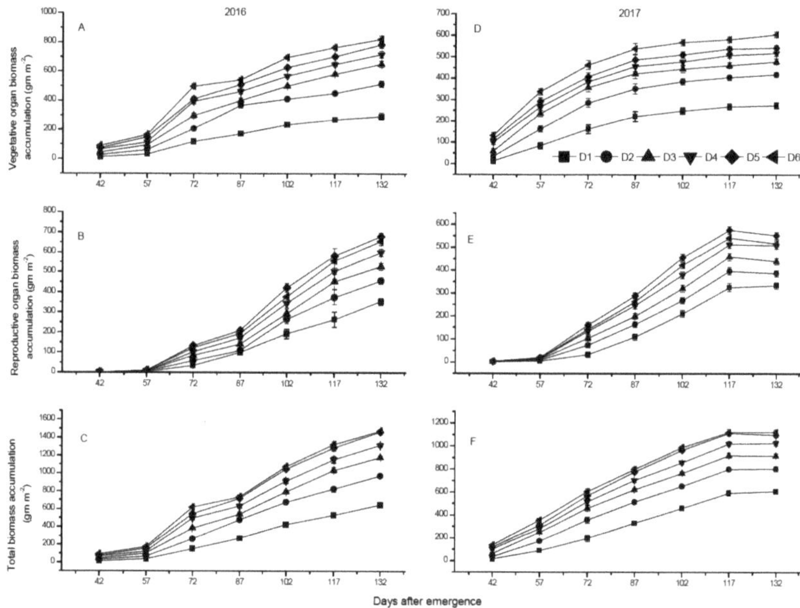

Figure 2. Vegetative organ biomass accumulation (**A,D**), reproductive organ biomass accumulation (**B,E**), and total organ biomass accumulation (**C,F**) of 2016 and 2017 cotton growing seasons.

Table 3. Analysis of variance for the effect of year (Y) and plant density (PD) on biomass accumulation.

Source	42 DAE		57 DAE		72 DAE		87 DAE		102 DAE		117 DAE		132 DAE	
	F	p-Value	F	p-value	F	p-Value	F	p-Value	F	p-Value	F	p-Value	F	p-Value
VOB														
Y	9.808	0.026	52.945	0.001	1.485	0.277	0.126	0.737	8.859	0.031	13.983	0.013	21.422	0.006
D	18.048	0.003	10.969	0.010	29.071	0.001	80.460	0.000	24.908	0.002	16.879	0.004	14.033	0.006
Y × D	4.421	0.007	14.299	<0.0001	5.652	0.002	1.710	0.178	14.539	<0.0001	27.077	<0.0001	21.498	<0.0001
ROB														
Y	39.283	0.002	14.100	0.013	15.211	0.011	31.227	0.003	19.709	0.007	1.360	0.296	27.305	0.003
D	5.136	0.048	3.505	0.097	54.730	0.000	20.054	0.003	140.225	<0.0001	61.440	0.000	25.162	0.001
Y × D	2.878	0.041	5.808	0.002	1.842	0.150	4.177	0.009	0.588	0.709	0.880	0.513	6.975	0.001
CPB														
Y	10.214	0.024	50.051	0.001	6.404	0.052	108.192	0.000	5.092	0.074	5.802	0.061	23.851	0.005
D	17.733	0.003	10.436	0.011	67.943	0.000	602.531	<0.0001	103.482	<0.0001	26.573	0.001	17.484	0.003
Y × D	4.492	0.007	13.345	<0.0001	3.051	0.033	0.501	0.772	3.956	0.012	11.936	<0.0001	33.702	<0.0001

Table 4. Regression of cotton plant biomass accumulation at growing seasons 2016 and 2017.

Items	Treatment	Regression Equation	R^2
Cotton plant biomass (2016)	D1	$Y = 697.81055/(1 + 348.84488e^{-0.06161t})$	0.9925 ***
	D2	$Y = 1011.12157/(1 + 310.52754e^{-0.06365t})$	0.9909 ***
	D3	$Y = 1302.14044/(1 + 174.95064e^{-0.05567t})$	0.9878 ***
	D4	$Y = 1435.81604/(1 + 134.01637e^{-0.05418t})$	0.9783 ***
	D5	$Y = 1575.89801/(1 + 138.95428e^{-0.05546t})$	0.9844 ***
	D6	$Y = 1594.84306/(1 + 111.53485e^{-0.05384t})$	0.9741 ***
Vegetative organ biomass	D1	$Y = 295.33909/(1 + 298.23496e^{-0.06989t})$	0.9879 ***
	D2	$Y = 484.50669/(1 + 743.27663e^{-0.08699t})$	0.9796 ***
	D3	$Y = 651.12066/(1 + 143.02059e^{-0.06253t})$	0.9782 ***
	D4	$Y = 701.83792/(1 + 131.19935e^{-0.06514t})$	0.9569 ***
	D5	$Y = 773.54645/(1 + 108.47096e^{-0.06237t})$	0.9738 ***
	D6	$Y = 807.8694/(1 + 122.95547e^{-0.06699t})$	0.9549 ***
Reproductive organ biomass	D1	$Y = 403.53388/(1 + 1096.07473e^{-0.06673t})$	0.9918 ***
	D2	$Y = 494.27754/(1 + 2706.97502e^{-0.07808t})$	0.9956 ***
	D3	$Y = 591.54947/(1 + 1594.68086e^{-0.07241t})$	0.9943 ***
	D4	$Y = 668.08731/(1 + 1195.61028e^{-0.07009t})$	0.9946 ***
	D5	$Y = 741.0549/(1 + 1078.67833e^{-0.07091t})$	0.9915 ***
	D6	$Y = 734.4757/(1 + 978.71309e^{-0.06826t})$	0.9929 ***
Cotton plant biomass (2017)	D1	$Y = 648.96068/(1 + 258.79492e^{-0.06442t})$	0.9936 ***
	D2	$Y = 838.50226/(1 + 150.01915e^{-0.06353t})$	0.9886 ***
	D3	$Y = 951.32332/(1 + 103.58331e^{-0.06163t})$	0.9857 ***
	D4	$Y = 1070.21853/(1 + 83.38347e^{-0.05897t})$	0.9918 ***
	D5	$Y = 1144.95086/(1 + 95.96115e^{-0.06213t})$	0.9936 ***
	D6	$Y = 1165.41818/(1 + 74.52055e^{-0.05987t})$	0.9938 ***
Vegetative organ biomass	D1	$Y = 268.83729/(1 + 327.895e^{-0.08594t})$	0.9911 ***
	D2	$Y = 404.49245/(1 + 299.20271e^{-0.0908t})$	0.9883 ***
	D3	$Y = 460.00848/(1 + 292.09854e^{-0.09749t})$	0.9855 ***
	D4	$Y = 505.9789/(1 + 101.20589e^{-0.08053t})$	0.9898 ***
	D5	$Y = 534.94333/(1 + 88.38951e^{-0.07954t})$	0.9911 ***
	D6	$Y = 586.38098/(1 + 97.73996e^{-0.08321t})$	0.9890 ***
Reproductive organ biomass	D1	$Y = 354.07086/(1 + 8856.53541e^{-0.09421t})$	0.9938 ***
	D2	$Y = 415.62152/(1 + 2397.24183e^{-0.08411t})$	0.9860 ***
	D3	$Y = 473.372/(1 + 1661.97624e^{-0.08182t})$	0.9825 ***
	D4	$Y = 539.06687/(1 + 1185.99181e^{-0.07956t})$	0.9877 ***
	D5	$Y = 582.96854/(1 + 1712.18326e^{-0.08643t})$	0.9869 ***
	D6	$Y = 548.10109/(1 + 1929.68342e^{-0.08696t})$	0.9867 ***

***, significant at the 0.001 probability level.

3.3. Simulation of Biomass Accumulation

Simulation of biomass accumulation based on Equation (1) followed the logistic function and all the biomass accumulation were significant. Calculation from Equations (2)–(7) based on Table 2 illustrates the day of starting and termination of cotton biomass fast accumulation period (FAP) during 2016 and 2017. The averaged highest speed for CPB in all plant densities were 68 and 114 DAE in 2016, and 56 and 98 in 2017, with the highest average (V_A = 16 and 14 gm m^{-2} d^{-1}) and maximum rate (V_M = 18 and 15 gm m^{-2} d^{-1}) (Tables 5 and 6).

Table 5. Eigen values of cotton biomass accumulation at growing season 2016.

Items	Treatment	Fast Accumulation Period				Fastest Accumulation Point	
		t_1 (DAE)	t_2 (DAE)	Δt (d)	V_A (gm m^{-2} d^{-1})	V_M (gm m^{-2} d^{-1})	at DAE
Cotton plant biomass	D1	73.7	116.4	42.8	9.4	10.7	95.0
	D2	69.5	110.8	41.4	14.1	16.1	90.2
	D3	69.1	116.4	47.3	15.9	18.1	92.8
	D4	66.1	114.7	48.6	17.1	19.4	90.4
	D5	65.2	112.7	47.5	19.2	21.8	89.0
	D6	63.1	112.0	48.9	18.8	21.5	87.6
	Average	67.8	113.9	46.1	15.7	18.0	90.8
Vegetative organ biomass	D1	62.7	100.4	37.7	4.5	5.2	81.5
	D2	60.9	91.1	30.3	9.2	10.5	76.0
	D3	58.3	100.4	42.1	8.9	10.2	79.4
	D4	54.6	95.1	40.4	10.0	11.4	74.9
	D5	54.0	96.3	42.2	10.6	12.1	75.1
	D6	52.2	91.5	39.3	11.9	13.5	71.8
	Average	57.1	95.8	38.7	9.2	10.5	76.5
Reproductive organ biomass	D1	85.2	124.6	39.5	5.9	6.7	104.9
	D2	84.4	118.1	33.7	8.5	9.6	101.2
	D3	83.7	120.0	36.4	9.4	10.7	101.8
	D4	82.3	119.9	37.6	10.3	11.7	101.1
	D5	79.9	117.1	37.1	11.5	13.1	98.5
	D6	81.6	120.2	38.6	11.0	12.5	100.9
	Average	82.8	120.0	37.1	9.4	10.7	101.4

t_1 is the starting and t_2 is the termination point of the fast accumulation period (FAP). Δt is the total duration of FAP. V_A is the average and V_M is the maximum rate of biomass accumulation during FAP. DAE represents days after emergence.

Table 6. Eigen values of cotton biomass accumulation at growing season 2017.

Items	Treatment	Fast Accumulation Period				Fastest Accumulation Point	
		t_1 (DAE)	t_2 (DAE)	Δt (d)	V_A (gm m^{-2} d^{-1})	V_M (gm m^{-2} d^{-1})	at DAE
Cotton plant biomass	D1	65.8	106.7	40.9	9.2	10.5	86.2
	D2	58.1	99.6	41.5	11.7	13.3	78.9
	D3	53.9	96.7	42.7	12.9	14.7	75.3
	D4	52.7	97.3	44.7	13.8	15.8	75.0
	D5	52.3	94.7	42.4	15.6	17.8	73.5
	D6	50.0	94.0	44.0	15.3	17.4	72.0
	Average	55.5	98.2	42.7	13.1	14.9	76.8
Vegetative organ biomass	D1	52.1	82.7	30.6	5.1	5.8	67.4
	D2	48.3	77.3	29.0	8.1	9.2	62.8
	D3	44.7	71.7	27.0	9.8	11.2	58.2
	D4	41.0	73.7	32.7	8.9	10.2	57.3
	D5	39.8	72.9	33.1	9.3	10.6	56.3
	D6	39.2	70.9	31.7	10.7	12.2	55.1
	Average	44.2	74.9	30.7	8.6	9.9	59.5
Reproductive organ biomass	D1	82.5	110.5	28.0	7.3	8.3	96.5
	D2	76.9	108.2	31.3	7.7	8.7	92.5
	D3	74.5	106.7	32.2	8.5	9.7	90.6
	D4	72.4	105.5	33.1	9.4	10.7	89.0
	D5	70.9	101.4	30.5	11.0	12.6	86.1
	D6	71.9	102.1	30.3	10.4	11.9	87.0
	Average	74.8	105.7	30.9	9.1	10.3	90.3

t_1 is the starting and t_2 is the termination point of fast accumulation period (FAP). Δt is the total duration of FAP. V_A is the average and V_M is the maximum rate of biomass accumulation during FAP. DAE represents days after emergence.

Cotton plant biomass accumulation was found significant among plant densities. In 2016, a fast accumulation period in D5 started at 65 DAE and terminated at 113 DAE, which lasts for 48 DAE with the ighest average (19.2 V_A gm m^{-2} d^{-1}) and maximum rate (22 V_M gm m^{-2} d^{-1}) at 89 DAE. The lengthiest fast accumulation period for CPB was noted in D6, which lasts for 49 DAE with the average rate of 18.8 V_A gm m^{-2} d^{-1} (Table 5).

The fast accumulation period of CPB in 2017 for D6 started earlier at 50 DAE and terminated at 94 DAE, while D1 FAP terminated last at 107 DAE. The highest average (15.6 V_A gm m^{-2} d^{-1}) and maximum rate (17.8 V_M gm m^{-2} d^{-1}) were noted in D5, followed by D4, D6, D3, D2, and D1 (Table 6).

Vegetative organ biomass responded positively to plant density. The earliest and highest FAP of VOB in both years was observed at D6 with the average rate (12 and 10.7 V_A gm m^{-2} d^{-1}), which lasts for 39 and 32 DAE, and maximum rate (13.5 and 12.2 V_M gm m^{-2} d^{-1}), which lasts for 72 and 55 DAE in 2016 and 2017, respectively. Both average and maximum VOB accumulation rates of D6 were 62%, 23%, 25%, 16%, and 11% higher than D1, D2, D3, D4, D5 in 2016 and 52%, 24%, 8%, 17%, and 13% higher than D1–D5 in 2017 (Tables 5 and 6).

The highest average rate (11.5 V_A gm m^{-2} d^{-1}) of reproductive structures biomass was observed in D5, which started at 80 DAE and terminated at 117 DAE and lasted for 37 DAE, with a maximum rate (13 V_M gm m^{-2} d^{-1}) at 99 DAE in 2016 (Table 4). Both average and maximum ROB accumulation rates of D5 were observed to be higher as compared to D1, D2, D3, D4, and D5. The earliest FAP in D5 began at 80 DAE, while the last terminated FAP was observed in D1, which ended at 125 DAE. In 2017, initial FAP of ROB began in D5 which lasted for 31 DAE and terminated at 101 DAE, with the highest average rate (11 V_A gm m^{-2} d^{-1}) and maximum rate (12.6 V_M gm m^{-2} d^{-1}) at 86 DAE, followed by D6, D4, D3, D2, and D1 (Tables 5 and 6).

3.4. Dry Matter Partitioning (DWRO/PB)

Dry matter partitioning, as indicated by the ratio of the dry weight of reproductive organs to plant biomass (DWRO/PB), increase slowly as the plant changes from one growth stage to another and peak stage of dry matter partitioning was observed at 120 DAE during 2016 and 2017 (Figure 3). During different growth stages, significant differences were observed between treatments (Table 7). The DWRO/PB of D1 was observed to be higher as compared to other treatments.

Table 7. Analysis of variance for the effect of year (Y) and plant density (PD) on dry matter partitioning.

Source	42 DAE		57 DAE		72 DAE		87 DAE		102 DAE		117 DAE		132 DAE	
	F	p-Value	F	p-Value	F	p-Value	F	p-Value	F	p-Value	F	p-Value	F	p-Value
Y	9.371	0.028	3.769	0.110	0.178	0.691	8.596	0.033	15.602	0.011	205.764	0.0000	13.113	0.015
D	64.200	0.0000	1.009	0.496	1.112	0.455	1.415	0.356	3.231	0.112	23.395	0.002	18.230	0.003
Y × D	0.591	0.707	26.356	0.000	5.720	0.002	6.189	0.001	1.904	0.139	0.249	0.935	1.498	0.235

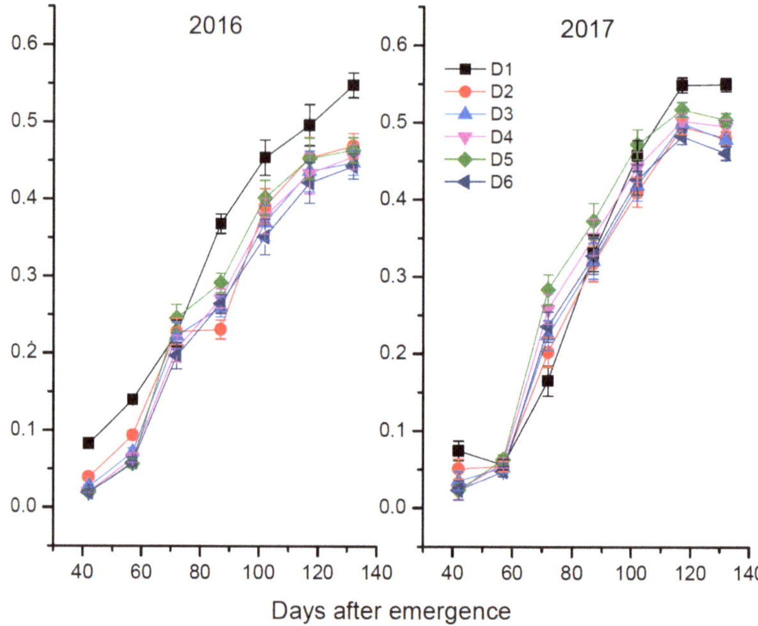

Figure 3. Ratio of dry weight of reproductive organs to plant biomass (DWRO/PB) in 2016 and 2017 growing seasons.

3.5. Boll Load (DWRO/LA)

Boll load, as indicated by the ratio of the dry weight of reproductive organs by leaf area (DWRO/LA), was found to be significantly higher in D6 as compared to other treatments and significant differences were observed between different treatments (Table 8). The DWRO/LA increased gradually with an increase in plant density and changing from one growth stage to another (Figure 4). At 132 DAE, DWRO/LA of D6 was 14%–82% and 4%–76% higher than treatment D1–D6 during 2016 and 2917, respectively.

Table 8. Analysis of variance for the effect of year (Y) and plant density (PD) on boll load.

Source	42 DAE		57 DAE		72 DAE		87 DAE		102 DAE		117 DAE		132 DAE	
	F	*p*-Value	*F*	*p*-Value	*F*	*p*-Value	*F*	*p*-Value	*F*	*p*-Value	*F*	*p*-Value	*F*	*p*-Value
Y	59.210	0.001	1.984	0.218	3.392	0.125	9.491	0.027	9.909	0.025	4.792	0.080	4.161	0.097
D	1.902	0.249	2.038	0.227	8.939	0.016	9.628	0.013	8.725	0.016	20.358	0.002	32.806	0.001
Y × D	5.208	0.003	1.565	0.215	6.946	0.001	7.186	0.001	5.631	0.002	4.300	0.008	1.138	0.373

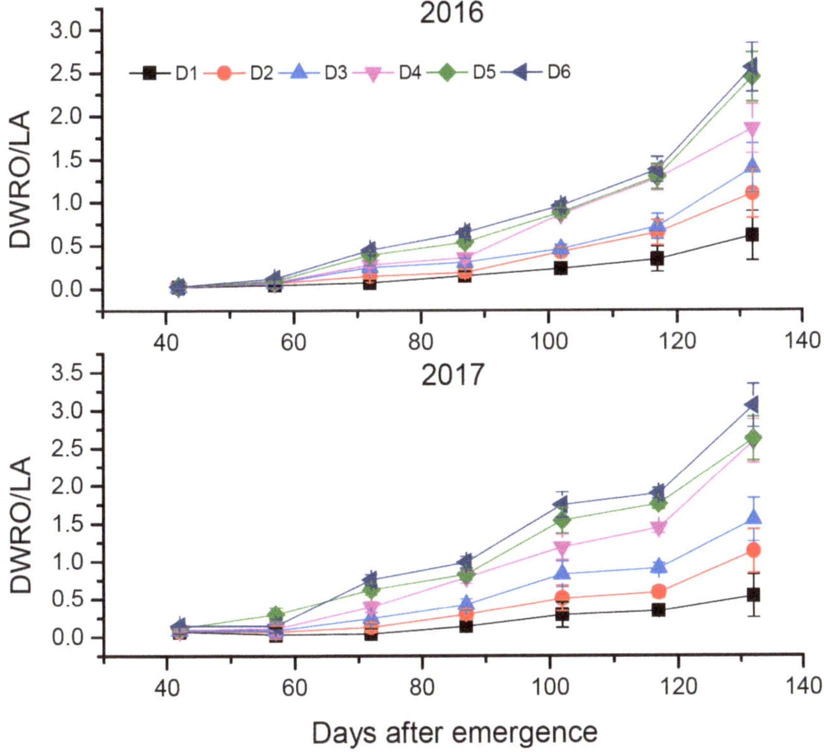

Figure 4. Ratio of dry weight of reproductive organs per leaf area (DWRO/LA) in 2016 and 2017 growing seasons.

3.6. Leaf Area Index

The leaf area index (LAI) at different days after emergence is shown in Figure 5. The LAI of D6 was higher during both years as compared to other treatments and increased linearly as plant density increased (Table 9). The LAI increased with the growth of the cotton plants and reached a peak at 102 DAE and then decreased linearly. LAI of high plant density reached 4.3 in 2016 and 4 in 2017, while in the case of lower plant density, it reached up to 1.3 in 2016 and 1.5 in 2017. In the last growth stages, no significant differences were observed in plant densities D4–D6.

Table 9. Analysis of variance for the effect of year (Y) and plant density (PD) on leaf area index.

Source	42 DAE		57 DAE		72 DAE		87 DAE		102 DAE		117 DAE		132 DAE	
	F	p-Value	F	p-Value	F	p-Value	F	p-Value	F	p-Value	F	p-Value	F	p-Value
Y	25.001	0.004	15.040	0.012	0.413	0.549	0.030	0.869	2.068	0.210	7.016	0.045	12.421	0.017
D	5.591	0.041	16.289	0.004	20.785	0.002	37.873	0.001	76.869	0.000	20.444	0.002	28.831	0.001
Y × D	7.678	0.000	1.573	0.213	1.237	0.329	0.896	0.503	1.032	0.426	3.175	0.029	0.635	0.676

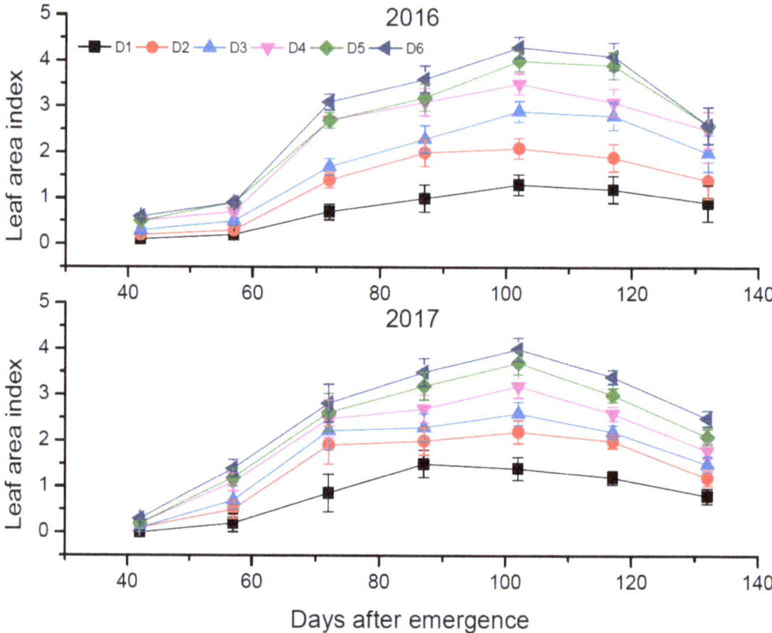

Figure 5. Leaf area index of cotton at different planting densities in 2016 and 2017.

3.7. Fiber Quality

Fiber quality parameters were significantly influenced by plant density in both years (Table 10). An increase in plant density led to longer fiber length while decreasing strength and micronaire value. Low planting density had low length and greater strength and micronaire value as compared to high treatments. The fiber length of high plant density D5 and D6 were statistically similar while the length of D5 was 5% and 7% longer than the lowest planting density in both years, respectively. Fiber strength of lowest plant density was recorded to be 9% and 10% higher, while micronaire was observed to be 15% and 9% higher as compared to the highest planting density in 2016 and 2017, respectively. Planting density had no significant effect on fiber elongation and uniformity index during both growing seasons.

Table 10. Comparison of fiber quality parameters of various densities of growing seasons 2016 and 2017.

Treatment	Length	Strength	Elongation	Uniformity Index	Microniare
	(mm)	(cN/tex)	(%)	(%)	
D1	28.86c	31.26a	6.40a	84.77b	5.33a
D2	29.40bc	30.97ab	6.28a	84.83b	4.93b
D3	29.52bc	30.10b	6.25a	85.17ab	4.83bc
D4	29.82b	29.10c	6.24a	85.47ab	4.80bc
D5	30.63a	28.80c	6.20a	86.07a	4.70c
D6	30.60a	28.33c	6.22a	85.53ab	4.63c
ANOVA					
Y	0.1170	0.0798	0.0000	<0.0001	0.0001
D	0.8030	0.0001	0.5164	0.0109	0.0007
Y × D	0.0614	0.9776	0.4447	0.9268	0.8639

Means followed by the same letters within the same category are statistically similar according to Duncan's multiple range test at $p < 0.05$.

3.8. Economic Analysis

The net returns were affected by different plant density. Net returns were determined on the basis of production cost and returns from the cotton crop. The highest net returns were obtained from D5 (1750 USD ha^{-1} and 1393 USD ha^{-1}) during 2016 and 2017, respectively, while the lowest was obtained from D1 (Table 11). Seed and labor cost mostly affected net returns. More labor was required for low plant densities due to more vegetative branches as compared to high plant densities.

Table 11. Cotton yield, production cost, and net returns of 2016 and 2017.

| Treatment | Lint Yield | | Returns from Cotton | | Production Cost | | Net Returns | |
	Kg ha^{-1}		USD ha^{-1}		USD ha^{-1}		USD ha^{-1}	
	2016	2017	2016	2017	2016	2017	2016	2017
D1	1459	1320	3040	2705	2429	2389	611	316
D2	1588	1391	3310	2850	2325	2257	985	593
D3	1685	1463	3510	2998	2297	2259	1213	739
D4	1762	1576	3671	3230	2269	2232	1402	998
D5	1843	1683	3840	3449	2090	2056	1750	1393
D6	1726	1493	3589	3061	2137	2102	1452	959

Production cost includes fertilizer, seed, and labor cost. Labor cost includes labor for planting, management, and harvesting. One labor unit per day cost was 6.02 USD in 2016 and 5.92 USD in 2017. Wholesale lint price of 1 kg was 2.08 USD in 2016 and 2.05 USD in 2017. Values were converted from Chinese Yuan to USD according to the official rate (USD 1 = 6.65 yuan in 2016 and 6.76 yuan in 2017).

4. Discussion

The main purpose of this study is to explore and compare different plant densities in response to cotton yield, leaf area index, dry matter partitioning, and fiber quality at different growth stages. Higher plant density is the key management practice for obtaining greater numbers of bolls per unit area, but in most cases, the yield enhances up to an optimum density, after which further increase in plant population decreases yield. Different regions of China have different optimum densities and lint production, which depends on climatic conditions along with other management practices. The Xinjiang autonomous region has the recommended PD of 21.0×10^4 to 24.0×10^4 plants ha^{-1} [27]; followed by Yellow River Valley with a PPD of 3.0×10^4, 4.5×10^4, and 6.0×10^4 plants ha^{-1} for hybrid Bt cotton, indigenous Bt cotton, and Bt cotton, respectively [28,29], and for late sowing, PD is 7.5×10^4 ha^{-1} [30]; while in the Yangtze River Valley where hybrid seeds are commonly used, it has the PD of 3.0×10^4 plants ha^{-1} [31]. Our results are consistent with previous studies that have shown that cotton yield increases up to a certain limit with increasing PD, while too low and too high plant density cause a reduction in yield [32]. In this study, yield and yield components were significantly affected by plant density, excluding boll weight and lint percentage. High yield and yield components were noted in plant density D5. Yield and number of bolls produced by a single plant of the treatment D5 was lower as compared to other treatments but was more based on per unit area. These results are consistent with Mao et al. [33], who reported that high plant population increase bolls m^{-2} while the weight of individual bolls decreases.

More biomass production is the foundation of high yield [34–36]. In this study, biomass accumulation was higher in 2016 as compared to 2017, which might be due to differences in environmental conditions. Total plant biomass and vegetative organ biomass accumulation were high in higher plant density while higher reproductive organ biomass was accumulated in D5 as compared to other treatments. In early growth stages of the cotton plant, plant density did not affect reproductive structure biomass accumulation, while after 87 DAE ROB, accumulation was influenced significantly. High biomass accumulation in high plant density was due to a greater number of plants per unit ground area with more vegetative growth. Our results are in line with other researchers who also reported that high plant density resulted in high biomass production [37,38]. Both high and low plant density lead to reductions in reproductive organ biomass. The less ROB production in high population might be due to less light penetration to the lower parts of plants, followed by a reduction in temperature and increased relative humidity in the cotton canopy, which enhanced fruit shedding as compared to other plant densities [39].

The ratio of dry weight of reproductive organs to plant biomass (DWRO/PB) also affected the yield of cotton [40]. In this study, the highest ratio was obtained in the lowest plant density D1 as compared to other treatments which showed less differences. Similar results were previously obtained by Dai et al. [37], who also reported high DWRO/PB in the lowest density. Boll load is also an important indicator of lint yield. In this study, a high and significant ratio of dry weight of reproductive organs per leaf area (DWRO/LA) was observed in high plant densities (D4–D6), mostly in the late growth stages. Our results are supported by Dong et al. [30], where high boll load led to an increase in leaf senescence and a decrease in cotton yield and quality. The high DWRO/LA in late growth stages is due to high competition for nutrients and assimilates between vegetative and reproductive growth after the bloom stage [40].

Leaf area index is an important factor that affects biomass production of cotton [41]. LAI is also one of the physiological parameters which determine crop yield and predict crop production up to some extent. For obtaining high yield, it is necessary to maintain optimum LAI for more light penetration and high light use efficiency, mostly at late growth stages: that is, the flowering and boll setting stages [22].

Cotton fiber is the extension of seed epidermal cells. Fiber quality indicators are affected by plant density and environmental factors [10]. In the present study, cotton fiber indicators were significantly affected by plant density. Low plant density had high strength and micronaire value as compared to high plant density, while the length of low plant density is shorter as compared to high and moderate plant density. Our results are in agreement with previous research that have reported high strength and micronaire and short fiber length at low planting density [5,33]. The lower fiber quality at high planting density may be due to less photosynthesis, which reduces carbohydrate supply for fiber formation. For obtaining good quality fiber, cultivar selection is of great importance, while managing plant populations to maintain genetic potential is the secondary part [42,43].

Economic benefit plays an important role in the success of agriculture business. In the Yellow River Valley, due to fast urbanization, high labor costs and a shortage of labor have become a challenge to traditional intensive cotton production [2]. Labor cost specifically affects the profitability of the cotton crop. High density has less vegetative branches as compared to high plant density, which needs less labor for vegetative branch removal and other field management.

5. Conclusions

In the present study, planting density positively affected yield, fiber quality, and dry matter accumulation and partitioning of cotton crop under equidistant row arrangement. Optimum or moderate plant density (8.7 plants m^{-2}) resulted in high reproductive organ biomass accumulation at later growth stages as compared to other treatments. More reproductive organ biomass accumulation in this density increased the yield of cotton. Good quality fiber was obtained at low and moderate plant densities as compared to higher ones. In conclusion, 8.7 plants m^{-2} is regarded to be an optimum plant density in term of high yield, uniform fiber quality, and dry matter accumulation. The finding of this research offers an alternative to cotton growers who use conventionally wider rows and lower plant population ha^{-1}.

Author Contributions: Y.L. and Z.W. designed the experiment. W.D., S.X., M.X. and H.C. conducted the experiment. Y.L., F.X. and G.W. helped and provided useful suggestions during experiment. N.K. processed and analyzed data and wrote first draft. Y.L. and L.F. revised and edited the manuscript. All authors have read and agreed to the published version of the manuscript.

Funding: We are thankful for the financial support by National Natural Science Foundation of China (31601264). The funder had no role in design, data collection and decision to publish or preparation of manuscript.

Acknowledgments: We highly acknowledge the help of technicians of the research station of institute of cotton research Chinese Academy of agricultural sciences.

Conflicts of Interest: The authors declare no conflict of interest.

References

1. Constable, G.A.; Bange, M.P. The yield potential of cotton (*Gossypium hirsutum* L.). *Field Crops Res.* **2015**, *182*, 98–106. [CrossRef]
2. Dai, J.; Dong, H. Field Crops Research Intensive cotton farming technologies in China: Achievements, challenges and countermeasures. *Field Crops Res.* **2014**, *155*, 99–110. [CrossRef]
3. Zhang, H.; Tian, W.; Zhao, J.; Jin, L.; Yang, J.; Liu, C.; Yang, Y.; Wu, S.; Wu, K.; Cui, J.; et al. Diverse genetic basis of field-evolved resistance to Bt cotton in cotton bollworm from China. *Proc. Natl. Acad. Sci. USA* **2012**, *109*, 10275–10280. [CrossRef] [PubMed]
4. Bednarz, C.W.; Nichols, R.L.; Brown, S.M. Plant density modifications of cotton within-boll yield components. *Crop Sci.* **2006**, *46*, 2076–2080. [CrossRef]
5. Zhi, X.Y.; Han, Y.C.; Li, Y.B.; Wang, G.P.; Du, W.L.; Li, X.X.; Mao, S.C.; Lu, F.E. Effects of plant density on cotton yield components and quality. *J. Integr. Agric.* **2016**, *15*, 1469–1479. [CrossRef]
6. Siebert, J.; Stewart, A. Influence of Plant Density on Cotton Response to Mepiquat Chloride Application. *Agron. J.* **2006**, *98*, 1634–1639. [CrossRef]
7. Bednarz, C.W.; Bridges, D.C.; Brown, S.M. Analysis of cotton yield stability across population densities. *Agron. J.* **2000**, *92*, 128–135. [CrossRef]
8. Dong, H.; Li, W.; Eneji, A.E.; Zhang, D. Nitrogen rate and plant density effects on yield and late-season leaf senescence of cotton raised on a saline field. *Field Crops Res.* **2012**, *126*, 137–144. [CrossRef]
9. Yang, G.; Tang, H.; Nie, Y.; Zhang, X. Responses of cotton growth, yield, and biomass to nitrogen split application ratio. *Eur. J. Agron.* **2011**, *35*, 164–170. [CrossRef]
10. Khan, A.; Kong, X.; Najeeb, U.; Zheng, J.; Kean, D.; Tan, Y.; Akhtar, K.; Munsif, F.; Zhou, R. Planting Density Induced Changes in Cotton Biomass Yield, Fiber Quality, and Phosphorus Distribution under Beta Growth Model. *Agronomy* **2019**, *9*, 500. [CrossRef]
11. Wingler, A.; Purdy, S.; MacLean, J.A.; Pourtau, N. The role of sugars in integrating environmental signals during the regulation of leaf senescence. *J. Exp. Bot.* **2006**, *57*, 391–399. [CrossRef]
12. Marcelis, L.F.M.; Heuvelink, E.; Baan Hofman-Eijer, L.R.; Den Bakker, J.; Xue, L.B. Flower and fruit abortion in sweet pepper in relation to source and sink strength. *J. Exp. Bot.* **2004**, *55*, 2261–2268. [CrossRef]
13. Khan, A.; Najeeb, U.; Wang, L.; Tan, D.K.Y.; Yang, G.; Munsif, F.; Ali, S.; Hafeez, A. Planting density and sowing date strongly influence growth and lint yield of cotton crops. *Field Crops Res.* **2017**, *209*, 129–135. [CrossRef]
14. Kaggwa-Asiimwe, R.; Andrade-Sanchez, P.; Wang, G. Plant architecture influences growth and yield response of upland cotton to population density. *Field Crops Res.* **2013**, *145*, 52–59. [CrossRef]
15. Mariscal, M.J.; Orgaz, F.; Villalobos, F.J. Modelling and measurement of radiation interception by olive canopies. *Agric. For. Meteorol.* **2000**, *100*, 183–197. [CrossRef]
16. Galanopoulou-Sendouka, S.; Sficas, A.G.; Fotiadis, N.A.; Gagianas, A.A.; Gerakis, P.A. Effect of Population Density, Planting Date, and Genotype on Plant Growth and Development of Cotton1. *Agron. J.* **1980**, *72*, 347. [CrossRef]
17. Sarlikioti, V.; De Visser, P.H.B.; Marcelis, L.F.M. Exploring the spatial distribution of light interception and photosynthesis of canopies by means of a functionalstructural plant model. *Ann. Bot.* **2011**, *107*, 875–883. [CrossRef] [PubMed]
18. Mao, L.; Zhang, L.; Zhao, X.; Liu, S.; van der Werf, W.; Zhang, S.; Spiertz, H.; Li, Z. Crop growth, light utilization and yield of relay intercropped cotton as affected by plant density and a plant growth regulator. *Field Crops Res.* **2014**, *155*, 67–76. [CrossRef]
19. Tetio-Kagho, F.; Gardner, F.P. Responses of Maize to Plant Population Density. I. Canopy Development, Light Relationships, and Vegetative Growth. *Agron. J.* **1988**, *80*, 930. [CrossRef]
20. Xu, W.; Liu, C.; Wang, K.; Xie, R.; Ming, B.; Wang, Y.; Zhang, G.; Liu, G.; Zhao, R.; Fan, P.; et al. Adjusting maize plant density to different climatic conditions across a large longitudinal distance in China. *Field Crops Res.* **2017**, *212*, 126–134. [CrossRef]

21. Srinivasan, V.; Kumar, P.; Long, S.P. Decreasing, not increasing, leaf area will raise crop yields under global atmospheric change. *Glob. Chang. Biol.* **2017**, *23*, 1626–1635. [CrossRef] [PubMed]
22. Flénet, F.; Kiniry, J.R.; Board, J.E.; Westgate, M.E.; Reicosky, D.C. Row spacing effects on light extinction coefficients of corn, sorghum, soybean, and sunflower. *Agron. J.* **1996**, *88*, 185–190. [CrossRef]
23. Chen, Z.; Niu, Y.; Zhao, R.; Han, C.; Han, H.; Luo, H. The combination of limited irrigation and high plant density optimizes canopy structure and improves the water use efficiency of cotton. *Agric. Water Manag.* **2019**, *218*, 139–148. [CrossRef]
24. Baytar, A.A.; Peynircioğlu, C.; Sezener, V.; Basal, H.; Frary, A.; Frary, A.; Doğanlar, S. Identification of stable QTLs for fiber quality and plant structure in Upland cotton (*G. hirsutum* L.) under drought stress. *Ind. Crops Prod.* **2018**, *124*, 776–786. [CrossRef]
25. Zhang, D.; Luo, Z.; Liu, S.; Li, W.; Dong, H. Effects of deficit irrigation and plant density on the growth, yield and fiber quality of irrigated cotton. *Field Crops Res.* **2016**, *197*, 1–9. [CrossRef]
26. Khan, N.; Han, Y.; Xing, F.; Feng, L.; Wang, Z.; Wang, G. Plant Density Influences Reproductive Growth, Lint Yield and Boll Spatial Distribution of Cotton. *Agronomy* **2019**, *10*, 14. [CrossRef]
27. Xia, Y.Q. A probe into planting density for cotton high yielding. *Xinjiang Agric. Sci.* **2008**, *45*, 70–71.
28. Dong, H.; Li, Z.; Tang, W.; Zhang, D. Evaluation of a Production System in China that Uses Reduced Plant Densities and Retention of Vegetation Branches. *J. Cotton Sci.* **2005**, *9*, 1–9.
29. Dong, H.Z.; Li, W.J.; Tang, W.; Li, Z.H.; Zhang, D.M. Crop/Stress Physiology Effects of Genotypes and Plant Density on Yield, Yield Components and Photosynthesis in Bt Transgenic Cotton. *J. Agron. Crop Sci.* **2006**, *139*, 132–139. [CrossRef]
30. Dong, H.; Li, W.; Tang, W.; Li, Z.; Zhang, D.; Niu, Y. Yield, quality and leaf senescence of cotton grown at varying planting dates and plant densities in the Yellow River Valley of China. *Field Crops Res.* **2006**, *98*, 106–115. [CrossRef]
31. Wang, R.Q.; Pang, C.Q.; Yang, G.Z. Effect of NPK fertilizer and density on cotton lint yield. *China Cott.* **2009**, *36*, 8–12.
32. Ali, M.; Ali, L.; Sattar, M. Response of seed cotton yield to various plant populations and planting methods. *J. Agric. Res.* **2010**, *48*, 163–169.
33. Mao, L.; Zhang, L.; Evers, J.B.; van der Werf, W.; Liu, S.; Zhang, S.; Wang, B.; Li, Z. Yield components and quality of intercropped cotton in response to mepiquat chloride and plant density. *Field Crops Res.* **2015**, *179*, 63–71. [CrossRef]
34. Tiwari, R.S.; Picchioni, G.A.; Steiner, R.L.; Jones, D.C.; Hughs, S.E.; Zhang, J. Genetic variation in salt tolerance at the seedling stage in an interspecific backcross inbred line population of cultivated tetraploid cotton. *Euphytica* **2013**, *194*, 1–11. [CrossRef]
35. Jones, M.A.; Wells, R. Dry matter allocation and fruiting patterns of cotton grown at two divergent plant populations. *Crop Sci.* **1997**, *37*, 797–802. [CrossRef]
36. Boquet, D. Cotton in Ultra-Narrow Row Spacing: Plant Density and Nitrogen Fertilizer Rates. *Agron. J.* **2005**, *97*, 279–287. [CrossRef]
37. Dai, J.; Li, W.; Tang, W.; Zhang, D.; Li, Z.; Lu, H.; Eneji, A.E.; Dong, H. Manipulation of dry matter accumulation and partitioning with plant density in relation to yield stability of cotton under intensive management. *Field Crops Res.* **2015**, *180*, 207–215. [CrossRef]
38. Brodrick, R.; Bange, M.P.; Milroy, S.P.; Hammer, G.L. Physiological determinants of high yielding ultra-narrow row cotton: Biomass accumulation and partitioning. *Field Crops Res.* **2012**, *134*, 122–129. [CrossRef]
39. Darawsheh, M.K.; Chachalis, D.; Aivalakis, G.; Khah, E.M. Cotton row spacing and plant density cropping systems. II. Effects on seedcotton yield, boll components and lint quality. *J. Food Agric. Environ.* **2009**, *7*, 262–265.
40. CRI (Cotton Research Institute Chinese Academy of Agricultural Sciences). *Cultivation of Cotton in China*; Shanghai Science and Technology Press: Shanghai, China, 2013.
41. Tsimba, R.; Edmeades, G.; Millner, J.; Kemp, P. The effect of planting date on maize grain yields and yield components. *Field Crops Res.* **2013**, *150*, 135–144. [CrossRef]

42. Bednarz, C.W.; Shurley, W.D.; Anthony, W.S.; Nichols, R. Yield, quality, and profitability of cotton produced at varying plant densities. *Agron. J.* **2005**, *97*, 235–240.

43. Ragsdale, P.I.; Smith, C.W. Germplasm potential for trait improvement in upland cotton: Diallel analysis of within-boll seed yield components. *Crop Sci.* **2007**, *47*, 1013–1017. [CrossRef]

 *agronomy*

Article

Phosphorus Application Improves the Cotton Yield by Enhancing Reproductive Organ Biomass and Nutrient Accumulation in Two Cotton Cultivars with Different Phosphorus Sensitivity

Babar Iqbal [1] , Fanxuan Kong [1], Inam Ullah [2], Saif Ali [1] , Huijie Li [1], Jiawei Wang [1], Wajid Ali Khattak [1] and Zhiguo Zhou [1,*]

[1] Key Laboratory of Crop Physiology and Ecology, Ministry of Agriculture, Nanjing Agricultural University, Nanjing 210095, China; agronomist19388@yahoo.com (B.I.); 2017101030@njau.edu.cn (F.K.); saifbhatti63@yahoo.com (S.A.); 2017201013@njau.edu.cn (H.L.); 2017101047@njau.edu.cn (J.W.); wajid_82@ymail.com (W.A.K.)

[2] Department of Agronomy, The University of Agriculture Peshawar-Pakistan; Peshawar 25000, Pakistan; inamullah72@yahoo.com

* Correspondence: giscott@njau.edu.cn; Tel.: +86-25-84396813; Fax: +86-25-84396813

Received: 30 November 2019; Accepted: 14 January 2020; Published: 21 January 2020

Abstract: Phosphorus (P) plays a pivotal role in cotton by enhancing the reproductive growth and yield formation. Cotton cultivars vary greatly in response to P availability, especially under P-deficient conditions. So, we hypothesized that the increasing P level promotes the reproductive growth in cotton cultivars varying with P sensitivity. For this, two cotton cultivars, Lu-54 (sensitive to low P) and Yuzaomian-9110 (tolerant to low P), in response to three different P levels (P0: 0 (control), P1: 100, and P2: 200 kg P_2O_5 ha^{-1}) were studied at 39, 52, 69, 83, and 99 days after transplanting during 2017 and 2018. The results revealed that the seed cotton yield was improved in P1 and P2 treatments by 23.9%–34.5% and 30.8%–52.3% in Lu-54, and 16.6%–25.6% and 20.6%–38.5% in Yuzaomian-9110 during 2017 and 2018, respectively. The accumulation of reproductive organ biomass was 21.0%–52.1% and 28.5%–56.8% higher in Lu-54 and 24.2%–56.8% and 34.8%–69.1% higher in Yuzaomian-9110 in P1 and P2 over the control, respectively. During the fast accumulation period, the average accumulation of N, P, K, and biomass across the years in P2 were recorded as 0.75, 0.6, 0.5, and 120.5 kg ha^{-1} d^{-1} in Lu-54, while they were 0.65, 0.5, 0.8, and 98.5 kg ha^{-1} d^{-1} in Yuzaomian-9110. Overall, a longer period, in terms of reproductive biomass accumulation, was recorded for Yuzaomian-9110 compared with Lu-54 in 2017 and vice versa across the 2018 growing season. The results suggested that increasing P rate improved yield, reproductive organ biomass, as well as nutrient accumulation in both cotton cultivars. However, low P-sensitive cultivar (Lu-54) was more responsive to P application compared with low P-tolerant cultivar.

Keywords: cotton; phosphorus sensitivity; phosphorus; reproductive organ biomass; nutrients accumulation; yield

---

1. Introduction

Cotton (*Gossypium hirsutum* L.) is globally considered as one of the most important commercial crops. Being a cash crop, it is grown worldwide for the purpose of oil, lint, and feed for animals [1]. About 30 million hectares of fertile land is engaged in cotton cultivation in almost 70 different countries of the world [2]. China is the world's largest cotton producer and consumer [3,4], with an average lint yield of 14.38 g m^{-2} during 2013 followed by the US, India, and Pakistan [5]. Due to its indeterminate growth habit, cotton exhibits morphological adaptation, such as modifying the canopy arrangement with phosphorus (P) application [6]. The morphological adaptations in terms of light capture, sink to source relationship, and photoassimilates distribution, are the main reasons of enhancing seed cotton yield [7,8]. For the last two decades, cotton yield per unit area remained stagnant in spite of the introduction of new high yielding cultivars [7,8]. Utilization of mineral fertilizers is sought as an effective strategy to improve soil nutrient and boost cotton yield.

The use of increased fertilization has influenced the crop production over the last several years because of its effect on soil nutritional status and fertility characteristics [9]. Phosphate fertilizer is central to crop productivity with a higher P requirement that is not in competition with soil P, especially at sub-optimal level compared with other nutrients [10]. Phosphorus is very important in crop production after nitrogen (N); however, its resources are limited worldwide [11]. The P application improves root architecture by increasing length, width, and diameter of root. Hence, P uptake by the plants is predominantly controlled by the availability and acquisition of P [12–14]. Therefore, the P deficiency inhibits cotton growth and development by declining the biomass accumulation, leading to lower seed cotton yield [15].

Crop growth requires nutrients' availability and constant supply throughout the growing season. Cotton yield responds positively to the availability of the nutrients, especially P [16]. The availability of P in the soil affects the nutrient accumulation and dry matter accumulation in the cotton plant parts [17]. The existence of varietal differences also fluctuated the accumulation of nutrient and biomass with greater tendency towards the vegetative organ [18]. Improving the cotton cultivars with better nutrient management to obtain higher economic yield is of great importance to minimize the environmental impact of inorganic fertilizers.

The cotton cultivars respond differently to the P availability and results in the increase of seed cotton yield [19,20]. Different plant cultivars show genetic diversity in the utilization and absorption of P ratios. Cotton cultivars showing sensitivity to low P may increase plant performance by the application of more P as compared to cultivars with low P tolerance [21]. The stunted growth and low yield resulted by P deficits in cotton [16] have been reported, but data regarding cotton cultivars with different P sensitivity under different P rates are still lacking. Screening and using P-efficient cotton cultivars with better uptake of P can provide a base to increase P utilization in plants [22,23]. The current study is aimed at evaluating the response of P application, P, N, and K relationship due to P fluctuation, and to estimate P's role in accumulating different nutrients in the reproductive organ (RO) of the plant. The evaluation was carried out at different days after transplanting (DAT) with 39, 52, 69, 83, and 99 at Squaring "SQ", first bloom "FB", peak bloom "PB", boll setting "BS", and boll opening "BO" in two different cotton cultivars having different response to inorganic P (i.e., sensitive versus tolerant to low P) and different P rates. The objectives of the present study were to assess the effects of different P regimes on different P-efficient cotton cultivars in their yield, nutrient, biomass accumulation, and allocation in the RO of cotton.

2. Materials and Methods

2.1. Experimental Site and Field Conditions

A 2-year field experiment was conducted at Pailou Research Station (118°50′ E, 32°02′ N), Nanjing Agricultural University, Jiangsu, P.R. China. The soil of the experimental field was mixed, acidic clay, thermic, and typic Alfisols (Udalfs; FAO Luvisol). The pre-planting soil samples were collected from 0–20 and 20–40 cm depth. Soil properties (Table 1) of the collected samples were determined by following Yang et al. [24]. Weather data of mean monthly air temperature and rainfall during 2017 and 2018 are given in Figure 1.

Table 1. Soil fertility status of the experimental site during 2017 and 2018. Data of soil samples collected from topsoil (0–20 cm) and subsoil (20–40 cm) depths before planting.

		pH (H$_2$O)	Bulk Density (g cm^3)	TN (g kg^{-1})	EC (μS cm^{-1})	SOM (g kg^{-1})	AN (NH$_4$$^+$)(mg kg^{-1})	AN (NO$_3$$^{-1}$)(mg kg^{-1})	AP (mg kg^{-1})	AK (mg kg^{-1})	TP (g kg^{-1})	TK (g kg^{-1})
Year 2017	topsoil (0–20 cm)	6.84	1.35	1.03	170	18.0	61.21	11.32	16.91	139.4	0.61	5.81
	subsoil (20–40 cm)	7.45	1.29	0.95	185	17.7	59.32	11.21	16.72	135.6	0.48	4.87
Year 2018	topsoil (0–20 cm)	6.94	1.44	1.11	183	19.1	64.92	13.91	17.85	144.3	0.69	5.87
	subsoil (20–40 cm)	7.54	1.32	1.06	190	18.4	61.01	12.1	17.02	140.6	0.52	4.21

TN: Total nitrogen; EC: Electrical conductivity; SOM: Soil organic matter; AN: Available nitrogen; AP: Available phosphorus; AK: Available potassium; TP: Total phosphorus; TK: Total potassium.

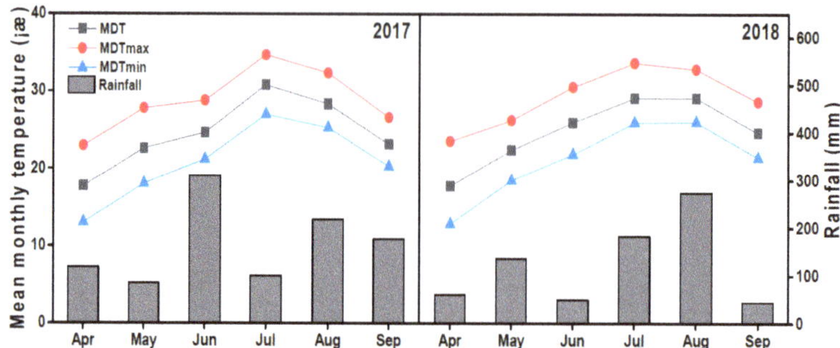

Figure 1. Mean monthly air temperature and rainfall during 2017 and 2018.

2.2. Experimental Design, Treatments, and Crop Management

The experiment was designed in a split-plot arrangement with two cotton cultivars (Lu-54: low-P-sensitive and Yuzaomian-9110: low-P-tolerant) [25] in main plots and three P rates (P0: 0, P1: 100, and P2: 200 kg P_2O_5 ha^{-1}) in subplots with three replications. Cotton seeds were sown in the middle of April which attained a three true leaf stage at 30 to 35 days. These seedlings were transplanted to the field on May 24, 2017 and 2018. Fertilizer, such as N and K, were applied at a rate of 225 kg ha^{-1} of each to every experimental plot. All of the K was applied at pre-planting while N was applied in four different splits as basal dose (20%), start of flowering (25%), bloom stage (40%), and end of flowering (15%), according to the stages described by Baker et al. [26]. Other field and plant management practices were adopted according to the local cotton production practices.

2.3. Soil and Plant Sampling

Before seedlings transplanting, soil samples (0–20 and 20–40 cm) were collected at three different locations from the field to make a composite sample from each plot. These sample were sealed and kept in an ice box immediately after collection. In the laboratory, the sample was divided into two equal parts. One portion was kept at –20 °C in a freezer and the other portion was kept outside in order to dry. The dried sample was meshed to make fine powder which was used for further analysis.

The plant samples were collected at five different times, i.e., squaring "SQ" (39 days after transplanting), first bloom "FB" (52 DAT), peak bloom "PB" (69 DAT), boll setting "BS" (83 DAT), and boll opening "BO" (99 DAT). Three plants from each plot were randomly collected and RO (bolls) were separated. The collected material was oven dried at 70 °C and dry matter was calculated on the base of per unit land area.

2.4. Nutrients and Biomass Accumulation

Total N, P, and K were determined by the H_2SO_4-H_2O_2 extraction method. The samples were weighed, put in a glass tube, and heated at 350 °C. H_2SO_4 was added followed by H_2O_2 in addition with pure water. Lastly, the solution was filtered, and filtrate was stored for further analysis.

Reproductive organ biomass was determined through destructive sampling by randomly selecting three cotton plants from each replication of treatments. Reproductive organs were separated from the plants and divided into three different parts (seed, bur, and lint). These parts were oven-dried first at 105 °C for 30 min and then at 80 °C until constant dry weight and expressed as kg ha^{-1}. The process of N, P, K, and biomass accumulation was described and calculated by the following logistics formulas [27].

$$Y = \frac{K}{1 + ae^{bt}} \tag{1}$$

where "t" denotes to the DAT, Y (g) represents the accumulation of K at t, K (g) signifies the higher accumulation biomass value for K, while "a" and "b" are constants.

From Equation (1), the following formulas can be derived;

$$t_1 = \frac{1}{b} \ln\left(2 + \sqrt{3}\right) \tag{2}$$

$$t_2 = \frac{1}{b} \ln\left(2 - \sqrt{3}\right) \tag{3}$$

$$t_m = -\ln\frac{a}{b} \tag{4}$$

The starting point denotes (t_1), ending point (t_2), and ($T = t_2 - t_1$) represents the difference of starting and ending time.

$$V_M = \frac{-bk}{4} \tag{5}$$

$$V_T = \frac{\Delta Y}{\Delta t} = (Y_2 - Y_1)(t_2 - t_1) \tag{6}$$

The fast accumulation period (FAP) can be explained as the period accumulating N, P, K, and biomass that starts and ends with an average speed of (V_T), average maximum speed (V_M) during FAP. Whereas Y_1 and Y_2 represent weight (N, P, K, and biomass) at t_1 and t_2, respectively, and can be calculated as above.

In each subplot, two rows were selected for seed cotton yield. Opened bolls were hand-picked from the two rows, seeds were removed to calculate seed yield, and expressed as kg ha^{-1}.

2.5. Data Analysis

Data were processed by using Microsoft Excel 2013. Statistical analysis was carried out by Statistix 8.1 (Analytical Software, Tallahassee, FL, USA). Mean difference between the treatments were separated by the least significant difference (LSD) test at the probability level of 0.05. Origin 9.1, Sigma plot 12.0 (Systat Software Inc., San Jose, CA, USA), and R 3.6.1 were employed to draw figures.

3. Results

3.1. Seed Cotton Yield

Phosphorus application expressively affected the seed cotton yield with a different trend in 2017 and 2018 (Figure 2). Seed cotton yield in 2017 was significantly higher than in 2018. The control treatment showed the lowest seed cotton yield compared with the other P application (Figure 2). Comparison of treatments showed that P1 and P2 increased the seed cotton yield by 23.9%–34.5% and 30.8%–52.3% for Lu-54 and 16.6%–25.6% and 20.6%–38.5% for Yuzaomian-9110 during 2017 and 2018, respectively.

3.2. Nitrogen Accumulation in Reproductive Organ

Nitrogen accumulation in RO of cotton plant represents a sigmoid curve with DAT. The early growth stage showed a quick average speed of accumulation of N and then gradually decreased in the later stage (Figure 3). The P application had a drastic effect on the accumulation of N content arrangement in the RO of cotton throughout the growing period after transplanting. Compared with the control, the final amount of N accumulation in P treatments (100 and 200 kg P$_2$O$_5$ ha^{-1}) increased by 24.0%–46.9% and 23.3%–47.1% in Lu-54 and 36.9%–67.9% and 21.8%–58.2% in Yuzaomian-9110 during 2017 and 2018, respectively. The RO N content was drastically increased with the increasing DAT and was highest at 83 DAT and continued to present the same accumulation at 99 DAT. During 2017, cotton plant accumulated comparatively more N in the RO than 2018. The difference of weather between the two years might be a big reason for change in N accumulation. Similarly, P2 showed higher N accumulated in the RO as compared with the control during 2017 and 2018, respectively (Figure 3).

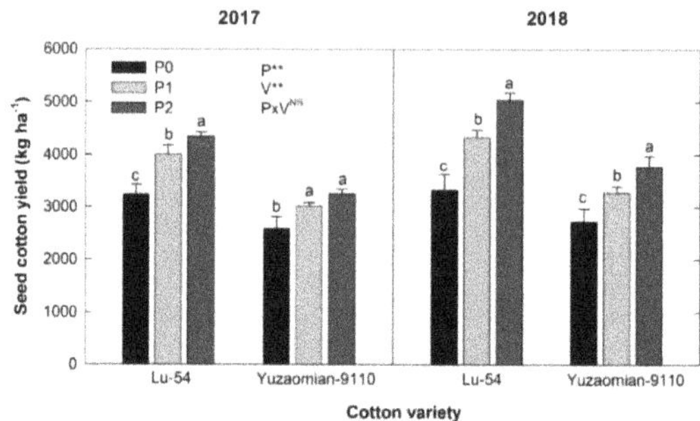

Figure 2. Seed cotton yield (kg ha^{-1}) as affected by phosphorus (P) levels during 2017 and 2018 in Lu-54 (low P sensitive) and Yuzaomian-9110 (low P tolerant). P0, P1, and P2 indicate P levels of 0, 100, and 200 kg P_2O_5 ha^{-1}, respectively. Whereas P, V, and P×V represent phosphorus, cultivar, and their interaction, respectively. While ** and NS show highly significant and non-significant, respectively. Vertical bars on the columns indicate standard errors of the mean (n = 3). The bars showing different letters are statistically significant at $p \le 0.05$.

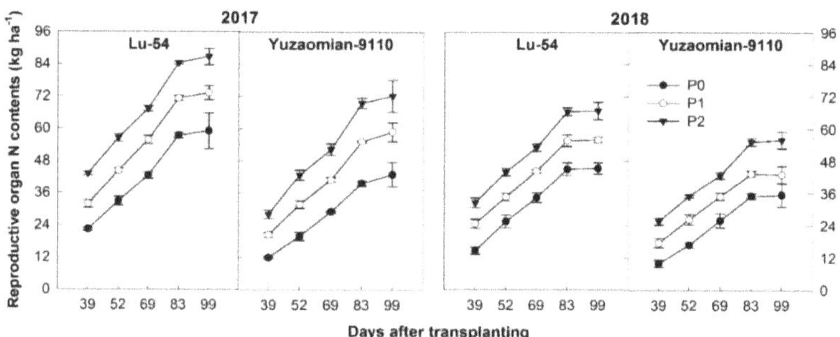

Figure 3. The accumulation of total N contents (kg ha^{-1}) in reproductive organ of cotton plant for Lu-54 and Yuzaomian-9110 in 2017 and 2018. The treatments P0, P1, and P2 represent the P levels, i.e., 0, 100, and 200 kg P_2O_5 ha^{-1}. The data are the means of three replications ± standard error.

3.3. Simulation of N Accumulation

Calculating overall data and fitting these into Formula (1) resulted in overall determination coefficients being high ($R^2 \ge 0.9383$**, $p < 0.01$, Table 2), concluding that logistic function was the most suitable for accumulation of N. The K in the P application treatments (P1 and P2) over control was increased by 14.0%–69.8% and 27.2%–66.8% for Lu-54 and 40.2%–39.0% and 20.3%–65.4% for Yuzaomian-9110 during 2017 and 2018, respectively. Calculations based on Formulas (2)–(6) indicated that the T was increased consistently by increasing P application in both cultivars during 2017 and 2018. The V_T and V_M calculated in the P application treatments were higher compared with the control treatment for both cultivars in both growing seasons. Based on the calculations from Formulas (2)–(6), it was indicated that initiation and termination day of 77 and 57, fast accumulation period of N accumulation in the RO was 26 and 22 and 104 and 79 DAT for Lu-54 and 65 and 55 d FAP for N accumulation was 35 and 27 and 101 and 82 DAT for Yuzaomian-9110, during 2017 and 2018, respectively. The fluctuating trend was observed across two years in different treatments with an average maximum speed (V_M) recorded higher than the average speed (V_T). The P0 began FAP last at 39 DAT and terminated at 119 DAT, stayed for 80 d with the maximum average speed of 0.8 kg ha^{-1}

d^{-1}, which was the same as the average speed of P1 with different starting and termination DAT in Lu-54 during 2017; however, the P2 began the fast accumulation period the earliest at 15 DAT and terminated at 84 DAT, and stayed for 69 d with the V_M of 0.6 kg ha^{-1} d^{-1} during 2018, which was recorded lower than in 2017. The same trend was observed for Yuzaomian-9110 with a lower average speed than Lu-54 during 2017 and 2018.

3.4. Phosphorus Accumulation in Reproductive Organ

Accumulation of P in RO of cotton cultivars showed a sigmoid curve with DAT (Figure 4). The average speed of P accumulation in RO increased in the early growth stages and then slowed a little at the late period. The P supply had a significant effect on the P accumulation pattern throughout the growth period after transplanting of cotton. Compared with the control, the amount of accumulated P in P application treatment (P1 and P2) increased by 43.5%–91.0% and 30.4%–65.3% for Lu-54 and 50.3%–108.7% and 52.9%–100.6% for Yuzaomian-9110 during 2017 and 2018, respectively. The P content in the RO of cotton was drastically increased with the increase of DAT and recorded the highest at 99 DAT in 2017; however, in 2018, it was recorded the highest at 83 DAT and continued with same the trend at 99 DAT. The P accumulated in the RO were recorded higher in 2017 compared with 2018. The P2 resulted in higher P in RO of cotton as compared to P0 in both cropping seasons (Figure 4).

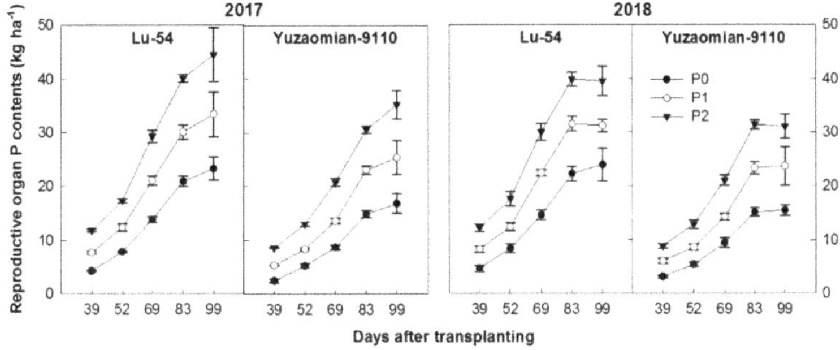

Figure 4. The accumulation of total P contents (kg ha^{-1}) in reproductive organ of cotton plant for Lu-54 and Yuzaomian-9110 in 2017 and 2018. The treatments P0, P1, and P2 represent the P levels, i.e., 0, 100 and 200 P$_2$O$_5$ kg ha^{-1}. The data are the means of three replications ± standard error.

3.5. Simulation of P Accumulation

The determination coefficient derived from the data fitting into Formula (1) indicated that it was low ($R^2 \geq 0.9495**$, $p < 0.05$, Table 3) and high ($R^2 \geq 0.9956**$, $p < 0.01$, Table 3), and confirmed the best indication of logistic function to define P accumulation in RO of cotton. The K in the P application-supply treatments was increased by 70.3%–202.7% and 14.2%–56.9% for Lu-54 and 181.1%–145.7% and 185.7%–136.1% for Yuzaomian 9110 in comparison with the control during 2017 and 2018, respectively. The calculation regarding T from Formulas (2)–(6) grounded in Table 3 confirmed the increased T with increasing P levels in 2017 and 2018 for both cultivars. The V_T and V_M were recorded as higher in the P application treatments compared with the control for both the cultivars (Table 3). On an average basis across all the treatments, the accumulation of P which initiated and terminated the fast accumulation period of 56 d were at 48 DAT and 104 DAT for Lu-54, while at 61 d, they were at 58 DAT and 119 DAT for Yuzaomian-9110 during 2017, respectively. During 2018, P accumulation starting and finishing the 46 d were 40 DAT and 86 DAT for Lu-54, while at 54 d, they were at 48 DAT and 103 DAT for Yuzaomian-9110 during 2017 and 2018, respectively. The increasing P application increased the V_M compared to V_T across the two years. The increased P application linearly increased the V_M and V_T during 2017 and 2018, respectively (Table 3).

Table 2. Dynamic models and the maximum accumulation rate of N (V_M), average speed of N accumulation (V_T), start time of the N rapid-accumulation period (t_1), termination time of the N rapid-accumulation period (t_2), duration of the N rapid-accumulation period (T), and occurrence time of maximum accumulation rate of N (t_m) in reproductive organ of cotton plants for Lu-54 and Yuzaomian-9110 in 2017 and 2018.

Year	Variety	P Levels (kg ha^{-1})	Regression Equation	R^2	Fast Accumulation Period		T (d)	V_T (kg ha^{-1} d^{-1})	Fastest Accumulation Point	
					t_1 (d)	t_2 (d)			V_m (kg ha^{-1} d^{-1})	t_m (d)
2017	Lu-54	P0	$W = 76.9/(1 + 13.41e^{-0.0327t})$	0.9856 **	39.1	119.5	80.4	0.8	0.87	79.3
		P1	$W = 87.72/(1 + 9.56e^{-0.0435t})$	0.9685 **	21.6	82.2	60.6	0.8	0.95	51.9
		P2	$W = 130.65/(1 + 6.28e^{-0.0290t})$	0.9766 **	17.9	108.7	90.8	0.9	0.95	63.3
		Mean			26.2	103.5	77.3	0.8	0.92	64.8
	Yuzaomian-9110	P0	$W = 65.54/(1 + 23.43e^{-0.0425t})$	0.9946 **	43.3	105.3	62.0	0.6	0.70	74.3
		P1	$W = 91.91/(1 + 14.97e^{-0.0374t})$	0.9806 **	37.2	107.7	70.5	0.8	0.86	72.5
		P2	$W = 91.11/(1 + 10.86e^{-0.0414t})$	0.9555 **	25.8	89.4	63.6	0.8	0.94	57.6
		Mean			35.4	100.8	65.4	0.7	0.83	68.1
2018	Lu-54	P0	$W = 48.89/(1 + 25.35e^{-0.0614t})$	0.9787 **	31.2	74.2	42.9	0.5	0.75	52.7
		P1	$W = 62.17/(1 + 8.54e^{-0.0446t})$	0.9720 **	18.6	77.6	59.1	0.5	0.69	48.1
		P2	$W = 81.53/(1 + 6.55e^{-0.0381t})$	0.9387 **	14.8	83.9	69.1	0.6	0.78	49.3
		Mean			21.5	78.6	57.0	0.6	0.74	50.0
	Yuzaomian-9110	P0	$W = 42.35/(1 + 37.06e^{-0.0616t})$	0.9901 **	37.2	80.0	42.7	0.5	0.65	58.6
		P1	$W = 50.94/(1 + 14.06e^{-0.0513t})$	0.9681 **	25.9	77.2	51.4	0.5	0.65	51.5
		P2	$W = 70.04/(1 + 6.83e^{-0.0366t})$	0.9383 **	16.5	88.4	71.9	0.5	0.64	52.5
		Mean			26.5	81.9	55.3	0.5	0.65	54.2

The treatments P0, P1, and P2 represent the P levels, i.e., 0, 100, and 200 kg P_2O_5 ha^{-1}. ** represents highly significant.

Table 3. Dynamic models and the maximum accumulation rate of P (V_M), average speed of P accumulation (V_T), start time of the P rapid-accumulation period (t_1), termination time of the P rapid-accumulation period (t_2), duration of the P rapid-accumulation period (T), and occurrence time of maximum accumulation rate of P (t_m) in reproductive organ of cotton plants for Lu-54 and Yuzaomian-9110 in 2017 and 2018.

Year	Variety	P Levels (kg ha^{-1})	Regression Equation	R^2	Fast Accumulation Period		T (d)	V_T (kg ha^{-1} d^{-1})	Fastest Accumulation Point	
					t_1 (d)	t_2 (d)			V_m (kg ha^{-1} d^{-1})	t_m (d)
2017	Lu-54	P0	$W = 27.92/(1+56.75e^{-0.0596t})$	0.9955 **	45.7	89.8	44.2	0.3	0.42	67.7
		P1	$W = 47.57/(1+33.64e^{-0.0481t})$	0.9956 **	45.8	100.7	54.9	0.5	0.57	73.3
		P2	$W = 84.52/(1+28.50e^{-0.0391t})$	0.9937 **	52.2	119.8	67.6	0.7	0.82	86.0
		Mean			47.9	103.5	55.6	0.5	0.60	75.7
	Yuzaomian-9110	P0	$W = 24.70/(1+68.20e^{-0.0538t})$	0.9636 **	54.0	102.9	48.9	0.3	0.33	78.4
		P1	$W = 69.44/(1+18.79e^{-0.0383t})$	0.9798 **	42.2	110.9	68.7	0.4	0.67	76.5
		P2	$W = 60.69/(1+28.89e^{-0.0401t})$	0.9933 **	51.0	116.7	65.7	0.5	0.61	83.9
		Mean			58.2	119.3	61.1	0.4	0.54	88.8
2018	Lu-54	P0	$W = 30.99/(1+53.32e^{-0.0569t})$	0.9846 **	46.7	92.9	46.2	0.4	0.44	69.8
		P1	$W = 35.41/(1+40.24e^{-0.0616t})$	0.9687 **	38.6	81.3	42.7	0.4	0.55	60.0
		P2	$W = 48.60/(1+27.26e^{-0.0555t})$	0.9737 **	35.8	83.2	47.4	0.5	0.68	59.5
		Mean			40.4	85.8	45.5	0.4	0.55	63.1
	Yuzaomian-9110	P0	$W = 18.44/(1+59.68e^{-0.0627t})$	0.9716 **	44.2	86.2	42.0	0.2	0.29	65.2
		P1	$W = 52.69/(1+28.02e^{-0.0388t})$	0.9495 **	51.9	119.7	67.8	0.4	0.51	85.8
		P2	$W = 43.54/(1+28.22e^{-0.0494t})$	0.9606 **	41.0	94.3	53.4	0.5	0.54	67.7
		Mean			48.1	102.5	54.4	0.4	0.45	75.3

The treatments P0, P1, and P2 represent the P levels, i.e., 0, 100, and 200 kg P_2O_5 ha^{-1}. ** represents highly significant.

3.6. Potassium Accumulation in Reproductive Organ

Cotton plant potassium uptake increased with the advancement in growth stages following a normal exponential growth curve with DAT (Figure 5). A momentous effect of phosphorus application on K status of plant at different growth stages after transplanting of cotton seedlings was observed. Compared with the control treatment, P1 and P2 increased K in RO by 27.0%–46.0% and 23.3%–47.1% in Lu-54 and 21.0%–40.9% and 21.8%–58.2% in Yuzaomian-9110 during 2017 and 2018, respectively. Along with the plant age, P application increased the accumulation of K until 83 DAT and then remained constant until 99 DAT.

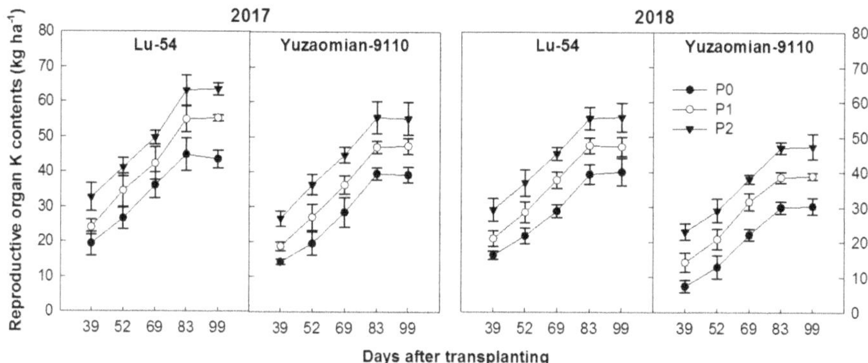

Figure 5. The accumulation of total K contents (kg ha^{-1}) in reproductive organ of cotton plant for Lu-54 and Yuzaomian-9110 in 2017 and 2018. The treatments P0, P1, and P2 represent the P levels, i.e., 0, 100, and 200 kg P$_2$O$_5$ ha^{-1}. The data are the means of three replications ± standard error.

3.7. Simulation of K

The experimental data were used in a Formula (1) to determine the simulation of K accumulation with cotton growth stages. The accumulation of K as a normal sigmoid curve were the best fitted in the logistic function since all the determination coefficients were high ($R^2 \geq 0.9408$ **, $p < 0.01$, Table 4). The K in the P1 and P2 treatments were increased by 26.5%–63.5% and 27.2%–66.8% in Lu-54 and 20.7%–43.1% and 20.3%–65.4% during 2017 and 2018 respectively, over the control. Data obtained from Formulas (2)–(6), exhibited that the starting and ending day of K uptake for cultivars and P application showed that T was higher with the application of P in 2017 while it was inconsistent across 2018 in Lu-54 and Yuzaomian-9110, respectively (Table 4). The V_T and V_M were relatively higher in P application over the control for both cultivars. Data obtained from Formulas (2)–(6) revealed that initiation and termination day of 58 d FAP for accumulation of K was 17 and 75 DAT averaged across the treatments for Lu-54, while 60 d was 15 DAT and 75 DAT averaged for Yuzaomian-9110 in 2017, respectively. In 2018, 66 d FAP of K was 19 and 85 DAT for Lu-54 while 48 d was 29 DAT and 77 DAT for 48 d was 29 DAT and 77 DAT for Yuzaomian-9110, respectively. The V_M was recorded as higher than V_T with different inclination between the treatments during two cropping seasons. The longer period (80–73 d) was observed in Lu-54, while Yuzaomian-9110 showed a relatively shorter period (60–69 d) at P2 treatment compared with the P0 and P1 during 2017 and 2018, respectively.

Table 4. Dynamic models and the maximum accumulation rate of K (V_M), average speed of K accumulation (V_T), start time of the K rapid-accumulation period (t_1), termination time of the K rapid-accumulation period (t_2), duration of the K rapid-accumulation period (T), and occurrence time of maximum accumulation rate of K (t_m) in reproductive organ of cotton plants for Lu-54 and Yuzaomian-9110 in 2017 and 2018.

Year	Variety	P Levels (kg ha^{-1})	Regression Equation	R^2	Fast Accumulation Period		T (d)	V_T (kg ha^{-1} d^{-1})	Fastest Accumulation Point	
					t_1 (d)	t_2 (d)			V_m (kg ha^{-1} d^{-1})	t_m (d)
2017	Lu-54	P0	$W = 46.13/(1 + 13.92e^{-0.0577t})$	0.9665 **	22.8	68.4	45.6	0.42	0.67	45.6
		P1	$W = 58.34/(1 + 12.09e^{-0.0546t})$	0.9865 **	21.5	69.8	48.2	0.52	0.80	45.6
		P2	$W = 75.43/(1 + 4.75e^{-0.0329t})$	0.9568 **	7.3	87.2	79.9	0.52	0.62	47.3
		Mean			17.2	75.1	57.9	0.49	0.69	46.2
	Yuzaomian-9110	P0	$W = 43.21/(1 + 21.62e^{-0.0602t})$	0.9758 **	29.1	72.9	43.7	0.45	0.65	51.0
		P1	$W = 52.13/(1 + 16.21e^{-0.0561t})$	0.9761 **	26.2	73.1	46.9	0.51	0.73	49.6
		P2	$W = 61.81/(1 + 7.21e^{-0.0437t})$	0.9698 **	15.0	75.2	60.2	0.49	0.68	45.1
		Mean			23.4	73.7	50.3	0.48	0.69	48.6
2018	Lu-54	P0	$W = 53.17/(1 + 9.58e^{-0.0364t})$	0.9615 **	25.9	98.3	72.3	0.44	0.48	62.1
		P1	$W = 52.57/(1 + 11.49e^{-0.0510t})$	0.9555 **	22.0	73.7	51.6	0.48	0.67	47.9
		P2	$W = 66.28/(1 + 5.25e^{-0.0359t})$	0.9408 **	9.5	82.9	73.3	0.48	0.59	46.2
		Mean			19.2	84.9	65.8	0.46	0.58	52.0
	Yuzaomian-9110	P0	$W = 33.54/(1 + 55.41e^{-0.0692t})$	0.9764 **	39.0	77.0	38.0	0.41	0.58	58.0
		P1	$W = 40.41/(1 + 34.96e^{-0.0724t})$	0.9702 **	30.9	67.3	36.4	0.44	0.73	49.1
		P2	$W = 59.06/(1 + 7.37e^{-0.0385t})$	0.9442 **	17.7	86.1	68.5	0.47	0.57	51.9
		Mean			29.2	76.8	47.6	0.44	0.63	53.0

The treatments P0, P1, and P2 represent the P levels, i.e., 0, 100, and 200 kg P_2O_5 ha^{-1}. ** represents highly significant.

3.8. Biomass Accumulation in Reproductive Organ

Cotton plant biomass accumulation followed a sigmoid curve with DAT (Figure 6). The average speed of biomass accumulation was the fastest at DAT and then recorded a similar accumulation with 83 DAT and 99 DAT in 2017, while in 2018, a higher biomass accumulation was recorded with the passage of the growth period. The significant effect by P application was observed on the accumulation of RO biomass throughout the growth period. The P1 and P2 improved RO biomass over the control, by 21.0%–52.1% and 28.5%–56.8% for Lu-54 and 24.2%–56.8% and 38.5%–69.1% for Yuzaomian-9110 during 2017 and 2018, respectively.

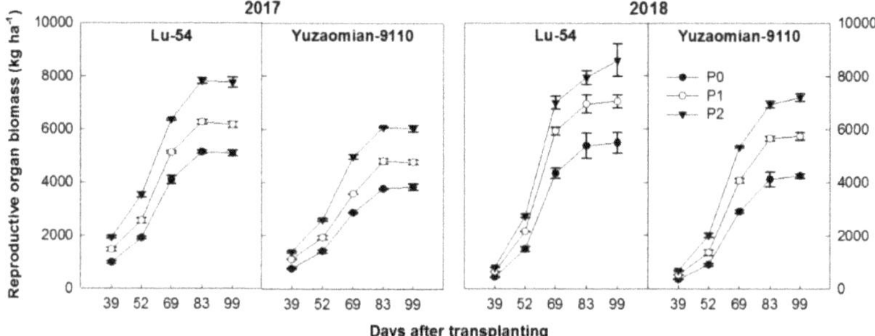

Figure 6. The accumulation of biomass in reproductive organ of cotton plant for Lu-54 and Yuzaomian-9110 in 2017 and 2018. The treatments P0, P1, and P2 represent the P levels, i.e., 0, 100, and 200 kg P_2O_5 ha^{-1}. The data are the means of three replications ± standard error.

3.9. Simulation of Biomass Accumulation

The experimental data fitting into Formula (1) showed a normal logistic organism growth pattern and revealed that all the coefficients were high ($R^2 \geq 0.9481^{**}$, $p < 0.01$, Table 5). The value of K in the P1 and P2 treatments over the control were improved by 11.3%–44.0% and 32.4%–53.3% for Lu-54 and 20.3%–60.9% and 38.5%–67.5% for Yuzaomian-9110 during 2017 and 2018, respectively. Total biomass accumulation calculated by putting the data into Formulas (2)–(6) grounded in Table 5 indicated that the T was unreliable across the years in both of the cultivars. The V_T and V_M were recorded as relatively higher in the P treatments compared with the control in both of the cultivars. Averaged across the treatments, the initiation and termination of the fast accumulation point of 34 d was 55 DAT and 88 DAT for Lu-54 and for 38 d was 55 DAT and 93 DAT for Yuzaomian-9110 in 2017, respectively. In 2018, RO biomass initiated and terminated the 28 d fast accumulation period at 65 DAT and 93 DAT for Lu-54, and the 27 d fast accumulation period of biomass at 63 DAT and 90 DAT for Yuzaomian-9110, respectively. Maximum and average speeds were apart from each other with higher maximum speed, followed by average speed averaged across the treatments in two cropping seasons. The P application increased the maximum and average speed by increasing the P level. The Lu-54 performance is better than Yuzaomian-9110 during 2017 and 2018, respectively (Table 5).

3.10. Allocation of N, P, K, and Biomass in Cotton

The total N, P, K, and biomass accumulation increased with the increase in P application in both cotton cultivars (Table 6). The accumulation of percentage of N in RO increased with the P treatments while cotton cultivars showed non-significant differences during 2017 and 2018. There were no understandable differences in the P and K accumulation of the RO recorded, neither with the application of P nor in cotton cultivars. Similarly, P treatment had no significant effect on the accumulation of biomass in the RO in 2017. However, in 2018, the cultivars showed a significant effect with a higher biomass accumulation in Lu-54 compared with Yuzaomian-9110 in P treatments over the control.

Table 5. Dynamic models and the maximum accumulation rate of biomass (V_M), average speed of biomass accumulation (V_T), start time of the biomass rapid-accumulation period (t_1), termination time of the biomass rapid-accumulation period (t_2), duration of the biomass rapid-accumulation period (T), and occurrence time of maximum accumulation rate of biomass (t_m) in reproductive organ of cotton plants for Lu-54 and Yuzaomian-9110 in 2017 and 2018.

Year	Variety	P Levels (kg ha⁻¹)	Regression Equation	R^2	Fast Accumulation Period		T (d)	V_T (kg ha⁻¹ d⁻¹)	Fastest Accumulation Point	
					t_1 (d)	t_2 (d)			V_m (kg ha⁻¹ d⁻¹)	t_m (d)
2017	Lu-54	P0	$W = 6029/(1 + 267e^{-0.072t})$	0.9616 **	59.4	96.0	36.6	77.3	108.41	77.7
		P1	$W = 6710/(1 + 393e^{-0.085t})$	0.9481 **	54.7	85.6	30.9	86.6	142.95	70.1
		P2	$W = 8681/(1 + 187e^{-0.075t})$	0.9796 **	52.3	87.4	35.2	105.9	162.60	69.8
		Mean			55.4	89.7	34.2	89.9	137.99	72.6
	Yuzaomian-9110	P0	$W = 4427/(1 + 234e^{-0.073t})$	0.9879 **	57.0	93.2	36.2	56.2	80.46	75.1
		P1	$W = 5328/(1 + 127e^{-0.067t})$	0.9925 **	52.7	92.0	39.3	63.4	89.26	72.3
		P2	$W = 7125/(1 + 176e^{-0.071t})$	0.9803 **	55.4	93.2	37.8	88.0	124.00	74.3
		Mean			55.0	92.8	37.8	69.2	97.91	73.9
2018	Lu-54	P0	$W = 6173/(1 + 2159e^{-0.098t})$	0.9540 **	64.9	91.8	26.9	89.6	151.16	78.4
		P1	$W = 8174/(1 + 1183e^{-0.088t})$	0.9756 **	65.4	95.3	29.9	115.3	179.90	80.4
		P2	$W = 9462/(1 + 1323e^{-0.092t})$	0.9729 **	63.6	92.1	28.5	134.5	218.39	77.9
		Mean			64.7	93.1	28.4	113.1	183.15	78.9
	Yuzaomian-9110	P0	$W = 4529/(1 + 1785e^{-0.095t})$	0.9692 **	64.7	92.3	27.6	65.2	108.07	78.5
		P1	$W = 6271/(1 + 1545e^{-0.10t})$	0.9892 **	63.4	91.1	27.7	89.7	149.02	77.3
		P2	$W = 7588/(1 + 1993e^{-0.10t})$	0.9935 **	61.9	87.8	25.9	109.1	192.56	74.9
		Mean			63.3	90.4	27.1	88.0	149.88	76.9

The treatments P0, P1, and P2 represent the P levels, i.e., 0, 100, and 200 kg P_2O_5 ha⁻¹. ** represents highly significant.

Table 6. Total N, P, K, and biomass percentage allocation in reproductive organ for Lu-54 and Yuzaomian-9110 in 2017 and 2018.

P Levels (kg ha^{-1})	2017				2018			
	RNR[1] (%)	RPR (%)	RKR (%)	RBR (%)	RNR (%)	RPR (%)	RKR (%)	RBR (%)
Lu-54								
P0	50.6b[2]	51.2	44.6	59.0	49.9	54.8	47.5	66.2abc
P1	51.2ab	51.6	46.1	56.5	50.4	53.7	46.9	69.9a
P2	52.9a	53.2	46.4	58.4	51.4	54.2	47.2	67.6ab
Mean	51.6	52.0	45.7	58.0	50.6	54.2	47.2	67.9
Yuzaomian-9110								
P0	49.3b	52.9	46.4	58.4	51.4	54.0	46.1	61.6d
P1	51.6ab	53.8	46.6	56.7	50.1	54.6	46.9	63.6cd
P2	53.4a	54.8	46.3	56.0	52.1	53.7	47.5	64.0bcd
Mean	51.4	53.8	46.4	57.1	51.2	54.1	46.8	63.1
Significance of factors								
Variety	NS[3]	NS	NS	NS	NS	NS	NS	**
P levels	**	NS	**	NS	NS	NS	NS	NS
Variety × P level	NS	NS	NS	NS	NS	NS	NS	NS

[1] RNR: the percentage of total N in reproductive organ; RPR: the percentage of total P in reproductive organ; RKR: the percentage of total K in reproductive organ; RBR: the percentage of biomass in reproductive organ. [2] For each cultivar, values followed by a different letter within the same column are significantly different at $p < 0.05$ probability level. Each value represents the mean of three replications. [3] NS means non-significant. ** Indicates significant difference at $p < 0.01$ probability level. The treatments P0, P1, and P2 represent the P levels i.e., 0, 100, and 200 P_2O_5 kg ha^{-1}.

3.11. Inter-Trait Relationship

The RO biomass, N, and K content showed a positively linear relationship with the P content accumulated in the organ. However, the slope of the fitted line varied for biomass ($R^2 = 0.9064$–0.9021), N content ($R^2 = 0.9430$–0.9474), and K content ($R^2 = 0.9081$–0.9261) in 2017 and 2018 respectively, as presented in Figure 7a–c. Similarly, the N, P, K, and biomass accumulation in the RO showed a positive correlation between each other (Figure 8), but the strength of this correlation varied between 2017 and 2018 cropping seasons.

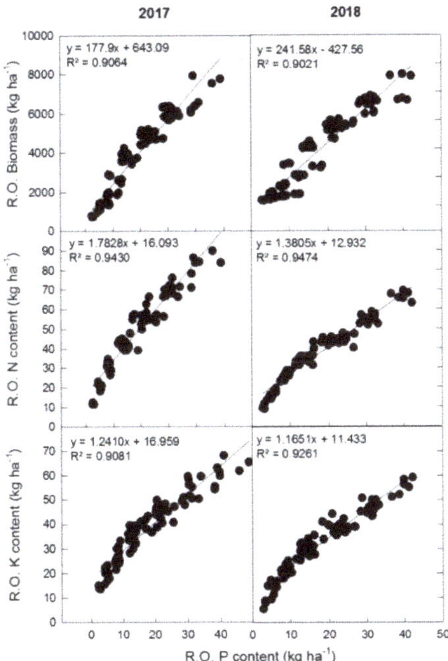

Figure 7. Relationship between accumulation of N, P, K, and biomass during 2017 and 2018 cropping seasons.

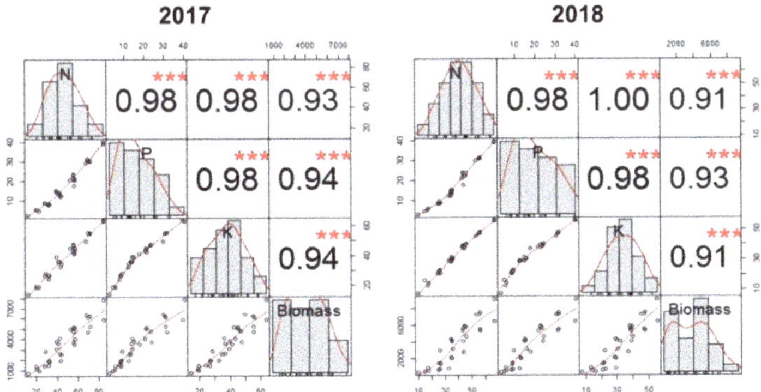

Figure 8. Correlation between N, P, K, and reproductive organ biomass during 2017 and 2018 cropping seasons. The bars and line, shown with observation, are regarded as correlation coefficients as 1 ($R = 1$). Other line graphs and correlation coefficients represent the comparative values to $R = 1$. Where *** showed very highly significant ($p \leq 0.001$).

4. Discussion

The present study was based on the effect of different P application rates on biomass and nutrients accumulation in RO of two cotton cultivars differing in P sensitivity. In the present study, P application had a significant effect on nutrients (N, P, and K) and biomass accumulation in the RO. Both cotton cultivars significantly varied for the aforementioned parameters during 2017 and 2018. This change in nutrients' accumulation might be due to a good balance between N, P, and K contents in the soil, resulting in a higher uptake of the nutrients by the roots. Similarly, higher P application treatments might increase the available P contents in the soil, leading to a well-established transport of water and nutrients towards the areal parts of the plant. These results are in line with the findings of Widowati et al. [28], who found that the proper and balanced inorganic nutrients promoted the fertility of the soil which enhanced the crop productivity. The availability of proper nutrients to the cotton plants might help in leaf growth and development followed by its proper functions with the application of P1 and P2 over the control. Li et al. [23] also reported similar results that a good nutrients balance improved the crop growth. In the present study, cotton cultivars showed a different behavior regarding nutrient uptake that might be due to their sensitivity and tolerance to P deficiency. The patterns of N, P, and K were similar with a slight change between two growing seasons. The increase in the nutrients at early 39 DAT, which reached to a peak at 83 DAT, was consistent. The possible reason for the trend might be that the plant demand for NPK was higher at later stages. These results are in a good agreement with the findings of Yang et al. [29] and Hu et al. [30]. The difference of NPK in the RO of different treatments was due to different levels of P supply from the soil to plants. Furthermore, higher P content in plant RO helped to improve the N and K, and a similar phenomenon was recorded in the present study (Figure 7b,c). The improved N, P, and K contents are also linked to the RO biomass; hence, it enhanced the boll biomass in the present study (Figure 7a). However, two cotton growing seasons (2017 and 2018) showed a slight variation that might be due to the difference of weather conditions during both years (Figure 1). Alterations in the nutrient application and weather conditions changed the nutrient accumulation, speed, initiation, and termination time during the fast accumulation period (FAP) and significantly affected the yield of cotton [31].

In cotton, RO biomass is considered as one of the most important factors directly influencing the yield. In the present study, the cultivar differences in P utilization showed a significant change in the RO biomass accumulation, which might create the yield difference (Figure 2). The RO biomass attained a peak at 83 DAT with a clear difference in P application treatments (Figure 6). After 83 DAT, the RO biomass did not change because at this stage, bolls reached near to a maturation stage. This biomass increase with the increase in P application might be attributed to the higher P supply that favored the nutrients (N and K) and photosynthates translocation toward the reproductive part rather than vegetative organs. These finding are in accord with the conclusion of Cao et al. [32] and Stewart et al. [33], that P application restricts the photoassimilates' translocation from vegetative organs and is increasingly directed towards the RO. Singh et al. [34] concluded that the accumulation of dry matter was higher due to better photosynthesis, which resulted in a better supply of photosynthates and assimilates to the fruit. In the present study, higher biomass accumulation was recorded in the RO of the cotton in Lu-54 during 2018 compared to Yuzaomian-9110 (Figure 6). These findings were similar to previous results presented by Wang et al. [35], that dry matter accumulation was different in different cotton cultivars with different partitioning to vegetative and reproductive parts. The main reason behind this phenomenon was that the P application might increase the N, P, and K contents in leaf and other parts which influence the nutrients' availability to the plants, establishing a better source–sink relationship. A well-established source–sink balance guarantees the better RO development, resulting in an increase in final yield [36].

The accumulation of nutrients in cotton plant parts and its distribution to the sink play a vital role in the production of seed cotton yield. In the present study, nutrients (N, P, and K) and biomass in cotton RO significantly increased with the P application and advancement of growth stages at higher speed compared to the control (Table 2, Table 3, Table 4, and Table 5). These results might be due to the proper P nutrient application which increased the N, P, K, and biomass accumulation average maximum speed (V_M) and average speed (V_T) during the FAP. A similar result was stated by Tung et al. [31] and Yang et al. [37], that the fluctuation in the nutrient application and rate along with management practices in cotton affected the accumulation of different nutrients and its speed to the RO.

The N and K are the highly absorbable nutrients in soil compared with the P [38]. However, P availability influenced the accumulation of N and K in the plants by increasing the pool of ATP [39].

The current study revealed a higher accumulation of N and K in the RO of cotton in the P2 treatment in Lu-54 during 2017 compared to the cultivar Yuzaomian-9110 (Figures 3 and 5). Similarly, the average speed (V_T) at FAP was observed to be higher at P2 treatment in Lu-54 during 2017 compared to other treatments (Tables 2 and 4). These results might be due to the availability and accumulation of N and K from the soil in the presence of a higher application of P compared to the control, which enhanced the performance of cultivar to respond better to P fertilizer. Similarly, the results presented by Yang et al. [37] revealed that increases in yield occurred with higher inorganic fertilizer, which could be interlinked with the extended growing season by promoting the growth period of cotton by getting more P compared to the control. The current results are also in line with the finding of Gebaly et al. [40], who reported that cotton accumulated a high rate of K from the soil. Plenty of studies have reported that cotton cultivars showed different variations in partitioning of dry matter between vegetative and reproductive parts [27,41–43]. Cultivars with a greater dry matter partitioning ability to their reproductive parts are usually considered as efficient in increasing seed cotton yield.

The P accumulation in cotton resulted higher in Lu-54 at P2 treatment at 99 DAT during 2017 compared to the control (Figure 4), which improved the biomass of RO and resulted in a higher yield. This might be due to the better availability of P and mobility of photoassimilates to RO of cotton with the advancement of growth stages. Similarly, increased P uptake improved the plants' growth and resulted in increased yield in other crops, such as maize [44]. In the current study, P application promoted the average speed (V_T) and average maximum speed (V_M) of P accumulated in Lu-54 during 2017 compared to Yuzaomian-9110 in the RO of cotton (Table 3), which might be due to the better availability of P nutrients and unexpected weather condition to prolong the growing period of cotton. Similarly, uptake of P and its distribution to the cotton boll resulted in a higher increase in boll biomass (Table 5), with higher V_T and V_M [45]. However, Gill et al. [46] reported that P deficiency promoted dry matter accumulation in cotton, which might be due to different soil and climatic conditions as well as difference of cultivar P sensitivity. Nitrogen and K concentration in the RO remained higher compared with the control because these nutrients are easily taken up by the plants. Similar results were reported by Blom-Zandstra et al. [47] and Hsiao et al. [48], showing that the concentration of N and K remained the same due to their ionic solution forms that played a pivotal role in osmotic adjustment. Further, P-deficient conditions resulted in a stagnate growth and increased the P level in petiole sap. On other hand, P application improved nutrients' concentration and biomass in the plants, leading to an increase in reproductive growth, such as improving flowerings [49] and to extend the growing season of cotton [36,37]. The P deficiency strongly constrains the translocation of nutrients; however, it mainly affects the biomass because cotton growth responded highly to ambient weather, better supply of nutrients, and availability of water in the soil [50,51]. This leads to a better equilibrium among vegetative and reproductive growth by supplying a higher P to establish a good and balanced source–sink relationship.

Nutrients are still limiting factors in agro-ecosystem, although human activities increased N, P, K [52–54], and micronutrients [55,56]. In the present study, the boll biomass, by the combined effect of all the three major nutrients, increased with the combination of N, P, and K, especially under P1 and P2 application in Lu-54 compared to Yuzaomian-9110 during the 2018 growing season (Figure 6). Initiation and termination days for biomass accumulation changed with the application of P, while V_M and V_T were increased with the application of P2 compared to the control in both the cultivars (Table 5). The percent allocation of N and K varied significantly with the application of P; however, the cultivar was not significantly affected in 2017 (Table 6). Contrary to this, for the P and biomass allocation in 2017, while N, P, and K allocation in 2018 did not vary significantly in both cultivars and P application, the biomass, however, was significantly affected by cultivars and showed no effect with the P application (Table 6). These results might be due to the synergistic effect of nutrients which increased the biomass and alternately affected the yield of cotton. Opposite to the P deficiency, the higher rate of P application negatively affected the biomass of RO in cotton [23]. This reduction is due to a disturbance in the source–sink relationship under P imbalance [57]. Marshner and Rengel [58] reported synergistic effects of major nutrients on cotton boll biomass. On the other hand, Zahreddine et al. [59] reported that the exceeded nutrients' concentration from optimal, reduced the biomass production. Extension in the growing period occurred due to a loss of fruiting structures, which resulted in leaf expansion, photosynthetic capacity, and increased carbon assimilation [60–62]. The photoassimilates were accumulated with the continued growth period, with a better supply, rate, and quantity of nutrient accumulation with differed growth periods [63]. Crops benefited from the fast N uptake during initial to peak flowering stage. Cotton yield and growth were strongly affected by uptake of nutrients and its speed at FAP [64]. Many reports, including Jenkins et al. [65], assumed that the change from vegetative to reproductive growth was earlier in new cultivars. Various studies also reported a different response of cultivars in biomass production and dry matter partitioning between vegetative and reproductive growth [63,66,67]. Due to indeterminate growth of cotton, the nutrients' deposition varies with the time and with the advancement of growth stages. A higher and faster accumulation of nutrients in the RO of cotton with higher biomass accumulation occurs. Nutrient accumulation and biomass production in RO of cotton data would be useful for growers to take management decisions for maximizing the seed cotton yield.

5. Conclusions

The present study evaluated the response of two cotton cultivars with different P sensitivity to P application. The results concluded that: higher accumulation of nutrients was reported in Lu-54 compared to Yuzaomian-9110. Higher total N, P, and K accumulation was found with the incorporation of high P level at 83 DAT, and then remained constant at 99 DAT in both used cultivars during 2017 and 2018, respectively. Higher nutrient accumulation was recorded in total N and K forms while P remained lower in all parts of the reproductive organ of cotton cultivars. Total biomass accumulation was recorded higher at 83 DAT and remained stable until 99 DAT in response to P application. Cotton cultivar Lu-54 produced higher reproductive organ biomass as compared with the Yuzaomian-9110. The increment in seed cotton yield was associated with higher N, P, K, and RO biomass. Conclusively, 200 kg P_2O_5 ha^{-1} (P2) application could be a sufficient level with a boll opening stage at 99 DAT for the better yield performance in Lu-54 compared to Yuzaomian-9110.

Author Contributions: Data curation and writing—original draft, B.I.; investigation, F.K., H.L., J.W., and W.A.K.; writing—review and editing, I.U. and S.A.; conceptualization and supervision, Z.Z. All authors have read and agreed to the published version of the manuscript.

Funding: This research was funded by the National Key Research and Development Program of China (2017YFD0201900), China Agriculture Research System (CARS-15-14) and Jiangsu Collaborative Innovation Center for Modern Crop Production (JCIC-MCP). The APC was funded by Z.Z.

Conflicts of Interest: There are no conflicts of interest to declare.

References

1.   Constable, G.A.; Bange, M.P. The yield potential of cotton (*Gossypium hirsutum* L.). *Field Crops Res.* **2015**, *182*, 98–106. [CrossRef]

2.   Saleem, M.; Maqsood, M.; Javaid, A.; Hassan, M.; Khaliq, T. Optimum irrigation and integrated nutrition improves the crop growth and net assimilation rate of cotton (*Gossipium hirsutum* L.). *Pak. J. Bot.* **2010**, *42*, 3659–3669.

3.   Mao, S.C. *Cotton Farming in China Shanghai*; Shanghai Scientific and Technical Press: Shanghai, China, 2013; pp. 66–91. (In Chinese)

4.   Dai, J.; Dong, H. Intensive cotton farming technologies in China: Achievements, challenges and countermeasures. *Field Crops Res.* **2014**, *155*, 99–110. [CrossRef]

5.   USDA. United States Department of Agriculture 2013 Cotton: World Markets and Trade. Available online: https://www.fas.usda.gov/data/cotton-world-markets-and-trade (accessed on 14 October 2016).

6.   Sarkar, B.; Patra, A.K.; Purakayastha, T.J. Transgenic Bt-Cotton Affects Enzyme Activity and Nutrient Availability in a Sub-Tropical Inceptisol. *J. Agron. Crop Sci.* **2008**, *194*, 289–296. [CrossRef]

7. Yang, G.; Jiao, L.X.; Chun, N.Y.; Long, Z.X. Effects of Plant Density on Yield and canopy micro environment in hybrid cotton. *J. Integr. Agric.* **2014**, *13*, 2154–2163. [CrossRef]

8. Wang, Y.; Li, X.L. Progress in adaptive mechanisms of different genotypic plants to low phosphorus stress. *Eco-Agr. Res.* **2000**, *8*, 34–36.

9. Geisseler, D.; Scow, K.M. Long-term effects of mineral fertilizers on soil microorganisms—A review. *Soil Biol. Biochem.* **2014**, *75*, 54–63. [CrossRef]

10. Parfitt, R.L.; Baisden, W.T.; Elliott, A.H. Phosphorus inputs and outputs for New Zealand in 2001 at national and regional scales. *J. R. Soc. N. Z.* **2008**, *38*, 37–50. [CrossRef]

11. Gilbert, N. The disappearing nutrient. *Nature* **2009**, *461*, 716–718. [CrossRef] [PubMed]

12. Lynch, J.P. Root phenes for enhanced soil exploration and phosphorus acquisition: Tools for future crops. *Plant Physiol.* **2011**, *156*, 1041–1049. [CrossRef]

13. Mai, W.; Xue, X.; Gu, F.; Rong, Y.; Tian, C. Can optimization of phosphorus input lead to high productivity and high phosphorus use efficiency of cotton through maximization of root/mycorrhizal efficiency in phosphorus acquisition? *Field Crops Res.* **2018**, *216*, 100–108. [CrossRef]

14. Vance, C.P.; Uhde-Stone, C.; Allan, D.L. Phosphorus acquisition and use: critical adaptations by plants for securing a nonrenewable resource. *New Phytol.* **2003**, *157*, 423–447. [CrossRef]

15. Singh, V.; Pallaghy, C.K.; Singh, D. Phosphorus nutrition and tolerance of cotton to water stress: I. Seed cotton yield and leaf morphology. *Field Crops Res.* **2006**, *96*, 191–198. [CrossRef]

16. Girma, K.; Teal, R.K.; Freeman, K.W.; Boman, R.K.; Raun, W.R. Cotton lint yield and quality as affected by applications of N, P, and K fertilizers. *J. Cotton Sci.* **2007**, *11*, 12–19.

17. Roy, R.N. Integrated plant nutrition systems—Conceptual overview. In *Proceedings of symposium on "Integrated plant nutrition management"*; National Fertilizer Development Center: Islamabad, Pakistan, 2000; pp. 45–48.

18. Unruh, B.L.; Silvertooth, J.C. Comparison between upland and a Pima cotton cultivars II. Nutrient uptake and partitioning. *Agron. J.* **1996**, *88*, 589–595. [CrossRef]

19. Sawan, Z.M. Mineral fertilizers and plant growth retardants: Its effects on cottonseed yield; its quality and contents. *Cogent Biol.* **2018**, *4*, 1459010. [CrossRef]

20. Gebaly, S.G.; El-Gabiery, A.E. Response of cotton Giza 86 to foliar application of phosphorus and mepiquat chloride under fertile soil condition. *J. Agric. Res.* **2012**, *90*, 191–205.

21. Wang, J.; Chen, Y.; Wang, P.; Li, Y.S.; Wang, G.; Liu, P.; Khan, A. Leaf gas exchange, phosphorus uptake, growth and yield responses of cotton cultivars to different phosphorus rates. *Photosynthetica* **2018**, *56*, 1414–1421. [CrossRef]

22. Luo, H.H.; Zhang, Y.L.; Zhang, W.F. Effects of water stress and rewatering on photosynthesis, root activity, and yield of cotton with drip irrigation under mulch. *Photosynthetica* **2016**, *54*, 65–73. [CrossRef]

23. Li, H.G.; Huang, G.; Meng, Q.; Ma, L.; Yuan, L.X.; Wang, F.; Zhang, W.; Cui, Z.L.; Shen, J.B.; Chen, X.P.; et al. Integrated soil and plant phosphorus management for crop and environment in China. A review. *Plant Soil* **2011**, *349*, 157–167. [CrossRef]

24. Yang, H.; Meng, Y.; Chen, B.; Zhang, X.; Wang, Y.; Zhao, W. How integrated management strategies promote protein quality of cotton embryos: High levels of soil available N, N assimilation and protein accumulation rate. *Front. Plant Sci.* **2016**, *7*, 1118. [CrossRef] [PubMed]

25. Li, H.J.; Wang, J.; Saif, A.; Babar, I.; He, Z.; Shanshan, W.; Binglin, C.; Zhiguo, Z. Agronomic traits at the seedling stage, yield and fiber quality in two cotton (*Gossypium hirsutum* L.) cultivars in response to phosphorus deficiency. *Soil Sci. Plant Nutr.* **2019**. [CrossRef]

26. Baker, D.A.; Young, D.L.; Huggins, D.R.; Pan, W.L. Economically optimal nitrogen fertilization for yield and protein in hard red spring wheat. *Agron. J.* **2004**, *96*, 116–123. [CrossRef]

27. Yang, G.; Tang, H.; Nie, Y.; Zhang, X. Responses of cotton growth, yield, and biomass to nitrogen split application ratio. *Eur. J. Agron.* **2011**, *35*, 164–170. [CrossRef]

28. Widowati, I.; Utomo, W.; Guritno, B.; Soehono, L. The effect of biochar on the growth and N fertilizer requirement of maize (Zea mays L) in greenhouse experiment. *J. Agric. Sci.* **2012**, *4*, 255–262.

29. Yang, M.H.; Jiang, Y.J.; Nie, W.L.; Zheng, D.M. Effect of phosphorus application on accumulating dynamic change of nitrogen, phosphorus and potassium nutrient of the hybrid cotton. *Xinjiang Agric. Sci.* **2010**, *47*, 1172–1177. (In Chinese)

30. Hu, W.; Nan, J.; Jiashuo, Y.; Yali, M.; Youhua, W.; Binglin, C.; Wenqing, Z.; Derrick, M.O.; Zhiguo, Z. Potassium (K) supply affects K accumulation and photosynthetic physiology in two cotton (*Gossypium hirsutum* L.) cultivars with different K sensitivities. *Field Crops Res.* **2016**, *196*, 51–63. [CrossRef]

31. Tung, S.A.; Huang, Y.; Hafeez, A.; Ali, S.; Khan, A.; Souliyanonh, B.; Song, X.; Liu, A.; Guozheng, Y. Mepiquat chloride effects on cotton yield and biomass accumulation under late sowing and high density. *Field Crops Res.* **2018**, *215*, 59–65. [CrossRef]

32. Cao, Y.; He, J.; Yan, Y.; Feng, Z.; Wang, Q.; Zhang, Y. Effect of water deficit on cotton characteristics. *China Cotton* **2003**, *30*, 29–30, (In Chinese with English Abstract).

33. Stewart, W.M.; Reiter, J.S.; Krieg, D.R. Cotton response to multiple application of phosphorus fertilizer. *Better Crops Plant Food.* **2005**, *89*, 18–20.

34. Singh, S.K.; Reddy, V.R.; Fleisher, D.H.; Timlin, D.J. Growth, nutrient dynamics, and efficiency responses to carbon dioxide and phosphorus nutrition in soybean. *J. Plant Interact.* **2014**, *9*, 838–849. [CrossRef]

35. Wang, S.; Han, X.; Yan, J.; Xiao, L.I.; Qiao, Y. Impact of phosphorus deficiency stress on root morphology, nitrogen concentration and phosphorus accumulation of soybean (*Glycine max* L.). *China J. Soil Sci.* **2010**, *41*, 644–650.

36. Saleem, M.F.; Shakeel, A.; Bilal, M.F.; Shahid, M.Q.; Anjum, S.A. Effect of different phosphorus levels on earliness and yield of cotton cultivars. *Soil Environ.* **2010**, *29*, 128–135.

37. Yang, G.; Tang, H.; Tong, J.; Nie, Y.; Zhang, X. Effect of fertilization frequency on cotton yield and biomass accumulation. *Field Crops Res.* **2012**, *125*, 161–166. [CrossRef]

38. Zhu, J.; Li, M.; Whelan, M. Phosphorus activators contribute to legacy phosphorus availability in agricultural soils: A review. *Sci. Total Environ.* **2018**, *612*, 512–537. [CrossRef]

39. Gautam, P.; Gustafson, D.M.; Wicks, Z. Phosphorus concentration, uptake and dry matter yield of corn hybrids. *World J. Agric. Sci.* **2011**, *7*, 418–424.

40. Gebaly, S.G. Physiological effects of potassium forms and methods of application on cotton variety Giza 80. *Egypt. J. Agric. Res.* **2012**, *90*, 1633–1646.

41. Meredith, W.R., Jr.; Wells, R. Potential for increasing cotton yields through enhanced partitioning to reproductive structures. *Crop Sci.* **1989**, *29*, 636–639. [CrossRef]

42. Unruh, B.L.; Silvertooth, J.C. Comparison between upland and a Pima cotton cultivar. I. Growth and yield. *Agron. J.* **1996**, *88*, 583–589. [CrossRef]

43. Mao, L.; Zhang, L.; Zhao, X.; Liu, S.; van der Werf, W.; Zhang, S.; Li, Z. Crop growth, light utilization and yield of relay intercropped cotton as affected by plant density and a plant growth regulator. *Field Crops Res.* **2014**, *155*, 67–76. [CrossRef]

44. Ma, Q.H.; Zhang, F.S.; Rengel, Z.; Shen, J.B. Localized application of NH_4^+-N plus P at the seedling and later growth stages enhances nutrient uptake and maize yield by inducing lateral root proliferation. *Plant Soil* **2013**, *372*, 65–80. [CrossRef]

45. Wang, X.X.; Liu, S.; Zhang, S.; Li, H.; Maimaitiaili, B.; Feng, G.; Rengel, Z. Localized ammonium and phosphorus fertilization can improve cotton lint yield by decreasing rhizosphere soil pH and salinity. *Field Crops Res.* **2018**, *217*, 75–81. [CrossRef]

46. Gill, M.A.; Sabir, M.; Ashraf, S.; Rahmatullah, A.T. Effect of P-stress on growth, phosphorus uptake and utilization efficiency of different cotton cultivars. *Pak. J. Agric. Sci.* **2005**, *42*, 42–47.

47. Blom-Zandstra, M.; Lampe, J.E.M. The effect of chloride and sulpate salts on nitrate content in lettuce plants (*Lactuca satica* L.). *J. Plant Nutr.* **1983**, *6*, 611–628. [CrossRef]

48. Hsiao, T.C.; Lauchli, L. Role of potassium in plant water relations. In *Advances in Plant Nutrition*; Tinker, P.B., Lauchli, A., Eds.; Praeger: New York, NY, USA, 1986; Volume 2, pp. 281–382.

49. Biles, S.P.; Cothren, J.T. Flowering and yield response of cotton to application of mepiquat chloride and PGR-IV. *Crop Sci.* **2001**, *41*, 1834–1837. [CrossRef]

50. Xu, X.; Taylor, H.M. Increase in drought resistance of cotton seedlings treated with mepiquat chloride. *Agron. J.* **1992**, *84*, 569–574. [CrossRef]

51. Song, C.J.; Ma, K.M.; Qu, L.Y.; Liu, Y.; Xu, X.L.; Fu, B.J.; Zhong, J.F. Interactive effects of water, nitrogen and phosphorus on the growth, biomass partitioning and water-use efficiency of Bauhinia faberi seedlings. *J. Arid Environ.* **2010**, *74*, 1003–1012. [CrossRef]

52. Sardans, J.; Rodà, F.; Peñuelas, J. Effects of water and a nutrient pulse supply on Rosmarinus officinalis growth, nutrient content and flowering in the field. *Environ. Exp. Bot.* **2005**, *53*, 1–11. [CrossRef]

53. Sardans, J.; Peñuelas, J.; Rodà, F. Changes in nutrient status, retranslocation and use efficiency in young post-fire regeneration Pinus halepensis in response to sudden N and P input, irrigation and removal of competing vegetation. *Trees* **2005**, *19*, 233–250. [CrossRef]

54. Makhdum, M.I.; Pervez, H.; Ashraf, M. Dry matter accumulation and partitioning in cotton (*Gossypium hirsutum* L.) as influenced by potassium fertilization. *Biol. Fertil. Soils* **2007**, *43*, 295–301. [CrossRef]

55. Kidron, G.J.; Zilberman, A. Low cotton yield is associated with micronutrient deficiency in West Africa. *Agron. J.* **2019**, *111*, 1977–1984. [CrossRef]

56. Lambrecht, I.; Vanlauwe, B.; Maertens, M. Integrated soil fertility management: From concept to practice in eastern DR Congo. *Int. J. Agric. Sustain.* **2016**, *14*, 100–118. [CrossRef]

57. Rosolem, C.A.; Oosterhuis, D.M.; Souza, F.S.D. Cotton response to mepiquat chloride and temperature. *Sci. Agric.* **2013**, *70*, 82–87. [CrossRef]

58. Marshner, P.; Rengel, Z. Plant-Soil Relationships. In *Mineral Nutrition, Yield and Source-Sink Relationships*, 3rd ed.; Academic Press: Orlando, FL, USA, 2012.

59. Zahreddine, H.G.; Struve, D.K.; Talhouk, S.N. Growth and nutrient partitioning of containerized *Cercis siliquastrum* L. under two fertilizer regimes. *Sci. Hortic.* **2007**, *112*, 80–88. [CrossRef]

60. Dong, H.Z.; Tang, W.; Li, W.J.; Li, Z.H.; Niu, Y.H.; Zhang, D.M. Yield, leaf senescence, and Cry1Ac expression in response to removal of early fruiting branches in transgenic Bt cotton. *Agric. Sci. China* **2008**, *7*, 692–702. [CrossRef]

61. Lu, B.Q.; Downes, S.; Wilson, L.; Gregg, P.; Knight, K.; Kauter, G.; McCorkell, B. Yield, development and quality response of dual-toxin Bt-cotton to manual simulation of damage by Helicoverpa spp. in Australia. *Crop Prot.* **2012**, *41*, 24–29. [CrossRef]

62. Wells, R. Leaf pigment and canopy photosynthetic response to early flower removal in cotton. *Crop Sci.* **2001**, *41*, 1522–1529. [CrossRef]

63. Gao, Y.; Lynch, J.P. Reduced crown root number improves water acquisition under water deficit stress in maize (*Zea mays* L.). *J. Exp. Bot.* **2016**, *67*, 4545–4557. [CrossRef]

64. Bange, M.P.; Milroy, S.P.; Thongbai, P. Growth and yield of cotton in response to waterlogging. *Field Crops Res.* **2004**, *88*, 129–142. [CrossRef]

65. Jenkins, J.N.; McCarty, J.C., Jr.; Parrott, W.L. Effectiveness of fruiting sites in cotton: Yield. *Crop Sci.* **1990**, *32*, 365–369. [CrossRef]

66. Ma, Q.; Rengel, Z.; Rose, T. The effectiveness of deep placement of fertilisers is determined by crop species and edaphic conditions in Mediterranean-type environments: A review. *Soil Res.* **2009**, *47*, 19–32. [CrossRef]

67. Chen, B.L.; Sheng, J.D.; Jiang, P.A.; Liu, Y.G. Effect of applying different forms and rates of phosphoric fertilizer on phosphorus efficiency and cotton yield. *Cotton Sci.* **2010**, *22*, 49–56.

 *agronomy*

Article

Using Digestate and Biochar as Fertilizers to Improve Processing Tomato Production Sustainability

Domenico Ronga [1,2,*] , Federica Caradonia [1], Mario Parisi [3], Guido Bezzi [4], Bruno Parisi [5], Giulio Allesina [6], Simone Pedrazzi [6] and Enrico Francia [1]

[1] Centre BIOGEST-SITEIA, Department of Life Sciences, University of Modena and Reggio Emilia, 42122 Reggio Emilia (RE), Italy; federica.caradonia@unimore.it (F.C.); enrico.francia@unimore.it (E.F.)

[2] CREA Research Centre for Animal Production and Aquaculture, 26900 Lodi (LO), Italy

[3] CREA Research Centre for Vegetable and Ornamental Crops, 84098 Pontecagnano Faiano (SA), Italy; mario.parisi@crea.gov.it

[4] CIB—Consorzio italiano biogas e gassificazione, 26900 Lodi (LO), Italy; guido.bezzi@gmail.com

[5] CREA Research Centre for Cereal and Industrial Crops, 40128 Bologna (BO), Italy; bruno.parisi@crea.gov.it

[6] Department of Engineering 'Enzo Ferrari', University of Modena and Reggio Emilia, 41125 Modena (MO), Italy; giulio.allesina@unimore.it (G.A.); simone.pedrazzi@unimore.it (S.P.)

* Correspondence: domenico.ronga@unimore.it; Tel.: +39-0522-522003

Received: 14 December 2019; Accepted: 15 January 2020; Published: 17 January 2020

Abstract: The principal goal of the organic farming system (OFS) is to develop enterprises that are sustainable and harmonious with the environment. Unfortunately, the OFS yields fewer products *per* land than the non-organic farming system in many agricultural products. The objective of our study was to assess the effects of digestate and biochar fertilizers on yield and fruit quality of processing tomato produced under the OFS. The experiment was carried out in Po Valley, during the 2017 and 2018 growing seasons. Liquid digestate (LD), LD + biochar (LD + BC) and pelleted digestate (PD) were evaluated and compared to biochar (BC) application and unfertilized control. The results showed that plants fertilized with LD + BC recorded the maximum marketable yield (72 t ha^{-1}), followed by BC (67 t ha^{-1}), PD (64 t ha^{-1}) and LD (59 t ha^{-1}); while the lowest production (47 t ha^{-1}) was recorded in unfertilized plants. Over the two cropping seasons, LD + BC, BC, PD, and LD, increased fruit number *per* plant (+15%), fruit weight (+24%), Brix t ha^{-1} (+41%) and reduced Bostwick index (−16%), if compared to the untreated control. Considering the overall agronomic performances, digestate and biochar can be useful options for increasing yield and quality of processing tomato production in the OFS. Hence, these fertilizers can be assessed in future research both on other crops and farming systems.

Keywords: organic farming system; yield; pH; soluble solid content; Bostwick viscosity

1. Introduction

Processing tomato (*Solanum lycopersicum* L.) is a globally important cash crop, grown under different environments and input regimes. In 2019, worldwide production was estimated at ~37 million tones [1].

In the last 20 years, agriculture challenge is to provide enough and nutritious food for the growing population, minimizing its environmental impact in order to meet the sustainable development goals [2]. The organic farming system (OFS) can be an alternative approach to improve agricultural sustainability compared to the conventional one. OFS emphasizes rotating crops, adopts animal and green manure or compost to fertilize the crops, managing abiotic and biotic stress naturally, and improving biodiversity, soil and water conservations [3].

Reviews and meta-analyses revealed that the OFS has greater soil carbon content and less soil erosion compared with conventional systems [4–6]. Different works also reported that the agrosystem biodiversity is improved in the OFS [5,7,8]. In addition, synthetic pesticides and fertilizers are not allowed, and there is a reduction of nitrate leaching and greenhouse gas emissions in comparison with the conventional farming system [3,8,9]. However, since the OFS has lower land-use efficiency than the conventional system, these positive effects are less pronounced and in sometime reversed when expressed *per* unit product [9,10]. Seufert et al. [11] found wheat and vegetables to be the lowest, yielding organic crops, 37% and 33% less than conventional systems, respectively. In addition, when crop production depends only on green manure crops, the crop yield might be further reduced [12]. Among vegetables processing tomato when cultivated in the OFS showed marketable yield reduction of ~50% compared to conventional systems [13–17].

Cavigelli et al. [18] reported that nitrogen availability and weed control are the two main factors influencing crop yield in the OFS. Particularly, nitrogen is one of the main essential nutrients for tomato growth [19]. Scholberg et al. [20] reported that nitrogen deficiency can reduce tomato leaf area index, biomass production and fruit yield within a range from 60% to 70%.

Digestate is a by-product of the anaerobic digestion coming from the biogas plant production. Digestate mainly derives from the digestion of different biomasses such as energy crops (e.g., corn silage, triticale silage, etc.), vegetable by-products and manure. The solid fraction of the digestate is rich in minerals (like nitrogen and phosphorus) and organic matter [21,22]. Therefore, digestate could be interesting as a sustainable fertilizer for crop production. Ronga et al. [23–25] suggested the use of digestate as innovative fertilizer and growing media for the production of basil, peppermint, baby leaf lettuce and grapevine in soilless cropping systems. Other researches highlighted improving in quality and yield of digestate-fertilized crops. In fact, Barzee et al. [19] reported that tomatoes fertilized with digestate had higher soluble solids contents then synthetically fertilized one, and Šimon et al. [26] reported increases in grain yield in *Triticum aestivum* L. to respect the untreated control.

Biochar (BC) is considered an inorganic carbon-rich matrix obtained from organic material in the total or partial absence of oxygen at temperatures below 700 °C [27]. Biochar may enhance the growth performance and yield of crops, modifying the chemical properties of soil [28]. Changes in soil properties can make available some mineral nutrients and improve microbial activity [28]. Xu et al. [29] stated that nitrogen leaching was reduced after the application of biochar. Contradictory reports on the effectiveness of biochar on crop production are found in the literature. Indeed, squash yield was increased by biochar applications in the OFS [30], while Gonzaga et al. [31] reported that maize biomass and its nutrient uptake was not improved by the application of pinewood chip biochar. Moreover, the same authors highlighted that the increase in soil pH may result in potentially greater nitrogen losses than unfertilized control. Finally, Hol et al. [32] suggested that biota from biochar-amended soil was less beneficial for legume plant growth and flowering was delayed.

Few pieces of research are focused on the use of digestate and biochar in the OFS and no one has yet assessed the effectiveness of these two products applied together in the OFS. In light of the revision of the European organic regulation (CE 889/2008) that in addition to the recent inclusion of the digestate could allow also the use of biochar as fertilizers, is fundamental to provide useful information both to farmers and policymakers. Hence, the objective of the present study was to assess the effectiveness of different digestate fertilizers, biochar, and digestate + biochar on yield and fruit technological characteristics (pH, soluble solid contents, Bostwick viscosity) of processing tomato under the OFS.

2. Materials and Methods

2.1. Field Experiments

The trial was carried out, during a two-year period (2017–2018), in an organic farm located in Po Valley (44°41′17.9″ N; 10°34′00.2″ E and altitude of 65 m a.s.l., Reggio Emilia, Italy), however on two different fields to follow the traditional crop rotation used by the farmer. The well-drained soil was classified as Alfisoil, according to the American classification of Soil Taxonomy [33]. Sampling up to 30 cm depth was done one month before the transplanting and were immediately analyzed for the main physical and chemical properties, reported in Table 1.

Table 1. Physical and chemical soil properties of two-year experiment. EC = electrical conductivity; TN = total nitrogen; CEC = cation exchange capacity.

Soil Characteristics	2017	2018
Sand (%)	5.8	11.2
Silt (%)	54.1	67.5
Clay (%)	40.1	21.3
pH (-) (Soil water suspension)	7.2	7.8
EC (dS m^{-1}) (1:5 soil-to-water)	0.1	0.2
CaCO3 eq (%)	2.8	9.4
Exchangeable K$_2$O (mg kg^{-1}) (Ammonium acetate method)	226.1	179.9
Available P$_2$O$_5$ (mg kg^{-1}) (Olsen method)	34.4	55.0
TN (‰) (Kjeldahl method)	1.5	1.3
Organic matter (%)	2.3	1.8
CEC (meq 100 g^{-1})	27.0	17.9

The climate is typical continental with cold winter and dry and warm summer. The mean maximum and minimum air temperatures and total rainfall recorded during the cropping cycles (May to September) were 29.6 °C, 18.1°C and 191.4 mm in 2017, and 28.8 °C, 18.7 °C and 279.2 mm, in 2018, respectively (Figure 1).

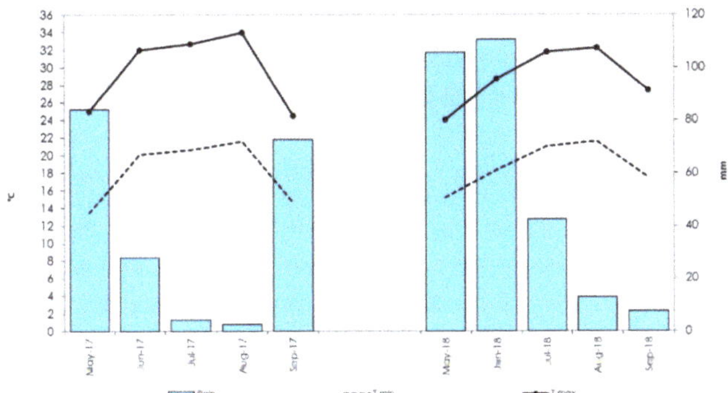

Figure 1. The mean maximum and minimum air temperatures and total rainfall recorded during the cropping cycles (May to September) in the two growing seasons (2017 and 2018).

The OFS comprising the two-year experiment provided for the following crop rotation: alfalfa (four years)—bread wheat (one crop cycle)—autumn—cover crop of *Vicia faba* L. and *Sinapis arvensis* (half and half mixture, one crop cycle)—processing tomato. The cover crop was used as green manure for the processing tomato production.

2.2. Growth Condition

In both years, tomato seedlings (Barone Rosso cultivar with blocky-shaped fruits provided by Tomato Colors, Bologna, Italy) were transplanted into the open field within the first week of May at 2.8

plants m^{-2} in a single row with a spacing of 1.60 m between each row and 0.22 m between plants in the row. This plant density and row spacing were typical for processing tomato cultivation in Northern Italy and were suitable for mechanical harvesting.

Three digestate fertilizers [liquid digestate (LD), liquid digestate + biochar (LD + BC) and pelleted digestate (PD)] were assessed in comparison with biochar (BC) application and unfertilized plants (null control). For treatment LD + BC, LD and BC were applied together using half rates of the dose adopted for LD and BC treatment, respectively.

A randomized complete block design with three replications was adopted and each plot measured 32 m^{-2} (6.4 m × 5.0 m).

The amounts of different fertilizers [LD, (LD + BC), BC and PD] applied were calculated considering a total of 150 kg ha^{-1} of N to supply for a tomato crop cycle. The control (CTRL) was untreated. All the tested fertilizers were manually applied (on the row afterward used to transplant seedlings) and buried using a disc harrow (one week before the manual transplant).

For irrigation scheduling, evapotranspiration of the crop (ETc) was calculated as ETc = ETo × Kc, where ETo (reference evapotranspiration) was determined according to Hargreaves and Samani [34], and Kc (crop coefficient) for tomato crop was adjusted for the environmental conditions and crop growth stage [35]. In each plot, 100% ETc was restored when 40% of the total available water was depleted, according to the evapotranspiration method of Doorenbos and Pruitt [36]. A total of 351.0 mm and 224.4 mm of irrigation water were applied in 2017 and 2018, respectively, by drip irrigation.

Weeds control and plant protection were done according to the OFS cultivation guidelines of the Emilia-Romagna Region (Italy). A single harvest was carried out at the end of the growing seasons in each year, i.e., within the first ten days of September 2017 and 2018, when the ripe fruits accounting for approximately 85% of the total.

2.3. Digestate Fertilizers and Biochar Productions

Digestate and pellets were produced in an anaerobic digester (AD) plant owned by CAT–Cooperativa Agroenergetica Territoriale (Correggio, Reggio Emilia, Italy) as described by Pulvirenti et al. [21]. The raw materials (ingestates) used in AD were maize (*Zea mais* L.) silage (43%), triticale (X *Triticosecale* Wittmack) silage (22%), cow slurry (27%), and grape stalks (of *Vitis vinifera* L.) (8%). Ingestate proportions were calculated according to their fresh weight [21]. After solid/liquid separation (using a dewatering machine) of the fresh digestate, the chemical parameters (on fresh weight basis) of the liquid phase were the following: total carbon (TC, 4.45%), total organic carbon (TOC, 3.74%), total nitrogen (TN, 0.34%), potassium (K_2O, 0.95%), (phosphorous as P_2O_5, was completely absent). Electrical conductivity (EC = 1.07 dS m^{-1}) and pH (8.03) were measured on wet material (1:5 ratio), using a pH-conductivity meter (FiveEasy™model, Mettler Toledo, Giessen, Germany). Conversely, solid-phase digestate was dried and pelleted accordingly to the description of Pulvirenti and collaborators [21]. This pellet (PD) was also analyzed and the results (expressed on a fresh weight basis) were here reported (TC 17.19%, TOC 16.32%, TN 1.5%, P_2O_5 2.5%, K_2O 2.0%, EC 4.17 dS m^{-1}, and pH 8.28).

The BC used, in our work, was produced as described by Ronga et al. [25], except that pine wood chips were used as feedstock in the gasifier (PP20 gasifier, manufactured by ALL Power Labs, Berkeley, CA, USA). The obtained BC displayed the following chemical characteristics: total inorganic carbon (TIC, 73.4%), TN 0.37%, K_2O 3.75%, EC 2.57 dS m^{-1}, and pH 10.1.

2.4. Recorded Parameters on Tomato Crops

At harvest time, five plants were sampled, and plant height was measured. Moreover, physiological parameters were detected on the youngest fully expanded leaf, using the portable Dualex 4 Scientific (FORCEA, Orsay, France) instrument: chlorophyll (CHL), flavonoid (FLAV), and anthocyanin (ANTH) contents were estimated. Finally, the nitrogen balance index (NBI) was calculated as the ratio between CHL and FLAV parameters.

At canopy level, Normalized Difference Vegetation Index (NDVI) was measured by SRS-NDVI (Decagon Device, Pullman, WA, USA), while, the photochemical reflectance index (PRI) were detected using Decagon SRS-PRI (Decagon Device, Pullman, WA, USA) instrument. The measurements of this physiological index were taken at a distance of 1 m above the canopy. Ten average spectra, each a mean of 10 spectra, were recorded *per* plot.

In order to compare N use across the treatments, the nitrogen applied efficiency (NAE) index was calculated, which was derived from the marketable fruit yield (t ha^{-1}) and the amount of N applied (kg ha^{-1}) and expressed as t yield kg^{-1} N [37].

Fruit water productivity (FWP) was also calculated as the ratio between the marketable yield (kg) and the total water used by plants (mm) during the growing season [38].

For destructive measurements, five plants were measured for the main stem length and then harvested dividing the fruits in ripe, unripe and affected by blossom-end rot (BER). Collected berries were counted and weighed, so total and marketable yield, and mean fruit weight were calculated. The above ground biomass was weighed, recorded and oven-dried at 65 °C until constant weight and total biomass dry weight was obtained.

For fruit quality, ~35 collected fruits *per* each harvest plot were ground and homogenized (under cold break preparation) and different parameters were then assessed. The pH was measured with a Basic 20 pH–meter (Crison, Instrument, Barcelona, Spain), while °Brix was determined using a digital refractometer (HI 96814, Hanna Instruments, Villafranca Padovana, Italy). Brix t ha^{-1} was calculated by multiplying the marketable yield (t ha^{-1}) by °Brix and dividing the result by 100. Finally, Bostwick test was carried out according to that described by Ranganna [39] and viscosity was expressed as distance (cm) a sample flows in each time interval (1 min). The experiment was performed at room temperature and repeated three times.

2.5. Data Analysis

Agronomic and physiological data were subjected to analysis of variance (one-way ANOVA) separately *per* each growing season due to unpredictable weather conditions under the Mediterranean basin [40]. Means were separated, when the 'F' test of ANOVA for treatment was significant at least at the 0.05 probability level. For statistical analysis GENSTAT 17th software package (VSN International, Hemel Hempstead, UK) was used.

3. Results

Three digestate fertilizers (LD, LD + BC and PD) were assessed and compared with biochar application and unfertilized plants (CTRL). In both years of the trial all fertilizers increased marketable yield respect to CTRL; moreover, the highest values resulted under LD + BC applications (+54% and +51%, respect to the CTRL, in 2017 and 2018, respectively) (Table 2). A similar trend was also recorded for total yield in both years, with LD + BC treatment that also displayed the highest values (+36% and +47%, respect to the CTRL, in 2017 and 2018, respectively). In addition, our results revealed that LD + BC, BC, PD and LD, significantly increased fruit number *per* plant compared to unfertilized plots in both the investigated years. Among treatments LD + BC displayed the highest values, and across years an average increment of +16% compared to the unfertilized plant was found. In addition, LD and BC when separately applied, resulted in a significantly lower number of fruits *per* plant than LD + BC combination.

The fertilizers assessed in the present study also affected plant morphological parameters in both cropping seasons. Significant lower values of plant height were found for LD + BC thesis in both years (34 cm and 31 cm, in 2017 and 2018, respectively). However, in the second year, plant height under untreated control did not differ from LD + BC.

Table 2. Effect of different digestate and biochar fertilizers on yield and its components. CTRL = unfertilized plants; BC = biochar; LD = liquid digestate; LD + BC = liquid digestate + biochar; PD = pelleted digestate. Different letters within each column and year indicate significant differences according to Duncan's test ($p \leq 0.05$).

TREATMENT	Total Yield (t ha^{-1})		Marketable Yield (t ha^{-1})		Fruit Number (no. plant^{-1})		Plant Height (cm)		Main Stem Lenght (cm)		Aboveground Fresh Weight (g plant^{-1})		Aboveground Dry Weight (g plant^{-1})	
YEAR 2017														
CTRL	52.1	d	45.2	e	24.7	d	43.0	b	69.0	e	667.0	d	22.6	c
BC	66.2	c	64.2	b	27.8	b	51.0	a	75.0	d	1114.0	bc	25.8	b
LD	71.1	b	57.0	d	26.1	c	50.0	a	86.0	b	994.0	c	23.3	c
LD + BC	77.6	a	69.6	a	28.8	a	34.0	c	95.0	a	1294.0	a	29.8	a
PD	70.7	b	61.4	c	27.7	b	51.0	a	80.0	c	1234.0	ab	23.1	c
YEAR 2018														
CTRL	54.6	c	49.5	e	25.3	d	32.0	c	70.0	bc	538.0	c	22.2	c
BC	71.0	b	69.6	b	28.4	b	40.0	b	69.0	c	914.0	ab	19.7	e
LD	71.7	b	61.2	d	25.9	c	39.0	b	72.0	b	821.0	b	21.5	d
LD + BC	80.1	a	74.5	a	29.2	a	31.0	c	94.0	a	1034.0	a	25.5	a
PD	73.1	b	66.8	c	28.2	b	50.0	a	70.0	bc	991.0	a	23.2	b

More consistent differences for the main stem length as effect of different treatments were noticed in the first year of the experiment than the second one. However, the LD + BC application always resulted in the highest values of main stem length (95 cm and 94 cm, in the 2017 and 2018, respectively) that significantly differed by unfertilized and fertilized thesis.

For the aboveground biomass fresh weight, the highest values were found for LD + BC treatment that did not significantly differ from PD (1294 and 1234 g *per* plant, respectively) in 2017 and from PD and BC fertilizers (1034, 991 and 914 g *per* plant, respectively) in 2018.

LD + BC fertilization also increased the aboveground biomass dry weight respect to the control both in 2017 (29.8 and 22.6 g *per* plant, respectively) and in 2018 (25.5 and 22.2 g *per* plant, respectively).

Considering the parameters related to tomato crop physiology (Table 3), all of them were statistically affected by fertilizer treatments. For the leaf measurements, our research highlighted that LD displayed the highest values of leaf chlorophyll content (CHL) (+19% and +9%, respect to the CTRL, in 2017 and 2018, respectively). Flavonoid contents measured in 2017 ranged from the lowest value of 2.22 for LD to the highest one of 2.83 for CTRL, and the same trend was observed in 2018 with the lowest value detected in LD-fertilized plants (2.27), while CTRL confirmed the highest flavonoid contents (3.29). For leaf anthocyanin content (ANTH), PD fertilizer resulted in the highest amounts of these metabolites in both years, without significant differences from BC treatment in 2017. For the nitrogen balance index (NBI) values, as ratio between CHL and FLAV, the highest values were recorded by LD treatment in both years (19.1 and 15.8, for 2017 and 2018 respectively); moreover, for this parameter increases of +52% and +56% (in 2017 and 2018, respectively) were detected comparing LD fertilizer with CTRL. Regarding measurements on the canopy, in both years the lowest values of NDVI were noticed, when LD + BC fertilizer was applied (0.54 and 0.53 in 2017 and 2018, respectively). Conversely, LD treatment and CTRL showed the highest values of NDVI, in the first and second years of the experiment, respectively. Considering the photochemical reflectance index (PRI), the highest values were displayed by BC treatment (+16% and +49% than the CTRL, in 2017 and 2018, respectively), while the lowest values resulted adopting LD + BC combination in both cropping seasons.

For fruit water productivity (FWP) and nitrogen applied efficiency (NAE), LD + BC displayed higher values than the other treatments, with average values of +25% and +13% (across treatments and years), respectively.

Regarding the effect of treatments on tomato fruit quality all the fertilizers significantly increased mean fruit weight (+24% across treatments and years) and the number of fruits affected by blossom-end rot (BER) *per* plant compared to the unfertilized plants (CTRL) (Table 4).

In both investigated years, LD induced the highest °Brix values (+11% and +13%, respect to the CTRL, in 2017 and 2018, respectively), while Brix t ha^{-1} was positively affected by LD + BC application in both years (+52% and +48%, respect to the CTRL, in 2017 and 2018, respectively). However, no significant differences were found between LD + BC combination and BC fertilizer in 2017. Other technological parameters of fresh fruit were also affected by fertilizers in comparison to untreated plants (CTRL). Indeed, LD + BC significantly increased pH (+10% and 7%, respect to the CTRL, in 2017 and 2018, respectively), and, on average, all fertilizers decreased the Bostwick index (≈ −16% respect to the CTRL, across treatments and years), hence increasing the juice consistency.

Table 3. Effect of different digestate and biochar fertilizers on physiological parameters. CTRL = unfertilized plants; BC = biochar; LD = liquid digestate; LD + BC = liquid digestate + biochar; PT = pelleted digestate; CHL = leaf chlorophyll content index; FLAV = leaf flavonoid contents index; ANTH = leaf anthocyanin contents index; NBI = nitrogen balance index (CHL/FLAV ratio); NDVI = normalized difference vegetation index; PRI = photochemical reflectance index; FWP = fruit water productivity; NAE = nitrogen applied efficiency; $_x$ = measured using Dualex 4 Scientific; $_y$ = using instrument SRS-NDVI; $_z$ = using instrument SRS-PRI. Different letters within each column and year indicate significant differences according to Duncan's test ($p \leq 0.05$).

TREATMENT	CHL_x (-)		$FLAV_x$ (-)		$ANTH_x$ (-)		NBI_x (-)		$NDVI_y$ (-)		PRI_z (-)		FWP (kg mm^{-1})		NAE (t kg^{-1})	
YEAR 2017																
CTRL	35.64	b	2.83	a	0.272	c	12.6	b	0.74	b	−0.005	b	0.0029	d	-	
BC	24.13	d	2.47	c	0.406	a	9.8	d	0.73	c	0.001	a	0.0042	b	0.428	b
LD	42.36	a	2.22	d	0.203	d	19.1	a	0.80	a	−0.041	d	0.0038	c	0.380	d
LD+BC	30.61	c	2.71	b	0.382	b	11.3	c	0.54	e	−0.289	e	0.0046	a	0.464	a
PD	23.01	e	2.75	b	0.408	a	8.4	e	0.68	d	−0.036	c	0.0040	b	0.410	c
YEAR 2018																
CTRL	32.80	b	3.29	a	0.334	d	10.1	b	0.79	a	−0.067	d	0.0035	d	-	
BC	22.41	d	3.04	b	0.254	e	7.4	d	0.68	d	−0.033	a	0.0049	b	0.464	b
LD	35.74	a	2.27	e	0.479	b	15.8	a	0.73	b	−0.036	b	0.0043	c	0.408	d
LD+BC	29.80	c	2.69	c	0.372	c	11.1	c	0.53	e	−0.288	e	0.0053	a	0.497	a
PD	15.66	e	2.34	d	0.488	a	6.7	d	0.71	c	−0.048	c	0.0047	b	0.445	c

Table 4. Effect of different digestate and biochar fertilizers on fruit quality parameters. CTRL = unfertilized plants; BC = biochar; LD = liquid digestate used as fertilizer; LD + BC = liquid digestate + biochar; PD = pelleted digestate; BER = blossom-end rot. Different letters within each column indicate significant differences according to Duncan's test ($p \leq 0.05$).

TREATMENT	Mean Fruit Weight (g)		Fruit Affected by BER (no. plant^{-1})		°Brix		Brix Yield (t ha^{-1})		pH		Bostwick (cm 30 s^{-1})	
YEAR 2017												
CTRL	65.33	d	7.0	d	6.1	c	2.75	d	4.32	d	5.5	a
BC	82.67	b	10.0	c	6.5	b	4.18	a	4.51	b	5.0	b
LD	78.00	c	43.0	a	6.8	a	3.87	b	4.30	d	4.5	c
LD + BC	86.30	a	15.0	b	6.0	c	4.18	a	4.74	a	4.0	d
PD	79.33	c	11.0	c	5.7	d	3.51	c	4.45	c	5.0	b
YEAR 2018												
CTRL	70.00	d	8.0	e	6.0	bc	2.97	d	4.27	c	5.5	a
BC	87.50	b	20.0	b	6.1	b	4.24	b	4.46	b	4.5	c
LD	84.30	c	39.0	a	6.8	a	4.16	b	4.29	c	4.5	c
LD + BC	91.10	a	13.0	c	5.9	cd	4.40	a	4.55	a	4.5	c
PD	84.70	c	11.0	d	5.8	d	3.87	c	4.41	b	5.0	b

4. Discussion

High marketable yield is the most important goal in processing tomato production [41]; however, a large amount of external inputs is required [42]. Processing tomato sustainability may be increased by adopting the OFS [43] and recently consumers are increasing the purchase of organic farming products [44]. However, as reported by Ronga et al. [13], lower marketable yields have been reported in tomato crops cultivated in Southern Italy under the OFS in comparison to conventional farming systems. Hence, it is of paramount importance for the OFS, that innovative agronomic strategy can be identified to reduce the current gap with the conventional farming system. In our work digestate coming from biogas plant and pinewood chip biochar were assessed as alternative fertilizers studying their effects on processing tomato physiology, yield and fruit quality.

In our study marketable yield was positively affected by different fertilizations showing an average of 66 t ha^{-1} across the thesis—furthermore, LD + BC combination resulted in the highest productivity in both years (2017–2018).

Comparisons with other studies on processing tomato produced under the OFS seems not to be easy in relation to the different environments, cultivars, and agronomic management adopted in other studies. Nonetheless, the marketable yield recorded in our research falls into the range reported by Ronga et al. [45,46], showing an average of 45 t ha^{-1} and 86 t ha^{-1} in Italian and Californian experiments, respectively.

Considering that the average marketable yield recorded under a conventional farming system in Italy in the last two years was ~50 t ha^{-1} [47], our results demonstrated that it is possible to reduce the production gap between the OFS and traditional management in processing tomato production. However, the average yield reported under conventional systems, considered both specialized and not specialized farms in processing tomato production, but also fields irrigated with different water distribution systems. Indeed, in the same area, where the present study was carried out and especially in specialized farms, processing tomato crops grown under the conventional system reached productions of ~100 t ha^{-1} [48].

The highest marketable yield displayed by LD + BC fertilization can be related to seven main parameters: fruit number *per* plant, fruit weight, plant height, main stem length, aboveground biomass production, FWP and NAE, according to the results showed by Barrios-Masias and Jackson [49] and, Ronga et al. [48], which investigated the main morphological and physiological parameter involved in increasing marketable yield under different environments (California and Italy, respectively).

Fruit number *per* plant and fruit weight are the two most important parameters contributing in tomato yield. Gains achieved in marketable yield for processing tomatoes were mainly ascribed to an increase in fruit number *per* plant [50], nevertheless, Hihashide and Heuvelink [51] reported the importance of fruit weight in increasing gains in fresh tomatoes for greenhouse productions. The highest values of fruit number *per* plant and fruit weight found using LD + BC fertilizer can be due to an increase in water (rainfall and irrigation) and nutrient (carrying by LD) retentions in the soil as effect of biochar administration, as already reported by Laird et al. [52] and Sun and Lu [53]. Furthermore, Scaglia et al. [54] reported that digestates, coming from the biogas plants, can contain phytohormones as well as other bioactive compounds able to improve plant growth. Nitrogen availability, different potassium, phosphorous and bioactive compounds contents of fertilizers investigated in our study, as well as the capacity of biochar to increase the nutrient retentions supplied by LD fertilization represent critical aspects to be investigated in a future study in order to explain the results obtained here.

Barrios-Masias et al. [49] reported a positive correlation between tomato yield and leaf photosynthetic activity and the same physiological behavior was also reported for other crops like wheat [55]. NBI, calculated as the ratio between chlorophyll and polyphenols leaf contents, is an index of the crop nitrogen status [56]. The highest values of NBI and leaf chlorophyll content showed by LD-fertilized plants, in both growing seasons, can suggest a low nitrogen utilization for increasing fruit production. This hypothesis was also confirmed by the lowest value of NAE for the same thesis.

NDVI and PRI, as physiological indexes linked to yield [57], highlighted the lowest values in tomatoes treated with LD + BC, suggesting a better utilization of the readily available nitrogen forms by plants [58], during the crop growth cycle, than the other fertilizers and the untreated control, putatively with a lower impact on nitrate leaching and nitrogen volatilization. However, further studies investigating these physiological and agronomic parameters in different growth stages and under environmental conditions seem to be necessary to corroborate this hypothesis.

While considering the importance of climatic conditions, genotype, soil properties on crop performances, water and nitrogen availability are the main factor affecting yield [59,60] and hence the profitability of the processing tomato production [61,62].

The highest values of FWP and NAE as resulted in LD + BC administration indicated a maximized use of water and nitrogen, in both the trial years. These findings agreed with previous investigations, indicating the positive correlation between marketable yield and water and nutrient use efficiencies in processing tomato production [49,63].

Regarding fruit quality, °Brix, a measure of total soluble solids, is a very impacting parameter for tomato canning companies [64] and is often negatively correlated to fruit yield [65,66]. Indeed, as shown in the present study, the highest yield as the effect of LD + BC application resulted in lower °Brix respect to BC and LD fertilizers separately applied to the plants. Conversely, for the last treatments, lower fruit yield and a higher °Brix than LD + BC administration were noticed. Nonetheless, the highest yield recorded for LD + BC treatments allowed to achieve the highest values of Brix t ha^{-1}, in both assessed years, resulting in the high profitability of processing tomato production. Indeed, tomato paste is produced and sold based on its total soluble solids content, thus, the total soluble solids dictate the factory yield [63].

BER and Bostwick viscosity are other two important quality parameters in processing tomatoes. The highest values of BER found in LD-fertilized plants, both 2017 and 2018, can be due to the high concentration of the ammonia and ammonium nitrogen forms contained in the LD [67,68]. According to this hypothesis, recently Hagassou et al. [69], reported an increment of BER incidences on tomato fruits when fertilizers containing ammonium nitrogen were applied. Among the fertilizers studied in this research, PD displayed the lowest values of fruit affected by BER, and LD + BC combination was also interesting because it did not increase this fruit physiological disorder. With regards to BER, more studies are necessary to clarify the effects of the investigated fertilizers on soil calcium availability as well as plant calcium uptake and its translocation to fruit. Finally, as expected high values of °Brix resulted in low values of Bostwick viscosity, as also suggested by May and Gonzales [70]. For fruit quality attributes, a usefully improved of Bostwick viscosity was found applying LD fertilizer alone or in combination with BC and these results can be related to better plant nutrition during fruit ripening. On the other hand, no information is available about the effect of the adopted fertilizers on the Bostwick index.

The highest yield of LD + BC thesis also resulted in a worsening of the pH of tomato juice. An inverse correlation between yield and pH was already reported by Parisi et al. [71] studying nitrogen fertilization in processing tomato grown in Southern Italy.

Our results demonstrated that the organic fertilizers assessed in our work improved different fruit quality attributes of processing tomato in agreement with the results reported by Asami et al. [72], on strawberry grown under the OFS.

Finally, an assessment of the carbon footprint and the economic impact of the fertilizers should be investigated in the next studies.

5. Conclusions

Fertilizers assessed in our work improved marketable yield and fruit quality of processing tomato cultivated in Northern Italy under the OFS. The highest values of total and marketable yields were obtained with LD + BC combination and these results were related to the highest plant growth and fertility in terms of fruit number *per* plant, fruit weight, main stem length, aboveground biomass, FWP and NAE. Moreover, LD + BC improved two important fruit quality parameters like Brix t ha^{-1} and Bostwick viscosity thus ensuring an improved fruit quality for tomato canning companies. Furthermore, LD + BC fertilization showed the lowest values of NDVI and PRI, suggesting more rapid nitrogen assimilation during the crop growth cycle and early plant senescence at fruit maturity, ultimately resulting in facilitated mechanical harvesting. Hence, our results can be considered in future research aiming to improve fruit yield and quality in other crops grown under the OFS, as well as for new precision agronomic strategies and facing the environmental uncertainties of climate change. However, further studies are needed to study the effects of the available macro- and micronutrients of the fertilizers assessed in the present study, as well as the presence of substances and microorganisms able to stimulate plant growth.

Author Contributions: D.R. and F.C.; methodology, D.R., F.C., G.B.; investigation, D.R., F.C., B.P., M.P., G.A., S.P.; resources, D.R. and E.F.; data curation, D.R., B.P., M.P.; writing—original draft preparation, D.R. and F.C.; writing—review and editing, D.R., F.C., E.F., M.P., B.P., G.A., S.P., G.B.; supervision, D.R.; funding acquisition, D.R. and E.F. All authors have read and agreed to the published version of the manuscript.

Funding: This research was partially funded by the EU-ERDF, Emilia-Romagna Regional Operational Programme POR-FESR 2014/2020, project GENBACCA (PG/2015/728079), and by BIOPRIME (Italian MiPAF project code 1.10.05).

Acknowledgments: We wish to thank Andrea Ferretti (La Collina, Reggio Emilia, RE, Italy) for providing the field trials and Massimo Zaghi (CAT, Correggio, RE, Italy) for providing digestate and pellets. The authors thanks Mariangela Cucino and Giulio Zaccagnini for their collaboration.

Conflicts of Interest: The authors declare no conflict of interest.

References

1. World Processing Tomato Council (WPTC). 2019. Available online: https://www.wptc.to (accessed on 8 December 2019).

2. Eyhorn, F.; Muller, A.; Reganold, J.P.; Frison, E.; Herren, H.R.; Luttikholt, L.; Mueller, A.; Sanders, J.; El-Hage Scialabba, N.; Seufert, V.; et al. Sustainability in global agriculture driven by organic farming. *Nature* **2019**, *2*, 253–255. [CrossRef]

3. Reganold, J.P.; Wachter, J.M. Organic agriculture in the twenty-first century. *Nat. Plants* **2016**, *2*, 1–8. [CrossRef] [PubMed]

4. Tuomisto, H.L.; Hodge, I.D.; Riordan, P.; Macdonald, D.W. Does organic farming reduce environmental impacts? A meta-analysis of European research. *J. Environ. Manag.* **2012**, *112*, 309–320. [CrossRef] [PubMed]

5. Pimentel, D.; Hepperly, P.; Hanson, J.; Douds, D.; Seidel, R. Environmental, energetic, and economic comparisons of organic and conventional farming systems. *BioScience* **2005**, *55*, 573–582. [CrossRef]

6. Gomiero, T.; Pimentel, D.; Paoletti, M.G. nvironmental impact of different agricultural management practices: Conventional vs. organic agriculture. *Crit. Rev. Plant Sci.* **2011**, *30*, 95–124. [CrossRef]

7. Crow der, D.W.; Northfield, T.D.; Strand, M.R.; Snyder, W.E. Organic agriculture promotes evenness and natural pest control. *Nature* **2010**, *466*, 109–112. [CrossRef]

8. Lynch, D.H.; Halberg, N.; Bhatta, G.D. Environmental impacts of organic agriculture in temperate regions. *CAB Rev.* **2012**, *7*, 1–17. [CrossRef]

9. Lee, K.S.; Choe, Y.C.; Park, S.H. Measuring the environmental effects of organic farming: A meta-analysis of structural variables in empirical research. *J. Environ. Manag.* **2015**, *162*, 263–274. [CrossRef]

10. Mondelaers, K.; Aertsens, J.; Van Huylenbroeck, G. A meta-analysis of the differences in environmental impacts between organic and conventional farming. *Br. Food J.* **2009**, *111*, 1098–1119. [CrossRef]

11. Seufert, V.; Ramankutty, N.; Foley, J.A. Comparing the yields of organic and conventional agriculture. *Nature* **2012**, *485*, 229. [CrossRef]

12. De Ponti, T.; Rijk, B.; van Ittersum, M.K. The crop yield gap between organic and conventional agriculture. *Agric. Syst.* **2012**, *108*, 1–9. [CrossRef]

13. Ronga, D.; Lovelli, S.; Zaccardelli, M.; Perrone, D.; Ulrici, A.; Francia, E.; Milc, J.; Pecchioni, N. Physiological responses of processing tomato in organic and conventional Mediterranean cropping systems. *Sci. Hortic.* **2015**, *190*, 161–172. [CrossRef]

14. Ronga, D.; Zaccardelli, M.; Lovelli, S.; Perrone, D.; Francia, E.; Milc, J.; Ulrici, A.; Pecchioni, N. Biomass production and dry matter partitioning of processing tomato under organic vs. conventional cropping systems in a Mediterranean environment. *Sci. Hortic.* **2017**, *224*, 163–170. [CrossRef]

15. Ronga, D.; Gallingani, T.; Zaccardelli, M.; Perrone, D.; Francia, E.; Milc, J.; Pecchioni, N. Carbon footprint and energetic analysis of tomato production in the organic vs. the conventional cropping systems in Southern Italy. *J. Clean. Prod.* **2019**, *220*, 836–845. [CrossRef]

16. Riahi, A.; Hdider, C.; Sanaa, M.; Tarchoun, N.; Ben Kheder, M.; Guezal, I. Effect of conventional and organic production systems on the yield and quality of field tomato cultivars grown in Tunisia. *J. Sci. Food Agric.* **2009**, *89*, 2275–2282. [CrossRef]

17. Bettiol, W.; Ghini, R.; Galvão, J.A.H.; Siloto, R.C. Organic and conventional tomato cropping systems. *Sci. Agric.* **2004**, *61*, 253–259. [CrossRef]

18. Cavigelli, M.A.; Teasdale, J.R.; Conklin, A.E. Long-term agronomic performance of organic and conventional field crops in the mid-Atlantic region. *Agrobiol. J.* **2008**, *100*, 785–794. [CrossRef]

19. Barzee, T.J.; Edalati, A.; El-Mashad, H.; Wang, D.; Scow, K.; Zhang, R. Digestate biofertilizers support similar or higher tomato yields and quality than mineral fertilizer in a subsurface drip fertigation system. *Front. Sustain. Food Syst.* **2019**, *3*, 58. [CrossRef]

20. Scholberg, J.; McNeal, B.L.; Boote, K.J.; Jones, J.W.; Locascio, S.J.; Olson, S.M. Nitrogen stress effects on growth and nitrogen accumulation by field-grown tomato. *Agron. J.* **2000**, *92*, 159–167. [CrossRef]

21. Pulvirenti, A.; Ronga, D.; Zaghi, M.; Tomasselli, A.R.; Mannella, L.; Pecchioni, N. Pelleting is a successful method to eliminate the presence of Clostridium spp. from the digestate of biogas plants. *Biomass Bioenerg.* **2015**, *81*, 479–482. [CrossRef]

22. Sapp, M.; Harrison, M.; Hany, U.; Charlton, A.; Thwaites, R. Comparing the effect of digestate and chemical fertilizer on soil bacteria. *Appl. Soil Ecol.* **2015**, *86*, 1–9. [CrossRef]

23. Ronga, D.; Pellati, F.; Brighenti, V.; Laudicella, K.; Laviano, L.; Fedailaine, M.; Benvenuti, S.; Pecchioni, N.; Francia, E. Testing the influence of digestate from biogas on growth and volatile compounds of basil (*Ocimum basilicum* L.) and peppermint (*Mentha* × *piperita* L.) in hydroponics. *J. Appl. Res. Med. Aromat. Plants* **2018**, *11*, 18–26. [CrossRef]

24. Ronga, D.; Setti, L.; Salvarani, C.; De Leo, R.; Bedin, E.; Pulvirenti, A.; Milc, J.; Pecchioni, N.; Francia, E. Effects of solid and liquid digestate for hydroponic baby leaf lettuce (*Lactuca sativa* L.) cultivation. *Sci. Hortic.* **2019**, *244*, 172–181. [CrossRef]

25. Ronga, D.; Francia, E.; Allesina, G.; Pedrazzi, S.; Zaccardelli, M.; Pane, C.; Tava, A.; Bignami, C. Valorization of vineyard by-products to obtain composted digestate and biochar suitable for nursery grapevine (*Vitis vinifera* L.) production. *Agronomy* **2019**, *9*, 420. [CrossRef]

26. Šimon, T.; Kunzová, E.; Friedlová, M. The effect of digestate. Cattle slurry and mineral fertilization on the winter wheat yield and soil quality parameters. *Plant Soil Environ.* **2015**, *61*, 522–527. [CrossRef]

27. Lehmann, J.; Joesph, S. *Biochar for Environmental Management: Science and Technology*; Earthscan Ltd.: London, UK, 2009.

28. Palansooriya, K.N.; Ok, Y.S.; Awad, Y.M.; Lee, S.S.; Sung, J.-K.; Koutsospyros, A.; Moon, D.H. Impacts of biochar application on upland agriculture: A review. *J. Environ. Manag.* **2019**, *234*, 52–64. [CrossRef]

29. Xu, N.; Tan, G.; Wang, H.; Gai, X. Effect of biochar additions to soil on nitrogenleaching, microbial biomass and bacterial community structure. *Eur. J. Soil Biol.* **2016**, *74*, 1–8. [CrossRef]

30. Gao, S.; Hoffman-Krull, K.; DeLuca, T.H. Soil biochemical properties and crop productivity following application of locally produced biochar at organic farms in Waldron Island, WA. *Biogeochemistry* **2017**, *136*, 31. [CrossRef]

31. Gonzaga, M.I.S.; Mackowiak, C.; de Almeida, A.Q.; Wisniewski, A., Jr.; de Souza, D.F.; da Silva Lima, I.; de Jesus, A.N. Assessing biochar applications and repeated Brassica juncea L. Production cycles to remediate Cu contaminated soil. *Chemosphere* **2018**, *201*, 278–285. [CrossRef]

32. Hol, W.G.; Vestergård, M.; ten Hooven, F.; Duyts, H.; van de Voorde, T.F.; Bezemer, T.M. Transient negative biochar effects on plant growth are strongest after microbial species loss. *Soil Biol. Biochem.* **2017**, *115*, 442–451. [CrossRef]

33. USDA NRCS. *Keys to Soil Taxonomy*, National Cooperative Soil Survey, 10th ed.; USDA NRCS: Madison, WI, USA, 2006.

34. Hargreaves, G.H.; Samani, Z.A. Reference crop evapotranspiration from tem-perature. *Appl. Eng. Agric.* **1985**, *1*, 96–99. [CrossRef]

35. Allen, R.G.; Pereira, L.S.; Raes, D.; Smith, M. *Crop Evapotranspiration. Guidelines for Computing Crop Water Requirements FAO Irrigation and Drainage Paper 56*; FAO: Rome, Italy, 1998; Paper No. 24 (review).

36. Doorenbos, J.; Pruitt, W.O. *Crop Water Requirement*; FAO Irrigation and Drainage; FAO: Rome, Italy, 1977.

37. Cabello, M.J.; Castellanos, M.T.; Romojaro, F.; Martinez-Madrid, C.; Ribas, F. Yield and quality of melon grown under different irrigation and nitrogen rates. *Agric. Water Manag.* **2009**, *96*, 866–874. [CrossRef]

38. Padilla-Díaz, C.M.; Rodriguez-Dominguez, C.M.; Hernandez-Santana, V.; Perez-Martin, A.; Fernandes, R.D.M.; Montero, A.; García, J.M.; Fernández, J.E. Water status, gas exchange and crop performance in a super high density olive orchard under deficit irrigation scheduled from leaf turgor measurements. *Agric. Water Manag.* **2018**, *202*, 241–252. [CrossRef]

39. Ranganna, S. *Handbook of Analysis and Quality Control for Fruits and Vegetable Products*; Tata McGraw Hill Publishing Company Ltd.: New Delhi, India, 2011.

40. Gomez, K.A.; Gomez, A.A. *Statistical Procedures for Agricultural Research*; John Wiley & Sons: New York, NY, USA, 1984.

41. Lovelli, S.; Potenza, G.; Castronuovo, D.; Perniola, M.; Candido, V. Yield, quality and water use efficiency of processing tomatoes produced under different irrigation regimes in Mediterranean environment. *Ital. J. Agron.* **2017**, *12*, 17–24. [CrossRef]

42. Clark, M.S.; Horwath, W.R.; Shennan, C.; Scow, K.M.; Lantni, W.T.; Ferris, H. Nitrogen, weeds and water as yield-limiting factors in conventional, low-input, and organic tomato systems. *Agric. Ecosyst. Environ.* **1999**, *73*, 257–270. [CrossRef]

43. Muller, A.; Schader, C.; Scialabba, N.E.H.; Brüggemann, J.; Isensee, A.; Erb, K.H.; Smith, P.; Klocke, P.; Leiber, F.; Stolze, M.; et al. Strategies for feeding the world more sustainably with organic agriculture. *Nat. Commun.* **2017**, *8*, 1290. [CrossRef]

44. Hempel, C.; Hamm, U. Local and/or organic: A study on consumer preferences for organic food and food from different origins. *Int. J. Consum. Stud.* **2016**, *40*, 732–741. [CrossRef]

45. Ronga, D.; Caradonia, F.; Setti, L.; Hagassou, D.; Giaretta Azevedo, C.V.; Milc, J.; Pedrazzi, S.; Allesina, G.; Arru, L.; Francia, E. Effects of innovative biofertilizers on yield of processing tomato cultivated in organic cropping systems in northern Italy. *Acta Hortic.* **2019**, *1233*, 129–136. [CrossRef]

46. Bustamante, S.C.; Hartz, T.K. Nitrogen management in organic processing tomato production: Nitrogen sufficiency prediction through early-season soil and plant monitoring. *HortScience* **2015**, *50*, 1055–1063. [CrossRef]

47. Stat, I. Available online: http://dati.istat.it/Index.aspx?DataSetCode=DCSP_COLTIVAZIONI (accessed on 8 December 2019).

48. Barrios-Masias, F.H.; Jackson, L.E. California processing tomatoes: Morphological, physiological and phenological traits associated with crop improvement during the last 80 years. *Eur. J. Agron.* **2014**, *53*, 45–55. [CrossRef]

49. Ronga, D.; Francia, E.; Rizza, F.; Badeck, F.W.; Caradonia, F.; Montevecchi, G.; Pecchioni, N. Changes in yield components, morphological, physiological and fruit quality traits in processing tomato cultivated in Italy since the 1930's. *Sci. Hortic.* **2019**, *257*, 108726. [CrossRef]

50. Grandillo, S.; Zamir, D.; Tanksley, S.D. Genetic improvement of processing tomatoes: A 20 years perspective. *Euphytica* **1999**, *110*, 85–97. [CrossRef]

51. Higashide, T.; Heuvelink, E. Physiological and morphological changes over the past 50 years in yield components in Tomato. *J. Am. Soc. Hortic. Sci.* **2009**, *134*, 460–465. [CrossRef]

52. Laird, D.; Fleming, P.; Wang, B.; Horton, R.; Karlen, D. Biochar impact on nutrient leaching from a Midwestern agricultural soil. *Geoderma* **2010**, *158*, 436–442. [CrossRef]

53. Sun, F.; Lu, S. Biochars improve aggregate stability, water retention, and pore-space properties of clayey soil. *J. Plant Nutr.* **2014**, *177*, 26–33. [CrossRef]

54. Scaglia, B.; Pognani, M.; Adani, F. The anaerobic digestion process capability to produce biostimulant: The case study of the dissolved organic matter (DOM) vs. auxin-like property. *Sci. Total Environ.* **2017**, *589*, 36–45. [CrossRef]

55. Fischer, R.A.; Rees, D.; Sayre, K.D.; Lu, Z.M.; Condon, A.G.; Saavedra, A.L. Wheat yield progress associated with higher stomatal conductance and photosynthetic rate, and cooler canopies. *Crop Sci.* **1998**, *38*, 1467–1475. [CrossRef]

56. Cerovic, Z.G.; Masdoumier, G.; Ghozlen, N.B.; Latouche, G. A new optical leaf-clip meter for simultaneous non-destructive assessment of leaf chlorophyll and epidermal flavonoids. *Physiol. Plant.* **2012**, *146*, 251–260. [CrossRef]

57. Fortes, R.; Prieto, M.H.; Terrón, J.M.; Blanco, J.; Millán, S.; Campillo, C. Using apparent electric conductivity and NDVI measurements for yield estimation of processing tomato crop. *Trans. ASABE* **2014**, *57*, 827–835.

58. Möller, K.; Müller, T. Effects of anaerobic digestion on digestate nutrient availability and crop growth: A review. *Eng. Life Sci.* **2012**, *12*, 242–257. [CrossRef]

59. Cammarano, D.; Ceccarelli, S.; Grando, S.; Romagosa, I.; Benbelkacem, A.; Akar, T.; Ronga, D. The impact of climate change on barley yield in the Mediterranean basin. *Eur. J. Agron.* **2019**, *106*, 1–11. [CrossRef]

60. Cammarano, D.; Hawes, C.; Squire, G.; Holland, J.; Rivington, M.; Murgia, T.; Ronga, D. Rainfall and temperature impacts on barley (*Hordeum vulgare* L.) yield and malting quality in Scotland. *Field Crop. Res.* **2019**, *241*, 107559. [CrossRef]

61. Giuliani, M.M.; Gatta, G.; Nardella, E.; Tarantino, E. Water saving strategies assessment on processing tomato cultivated in Mediterranean region. *Ital. J. Agron.* **2016**, *11*, 69–76. [CrossRef]

62. Farneselli, M.; Benincasa, P.; Tosti, G.; Pace, R.; Tei, F.; Guiducci, M. Nine-year results on maize and processing tomato cultivation in an organic and in a conventional low input cropping system. *Ital. J. Agron.* **2013**, *8*, e2. [CrossRef]

63. Di Cesare, L.F.; Migliori, C.; Viscardi, D.; Parisi, M. Quality of tomato fertilized with nitrogen and phosphorous. *Ital. J. Food Sci.* **2010**, *22*, 186–191.

64. Ronga, D.; Parisi, M.; Pentangelo, A.; Mori, M.; Di Mola, I. Effects of nitrogen management on biomass production and dry matter distribution of processing tomato cropped in southern Italy. *Agronomy* **2019**, *9*, 855. [CrossRef]

65. Fulton, T.M.; Beck-Bunn, T.; Emmatty, D.; Eshed, Y.; Lopez, J.; Petiard, V.; Uhlig, J.; Zamir, D.; Tanksley, S.D. QTL analysis of an advanced backcross of *Lycopersicon peruvianum* to the cultivated tomato and comparisons with QTLs found in other wild species. *Theor. Appl. Genet.* **1997**, *95*, 881–894. [CrossRef]

66. Schauer, N.; Semel, Y.; Roessner, U.; Gur, A.; Balbo, I.; Carrari, F.; Pleban, T.; Perez-Melis, A.; Bruedigam, C.; Kopka, J.; et al. Comprehensive metabolic profiling and phenotyping of interspecific introgression lines for tomato improvement. *Nat. Biotechnol.* **2006**, *24*, 447–454. [CrossRef]

67. Xia, A.; Murphy, J.D. Microalgal cultivation in treating liquid digestate from biogas systems. *Trends Biotechnol.* **2016**, *34*, 264–275. [CrossRef]

68. Törnwall, E.; Pettersson, H.; Thorin, E.; Schwede, S. Post-treatment of biogas digestate—An evaluation of ammonium recovery, energy use and sanitation. *Energy Procedia* **2017**, *142*, 957–963. [CrossRef]

69. Hagassou, D.; Francia, E.; Ronga, D.; Buti, M. Blossom end-rot in tomato (*Solanum lycopersicum* L.): A multidisciplinary overview of inducing factors and control strategies. *Sci. Hortic.* **2019**, *249*, 49–58. [CrossRef]

70. May, D.M.; Gonzales, J. Irrigation and nitrogen management as they affect fruit quality and yield of processing tomatoes. In Proceedings of the Acta Horticulturae, V International Symposium on the Processing Tomato, Sorrento (SA), Italy, 23–27 November 1993; Volume 376, pp. 227–234.

71. Parisi, M.; Giordano, I.; Pentangelo, A.; D'Onofrio, B.; Villari, G. Effects of different levels of nitrogen fertilization on yield and fruit quality in processing tomato. *Acta Hortic.* **2006**, *700*, 129–132. [CrossRef]

72. Asami, D.K.; Hong, Y.J.; Barrett, D.M.; Mitchell, A.E. Comparison of the total phenolic and ascorbic acid content of freeze-dried and air-dried marionberry, strawberry, and corn grown using conventional, organic, and sustainable agricultural practices. *J. Agric. Food Chem.* **2003**, *51*, 1237–1241. [CrossRef] [PubMed]

 *agronomy*

Article

Equal K Amounts to N Achieved Optimal Biomass and Better Fiber Quality of Late Sown Cotton in Yangtze River Valley

Xiaolei Ma, Saif Ali, Abdul Hafeez, Anda Liu, Jiahao Liu, Zhao Zhang, Dan Luo, Adnan Noor Shah and Guozheng Yang *

MOA Key Laboratory of Crop Eco-physiology and Farming system in the Middle Reaches of Yangtze River, College of Plant Science and Technology, Huazhong Agricultural University, Wuhan 430000, China; xiaoleima@webmail.hzau.edu.cn (X.M.); drsaifbhatti@webmail.hzau.edu.cn (S.A.); ahafeez1226@webmail.hzau.edu.cn (A.H.); lada199@163.com (A.L.); liujiahao@webmail.hzau.edu.cn (J.L.); zzhang@webmail.hzau.edu.cn (Z.Z.); ld17191091496@126.com (D.L.); ans.786@yahoo.com (A.N.S.)
* Correspondence: ygzh9999@mail.hzau.edu.cn; Tel.: +139-9555-3884

Received: 24 November 2019; Accepted: 8 January 2020; Published: 13 January 2020

Abstract: Potassium (K) fertilizer plays a crucial role in the formation of the biological and economic yield of cotton (*Gossypium hirsutum* L.). Here we investigated the effects of the amount of K on biomass accumulation and cotton fiber quality with lowered N amounts (210 kg ha^{-1}) under late sowing, high density and fertilization once at 2 weeks after squaring. A 2-year field experiment was performed with three K fertilizer amounts (168 kg ha^{-1} (K$_1$), 210 kg ha^{-1} (K$_2$), and 252 kg ha^{-1} (K$_3$)) using a randomized complete block design in 2016 and 2017. The results showed correspondingly, K$_3$ accumulated cotton plant biomass of 7913.0 kg ha^{-1}, next to K$_2$ (7384.9 kg ha^{-1}) but followed by K$_1$ (6985.1 kg ha^{-1}) averaged across two growing seasons. Higher K amounts (K$_2$, K$_3$) increased biomass primarily due to a higher accumulation rate (32.68%–74.02% higher than K$_1$) during the fast accumulation period (FAP). Cotton fiber length, micronaire, and fiber strength in K$_2$ were as well as K$_3$ and significantly better than K$_1$. These results suggest that K fertilizer of 210 kg ha^{-1} should be optimal to obtain a promising benefit both in cotton biomass and fiber quality and profit for the new cotton planting model in the Yangtze River Valley, China and similar climate regions.

Keywords: cotton; potassium; fertilizer; biomass accumulation; fiber quality

1. Introduction

Cotton is one of the most important fiber crops grown not only for fiber but also for the paper and oil industries [1,2]. China is one of the leading countries for cotton production. The Yangtze River Valley is one of the three cotton-growing regions in China where seedlings are transplanted after wheat or rapeseed is harvested and more than 300 kg ha^{-1} N is applied in three splits (30% at pre-plant, 40% at first bloom, and 30% at peak bloom) [3,4]. However, the arduous procedure and excess fertilizer input are depleting cotton production profits [5]. To improve production benefit, a new planting model with late sowing (mid-May) [6], high density (9–10 plants m^{-2}) [6,7], low N amounts (180–225 kg ha^{-1}) [7], and once fertilization [3,8] has been practiced as an effective way to fight the challenge of high cost in cotton production in the region. The new planting model harvested similar yield to the conventional practice [9] but greatly reduced the cost resulted from less manual work, low N fertilizer amount and less application of chemicals, due to the short cotton growing season with high planting density.

Numerous studies have demonstrated that K is a fundamental element for plant growth which markedly affects biomass accumulation and biomass partitioning [10–12]. Applying potassium fertilizer improved cotton plant biomass [13], especially the biomass of cotton bolls [14], and it increased the reproductive parts biomass per unit area [15–17]. On the contrary, K deficiency reduced not only the production but also the transportation of dry matter, leading to poor growth and reduced biomass accumulation in bolls [18]. Excessive K fertilizer has increased not only luxurious consumption and environmental concern [19] but also canopy closure, leading to rotten bolls and delayed maturation [20]. Tsialtas et al. [21] revealed that 80 kg K_2O ha^{-1} was sufficient for cotton growth to achieve considerable yields in Australia. However, it remains to study how much K has to be applied to ensure enough cotton products for the new planting model. Previous studies have indicated that the cotton plant could produce a considerable yield of 2691 kg ha^{-1} seed cotton when the K amount was in line with N amount. It is hypothesized that K could also be reduced in accordance with N because the plant should keep in balance in nutrients accumulation for normal growth and fruits.

Cotton fiber quality is an important standard in cotton production based on high yield. Many studies focused on the effect of K on cotton fiber quality traits but the results had many differences. Some studies showed that the K amount significantly affected the fiber length [21,22], strength, micronaire, uniformity, and elongation of the cotton [23]. However, some studies indicated that fiber properties were not significantly affected by the K amount [16,24,25].

The study aimed to (1) determine the effects of K fertilizer amount (ranging from 168–252 kg ha^{-1} K_2O) on cotton phenology, biomass accumulation (duration and rate of FAP and distribution) and fiber quality; (2) find the optimal K amount to achieve high productivity and fiber quality of cotton in the new planting model.

2. Materials and Methods

2.1. Experimental Site and Cultivar

The field experiment was conducted in 2016 and 2017 with Huamian 3109 (*G. hirsutum* L.) on the experimental farm of Huazhong Agricultural University, Wuhan, China (30°37′ N latitude, 114°21′ E longitude, 23 m elevation). The soil of the experimental field was yellowish-brown and clay loam comprising of 89.3 mg kg^{-1} alkaline N, 26.4 mg kg^{-1} P_2O_5, and 177.0 mg kg^{-1} K_2O.

2.2. Climate

The mean air temperatures from May to October in 2016 were 25.4 °C with 0.1 °C lower than that in 2017, and from June to September, air temperatures in 2016 were 0.3–1.6 °C lower than that in early 2017. The total rainfall from May to October in 2016 was 1311.6 mm with 925 mm more than that in 2017, and rainfall was mainly concentrated on June and July in 2016 with 823 mm more than that in 2017, but 107 mm less from August to September in 2016 than in 2017 [26].

2.3. Experiment Design

A randomized complete block design was employed with four replicates. Three K fertilizer amounts were 168 kg ha^{-1} (K_1), 210 kg ha^{-1} (K_2), and 252 kg ha^{-1} (K_3).

Fertilizers, as provided by urea (46.3% N) for 210 kg N ha^{-1}, calcium superphosphate (12% P_2O_5) for 63 kg P_2O_5 ha^{-1}, potassium chloride (59% K_2O) for three amounts, and borate (10% B) for 1.5 kg B ha^{-1}, were mixed evenly and buried in 10 cm deep between cotton rows in bed 2 weeks after squaring.

2.4. Field Management

The plant density was 9×10^4 plants ha^{-1} with a row to row space of 76 cm. The plot size was 36.48 m^2 (12 m × 3.04 m) with four rows in two beds. Cotton seeds were sown directly on 18 May 2016 and 10 May 2017. Seedlings were thinned at the three leave stage to the target planting density. Other field managements were carried out according to conventional practice.

2.5. Data Collection

2.5.1. Cotton Phenology

Fifteen successive and uniform plants in one row from each plot were fixed for the investigation of plant growth stages, such as squaring (50% plant bearing squares), first bloom (50% plants showing flowers), peak bloom (normally 15 d after first bloom), boll opening (50% plants showing open boll), and plant senescence. The specific growth period in days were identified as the duration from the day of the first stage to the day of the next stage, such as seedling, from emergence to squaring; squaring, from squaring to first bloom; flowering, from first bloom to peak bloom; boll setting, from peak bloom to boll opening; flowering and boll setting, from first bloom to boll opening; boll opening (or maturation), from boll opening to plant senescence.

2.5.2. Cotton Biomass Accumulation

Cotton biomass was measured five times (squaring, first bloom, peak bloom, boll opening, and plant senescence) in the fourth replication. Nine (eighteen at squaring stage) successive plants were carefully uprooted and grouped randomly but equally in number into 3 as replicates from each plot at each stage. Plants were separated into vegetative parts (root, stem, and leaves) and reproductive parts (square, flower, and boll). Sub-samples were packed separately and dried in an electric fan-assisted oven at 105 °C for 30 min, at 80 °C for constant weight, and then weighted. Vegetative part biomass (VPB) is the total biomass of root, stem, and leaves, and reproductive part biomass (RPB) is the total biomass of squares, flowers, and bolls, and cotton plant biomass (CPB) is the sum of VPB and RPB.

Cotton plant biomass accumulation progress was described by a logistic regression model [3],

$$W = \frac{W_M}{1 + ae^{bt}},\tag{1}$$

where a and b are constants to be found, t is the time as the days after emergence (DAE), W is the biomass (g) at t, and W_M is the maximum biomass (g).

According to Equation (1), the following equations will be calculated:

$$t_1 = \frac{1}{b}\ln\left(\frac{2 + \sqrt{3}}{a}\right),\tag{2}$$

$$t_2 = \frac{1}{b}\ln\left(\frac{2 - \sqrt{3}}{a}\right),\tag{3}$$

$$T = -\frac{\ln a}{b},\tag{4}$$

$$V_T = \frac{W_1 - W_2}{t_1 - t_2},\tag{5}$$

$$V_M = -\frac{bW_M}{4},\tag{6}$$

where t_1 and t_2 (DAE) are the initiation and termination of FAP (fast accumulation period), respectively; T (d) is the duration of FAP; V_T and V_M (g d^{-1}) are the average and the highest biomass accumulation rate during FAP, respectively; W_1 and W_2 are the biomass at t_1 and t_2, respectively.

The accumulation rate (AR) of cotton plant biomass during each period was calculated by the following formula:

$$AR\left(\text{kg ha}^{-1}\text{d}^{-1}\right) = \frac{W_T - W_I}{\text{period length}},\tag{7}$$

where W_I and W_T (kg ha^{-1}) are the biomasses on the first day and the last day of the period, respectively, and the period length (d) is the duration in days of this period.

2.5.3. Cotton Fiber Quality

One hundred maturated bolls were picked from each plot before harvest to get the fiber samples. High volume instrumentation (HVI) was used to analyze fiber quality parameters for each fiber sample, as described by [15]. The reports of five important quality parameters describing the fiber length, strength, fineness, elongation, uniformity was provided by HVI.

2.6. Statistical Analysis

Data are processed with Microsoft Excel 2010; ANOVA was performed with SPSS 21.0 (IBM Company, Chicago, IL, USA) and figures were drawn with Sigma Plot 12.5 (Systat Software Inc., San Jose, CA, USA). Least Significant Difference (LSD) among the treatments was conducted with Duncan at a 5% probability level ($p = 0.05$).

Higher K fertilizer amounts (K_2 and K_3) increased 10.34%–20.03% seed cotton yield over K_1 due to higher boll density in 2016 and boll weight in both years, although differences existed between years in yield and its components [26,27].

3. Results

3.1. Cotton Plant Phenology

Cotton flowering and boll setting period took the longest while squaring the shortest, although differences existed between years in each specific cotton growth period (seedling, squaring and flowering, and boll setting) (Table 1).

Table 1. Cotton growth stages and periods influenced by K fertilizer amounts.

Year	Treatment	Growing Stage (m/d) #				Growth Period (d) #			
		Emergence	Squaring	First Bloom *	Opening	Seedling	Squaring	Flowering and Boll Setting	Total
2016	K_1	5/28	7/15	8/1	9/23	48a *	17a	53a	118a
	K_2	5/28	7/15	8/1	9/22	48a	17a	52a	117a
	K_3	5/28	7/15	8/1	9/22	48a	17a	52a	117a
2017	K_1	5/18	6/20	7/15	8/24	33a	25a	40a	98a
	K_2	5/18	6/20	7/15	8/25	33a	25a	41a	99a
	K_3	5/18	6/20	7/15	8/25	33a	25a	41a	99a

m/d shows month/date, d means days. * Values followed by different letters within the same column in the same year are significantly different at probability levels ($p < 0.05$) according to the Least Significant Difference (LSD) test.

None of the specific cotton growth periods were affected by the K fertilizer amounts within the same year. However, the cotton growth period in 2016 was 18 d longer than that in 2017, due to 15 d longer in seedling and 11 d longer in boll setting, but 8 d shorter in squaring.

3.2. Cotton Plant Biomass Accumulation

Cotton plant biomass (CPB) was significantly increased with increased K amounts in both years (Table 2). The same trends were observed in root and stem biomass. Compared with K_1, K_2 increased root and stem 11.55% and 2.11% in 2016, respectively. However, the root biomass in K_2 was lower than that in K_1 in 2017 with no significant difference and stem biomass in K_2 was 13.87% higher than that in K_1. Cotton plants in K_3 produced 24.71% (2016) and 0.65% (2017) more root biomass and 27.36% (2016) and 26.40% (2017) more stem biomass compared with K_1. Leaves and reproductive parts accumulated higher in K_2 and K_3, and significantly lower in K_1. The ratios of RPB to CPB had no significant difference among the three K amounts in 2016, but that is significantly higher in K_2 and K_3 than K_1. There were no significant differences between K_2 and K_3 for Leaves and RPB and the ratios of RPB to CPB. Furthermore, the ratios of RPB/CPB in K_2 were higher than other treatments.

Table 2. Cotton and each part biomass accumulation influenced by K fertilizer amounts.

Year	Treatment	Biomass Accumulation (kg ha^{-1})					RPB/CPB (%)
		Root	Stem	Leaves	Reproductive Parts	Total	
2016	K$_1$	815.6c *	1950.1b	406.0b	3258.8a	6603.9b	48.35a
	K$_2$	909.8b	1991.3b	494.6a	3338.6a	6759.6b	49.39a
	K$_3$	1017.1a	2206.3a	517.1a	3464.0a	7086.8a	48.86a
	Average	914.2	2049.2	499.6	3353.8	6816.8	48.87
2017	K$_1$	1169.1a	2374.0c	406.0b	3417.1b	7366.2c	44.18b
	K$_2$	1103.1a	2703.3b	494.6a	3709.1ab	8010.1b	46.32a
	K$_3$	1176.7a	3000.8a	517.1a	4044.6a	8739.2a	46.29a
	Average	1149.6	2692.7	472.6	3723.6	8038.5	45.60

* Values followed by different letters within the same column in the same year are significantly different at probability level ($p < 0.05$) according to Least Significant Difference (LSD) test.

The growth curves of CPB, VPB, and RPB increased along with the cotton growth stage following a sigmoid curve with different slopes from K fertilizer amounts (Figure 1).

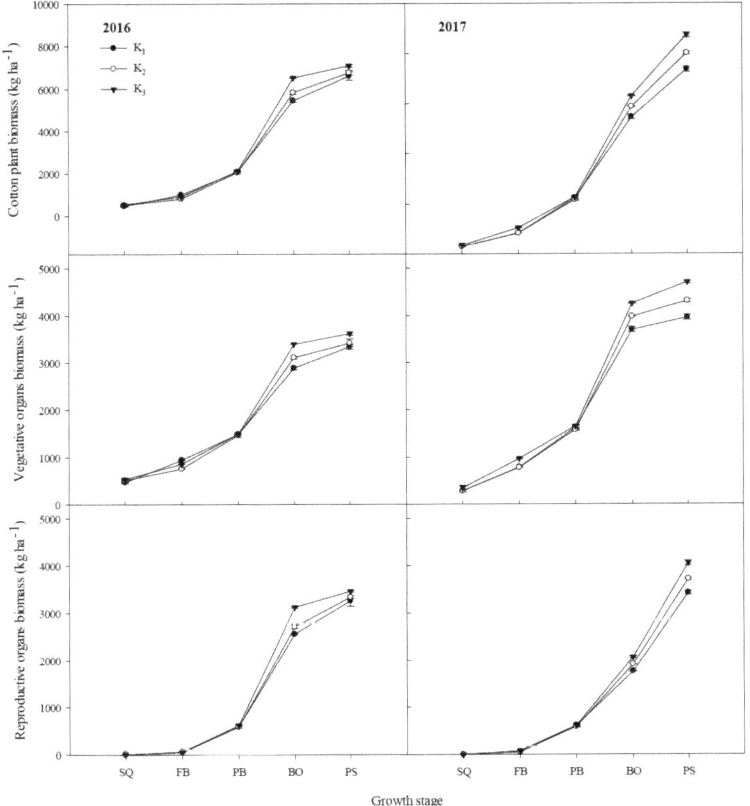

Figure 1. Cotton plant, vegetative parts, reproductive parts, and reproductive related parts biomass of field-grown cotton influenced by K fertilizer amounts at different growth stages in 2016. SQ, FB, PB, BO, and PS indicate squaring (51 days after emergence (DAE)), first bloom (66 DAE), peak bloom (81 DAE), boll opening (128 DAE), and plant senescence (168 DAE) stage, respectively. Error bar plus shows SEMs.

The growth curve slopes of CPB, VPB, and RPB were gradually increased until the boll opening stage and then decreased. Compared with RPB, the slopes of VPB curves were higher before peak bloom stage, and 41.04%–44.47% and 35.05%–40.90% VPB was produced in 2016 and 2017, respectively. However, RPB grew faster after peak bloom stage, and 81.75%–82.40% and 82.09%–84.82% RPB were produced in 2016 and 2017, respectively. The growth curves of CPB, VPB, and RPB in different K amounts gradually diverged from peak bloom. At the plant senescence stage, K3 plants produced 4.84% CPB, 5.90% VPB, and 3.76% RPB more than K2 plants and 7.31% CPB, 8.30% VPB, and 6.30% RRB more than K1 plants, respectively, in 2016. K3 plants produced 9.10% CPB, 9.15% VPB, and 9.04% RPB more than K2 plants and 18.64% CPB, 18.88% VPB, and 18.36% RRB more than K1 plants, respectively, in 2017.

3.3. Simulation of Biomass Accumulation

The biomass of cotton plants and each plant part accumulated following the logistic regression Equation (1) ($p < 0.01$) and showed different accumulation characteristics during FAP in both years (Table 3). Eigenvalues of cotton plant biomass accumulation were calculated by Equations (2)–(6) using the coefficient of Equation (1).

Table 3. Regression equation and Eigenvalues of cotton plant biomass accumulation of field-grown cotton influenced by K fertilizer amount in 2016 and 2017.

Year	Treatment	Regression Eqs. $W =$ kg/ha, $t =$ DAE	p-Value	t_1 (DAE)	t_2 (DAE)	Δt	VT (kg/ha d^{-1})	VM (kg/ha d^{-1})
Cotton plant								
2016	K1	$W=6771.764/(1+4.98e^{-0.050t})$	0.0006	72.7	125.0	52.3	74.7	85.2
	K2	$W=6850.748/(1+5.58e^{-0.057t})$	0.0007	74.1	120.0	45.8	86.3	98.4
	K3	$W=7170.924/(1+6.19e^{-0.066t})$	0.0004	74.0	114.0	40.0	103.5	118.0
	Average			73.6	119.7	46.0	88.2	100.6
2017	K1	$W=7387.145/(1+6.18e^{-0.073t})$	0.0003	66.7	102.9	36.1	118.0	134.6
	K2	$W=8036.283/(1+6.54e^{-0.076t})$	0.0003	68.7	103.3	34.6	134.0	152.8
	K3	$W=8785.985/(1+6.33e^{-0.073t})$	0.0005	68.6	104.6	36.0	140.8	160.6
	Average			68.0	103.6	35.6	130.9	149.3
Vegetative parts								
2016	K1	$W=3399.527/(1+4.00e^{-0.046t})$	0.0008	58.7	116.4	57.7	34.0	38.8
	K2	$W=3477.372/(1+4.61e^{-0.053t})$	0.0010	62.6	112.7	50.0	40.1	45.8
	K3	$W=3714.961/(1+4.71e^{-0.054t})$	0.0010	63.2	112.2	49.0	43.8	49.9
	Average			61.5	113.7	52.3	39.3	44.8
2017	K1	$W=4036.745/(1+6.64e^{-0.087t})$	0.0041	61.0	91.2	30.2	77.2	88.1
	K2	$W=4390.263/(1+7.01e^{-0.090t})$	0.0044	63.2	92.4	29.2	86.7	98.9
	K3	$W=4801.295/(1+6.43e^{-0.082t})$	0.0066	62.3	94.3	32.1	86.5	98.6
	Average			62.1	92.6	30.5	83.5	95.2
Reproductive parts								
2016	K1	$W=3315.932/(1+6.97e^{-0.064t})$	0.0021	88.0	129.0	41.0	46.7	53.3
	K2	$W=3374.119/(1+7.34e^{-0.069t})$	0.0017	87.7	126.0	38.4	50.8	57.9
	K3	$W=3450.948/(1+8.52e^{-0.085t})$	0.0009	84.8	115.8	31.0	64.3	73.3
	Average			86.8	123.6	36.8	53.9	61.5
2017	K1	$W=3445.530/(1+6.85e^{-0.070t})$	0.0020	79.3	117.0	37.7	52.7	60.1
	K2	$W=3733.200/(1+7.25e^{-0.074t})$	0.0012	80.4	116.1	35.7	60.4	68.9
	K3	$W=4072.099/(1+7.27e^{-0.074t})$	0.0009	80.7	116.4	35.7	65.9	75.1
	Average			80.1	116.5	36.4	59.7	68.0

Where t_1 and t_2 (DAE) mean the initiation and termination, respectively, of the fast accumulation period (FAP); Δt (d) means the duration of FAP; V_T and V_M (g d^{-1}) mean the average, and the highest biomass accumulation rate, respectively, during FAP.

The W_M values of the logistic regression equation in CPB, VPB, and RPB were higher with increased K amounts (Table 2). The average and the highest biomass accumulation rates of CPB and each part biomass during FAP showed higher values in higher K amounts in both years.

Compared with RPB, VPB initiated FAP 25 d earlier in 2016 and 18 d in 2017 and terminated FAP 10 d earlier in 2016 and 24 d in 2017 with 15.5 d longer duration in 2016 and 6 d shorter in 2017. CPB initiated FAP 13 d (2016) and 12 d (2017) earlier than RPB and terminated 4 d (2016) and 13 d (2017) earlier with 9 d longer duration in 2016 and 1 d shorter in 2017. Compared with the average accumulation rates of FAP in RPB, the rates in CPB was 34.3 kg ha^{-1} d^{-1} (2016) and 71.2 kg ha^{-1} d^{-1} (2017) faster, and the rates in VPB was 14.6 kg ha^{-1} d^{-1} slower in 2016 and 23.8 kg ha^{-1} d^{-1} faster in 2017.

CPB initiated FAP in flowering and boll setting period (74 DAE in 2016 and 68 DAE in 2017) and terminated at 120 DAE (2016) and 104 DAE (2017) with the duration of 46 d and 36 d averaged across three K fertilizer amount in 2016 and 2017, respectively. The FAP initiations in K$_2$ and K$_3$ were similar but later than that in K$_1$ in both years. The FAP termination was earlier with increased K amounts in 2016, but later in 2017. The FAP duration was decreased with increased K amounts in 2016, but no similar result was observed in 2017.

VPB initiated FAP at 62 DAE in both years and terminated in flowering and boll setting period at 114 DAE (2016) and 93 DAE (2017), respectively. Many differences existed in FAP durations and biomass accumulation rates of VPB between both years which reflected the VPB were accumulated more slowly in 2016 than in 2017. With the increase in K amount, the duration of FAP become shorter and the accumulation rates were higher in 2016, but the shortest duration and highest accumulation of FAP in VPB were observed in K$_2$ in 2017.

RPB initiated and terminated FAP in the flowering and boll setting period with 37 d FAP duration in both years. The accumulation rates were higher in 2017 than in 2016. With increased K amounts, FAP durations decreased and the accumulation rates increased.

3.4. Fiber Quality

K amount significantly affected the fiber length, micronaire, and fiber strength. With increased K amount, the fiber length and the fiber strength increased significantly but there was no significant difference between that in K$_2$ and K$_3$. The micronaire values in different K amounts were no significant difference in 2016, but significantly lower in K$_1$ in 2017 with no significant difference between that in K$_2$ and K$_3$.

4. Discussion

K fertilizer is one of the main cotton fertilizers and has great correlations with cotton growth and the economic benefits of cotton production.

Many studies also reported that K deficiency could accelerate the growth process of cotton and result in premature senescence [28–31]. However, another study showed that K deficiency elicited similar effects on cotton earliness with late sowing which delayed flowering and boll development [32]. In the present study, the growth period was not affected by K amount, although apparent differences existed between the two growing seasons (Table 1). This might be due to the closeness of the K amount range in this study which was not sufficient to bring significant differences in growth period among different K amounts in the same year. Furthermore, the K amounts of three treatments were within the appropriate range for cotton growth under medium fertility, ensuring no K-deficiency in this study. The big differences between the two growing seasons were possibly due to a large amount of precipitation during the early cotton growth period but draught occurred in the flowering and boll setting periods in the 2016 growing season.

The biological yield was the basis of economic yield. Biomass accumulation could be explained in the context of plant photosynthesis, photo-assimilate translocation from vegetative to reproductive parts. K fertilizer affected the photo-assimilate export from leaves to sink parts and regulated the sugar signaling in reproductive parts [24]. Potassium deficiency led to a reduction of main stem length, nodes and bolls, and also leaf photosynthesis and stomatal conductance [18,23,33,34], resulting in less carbohydrate production, a small sink and an in-balanced source-sink ratio. A previous study also revealed that a linear effect between K amounts and the growth efficiency of the reproductive part [35]. In this study, the RPB and the ratios of RPB to CPB were significantly higher in K_2 and K_3 with no significant difference between the two treatments (Table 2). This indicated that K_2 can benefit the carbohydrate production transportation from vegetative parts to reproductive parts and get the approximate RPB with K_3. Higher carbohydrate production in the reproductive parts can result in higher yield [36]. In the present study, the biomass of each cotton part was increased along with the increase in K amount (Table 2 and Figure 1). Similar studies revealed that increasing the K amount can increase the biomass of total plant and cotton bolls [14,18,37]. Furthermore, in this study, the FAP of CPB and RPB were initiated in the flowering and boll setting period and the FAP durations of RPB in both years were the same but the FAP accumulation rates were higher in 2017 with higher RPB (Table 3). With the increased K amount, the FAP accumulation rates were higher with higher biomass accumulation, but the duration of FAP shortened (Table 3). Similar results were also observed in Khan et al. [6] and Tung et al. [38]. This indicated that higher K amounts increased the cotton plant biomass mainly by higher accumulation rate during flowering and boll setting period.

In this study, fiber length, fiber strength, and micronaire were significantly affected by K amounts and better fiber quality traits were observed in K_2 and K_3 with no significant difference between K_2 and K_3 (Table 4). That indicated the fiber quality in K_2 was as well as that in K_3 and significantly better than that in K_1.

Table 4. Cotton fiber quality influenced by K fertilizer amounts.

Year	Treatment	Length (mm)	Uniformity (%)	Micronaire	Strength (g/tex)	Elongation (%)
2016	K_1	23.3b *	83.8a	4.6a	25.8b	6.57a
	K_2	25.0ab	83.8a	4.8a	27.6ab	6.60a
	K_3	25.4a	84.1a	4.5a	28.7a	6.60a
	Average	24.6	83.9	4.6	27.4	6.6
2017	K_1	22.2b	84.6a	5.2a	24.6b	6.57a
	K_2	23.1a	85.1a	4.4b	26.0a	6.60a
	K_3	23.6a	84.9a	4.3b	26.4a	6.60a
	Average	23.0	84.9	4.6	25.7	6.6

* Values followed by different letters within the same column in the same year are significantly different at probability level ($p < 0.05$) according to Least Significant Difference (LSD) test.

5. Conclusions

K amounts ranging from 168–252 kg K_2O ha^{-1} have not altered the cotton growth period. Higher K increased cotton biomass due to a higher accumulation rate during FAP. Nevertheless, K_2 had similar fiber qualities, biomass accumulation, and partitioning as K_3.

The results suggest that an equal K amount to lowed N of 210 kg ha^{-1} should be the optimal strategy under this new planting model in Yangtze River Valley, China, and similar regions.

Author Contributions: Conceptualization, X.M. and G.Y.; methodology, X.M.; data curation, X.M., S.A., A.H., A.L., J.L., Z.Z., D.L., A.N.S. and G.Y.; writing—original draft preparation, X.M.; writing—review and editing, X.M. and G.Y.; funding acquisition, G.Y. All authors have read and agreed to the published version of the manuscript.

Funding: This research was funded by the National Natural Science Foundation of China (31271665, 31771708).

Conflicts of Interest: The authors declare no conflict of interest.

References

1. Reddy, N.; Yang, Y. Properties and potential applications of natural cellulose fibers from the bark of cotton stalks. *Bioresour. Technol.* **2009**, *100*, 3563–3569. [CrossRef]

2. Esteve-Turrillas, F.A.; Guardia, M.D.L. Environmental impact of recover cotton in textile industry. *Resour. Conserv. Recycl.* **2017**, *116*, 107–115. [CrossRef]

3. Yang, G.; Tang, H.; Nie, Y.; Zhang, X. Responses of cotton growth, yield, and biomass to nitrogen split application ratio. *Eur. J. Agron.* **2011**, *35*, 164–170. [CrossRef]

4. Yang, G.; Luo, X.; Nie, Y.; Zhang, X. Effects of plant density on yield and canopy micro environment in hybrid cotton. *J. Integr. Agric.* **2014**, *13*, 2154–2163. [CrossRef]

5. Yang, G.; Zhou, M.Y. Multi-location investigation of optimum planting density and boll distribution of high-yielding cotton (*Gossypium hirsutum* L.) in Hubei province, China. *Agric. Sci. China* **2010**, *9*, 1749–1757. [CrossRef]

6. Khan, A.; Najeeb, U.; Wang, L.; Tan, D.K.Y.; Yang, G.; Munsif, F.; Ali, S.; Hafeez, A. Planting density and sowing date strongly influence growth and lint yield of cotton crops. *Field Crops Res.* **2017**, *209*, 129–135. [CrossRef]

7. Shah, A.N.; Yang, G.; Tanveer, M.; Iqbal, J. Leaf gas exchange, source-sink relationship, and growth response of cotton to the interactive effects of nitrogen rate and planting density. *Acta Physiol. Plant.* **2017**, *39*, 119. [CrossRef]

8. Yang, G.; Tang, H.; Tong, J.; Nie, Y.; Zhang, X. Effect of fertilization frequency on cotton yield and biomass accumulation. *Field Crops Res.* **2012**, *125*, 161–166. [CrossRef]

9. Hafeez, A.; Ali, S.; Ma, X.; Tung, S.A.; Shah, A.N.; Liu, A.; Zhang, Z.; Liu, J.; Yang, G. Sucrose metabolism in cotton subtending leaves influencedby potassium-to-nitrogen ratios. *Nutr. Cycl. Agroecosyst.* **2019**, *113*, 201–216. [CrossRef]

10. Reddy, K.R.; Zhao, D. Interactive effects of elevated CO2 and potassium deficiency on photosynthesis, growth, and biomass partitioning of cotton. *Field Crops Res.* **2005**, *94*, 201–213. [CrossRef]

11. Clementbailey, J.; Gwathmey, C.O. Potassium effects on partitioning, yield, and earliness of contrasting cotton cultivars. *Agron. J.* **2007**, *99*, 1130–1136. [CrossRef]

12. Gerardeaux, E.; Jordan-Meille, L. Effect of carbon assimilation on dry weight production and partitioning during vegetative growth. *Plant Soil.* **2009**, *324*, 329–343. [CrossRef]

13. Hu, W.; Yang, J.; Meng, Y.; Wang, Y.; Chen, B.; Zhao, W.; Oosterhuis, D.M.; Zhou, Z. Potassium application affects carbohydrate metabolism in the leaf subtending the cotton (*Gossypium hirsutum* L.) boll and its relationship with boll biomass. *Field Crops Res.* **2015**, *179*, 120–131. [CrossRef]

14. Hu, W.; Zhao, W.; Yang, J.; Oosterhuis, D.M.; Loka, D.A.; Zhou, Z. Relationship between potassium fertilization and nitrogen metabolism in the leaf subtending the cotton (*Gossypium hirsutum* L.) boll during the boll development stage. *Plant Physiol. Biochem.* **2016**, *101*, 113–123. [CrossRef] [PubMed]

15. Gormus, O. Effects of rate and time of potassium application on cotton yield and quality in Turkey. *J. Agron. Crop Sci.* **2002**, *188*, 382–388. [CrossRef]

16. Hu, W.; Coomer, T.D.; Loka, D.A.; Oosterhuis, D.M.; Zhou, Z. Potassium deficiency affects the carbon-nitrogen balance in cotton leaves. *Plant Physiol. Biochem.* **2017**, *115*, 408–417. [CrossRef]

17. Hu, W.; Dai, Z.; Yang, J.; Snider, J.L.; Wang, S.; Meng, Y.; Wang, Y.; Chen, B.; Zhao, W.; Zhou, Z. Cultivar sensitivity of cotton seed yield to potassium availability is associated with differences in carbohydrate metabolism in the developing embryo. *Field Crops Res.* **2017**, *214*, 301–309. [CrossRef]

18. Gerardeaux, E.; Jordan-Meille, L.; Constantin, J.; Pellerin, S.; Dingkuhn, M. Changes in plant morphology and dry matter partitioning caused by potassium deficiency in *Gossypium hirsutum* (L.). *Environ. Exp. Bot.* **2010**, *67*, 451–459. [CrossRef]

19. Onanuga, A.O.; Jiang, P.; Sina, A. Residual level of phosphorus and potassium nutrients in hydroponically grown cotton (*Gossypium hirsutum* L.). *J. Agric. Sci.* **2012**, *4*, 149–160. [CrossRef]

20. Oosterhuis, D.M.; Loka, D.A.; Raper, T.B. Potassium and stress alleviation: Physiological functions and management of cotton. *J. Plant Nutr. Soil Sci.* **2013**, *176*, 331–343. [CrossRef]

21. Tsialtas, I.T.; Shabala, S.; Baxevanos, D.; Matsi, T. Effect of potassium fertilization on leaf physiology, fiber yield and quality in cotton (*Gossypium hirsutum* L.) under irrigated mediterranean conditions. *Field Crops Res.* **2016**, *193*, 94–103. [CrossRef]

22. Chen, Y.; Li, Y.; Hu, D.; Zhang, X.; Wen, Y.; Chen, D. Spatial distribution of potassium uptake across the cotton plant affects fiber length. *Field Crops Res.* **2016**, *192*, 126–133. [CrossRef]

23. Lokhande, S.; Reddy, K.R. Reproductive performance and fiber quality responses of cotton to potassium nutrition. *Am. J. Plant Sci.* **2015**, *6*, 911–924. [CrossRef]

24. Gormus, O.; Yucel, C. Different planting date and potassium fertility effects on cotton yield and fiber properties in the Çukurova region, Turkey. *Field Crops Res.* **2002**, *78*, 141–149. [CrossRef]

25. Tariq, M.; Afzal, M.N.; Muhammad, D.; Ahmad, S.; Shahzad, A.N.; Kiran, A.; Wakeel, A. Relationship of tissue potassium content with yield and fiber quality components of *Bt* cotton as influenced by potassium application methods. *Field Crops Res.* **2008**, *229*, 37–43. [CrossRef]

26. Ali, S.; Hafeez, A.; Ma, X.; Tung, S.A.; Chattha, M.S.; Shah, A.N.; Luo, D.; Ahmad, S.; Liu, J.; Yang, G. Equal potassium-nitrogen ratio regulated the nitrogen metabolism and yield of high-density late-planted cotton (*Gossypium hirsutum* L.) in Yangtze River valley of China. *Ind. Crops Prod.* **2019**, *129*, 231–241. [CrossRef]

27. Hafeez, A.; Ali, S.; Ma, X.; Tung, S.A.; Shah, A.N.; Liu, A.; Ahmed, S.; Chattha, M.S.; Yang, G. Potassium to nitrogen ratio favors photosynthesis in late-planted cotton at high planting density. *Ind. Crops Prod.* **2018**, *124*, 369–381. [CrossRef]

28. Ali, S.; Hafeez, A.; Ma, X.; Tung, S.A.; Liu, A.; Shah, A.N.; Chattha, M.S.; Zhang, Z.; Yang, G. Potassium relative ratio to nitrogen considerably favors carbon metabolism in late-planted cotton at high planting density. *Field Crops Res.* **2018**, *223*, 48–56. [CrossRef]

29. Wright, P.R. Premature senescence of cotton (*Gossypium hirsutum* L.)-Predominantly a potassium disorder caused by an imbalance of source and sink. *Plant Soil.* **1999**, *211*, 231–239. [CrossRef]

30. Bedrossian, S.; Singh, B. Potassium adsorption characteristics and potassium forms in some New South Wales soils in relation to early senescence in cotton. *Aust. J. Soil Res.* **2004**, *42*, 747–753. [CrossRef]

31. Li, B.; Wang, Y.; Zhang, Z.; Wang, B.; Eneji, A.E.; Duan, L.; Li, Z.; Tian, X. Cotton shoot plays a major role in mediating senescence induced by potassium deficiency. *J. Plant Physiol.* **2012**, *169*, 327–335. [CrossRef] [PubMed]

32. Hu, W.; Lv, X.; Yang, J.; Chen, B.; Zhao, W.; Meng, Y.; Wang, Y.; Zhou, Z. Effects of potassium deficiency on antioxidant metabolism related to leaf senescence in cotton (*Gossypium hirsutum* L.). *Field Crops Res.* **2016**, *191*, 139–149. [CrossRef]

33. Read, J.J.; Reddy, K.R.; Jenkins, J.N. Yield and fiber quality of upland cotton as influenced by nitrogen and potassium nutrition. *Eur. J. Agron.* **2006**, *24*, 282–290. [CrossRef]

34. Zahoor, R.; Dong, H.; Abid, M.; Zhao, W.; Wang, Y.; Zhou, Z. Potassium fertilizer improves drought stress alleviation potential in cotton by enhancing photosynthesis and carbohydrate metabolism. *Environ. Exp. Bot.* **2017**, *137*, 73–83. [CrossRef]

35. Makhdum, M.I.; Pervez, H.; Ashraf, M. Dry matter accumulation and partitioning in cotton (*Gossypium hirsutum* L.) as influenced by potassium fertilization. *Biol. Fertil. Soils.* **2007**, *43*, 295–301. [CrossRef]

36. Tang, F.; Wang, T.; Zhu, J. Carbohydrate profiles during cotton (*Gossypium hirsutum* L.) boll development and their relationships to boll characters. *Field Crops Res.* **2014**, *164*, 98–106. [CrossRef]

37. Pettigrew, W.T.; Meredithjr, W.R. Dry matter production, nutrient uptake, and growth of cotton as affected by potassium fertilization. *J. Plant Nutr.* **1997**, *20*, 531–548. [CrossRef]

38. Tung, S.A.; Huang, Y.; Hafeez, A.; Ali, S.; Khan, A.; Souliyanonh, B.; Song, X.; Liu, A.; Yang, G. Mepiquat chloride effects on cotton yield and biomass accumulation under late sowing and high density. *Field Crops Res.* **2018**, *215*, 59–65. [CrossRef]

 *agronomy*

Article

The Use of Appropriate Cultivar of Basil (*Ocimum basilicum*) Can Increase Water Use Efficiency under Water Stress

Iakovos Kalamartzis [1], Christos Dordas [1,*], Pantazis Georgiou [2] and George Menexes [1]

[1] Laboratory of Agronomy, School of Agriculture, Aristotle University of Thessaloniki, 54124 Thessaloniki, Greece; zisvos@yahoo.gr (I.K.); gmenexes@agro.auth.gr (G.M.)

[2] Laboratory of General and Agricultural Hydraulics and Land Reclamation, School of Agriculture, Aristotle University of Thessaloniki, 54124 Thessaloniki, Greece; pantaz@agro.auth.gr

* Correspondence: chdordas@agro.auth.gr; Tel.: +30-231-099-8602; Fax: +30-231-099-8634

Received: 3 November 2019; Accepted: 25 December 2019; Published: 3 January 2020

Abstract: Drought is one of the major yield constraints of crop productivity for many crops. In addition, nowadays, climate change creates new challenges for crop adaptation in stressful environments. The objective of the present study was to determine the effect of water stress on five cultivars of basil (Mrs Burns, Cinnamon, Sweet, Red Rubin, Thai) and whether water use efficiency (WUE) can be increased by using the appropriate cultivar. Water stress affected the fresh and dry weight and also the partitioning of dry matter to leaves, flowers, and stems. Also, there are cultivars, such as Mrs Burns and Sweet, which were not affected by the limited amount of water and continued to produce a high amount of dry matter and also showed high essential oil yield. Essential oil content was not affected by the irrigation; however, essential oil yield was affected by the irrigation, and the highest values were found at Mrs Burns. The water use efficiency was affected by the cultivar and irrigation level, and the highest was found at Mrs Burns. The results show that using appropriate cultivars basil can achieve higher WUE and allow saving water resources and utilizing fields in areas with limited water resources for irrigation.

Keywords: drought tolerance; dry weight yield; essential oil content; leaf area index; *Ocimum basilicum*

1. Introduction

Water deficit is one of the most important yield-limiting factors for agricultural crops worldwide and especially in the Mediterranean area. In addition, climate change and the scenarios that are proposed indicate that water availability will be a limiting factor for many countries in the following years [1]. Therefore, it is important to use water resources more efficiently, which will help us in preserving water resources. One of the ways to conserve water is by using the appropriate crop species and cultivars that have low requirements for water [2].

Aromatic and medicinal plants have received great attention in the last few years because of their multiple uses, such as basil (*Ocimum basilicum* L.) as it can be used for its essential oil, dry leaves, and flowers and also as an ornamental plant [3–5]. Basil has more than 60 different species that were reported throughout the world, and some of them can have important uses [6–8]. Some of the medicinal properties that basil has are that it can be used to cure coughs, headaches, abdominal aches, and kidney diseases [3]. Despite the medicinal properties that basil has, there are also a number of other uses, such as it is used in foods and beverages and can be used as insect repellent [3,9]. Another important characteristic is that basil can be used to produce essential oil with high economic value because it contains important components, such as eugenol, chavicol, and their derivates, and terpenoids, like monoterpene alcohol linalool, methyl cinnamate, and limonite. In addition, different chemotypes of *Ocimum basilicum* L. with a specific chemical composition of essential oils were found [10,11].

Basil is an underutilized crop species with great potential for using it as an alternative crop in many countries because of the many different uses. There are a few studies about the genotypes, the essential oil content, the composition of the essential oil, the effect of fertilization, and growing conditions, such as density, but the effect of water availability was not adequately determined, especially under field conditions. Ekren et al. (2012) [12] found that purple basil was very sensitive to water stress, leading to a significant reduction in the dry matter yield. Despite the fact that there are only a few studies that determined the effect of water stress on basil under field conditions, there are also other studies that were conducted in pots and determined the effect of water stress [13–17]. When *O. basilicum* L. and *O. americanum* L. were exposed to different levels of water stress in pot experiments, they showed significant differences in fresh and dry weight, essential oil content, the main components of the essential oil, proline content, total carbohydrate content, content of major nutrients, such as N, P, K, and protein content decreased [16]. In addition, Yassen et al. (2003) [13] reported that the irrigation of basil at 0.6 and 0.8 irrigation water/cumulative pan evaporation ratio showed the highest herb yield. Under water stress, there was an increase in essential oil content [14]. However, others found that irrigation levels did not affect essential oil content and essential oil components [15]. Basil is characterized by a large leaf area [18], and also, the water consumed per area can be up to 849 mm [12,19].

There are no studies that show the effect of water stress on different basil cultivars in the dry weight yield, on the essential oil yield and agronomic and morphological characteristics, and on water use efficiency (WUE) under field conditions. The objectives of the present study were: 1. To determine the effect of water stress on dry matter and essential oil yield of five basil cultivars under field conditions. 2. To determine the water use efficiency (WUE) of the different basil cultivars and different irrigation treatments and also the WUE of the different growth stages of basil under field conditions.

2. Materials and Methods

2.1. Study Site

A field experiment over two years (2015 and 2016) was established in Northern Greece at the University farm of Aristotle University of Thessaloniki (40°32′9″ N 22°59′18″ E, 0 m) in a clay loam soil with pH (1:1 H_2O) 7.77, $CaCO_3$ 11.3%, EC (dS m^{-1}) 1.07, organic matter 12.40 g kg^{-1}. The pressure plate extractor method [20] was used to obtain the values of physical and hydraulic properties in undisturbed soil samples at 0–30 cm: bulk density (Mg m^{-3}) 1.3, field capacity (at 10 kPa, m^3 m^{-3}) 0.373 and wilting point (at 1500 kPa, m^3 m^{-3}) 0.132. Irrigation water had the following characteristics of pH = 7.0, electrical conductivity EC = 0.6 dS m^{-1}, sodium adsorption ratio (SAR) SAR = 2.00. Durum wheat (*Triticum turgidum subsp. durum* L.) was the previous crop, and wheat straw was baled and removed after harvest. The cultivation area was prepared for seeding by ploughing and harrowing with the use of a cultivator. Nitrogen and P fertilizer were applied at the rates of 100 and 50 kg ha^{-1}, respectively, before planting. Weed control was achieved by tilling and hand weeding. The weather data (rainfall, temperature, relative humidity, solar radiation, and wind speed) were recorded daily with an automatic weather station, which was close to the experimental site, and are reported as mean monthly data for both years (Table 1). The automatic weather station consisted of a data acquisition system and a set of sensors for the measurement of the above-mentioned variables. The data acquisition system used was ZENO®-3200 (Coastal Environmental Systems Inc., Seattle, WA, USA), which is a versatile, low-power, 32-bit data acquisition system, designed to collect, process, store, and transmit data from multiple sensors. The humidity and air temperature sensor used was a Delta OHM HD9009TR (Delta OHM S.r.l., Caselle di Selvazzano (PD), Italy), the wind speed sensor used was the Thies CLIMA Small Wind Transmitter 4.3515.30.000 (THIES CLIMA, Gottingen, Germany), the solar radiation sensor used was the LP PYRA 02 pyranometer (Delta OHM S.r.l., Caselle di Selvazzano (PD) Italy), and the rainfall sensor used was the aerodynamic rain gauge ARG100 (Campbell Scientific Ltd, Loughborough, UK) and RGB1 levelling base plate (Campbell Scientific Ltd, Loughborough, UK) [21]. Weather conditions during the two years were different as the temperature was lower in May of 2015, and the rainfall was higher compared with 2016 (Table 1). In addition, during the irrigation period, the other parameters were similar to the 30-year average values (Table 1).

Table 1. The main weather parameters (mean relative humidity (RH_{mean}), rainfall, maximum (T_{max}), minimum (T_{min}), and mean (T_{mean}) temperature), and reference evapotranspiration (ET_o), for the two years and its comparison to the 30-year average. The weather data were recorded with an automatic weather station close to the experimental site.

	Year 2015			Year 2016			30-Year Average		
	June	July	August	June	July	August	June	July	August
T_{max} (°C)	29.8	34.3	33.8	32.4	34.5	33.8	30.2	32.5	32.2
T_{min} (°C)	17.1	20.5	20.4	18.7	21.2	20.4	15.9	18.2	18.0
T_{mean} (°C)	23.2	27.5	27.1	25.9	27.8	27.1	24.5	26.7	26.0
RH_{mean} (%)	66.7	62.7	63.9	62.3	58.9	62.1	60	58	62
Rainfall (mm)	96.2	8.2	1.1	15.2	1.2	0.8	32	31	24
ET_o (mm/day)	4.5	5	4.5	4.8	5	5	4	5	5

2.2. Plant Cultivars Used in the Study

During 2013 and 2014, twenty basil cultivars were evaluated under field conditions for their agronomic characteristics and also for their essential oil yield [22]. From the twenty basil cultivars, five different basil cultivars were used in this study that had differences in earliness and essential oil content. The cultivars were Cinnamon (early vigorous plant with distinctive cinnamon scent), Mrs Burns Lemon (early, vigorous plant with distinctive lemon scent), Sweet (medium maturity, new hybrid of Genovese type cultivar, more pointed leaf, vigorous with good uniformity, slow to flower, and has broad-spectrum tolerance to *Fusarium* wilt.), Thai (late, mild anise or liquorice flavour, with attractive purple stems and dark green leaves tinged purple), and Red Rubin (late, a good red for cut leaf or pot production).

2.3. Crop Management and Experimental Design

The experiment was based on the randomized complete block design (RCBD) in a split-split plot arrangement. Irrigation levels were the main plots, cultivars were the sub-plots, and the repeated measures on the three different growth stages were the sub-sub plots. There were four replications (blocks) per treatment combination. Each block was divided into three strips corresponding to the three irrigation levels, and within each strip, the five cultivars were randomized. Every plot had five rows; the length of each row was 5 m, the rows were 50 cm apart, and the total size of each plot was 12.5 m^2. Seeds were sown in a mixture of peat and perlite (9:1) on 4 April 2015 and 19 March 2016. When the basil seedlings reached 10 cm in plant height, they were hand-planted on 16 May 2015 and on 25 April 2016 at a rate of 8 plants m^{-2}.

Irrigation treatments that were applied were 100%, 70%, and 40% of the net irrigation requirements (IR_n) and are presented as d_{100}, d_{70}, and d_{40}, respectively. IR_n was calculated from the equation:

$$IR_n = ET_c - P_e - CR + D_p + R_{off}$$ (1)

where ET_c is the crop evapotranspiration, P_e is the effective rainfall, which was taken into account only when it was higher than 4 mm on any day, and entire rainfall was considered as effective rainfall, CR is the capillary rise from the groundwater table, D_p is the deep percolation, and R_{off} is the runoff. In this study, the CR, D_p, and R_{off} were negligible because (a) there was no shallow water table problem in the experimental area; thus, the CR value was assumed to be zero, (b) D_p was not assumed since the amount of irrigation water was equal to the deficit amount in the root zone, and (c) irrigation was performed with drip irrigation, and there was no runoff.

Reference evapotranspiration (ET_o) was calculated with the FAO Penman-Monteith method [23] with the following equation:

$$ET_o = [0.408\Delta(R_n - G) + \gamma[900/(T + 273)]u_2(e_s - e_a)]/[\Delta + \gamma(1 + 0.34u_2)] \tag{2}$$

where ET_o is the reference evapotranspiration (mm day^{-1}), R_n is net radiation at the crop surface (MJ m^{-2} day^{-1}), G is soil heat flux density (MJ m^{-2} day^{-1}), T is mean daily air temperature at 2 m height (°C), u_2 is wind speed at 2 m height (m s^{-1}), e_s is saturation vapour pressure (kP$_a$), e_a is actual vapour pressure (kP$_a$), $e_s - e_a$ is the saturation vapour pressure deficit (kP$_a$), Δ is the slope vapour pressure curve (kP$_a$ °C^{-1}), γ is the psychrometric constant (kP$_a$ °C^{-1}).

Crop evapotranspiration (ET_c) was calculated with the following equation:

$$ET_c = k_c \times ET_o \tag{3}$$

where k_c is the crop coefficient.

The following values of crop coefficient (k_c) were used: for the beginning of flowering, 0.9; for full bloom, 1.1; for the end of flowering, 1.0 [19].

After transplantation, 30 mm of irrigation water was applied in order to promote the establishment of the newly transplanted plants. The differentiation of irrigation levels started when the plants were at the vegetative stage and 40 days after transplantation and 30 days before anthesis. The irrigation date was the same for the three treatments and was determined when the soil moisture was at 70% of field capacity of the full irrigation treatment (d_{100}), which is considered adequate for plant growth in all growth stages (Table 2). For the two deficit irrigation treatments (d_{70} and d_{40}), the water amount was determined according to the full irrigation treatment, which was 70% and 40% of the full irrigation (Table 2). The water was applied with a drip irrigation system; after transplanting with the drippers spaced at 50 cm intervals, the water supply of the drippers was 4 L h^{-1}. The drip irrigation lines were placed every other row. The same irrigation system was extensively used in other experiments [24,25].

Table 2. Date and amount (mm) of applied irrigation water during the two years of the study and the three treatments (d_{100}, d_{70}, and d_{40}).

| | 2015 | | | | | |
| | Date (DD/MM/YEAR) | | | | | |
Treatment	16/7/2015	21/7/2015	28/7/2015	31/7/2015	13/8/2015	Total water applied
d_{100}	74.9	42.5	42.5	21.3	21.3	202.5
d_{70}	52.4	29.8	29.8	14.9	14.9	141.8
d_{40}	30	17.0	17.0	8.5	8.5	81
	2016					
	Date (DD/MM/YEAR)					
	14/6/2016	18/6/2016	22/6/2016	1/7/2016	12/7/2016	
d_{100}	38.3	51.1	42.5	46.8	65.9	244.6
d_{70}	26.8	35.7	29.8	32.8	46.2	171.3
d_{40}	15.3	20.4	17.0	18.7	26.4	97.8

2.4. Morphological Parameters

The following morphological parameters were determined: plant height, leaf area index (LAI), and number of branches. Plant height was determined three times before each sampling by measuring the height of five randomly selected plants per plot from the soil to the top of the plant and getting an average value for each plot. LAI was recorded three times before each sampling using the *AccuPAR* system (model LP-80, *PAR/LAI* Ceptometer, Decagon Devices, Inc., Pullman, WA, USA). For the determination of LAI, one measurement of photosynthetically active radiation (PAR) was taken above the canopy, and three measurements of PAR were taken at the soil level following the manufacturer's recommendation. The number of branches was determined by measuring the number of main branches from eight plants from each plot at the three samplings.

2.5. Crop Sampling and Essential Oil Determination

The crop was sampled three times during the growing season. The first was at the beginning of flowering, the second at full bloom, and the third at the end of flowering, and started from the first week of July until the first week of August in both years. At each sampling, 1 m^2 of the inner row was randomly selected and cut at the ground level. It was then weighted to obtain the fresh weight (kg ha^{-1}) and left to dry for a week at room temperature; when the samples reached a constant weight, they were weighted to obtain the final dry weight. A subsample of 0.5 kg biomass was obtained and dried at 65 °C to a constant weight to determine the relative water content and the dry weight yield. The leaves and flowers were separated from stems by hand and weighed. The essential oil content was determined by using 40 g of dry leaf materials subjected to water distillation for 3 h using a Clevenger apparatus. The essential oil yield was determined by multiplying the essential oil content by the dry weight.

2.6. Water Use Efficiency

The water use efficiency (WUE) for the different cultivars and harvests was determined by dividing the dry weight yield by the total water (rainfall and irrigation) that each treatment received [26].

2.7. Statistical Analysis

The data were analyzed with the ANOVA method according to a split-split-split-plot design combined over years (years × irrigation levels × cultivars × growth stages) with four replications (blocks) per treatment combination. The years were considered as the main plots, irrigation levels were the sub-plots, cultivars were the sub-sub-plots, and growth stages were the sub-sub-sub plots [27]. A combined analysis over years was carried out according to the aforementioned design. It must be noted that, within each year, the basic experimental design was based on the RCBD in a split-split plot arrangement, as described previously. The combined analysis over the years is statistically equivalent to a split-split-split plot arrangement and analysis, where the years are now considered as the main plots and, consequently, the main plot factor (irrigation levels) previously specified now becomes a split factor and so on.

The least significant difference (LSD) criterion was used to test the differences between treatment means, and the significance level of all hypotheses tested was preset at $p < 0.05$. All statistical analyses were performed using the SPSS software package (ver. 17, SPSS Inc., Chicago, IL, USA). The statistical analysis (ANOVA) with SPSS was done within the frame of mixed linear models using an SPSS syntax code developed and programmed by the authors.

3. Results

The rainfall was different between the two years. The first year, 2015, rainfall was higher during June and low during July and August. In contrast, during the second year, 2016, there was a different trend as there was lower rainfall during the summer months (June, July, and August). The other weather parameters were similar in both years (Table 1). The irrigation water applied in 2016 was 20% higher than that applied in 2015 because in June 2015 there was higher rainfall of total 96.2 mm and this decreased the need for irrigation in 2015. Most of the characteristics were affected by the main effect of year (Y), irrigation (W), cultivar (C), and growth stages (S) and also by some of their two way and higher order interactions (Table 3). Ratio of leaves and flowers per stems was affected only by the main effect of year and growth stages (Table 4). The number of branches was affected only by the main effect of irrigation and cultivar (Table 4). The two-way interaction "cultivar × year" had a statistically significant effect on all plant characteristics except on LAI. The effect of the interaction "irrigation × year" was only statistically significant for the ratio of leaves and flowers per stems and LAI. The interaction "growth stages × year" had a statistically significant effect on all characteristics except on dry weight of stems. The effect of the interaction "cultivar × growth stages" was statistically significant for all the characteristics except the ratio of leaves and flowers per stems. The interaction "irrigation × growth stages" had a statistically significant effect only on plant height. The interaction "cultivar × irrigation" effect was only statistically significant for the ratio of leaves and flowers per stems. The three-way interaction "cultivar × year × irrigation" was statistically significant only for plant height and LAI. The interaction "cultivar × irrigation × growth stages" effect was statistically significant for the ratio of leaves and flowers per stems, the plant height, the LAI, and essential oil content. The interaction "irrigation × year × growth stage" had a statistically significant effect only on the ratio of leaves and flowers per stems. The interaction "cultivar × year × growth stages" effect was statistically significant for all the measured plant characteristics. Finally, the four-way interaction "cultivar × year × irrigation × growth stages" had a statistically significant effect on the ratio of leaves and flowers per stems, LAI, essential oil content, and essential oil yield. Based on Table 3, for all measured plant characteristics, there are significant two-way and three-way interactions (and in three cases, there are significant four-way interactions) that involve the combination of the four factors, in some cases, in pairs, and in others, in triplets. Consequently, there is a point to present the synergistic effect of cultivar, irrigation, year, and growing stage; that is, to present the mean values for all treatments' combinations in Table 5 and in Figures 1–4.

3.1. Fresh and Dry Weight

The fresh weight of the different cultivars was affected by the irrigation treatments, growth stages, years, and cultivars. In addition, the fresh weight was affected by the interactions between cultivars and years, growth stages and years, growth stages and cultivars, and by the three-way interactions of cultivars, years, and growth stages (Table 4). There was a much higher reduction in the Thai cultivar compared with the Mrs Burns cultivar in the fresh weight between the first sampling (beginning of flowering) and in the other growth samplings. The fresh weight was reduced by 25%, 36%, and 34% at d_{40} compared with the d_{100} at the beginning of flowering, full bloom, and at the end of flowering, respectively. In addition, during 2015, the fresh weight overall was much higher than in 2016 (Table 4).

Similarly, the dry weight was affected by the cultivar and also by the irrigation treatments, growth stage, year, cultivars, and by the interactions of cultivars with years and growth stages (Tables 3 and 4). In addition, there was an increase in dry weight for Mrs Burns and Cinnamon from the initiation of flowering to full bloom. In 2016, both 'Mrs Burns' and 'Cinnamon' showed a growing pattern not just from the initiation to flowering to full bloom but from beginning of flowering to end of flowering. On the other hand, in 2015, the dry weight of 'Mrs Burns' remained stable from full bloom to the end of flowering while the dry weight of 'Cinnamon' increased. For the cultivar Sweet, there was an increase in the dry weight from the first stage to the second, and then there was a decrease from the second to the third. For the cultivar Red Rubin in 2016, there was also an increase between the first and the second growth stage, but there was no difference between the second and the third growth stage. Thai cultivar showed no significant response between the three growth stages.

The dry weight of leaves and flowers was also affected by all factors that were determined and also by their interactions (Tables 3 and 4). It was followed by a similar response to the total dry weight, and it was higher in the second growth stages. The cultivar that showed the highest dry weight of leaves and flowers was Mrs Burns in both years. The lowest was found at the Red Rubin for both years. Red Rubin and Thai had a higher dry weight of leaves and flowers compared with the stems.

Table 3. Analysis of variance results (significance of the effects) for testing the effects (main and interactions) of Year (Y), Irrigation (W), Cultivar (C), and Growth Stages (S) on the measured plant characteristics. Where LAI is leaf area index, and WUE is water use efficiency.

Plant Characteristics	Year (Y)	Irrigation (W)	Cultivar (C)	Growth Stages (S)	C×Y	W×Y	S×Y	C×S	W×S	C×W	C×Y×W	C×W×S	W×Y×S	C×Y×S	C×Y×W×S
Fresh weight	***	***	***	***	**	NS	***	***	NS	NS	NS	NS	NS	***	NS
Dry weight	***	***	***	***	***	NS	***	***	NS	NS	NS	NS	NS	***	NS
Dry weight of leaves and flowers	***	**	***	***	**	NS	***	***	NS	NS	NS	NS	NS	***	NS
Dry weight of stems	***	***	***	***	***	NS	NS	***	NS	NS	NS	NS	NS	***	NS
Ratio of leaves and flowers/stems	*	NS	NS	*	**	**	***	NS	NS	*	NS	*	**	**	*
Plant height	***	***	***	***	***	NS	***	***	***	NS	*	**	NS	***	NS
Number of branches	NS	*	***	NS	*	NS	***	NS	NS	NS	NS	NS	NS	**	NS
LAI	***	***	***	***	NS	*	***	***	NS	NS	*	***	NS	***	**
Essential oil yield	***	**	***	**	NS	NS	***	***	NS	NS	NS	NS	NS	***	NS
Essential oil content	***	NS	**	***	***	NS	***	***	NS	NS	NS	*	NS	*	**
WUE$_{DW}$	***	***	***	***	*	NS	***	***	NS	NS	NS	NS	NS	***	NS

* Significant at 0.05 significance level according to ANOVA; ** Significant at 0.01 significance level according to ANOVA; *** Significant at 0.001 significance level according to ANOVA; NS, non-significant ($p > 0.05$).

Table 4. Combined effect of Cultivar, Year (2015, 2016), and Growth Stages, and the main effect of Irrigation on fresh and dry weight, dry weight of leaves, flowers, and stems, and number of branches. Data presented are mean values, where LSD is the least significant difference at the 0.05 significance level.

Cultivars	Fresh Weight (g/m^2)	Dry Weight (g/m^2)	Dry Weight of Leaves and Flowers (g/m^2)	Dry Weight of Stems (g/m^2)	Number of Branches
			2015 beginning of flowering		
Mrs Burns	3285.4	569.2	246.3	322.9	8.0
Cinnamon	2730.4	513.1	213.6	299.5	9.5
Sweet	3965.8	898.3	441.8	456.6	10.1
Red Rubin	2028.3	405.8	197.6	208.3	8.8
Thai	3081.7	594.2	303.3	290.8	8.0
			2016 beginning of flowering		
Mrs Burns	2223.3	374.9	192.3	182.6	10.6
Cinnamon	1440.3	278.1	173.4	104.7	11.3
Sweet	886.5	213.3	107.7	105.6	9.9
Red Rubin	466.0	74.8	43.1	31.7	8.2
Thai	1029.4	191.2	111.4	79.7	10.4

Table 4. *Cont.*

Cultivars	Fresh Weight (g/m^2)	Dry Weight (g/m^2)	Dry Weight of Leaves and Flowers (g/m^2)	Dry Weight of Stems (g/m^2)	Number of Branches
			2015 full bloom		
Mrs Burns	4651.7	1117.5	572.7	544.8	8.5
Cinnamon	3755.8	880.8	456.0	424.8	10.7
Sweet	2315.8	966.7	420.9	545.8	8.6
Red Rubin	1632.5	531.7	244.0	287.7	8.9
Thai	1810.8	669.2	356.6	312.6	8.8
			2016 full bloom		
Mrs Burns	2970.4	561.9	284.8	277.1	9.6
Cinnamon	1640.8	338.7	184.3	154.3	10.5
Sweet	1682.5	442.8	191.8	250.9	8.9
Red Rubin	963.3	218.8	89.8	128.9	8.3
Thai	976.6	243.6	130.1	113.6	8.2
			2015 end of flowering		
Mrs Burns	3088.3	1141.7	546.3	595.4	10.2
Cinnamon	3265.8	1080.8	541.5	539.3	10.3
Sweet	1912.5	819.2	294.3	524.9	9.4
Red Rubin	1139.2	426.7	199.8	226.8	9.9
Thai	1662.5	668.3	384.6	283.8	8.9
			2016 end of flowering		
Mrs Burns	3110.1	876.9	372.8	504.1	8.7
Cinnamon	2157.0	629.8	302.3	327.4	9.0
Sweet	1282.0	330.8	168.1	162.7	9.0
Red Rubin	1233.8	214.6	122.3	92.3	7.9
Thai	1062.2	229.9	144.8	85.1	9.1
LSD$_{0.05}$	376.7	85.7	49.6	52.9	1.3
Irrigation treatment	**Fresh weight (g/m^2)**	**Dry weight (g/m^2)**	**Dry weight of leaves and flowers (g/m^2)**	**Dry weight of stems (g/m^2)**	**Number of branches**
d$_{40}$	1716.5	500.7	254.1	246.7	9.6
d$_{70}$	2112.0	547.8	270.4	277.4	9.4
d$_{100}$	2516.5	601.7	279.3	322.4	8.8
LSD$_{0.05}$	173.5	38.6	16.3	27.8	0.6

3.2. Ratio of Leaves and Flowers to Stems, Plant Height, Number of Branches, and Leaf Area Index

The ratio of leaves and flowers by the stems was also affected by the growth stages, year, cultivars, and also by their interactions (Tables 3 and 5). The cultivar that showed the highest ratio was Mrs Burns and Cinnamon, and the lowest was found at Red Rubin and Thai. In addition, the water level affected the ratio, and the highest was found at the d_{100} and d_{70}, and the lowest at the d_{40} overall in the five cultivars and the two years of the experiments.

Plant height was affected by all the factors that were determined and more specifically by growth stages, year, irrigation, cultivar, and also by their interactions. Plant height increased from the first to the third measurement. The tallest cultivar was Sweet, followed by the cultivar Mrs Burns. The shortest variety was Thai, and the following cultivar was Red Rubin (Table 5). There were differences between the three water levels, and the highest plant height was at d_{100}, which was 16% higher compared with the d_{40}.

Table 5. Combined effect of Cultivar, Irrigation, Year (2015, 2016), and Growth Stages on plant height and ratio of leaves and flowers/stems. Data presented are mean values, where LSD is the least significant difference at the 0.05 significance level.

Cultivars	Irrigation Treatment	Plant Height (cm)	Ratio of Leaves and Flowers/Stems (g/m^2)	Plant Height (cm)	Ratio of Leaves and Flowers/Stems (g/m^2)	Plant Height (cm)	Ratio of Leaves and Flowers/Stems (g/m^2)
		Beginning of Flowering		Full Bloom		End of Flowering	
2015							
Mrs Burns	d_{40}	66.4	0.88	75.4	1.02	78.8	0.72
	d_{70}	60.9	1.19	82.0	0.92	83.6	1.01
	d_{100}	60.3	0.85	80.3	0.66	85.2	1.04
Cinnamon	d_{40}	57.5	1.05	63.3	0.68	65.9	1.01
	d_{70}	57.1	1.18	67.6	1.10	74.9	1.10
	d_{100}	52.2	1.06	66.2	1.02	78.6	0.86
Sweet	d_{40}	71.7	0.74	73.5	1.94	76.7	1.25
	d_{70}	75.6	1.19	84.4	1.30	86.4	0.92
	d_{100}	75.8	0.91	95.6	1.08	93.3	0.55
Red Rubin	d_{40}	44.7	1.16	47.9	0.73	48.2	1.09
	d_{70}	47.7	1.07	55.1	1.14	57.1	0.95
	d_{100}	54.3	0.74	61.9	0.95	62.3	1.18
Thai	d_{40}	37.0	0.74	43.2	1.50	40.9	0.96
	d_{70}	38.7	0.95	43.1	1.41	44.8	0.77
	d_{100}	43.7	0.99	50.4	0.97	53.7	0.57
2016							
Mrs Burns	d_{40}	44.1	1.28	54.4	0.91	65.1	1.70
	d_{70}	44.1	0.82	57.4	0.82	71.0	1.46
	d_{100}	51.0	1.17	68.5	0.89	84.2	2.21
Cinnamon	d_{40}	39.2	0.75	45.7	1.25	53.2	0.97
	d_{70}	39.9	0.64	49.9	0.70	57.2	1.09
	d_{100}	44.4	0.58	54.0	1.41	62.4	2.89
Sweet	d_{40}	50.6	0.95	57.9	0.95	64.1	1.63
	d_{70}	48.5	0.68	63.3	0.69	67.1	1.18
	d_{100}	50.2	0.86	70.2	0.77	73.5	1.71
Red Rubin	d_{40}	34.9	0.96	42.8	1.29	49.0	1.01
	d_{70}	31.7	1.16	45.7	0.57	50.1	1.31
	d_{100}	37.7	0.96	46.8	1.37	54.1	1.23
Thai	d_{40}	25.8	1.53	29.2	1.22	30.9	1.03
	d_{70}	28.2	1.10	26.4	1.19	33.5	0.98
	d_{100}	30.0	1.24	32.6	1.21	40.8	1.50
LSD$_{0.05}$		4.8	0.8	4.8	0.8	4.8	0.8

The number of branches was affected by irrigation, cultivar, and by the interaction of year with cultivar and growth stages. The cultivar with the more branches was Mrs Burns, followed by the cultivar Cinnamon. The cultivar with the lowest number of branches was Sweet (Table 5).

Leaf area index was affected by growth stages, year, irrigation, and cultivar (Figure 1). In addition, LAI showed significant interaction of the different factors that were studied. LAI was increased from the first to the second measurement in all treatments and cultivars and showed a decrease from the second to the third measurement. The cultivar with the highest LAI was Mrs Burns, followed by the cultivar Cinnamon. The cultivar with the lowest LAI was Thai (Figure 1). There were differences between the three water levels, and the highest LAI in most cultivars was at d_{100}.

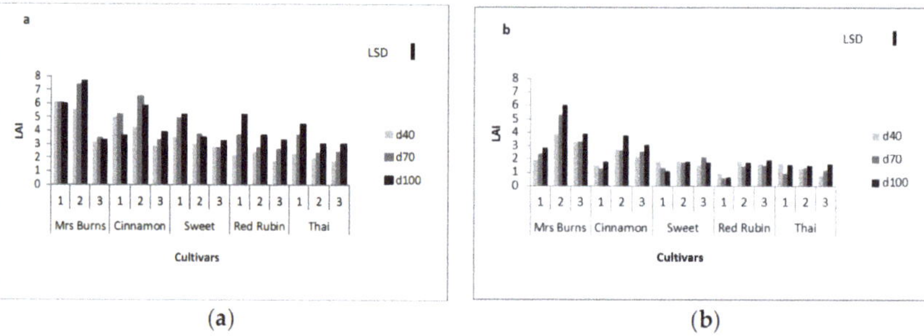

Figure 1. Leaf area index (LAI) of the five basil cultivars during the three growing stages (where 1 is for the initiation of flowering, 2 is for the full bloom, 3 is the end of flowering) at the three irrigation levels (d_{40}, d_{70}, and d_{100}) for the two growing seasons 2015 (**a**) and 2016 (**b**). Data presented are mean values; vertical bar corresponds to the least significant difference (LSD).

3.3. Essential Oil Content and Yield

Essential oil content was affected by growth stages, year, cultivar, and also by their interactions (Figure 2). However, it was not affected by irrigation and the two-way interactions of irrigation with other factors. The highest essential oil content was found at Mrs Burns cultivar, followed by Cinnamon and Thai. The lowest essential oil content was found at Red Rubin (Figure 2).

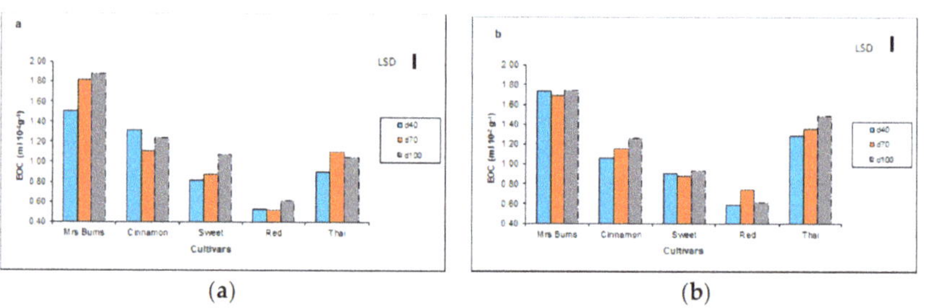

Figure 2. Essential oil content (EOC) of the five basil cultivars at the three irrigation levels (d_{40}, d_{70}, and d_{100}) for 2015 (**a**) and 2016 (**b**). Data presented are mean values of the three growth stages; vertical bar corresponds to the least significant difference (LSD).

Essential oil yield was affected by growth stages, year, irrigation, cultivar, and also by their interactions (Figure 3). Essential oil yield was different for each cultivar, and it was also different between the two years of the study. The cultivar with the highest essential oil yield was Mrs Burns, followed by the cultivar Cinnamon. The cultivar with the lowest essential oil yield was Red Rubin (Figure 3). There were no differences between the d_{100} and d_{70} treatments in the essential oil yield in most cultivars, and the lowest essential oil yield was found at the d_{40}.

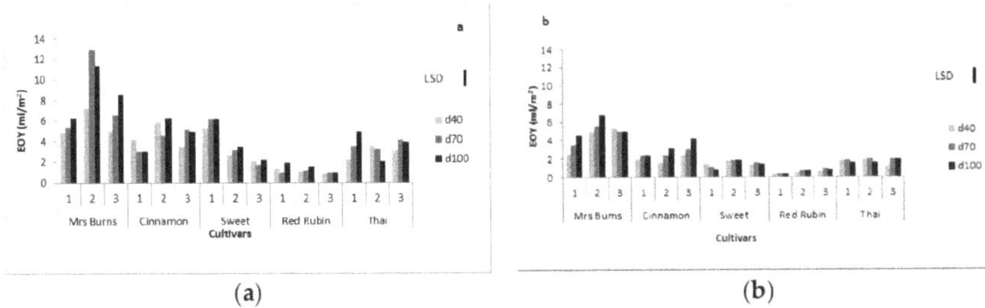

Figure 3. Essential oil yield (EOY) of the five basil cultivars during the three growing stages (where 1 is for the initiation of flowering, 2 is for the full bloom, 3 is the end of flowering) at the three irrigation levels (d_{40}, d_{70}, and d_{100}) for the two growing seasons 2015 (**a**) and 2016 (**b**). Data presented are mean values; vertical bar corresponds to the least significant difference (LSD).

3.4. Water Use Efficiency

Water use efficiency (WUE) was affected by growth stages, year, irrigation, cultivar, and by the interactions between cultivars and years, growth stages and years, cultivars and growth stages, and by the interaction of cultivars, years and growth stages. WUE was higher at the d_{40} treatment and lower at the d_{100} treatment (Figure 4). The highest WUE was found at Mrs Burns cultivar, followed by Cinnamon, and the lowest was found at Red Rubin. The trend was similar in all cultivars, and the lowest WUE was found at the d_{100} treatment.

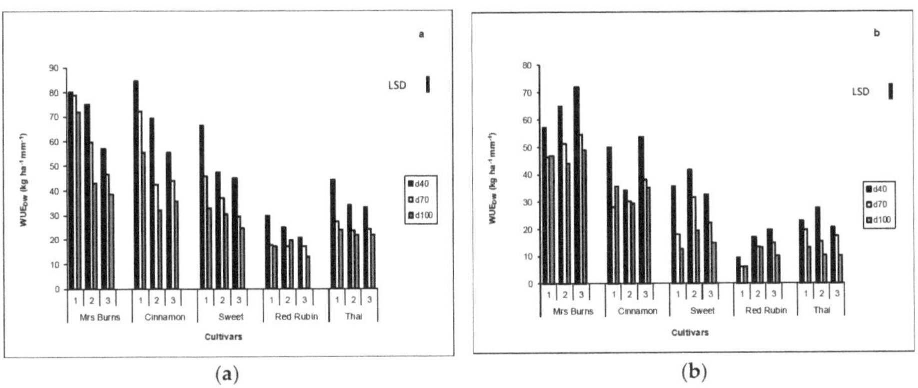

Figure 4. Water use efficiency (WUE) of the five basil cultivars at the three irrigation levels (d_{40}, d_{70}, and d_{100}), at the three growing stages of the two growing seasons 2015 (**a**) and 2016 (**b**). Data presented are mean values; vertical bar corresponds to the least significant difference (LSD).

4. Discussion

4.1. Fresh and Dry Weight

Basil is a species that was not studied extensively, and there is a need to determine the effect of water availability on the productivity of the different basil cultivars with economic importance under field conditions and especially in a Mediterranean environment. It was found that the water level affected the fresh and dry weight of the crop and also the partitioning of dry weight to leaves and flowers, which is an important commercial characteristic. Water stress can affect the fresh and dry weight of basil [12]; however, it is not known how the different basil cultivars can be affected by water stress and whether using the appropriate cultivar can help conserve water and at the same time maintain high productivity and quality [28,29]. Some cultivars showed a much higher reduction in fresh and dry weight, like the Thai cultivar, compared with others, like Mrs Burns, which were affected less by water stress (Tables S1–S3). This could be because Mrs Burns has a much deeper root system and can take up water from deep in the soil. In addition, Mrs Burns was a cultivar with higher growth rate and higher biomass, and the dry matter of the cultivar was much higher than the others, even under stress conditions. This could also be because this cultivar has other adaptations compared with bigger root systems, such as cuticles, hairs, and better control of stomata, which can reduce the amount of water that is lost from the plant and can maintain its growth under a limited water status [30,31]. From the five cultivars that were tested, Mrs Burns showed the lowest reduction in dry weight, followed by Cinnamon and Sweet, which indicates that these cultivars have better adaptability to water stress. In addition, the cultivars that showed the highest reduction in dry weight were Thai and Red Rubin, and they probably have lower adaptability to water stress. The water stress treatment d_{70} showed that it did not reduce the yield significantly, but, on the other hand, it can help us to conserve water for other crops.

Also, in some cultivars, as the water level increased, there was a decrease in the fresh weight. This can be because of the higher soil water level, which can cause extensive leaf area sensitive to leaf diseases and also show premature leaf senescence and leaf drop [32].

In addition, there was an increase in dry weight for Mrs Burns and Cinnamon from the initiation of flowering to full bloom. Cultivar Sweet showed an increase in the dry weight from the first growth stage to the second growth stage, and then there was a decrease from the second to the third; this decrease could be because of leaf senescence and leaf drop [32]. The fresh and dry weight was much higher than other studies, as, in most studies, the fresh weight of different basil cultivars grown in the field was in the range of 240.2–1105.9 g m^{-2} and dry weight was in the range of 47.9–202.8 g m^{-2} [12,18,20], and in our study, fresh weight was in the range of 378.5–4357.5 g m^{-2} and dry weight in the range of 65.8–922.5 g m^{-2}. It is known that the fresh and dry weight of basil are affected by a number of factors such as irrigation, fertilization, sowing time, plant density, weather conditions (temperature, humidity), and genotype [12,18,33,34]. The fresh and dry weight that was found in the present study was higher than other studies because of the higher growth of the basil plants as the plants reached 90 cm in height, possibly due to better growth conditions.

The dry weight of leaves and flowers was also affected by the irrigation treatments, growth stages, year, and cultivars. The cultivar that showed the highest dry weight of leaves and flowers was Mrs Burns in both years. The lowest was found at Red Rubin for both years. It was reported in several plant species that except from the environmental factors, genetic differences may also affect the productivity of the aromatic and medicinal plants and also their essential oil yield. Dry weight was reduced as water level was reduced, and this could be because of the reduction in leaf area index and, consequently, photosynthesis [16,35–38]. The ontogenetic stage in which water stress had the highest effect was the end of flowering as the stress was more pronounced and followed by the full bloom.

The ratio of leaves and flowers by the stems is also an important commercial characteristic for basil and also other aromatic and medicinal plants since the most important characteristic is the yield of dry leaves and flowers. The cultivar that showed the highest ratio was Cinnamon, and the cultivar with the lowest was Thai; this trend was similar to the dry matter accumulation. In addition, the water level affected the ratio differently of the five cultivars that were tested. Red Rubin was quite sensitive to water stress, and a similar response was found by others [12].

4.2. Plant Height, Number of Branches, and Leaf Area Index

The plant height of the basil cultivars that were tested was affected by all the factors that were studied. There were significant differences between the two years of the experiments, and this can be because of the weather conditions as in 2016 the temperature was higher in June and in July and the rainfall was much lower [4]. Similar trends between the years of the experiments were reported by others [12,22]. In addition, the water level affected the plant height and the highest difference between d_{100} and d_{40} was found at the end of flowering, which was 16%. This is a common response, as when plants are exposed to water stress, there is a decrease in growth rate and also in plant height [12,14,39]. In addition, in several other aromatic and medicinal plants, there was a decrease in plant height, like *Origanum majorana* [40] and *Mentha arvensis* [41] under water stress. The plant height that was found in the present study was much higher than other studies [12,42], which is because of the better conditions, and this was similar to other studies [35,36].

The number of branches was affected by the irrigation, cultivar, and the interaction of cultivar and year, growth stage and year, and the interaction of cultivar, year, and growth stage. Morphological characteristics, such as number of branches, are affected by irrigation, fertilization, and cultivar [43,44].

One of the plant adaptations of leafy plants is to reduce the leaf area [31]. Leaf area index is an important characteristic for dry matter yield, and the commercial products of the aromatic and medicinal plants, and also affects the photosynthesis and the dry matter production. Irrigation can affect the development of the leaf area and also the production of dry weight [18]. Leaf area index was affected by different treatments and also their interactions, and when there was a reduction in water availability, there was a significant reduction in leaf area index by an average of 22% and reached 59% in some cultivars, like in Red Rubin in 2015 at the first measurement.

4.3. Essential Oil Content and Yield

Despite the fact that essential oil content was not affected by the irrigation level, essential oil yield was affected by year, irrigation, cultivar, and growth stages as it was increased from the first to the third stage. This is because there was an increase in dry weight, and the same response was found in other species [17,40,41,45]. In addition, in some species, it was found that water stress can increase the essential oil content, but in other species, there was no effect [17,40,41,45]. Also, essential oil yield can be affected in the same way [17,40,41,45]. In the present study, the highest essential oil yield was found at Mrs Burns and followed by the cultivar Cinnamon because these cultivars had the highest dry matter and the essential oil content was not affected much by water stress. In contrast, the cultivars that showed the lowest essential oil yield was Thai, which was affected more by water stress and had the lowest dry matter yield. Similar responses were reported for other species [17,40,41,45].

4.4. Water Use Efficiency

The highest WUE was found at Mrs Burns cultivar and followed by Cinnamon and the lowest at Red Rubin. The trend was similar in all cultivars, and the lowest WUE was found at the d_{100} treatment. WUE was affected by the irrigation level and also by the cultivar, which is very important, as, in limited water supply, it is better to find cultivars tolerant to water stress that can efficiently use water [24,25,45–48]. In addition, the WUE that was found was much higher than in other species such as maize, and this is because basil and also other aromatic and medicinal plants are not harvested for grains, but they are harvested at full anthesis and do not require water for their whole growth period [46–48]. Therefore, basil can help in conserving water resources that can be used for other crops.

5. Conclusions

This study describes the effect of water stress on five different basil cultivars, and it was found that water affects the fresh and dry weight and also the partitioning of dry matter to leaves, flowers, and stems. Basil does not seem very sensitive to water stress, and a reduction of water by 60% compared with the full irrigation was not great to for significantly affecting the dry weight as it was lower by 34% compared with the full irrigation. Also, it was found that some cultivars, like Mrs Burns, were not affected by the limited amount of water and continue to show high dry weight accumulation even at d_{40} and also have high essential oil yield. These cultivars can be used in water limited environments and help to conserve our water resources, and also can be used by the farmers for higher yield under water limited environments. In addition, a significant increase in WUE can be achieved by the selection of appropriate cultivars and water management systems and can be used to conserve water resources.

Supplementary Materials: The following are available online at http://www.mdpi.com/2073-4395/10/1/70/s1, Table S1: Effect of cultivar and irrigation on fresh and dry weight, dry weight of leaves, flowers and stems and ratio of leaves and flowers to stems at the beginning of flowering during the two years (2015 and 2016). Table S2. Effect of cultivar and irrigation on fresh and dry weight, dry weight of leaves and flowers, stems and ratio of leaves and flowers to stems at full bloom during the two years (2015 and 2016). Table S3. Effect of cultivar and irrigation on fresh and dry weight, dry weight of leaves and flowers, stems and ratio of leaves and flowers to stems at the end of flowering during the two years (2015 and 2016).

Author Contributions: All the authors have contributed to the manuscript significantly. I.K. conducted the experiments. P.G. took care of the water treatments and analysis of results related to the water treatments. G.M. was responsible for the statistical analysis. C.D. was responsible for conducting the experiment and also writing the manuscript. All authors have read and agreed to the published version of the manuscript.

Funding: This research received no external funding.

Acknowledgments: We are grateful to Anastasios Lithourgidis and the personnel of the University Farm of the Aristotle University of Thessaloniki for assistance with the field experiments.

Conflicts of Interest: The authors declare no conflict of interest.

References

1. Parry, M.L.; Canziani, O.F.; Palutikof, J.P.; van der Linden Hanson, C.E. *Climate Change 2007: Impacts, Adaptation and Vulnerability. Contribution of Working Group II to the Fourth Assessment Report of the Intergovernmental Panel on Climate Change (IPCC)*; Cambridge University Press: Cambridge, UK, 2007; p. 976.

2. Nemeskéri, E.; Helyes, L. Physiological Responses of Selected Vegetable Crop Species to Water Stress. *Agronomy* **2019**, *9*, 447. [CrossRef]

3. Simon, J.E.; Quinn, J.; Murray, R.G. Basil: A source of essential oils. In *Advances in New Crops*; Janick, J., Simon, J.E., Eds.; Timber Press: Portland, OR, USA, 1990; pp. 484–489.

4. Makri, O.; Kintzios, S. *Ocimum sp.* (basil): Botany, cultivation, pharmaceutical properties, and biotechnology. *J. Herbs. Spices Med. Plants* **2007**, *13*, 123–150. [CrossRef]

5. Bączek, K.; Kosakowska, O.; Gniewosz, M.; Gientka, I.; Węglarz, Z. Sweet Basil (*Ocimum basilicum* L.) Productivity and Raw Material Quality from Organic Cultivation. *Agronomy* **2019**, *9*, 279. [CrossRef]

6. Darrah, H.H. *The Cultivated Basils*; Buckeye Printing Company: Independence, MO, USA, 1988.

7. Carovic-Stanko, K.; Liber, Z.; Besendorfer, V.; Javornik, B.; Bohanec, B.; Kolak, I.; Satovic, Z. Genetic relations among basil taxa (*Ocimum* L.) based on molecular markers, nuclear DNA content, and chromosome number. *Plant Syst. Evol.* **2010**, *285*, 13–22. [CrossRef]

8. Rewers, M.; Jędrzejczyk, I. Genetic characterization of *Ocimum* genus using flow cytometry and inter-simple sequence repeat markers. *Ind. Crop. Prod.* **2016**, *91*, 142–151. [CrossRef]

9. Juliani, H.R.; Simon, J.E. Antioxidant activity of basil. In *Trends in New Crops and New Uses*; Janic, J., Whipkey, A., Eds.; ASHS Press: Alexandria, VA, USA, 2002; pp. 575–579.

10. Grayer, R.G.; Kite, G.C.; Goldstone, F.J.; Bryan, S.E.; Paton, A.; Putievsky, E. Infraspecific taxonomy and essential oil chemotypes in basil. *Ocimum. Basilicum. Phytochem.* **1996**, *43*, 1033–1039. [CrossRef]

11. Nacar, S.; Tansi, S. Chemical components of different basil (*Ocimum basilicum* L.) cultivars grown in Mediterranean regions in Turkey. *Israel J. Plant Sci.* **2000**, *48*, 109–112. [CrossRef]

12. Ekren, S.; Sönmez, C.; Özcakal, E.; Kurttas, Y.S.K.; Bayram, E.; Gürgülü, H. The effect of different irrigation water levels on yield and quality characteristics of purple basil (*Ocimum basilicum* L.). *Agric. Water Manag.* **2012**, *109*, 155–161. [CrossRef]

13. Yassen, M.; Ram, P.; Anju, Y.; Singh, K. Response of Indian basil (*Ocimum basilicum* L.) to irrigation and nitrogen schedule in Central Uttar Pradesh. Ann. *Plant Physiol.* **2003**, *17*, 177–181.

14. Omidbaigi, R.; Hassani, A.; Sefidkon, F. Essential oil content and composition of sweet basil (*Ocimum basilicum* L.) at different irrigation regimes. *J. Essent. Oil-Bear Plants* **2003**, *6*, 104–108. [CrossRef]

15. Singh, M. Effect of nitrogen and irrigation on the yield and quality of sweet basil (*Ocimum basilicum* L.). *J. Spices Aromat. Crop.* **2003**, *11*, 151–154.

16. Khalid, K.A. Influence of water stress on growth, essential oil and chemical composition of herbs (*Ocimum* sp.). *Int. Agrophys.* **2006**, *20*, 289–296.

17. Asadollahi, A.; Mirza, M.; Abbaszadeh, B.; Azizpour, S.; Keshavarzi, A. Comparison of Essential oil from Leaves and Inflorescence of three Basil (*Ocimum basilicum* L.). Populations under Drought Stress. *Int. J. Agron. Plant Prod.* **2013**, *4*, 2764–2767.

18. Bekhradi, F.; Luna, M.C.; Delshad, M.; Jordan, M.J.; Sotomayor, J.A.; Martínez-Conesa, C.; Gil, M.I. Effect of deficit irrigation on the postharvest quality of different genotypes of basil including purple and green Iranian cultivars and a Genovese variety. *Postharvest Biol. Technol.* **2015**, *100*, 127–135. [CrossRef]

19. Ghamarnia, H.; Amirkhani, D.; Issa Arji, I. Basil (*Ocimum basilicum* L.) Water Use, Crop Coefficients and SIMDualKc Model Implementing in a Semi-Arid Climate. *Int. J. Plant Soil* **2015**, *4*, 535–547. [CrossRef]

20. Dane, J.H.; Hopmans, J.W. Pressure Plate Extractor. In *Methods of Soil Analysis*; Part 4: Physical Methods. SSSA Book Ser. 5; Dane, J.H., Topp, E.C., Eds.; SSSA: Madison, WI, USA, 2002; pp. 688–690.

21. Kavalieratou, S.; Karpouzos, D.K.; Babajimopoulos, C. *Monitoring Equipment Installation*; Technical report in the Program INTERREG IIIB: "Integrated Water Resources Management, Development and Confrontation of Common and Transnational Methodologies for Combating Drought within the MEDOCC Region"; EU: Thessaloniki, Greece, 2008; p. 15.

22. Karagiannioy, I.; Dordas, C. Evaluation of basil genotypes using physiological and agronomic characteristics. In Proceedings of the NAROSSA®2016, International Conference for Renewable Resources and Plant Biotechnology, Magdeburg, Germany, 13 June 2016.

23. Allen, R.G.; Pereira, L.S.; Raes, D.; Smith, M. Crop evapotranspiration. Guidelines for computing crop water requirements. In *FAO Irrigation and Drainage Paper*; FAO—Food and Agriculture Organization of the United Nations: Rome, Italy, 1998.

24. Dordas, C.; Papathanasiou, F.; Lithourgidis, A.; Petrevska, J.-K.; Papadopoulos, I.; Pankou, C.; Gekas, F.; Ninou, E.; Mylonas, I.; Sistanis, I.; et al. Evaluation of physiological characteristics as selection criteria for drought tolerance in maize inbred lines and their hybrids. *Maydica* **2018**, *63*, 1–14.

25. Tokatlidis, I.S.; Dordas, C.; Papathanasiou, F.; Papadopoulos, I.; Pankou, C.; Gekas, F.; Ninou, E.; Mylonas, I.; Tzantarmas, C.; Petrevska, J.K.; et al. Improved Plant Yield Efficiency is Essential for Maize Rainfed Production. *Agron. J.* **2015**, *107*, 1011–1018. [CrossRef]

26. Howell, T.A. Enhancing water use efficiency in irrigated agriculture. *Agron. J.* **2001**, *93*, 281–289. [CrossRef]

27. Steel, R.G.D.; Torrie, J.H.; Dickey, D.A. *Principles and Procedures of Statistics: A Biometrical Approach*, 2nd ed.; McGraw-Hill: New York, NY, USA, 1997.

28. Debaeke, P.; Aboudrare, A. Adaptation of crop management to water-limited environments. *Eur. J. Agron.* **2004**, *21*, 433–446. [CrossRef]

29. Jacobsen, S.-E.; Jensen, C.R.; Liu, F. Improving crop production in the arid Mediterranean climate. *Field Crops Res.* **2012**, *128*, 34–47. [CrossRef]

30. Blum, A. Effective use of water (EUW) and not water-use efficiency (WUE) is the target of crop yield improvement under drought stress. *Field Crop. Res.* **2009**, *112*, 119–123. [CrossRef]

31. Blum, A. *Plant Breeding for Water-Limited Environments*; Springer: New York, NY, USA, 2011.

32. Buchanan, B.B.; Gruissem, W.; Jones, R.L. *Biochemistry Molecular Biology of Plants*, 2nd ed.; Wiley: New York, NY, USA, 2015.

33. Arabaci, O.; Bayram, E. The Effect of Nitrogen Fertilization and Different Plant Densities on Some Agronomic and Technologic Characteristic of (*Ocimum basilicum* L.) Basil. *J. Agron.* **2004**, *3*, 255–262.

34. Daneshnia, F.; Amini, A.; Chaichi, M.R. Surfactant effect on forage yield and water use efficiency for berseem clover and basil in intercropping and limited irrigation treatments. *Agric. Water Manag.* **2015**, *160*, 57–63. [CrossRef]

35. Abdul-Hamid, A.F.; Kubota, F.A.; Morokuma, M. Photosynthesis, transpiration, dry matter accumulation and yield performance of mungbean plant in response to water stress. *J. Fac. Agric. Kyushu Univ.* **1990**, *1–2*, 81–92.

36. Castonguay, Y.; Markhart, A.H. Saturated rates of photosynthesis in water stressed leaves of common bean and tepary bean. *Crop. Sci.* **1991**, *31*, 1605–1611. [CrossRef]

37. Nunez-Barrios, A. Effect of Soil Water Deficits on the Growth and Development of Dry Bean at Different Stages of Growth (1991). Ph.D. Thesis, Michigan State University, East Lansing, MI, USA, 1992.

38. Viera, H.J.; Bergamaschi, H.; Angelocci, L.R.; Libardi, P.L. Performance of two bean cultivars under two water availability regimes. II. Stomatal resistance to vapour diffusion, transpiration flux density and water potential in the plant (in Portugal). *Pesqui. Agropeularia Bras.* **1991**, *9*, 1035–1045.

39. Durigon, A.; Evers, J.; Metselaar, K.; de Jong van Lier, Q. Water Stress Permanently Alters Shoot Architecture in Common Bean Plants. *Agronomy* **2019**, *9*, 160. [CrossRef]

40. Rhizopoulou, S.; Diamantoglou, S. Water stress induced diurnal variations in leaf water relations, stomatal conductance, soluble sugars, lipids and essential oil content of *Origanum majorana* L. *J. Hortic. Sci.* **1991**, *66*, 119–125. [CrossRef]

41. Misra, A.; Srivastava, N.K. Influence of water stress on Japanese mint. *J. Herbs. Spices Med. Plants* **2000**, *7*, 51–58. [CrossRef]

42. Nurzyńska-Wierdak, R.; Bogucka-Kocka, A.; Kowalski, R.; Borowsk, B. Changes in the chemical composition of the essential oil of sweet basil (*Ocimum basilicum* L.) depending on the plant growth stage. *Chemija* **2012**, *23*, 216–222.

43. Sirousmehrm, A.; Arbabi, J.; Asharipour, M.R. Effect of drought stress levels and organic manures on yield, essential oil content and some morphological characteristics of sweet basil (*Ocimum basilicum* L.). *Adv. Environ. Biol.* **2014**, *8*, 880–885.

44. Pirbalouti, A.G.; Malekpoor, F.; Salimi, A.; Golparvar, A. Exogenous application of chitosan on biochemical and physiological characteristics, phenolic content and antioxidant activity of two species of basil (*Ocimum ciliatum* and *Ocimum basilicum*) under reduced irrigation. *Sci. Hortic.* **2017**, *217*, 114–122. [CrossRef]

45. Kulaka, M.; Ozkanc, A.; Bindakd, R. A bibliometric analysis of the essential oil-bearing plants exposed to the water stress: How long way we have come and how much further? *Sci. Hortic.* **2019**, *246*, 418–436. [CrossRef]

46. Barideh, R.; Besharat, S.; Morteza, M.; Rezaverdinejad, V. Effects of Partial Root-Zone Irrigation on the Water Use Efficiency and Root Water and Nitrate Uptake of Corn. *Water* **2018**, *10*, 526. [CrossRef]

47. Al-Ghzawi, A.L.A.; Khalaf, Y.B.; Al-Ajlouni, Z.I.; AL-Quraan, N.A.; Musallam, I.; Hani, N.B. The Effect of Supplemental Irrigation on Canopy Temperature Depression, Chlorophyll Content, and Water Use Efficiency in Three Wheat (*Triticum aestivum* L. and *T. durum* Desf.) Varieties Grown in Dry Regions of Jordan. *Agriculture* **2018**, *8*, 67. [CrossRef]

48. Wei, Y.; Jin, J.; Jiang, S.; Ning, S.; Liu, L. Quantitative Response of Soybean Development and Yield to Drought Stress during Different Growth Stages in the Huaibei Plain, China. *Agronomy* **2018**, *8*, 97. [CrossRef]

 *agronomy*

Article

Changes in Leaf Structural and Functional Characteristics when Changing Planting Density at Different Growth Stages Alters Cotton Lint Yield under a New Planting Model

Aziz Khan [1], Jie Zheng [1], Daniel Kean Yuen Tan [2], Ahmad Khan [3], Kashif Akhtar [4], Xiangjun Kong [1], Fazal Munsif [1,3], Anas Iqbal [1], Muhammad Zahir Afridi [3], Abid Ullah [5], Shah Fahad [6] and Ruiyang Zhou [1,*]

[1] Key Laboratory of Plant Genetics and Breeding, College of Agriculture, Guangxi University, Nanning 530005, China; azizkhanturlandi@gmail.com (A.K.); summerjackk@gmail.com (J.Z.); kongxiangjun201010@163.com (X.K.); munsiffazal@yahoo.com (F.M.); anasiqbalagr@gmail.com (A.I.)

[2] Plant Breeding Institute, Sydney Institute of Agriculture, School of Life and Environmental Sciences, Faculty of Science, The University of Sydney, Sydney, NSW 2006, Australia; daniel.tan@sydney.edu.au

[3] Department of Agronomy, University of Agriculture, Peshawar 25000, Khyber Pakhtunkhwa, Pakistan; ahmad0936@yahoo.com (A.K.); drmzahir@aup.edu.pk (M.Z.A.)

[4] Institute of Nuclear Agricultural Sciences, College of Agriculture and Biotechnology, Zhejiang University, Hangzhou 310058, China; kashif@zju.edu.cn

[5] Department of Botany, University of Malakand, Chakdara Dir Lower, Malakand, Khyber Pakhtunkhwa 18800, Pakistan; abid.ullah@uom.edu.pk

[6] Department of Agriculture, University of Swabi, Swabi 23561, Khyber Pakhtunkhwa, Pakistan; shah_fahad80@yahoo.com

* Correspondence: ruiyangzhou@aliyun.com or ruiyangzh@gmail.com

Received: 22 October 2019; Accepted: 3 December 2019; Published: 7 December 2019

Abstract: Manipulation of planting density and choice of variety are effective management components in any cropping system that aims to enhance the balance between environmental resource availability and crop requirements. One-time fertilization at first flower with a medium plant stand under late sowing has not yet been attempted. To fill this knowledge gap, changes in leaf structural (stomatal density, stomatal length, stomata width, stomatal pore perimeter, and leaf thickness), leaf gas exchange, and chlorophyll fluorescence attributes of different cotton varieties were made in order to change the planting densities to improve lint yield under a new planting model. A two-year field evaluation was carried out on cotton varieties—V_1 (Zhongmian-16) and V_2 (J-4B)—to examine the effect of changing the planting density (D_1, low, 3×10^4; D_2, moderate, 6×10^4; and D_3, dense, 9×10^4) on cotton lint yield, leaf structure, chlorophyll fluorescence, and leaf gas exchange attribute responses. Across these varieties, J-4B had higher lint yield compared with Zhongmian-16 in both years. Plants at high density had depressed leaf structural traits, net photosynthetic rate, stomatal conductance, intercellular CO_2 uptake, quenching (qP), actual quantum yield of photosystem II (ΦPSII), and maximum quantum yield of PSII (F_v/F_m) in both years. Crops at moderate density had improved leaf gas exchange traits, stomatal density, number of stomata, pore perimeter, length, and width, as well as increased qP, ΦPSII, and F_v/F_m compared with low- and high-density plants. Improvement in leaf structural and functional traits contributed to 15.9%–10.7% and 12.3%–10.5% more boll m^{-2}, with 20.6%–13.4% and 28.9%–24.1% higher lint yield averaged across both years, respectively, under moderate planting density compared with low and high density. In conclusion, the data underscore the importance of proper agronomic methods for cotton production, and that J-4B and Zhongmian-16 varieties, grown under moderate and lower densities, could be a promising option based on improved lint yield in subtropical regions.

Keywords: leaf chlorophyll fluorescence; fiber yield; leaf gas exchange; leaf structure

---

1. Introduction

Cotton (*Gossypium hirsutum* L.) is a natural white fiber and cash crop that is grown globally [1]. The cotton plant is characterized by indeterminate growth habits and shows morphological and physiological adaptation to a wide range of environmental and management practices, including planting density and cultivar. An expanding population necessitates global efforts to increase crop production, especially those fulfilling food and fiber needs. Currently, numerous management practices have been introduced for cotton production systems, but lint production per unit area has remained stagnant [2]. High input costs combined with multiple management and material inputs have threatened cotton productivity. [2]. An efficient agricultural production system characterized by moderate planting density with one-time fertilization under a short growing season can reduce inputs without yield loss [3,4].

Planting density and choice of cultivar are important agronomic practices that have the potential to optimize the canopy photosynthetic rate and crop productivity of any cropping system [5]. Changes in plant architecture and canopy dynamics in response to planting density can have impacts on disease incidence, water use, canopy temperature, and enzymatic activity of assimilate metabolism [6]. Manipulations of planting density in cotton have remarkable impacts on biomass partitioning, nutrient uptake, boll distribution, changes in the light spectrum, and crop production [3,5,7,8], which can influence yield and profits for producers [9]. Plants at high density can minimize evaporation and irrigation frequency, as well as increase the utilization of irrigation water [10,11]. In contrast, high-density planting can slow down leaf appearance and reduce open boll density [12], boll weight, and boll number [7]. It also delays leaf senescence [13] and decreases nitrogen use efficiency and nitrogen recovery efficiency [14]. A planting density of up to nine plants m^{-2} has been reported to sustain leaf photosynthetic rate and reproductive organ biomass formation by increasing plant potassium uptake at various developmental stages. However, a sowing density of >10 plants m^{-2} and subsequent shading can result in disease infestation, small boll size, fruit shedding, delayed maturity, and decreased individual plant development [4]. Dense planting can also delay crop maturity by promoting vegetative growth and can substantially depress net photosynthetic rate [4] due to decreased RuBP carboxylase activity and chlorophyll content [15]. High planting density can increase the auxin (IAA) content and enhance auxin polar transport by increasing the expression of the auxin biosynthesis gene (Gh*YUC5*) and the auxin polar transport gene (Gh*PIN1*). It can also inhibit vegetative branching by decreasing IAA, cytokinin, gibberellic acid, and brassinosteroid contents, followed by increased strigolactone content due to differential expression of hormone-associated genes in the tips of vegetative branches [15]. Optimal plant density can ensure healthy plant development by maintaining a core population of plants synchronizing boll number and fiber quality to achieve optimal yield [16].

Leaf morphological and physiological attributes are important players in photosynthetic regulation [4] and can provide a structural framework for gas exchange as well as optimize the photosynthetic function [17]. Cotton leaf surface characteristics, including cuticular thickening, wax layer, and trichomes, play critical roles in the variability of optical properties [18]. Generally, leaves developed under high sunlight can have thicker and smaller leaves with well-developed plastid tissues, greater stomatal density, and smaller granal stacks than shade leaves [19]. Plants under low density planting have a lower chlorophyll content and a higher electron transfer rate and ribulose-1,5-bi-phosphate carboxylase/oxygenase compared with high-density planting [20,21]. Leaves developed under lower density (sun leaves) are tolerant to strong light; conversely, shade leaves have weak photoprotection potential and are more sensitive to high light [17,21].

Studies regarding cotton growth and lint yield in response to diverse populations are common [22–24]. However, we are the first to report the effects of changing the planting density on cotton lint yield, leaf structure, chlorophyll fluorescence, and leaf gas exchange characteristics in subtropical regions. The objectives of this study were to investigate leaf structural and functional characteristics in response to different planting densities and varieties. It also explored optimal plant density and variety for improved lint yield in subtropical regions.

2. Materials and Methods

2.1. Plant Material and Experimental Site

Seeds of two cotton cultivars—V_1 (Zhongmian-16) and V_2 (J-4B)—were procured from the Cotton Research Institute, Chinese Academy of Agricultural Sciences and were grown under field conditions for two years. A replicated two-year (2017 and 2018 growing seasons) field experiment was conducted at Guangxi University, Nanning, China. The soil properties of the experimental field were sandy loam and yellowish, having a pH of 6.5; organic matter of 23.37 mg kg^{-1}; and available nitrogen, phosphorus, and potassium content of 53.24, 77.58, and 6.30 mg kg^{-1}, respectively. The experimental design layout was a balanced split plot with three replications.

2.2. Crop Management and Experimental Design

Before sowing, the experimental field was ploughed, laser leveled approximately three weeks prior, and covered with plastic film to conserve moisture and suppress weed germination. The experiments were designed in a split plot arrangement with three replications of each of the six treatment combinations. Two cotton varieties (V_1, Zhongmian-16; V_2, J-4B), were randomly allocated to the main plots and three plant population levels (D_1, low, 3×10^4; D_2, medium, 6×10^4; and D_3, dense, 9×10^4 ha^{-1}) were randomized in subplots. By increasing the precision of comparisons, split plot arrangements were adopted. Seeds were sown on 5 June in double rows on each raised plot (3.0 m wide and 11 m long), with a total plot size of 33.0 m^2. Each subplot was 11 m long and 1.5 m wide, consisting of four rows with narrow (10 cm) and wide (66 cm) row spaces for a total of eight rows on each main plot. Plant-to-plant spacing was controlled according to the corresponding population level. Crops were irrigated one day after sowing to ensure uniform germination. Cotton seedlings were hand-thinned at the third leaf stage to the target population level for each plot. A basal application of phosphorus (P_2O_5) at 66 kg ha^{-1}, nitrogen (N) at 170 N kg ha^{-1}, and potash (K_2O) at 190 kg ha^{-1} was applied using superphosphate (12% P_2O_5), urea (46% N), and potassium chloride (59% K_2O), respectively, during the pinhead stage. A plant growth regulator (i.e., mepiquat chloride) at the rate of 0.057 active ingredient ha^{-1} was sprayed to control vegetative growth. All the necessary field management practices were performed according to crop requirements during the whole crop cycle.

2.3. Data Collection

Data on leaf structure, chlorophyll fluorescence, leaf gas exchange attributes, cotton yield, and yield contributors were recorded for each treatment in three replications. The details of each measurement are given below.

2.4. Yield and Yield Components

To assess cotton yield, fully opened bolls were hand-picked at three times in each treatment. The harvested seed cotton was sun-dried to ≤11% moisture content [16]. The seed cotton was ginned to calculate seed cotton and lint yield. During the second picking, 100 mature bolls were manually picked to calculate single boll mass and lint percentage. Seed cotton yield of 100 bolls was divided by the number of bolls to assess individual boll weight. Lint % was determined using the lint yield of 100 bolls and divided by seed cotton mass.

2.5. Cotton Leaf Structure Attributes

Ten plants in each plot were randomly tagged to measure leaf structure and plant growth characteristics at the boll setting stage. Leaf thickness was determined on 10 fully expanded leaves from the upper part of three plants (functional leaves, i.e., upper fourth leaf). A hand-held micrometer (Mitutoyo Digital Micrometer Model 293-185, Kawasaki, Japan) with a digital display and a clutch that ensured uniform pressure [25] was used for leaf thickness assessment. A 5 × 8 mm leaf section was removed for each treatment. Samples were then added into 10 mL tubes containing 50%, 5%, and 5% alcohol solution, formaldehyde, and glacial acetic acid, respectively. Scanning electron microscopy was performed at Guanxi Medical University using a SUPRA 55VP (Carl Zeiss AG, Oberkochen, Germany). Image software was used to assess cotton leaf stomatal length, width, density, and pore perimeter according to the method reported in [26].

2.6. Chlorophyll Fluorescence Traits

Cotton leaf chlorophyll fluorescence attributes were measured on a fully expanded functional leaf (upper fourth leaf on the main stem) on a sunny day (between 1000 and 1200 h) via a portable mini PAM-2100 fluorometer coupled with a 2030-B leaf (Walz, Germany). Maximum (F_m) and minimum (F_o) fluorescence values of dark-adapted leaves (photosystem II (PSII) centers open) were measured using leaf clips. The maximum and minimum fluorescence values were assessed at 0.5 μmol m^{-2} s^{-1} with a frequency of 0.6 kHz and a 0.8 saturating pulse at >8000 μmol m^{-2} s^{-1}, respectively. Maximum quantum yield of PSII photochemistry (F_v/F_m) was calculated as $F_v/F_m = 1 - (F_o/F_m)$ [27]. The effective quantum yield of PSII photochemistry of light-adapted leaves was determined by Φ_{PSII} ($F_m' - F)/F_m'$ [28]. Coefficient of photochemical quenching (qP) was assessed using the formula qP = ($F_m' - F_s)/(F_m' - F_o'$) [29]. Minimal fluorescence of light-adapted leaves (F_o') was calculated according to the equation $F_o' = F_o/(F_v/F_m + F_o/F_m')$ done by [28]. Nonphotochemical quenching (NPQ) was recorded according to [6] as NPQ = ($F_m - F_m')/F_m'$, where F_m represents the value of the predawn observations. The electron transport rate (ETR) was assessed using a leaf absorptance of 0.85 and half of the absorbed light was partitioned to each photosystem: ETR = PSII × PPFD × 0.85 × 0.5 [30].

2.7. Leaf Gas Exchange Attributes

At squaring, flowering, peak bloom, and boll setting stages, fully expanded leaves from the upper part of three plants (functional leaves, i.e., upper fourth leaf) were chosen to assess net photosynthetic rate (Pn), stomatal conductance (g_s), intercellular CO_2 concentration (C_i), and transpiration rate (E). Net rate of photosynthesis was measured from the six functional leaves of three plants in each treatment using a portable infrared gas exchange analyzer (Li-6400, Li-Cor, Lincoln, NE, USA). These observations were made on a clear day between 10:00 a.m. and 12:00 p.m. Beijing time in each experimental unit of four replications. Leaves in each plot followed the following adjustments: PAR, 1800 μmol m^{-2} s^{-2}; air flow, 389.42 mmol^{-1} m^{-2} s^{-1}; water vapor pressure into leaf chamber, 3.13 mbar; leaf temperature, 30 °C; ambient temperature, 33.69 °C; and ambient carbon dioxide concentration, 330–350 mol mol^{-1}.

2.8. Statistical Analysis

All the data were processed using Microsoft Excel 2016. Figures were plotted using Sigma Plot 14.00 software. Analysis of variance was implemented using SAS software (version 8.1, SAS Institute, Cary, NC, USA). The initial combined data showed no interactions with years. Therefore, the data were pooled and presented across the two years. Means of planting density were separated using the least significant difference (LSD) test at the 5% probability level. Both planting densities and cultivars were taken as main factors and fixed effects with cropping season as the repetitive measured factor with a fixed effect. Similarly, the interaction was taken as fixed effects and treatment × replication interaction, which was taken as a random effect. Differences among treatments imply statistical difference ($p = 0.05$).

3. Results

3.1. Yield and Yield Components

The analysis of variance (Table 1) showed that effects of year, planting density, variety, and their interaction on cotton yield and yield contributors. The year effect was statistically significant, but the differences were not large. Planting density and variety did not affect lint percentage and boll weight. There were 14.5% and 7.1% more bolls m^{-2} with a 19% and 11.5% higher lint yield in moderate-density crops compared with low- and high-density crops, respectively. Under high-density conditions, a reduction of 9.6% and 2.3% was noted in boll weight and lint percentage, respectively, compared with low- and moderate-density crops. Across the varieties, J-4B produced 6% and 7.8% greater bolls m^{-2} and lint yield, respectively, compared with the Zhongmian-16 variety (Figure 1A–D). Interaction was significant for density × variety across two years. Cotton plant individual boll weight, boll density m^{-2}, and lint yield were highest under moderate-to-high planting density for J-4B, while under low-density conditions, Zhongmian-16 had a higher boll weight.

Table 1. Summary of mean square (MS) values from analysis of variance (ANOVA) for cotton yield and yield contributors.

Source of Variance	Year	Density	Variety	Density × Variety
Bolls number (m^{-2})	20.59 *	170.7 **	96.13 **	2.995 **
Boll weight (g)	2.402 **	0.640 **	0.003 ns	0.190 **
Lint (%)	121.5 **	3.001 ns	4.448 ns	4.749 ns
Lint yield (kg ha^{-1})	50,400 **	123,003 **	65,451 **	3561 **

Different values obtained from ANOVA represent * significant at $p < 0.05$, ** significant at $p < 0.01$ and ns: nonsignificant.

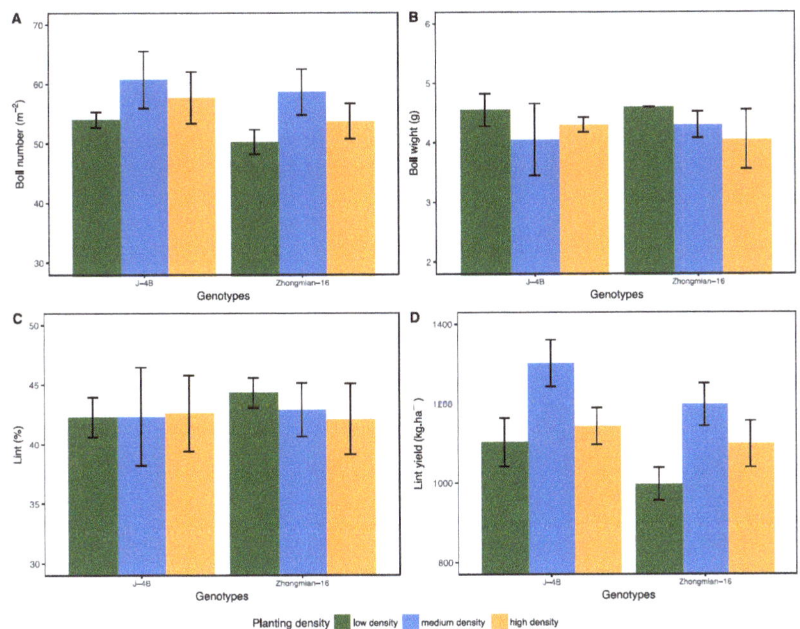

Figure 1. Cotton (in response to different planting densities and cultivars: (**a**) boll number (m^{-2}); (**b**) boll weight (g); (**c**) lint (%); (**d**) lint yield (kg.ha^{-1}) Values are the sum of three independent replicates. Error bars correspond to confidence interval at $p = 0.05$.

3.2. Leaf Structure Attributes

Cotton leaf structural characteristics (e.g., stomatal density, length, width, pore perimeters, and leaf thickness) significantly influenced by planting density and cultivar (Table 2). Under dense crops, leaf stomatal density, length, width, and pores were reduced by 7.1% and 11.7%; 3.3% and 9.3%; and 11.2%, 2.2%, and 7.9% compared with lower- and medium-density crops, respectively. Likewise, J-4B had improved stomatal density, length, width, pores, and leaf thickness by 10.3%, 13.7%, 1.1%, 9.9%, and 10.7%, respectively, compared with the Zhongmian-16 variety. Significant density × variety interaction revealed that, unlike J-4B, increasing planting density reduced stomatal density, length, width, and pore perimeters in Zhongmian-16 during both growing seasons.

Table 2. Cotton leaf structural attributes as influenced by planting density and cultivars.

Treatment	Plant Height (cm)	Stomatal Density (mm^{-2})	Stomata Length (μm)	Stomata Width (μm)	Stomatal Pore Perimeter (μm)	Leaf Thickness (μm)
Year (Y)						
Year 2017	66.9a	28.8a	146.3a	20.8a	28.3a	143.0a
Year 2018	45.6b	20.1b	125.3b	14.3b	20.9b	106.6b
Density (D)						
D_1 (low)	57.0a	25.3a	144.9a	18.3a	25.6a	128.9a
D_2 (moderate)	56.1ab	24.7a	134.6b	18.1a	24.8b	124.5b
D_3 (high)	57.0a	23.3b	127.9c	16.3a	23.4c	121.0c
Varity (V)						
V_1 (Zhongmian-16)	59.1a	23.3b	129.2b	17.7a	23.1b	117.9b
V_2 (J-4B)	53.5b	25.6a	142.4a	17.5a	26.2a	131.7a
Source of variance						
Y	4091 **	689.8 **	4001 **	381.1 *	485.47 **	11916.8 *
D	5.53 *	12.83 **	878.3 **	15.55 ns	0.422 **	3.10 **
V	276.39 **	48.22 *	1579 **	0.358 *	89.30 *	1717.6 **
D × V	744.18 **	256.9 **	8971 **	208.6 **	201.59 *	6219.4 **

Values within columns followed by the same letter are statistically insignificant at the 0.05 level. ** significant at $p < 0.01$ and * significant at $p < 0.05$. ns: nonsignificant.

3.3. Chlorophyll Fluorescence Traits

During both years, planting densities, varieties, and their interaction had significant impacts on chlorophyll fluorescence traits in different growth stages (Tables 3–5). Except the squaring stage, ΦPSII at first bloom, peak bloom, and boll setting stages were increased by moderate-density compared with low- and high-density crops, while the F_v/F_m yield was greater at all growth stages (Table 3). Across the varieties, J-4B had higher ΦPSII and F_v/F_m at peak bloom and boll setting stages than Zhongmian-16, respectively. The interaction between density × variety remained significant for ΦPSII and F_v/F_m at different growth stages. The J-4B variety with moderate crops had greater ΦPSII and F_v/F_m across the years.

Significant variation between planting densities, varieties, and years was found for photochemical quenching (qP) and nonphotochemical quenching (NPQ) of cotton at all growth stages (Table 4). Across densities, medium competitive plants yielded higher qP and NPQ rates. The variety J-4B resulted in higher qP at squaring and boll setting stages, while Zhongmian-16 had higher NPQ rates at first and full bloom stages. The interaction showed that J-4B had a higher qP under moderate density at different growth stages. J-4B had higher values for NPQ at low density compared with Zhongmian-16, followed by moderate density for the same variety at the peak bloom stage.

Significant differences existed between years, densities, and varieties for the ETR at four growth stages (Table 5). Interaction between density × variety revealed substantial variation between varieties to planting density at all growth stages. Increased planting density substantially reduced ETR at all growth stages in both years (Table 5). The low-density plants improved ETR at squaring, first, peak bloom, and boll setting stages, followed by moderate density, while there was a lower ETR in high-density crops. A higher ETR was noted for the variety Zhongmian-16 at squaring, first, and peak bloom stages compared with J-4B; however, J-4B had a higher ETR at the boll setting stage than

Zhongmian-16. ETR values were substantially reduced under high density for both varieties. Lower planting density had higher ETR values for Zhongmian-16 or J-4B during both years.

Table 3. Quantum and maximum quantum yield of photosystem II (PSII) of cotton cultivars under varied planting densities.

Treatment	Squaring	First Bloom	Peak Bloom	Boll Set
Quantum yield (ΦPSII)				
Year				
Year 2017	0.37b	0.57a	0.57a	0.55a
Year 2018	0.43a	0.51b	0.48b	0.38b
Density (D)				
D_1 (low)	0.40a	0.53b	0.49b	0.46b
D_2 (moderate)	0.40a	0.57a	0.58a	0.52a
D_3 (high)	0.40a	0.52b	0.50b	0.43c
Variety (V)				
V_1 (Zhongmian-16)	0.38a	0.52a	0.48b	0.52a
V_2 (J-4B)	0.41a	0.56a	0.56a	0.56a
Source of variance				
Y	0.034 **	0.029 **	0.07 **	0.260 **
D	0.008 ns	0.010 *	0.025 **	0.038 **
V	0.007 ns	0.009 ns	0.053 **	0.072 **
D × V	0.010 ns	0.007 ns	0.038 **	0.017 *
Maximal quantum yield (F_v/F_m)				
Year				
Year 2017	0.79a	0.78a	0.78a	0.73a
Year 2018	0.44b	0.70b	0.76b	0.59b
Density (D)				
D_1 (low)	0.62b	0.76a	0.78a	0.67b
D_2 (moderate)	0.63a	0.76a	0.78a	0.70a
D_3 (high)	0.59c	0.71b	0.75b	0.62c
Variety (V)				
V1 (Zhongmian-16)	0.59a	0.74a	0.77a	0.63b
V_2 (J-4B)	0.62a	0.74a	0.77a	0.69a
Source of variance				
Y	1.123 **	0.057 **	0.004 **	0.161 **
D	0.006 **	0.008 **	0.004 **	0.023 **
V	0.008 ns	0.001 ns	0.003 ns	0.036 *
D × V	0.008 ns	0.017 *	0.004 ns	0.016 ns

Values within columns followed by the same letter are statistically insignificant at the 0.05 level. * indicate significant at $p < 0.05$, ** significant at $p < 0.01$ and ns: nonsignificant.

Table 4. Photochemical and nonphotochemical quenching of cotton cultivars under varied planting densities.

Treatment	Squaring	First Bloom	Peak Bloom	Boll Set
Photochemical quenching (qP)				
Year				
Year 2017	0.63a	0.78a	0.75a	0.7a
Year 2018	0.64a	0.69b	0.61b	0.6b
Density (D)				
D_1 (low)	0.64b	0.74b	0.65b	0.63b
D_2 (moderate)	0.60c	0.83a	0.73a	0.79a
D_3 (high)	0.66a	0.64c	0.65b	0.58c

Table 4. *Cont.*

Treatment	Squaring	First Bloom	Peak Bloom	Boll Set
Photochemical quenching (qP)				
Variety (V)				
V1 (Zhongmian-16)	0.62a	0.70b	0.68a	0.63b
V_2 (J-4B)	0.65a	0.77a	0.67a	0.70a
Variance				
Y	0.002 ns	0.078 **	0.156 **	0.137 **
D	0.009 **	0.109 **	0.023 **	0.148 **
V	0.008 ns	0.036 **	0.002 ns	0.048 **
D × V	0.001 ns	0.029 **	0.016 *	0.034 **
Nonphotochemical quenching (NPQ)				
Year				
Year 2017	1.07a	1.78a	1.86a	1.33a
Year 2018	0.82b	0.64b	0.97b	0.95b
Density (D)				
D_1 (low)	1.06a	1.39a	1.91a	1.24a
D_2 (moderate)	0.98b	1.21b	1.26b	1.15b
D_3 (high)	0.77c	1.04c	1.08c	1.04c
Variety (V)				
V_1 (Zhongmian-16)	0.94a	1.25a	1.55a	1.18a
V_2 (J-4B)	0.94a	1.17b	1.28b	1.10a
Source of variance				
Y	0.555 **	10.856 **	7.124 **	1.355 **
D	0.257 **	0.374 **	2.277 **	0.128 **
V	0.001 ns	0.051 **	0.699 **	0.049 ns
D × V	0.007 ns	0.002 ns	0.613 **	0.004 ns

Values within columns followed by the same letter are statistically insignificant at the 0.05 level. ** significant at $p <$ 0.01 and * significant at $p < 0.05$. ns: nonsignificant.

Table 5. Electron transport rate (ETR) of cotton cultivars at different planting densities.

Treatment	Squaring	First Bloom	Peak Bloom	Boll Set
Year				
Year 2017	118.7b	168.0a	167.6a	166.3a
Year 2018	136.3a	162.2b	130.9b	109.9b
Density (D)				
D1 (low)	140.7a	172.8a	156.2a	156.0a
D_2 (moderate)	125.0b	162.3b	147.7b	131.4b
D_3 (high)	116.8c	160.3c	143.8c	126.9c
Variety (V)				
V_1 (Zhongmian-16)	129.9a	167.1a	151.9a	133.9b
V_2 (J-4B)	125.0b	163.2b	146.6b	142.3a
Source of variance				
Y	2770 **	301.6 **	12115 **	28685 **
D	1769 **	538.1 **	478.7 **	2940
V	216.1 **	126.2 **	261.4 **	641.8 **
D × V	34.08 **	113.9 **	50.60 **	930.1 **

Values within columns followed by the same letter are statistically insignificant at the 0.05 level. ** significant at $p <$ 0.01 and ns: nonsignificant.

3.4. Leaf Gas Exchange Attributes

Cotton leaf gas attributes were significantly influenced by plant density, variety, and growing year (Tables 6 and 7). Under moderate-density conditions, net photosynthetic rate (Pn) was increased at all growth stages except squaring, while stomatal conductance (g_s) was higher at the first bloom and boll setting stages. Plants under high density had significantly lower Pn and g_s compared with low and moderate density (Table 6). J-4B had higher Pn and g_s compared with Zhongmian-16 under moderate density. Interaction between density × variety was significant only at full bloom and boll setting for Pn and at the peak bloom stage for g_s. J-4B under low-to-moderate planting density had a higher Pn at squaring and first bloom stages, while it was higher in Zhongmian-16 at the peak bloom and boll set stages. A higher g_s under moderate planting density was noted in J-4B at the peak stage than Zhongmian-16 at low or high density.

Table 6. Net photosynthetic rate (Pn) and stomatal conductance (g_s) of cotton cultivars at varied planting densities.

Treatment	Squaring	First Bloom	Peak Bloom	Boll Set
Photosynthesis (Pn (μmol (CO_2) m^{-2} s^{-1}))				
Year				
Year 2017	25.5a	27.0a	32.3b	35.9a
Year 2018	26.0a	26.8a	32.5a	35.7b
Density (D)				
D$_1$ (low)	25.8a	26.4b	31.9c	34.8c
D$_2$ (moderate)	25.5b	27.7a	33.3a	36.8a
D$_3$ (high)	25.9a	26.2b	32.1b	35.8b
Variety (V)				
V$_1$ (Zhongmian-16)	25.6a	26.6b	32.1b	35.6b
V$_2$ (J-4B)	25.9a	27.2a	32.7a	36.1a
Source of variance				
Y	2.576 ns	0.276 ns	0.681 **	0.123 **
D	0.735 **	5.623 **	6.544 **	11.56 **
V	0.664 ns	3.453 *	3.901 **	1.823 **
D × V	0.323 ns	0.948 ns	7.696 **	3.399 **
Stomatal conductance (g_s (mol (H_2O) m^{-2} s^{-1}))				
Year				
Year 2017	0.49a	0.58a	0.45a	0.33a
Year 2018	0.49a	0.55a	0.44a	0.32a
Density (D)				
D$_1$ (low)	0.48a	0.54b	0.46a	0.31b
D$_2$ (moderate)	0.49a	0.61a	0.46a	0.35a
D$_3$ (high)	0.49a	0.55b	0.42a	0.33ab
Variety (V)				
V$_1$ (Zhongmian-16)	0.45a	0.55b	0.43b	0.32b
V$_2$ (J-4B)	0.48a	0.58a	0.46a	0.34a
Source of variance				
Y	0.001 ns	0.005 ns	0.003 ns	0.006 ns
D	0.003 ns	0.027 **	0.006 ns	0.044 *
V	0.004 ns	0.017 **	0.019 ns	0.064 **
D × V	0.004 ns	0.016 ns	0.008 **	0.006 ns

Values within columns followed by the same letter are statistically insignificant at the 0.05 level. ** significant at $p < 0.01$ and * significant at $p < 0.05$. ns: nonsignificant.

Table 7. Intercellular CO_2 concentration (C_i) and transpiration rate (E) of cotton cultivars under different planting densities.

Treatment	Squaring	First Bloom	Peak Bloom	Boll Setting
Intercellular CO_2 concentration (C_i (μmol (CO_2) m^{-2} s^{-1}))				
Year				
Year 2017	246.0a	274.6a	167.6a	243.2a
Year 2018	243.8b	271.6b	164.9b	239.3b
Density (D)				
D_1 (low)	248.7a	272.2c	164.8b	242.2a
D_2 (moderate)	244.0b	274.7a	167.9a	242.1a
D_3 (high)	241.9c	272.4b	165.9b	239.4b
Variety (V)				
V_1 (J-4B)	245.9a	274.6a	166.9a	241.5a
V_2 (Zhongmian-16)	243.9b	271.6b	165.5b	240.9b
Source of variance				
Y	42.25 **	81.00 **	64.00 **	132.25 **
D	145.63 **	21.948 **	28.89 **	29.84 **
V	37.21 **	0.0004 **	17.64 **	2.89 **
D × V	16.74 **	8.703 **	25.40 **	56.12 **
Transpiration rate (E (mmol (H_2O) m^{-2} s^{-1}))				
Year				
Year 2017	6.8a	9.1a	6.6a	4.6a
Year 2018	6.7b	9.1b	6.5b	4.4b
Density (D)				
D_1 (low)	6.6c	9.3a	6.7a	4.4a
D_2 (moderate)	6.7b	9.1b	6.5b	4.4b
D_3 (high)	6.8a	9.0c	6.1c	4.4b
Variety (V)				
V_1 (J-4B)	6.68b	9.21a	6.48b	4.52a
V_2 (Zhongmian-16)	6.82a	9.05b	6.68a	4.29b
Source of variance				
Y	2770 **	0.007 **	301.6 **	28685 **
D	1769 **	0.203 **	538.1 **	2940 **
V	216.0 **	0.226 **	126.2 **	641.8 **
D × V	34.10 **	0.139 **	113.9 **	930.1 **

Values within columns followed by the same letter are statistically insignificant at the 0.05 level. ** significant at $p <$ 0.01 and ns: nonsignificant.

Increasing planting density significantly reduced C_i in cotton leaves for both varieties. Plants with moderate density had higher C_i uptake at first bloom and peak bloom stages compared with low- and high-density crops, respectively (Table 7). Plants under low density resulted in a higher rate of E during first bloom, peak bloom, and boll set stages compared with moderate- and high-density crops, respectively (Table 7). Across the varieties, Zhongmian-16 yielded higher for both C_i uptake and E rates compared with J-4B. Interaction between density × variety remained significant at all growth stages for C_i. The transpiration rate was decreased in both varieties when the planting density increased.

4. Discussion

The current study has provided new data on the common perception that high planting density significantly decreases leaf structural characteristics, such as stomatal density, length, width, pore perimeter, and leaf thickness, as well as functional traits (leaf gas exchange and chlorophyll fluorescence traits), which leads to lint yield loss. However, we found that improved leaf functional and structural traits for J-4B under moderate density had a higher lint yield. Under high-density treatment, reductions in lint yield for Zhongmian-16 were associated with repression in leaf structural and functional attributes, which in turn caused depression in leaf photosynthetic capacity due to nutrient competition. The difference between varieties from changing planting density might be associated with canopy architecture and genetic variation. Therefore, these changes in varieties might have significant impacts on leaf structural and functional attributes and, ultimately, on yield formation.

High planting density responses to cotton lint yield, growth, biomass production, nutrient uptake, and fiber quality have been extensively investigated [3,4,13,22]. The mechanisms of interplant competitiveness under low-to-high planting density on leaf structure, chlorophyll fluorescence, and leaf gas exchange attributes for optimal cotton lint yield have not yet been reported. Across densities, the moderate population had a higher boll number m^{-2} with improved lint yield for J-4B compared with Zhongmian-16 across two years. High-density plants substantially reduced yield and yield components in both years, probably due to competition for nutrients. The phenomenon of increased lint yield under moderate density can be associated with improved leaf structural and chlorophyll fluorescence traits and higher leaf photosynthetic capacity, which resulted in higher boll density m^{-2} compared with other densities.

Moderate density favors dry matter partitioning to the reproductive structures rather than vegetative organs [31] and less fruit shedding compared with denser plants. The reductions in lint yield under high density can be attributed to decreased leaf structural and physiological traits, which were observed in this study. The differences that existed between varieties for yield when changing planting density might be attributable to canopy architecture. Differences in plant canopy architectural traits among varieties have an impact on growth characteristics and lint yield. These data further confirmed that an appropriate selection of variety and optimal density can contribute to successful cotton production. Reducing population density may also have other implications, such as decreased frequency and insecticide inputs per season without any yield loss to increase profit. Moreover, high plant density can substantially depress leaf structural and physiological attributes, which in turn cause a severe yield penalty.

Plants respond to ambient and management interventions via architectural and structural changes. Plant growth and leaf morphological attributes, including stomatal density, size, number of pores, width, length, and leaf thickness features, are pivotal windows regulating leaf photosynthetic capacity [10,25] and offer a structural framework for CO_2 exchange and optimization of photosynthetic activities, which in turn can improve crop yield [17]. In this study, high planting density substantially decreased leaf thickness, stomatal density, leaf length, width, and number of stomatal pores. Limitations in these attributes disrupted the photosynthetic capacity of plants by restricting entry of CO_2 to the mesophyll through the stomata of leaves, which is extremely responsive to light environments. Thus, the exchange of CO_2 by means of stomata might be restricted [32]. Higher stomatal density, thicker leaves, and rapid metabolite transfer between the mesophyll and bundle sheet cells can favor higher leaf photosynthetic capacity [33]. Increasing planting density has been proposed to decrease the stomatal density of wheat leaves [34]. A greater stomatal size can facilitate CO_2 distribution into the leaf due to its conductance being proportional to the square of the effective radius of the stomatal pore, resulting in increased stomatal conductance [35]. However, the responses of leaf structural attributes vary under different abiotic stresses in different plant species or varieties [36]. These data suggest that plants under high-density conditions have significantly decreased leaf morphological characteristics, which might be particularly responsible for depressing leaf photosynthetic capacity.

Chlorophyll fluorescence is a nondestructive evaluation of PSII activity. In plant physiology, this technique is commonly used and has become a classical method for crop improvement, screening of beneficial traits, and linking genomic knowledge to phenological response. Due to the sensitivity of PSII to undesirable ambient conditions, this is a useful method for understanding photosynthetic mechanisms and a good indicator of how plants respond to ambient change [37,38].

ΦPSII is a measure of light energy capture efficiency, which reflects the actual primary sunlight energy conversion efficiency of the PSII reaction center [15]. In this study, ΦPSII substantially declined under high-density conditions. Probably, a lower ΦPSII value under high-density conditions did not efficiently convert photon energy to chemical energy; however, this phenomenon needs further exploration. Under shading conditions, a low ΦPSII may be responsible for depressing Pn due to the adjustment in photochemical reaction centers [39], which was observed in our study. The efficient use of limited light energy and the degree of the PSII reaction center openness can increase, resulting in improved energy conversion efficiency. This is associated with the increase of F_v/F_m, ΦPSII, and qP at early shading [40]. The maximal photochemical efficiency of PSII (F_v/F_m) determines the potential quantum efficiency of PSII [41]. In this study, F_v/F_m had higher values under low rather than high planting density, which is consistent with [32], and reductions in F_v/F_m values might be due to the lower values of F_m and increased values of F_o. The ETR is an important chlorophyll fluorescence attribute affected by the external light environment. The rate of ETR declined from low to high density in this study, which corresponds with [40], and shading can significantly decrease ETR values by affecting PSII photochemical reaction centers and consequently diminish the primary stable quinine acceptor of PSII, leading to a decrease in the activity of photosynthetic electron transport efficiency via PSII [27,33]. NPQ can have critical roles in the nonradiative dissipation of surplus light energy [42]. A low-light environment can cause a reduction in NPQ, possibly associated with reduced light energy [32,43]. In this study, a severe decline in NPQ values was noted under high-density compared with low-density crops. This can be explained as the decreased NPQ being associated with the decreased efficiency of photochemical reactions through the reduced fraction of incident light in photochemical energy utilization, which resulted in lower thermal dissipation in PSII [44]. The rate of photochemical quenching (qP) under dense crops showed a substantial reduction compared with low and moderate densities. Probably, a low-light environment can cause reductions in the amount of pigment and the efficiency of photochemical energy conversion, resulting in the depressed quantum yield of PSII and decreased qP. The qP reflects the efficiency of light quantum harvested by PSII to chemical energy and represents the openness degree of the PSII reaction center, and a greater qP results in greater activity of electron transfer in PSII.

Leaf gas exchange traits can play a central role in biomass formation and the prime determination of cotton lint yield [45]. High planting density results in rapid canopy closure and an increase in radiation interception, which reduces weed competition [46], but this impedes leaf gas exchange traits, leading to yield loss [47]. In the current study, cotton leaf gas exchange parameters were substantially depressed under close planting at different growth stages. Accordingly, high-density conditions resulted in reductions in leaf stomatal density, length, width, pores, and leaf thickness, probably due to mutual shading, which may be responsible for depressing stomatal conductance (g_s) and CO_2 uptake through the stomata, which in turn suppressed the photosynthetic capacity. Plants under high-density conditions can significantly decrease g_s and Ci, which can negatively influence the photosynthetic system [38]. The CO_2 concentration plays a central role in net photosynthetic rate (Pn), but this varies across species and ambient conditions [48,49]. The g_s might respond to alterations in Pn and thus prevents C_i near saturation. The primary function of stomata is to avoid desiccation and enable the passage of CO_2. Stomata induce a substantial disruption in the CO_2 assimilation rate, which reduces more in C_4 than C_3 plants. The stomatal limitation of Pn is the role of stomatal resistance to contribute to "resistance" to CO_2 uptake and stomatal limitation in spite of a decline in C_i [50]. The higher transpiration (E) rates in low-density conditions may have been due to low mutual shading, which allowed rapid stomata opening. Our data showed that high plant density substantially decreased leaf thickness, stomatal density, width, length, and stomatal pores and resulted in lower C_i and g_s, which in turn depressed leaf photosynthetic capacity.

5. Conclusions

In the present study, planting densities and varieties significantly influenced lint yield by affecting leaf stomatal density, thickness, width, length, pore perimeter, leaf gas exchange, and chlorophyll fluorescence characteristics. The J-4B variety in the moderate-density condition produced a higher lint yield due to improved leaf structure, leaf gas exchange, and chlorophyll fluorescence attributes compared with low or high planting densities. Plants at high density substantially depressed leaf stomatal density, thickness, width, length, and pore perimeter, probably due to more competition for nutrients compared with low and moderate planting densities in both varieties. The offset in these attributes further disrupted ΦPSII, F_v/F_m, ETR, and NPQ, which in turn reduced leaf photosynthetic capacity and consequently, lint yield loss. Conclusively, J-4B and Zhongmian-16 grown under medium- and lower-density conditions may be a promising option based on improved leaf structural and functional traits in subtropical regions. Our data will substantially contribute to cotton breeding programs in subtropical environments in the future.

Author Contributions: Conceptualization, A.K. (Aziz Khan) and R.Z.; methodology, A.K. (Aziz Khan); investigation, J.Z., Z.Z., X.K. and A.I.; review and editing, D.K.Y.T. and A.U.; formal analysis, A.K. (Ahmad Khan), K.A. and F.M.; software, M.Z.A., A.B. and S.F.

Funding: We are thankful for the financial supported by National Natural Science Foundation of China (Grant No. 31360348). The supporters did not play any role in the design, analysis, or interpretation of this work and the relevant data.

Conflicts of Interest: The authors declare no conflict of interest.

Abbreviations

PSII, photosystem II; ΦPSII, actual quantum yield of PSII; F_v/F_m, maximal photochemical efficiency of PSII; ETR, electron transport rate; NPQ, nonphotochemical quenching; qP, photochemical quenching; Pn, net photosynthetic rate; g_s, stomatal conductance; C_i, intercellular CO_2 concentration; E, transpiration rate; HNR, height-to-node ratio; D_1, low; D_2, moderate; D_3, high density; V_1, Zhongmian-16; V_2, J-4B.

References

1. Constable, G.A.; Bange, M.P. The yield potential of cotton (*Gossypium hirsutum* L.). *Field Crops Res.* **2015**, *182*, 98–106. [CrossRef]

2. Dai, J.; Dong, H. Intensive cotton farming technologies in China: Achievements, challenges and countermeasures. *Field Crops Res.* **2014**, *155*, 99–110. [CrossRef]

3. Khan, A.; Najeeb, U.; Wang, L. Planting density and sowing date strongly influence growth and lint yield of cotton crops. *Field Crops Res.* **209**, 129–135. [CrossRef]

4. Khan, A.; Wang, L.; Ali, S. Optimal planting density and sowing date can improve cotton yield by maintaining reproductive organ biomass and enhancing potassium uptake. *Field Crops Res.* **2017**, *214*, 164–174. [CrossRef]

5. Yao, H.S.; Zhang, Y.L.; Yi, X.P.; Zhang, X.J.; Zhang, W.F. Cotton responds to different plant population densities by adjusting specific leaf area to optimize canopy photosynthetic use efficiency of light and nitrogen. *Field Crops Res.* **2016**, *188*, 10–16. [CrossRef]

6. Kalaji, H.M.; Schansker, G.; Brestic, M.; Bussotti, F.; Calatayud, A.; Ferroni, L.; Losciale, P. Frequently asked questions about chlorophyll fluorescence, the sequel. *Photosynth. Res.* **2017**, *132*, 13–66. [CrossRef] [PubMed]

7. Dong, H.; Li, W.; Tang, W.; Li, Z.; Zhang, D.; Niu, Y. Yield, quality and leaf senescence of cotton grown at varying planting dates and plant densities in the Yellow River Valley of China. *Field Crops Res.* **2006**, *98*, 106–115. [CrossRef]

8. Wherley, B.G.; Gardner, D.S.; Metzger, J.D. Tall fescue photomorphogenesis as influenced by changes in the spectral composition and light intensity. *Crop Sci.* **2005**, *45*, 562–568. [CrossRef]

9. Adams, C.; Thapa, S.; Kimura, E. Determination of a plant population density threshold for optimizing cotton lint yield: A synthesis. *Field Crops Res.* **2019**, *230*, 11–16. [CrossRef]

10. Yao, H.; Zhang, Y.; Yi, X. Plant density alters nitrogen partitioning among photosynthetic components, leaf photosynthetic capacity and photosynthetic nitrogen use efficiency in field-grown cotton. *Field Crop Res.* **2015**, *184*, 39–49. [CrossRef]

11. Antonietta, M.; Fanello, D.D.; Acciaresi, H.A.; Guiamet, J.J. Senescence and yield responses to plant density in stay green and earlier-senescing maize hybrids from Argentina. *Field Crop Res.* **2014**, *155*, 111–119. [CrossRef]

12. Sawan, Z.M. Plant density; plant growth retardants: Its direct and residual effects on cotton yield and fiber properties. *Cogent Biol.* **2016**, *2*, 1234959. [CrossRef]

13. Luo, Z.; Liu, H.; Li, W.; Zhao, Q.; Dai, J.; Tian, L.; Dong, H. Effects of reduced nitrogen rate on cotton yield and nitrogen use efficiency as mediated by application mode or plant density. *Field Crops Res.* **2018**, *218*, 150–157. [CrossRef]

14. Li, P.; Dong, H.; Zheng, C.; Sun, M.; Liu, A.; Wang, G.; Pang, C. Optimizing nitrogen application rate and plant density for improving cotton yield and nitrogen use efficiency in the North China Plain. *PLoS ONE* **2017**, *12*, e0185550. [CrossRef] [PubMed]

15. Li, T.; Zhang, Y.; Dai, J.; Dong, H.; Kong, X. High plant density inhibits vegetative branching in cotton by altering hormone contents and photosynthetic production. *Field Crops Res* **2019**, *230*, 121–131. [CrossRef]

16. Dong, H.Z.; Kong, X.Q.; Li, W.J.; Tang, W.; Zhang, D.M. Effects of plant density and nitrogen and potassium fertilization on cotton yield and uptake of major nutrients in two fields with varying fertility. *Field Crops Res.* **2010**, *119*, 106–113. [CrossRef]

17. Jiang, C.D.; Wang, X.; Gao, H.Y.; Shi, L.; Chow, W.S. Systemic regulation of leaf anatomical structure, photosynthetic performance, and high-light tolerance in sorghum. *Plant Physiol.* **2011**, *155*, 1416–1424. [CrossRef]

18. Bondada, B.R.; Oosterhuis, D.M. Comparative Epidermal Ultrastructure of Cotton (Gossypium hirsutum L.) Leaf, Bract and Capsule Wall. *Ann. Bot.* **2000**, *86*, 1143–1152. [CrossRef]

19. Anderson, J.M. Photo regulation of the composition, function, and structure of thylakoid membranes. *Ann. Rev. Plant Physiol.* **1986**, *37*, 93–136. [CrossRef]

20. Marchiori, P.E.R.; Machado, E.C.; Ribeiro, R.V. Photosynthetic limitations imposed by self-shading in field-grown sugarcane varieties. *Field Crops Res.* **2014**, *155*, 30–37. [CrossRef]

21. Naramoto, M.; Katahata, S.I.; Mukai, Y.; Kakubari, Y. Photosynthetic acclimation and photoinhibition on exposure to high light in shade-developed leaves of Fagus crenata seedlings. *Flora* **2006**, *201*, 120–126. [CrossRef]

22. Tung, S.A.; Huang, Y.; Hafeez, A.; Ali, S.; Khan, A.; Souliyanonh, B.; Yang, G. Mepiquat chloride effects on cotton yield and biomass accumulation under late sowing and high density. *Field Crops Res.* **2018**, *215*, 59–65. [CrossRef]

23. Mao, L.; Zhang, L.; Zhao, X.; Liu, S.; Zhang, S.; Li, Z. Crop growth, light utilization and yield of relay intercropped cotton as affected by plant density and a plant growth regulator. *Field Crop Res.* **2014**, *155*, 67–76. [CrossRef]

24. Liao, J.; Ma, F.Y.; Fan, H. Effects of sowing rate on population growth, canopy light distribution and yield of drip irrigated spring wheat. *J. Triticeae Crops* **2012**, *32*, 739–742.

25. Pauli, D.; White, J.W.; Andrade-Sanchez, P.; Conley, M.M.; Heun, J.; Thorp, K.R.; Gore, M.A. Investigation of the influence of leaf thickness on canopy reflectance and physiological traits in upland and Pima cotton populations. *Front. Plant Sci.* **2017**, *8*, 1405. [CrossRef] [PubMed]

26. Zhou, C.B.; Xie, C. A simple method to quantify the size and shape of stomatal pore. In Proceedings of the 17th International Congress on Photosynthesis Research, Maastricht, The Netherlands, 7–12 August 2016.

27. Genty, B.; Briantais, J.M.; Baker, N.R. The relationship between the quantum yield of photosynthetic electron transport and quenching of chlorophyll fluorescence. *Biochim. Biophys. Acta* **1989**, *1*, 87–92. [CrossRef]

28. Oxborough, K.; Baker, N.R. Resolving chlorophyll a fluorescence images of photosynthetic efficiency into photochemical and non- photochemical components-calculation of qP and Fv/Fm without measuring Fo. *Photosynth. Res.* **1997**, *54*, 135–142. [CrossRef]

29. Krause, G.H.; Weis, E. Chlorophyll fluorescence and photosynthesis: The basics. *Annu. Rev. Plant Biol.* **1991**, *42*, 313–349. [CrossRef]

30. Kromkamp, J.; Barranguet, C.; Penne, J. Determination of microphytobenthos quantum efficiency and photosynthetic activity by means of variable chlrophyll fluorescence. *Mar. Ecol. Prog. Ser.* **1998**, *162*, 45–55. [CrossRef]

31. Pettigrew, W.T.; Gerik, T.J. Cotton leaf photosynthesis and carbon metabolism. *Adv. Agron.* **2007**, *94*, 209–236.

32. Li, T.; Liu, L.N.; Jiang, C.D.; Liu, Y.J.; Shi, L. Effects of mutual shading on the regulation of photosynthesis in field-grown sorghum. *J. Photochem. Photobiol. B Biol.* **2014**, *137*, 31–38. [CrossRef] [PubMed]

33. Szczepanik, J.; Minchin, P.E.H.; Sowiński, P. On the mechanism of C4 photosynthesis intermediate exchange between Kranz mesophyll and bundle sheath cells in grasses. *J. Exp. Bot.* **2008**, *59*, 1137–1147.

34. Xiao, Y.; Tholen, D.; Zhu, X.G. The influence of leaf anatomy on the internal light environment and photosynthetic electron transport rate: Exploration with a new leaf ray tracing model. *J. Exp. Bot.* **2016**, *67*, 6021–6035. [CrossRef] [PubMed]

35. Xu, Z.; Zhou, G. Responses of leaf stomatal density to water status and its relationship with photosynthesis in a grass. *J. Exp. Bot.* **2008**, *59*, 3317–3325. [CrossRef]

36. Liu, S.; Liu, J.; Cao, J.; Bai, C.; Shi, R. Stomatal distribution and character analysis of leaf epidermis of jujube under drought stress. *J. Anhui Agric. Sci.* **2006**, *34*, 1315–1318.

37. Stirbet, A.; Lazár, D.; Kromdijk, J. Chlorophyll a fluorescence induction: Can just a one-second measurement be used to quantify abiotic stress responses. *Photosynthetica* **2018**, *56*, 86–104. [CrossRef]

38. Murchie, E.H.; Lawson, T. Chlorophyll fluorescence analysis: A guide to good practice and understanding some new applications. *J. Exp. Bot.* **2013**, *64*, 3983–3998. [CrossRef]

39. Chen, B.L.; Yang, H.K.; Ma, Y.N.; Liu, J.R.; Lv, F.J.; Chen, J.; Zhou, Z.G. Effect of shading on yield, fiber quality and physiological characteristics of cotton subtending leaves on different fruiting positions. *Photosynthetica* **2017**, *55*, 240–250. [CrossRef]

40. Zhong, X.M.; Shi, Z.S.; Li, F.H.; Huang, H.J. Photosynthesis and chlorophyll fluorescence of infertile and fertile stalks of paired near-isogenic lines in maize (*Zea mays* L.) under shade conditions. *Photosynthetica* **2014**, *52*, 597–603. [CrossRef]

41. Singh, S.K.; Badgujar, G.; Reddy, V.R.; Fleisher, D.H.; Bunce, J.A. Carbon dioxide diffusion across stomata and mesophyll and photo-biochemical processes as affected by growth CO_2 and phosphorus nutrition in cotton. *J. Plant Physiol.* **2013**, *170*, 801–813. [CrossRef]

42. Zhao, W.Q.; Meng, Y.L.; Chen, B.L.; Wang, Y.H.; Li, W.F.; Zhou, Z.G. Effects of fruiting-branch position, temperature-light factors and nitrogen rates on cotton (*Gossypium hirsutum* L.) fiber strength formation. *Sci. Agric. Sin.* **2011**, *48*, 3721–3732.

43. Dai, Y.; Shen, Z.; Liu, Y.; Wang, L.; Hannaway, D.; Lu, H. Effects of shade treatments on the photosynthetic capacity, chlorophyll fluorescence, and chlorophyll content of *Tetrastigmahemsleyanum* Diels Gilg. *Environ and Exp. Bot.* **2009**, *65*, 177–182. [CrossRef]

44. Guo, C.J.; Qi, W.M.; Yi, Z.; Long, W.Y. Effects of shading on photosynthetic characteristics and chlorophyll fluorescence parameters in leaves of *Hydrangea macrophylla*. *Chin. J. Plant Ecol.* **2017**, *41*, 570–576.

45. Zahoor, R.; Dong, H.; Abid, M.; Zhao, W.; Wang Zhou, Y.Z. Potassium fertilizer improves drought stress alleviation potential in cotton by enhancing photosynthesis and carbohydrate metabolism. *Environ. Exp. Bot.* **2017**, *137*, 73–83. [CrossRef]

46. Yao, H.; Zhang, Y.; Yi, X.; Zuo, W.; Lei, Z.; Sui, L.; Zhang, W. Characters in light-response curves of canopy photosynthetic use efficiency of light and N in responses to plant density in field-grown cotton. *Field Crops Res.* **2016**, *203*, 192–200. [CrossRef]

47. Riar, R.; Wells, R.; Edmisten, K.; Jordan, D.; Bacheler, J. Cotton yield and canopy closure in North Carolina as influenced by row width, plant population, and leaf morphology. *Crop Sci.* **2013**, *53*, 1704–1711. [CrossRef]

48. Wang, J.; Chen, Y.; Wang, P.; Li, Y.S.; Wang, G.; Liu, P.; Khan, A. Leaf gas exchange, phosphorus uptake, growth and yield responses of cotton cultivars to different phosphorus rates. *Photosynthetica* **2018**, *56*, 1414–1421. [CrossRef]

49. Ku, S.; Edwards, G. Oxygen inhibition of photosynthesis. II. Kinetic characteristics as affected by temperature. *Plant Physiol.* **1977**, *59*, 991–999. [CrossRef]

50. Farquhar, G.D.; Sharkey, T.D. Stomatal conductance and photo- synthesis. *Annu. Rev. Plant Physiol.* **1982**, *33*, 317–345. [CrossRef]

 *agronomy*

Article

Moderate Drip Irrigation Level with Low Mepiquat Chloride Application Increases Cotton Lint Yield by Improving Leaf Photosynthetic Rate and Reproductive Organ Biomass Accumulation in Arid Region

Hongyun Gao [1,†], Hui Ma [1,†], Aziz Khan [1], Jun Xia [1], Xianzhe Hao [1], Fangyong Wang [2,*] and Honghai Luo [1,*]

[1] Key Laboratory of Oasis Eco-Agriculture, Xinjiang Production and Construction Corps, Shihezi University, Shihezi, Xinjiang 832003, China; gaohongyun1995@163.com (H.G.); Mahui3219@163.com (H.M.); azizkhanturlandi@gmail.com (A.K.); xiajun9505@163.com (J.X.); haoxz386@163.com (X.H.)

[2] Cotton Institute, Xinjiang Academy Agricultural and Reclamation Science, Shihezi, Xinjiang 832003, China

* Correspondence: fangywang425@163.com (F.W.); luohonghai@shzu.edu.cn (H.L.); Tel.: +139-9953-8919 (F.W.); Fax: +133-4546-670 (H.L.)

† These authors contributed equally to this work.

Received: 15 October 2019; Accepted: 28 November 2019; Published: 2 December 2019

Abstract: Due to the changing climate, frequent episodes of drought have threatened cotton lint yield by offsetting their physiological and biochemical functioning. An efficient use of irrigation water can help to produce more crops per drop in cotton production systems. We assume that an optimal drip irrigation with low mepiquat chloride application could increase water productivity (WP) and maintain lint yields by enhancing leaf functional characteristics. A 2-year field experiment determines the response of irrigation regimes (600 (W_1), 540 (W_2), 480 (W_3), 420 (W_4) 360 (W_5) m^3 ha^{-1}) on cotton growth, photosynthesis, fiber quality, biomass accumulation and yield. Mepiquat chloride was sprayed in different concentration at various growth phases (see material section). Result showed that W_1 increased leaf area index (LAI) by 5.3–36.0%, net photosynthetic rate (Pn) by 3.4–23.2%, chlorophyll content (Chl) by 1.3–12.0% than other treatments. Improvements in these attributes led to higher lint yield. However, no differences were observed between W_1 and W_2 in terms of lint and seed cotton yield, but W_2 increased WP by 3.7% in both years. Compared with other counterparts, W_2 had the largest LAI (4.3–32.1%) at the full boll stage and prolonged reproductive organ biomass (ROB) accumulation by 30–35 d during the fast accumulation period (FAP). LAI, the average (V_T) and maximum (V_M) biomass accumulation rates of ROB were positively correlated with lint yield. In conclusion, the drip irrigation level of 540–600 m^3 ha^{-1} with reduced MC application is a good strategy to achieve higher WP and lint yield by improving leaf photosynthetic traits and more reproductive organ biomass accumulation.

Keywords: drip irrigation quota; cotton; lint yield; water productivity; biomass

1. Introduction

Cotton (*Gossypium hirsutum* L.) is an important fiber crop and oil seed crop worldwide [1]. China produces an average lint yield of 1200 kg ha^{-1}, which is higher than India, Pakistan and USA [2]. With the increasing population comes an increased demand for fiber, and changes in climatic conditions are threating cotton productivity [3]. Crop intensification to produce more food, fiber and feed requires more water, but water resources are limited. Although, cotton is considered drought-resistant crop and its productivity is negatively affected by drought stress. This can lead to reduced growth by negatively influencing plant physiological, biochemical and molecular events [2]. Drought stress can cause 50% to 73% reductions in cotton yields [4]. Limited water availability has threatened irrigated cotton production. Oh the other hand, sufficient fertilizer and irrigation supply results in luxury vegetative growth and increase insect pest incidence which lead to yield penalty [5]. In this context, there is a need to develop water conservation strategy to achieve more crops per drop [6].

Photosynthesis is the prerequisite for lint yield formation. The crops photosynthetic ability can be improved by regulating plant function and irrigation conditions, which in turn affect lint yield [7,8]. Deficit water affects biomass distribution and facilitates assimilate transfer to reproductive organs [9]. A short period of mild drought may stimulate the compensatory effect of photosynthesis [10]. These compensations favor the translocation of assimilate to reproductive organs and the improvement of WP (water productivity) without sacrificing yield [11]. Hence, these compensatory effects represent a self-regulatory mechanism that helps crop to adapt stressful environment by efficient utilization of limited water resources [1].

Xinjiang is the major cotton-growing provinces in China, contributing 67% of the total national lint production [12]. However, low water availability has imposed a great challenge to cotton production in this area. Currently, mulch drip irrigation is widely adopted to increase cotton lint yield and WP in Xinjiang [13,14]. To conserve water and produce high yields under irrigation systems, cotton growers have adopted the concept of regulated deficit irrigation (RDI) [10,13,15]. RDI can increase WP rapidly, this reduces leaf area and leading to lower photosynthesis rate [15,16], it was not conducive to biomass accumulation, resulting in failure to increase production [17,18]. Therefore, understanding the changes in photosynthetic characteristics and dry matter accumulation are needed to achieve optimal lint yield and WP under mulch drip irrigation system.

The main purpose of mulch drip irrigation technology is to conserve soil water and achieve high crop yield, but yield and productivity do not always increase with increased irrigation quota [19,20]. Hence, a reasonable control of drip irrigation quotas is essential to identify the optimal combination of water conservation and high yield. Mepiquat chloride (MC) (N,N-dimethylpiperidinium chloride) is a growth regulator used worldwide to control plant geometry. MC can be absorbed by leaves and is distributed throughout plants [21]. MC applications induce reductions in leaf expansion, stems, petiole length, and node number and enhance the maturity of cotton crops, with variable yield responses [22–24]. Therefore, it is hypothesized that moderately reduced irrigation quotas in conjunction with a low application rate of MC in the field can improve WP and maintain lint yields by utilizing the compensatory effects of photosynthesis under deficient drip irrigation. The objectives of this research were to explore the effects of various drip irrigation quotas on the photosynthesis capability, biomass accumulation, yield and WP using different concentration of MC under mulch drip irrigation systems. It also determines the quantitative relationships among these factors.

2. Materials and Methods

2.1. Experimental Site and Cultivar

Field experiments were conducted in 2016 and 2017 at the experimental farm of Shihezi University (45°19′ N latitude, 86°03′ E longitude). A Xinluzao 45 (*Gossypium hirsutum* L.) cultivar was used in this study. This cultivar was developed by the Xinjiang Academy of Agricultural and Reclamation Science and is officially registered and released by the Xinjiang Crop Cultivar Registration Committee. The total growth period from emergence to initial boll opening (BO) is 122 days. The soil was a purple clay loam with a pH of 7.65 and contained 15.3 g kg^{-1} organic matter, 1.1 g kg^{-1} total N, 54.9 mg kg^{-1} available N, 23.0 mg kg^{-1} available P and 194 mg kg^{-1} available K within the 0–20 cm soil layer. The evapotranspiration, temperature and precipitation data from April to October are shown in Figure 1.

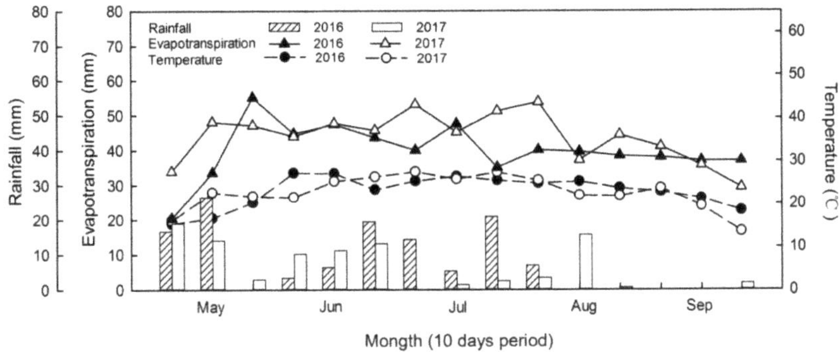

Figure 1. Monthly average evapotranspiration, temperature and rainfall of Shihezi (2016–2017).

2.2. Experimental Design

A randomized complete block design with four replications was used in this study. Generally, a total irrigation amount of 4800–5000 m^3 ha^{-1} is required to achieve more than 6000 kg ha^{-1} of seed cotton yield in northern Xinjiang region [10,18]. Five drip irrigation treatments were targeted i.e., W_1 (600 m^3 ha^{-1} of water each time with the total amount of irrigation 4800 m^3 ha^{-1}; control), W_2 (540 m^3 ha^{-1} of water each time with the total amount of irrigation 4320 m^3 ha^{-1}), W_3 (480 m^3 ha^{-1} of water each time with the total amount of irrigation 3840 m^3 ha^{-1}), W_4 (420 m^3 ha^{-1} of water each time, with the total amount of irrigation 3360 m^3 ha^{-1}), and W_5 (360 m^3 ha^{-1} of water each time with the total amount of irrigation 2880 m^3 ha^{-1}). The drip irrigation rates were controlled by water meter and switch ball valve. The irrigation was applied in the same dates for all the treatments, and the duration was approximately 10–14 h (07:30 AM–21:30 PM).

2.3. Field Management

Prior to sowing, the experimental field was covered with a plastic film. Two drip irrigation lines (Beijing Luckrain Inc., China) were installed under each plastic film. The drip irrigation line had an inner diameter of 2.5 cm with emitter distance of 50 cm, and a flow rate of 2.7 L h^{-1}. Cotton seeds were sown on both sides of the drip irrigation belt at a distance of 13.5 cm on 21 and 23 April in 2016 and 2017, respectively. The plots were randomly arranged with the total area of 56 m^2 (7.0 × 8.0 m^2). After two weeks, seedlings were thinned to maintain the desired planting density. The row spacing was maintained as 12 cm with a planting density of 18,000 plants ha^{-1} which was commonly practiced in this region. Fertilizer was applied with water by 8 times (first via drip irrigation for half an hour, then via fertigation). Thereafter, the field was fertilized with 4500 kg ha^{-1} of oil residue (with 13% N, 2% P_2O_5 and 16% K_2O) as a basal fertilizer. In addition, 72 kg ha^{-1} of urea (comprising 46% N) and 225 kg ha^{-1} of triple superphosphate (comprising 45% P_2O_5) were applied throughout the growth period. The amount and time of the drip irrigation was controlled to maintain to distribute equal amount of fertilizer for each treatment. MC was applied to control vegetative growth. MC solution at 208 g hm^{-2} concentration was sprayed 5 times in the W_1 treatment. A 6 g hm^{-2} MC solution was sprayed from cotyledon stage to the two-leaf stage and 11 g hm^{-2} was sprayed at the 5–7-leaf stage. Moreover, 26 g hm^{-2}, 45 g hm^{-2}, 120 g hm^{-2} was sprayed 2 days before the first irrigation, 2 days before the second irrigation and 5–7 days after topping, respectively. The first and second spray was similar to W_1 treatment. However, an MC solution at a concentration of 137 g hm^{-2} was sprayed on W_2, W_3, W_4 and W_5 treatments. To hasten the crop maturity, a defoliant at 450 g ha^{-1} tribenuron combined with 1350 mL ha^{-1} ethephon was used in both years. Artificial topping was carried out on 3 and 8 July in 2016 and 2017, respectively. Other management practices such as insect and weed control were conducted according to the local agronomic practices.

2.4. Net Photosynthetic Rate

Net photosynthetic rate (Pn) was assessed from functional leaf at the full squaring (FS), initial flowering (IF), full flowering (FF), full boll setting (FB), late boll setting (LFB) and BO stages between 10:00 a.m. and 12:00 p.m. on sunny day using an open-type photosynthesis system (LI-6400, LI-COR, Lincoln, NE, USA) equipped with a red/blue light source chamber. The machine was configured at light intensity of 1800 μmol m^{-2} s^{-1}, temperature, 32 °C and at 20% relative humidity. Four plants in each plot were selected for measurement.

2.5. Water Productivity (WP)

The WP was determined according to [6],

$$WP = Y/I \tag{1}$$

WP, Y and I represent the water productivity (kg m^{-3}), seed cotton yield (kg ha^{-1}) and the amount of drip irrigation (m^{-3}) during the whole cotton growth period, respectively.

2.6. Chlorophyll Content

To measure chlorophyll content leaves were removed and the petiole was wrapped in wet gauze. A 0.1 g leaf sample was used to determine the Chl content. Leaves were placed in a 25 mL test tube and the pigment was extracted with 13 mL of 80% acetone. Tubes were wrapped with a black cloth and placed in dark conditions. Tubes were shaken at regular intervals and incubated for 72 h until the leaves become white color. The optical density (OD) value was measured at 470 nm, 663 nm, and 645 nm wavelength using a UV-2041 spectrophotometer (Shimadzu, Kyoto, Japan).

2.7. Biomass Accumulation

Four successive plants at the FS, IF, FF, FB, LFB and BO stages from each plot of the fourth replicates were carefully uprooted, divided into vegetative organs (roots, stem, leaves and branches) and reproductive organs (buds, flowers, boll shells and bolls). Samples were enveloped separately and placed into an electric fan-amended oven at 105 °C for 30 min then dried at 80 °C to a constant weight. The leaf area was measured using an LI-3000 area meter (LI-COR, Lincoln, NE, USA). Leaf area index (LAI) was calculated by multiplying the total leaf area of single plant (m^2 plant^{-1}) $\times$ plant density (plants m^{-2}). The declining rate of LAI, Pn and Chl content at the FB and BO stages was determined as follows:

$$\text{Declining Rate (\%)} = -(V_{BO} - V_{FB})/V_{BO} \tag{2}$$

Of which, V_{BO} and V_{FB} represent the LAI, Pn and Chl content parameters at the FB and BO stages, respectively.

A logistic formula was used to describe the progress of biomass accumulation [25]:

$$Y = \frac{K}{1 + ae^{bt}} \tag{3}$$

Of which, t (day) indicates days after emergence (DAE), Y (g) indicates the biomass at t, K (g) is the maximum biomass, and a and b are the constants to be found.

$$t_0 = \frac{\ln a}{b} \tag{4}$$

$$t_1 = \frac{1}{b} \ln\left(\frac{2 + \sqrt{3}}{a}\right) \tag{5}$$

$$t_2 = \frac{1}{b} \ln\left(\frac{2 - \sqrt{3}}{a}\right) \tag{6}$$

At $t = t_0$, the biomass accumulation reaches a maximum speed defined as follows:

$$V_M = \frac{-bk}{4} \tag{7}$$

The period at which 58% of the biomass accumulated is defined as the biomass fast accumulation period (FAP), which begins at t_1 and terminates at t_2. During the FAP, Y is linearly correlated with t and the average speed, defined as follows:

$$V_T = \frac{Y_2 - Y_1}{t_2 - t_1} \tag{8}$$

2.8. Yield, Yield Contributors and Fiber Quality

Seed cotton from each plot was hand-picked (on 3 and 15 October in 2016; 30 September and 15 October in 2017). Seed cotton was sun dried and weighed. One hundred fully opened bolls were sampled to calculate individual boll weight and lint percentage lint percentage. Boll number were determined by counting bolls (>2 cm in diameter) of each plant on 15 September and 20 September in 2016 and 2017, respectively.

To assess fiber quality attributes (length, strength, micronaire, and uniformity) lint samples were sent to the Chinese Academy of Agricultural Sciences for high-volume instrumentation analysis.

2.9. Data Analysis

SPSS 19.0 software (SPSS Institute Inc., Chicago, IL, USA) was used for analysis of variance. Means were tested by Duncan multiple comparison at a level of 0.05. Sigma Plot 12.5 (Aspire Software Intl., Ashburn, VA, USA) was used for data processing and figures as well as linear regression.

3. Results

3.1. Leaf Area Index

The LAI decreased with the decreasing drip irrigation level (except the FB stage) in both years (Figure 2). Under W1, the LAI was 5.6 at the LFB stage, 3.6 to 5.3 at the FB stage under for W_5, W_4, W_3 and W_2 treatments. At the FB stage, the LAI was 3.1–5.9% higher in W_2 compared with W_1. Moreover, W_1, W_2, W_3, W_4 and W_5 decreased the LAI by 5.7%, 14.6%, 18.6%, 18.6% and 18.7%, respectively, at all growth stages.

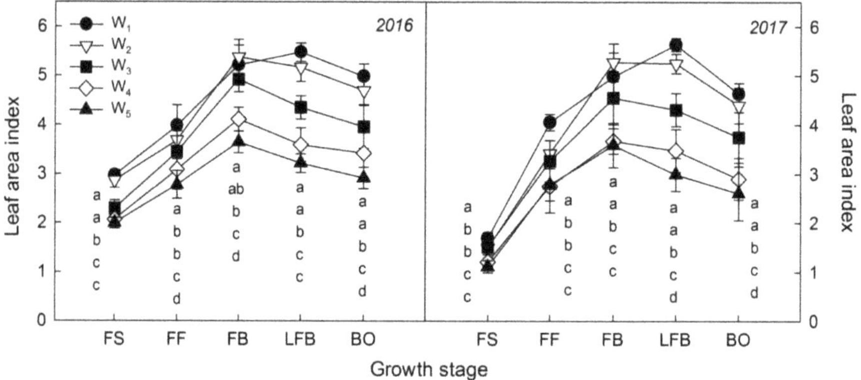

Figure 2. Effect of different drip irrigation quotas on leaf area index (LAI) of cotton at full squaring (FS), full flowering (FF), full boll (FB), later full boll (LFB) and boll opening stage (BO) in 2016 and 2017. Error bar shows standard error (SE) of means.

3.2. Chlorophyll Content

Cotton leaf Chl content was significantly influenced by irrigation levels at different growth stages (Figure 3). with the decreased in the irrigation level Chl content was significantly decreased at various growth stages. W_1 and W_2 had higher leaf Chl contents at FB stage and then decreased later in season. W_2–W_5 decreased the Chl by 0.4–5.2% at the FS stage, −0.11–9.7% at the FF stage, 7–16.6% at the FB stage, 1.0–6.3% at the LFB stage and 3.4–21.7% at the BO stage.

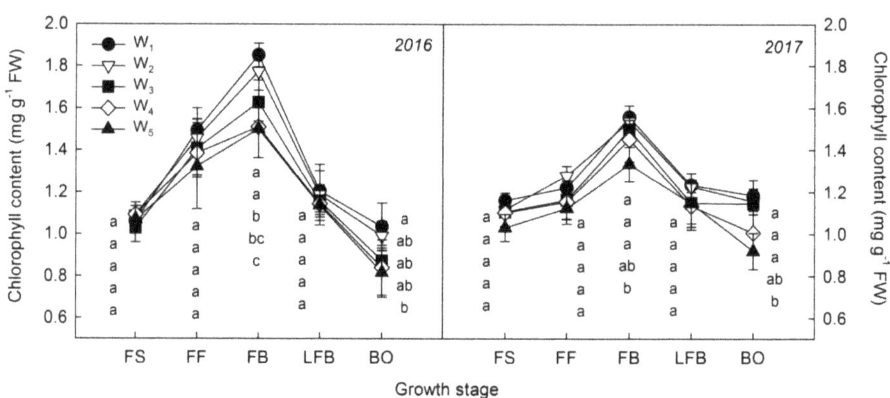

Figure 3. Effect of different drip irrigation quotas on chlorophyll (Chl) contents of cotton at full squaring (FS), full flowering (FF), full boll (FB), later full boll (LFB) and boll opening stage (BO) in 2016 and 2017. Error bar shows standard error (SE) of means.

3.3. Net Photosynthetic Rate

The Pn rate was substantially influenced by irrigation levels. With the crop development the Pn rate was increased and then decreased (Figure 4). The rate of Pn was lowered with the decreased of irrigation level during the whole growth stages. W_1 and W_2 resulted in higher net Pn compared with other counterparts. Across the years, the Pn was higher in 2016 compared with 2017.

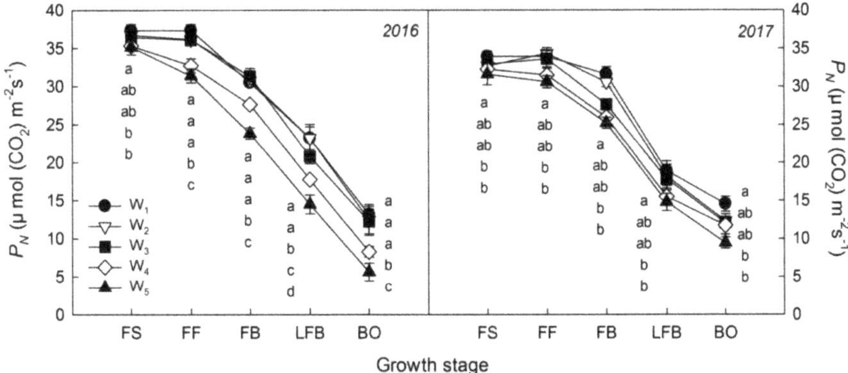

Figure 4. Effect of different drip irrigation quotas on net photosynthetic rate (Pn) of cotton leaf at full squaring (FS), full flowering (FF), full boll (FB), later full boll (LFB) and boll opening stage (BO) in 2016 and 2017. Error bar shows standard error (SE) of means.

3.4. Cotton Plant Biomass Accumulation

Cotton plant biomass (CPB) accumulation was increased rapidly and then decreased with decreasing drip irrigation (Figure 5). W_1 resulted in 7.84%, 13.13%, 22.17%, and 28.09% at the FB stage; 8.24%, 15.80%, 27.82%, and 33.91% at the LFB stage; and 9.72%, 15.60%, 27.54%, and 34.09% higher biomass at the BO stage averaged across both years. Vegetative organ biomass (VOB) accumulation increased sharply before the FF stage and then decreased with decreasing drip irrigation (Figure 5). W_1 increased the VOB by 6.97–32.97% at the FB stage and 5.86–34.20% at the LFB stage than other treatment. Reproductive organ biomass (ROB) accumulation decreased with decreasing drip irrigation (Figure 5). W_1 had higher ROB by 15.32% and 11.04% at the LFB stage and 17.21% and 18.74% at the BO stage, respectively.

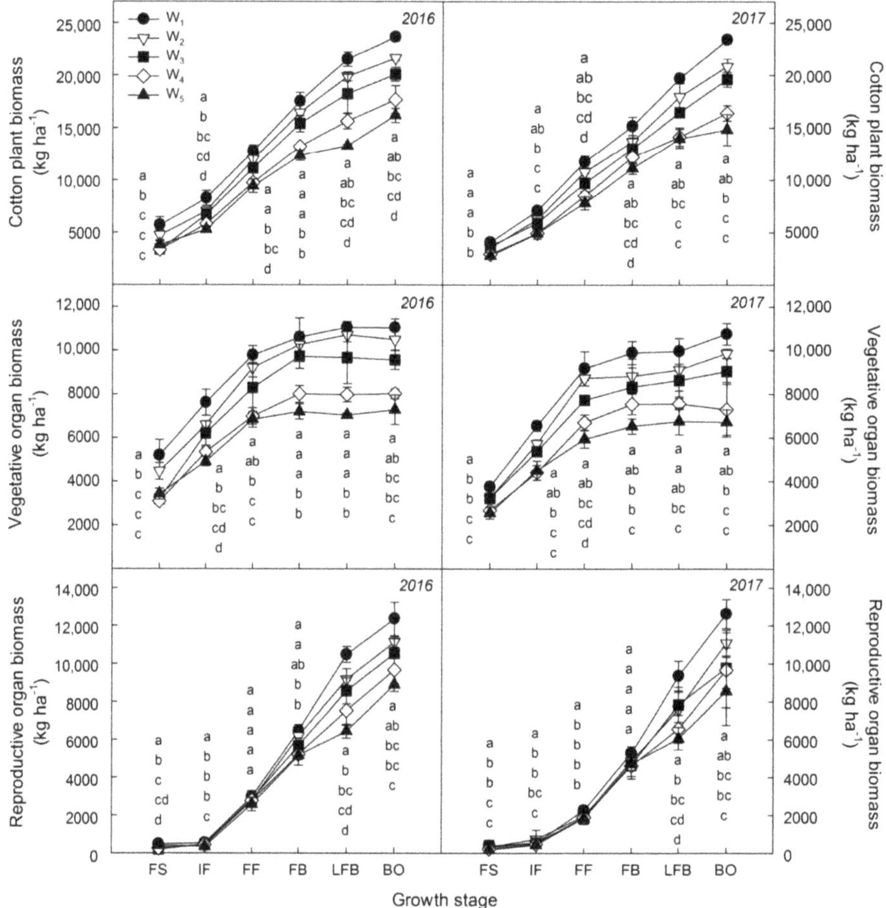

Figure 5. Response of cotton plant biomass (CPB), vegetative organs biomass (VOB), and reproductive organ biomass (ROB) at full squaring (FS), initial flowering (IF), full flowering (FF), full boll (FB), later full boll (LFB) and boll opening stage (BO) under drip irrigation quotas in 2016 and 2017. Error bar shows standard error (SE) of means.

3.5. Characteristics of Biomass Accumulation

The simulation of biomass as a function of DAE was assessed via equation (2). The logistic function of the biomass accumulation was followed by a sigmoidal growth pattern. All the coefficients of determination were significant, although the equation coefficients differed among the treatments (Table 1). Calculations by Equations (3)–(8) based on Table 1 revealed the beginning and end day of the FAP for CPB accumulation during both years. W_1 and W_2 begins and ends at 68 and 119 DAE and 69 and 113 DAE, respectively, in 2016 and 64 and 124 DAE and 77 and 115 DAE, respectively, in 2017, with greater average and maximum rates of biomass over other treatments.

Table 1. Eigen parameters of cotton biomass accumulations for different drip irrigation quotas.

Year	Treatment	Regression Equations	p	Fast Accumulation Period				Fast Accumulation Point		
				t_1 (DAE)	t_2 (DAE)	T (day)	V_T (kg ha^{-1} day^{-1})	V_M (kg ha^{-1} day^{-1})	t_m (DAE)
	Cotton plant biomass									
	W_1	$Y = 33258.5934/(1 + 119.3122e^{-0.051359t})$	0.0011	67.5	118.8	51.3	374.4	427.0	93.1
	W_2	$Y = 29446.6185/(1 + 231.5378e^{-0.05993t})$	0.0016	68.9	112.8	44.0	386.8	441.2	90.9
	W_3	$Y = 25670.201/(1 + 403.4676e^{-0.067762t})$	0.0020	69.1	108.0	38.9	381.3	434.9	88.5
	W_4	$Y = 22840.8431/(1 + 299.7992e^{-0.063716t})$	0.0026	68.8	110.2	41.3	319.0	363.8	89.5
	W_5	$Y = 20179.7177/(1 + 237.2227e^{-0.062861t})$	0.0059	66.1	108.0	41.9	278.1	317.1	87.0
	Vegetative organ biomass									
2016	W_1	$Y = 13417.5274/(1 + 2823.3601e^{-0.114252t})$	0.0007	58.0	81.1	23.1	336.0	383.2	69.5
	W_2	$Y = 13262.6927/(1 + 1995.0725e^{-0.105198t})$	0.0007	59.7	84.8	25.0	305.8	348.8	72.2
	W_3	$Y = 11790.2110/(1 + 14040.0792e^{-0.131561t})$	0.0019	62.6	82.6	20.0	340.0	387.8	72.6
	W_4	$Y = 9872.6580/(1 + 11420.6305e^{-0.129823t})$	0.0019	61.8	82.1	20.3	280.9	320.4	72.0
	W_5	$Y = 8836.4909/(1 + 14624.8702e^{-0.136997t})$	0.0007	60.4	79.6	19.2	265.4	302.6	70.0
	Reproductive organ biomass									
	W_1	$Y = 17786.1982/(1 + 7812.2131e^{-0.083488t})$	0.0012	91.6	123.1	31.6	325.5	371.2	107.4
	W_2	$Y = 15248.9874/(1 + 9516.8349e^{-0.086831t})$	0.0024	90.3	120.7	30.3	290.2	331.0	105.5
	W_3	$Y = 14945.7964/(1 + 5473.7872e^{-0.080545t})$	0.0033	90.5	123.2	32.7	263.9	301.0	106.9
	W_4	$Y = 14135.8253/(1 + 3096.8490e^{-0.074301t})$	0.0053	90.5	125.9	35.5	230.2	262.6	108.2
	W_5	$Y = 12496.6137/(1 + 2551.9707e^{-0.073465t})$	0.0109	88.9	124.7	35.9	201.2	229.5	106.8

Table 1. *Cont.*

Year	Treatment	Regression Equations	p	Fast Accumulation Period				Fast Accumulation Point	
				t_1 (DAE)	t_2 (DAE)	T (day)	V_T (kg ha^{-1} day^{-1})	V_M (kg ha^{-1} day^{-1})	t_m (DAE)
	Cotton plant biomass								
	W_1	$Y = 35207.8503/(1 + 60.9476e^{-0.043942t})$	0.0017	63.6	123.5	59.9	339.1	386.8	93.5
	W_2	$Y = 30621.6718/(1 + 67.5922e^{-0.046357t})$	0.0017	62.5	119.3	56.8	311.2	354.9	90.9
	W_3	$Y = 28700.9055/(1 + 60.3377e^{-0.044661t})$	0.0015	62.3	121.3	59.0	281.0	320.5	91.8
	W_4	$Y = 22741.8283/(1 + 106.6472e^{-0.054536t})$	0.0013	61.5	109.8	48.3	271.9	310.1	85.6
	W_5	$Y = 20470.5648/(1 + 88.5934e^{-0.053107t})$	0.0021	59.6	109.2	49.6	238.3	271.8	84.4
	Vegetative organ biomass								
2017	W_1	$Y = 12345.7473/(1 + 701.2048e^{-0.108047t})$	0.0003	48.5	72.8	24.4	292.4	333.5	60.6
	W_2	$Y = 11733.5035/(1 + 613.4740e^{-0.102951t})$	0.0020	49.6	75.1	25.6	264.8	302.0	62.4
	W_3	$Y = 10606.1122/(1 + 356.4513e^{-0.096032t})$	0.0022	47.5	74.9	27.4	223.3	254.6	61.2
	W_4	$Y = 9074.6476/(1 + 718.0429e^{-0.106314t})$	0.0016	49.5	74.2	24.8	211.5	241.2	61.9
	W_5	$Y = 8327.4903/(1 + 370.5410e^{-0.097825t})$	0.0019	47.0	73.9	26.9	178.6	203.7	60.5
	Reproductive organ biomass								
	W_1	$Y = 18820.0391/(1 + 3123.6047e^{-0.076271t})$	0.0004	88.2	122.8	34.5	314.6	358.9	105.5
	W_2	$Y = 16606.3387/(1 + 2573.79391e^{-0.074759t})$	0.0027	87.4	122.7	35.2	272.1	310.4	105.0
	W_3	$Y = 15124.7130/(1 + 2981.4888e^{-0.077040t})$	0.0010	86.7	120.9	34.2	255.4	291.3	103.8
	W_4	$Y = 13679.9428/(1 + 2689.8974e^{-0.077117t})$	0.0056	85.3	119.5	34.2	231.2	263.7	102.4
	W_5	$Y = 9488.9482/(1 + 7536.0818e^{-0.09532t})$	0.0050	79.8	107.5	27.6	198.3	226.1	93.7

DAE means days after emergence (day); t_1 and t_2 are the beginning and terminating day after the fast accumulation period; T indicates the duration of FAP, T = t_1 − t_2; V_T and V_M are the average and maximum biomass accumulation rates during the FAP, respectively.

Differences in CPB accumulation were noticed among the treatments in both years (Table 1). W_1 had the greatest average (356.8 kg ha^{-1} day^{-1} V_T) and maximum (406.9 kg ha^{-1} day^{-1} V_M) biomass accumulation rates followed by W_2 (349.0 kg ha^{-1} day^{-1} V_T, 398.1 kg ha^{-1} day^{-1} V_M) and W_3 (331.0 kg ha^{-1} day^{-1} V_T, 377.7 kg ha^{-1} day^{-1} V_M). The FAP of the CPB accumulation under W_2 and W_3 initiated at the same time (65.7 DAE), and they terminated 5 days and 7 days sooner than that under W_1. W_5 had a shorter duration of CPB at FAP which showed lowest average (258.2 kg ha^{-1} day^{-1} V_T) and maximum (294.5 kg ha^{-1} day^{-1} V_M) biomass accumulation rates.

The VOB accumulation was also affected by the irrigation quota during both years (Table 1). The V_T and V_M increased with increasing drip irrigation in both years. VOB accumulation at FAP under W_2 and W_3 begun and terminated almost at the same time, which was 1–3 days delayed than W_1. W_2 had longer VOB duration at FAP (25.3 days) with both average (V_T 285.3 kg ha^{-1} day^{-1}) and maximum (V_M 325.4 kg ha^{-1} day^{-1}) biomass formation rates.

The drip irrigation quotas significantly altered cotton plant ROB organ biomass formation in both years (Table 1). Under, the FAP of W_2 ROB accumulation at FAP began at 89 DAE and terminated at 122 DAE, both of which 1 day delayed than W_1. Moreover, W_2 and W_3 had similar FAP time. W_1 had higher both average and maximum biomass accumulation rates at FAP followed by W_2.

3.6. Yield, Water Productivity and Fiber Quality

Cotton yield and yield components were significantly influenced by irrigation levels (Table 2). Among irrigation levels W_1 produced the highest seed cotton and lint yield in both years compared with other treatments. Compared with W_1, W_2 slightly influenced the cotton yield in 2016 and 2017. W_2 resulted in 5.3–7.7% lower seed cotton yield and a 5.0–5.7% lower lint yield. Compared with W_1, individual boll weight was decreased by 1.6%, 2.4%, 4.1%, and 5.3% W_2, W_3, W_4, and W_5 respectively. Similarly boll numbers per unit area decreased by 5.4%, 6.2%, 9.9%, and 18.4%, respectively. However the lint % increased. In addition, WP under W_2, W_3, W_4, and W_5 was 3.9%, 13.0%, 23.6%, and 29.8% greater than over W_1. The differences were minor between both years.

Table 2. Cotton yield and WP under different drip irrigation quotas.

Year	Treatment	Seed Yield (kg ha^{-1})	Lint Yield (kg ha^{-1})	Boll Weight (g)	Bolls Per Unit Area (10^4 ha^{-1})	Lint Percentage (%)	Water Productivity (kg m^{-3})
	W_1	6607 ± 392 a	2739 ± 194 a	4.78 ± 0.12 a	140.2 ± 11.4 a	41.45 ± 1.43 b	1.38 ± 0.08 b
	W_2	6099 ± 305 ab	2581 ± 109 ab	4.72 ± 0.07 ab	129.2 ± 5.9 ab	42.32 ± 1.83 ab	1.41 ± 0.07 b
2016	W_3	5968 ± 286 ab	2531 ± 212 ab	4.69 ± 0.14 ab	129.8 ± 9.0 ab	42.41 ± 1.84 ab	1.55 ± 0.07 ab
	W_4	5694 ± 340 b	2393 ± 235 bc	4.62 ± 0.19 bc	123.1 ± 9.5 bc	42.39 ± 1.82 ab	1.68 ± 0.10 a
	W_5	5013 ± 260 c	2147 ± 85 c	4.55 ± 0.17 c	110.2 ± 4.7 c	42.84 ± 1.85 a	1.74 ± 0.09 a
	W_1	6492 ± 466 a	2615 ± 199 a	4.76 ± 0.16 a	135.9 ± 12.1 a	40.28 ± 1.39 b	1.35 ± 0.10 c
	W_2	6151 ± 341 ab	2484 ± 96 ab	4.67 ± 0.26 ab	132.0 ± 10.7 ab	40.39 ± 1.75 b	1.42 ± 0.08 bc
2017	W_3	5874 ± 441 bc	2397 ± 248 b	4.62 ± 0.15 ab	129.3 ± 11.1 ab	40.63 ± 1.76 ab	1.53 ± 0.11 b
	W_4	5689 ± 342 c	2358 ± 233 b	4.53 ± 0.09 b	125.6 ± 7.5 b	41.45 ± 1.79 a	1.69 ± 0.10 a
	W_5	5184 ± 533 d	2132 ± 165 c	4.49 ± 0.11 b	115.2 ± 14.4 c	41.13 ± 1.78 ab	1.80 ± 0.18 a
	Year	ns	*	ns	ns	**	ns
	Year × Treatment	ns	ns	ns	ns	ns	ns

Means within a column of the same year followed by a different letter are significantly different ($p < 0.05$) according to the Duncan multiple range test. The same letters in the same column indicated no significant difference at 0.05 level in Duncan's analysis in the same year. "*", "**" means significance at the 0.05, 0.01 level, respectively. "ns" indicates non-significant.

Fiber quality parameters were substantially influenced by irrigation levels (Table 3). The fiber length and uniformity increased as the drip irrigation quota increased. Compared with W_1, W_2 and W_3 had higher fiber lengths and fiber uniformity compared with other treatment. W_4 and W_5 treatment resulted in significantly lower fiber length and uniformity during both years. The uniformity was significantly greater in 2016 than in 2017, while the fiber length, specific strength, and micronaire values remained similar.

Table 3. Change of the fiber quality attributes under different drip irrigation quotas.

Year	Treatment	Fiber Length (mm)	Fiber Uniformity (%)	Specific Strength (CN tex^{-1})	Micronaire Value
2016	W$_1$	30.4 ± 0.38 a	87.3 ± 0.35 a	30.5 ± 1.5 a	4.0 ± 0.11 a
	W$_2$	30.0 ± 0.06 ab	87.0 ± 0.69 a	30.7 ± 1.5 a	4.2 ± 0.15 a
	W$_3$	29.9 ± 0.38 ab	86.8 ± 0.33 a	31.2 ± 1.2 a	4.1 ± 0.10 a
	W$_4$	29.6 ± 0.33 b	85.9 ± 0.31 b	30.8 ± 0.7 a	4.2 ± 0.17 a
	W$_5$	29.5 ± 0.35 b	85.8 ± 0.15 b	30.7 ± 1.3 a	4.2 ± 0.15 a
2017	W$_1$	30.6 ± 0.48 a	85.1 ± 0.78 a	30.7 ± 0.20 a	4.0 ± 0.31 a
	W$_2$	30.2 ± 0.53 ab	84.7 ± 0.50 ab	30.8 ± 0.10 a	4.0 ± 0.14 a
	W$_3$	30.2 ± 0.10 ab	84.6 ± 0.62 ab	30.7 ± 0.72 a	4.1 ± 0.07 a
	W$_4$	29.5 ± 0.21 bc	84.4 ± 0.19 ab	30.7 ± 0.40 a	4.0 ± 0.31 a
	W$_5$	29.0 ± 0.84 c	84.2 ± 0.47 b	30.6 ± 0.62 a	4.1 ± 0.26 a
	Year	ns	**	ns	ns
	Year × Treatment	*	ns	ns	ns

Means within a column of the same year followed by a different letter are significantly different ($p < 0.05$) according to the Duncan multiple range test. The same letters in the same column indicated no significant difference at 0.05 level in Duncan's analysis in the same year. "*", "**" means significance at the 0.05, 0.01 level, respectively. "ns" indicates non-significant.

3.7. Correlation Analysis and Regression Analysis

The relationships between photosynthetic characteristic parameters (the LAI, Pn and Chl), biomass accumulation (CPB, VOB and ROB) and lint yield were analyzed at different growth stages in both years (Table 4). The correlation intensity between the photosynthetic characteristic and lint yield was determined i.e., Pn (0.915) > LAI (0.896) > Chl (0.840) at the FB stage as well as LAI (0.916) > Pn (0.901) > Chl (0.727) at the LFB stage. During the FB to BO stages, the CPB, VOB and ROB were highly significantly ($p < 0.001$) correlated with lint yield at especially at the LFB stage. In addition, the correlation association between LAI and ROB and lint yield was gradually increased from FS to BO stage.

Table 4. Correlation between physiological parameters and lint yield and at different growth stages.

Growth Stages	LAI	Pn	Chl	CPB	VOB	ROB
FS	0.637*	0.577	0.324	0.722*	0.748*	0.291
FF	0.918**	0.907**	0.592	0.919**	0.940**	0.47
FB	0.896**	0.915**	0.840**	0.924**	0.964**	0.704*
LFB	0.916**	0.901**	0.727*	0.959**	0.964**	0.944**
BO	0.946**	0.806**	0.498	0.946**	0.946**	0.937**

"*" and "**" means significance at the 0.05, 0.01 level, respectively (both sides).

The relationships from regression analysis showed a declining rate of photosynthetic characteristics traits (LAI, Pn and Chl) from FB to BO stages. The lint yield has been shown in Figure 6. Cotton biomass accumulation (CPB, VOB and ROB), simulation (T, V$_T$, V$_M$ and t$_m$) and lint yield (Figure 7) were described using linear functions during different growth stages in both years. The declined rate of Pn and Chl content during the FB to BO stages, the T of CPB, VOB and ROB accumulation and t$_m$ of VOB and ROB accumulation were not significantly linearly correlated to lint yield. A negative correlation was observed between the declining rate of LAI ($R^2 = 0.7429$, $p < 0.001$) and lint yield. A positive correlation were noticed between V$_T$ ($R^2 = 0.7422$, $p < 0.001$) and V$_M$ ($R^2 = 0.7424$, $p < 0.001$) of CPB, VOB and ROB i.e., V$_T$ ($R^2 = 0.5791$, $p = 0.0106$), V$_M$ ($R^2 = 0.5791$, $p = 0.0106$), V$_T$ ($R^2 = 0.9354$, $p < 0.001$), V$_M$ ($R^2 = 0.9354$, $p < 0.001$) and lint yield. In addition, CPB t$_m$ was also correlated ($R^2 = 0.6702$, $p = 0.0038$) with lint yield.

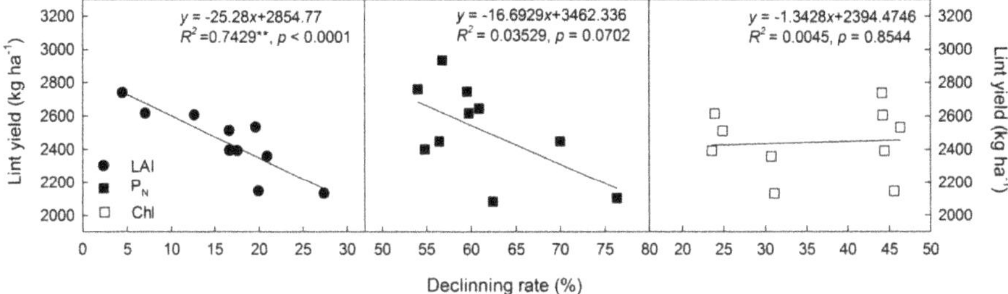

Figure 6. Regression analysis between the declining rate of photosynthetic capacity values (LAI, Pn and Chl) during FB to BO stage and lint yield. R^2 represents the coefficient of determination in linear regression. "**" means significance at the 0.01 level (both sides).

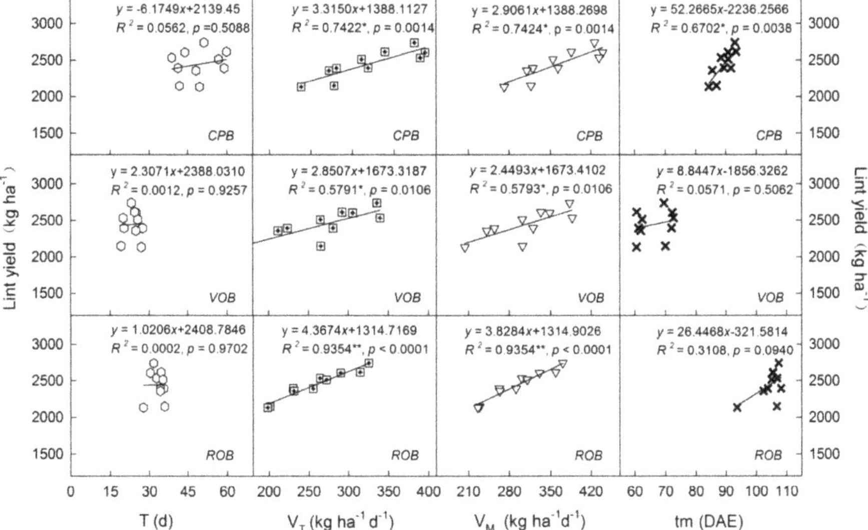

Figure 7. Regression analysis of the characteristic values (T, VT, VM and tm) of cotton biomass accumulation (CPB, VOB and ROB) and lint yield. T indicates the duration of FAP; V_T and V_M are the average and maximum biomass accumulation rates during the FAP, t_m (DAE) means days after emergence (day) respectively. R^2 represents the coefficient of determination in linear regression. "*" and "**" means significance at the 0.05, 0.01 level, respectively (both sides).

4. Discussion

Crop production is positively allied with photosynthesis capacity (i.e., photosynthetic area, Pn, and photosynthetic pigments) [26,27] and is significantly influenced by soil water content [28]. In this study, LAI, Pn and Chl content were positively correlated with lint yield from full boll to boll opening stages. The LAI was strongly correlated with lint yield during the later full boll to boll opening stages and the declining rate was negatively correlated with lint yield. This show that duration of LAI during late growth late stages and the leaf photosynthetic capacity important players for increasing cotton yield. These data are in line with [29] that the absorption of photosynthetically active radiation was not significantly affected by mild water deficit. Plants can respond to drought by reducing nonstomatal transpiration (soil evaporation) [30] and increasing stomatal resistance (reducing evaporation) and osmotic adjustment substances [31]. An optimistic growth i.e., Pn, root growth, the LAI, plant height and biomass accumulation maintain high values in under short term water deficit which in turn increase yield [10,32]. These physiological adjustments can be explained by the compensatory growth of cotton under moderate drought stress [33]. Hence, irrigation strategies can be used to alter leaf area expansion, the absorption of photosynthetically active radiation and carbohydrate production to enhance photosynthesis capacity, water conservation and consequently yield [27,29].

Photosynthesis is the basis of crop biomass accumulation and yield formation under drought conditions [34,35]. Chl affects electron transport and determines the photosynthesis capacity of crop plants as well plays a key role in the absorption, transmission and transformation of light energy [36]. In the present study, reductions in the Pn under water deficit conditions occurred due to Chl degradation [37]. This degradation may associate with low drip irrigation quota, increased stomatal resistance and low CO_2 supply to the chloroplast [38].

Leaf area is more sensitive to moisture stress compared with Pn and Chl [29]. A moderate reduction in drip irrigation quota is beneficial for low Chl and can delay leaf senescence. In this study, W_2 and W_3 had a negative effect on photosynthetic apparatus in the chloroplasts. This might be due to the change in the photosynthetic pigments or protection of the photosynthetic apparatus from photoinhibitory damage in the leaves [39]. However, the Calvin Cycle enzyme (ribulose-1,5-bisphosphate carboxylate/oxygenase, Rubisco) activity was maintained due to higher Pn and Chl content for biomass accumulation. The difference in dry matter was the result of size and duration of the photosynthetic area. W_1 delayed LAI which may be related to the exuberant development of leaves and the longer CPB accumulation time. W_2 maintained a relatively high LAI (>5.0) and a sufficient photosynthetic area resulted in assimilate formation and water conservation [30,40]. This phenomenon might be due to the growth compensatory effects of plants under slightly reduced irrigation quotas [41,42]. Plants can adapt to mild drought through various physiological activities such as increasing leaf area to maintain a favorable water content [43]. An optimal LAI of cotton plants lead to the absorption of sufficient light energy. This absorption improves both the population structure and canopy photosynthesis, thereby improving the light energy utilization and consequently high yield [44,45]. Furthermore, under a low level of MC application under W_2 treatment before the first and second irrigation events maintained a reasonable LAI and could create a reasonable population structure to guarantee a greater and more efficient photosynthetic system. An expansion of cotton leaves are considered more sensitive to drought than Pn [46]; this sensitivity could explain the significant decrease in photosynthetic area under W_3, W_4 and W_5 during the late growth period. Although, a lower LAI is conducive to light absorption within a lower canopy, it also decrease light energy and reduces yield [47]. This might be due to the lower irrigation quotas which did not provide suitable leaf moisture conditions. This further reduced LAI, increased the degradation rate of Chl and increased leaf senescence [48]. These alterations may affect integrity of the photosynthesis and reduced the photosynthetic efficiency [49].

Biomass accumulation is the final product of plant photosynthesis and more distribution of biomass to the reproductive organs are essential for high cotton yield [50]. More biomass accumulations are important to maintain high crop yields [51]. A significant or extremely significant positive correlation between biomass accumulation (VOB, VOB and ROB) and lint yield in the present study. Based on regression analysis, CPB, VOB and ROB biomass accumulation in both V_T and V_M were positively correlated to lint yield. Conversely, reductions in crop yield caused by irrigation have been attributed to decreased biomass formation [52]. Biomass accumulation at FAP was associated with increased water uptake. An appropriate irrigation quota could increase both average and maximum rates of CPB and can lead to increased biomass accumulation and consequently high yield [3]. In this study, W_2 and W_3 shortened CPB accumulation duration and facilitated maximum rates of CPB at FAP. However, W_2 had a longer duration of VOB accumulation. These conditions increased the distribution of assimilate to the reproductive organs. A moderate reduction in irrigation can maintain a high LAI to ensure high biomass accumulation [53]. This further transitioned more vegetative growth to reproductive growth and reduced evaporation during vegetative development [54]. Conversely, W_4 and W_5 significantly shortened the duration for biomass formation, thus reduced the maximum rates for VOB and ROB accumulation. This finding indicated that relatively low soil water contents are not good for growth and development of the aboveground parts during the vegetative growth stage. These adverse effects may involve physiological responses [55], leaf area expansion [48], root growth [56] resulting in a decreased plant VOB to reproductive biomass and ultimately reduced yield [11]. Luxury, vegetative growth can consume excessive amount of nutrients and increases competition between vegetative and reproductive growth and consequently fruit shedding [57]. Together, these data showed that reducing irrigation quotas are not conducive to cotton growth or yield formation.

An appropriate irrigation level is important for sustainable cotton production in arid regions. Different irrigation amount can lead to significant differences in crop growth and both the accumulation and redistribution of photosynthesis assimilate [49], which in turn affects crop yield, water use efficiency and fiber quality [9,58]. A 15% reduction in the total irrigation amount can save irrigation water and reduced yield losses. However, further reduction up to 25% conserve more irrigation water but can lead high yield penalty [20]. Interestingly, the yield under W_2 was not significantly different from W_1 and the WP also increased in the present study. These results are consistent with those of previous research [30,59] who also reported that a slight reduction in drip irrigation can cause physiologically relevant adaptations in cotton, such as improved photosynthesis capabilities (leaf area and Chl content per unit area) [29] and growth promotion of vegetative organs [60]. Another possible reason might be due to reduced application of MC under W_2. The reduced use of MC in this treatment may facilitated vegetative growth [54] and increased the balance between vegetative and reproductive growth [61]. These phenomena were also beneficial for cotton plants in terms of maintaining a self-adjustment ability via the relationship between boll number and single boll weight [62]. Although, W_4 and W_5 presented relatively high WP but did not increase yield. W_4 and W_5 significantly reduced boll weight and bolls per unit area. This reduction in boll number under reduced irrigation further decreased lint yield. However, W_2 slightly reduced individual boll weight and number of bolls per unit area. This increment in yield maight be associated with a moderate reduction in the drip irrigation quota [9].

Cotton fiber length, fiber uniformity, specific strength and micronaire value are the important fiber quality parameters. In this study, moderately reduced drip irrigation (W_2 and W_3) quotas did not significantly affect cotton fiber quality parameters. These data are consistent with the results of previous studies [59,63]. However, the extremely low drip irrigation quota (W_5) significantly reduced fiber length and uniformity. The difference in fiber length may be due to moisture effects on fiber length which influence fiber elongation phase. The micronaire value is a measure of fiber fineness and maturity [64]. No significant differences in micronaire value or specific strength among the different treatments were observed. This might be related to the time interval of irrigation, which influenced cotton boll development.

5. Conclusions

In this study, irrigation quota and MC application had a significant effect on leaf photosynthetic performance and biomass accumulation, cotton yield, fiber quality and water productivity. Compared with W_1, W_2 had higher Pn and Chl content during all growth period. Moreover, W_2 combined with reduced MC application resulted in greater LAI at the full boll stage, which ensured a sufficient photosynthetic area and prolonged ROB accumulation duration and yield formation. In conclusion, the drip irrigation level of 540–600 m^3 ha^{-1} with reduced MC application is a good strategy to maintain higher WP and achieve high lint yield as well as better fiber quality.

Author Contributions: H.L. and F.W. conceived and designed the experiments; J.X. and X.H. performed the experiments; H.G. and H.M. analyzed the data; A.K. edited and revised the paper; H.G. wrote the paper.

Funding: This project was supported by the National Key R&D Program of China (2017YFD0201900), National Natural Science Foundation of China (31760355), and Program of Youth Science and Technology Innovation Leader of The Xinjiang Production and Construction Corps (2017CB005).

Conflicts of Interest: The authors declare no conflict of interest. The funders had no role in the design of the study; in data collection, analyses, or interpretation of data; in the writing of the manuscript, or in the decision to publish the results.

References

1. Constable, G.A.; Bange, M.P. The yield potential of cotton (Gossypium hirsutum L.). *Field Crops Res.* **2015**, *182*, 98–106. [CrossRef]
2. Khan, A.; Wang, L.; Ali, S.; Tung, S.A.; Hafeez, A.; Yang, G. Optimal planting density and sowing date can improve cotton yield by maintaining reproductive organ biomass and enhancing potassium uptake. *Field Crops Res.* **2017**, *214*, 164–174. [CrossRef]
3. Feng, L.; Dai, J.; Tian, L.; Zhang, H.; Li, W.; Dong, H. Review of the technology for high-yielding and efficient cotton cultivation in the northwest inland cotton-growing region of China. *Field Crops Res.* **2017**, *208*, 18–26. [CrossRef]
4. Berry, P.; Ramirezvillegas, J.; Bramley, H.; Mgonja, M.A.; Mohanty, S. Regional impacts of climate change on agriculture and the role of adaptation. In *Plant Genetic Resources and Climate Change*; Jackson, M., Ford-Lloyd, B., Parry, M., Eds.; CAB International: Wallingford, UK, 2014; pp. 78–97.
5. Cathey, G.W.; Meredith, W.R. Cotton Response to Planting Date and Mepiquat Chloride. *Agron. J.* **1988**, *80*, 463–466. [CrossRef]
6. Choudhury, B.U.; Bouman, B.A.M.; Singh, A.K. Yield and water productivity of rice–wheat on raised beds at New Delhi, India. *Field Crops Res.* **2007**, *100*, 229–239. [CrossRef]
7. Long, S.P.; Zhu, X.G.; Naidu, S.L.; Ort, D.R. Can improvement in photosynthesis increase crop yields? *Plant Cell Environ.* **2006**, *29*, 315–330. [CrossRef]
8. Zhu, X.; Long, S.; Ort, D. What is the maximum efficiency with which photosynthesis can convert solar energy into biomass? *Curr. Opin. Biotechnol.* **2008**, *19*, 153–159. [CrossRef]
9. Basal, H.; Dagdelen, N.; Unay, A.; Yilmaz, E. Effects of deficit drip irrigation ratios on cotton (Gossypium hirsutumL.) yield and fibre quality. *J. Agron. Crop. Sci.* **2009**, *195*, 19–29. [CrossRef]
10. Luo, H.H.; Zhang, Y.L.; Zhang, W.F. Effects of water stress and rewatering on photosynthesis, root activity, and yield of cotton with drip irrigation under mulch. *Photosynthetica* **2016**, *54*, 65–73. [CrossRef]
11. Yan, M.; Zheng, J.; Zhang, J.; Shi, H.; Tian, L.; Guo, R.; Lin, T. Effects of regulated deficit irrigation on accumulation and distribution of biomass and nitrogen, and yield of island cotton. *Chin. J. Eco Agric.* **2015**, *23*, 841–850.
12. Lu, X.R.; Jia, X.Y.; Niu, J.H. The present Situation and prospects of cotton industry development in China. *Sci. Agric. Sin.* **2018**, *51*, 26–36.
13. Kang, S.Z.; Shi, W.J.; Zhang, J.H. An improved water-use efficiency for maize grown under regulated deficit irrigation. *Field Crops Res.* **2000**, *67*, 207–214. [CrossRef]
14. Zhang, G.Q.; Liu, C.W.; Xiao, C.H.; Xie, R.Z.; Ming, B.; Hou, P.; Liu, G.Z.; Xu, W.J.; Shen, D.P.; Wang, K.R.; et al. Optimizing water use efficiency and economic return of super high yield spring maize under drip irrigation and plastic mulching in arid areas of China. *Field Crops Res.* **2017**, *211*, 137–146. [CrossRef]
15. Tang, L.S.; Li, Y.; Zhang, J.H. Partial rootzone irrigation increases water use efficiency, maintains yield and enhances economic profit of cotton in arid area. *Agric. Water Manag.* **2010**, *97*, 1527–1533. [CrossRef]
16. Zhang, Y.; Zhang, Y.; Wang, Z.; Wang, Z. Characteristics of canopy structure and contributions of non-leaf organs to yield in winter wheat under different irrigated conditions. *Field Crops Res.* **2011**, *123*, 187–195. [CrossRef]
17. Yang, G.H. Research advance in the theory and technology of crop regulated deficit irrigation. *J. Anhui Agric. Sci.* **2008**, *36*, 2514–2516.
18. Cai, H.J.; Kang, S.Z.; Zhang, Z.H.; Chai, H.M.; Hu, X.T.; Wang, J. Proper growth stages and deficit degree of crop regulated deficit irrigation. *Trans. Chin. Soc. Agric. Eng.* **2000**, *16*, 24–27.
19. Qian, C.; Zhang, Y.; Sun, Z.; Zheng, J.; Wei, B.; Yang, L.; Feng, L.; Chen, F.; Zhe, Z.; Ning, Y. Morphological plasticity of root growth under mild water stress increases water use efficiency without reducing yield in maize. *Biogeosciences* **2017**, *14*, 1–24.
20. Yang, C.J.; Luo, Y.; Sun, L.; Wu, N. Effect of Deficit Irrigation on the Growth, Water Use Characteristics and Yield of Cotton in Arid Northwest China. *Pedosphere* **2015**, *25*, 910–924. [CrossRef]
21. Tung, S.A.; Huang, Y.; Ali, S.; Hafeez, A.; Shah, A.N.; Song, X.; Ma, X.; Luo, D.; Yang, G. Mepiquat chloride application does not favor leaf photosynthesis and carbohydrate metabolism as well as lint yield in late-planted cotton at high plant density. *Field Crops Res.* **2018**, *221*, 108–118. [CrossRef]
22. Bogiani, J.C.; Rosolem, C.A. Sensibility of cotton cultivars to mepiquat chloride. *Pesqui. Agropecu. Bras.* **2009**, *44*, 1246–1253. [CrossRef]

23. Cook, D.R.; Kennedy, C.W. Early flower bud loss and mepiquat chloride effects on cotton yield distribution. *Crop Sci.* **2000**, *40*, 1678–1684. [CrossRef]

24. Mao, L.; Zhang, L.; Evers, J.B.; Werf, W.V.D.; Liu, S.; Zhang, S.; Wang, B.; Li, Z. Yield components and quality of intercropped cotton in response to mepiquat chloride and plant density. *Field Crops Res.* **2015**, *179*, 63–71. [CrossRef]

25. Yang, G.Z.; Tang, H.Y.; Nie, Y.C.; Zhang, X.L. Responses of cotton growth, yield, and biomass to nitrogen split application ratio. *Eur. J. Agron.* **2011**, *35*, 164–170. [CrossRef]

26. Gu, J.F.; Zhou, Z.X.; Li, Z.K.; Chen, Y.; Wang, Z.Q.; Zhang, H. Rice (Oryza sativa L.) with reduced chlorophyll content exhibit higher photosynthetic rate and efficiency, improved canopy light distribution, and greater yields than normally pigmented plants. *Field Crops Res.* **2017**, *200*, 58–70. [CrossRef]

27. Yi, X.P.; Zhang, Y.L.; Yao, H.S.; Luo, H.H.; Gou, L.; Chow, W.S.; Zhang, W.F. Rapid recovery of photosynthetic rate following soil water deficit and re-watering in cotton plants (Gossypium herbaceum L.) is related to the stability of the photosystems. *J. Plant Physiol.* **2016**, *194*, 23–34. [CrossRef]

28. Chaves, M.M.; Flexas, J.; Pinheiro, C. Photosynthesis under drought and salt stress: Regulation mechanisms from whole plant to cell. *Ann. Bot.* **2009**, *103*, 551. [CrossRef]

29. Ennahli, S.; Earl, H.J. Physiological limitations to photosynthetic carbon assimilation in cotton under water stress. *Crop Sci.* **2005**, *45*, 2374–2382. [CrossRef]

30. Shareef, M.; Gui, D.W.; Zeng, F.J.; Waqas, M.; Zhang, B.; Iqbal, H. Water productivity, growth, and physiological assessment of deficit irrigated cotton on hyperarid desert-oases in northwest China. *Agric. Water Manag.* **2018**, *206*, 1–10. [CrossRef]

31. Xin, L.; Zheng, H.; Yang, Z.; Guo, J.; Liu, T.; Sun, L.; Xiao, Y.; Yang, J.; Yang, Q.; Guo, L. Physiological and proteomic analysis of maize seedling response to water deficiency stress. *J. Plant Physiol.* **2018**, *228*, 29–38. [CrossRef]

32. Niu, J.; Zhang, S.; Liu, S.; Ma, H.; Chen, J.; Shen, Q.; Ge, C.; Zhang, X.; Pang, C.; Zhao, X. The compensation effects of physiology and yield in cotton after drought stress. *J. Plant Physiol.* **2018**, *224*, 30–48. [CrossRef] [PubMed]

33. Kozlowski, T.T. Water Deficits and Plant Growth. *Water Relat. Plants* **1983**, 342–389.

34. Ullah, I.; Ashraf, M.; Zafar, Y. Genotypic variation for drought tolerance in cotton (Gossypium hirsutum L.): Leaf gas exchange and productivity. *Flora* **2008**, *203*, 105–115. [CrossRef]

35. Luo, H.; Li, J.; Gou, L. Regulation of under-mulch-drip irrigation on production and distribution of photosynthetic assimilate and cotton yield under different soil moisture contents during cotton flowering and boll-setting stage. *Sci. Agric. Sin.* **2008**, *41*, 1955–1962.

36. Santos, E.F.D.; Zanchim, B.J.; Campos, A.G.d.; Garrone, R.F.; Lavres Junior, J. Photosynthesis rate, chlorophyll content and initial development of physic nut without micronutrient fertilization. *Rev. Bras. Cienc. Solo.* **2013**, *37*, 1334–1342. [CrossRef]

37. Horvat, T.; Poljak, M.; Lazarevic, B.; Svecnjak, Z.; Karazlja, T.; Marietta, H. Effect of foliar fertilizers on chlorophyll content index and yield of potato crop grown under water stress conditions. *Növénytermelés* **2010**, *59*, 215–218.

38. Sobhkhizi, A.; Rayni, M.F.; Barzin, H.B.; Noori, M. Influence of drought stress on photosynthetic enzymes, chlorophyll, protein and relative water content in crop plants. *Int. J. Biosci.* **2014**, *5*, 89–100.

39. Gilmore, A.M. Mechanistic aspects of xanthophyll cycle-dependent photoprotection in higher plant chloroplasts and leaves. *Physiol. Plantarum* **2010**, *99*, 197–209. [CrossRef]

40. Koler, P.; Patil, B.C.; Chetti, M.B.; Hiremath, S.M. Influence of plant growth regulators on total dry matter, leaf area index and yield, yield components in hybrid cotton. *Karnataka J. Agric. Sci.* **2010**, *23*, 503–505.

41. Shan, L.; Zhang, S.Q. Is possible to save large irrigation water? The situation and prospect of water-saving agriculture in China. *Chin. J. Nature* **2006**, *2*, 71–74.

42. Zhao, L.Y.; Deng, X.P.; Shan, L. A review on types and mechanisms of compensation effect of crops under water deficit. *J. Appl. Ecol.* **2004**, *15*, 523–526.

43. Nepomuceno, A.L.; Oosterhuis, D.M.; Stewart, J.M. Physiological responses of cotton leaves and roots to water deficit induced by polyethylene glycol. *Environ. Exp. Bot.* **1998**, *40*, 29–41. [CrossRef]

44. Zhang, W.F.; Wang, Z.L.; Yu, S.J.; Li, S.K.; Cao, L.P.; Ren, L.T. Effect of under-mulch-drip irrigation on canopy apparent photosynthesis, canopy structure and yield formation in high-yield cotton of Xinjiang. *Sci. Agric. Sin.* **2002**, *35*, 632–637.

45. Zhan, D.X.; Zhang, C.; Yang, Y.; Luo, H.H.; Zhang, Y.L.; Zhang, W.F. Water deficit alters cotton canopy structure and increases photosynthesis in the mid-canopy layer. *Agron. J.* **2015**, *107*, 1947–1957. [CrossRef]

46. Constable, G.A.; Rawson, H.M. Distribution of 14C label from cotton leaves: Consequences of changed water and Nitrogen status. *Funct. Plant Biol.* **1982**, *9*, 735–747. [CrossRef]

47. Su, B.Y.; Song, Y.X.; Song, C.; Cui, L.; Yong, T.W.; Yang, W.Y. Growth and photosynthetic responses of soybean seedlings to maize shading in relay intercropping system in Southwest China. *Photosynthetica* **2015**, *52*, 332–340. [CrossRef]

48. Salah, H.; Tardieu, F. Control of leaf expansion rate of droughted maize plants under fluctuating evaporative demand (A superposition of hydraulic and chemical messages?). *Plant Physiol.* **1997**, *114*, 893–900. [CrossRef]

49. Kaiser, W.M. Effects of water deficit on photosynthetic capacity. *Physiol. Plantarum* **2010**, *71*, 142–149. [CrossRef]

50. Gao, X.L.; Sun, J.M.; Gao, J.F.; Feng, B.L.; Wang, P.K.; Chai, Y. Accumulation and transportation characteristics of dry matter after anthesis in different mung bean cultivars. *Acta Agron. Sin.* **2009**, *35*, 1715–1721. [CrossRef]

51. Plénet, D.; Mollier, A.; Pellerin, S. Growth analysis of maize field crops under phosphorus deficiency. II. Radiation-use efficiency, biomass accumulation and yield components. *Plant Soil* **2000**, *224*, 259–272. [CrossRef]

52. Fereres, E.; Soriano, M.A. Deficit irrigation for reducing agricultural water use. *J. Exp. Bot.* **2007**, *58*, 147–159. [CrossRef] [PubMed]

53. Xu, R.B.; Sun, H.C.; Liu, L.T.; Zhang, Y.J.; Liu, Y.C.; Bai, Z.Y.; Li, C.D.; Sun, G.J. Effect of irrigation patterns on accumulation and distribution of dry matter, yield and water use efficiency of cotton in southern Hebei. *Cotton Sci.* **2018**, *30*, 386–394.

54. Bell, J.M.; Schwartz, R.; McInnes, K.J.; Howell, T.; Morgan, C.L.S. Deficit irrigation effects on yield and yield components of grain sorghum. *Agric. Water Manag.* **2018**, *203*, 289–296. [CrossRef]

55. Chastain, D.R.; Snider, J.L.; Collins, G.D.; Perry, C.D.; Whitaker, J.; Byrd, S.A. Water deficit in field-grown Gossypium hirsutum primarily limits net photosynthesis by decreasing stomatal conductance, increasing photorespiration, and increasing the ratio of dark respiration to gross photosynthesis. *J. Plant Physiol.* **2014**, *171*, 1576–1585. [CrossRef] [PubMed]

56. Zhang, H.Z.; Khan, A.; Tan, D.K.Y.; Luo, H.H. Rational water and Nitrogen management improves root growth, increases yield and maintains water use efficiency of cotton under mulch drip irrigation. *Front. Plant Sci.* **2017**, *8*, 912. [CrossRef] [PubMed]

57. Dai, J.L.; Dong, H.Z. Intensive cotton farming technologies in China: Achievements, challenges and countermeasures. *Field Crops Res.* **2014**, *155*, 99–110. [CrossRef]

58. Balkcom, K.S.; Reeves, D.W.; Shaw, J.N.; Burmester, C.H.; Curtis, L.M. Cotton yield and fiber quality from irrigated tillage systems in the Tennessee valley. *Agron. J.* **2006**, *98*, 596. [CrossRef]

59. Ünlü, M.; Kanber, R.; Koç, D.L.; Tekin, S.; Kapur, B. Effects of deficit irrigation on the yield and yield components of drip irrigated cotton in a mediterranean environment. *Agric. Water Manag.* **2011**, *98*, 597–605.

60. Zhou, L.; Gan, Y.; Ou, X.B.; Wang, G.X. Progress in molecular and physiological mechanisms of water-saving by compensation for water deficit of crop and how they relate to crop production. *Chin. J. Eco Agric.* **2011**, *19*, 217–225. [CrossRef]

61. Tung, S.A.; Huang, Y.; Hafeez, A.; Ali, S.; Khan, A.; Souliyanonh, B.; Song, X.H.; Liu, A.D.; Yang, G.Z. Mepiquat chloride effects on cotton yield and biomass accumulation under late sowing and high density. *Field Crops Res.* **2018**, *215*, 59–65. [CrossRef]

62. Zhang, D.M.; Luo, Z.; Liu, S.H.; Li, W.J.; Wei, T.; Dong, H.Z. Effects of deficit irrigation and plant density on the growth, yield and fiber quality of irrigated cotton. *Field Crops Res.* **2016**, *197*, 1–9. [CrossRef]

63. Hussein, F.; Janat, M.; Yakoub, A. Assessment of yield and water use efficiency of drip-irrigated cotton (Gossypium hirsutum L.) as affected by deficit irrigation. *Turk. J. Agric. For.* **2011**, *35*, 611–621.

64. Bradow, J.M.; Davidonis, G.H. Quantitation of fiber quality and the cotton production-processing interface: A physiologist's perspective. *J. Cotton Sci.* **2000**, *1*, 34–64.

Article

Pre-Sowing Irrigation Plus Surface Fertilization Improves Morpho-Physiological Traits and Sustaining Water-Nitrogen Productivity of Cotton

Zongkui Chen [†], Hongyun Gao [†], Fei Hou, Aziz Khan * and Honghai Luo *

Key Laboratory of Oasis Eco-Agriculture, Xinjiang Production and Construction Group, Shihezi University, Shihezi 832003, China; chenzongkui90@foxmail.com (Z.C.); gaohongyun1995@163.com (H.G.); houfeishzu@163.com (F.H.)
* Correspondence: azizkhanturlandi@gmail.com (A.K.); luohonghai79@163.com (H.L.);
  Tel.: +156-7670-1561 (A.K.); +133-4546-6760 (H.L.)
† The authors contributed equally to this article.

Received: 23 September 2019; Accepted: 15 November 2019; Published: 19 November 2019

Abstract: The changing climatic conditions are causing erratic rains and frequent episodes of moisture stress; these impose a great challenge to cotton productivity by negatively affecting plant physiological, biochemical and molecular processes. This situation requires an efficient management of water-nutrient to achieve optimal crop production. Wise use of water-nutrient in cotton production and improved water use-efficiency may help to produce more crop per drop. We hypothesized that the application of nitrogen into deep soil layers can improve water-nitrogen productivity by promoting root growth and functional attributes of cotton crop. To test this hypothesis, a two-year pot experiment under field conditions was conducted to explore the effects of two irrigation levels (i.e., pre-sowing irrigation (W_{80}) and no pre-sowing irrigation (W_0)) combined with different fertilization methods (i.e., surface application (F_{10}) and deep application (F_{30})) on soil water content, soil available nitrogen, roots morpho-physiological attributes, dry mass and water-nitrogen productivity of cotton. W_{80} treatment increased root length by 3.1%–17.5% in the 0–40 cm soil layer compared with W_0. W_{80} had 11.3%–52.9% higher root nitrate reductase activity in the 10–30 cm soil layer and 18.8%–67.9% in the 60–80 cm soil layer compared with W_0. The $W_{80}F_{10}$ resulted in 4.3%–44.1% greater root nitrate reductase activity compared with other treatments in the 0–30 cm soil layer at 54–84 days after emergence. Water-nitrogen productivity was positively associated with dry mass, water consumption, root length and root nitrate reductase activity. Our data highlighted that pre-sowing irrigation coupled with basal surface fertilization is a promising option in terms of improved cotton root growth. Functioning in the surface soil profile led to a higher reproductive organ biomass production and water-nitrogen productivity.

Keywords: cotton; dry matter yield; root growth; root physiology; water productivity; nitrogen productivity

---

1. Introduction

Cotton is a commercial cash crop providing fiber, oil, and animal feed globally [1]. With the increasing population comes an increased demand for food and fiber, but the threats of climate change are challenging crop production. Crop intensification to produce more food, fiber and feed needs more water, but water resources are limited. Although cotton is considered a drought resistance crop, its productivity is negatively affected by drought stress and nutrient deficiency which results in reduced growth, physiological, biochemical and molecular events [2,3]. Drought stress causes a 50% to

73% reduction in cotton yield [4]. Transgenic cotton cultivars are more susceptible to moisture deficit conditions [3]. Therefore, lower water availability has threatened the productivity of irrigated cotton ecosystem. Hence, strategy to increase water conservation and nutrient uptake are needed to achieve optimal cotton yield [5,6].

Water-nitrogen productivity and cotton production can be improved by application of water-nutrient at the proper growth period of cotton crop [7,8]. However, many water-nutrient conservations strategies can lead to unbalanced organs development such as, the competition between root and aerial plant part (mainly reproductive organs), thus vegetative organs growth surpass reproductive organs development, which in turn decreased water productivity and yield. Moreover, greater above ground dry matter accumulation, especially in reproductive organs can drive cotton yield [9]. An excessive root expansion can reduce growth of aerial plant parts [10,11], but lower root dry matter accumulation affects root distribution and physiological activity in the soil [12,13]. Therefore, it is essential to enhance cotton root activity and distribution in the soil to achieve higher water-nutrient productivity via balancing the growth and development between aerial and underground parts of cotton plant.

Root morphology and physiology are closely associated with the growth and development of aboveground plants. The rates and modes of water and nutrient application influences crop growth and water-nutrient productivity [11,14,15] by affecting root morphological and physiological activity [16]. Poor irrigation practices can develop a large root system and induce aging signals (such as, ABA) that can lead to low dry matter accumulation and water-nutrient productivity [17–19]. An efficient water-nitrogen management can enhance root functioning, increases water-nitrogen absorption, which in turn promote reproductive organ dry matter accumulation and water-nutrient productivity [16,20]. Hence, facilitating the relationship between root and water-nutrient in the root zone is essential for improving water-nutrient productive potential of reproductive organs to achieve higher water-nutrient use efficiency.

Xinjiang is the major cotton growing province in China, contributing 67% to the total national lint production [21], where low water availability and poor nutrient management have imposed a great challenge to cotton production. In cotton, root development occurs before full flowering stage and is mainly affected by soil moisture and basal fertilization. Post-sowing irrigation and snow melt can enrich deep water layer (important soil moisture storage) in the soil. This can lead to a deeper root growth, enhance water uptake, improve photosynthetic capacity and reduces irrigation frequency [20,22]. Basal fertilization can promote root growth and increase nutrient availability [23,24]. Single effects of deep water layer [20] and basal fertilization [23] on cotton root have been documented, but the effect of combine application on cotton root growth and physiology in different soil profile to regulate water-nitrogen productivity is elusive. The aim of this study was firstly to determine the effects of pre-sowing irrigation and basal fertilization on soil water content, available nitrogen, root morpho-physiological traits and above dry mass production and secondly to analyze the relationship between root growth and water-nutrient productivity in the root zone of cotton crop.

2. Materials and Methods

2.1. Details of Experimental Site

A two-year pot experiment under field conditions was conducted at the research station of Shihezi University Xinjiang, China (45°19′ N, 74°56′ E) during 2015 and 2016 growing seasons. In the region, evapotranspiration was 1425 mm. The mean rainfall and temperatures in both years are presented in Figure 1. Cotton cultivar Xinluzao 45 seeds were sown in polyvinyl chloride (PVC) tubes (diameter, 30 cm; the tubes consisted of three stacked sections; each section was 40 cm high with 120 cm height). The bottom of the tube was covered with a wire to hold soil. The soil was clay loam comprised of 1.43 g m^{-3} bulk density, 24.6% field capacity, 7.6 pH, 54.9 mg kg^{-1} alkali hydrolysable N, 16.8 mg kg^{-1} Olsen-P, 196 mg kg^{-1} exchangeable K and 12.5 g kg^{-1} organic matter.

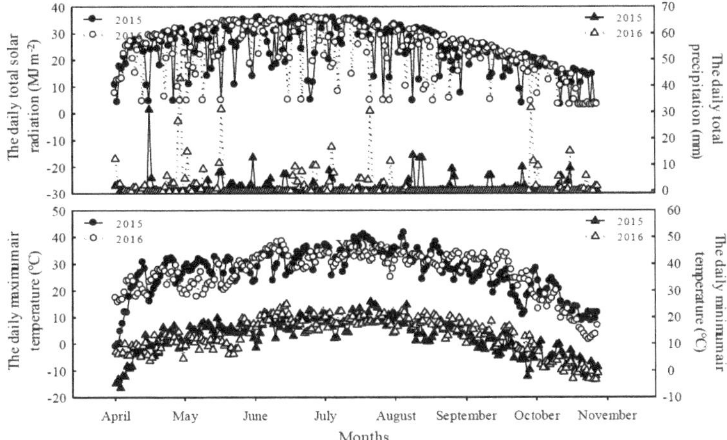

Figure 1. The daily total solar radiation (MJ m^{-2}), total precipitation (mm), maximum and minimum air temperature (°C) during cotton growing season in Shihezi (2015–2016).

2.2. Experimental Design and Crop Management

A randomized complete block design was employed with four treatments with 32 tubes per treatment. Irrigation treatments were: pre-plant irrigation (W$_{80}$, watered with 0.28 m^3 (80 ± 5% of field capacity) per tube before sowing), no pre-plant irrigation (W$_0$, no water was applied over the entire depth of the tube) with two fertilization depth (i.e., surface application (F$_{10}$, sufficient basal fertilizer in the 10–20 cm soil layer before sowing and deep fertilization (F$_{30}$, sufficient basal fertilizer in the 30–40 cm layer before sowing)) in each tube. Nitrogen (N) was applied at the ratio of 1:4 as basal fertilizer by topdressing method. Phosphorus (P$_2$O$_5$) and potassium (K$_2$O) were supplemented as basal fertilization. Urea (CO(NH$_2$)$_2$, 46.0% N) at the rate of 13.8 g per tube was used for N and mono-potassium phosphate at the rate of 18 g ((KH$_2$PO$_4$) 52.0% P$_2$O$_5$ and 35.4% K$_2$O) was used per tube as P$_2$O$_5$ and K$_2$O.

Four seeds per tube were sown at a depth of 3 cm on April 25th and May 1st in 2015 and 2016 growing season. Seeds were placed 10 cm apart in one direction and 20 cm apart in another direction. Four seedlings were left per tube. Drip laterals (Beijing Lvyuan Inc., Beijing, China) were installed on the top of each tube with a single emitter. The top of the tube was covered with a polyethylene film to reduce evaporation. Each pot was drip-irrigated each after four days. The total amount of water supplied to the plants was 434 mm each year. Standard local pest control measures were adopted in both cropping seasons.

2.3. Observations

During both years, soil water content, available N, dry matter accumulation, root morphological and physiological traits were assessed at 39, 54, 69, 84 and 99 days after emergence (DAE).

2.4. Soil Water Content and Available Nitrogen

The irrigation amount during growth period was based on measurement of the soil moisture content in the 0–40 cm soil layer using the Time Domain Reflectometry (TDR). Water supplied to the crop can be defined as:

$$A = (W_p - W_a) \times H \tag{1}$$

where A is the volume of water supplied (mm) and W_p is the field capacity in the 0–40 cm soil profile. W_a is the average relative soil moisture content in the 0–40 cm soil profile that was measured by TDR and H is the thickness of the soil layers using drip irrigation system (mm). Changes in soil moisture content in the 0–120 cm soil profile was determined by the stoving method. During root sampling, fresh soil samples were immediately collected from each soil layer (i.e., 20 cm or 10 cm) in each tube of three replications in 2015 and 2016 growing seasons. Soil was weighted then dried at 85 °C for constant weight. Soil moisture content was expressed as moisture content (g) per dry soil (g). Soil available N was determined by the alkaline hydrolysis diffusion method [25] and was expressed in mg kg^{-1} dry soil.

2.5. Root Growth Traits

Three tubes (each treatment) were carefully dug out from the ground level and cut down into 20 cm segments in 2015 and 10 cm segments in 2016 growing season. The segments were immersed in the water for 1 h; roots were placed on a 0.5 mm sieve and rinsed with running water. Plant debris such as weeds and dead roots were separated from 'living' roots according to Gwenzi et al. [26]. The live roots were placed in denoised water and stored in a refrigerator at 4°C for further analysis. Live roots were evenly spread on a plastic tray with deionized water and scanned using a flatbed scanner (300 dpi). Root images were obtained using WinRhizo image analysis software (Regent Instruments, Quebec, Canada). The software was configured to measure root length and then roots were oven-dried at 85 °C for 48 h and weighed.

2.6. Root Nitrate Reductase Activity (NR)

Nitrate reductase activity was measured according to Zhou et al. [27] method. Roots were homogenized in extraction buffer and centrifuged for 15 min at 4000 rpm. The supernatants were collected and added to the reaction buffer. After incubation at 25 °C for 30 mins, the reaction was suspended by 1 mL 1% sulphanilamide. The mixture was further centrifuged for 5 mins at 5000 rpm and N-(1-naphthyl) ethylenediamine dihydrochloride was added; the supernatant was used to assess nitrite production at 540 nm after. Root NR activity was expressed as nitrite production (µg) 1 g fresh root per hour.

2.7. Biomass Accumulation

To determine cotton biomass accumulation, three tubes (12 plants), in each treatment were chosen and cut down at the cotyledon node during each sampling day. Plant samples were dissected into leaves, stems, buds, flowers, bolls and roots. These samples were oven-dried at 85 °C for 48 h and weighed to a constant weight. A logistic function was used to describe the progress of biomass accumulation [28,29]:

$$Y = \frac{K}{1 + ae^{bt}} \tag{2}$$

In the formula, t (d) is the number of days after emergence (DAE), Y (g) is the biomass at t, K (g) is the maximum biomass while a and b are the constants.

Based on Formula (2), we could calculate:

$$t_1 = \frac{1}{b}\ln\left(\frac{2 + \sqrt{3}}{a}\right) \tag{3}$$

$$t_2 = \frac{1}{b}\ln\left(\frac{2 - \sqrt{3}}{a}\right) \tag{4}$$

$$t_m = -\frac{\ln a}{b} \tag{5}$$

$$V_m = -\frac{bK}{4} \tag{6}$$

$$V_t = \frac{Y_2 - Y_1}{t_2 - t_1} \tag{7}$$

where V_m (g d^{-1}) is the highest biomass accumulation rate; t_m (d) is the largest biomass accumulation period, beginning at t_1 and terminating at t_2. The factors Y_1 and Y_2 represent biomass at t_1 and t_2; V_t is the average biomass accumulation from t_1 to t_2.

2.8. Water-Nitrogen Productivity

Nitrogen productivity was defined as the total biomass (g plant^{-1}) or the biomass of each plant organ (root, stem and leaf, bud and boll) per unit of applied fertilizer-nitrogen (g plant^{-1}) at different growth stages [30]. In this study, nitrogen productivity was assessed at 39, 54, 69, 84 and 99 DAE.

Water productivity and soil moisture consumption rates were calculated at 39, 54, 69, 84 or 99 DAE according to the method described by Luo et al. [20]. Water productivity is the total biomass (g plant^{-1}) or the biomass of each organ (root, stem and leaf, bud and boll) per unit water consumption (cm^3 plant^{-1}). Moisture consumption rate was calculated according to Luo et al. [20].

2.9. Statistical Analysis

Analysis of variance (ANOVA), path analysis was performed using SPSS software version 16.0 (SPSS Inc., Chicago, IL, USA). Correlation analysis was performed using the "heatmap" package in R version 3.5.2. Treatments were separated using the least significant difference (*LSD*) tests at $p \leq 0.05$. Figures were constructed using Sigma Plot software version 10.0 (Systat Software Inc., San Jose, CA, USA). Data represent means ± SD.

3. Results

3.1. Soil Water Content and Available Nitrogen

Soil moisture content increased by 30.8%–53.1% for W_{80} treatment compared with W_0 in the 40–120 cm soil layer prior to 84 DAE (Figure 2). Water consumption of W_{80} was 28.1% more than that of W_0 in the 0–40 cm soil profile during whole growth period. No significant differences were observed between F_{10} and F_{30} treatment.

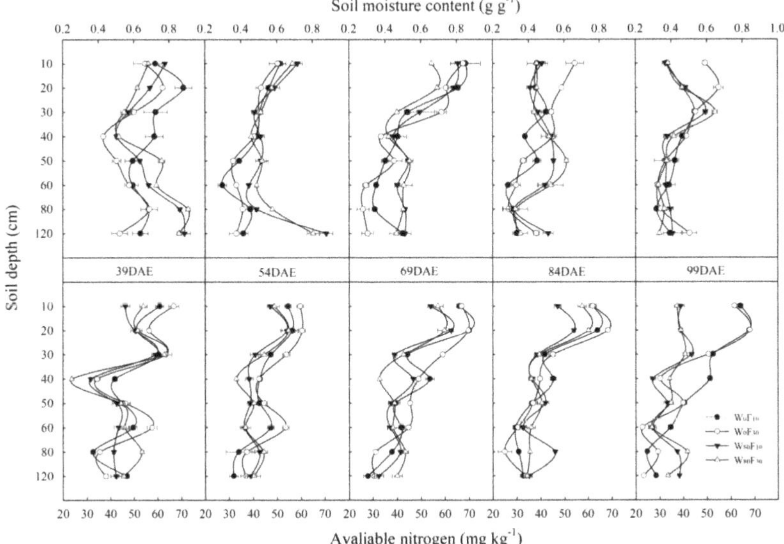

Figure 2. Changes in the soil moisture and the available nitrogen in different vertical soil layer (from 0 to 120 cm soil layer) at pre-sowing irrigation (W_{80}) or no pre-sowing irrigation (W_0) and base fertilizer surface (F_{10}) or deep (F_{30}) application with the days after emergence in 2016. Bars indicate SD ($n = 3$).

Under W_{80} treatment soil available N decreased by 22% in the 0–40 cm soil layer throughout the growth period (Figure 2) but increased by 7.6% in the 60–120 cm soil layer compared with W_0 treatment. No significant differences were observed in the 40–60 cm soil layer. F_{10} treatment had 0.8% and 13.0% lower soil available N compared with F_{30} treatment in the 0–30 cm and 60–80 cm soil layer before 84 DAE, while other soil layer remained unaffected.

3.2. Root Length

Cotton plant root length was significantly affected by irrigation levels and fertilization during the whole growth period. Root length gradually increased with the plant development but decreased later in the season (Figure 3). W_{80} treatment increased root length by 3.1–17.5% in the 0–40 cm soil layer but decreased by 7.7–66.1% in the 40–120 cm soil layer after 54 DAE than W_0 treatment. W_{80} F_{10} treatment had 3.5%–29.5% higher root length in the 0–40 cm soil layer, but 1.2%–10.5% lowered root length was observed in the 40–120 cm soil layer after 54 DAE compared with W_0 F_{30}.

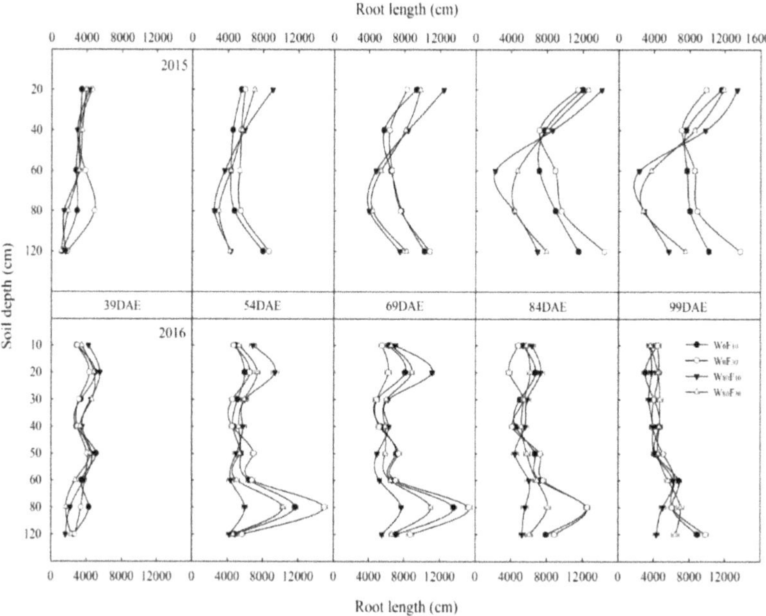

Figure 3. Changes in the root length in different vertical soil layer (from 0 to 120 cm soil layer) at pre-sowing irrigation (W_{80}) or no pre-sowing irrigation (W_0) and basal surface fertilization (F_{10}) or deep (F_{30}) application with the days after emergence (DAE).

3.3. Root Nitrate Reductase Activity

Nitrate reductase activity was rose with the plant development but gradually decreased in the 0–10 cm layer (Figure 4). Compared with W_{80} treatment, nitrate reductase activity in W_0 increased by 11.3%–52.9% and 18.8%–67.9%, respectively in the 0–40 and 60–120 cm soil depth at each growth stage but decreased by 13.5%–24.0% in the 40–60 cm soil profile. F_{10} treatment had 4.3%–44.1% and 7.2–18.3% higher nitrate reductase activity in the 10–20 cm and 40–60 cm soil layer soil profile at 54 to 84 and prior 69 DAE over F_{30} fertilization.

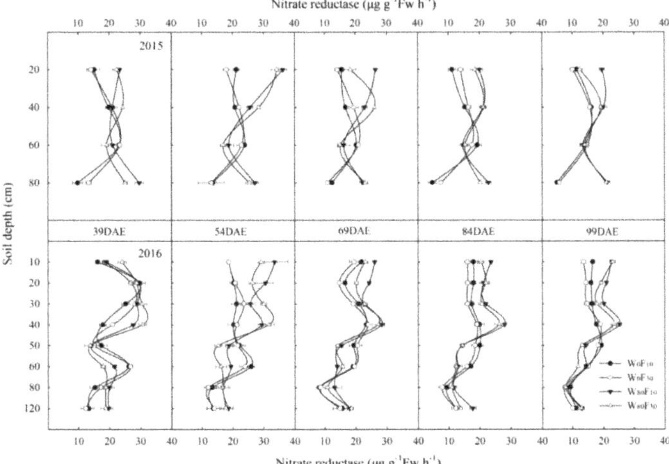

Figure 4. Changes in the nitrate reductase activity (μg g^{-1}FW h^{-1}) in different vertical soil layer (from 0 to 120 cm soil layer) at pre-sowing irrigation (W_{80}) or no pre-sowing irrigation (W_0) and basal surface fertilization (F_{10}) or deep (F_{30}) application with the days after emergence (DAE) in 2015 and 2016. Bars indicate SD (n = 3).

3.4. Cotton Plant Biomass Accumulation

Nitrogen and irrigation application method significantly altered cotton plant vegetative and reproductive organs biomass accumulation during both years (Table 1). Root, stem plus leaf, bolls and total plant biomass accumulation increased by 11.6%, 30.5%, 48.2% and 22.4%, respectively, in W_{80} over W_0 treatment. A 10% and 2% higher root and reproductive organs biomass produced in W_{80} F_{10} treatment compared with W_{80} F_{30} during both growing seasons.

Table 1. Changes in vegetative and reproductive and total organ biomass accumulation under different irrigation and fertilization during 2015 and 2016.

Years	Treatments	Root Dry Matter (g plant^{-1})	Stem and Leaf Dry Matter (g plant^{-1})	Bud and Boll Dry Matter (g plant^{-1})	The Total Dry Matter (g plant^{-1})
2015	W_0F_{10}	19.8 ± 0.42 c	19.6 ± 0.05 c	17.4 ± 0.59 c	62.06 ± 0.48 b
	W_0F_{30}	19.9 ± 0.29 c	16.2 ± 0.27 d	13.1 ± 0.9 d	53.47 ± 1.53 c
	$W_{80}F_{10}$	23.2 ± 0.19 a	22.1 ± 0.10 a	23.9 ± 0.01 a	70.02 ± 0.82 a
	$W_{80}F_{30}$	21.1 ± 0.31 b	24.6 ± 0.04 b	21.3 ± 1.37 b	71.4 ± 1.49 a
2016	W_0F_{10}	21.5 ± 0.79 c	12.2 ± 0.09 c	12.2 ± 0.04 c	45.87 ± 0.92 c
	W_0F_{30}	21.0 ± 0.61 c	10.0 ± 0.07 d	9.0 ± 0.04 d	40.106 ± 0.83 d
	$W_{80}F_{10}$	26.5 ± 0.95 a	16.9 ± 0.02 a	16.2 ± 0.04 a	59.54 ± 1.90 a
	$W_{80}F_{30}$	24.1 ± 0.82 b	15.2 ± 0.06 b	13.0 ± 0.04 b	52.28 ± 0.716 b

Note: pre-sowing irrigation (W_{80}) or no pre-sowing irrigation (W_{80}) and surface (F_{10}) or deep (F_{30}) fertilization. Data are the means of three replicates with standard errors and bars. Different letters indicate a significant difference at p = 0.05 according to Duncan's range test.

Simulation of biomass accumulation with respect to DAE was determined by formulas 2, 3, 4, 5, 6 and 7 (Table 2). In W_{80} treatments, total biomass and above biomass fast accumulation period was prolonged by 2–7 d and 5–10 d, root and boll biomass accumulation at fastest accumulation period was shortened by 2 d and 4–5 d, respectively, compared with W_0 treatment. W_{80} treatment had higher both total reproductive and vegetative organ biomass accumulation for maximum and average biomass accumulation rates during the fastest accumulation period than W_0 treatment. Under $W_{80}F_{10}$ total, stem, leaf and root biomass accumulation were extended by 1, 2, 10 and 1 d at fastest accumulation period compared with F_{30}. $W_{80}F_{10}$ had 13.9%, 12.5%, 10.9%, 15.0%, 17.5% 13.9%, 10.0% and 28.6%

higher maximum and average accumulation rates of total, aerial plant parts, boll, stem plus leaf and root biomass accumulation compared with $W_{80}F_{30}$ treatment.

3.5. Water-Nitrogen Productivity

Moisture consumption rate remained unaffected under both W_{80} and W_0 treatment (Table 3). Compared with W_0, 8.1%, 31.1%, 52.6% and 39.2% greater root, stem plus leaf, bud plus boll and total biomass water productivity resulted in W_{80} in all growth stages (Figure 5). $W_{80}F_{10}$ resulted in 32.0% and 15.2% higher total water and reproductive organs productivity respectively, compared with F_{30} after 84 DAE.

Root nitrogen productivity had no significant difference under both W_{80} treatment and W_0 treatment (Table 4, Figure 6). Nitrogen productivity of stem plus leaves, reproductive organs and total productivity increased by 31.3%, 42.9% and 23.1% in W_{80} compared with W_0 at 54 to 99 DAE. F_{10} produced 18.2%, 22.2% and 6.5% greater root, reproductive organs and total N productivity compared with F_{30} from 54 DAE to 99 DAE.

3.6. Factors Affecting Productivity

Soil moisture content was positively related to nitrate reductase activity and available N, but had a negative relationship with root length, root dry matter, vegetative and reproductive organs dry matter accumulation (Figure 6). Water productivity of stem, leaf, bud and boll were negatively associated with soil moisture content and available N, but had a positive relationship with root dry matter, stem plus leaf dry matter and bud plus boll dry matter production. Root, stem and leaf water-N were positively related with bud plus boll water and N productivity.

Pathway analysis showed that root length, nitrate reductase activity had a strong direct effect on boll water-nitrogen productivity (Table 5). Nitrate reductase activity had higher indirect effect on bud plus boll water productivity through soil moisture content. Nitrate reductase activity had significantly indirect effect on bud plus boll nitrogen productivity through available nitrogen than root length. This shows that improved root distribution and physiological activities could directly, or indirectly enhance water-nitrogen productivity.

Table 2. Equation of cotton plant biomass accumulation under different irrigation and fertilization during 2015 and 2016.

	Treatments	R^2	t_1 (DAE)	t_2 (DAE)	T (DAE)	t_m (DAE)	V_m (g plant^{-1} d^{-1})	V_t (g plant^{-1} d^{-1})	W_0 (g plant^{-1})
	W_0F_{10}	0.975	32.32	62.11	29.79	46.62	0.97	0.85	45.47
	W_0F_{30}	0.966	30.18	58.64	28.45	43.84	0.87	0.77	39.31
Total Biomass of Cotton Plant	$W_{80}F_{10}$	0.99	34.20	65.46	31.26	49.21	1.23	1.08	60.54
	$W_{80}F_{30}$	0.998	34.04	64.55	30.51	48.69	1.08	0.96	52.25
	W_0F_{10}	0.986	35.92	75.49	39.57	54.91	0.42	0.37	26.4
	W_0F_{30}	0.962	32.25	68.94	36.70	49.86	0.41	0.36	23.9
Aerial Part Biomass of Cotton Plant	$W_{80}F_{10}$	0.987	42.34	86.00	43.66	63.30	0.51	0.46	35.50
	$W_{80}F_{30}$	0.996	40.75	82.26	41.51	60.68	0.46	0.40	30.00
	W_0F_{10}	0.954	30.19	73.85	43.66	51.14	0.35	0.31	24.09
	W_0F_{30}	0.984	30.43	74.86	44.43	51.76	0.29	0.25	20.17
Cotton Bud and Boll Biomass	$W_{80}F_{10}$	0.997	28.46	69.3	40.84	48.07	0.47	0.41	30.22
	$W_{80}F_{30}$	0.997	28.84	70.35	41.51	48.76	0.40	0.36	26.33
	W_0F_{10}	0.999	28.96	66.20	37.24	46.84	0.52	0.46	30.63
	W_0F_{30}	0.988	34.17	70.87	36.70	51.79	0.36	0.32	20.70
Cotton Stem and Leaf Biomass	$W_{80}F_{10}$	0.997	27.37	66.94	39.57	46.36	0.59	0.52	36.84
	$W_{80}F_{30}$	0.996	25.85	54.96	29.11	39.83	0.59	0.52	26.90
	W_0F_{10}	0.997	30.86	43.24	12.38	58.70	0.29	0.21	22.03
	W_0F_{30}	0.999	29.58	38.96	9.38	61.30	0.28	0.17	20.80
Cotton Root Biomass	$W_{80}F_{10}$	0.99	26.57	34.42	7.86	62.88	0.33	0.18	25.56
	$W_{80}F_{30}$	0.999	23.97	30.69	6.72	63.92	0.30	0.14	23.81

Note: pre-sowing irrigation (W_{80}) or no pre-sowing irrigation (W_{80}) and base fertilization (F_{10}) or deep (F_{30}) application. DAE, indicates days after emergence (d). T1 and t2 are the beginning and termination days of the fast accumulation period, respectively. T indicates the duration of fast accumulation period. $T = t_2 - t_1$. t_m is the after-emergence days of the maximum biomass accumulation speeds. Vm and Vt are the maximum and average biomass accumulation speeds during the fast accumulation period, respectively. W_0 is the maximum biomass accumulation.

Table 3. Change of total moisture consumption rate, total moisture consumption ratio, water productivity of root, water productivity of stem, leaf, water productivity of bud plus boll and water productivity of total dry matter under different irrigation and fertilization during 2015 and 2016.

Year	Treatments	Total Moisture Consumption Rate (cm^3)	Total Moisture Consumption Ratio	Water Productivity of Root (g DM cm^{-3})	Water Productivity of Stem and Leaf (g DM cm^{-3})	Water Productivity of Bud and Boll (g DM cm^{-3})	Water Productivity of Total Dry Matter (g DM cm^{-3})
2015	W_0F_{10}	799.2 ± 14.62 b	0.62 ± 0.01 b	0.33 ± 0.02 a	0.26 ± 0.00 c	0.23 ± 0.01 b	0.60 ± 0.03 c
	W_0F_{30}	867.5 ± 13.98 a	0.69 ± 0.01 a	0.29 ± 0.01 b	0.19 ± 0.01 d	0.15 ± 0.01 c	0.42 ± 0.03 b
	$W_{80}F_{10}$	846.1 ± 16.39 a	0.46 ± 0.01 d	0.33 ± 0.01 a	0.31 ± 0.02 a	0.33 ± 0.01 a	0.76 ± 0.03 a
	$W_{80}F_{30}$	859.1 ± 19.91 a	0.55 ± 0.01 c	0.34 ± 0.01 a	0.28 ± 0.01 b	0.25 ± 0.02 b	0.66 ± 0.032 b
2016	W_0F_{10}	762.1 ± 13.42 b	0.65 ± 0.01 b	0.27 ± 0.01 bc	0.15 ± 0.00 c	0.15 ± 0.00 b	0.57 ± 0.01 c
	W_0F_{30}	849.1 ± 28.84 a	0.71 ± 0.02 a	0.24 ± 0.00 c	0.12 ± 0.00 d	0.11 ± 0.00 c	0.46 ± 0.02 d
	$W_{80}F_{10}$	822.7 ± 37.77 a	0.54 ± 0.00 c	0.31 ± 0.00 a	0.20 ± 0.00 a	0.19 ± 0.01 a	0.69 ± 0.02 a
	$W_{80}F_{30}$	872.2 ± 36.93 a	0.54 ± 0.02 c	0.29 ± 0.00 ab	0.18 ± 0.01 b	0.16 ± 0.01 b	0.62 ± 0.03 b

Note: pre-sowing irrigation (W_{80}) or no pre-sowing irrigation (W_{80}) and surface (F_{10}) or deep (F_{30}) fertilization. Data are the means of three replicates with standard errors and bars. Different letters indicate a significant difference at p = 0.05 according to Duncan's range test.

Table 4. Changes in root nitrogen, stem plus leaf, nitrogen bud plus boll and total dry matter nitrogen productivity under different irrigation and fertilization during 2015 and 2016.

Year	Treatments	Nitrogen Productivity of Root (g DM mg^{-1})	Nitrogen Productivity of Stem and Leaf (g DM mg^{-1})	Nitrogen Productivity of Bud and Boll (g DM mg^{-1})	Nitrogen Productivity of Total Dry Matter (g DM mg^{-1})
2015	W_0F_{10}	0.11 ± 0.004 b	0.09 ± 0.000 b	0.08 ± 0.003 c	0.28 ± 0.007 c
	W_0F_{30}	0.11 ± 0.002 b	0.07 ± 0.001 c	0.06 ± 0.004 d	0.24 ± 0.007 d
	$W_{80}F_{10}$	0.13 ± 0.000 a	0.10 ± 0.005 a	0.11 ± 0.001 a	0.33 ± 0.005 a
	$W_{80}F_{30}$	0.11 ± 0.001 b	0.11 ± 0.001 a	0.09 ± 0.006 b	0.31 ± 0.009 b
2016	W_0F_{10}	0.10 ± 0.003 b	0.06 ± 0.000 c	0.06 ± 0.000 c	0.21 ± 0.004 c
	W_0F_{30}	0.09 ± 0.003 b	0.05 ± 0.000 d	0.04 ± 0.000 d	0.18 ± 0.004 d
	$W_{80}F_{10}$	0.12 ± 0.00 a	0.07 ± 0.000 a	0.07 ± 0.000 a	0.26 ± 0.008 a
	$W_{80}F_{30}$	0.10 ± 0.003 b	0.07 ± 0.000 b	0.07 ± 0.000 b	0.23 ± 0.004 b

Note: pre-sowing irrigation (W_{80}) or no pre-sowing irrigation (W_{80}) and surface (F_{10}) or deep (F_{30}) fertilization. Data are the means of three replicates with standard errors and bars. Different letters indicate a significant difference at p = 0.05 according to Duncan's range test.

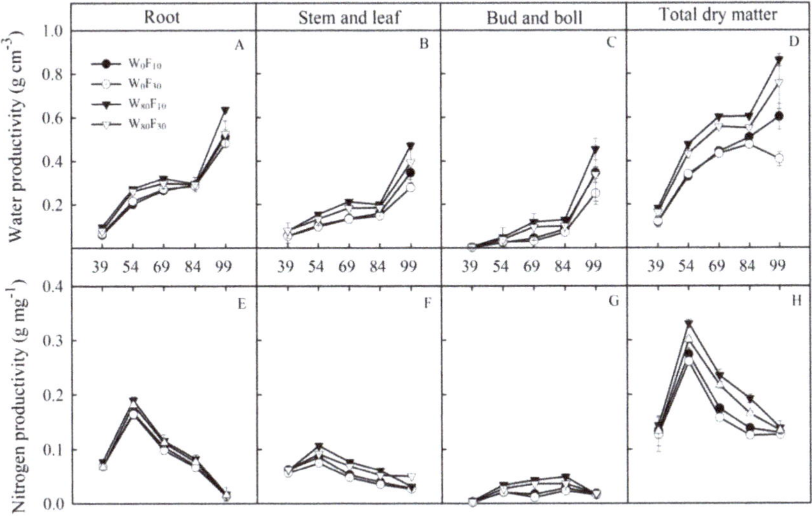

Figure 5. Dynamic change of root (**A**), stem and leaf (**B**), bud and boll (**C**) and total dry matter (**D**) water productivity and root (**E**), stem and leaf (**F**), bud and boll (**G**) and total dry matter (**H**) nitrogen productivity plant^{-1} at pre-sowing irrigation (W_{80}) or no pre-sowing irrigation and basal surface fertilization (F_{10}) or deep (F_{30}) application with the days after emergence (DAE) in 2016. Bars indicate SD ($n = 3$).

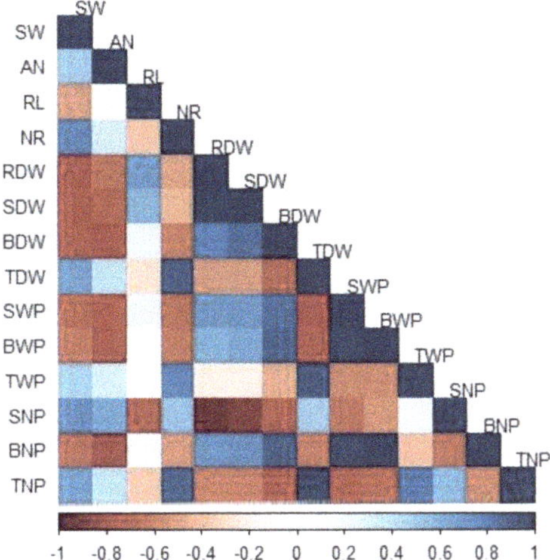

Figure 6. Correlation analysis ($n = 30$) of soil water content (SM), available nitrogen (AN, mg kg^{-1}), the root length (RL, cm), nitrate reductase (NR, μg g^{-1}FW h^{-1}), root dry matter (RDM, g plant^{-1}), stem and leaf dry matter (SDM, g plant^{-1}), bud and boll dry matter (BDM, g plant^{-1}) and total dry matter (TDM, g plant^{-1}) with nitrogen productivity and water productivity at pre-sowing irrigation (W_{80}) or no pre-sowing irrigation and base fertilizer surface (F_{10}) or deep (F_{30}) application with 2016. RWP, root water productivity (g DM cm^{-3}); SWP, stem and leaf water productivity (g DM cm^{-3}); BWP, bud and boll water productivity (g DM cm^{-3}); TWP, total water productivity (g DM cm^{-3}); RNP, root nitrogen productivity (g DM mg^{-1}); SNP, stem and leaf nitrogen productivity (g DM mg^{-1}); BNP, bud and boll nitrogen productivity (g DM mg^{-1}); TWP, total nitrogen productivity (g DM mg^{-1}).

Table 5. Path analysis (n = 60) for the direct or indirect effect on square and boll water-nitrogen productivity by soil moisture content, available nitrogen (mg kg^{-1}), the root length (cm), nitrate reductase (μg g^{-1}FW h^{-1}), root dry matter (g plant^{-1}), stem and leaf dry matter (g plant^{-1}) and bud and boll organ dry matter (g plant^{-1}) at pre-sowing irrigation or no pre-sowing irrigation and basal surface or deep fertilization in 2015 and 2016.

	Square and Boll Water Productivity						Square and Boll Nitrogen Productivity					
	x1-1	x2	x3	x4	x5	x6	x1-2	x2	x3	x4	x5	x6
Soil moisture content (x1-1)	0.058	0.41	0.455	−0.183	0.943	0.121	-	-	-	-	-	-
Available nitrogen (x1-2)	-	-	-	-	-	-	−0.027	0.211	0.375	−0.947	−1.221	−0.062
Root length (x2)	−0.177	0.382	−0.128	0.735	1.548	0.031	−0.013	0.312	−0.117	0.616	0.638	0.274
Nitrate reductase (x3)	0.37	0.031	0.339	0.25	1.18	−0.037	−0.069	−0.073	0.572	0.732	0.791	−0.007
Root dry matter (x4)	−0.277	−0.056	−0.165	0.564	1.633	0.226	0.143	0.13	0.179	0.201	1.765	0.055
Stem and leaf dry matter (x5)	−0.259	−0.047	−0.149	0.512	2.688	0.033	0.154	0.111	−0.151	0.471	1.801	0.152
Bud and boll dry matter (x6)	−0.254	−0.017	−0.2	0.183	0.27	0.585	0.177	0.039	−0.138	0.161	0.221	0.364

Note: "-" means no value.

4. Discussion

Water-nutrient application is an efficient strategy for improving plant performance under harsh environmental conditions i.e., drought stress, which ensures high cotton yield due to optimize root growth and activity in the soil [16,31]. In the present study, pre-sowing irrigation and basal surface fertilization significantly increased root distribution and physiological activity in the surface or deep soil profile at the boll setting stage. Improvement in these root traits contributed to greater shoot biomass and higher reproductive organ biomass accumulation led to greater water-nitrogen productivity.

Deeper root penetration can maximize soil moisture and nutrients uptake that can lead to maintain a high plant water and nutrient status [16,32,33]. We observed that pre-sowing irrigation and surface fertilization significantly increased root distribution and physiological activity in the surface soil (0–30 cm), indicating that improved the absorption and utility of water-nitrogen [34]. Because basal surface fertilizer application increased available water-N in the surface soil layer, which promoted cotton root distribution and physiological activity in the surface soil. This improved the absorption and utility of water-N and reduced the residual water-N in the surface soil profile. Moreover, root nitrate reductase activity in the deep soil profile (60–120 cm) enhanced, which indicated that decreasing root distribution regardless of improved root physiological activity in deep soil profile [20]. It is suggested that higher root distribution and physiological activity in both surface and deep soil profile could facilitate root and water-nutrient environment in the root zone, which can lead to higher root water-nitrogen absorption in cotton.

A strong relationship existed between root and shoot; shoots supply sufficient carbohydrates to roots that can develop and maintain root functioning which in turn can improve shoot growth by supplying a sufficient amount of nutrients, water and phytohormone. This further ensures crop productivity [5,35,36]. In this study, we observed that greater dry matter accumulated and allocation to the aerial parts has led to lower dry matter production in root and its physiological activity later in the season. The reason might be due to functional period of root (within 54–84 DAE) and the root biomass fast accumulation period (28–40 DAE) under different water-nitrogen management did not correspond. Root proliferation and physiological activity are positively associated with the root zone environment [32,33]. Therefore, pre-sowing irrigation and basal surface fertilization coordinates the relationship between root and water-nutrient in the soil. This in turn increased root absorptive capacity of water-nitrogen.

It is noteworthy that optimal water-nitrogen application could change the distance between water-nitrogen and root in the soil [16,37] as well as root physiological activity [33] to enhance the absorption of water-nitrogen. However, our data across the two years demonstrated that the water-nitrogen is an important management practice that can adjust the water-nitrogen productive ability in different plant organs which could result in greater water-nitrogen uptake. The possible reason might be improved root distribution and physiological activity in the surface soil and root physiological activity in the deep soil profile at 54–84 DAE promoted absorption of water-nitrogen from irrigation and deep layer water. This resulted in higher water-nitrogen productivity of reproductive organs at the boll setting stage. Secondly, root distribution and physiological activity could ensure the application of water-nitrogen, which increased leaf photosynthetic efficiency and leaf gas change parameters led to greater dry matter accumulation [22]. An adequate water-nitrogen in soil may decrease root distribution [15,20] and root dry matter at the fast accumulation period before 40 DAE. These phenomenon in turn decreased root dry matter accumulation and increased dry matter accumulation above ground parts at the boll setting stage (within 69–84 DAE).

Water use efficiency in terms of physiology is defined as the ratio transpiration and photosynthesis [38]. Lower dry mass accumulation in the aerial part can lead to a higher water-nitrogen use efficiency, but reducing water-nitrogen productive ability [10,32]. Interestingly, we observed that increasing root distribution and physiological activity in the surface soil layer and root physiological activity in deep soil layer at the boll setting stage can directly or indirectly promote dry mass accumulation and water-nitrogen productivity of the reproductive organs. More root distribution can boost water use efficiency and drought resistance and consequently greater crop yield [33,39]. Higher root distribution led to a lower biomass accumulation in the aerial parts of crop plants [10,32]. We speculated that increasing root distribution and physiological activity can drive reproductive organs dry mass accumulation which results in higher water-nitrogen productivity of cotton crop.

5. Conclusions

Pre-sowing irrigation and surface basal fertilization could significantly promote reproductive organ biomass accumulation and productive ability of water-nitrogen. Pre-sowing irrigation combined with basal surface fertilization favored root morphological and physiological performance i.e., greater root biomass, longer root length in the surface soil profile (0–30 cm), higher root nitrate reductase activity in the surface or deep soil profile (60–80 cm) at the boll setting stage. Improvements in these root traits led to a higher water-nitrogen consumption, accumulation and allocation of reproductive structures of cotton plant. This in turn contributed to a higher water-nitrogen productive ability of the reproductive organ at the boll setting stage. These data highlighted that pre-sowing irrigation combined with basal surface fertilization is a promising option in terms of higher root morphological and physiological activity and water-nitrogen productivity of cotton crop in the arid region.

Author Contributions: Writing & conceptualization, Z.C.; data curation, H.G.; investigation, F.H.; writing—review & editing, A.K. and H.L.

Funding: This research was funded by the National Natural Science Foundation of China (31760355) and Program of Youth Science and Technology Innovation Leader of The Xinjiang Production and Construction Corps (2017CB005).

Acknowledgments: Shihezi University and the National Natural Science Foundation of China are acknowledged for their financial support to the study.

Conflicts of Interest: All the authors have approved the manuscript and agree with submission to your esteemed journal. There are no conflicts of interest to declare.

References

1. Constable, G.A.; Bange, M.P. The yield potential of cotton (Gossypium hirsutum L.). *Field Crops Res.* **2015**, *182*, 98–106. [CrossRef]

2. Khan, A.; Tan, D.K.Y.; Afridi, M.Z.; Luo, H.; Tung, S.A.; Ajab, M.; Fahad, S. Nitrogen fertility and abiotic stresses management in cotton crop: A review. *Environ. Sci. Pollut. R* **2017**, *24*, 14551–14566. [CrossRef] [PubMed]

3. Khan, A.; Pan, X.D.; Najeeb, U.; Tan, D.K.Y.; Fahad, S.; Zahoor, R.; Luo, H.H. Coping with drought: Stress and adaptive mechanisms, and management through cultural and molecular alternatives in cotton as vital constituents for plant stress resilience and fitness. *Biol. Res.* **2018**, *51*, 47. [CrossRef] [PubMed]

4. Berry, P.; Ramirezvillegas, J.; Bramley, H.; Mgonja, M.A.; Mohanty, S.; Jackson, M.; Fordlloyd, B.; Parry, M. Regional impacts of climate change on agriculture and the role of adaptation. In *Plant Genetic Resources and Climate Change*; CABI: London, UK, 2014; Volume 2, pp. 78–97.

5. Zhang, X.Y.; Chen, S.Y.; Sun, H.Y.; Wang, Y.M.; Shao, L.W. Root size, distribution and soil water depletion as affected by cultivars and environmental factors. *Field Crops Res.* **2009**, *114*, 75–83. [CrossRef]

6. Bouman, B.A.M. A conceptual framework for the improvement of crop water productivity at different spatial scales. *Agric. Syst.* **2007**, *93*, 43–60. [CrossRef]

7. Afridi, M.Z.; Jan, M.T.; Arif, M.; Jan, A. Wheat yielding components response to different levels of fertilizer-N application time and decapitation stress. *Sarhad J. Agric.* **2010**, *26*, 499–506.

8. Liao, N.; Hou, Z.; Qi, L.I.; Ru, S.; Huijuan, B.O. Increase effect of biochar on cotton yield and nitrogen use efficiency under different nitrogen application levels. *Plant Nutr. Fertil. Sci.* **2015**, *21*, 782–791.

9. Dai, J.L.; Li, W.J.; Tang, W.; Zhang, D.M.; Li, Z.H.; Lu, H.Q.; Eneji, A.E.; Dong, H.Z. Manipulation of dry matter accumulation and partitioning with plant density in relation to yield stability of cotton under intensive management. *Field Crops Res.* **2015**, *180*, 207–215. [CrossRef]

10. Rao, S.S.; Tanwar, S.P.S.; Regar, P.L. Effect of deficit irrigation, phosphorous inoculation and cycocel spray on root growth, seed cotton yield and water productivity of drip irrigated cotton in arid environment. *Agric. Water Manag.* **2016**, *169*, 14–25. [CrossRef]

11. Han, H.M.; Tian, Z.W.; Fan, Y.H.; Cui, Y.K.; Cai, J.; Jiang, D.; Cao, W.X.; Dai, T.B. Water-deficit treatment followed by re-watering stimulates seminal root growth associated with hormone balance and photosynthesis in wheat (Triticum aestivum L.) seedlings. *Plant Growth Regul.* **2015**, *77*, 201–210. [CrossRef]

12. Song, H.; Feng, B.L.; Gao, X.L.; Dai, H.P.; Zhang, P.P. Relationship between root activity and leaf Senescence in different adzuki bean cultivars (lines). *Acta Bot. Boreali-Occident. Sin.* **2011**, *31*, 2270–2275.

13. Amanullah. Effects of NPK source on the dry matter partitioning in cool season C3-cereals (wheat, rye, barley, and oats) at various growth stages. *J. Plant Nutr.* **2017**, *40*, 352–364. [CrossRef]

14. Flexas, J.; Bota, J.; Galmes, J.; Medrano, H.; Ribas-Carbo, M. Keeping a positive carbon balance under adverse conditions: Responses of photosynthesis and respiration to water stress. *Physiol. Plant.* **2006**, *127*, 343–352. [CrossRef]

15. Li, K.F.; Zhang, F.C.; Qi, Y.L. Effects of water and fertility spatial coupling in root zone soil of winter wheat on root growth and activity. *Agric. Res. Arid Areas* **2009**, *27*, 48–52.

16. Wasson, A.P.; Richards, R.A.; Chatrath, R.; Misra, S.C.; Prasad, S.V.; Rebetzke, G.J.; Kirkegaard, J.A.; Christopher, J.; Watt, M. Traits and selection strategies to improve root systems and water uptake in water-limited wheat crops. *J. Exp. Bot.* **2012**, *63*, 3485–3498. [CrossRef]

17. Liu, R.X.; Zhou, Z.G.; Guo, W.Q. Effects of N fertilization on root development and activity of water-stressed cotton (Gossypium hirsutum L.) plants. *Agric. Water Manag.* **2008**, *95*, 1261–1270. [CrossRef]

18. Chen, Y.L.; Zhang, J.; Li, Q.; He, X.L.; Su, X.P.; Chen, F.J.; Yuan, L.X.; Mi, G.H. Effects of nitrogen application on post-silking root senescence and yield of maize. *Agron. J.* **2015**, *107*, 835–842. [CrossRef]

19. Klimesova, J.; Nobis, M.P.; Herben, T. Senescence, ageing and death of the whole plant: Morphological prerequisites and constraints of plant immortality. *New Phytol.* **2015**, *206*, 14–18. [CrossRef]

20. Luo, H.; Zhang, H.; Han, H.; Hu, Y.; Zhang, Y.; Zhang, W. Effects of water storage in deeper soil layers on growth, yield, and water productivity of cotton (Gossypium Hirsutum L.) in arid areas of northwestern China. *Irrig. Drain.* **2014**, *63*, 59–70. [CrossRef]

21. Lu, X.R.; Jia, X.Y.; Niu, J.H. The present situation and prospects of cotton industry development in China. *Sci. Agric. Sin.* **2018**, *51*, 26–36.

22. Chen, Z.K.; Niu, Y.P.; Ma, H.; Hafeez, A.; Luo, H.H.; Zhang, W.F. Photosynthesis and biomass allocation of cotton as affected by deep-layer water and fertilizer application depth. *Photosynthetica* **2017**, *55*, 638–647. [CrossRef]

23. Shen, Y.F.; Li, S.Q. Effects of the spatial coupling of water and fertilizer on the chlorophyll fluorescence parameters of winter wheat leaves. *Agric. Sci. Chin.* **2011**, *10*, 1923–1931. [CrossRef]

24. Garrido-Lestache, E.; Lopez-Bellido, R.J.; Lopez-Bellido, L. Effect of N rate, timing and splitting and N type on bread-making quality in hard red spring wheat under rainfed Mediterranean conditions. *Field Crops Res.* **2004**, *85*, 213–236. [CrossRef]

25. Baethgen, W.E.; Alley, M.M. Optimizing soil and fertilizer nitrogen use by intensively managed winter wheat. I. crop nitrogen uptake. *Agron. J.* **1989**, *81*, 120–125. [CrossRef]

26. Gwenzi, W.; Veneklaas, E.J.; Holmes, K.W.; Bleby, T.M.; Phillips, I.R.; Hinz, C. Spatial analysis of fine root distribution on a recently constructed ecosystem in a water-limited environment. *Plant Soil* **2011**, *344*, 255–272. [CrossRef]

27. Zhou, Z.; He, H.; Ma, L.; Yu, X.; Mi, Q.; Pang, J.; Tang, G.; Liu, B. Overexpression of a GmCnx1 gene enhanced activity of nitrate reductase and aldehyde oxidase, and boosted mosaic virus resistance in soybean. *PLoS ONE* **2015**, *10*, e0124273. [CrossRef]

28. Yang, G.Z.; Tang, H.Y.; Nie, Y.C.; Zhang, X.L. Responses of cotton growth, yield, and biomass to nitrogen split application ratio. *Eur. J. Agron.* **2011**, *35*, 164–170. [CrossRef]

29. Wang, Z.S.; Min, X.U.; Liu, R.X.; Xiao-Dong, W.U.; Zhu, H.; Chen, B.L.; Zhou, Z.G. Effects of nitrogen rates on biomass and nitrogen accumulation of cotton with different varieties in growth duration. *Cotton Sci.* **2011**, *23*, 537–544.

30. Xu, G.H.; Fan, X.R.; Miller, A.J. Plant nitrogen assimilation and use efficiency. *Annu. Rev. Plant Biol.* **2012**, *63*, 153–182. [CrossRef]

31. Hu, X.T.; Chen, H.; Wang, J.; Meng, X.B.; Chen, F.H. Effects of soil water content on cotton root growth and distribution under mulched drip irrigation. *Agric. Sci. Chin.* **2009**, *8*, 709–716. [CrossRef]

32. Palta, J.A.; Chen, X.; Milroy, S.P.; Rebetzke, G.J.; Dreccer, M.F.; Watt, M. Large root systems: Are they useful in adapting wheat to dry environments? *Funct. Plant Biol.* **2011**, *38*, 347–354. [CrossRef]

33. Lynch, J.P. Steep, cheap and deep: An ideotype to optimize water and N acquisition by maize root systems. *Ann. Bot.* **2013**, *112*, 347–357. [CrossRef] [PubMed]

34. Zwart, S.J.; Bastiaanssen, W.G.M. Review of measured crop water productivity values for irrigated wheat, rice, cotton and maize. *Agric. Water Manag.* **2004**, *69*, 115–133. [CrossRef]

35. Osaki, M.; Shinano, T.; Matsumoto, M.; Zheng, T.; Tadano, T. A root-shoot interaction hypothesis for high productivity of field crops. *Soil Sci. Plant Nutr.* **1997**, *43*, 445–454. [CrossRef]

36. Yang, L.; Zheng, B.; Mao, C.; Qi, X.; Liu, F.; Wu, P. Analysis of transcripts that are differentially expressed in three sectors of the rice root system under water deficit. *Mol. Genet. Genom.* **2004**, *272*, 433–442. [CrossRef]

37. Wu, H.X.; Ma, Y.Z.; Xiao, J.P.; Zhang, Z.H.; Shi, Z.H. Photosynthesis and root characteristics of rice (Oryza sativa L.) in floating culture. *Photosynthetica* **2013**, *51*, 231–237. [CrossRef]

38. Blum, A. Drought resistance, water-use efficiency, and yield potential—Are they compatible, dissonant, or mutually exclusive? *Aust. J. Agric. Res.* **2005**, *56*, 1159–1168. [CrossRef]

39. Lilley, J.M.; Kirkegaard, J.A. Benefits of increased soil exploration by wheat roots. *Field Crops Res.* **2011**, *122*, 118–130. [CrossRef]

 *agronomy*

Article

Growth, Critical N Concentration and Crop N Demand in Butterhead and Crisphead Lettuce Grown under Mediterranean Conditions

Giulia Conversa * and Antonio Elia

Department of the Science of Agriculture, Food and Environment (SAFE), University of Foggia, via Napoli 25, 71122 Foggia, Italy; antonio.elia@unifg.it
* Correspondence: giulia.conversa@unifg.it

Received: 3 October 2019; Accepted: 23 October 2019; Published: 25 October 2019

Abstract: Excessive nitrogen (N) fertilizers are applied in lettuce causing both environmental issues and N crop luxury consumption. In order to improve the N use efficiency (NUE) by defining optimal crop growth and N requirements of butterhead and crisphead lettuce, two field experiments were conducted using 0, 50, and 100 kg ha^{-1} of N fertilizer to study (i) the growth and productivity, (ii) the NUE, (iii) the critical N dilution curve, and (iv) the N demand. Nitrogen supply enhanced dry weight (DW) accumulation in the butterhead (from 295 to 410 g m^{-2}), but not in the crisphead type (251 g m^{-2}). The NUE indices underlined the poor ability of the crisphead type in absorbing soil N and also in the utilization of the absorbed N for producing DW. The critical N dilution curves %Nc = 3.96 DW$^{-0.205}$ and %Nc = 3.65 DW$^{-0.115}$ were determined for crisphead and butterhead lettuce, respectively. Based on these type-specific %Nc curves, the estimated N demand was 125 kg ha^{-1} in the butterhead and 80 kg ha^{-1} in the crisphead lettuce for producing 4.3 and 2.5 Mg ha^{-1} of DW, respectively, under Mediterranean climate. Neither N fertilization nor genotype affected crop productivity.

Keywords: growth; specific leaf nitrogen; nitrogen use efficiency; critical nitrogen uptake

1. Introduction

Among leafy vegetables, lettuce (*Lactuca sativa* L.) is the most important species, grown on over 1.86 million hectares around the world [1]. The crop is characterized by low efficiency in nitrogen (N) recovery, with the highest N absorption occurring during the last phase of the growing cycle [2], when a sub-optimal N availability may result in a decrease in head yield and quality. In order to avoid N deficiency, N fertilizers are frequently applied in excess compared with the crop demand [3], causing both environmental issues (e.g., nitrate contamination of aquifers and eutrophication of surface water) and N crop luxury consumption, the latter resulting in excessive nitrate accumulation in the plants. Nitrate-contaminated freshwater and leafy vegetables with a high concentration of nitrate are considered potentially dangerous for human health [4].

In order to improve N use efficiency (NUE) the determination of the patterns of crop growth and of N demand during the crop cycle are at the basis of optimal N fertilization planning, timing, and management [5–7].

The study employing a growth analysis of the effects of different N rates on dry matter production and N accumulation during the crop cycle may be useful to define optimal crop growth and N requirements [8]. The N crop demand, which can be defined as the amount of N necessary to sustain the potential growth of a crop at any time of the cycle [9], is frequently modeled using the concept of the critical N plant concentration. It is the minimum N concentration in dry biomass (critical N—%Nc) required for maximum dry weight (DW) accumulation under specific climatic conditions and agronomic practices [10]. Critical N concentration decreases during the crop cycle according to the equation %Nc = a DW^{-b} [11] (%Nc dilution curve) with a crop-specific pattern. Using the critical N curve, the crop's N demand can be obtained from the crop biomass accumulation during the crop cycle, which is in turn determined by environment and genotype [12]. The availability of a crop-specific critical N dilution curve is therefore of key importance for deriving N uptake and improving the dynamic assessment of N status in a crop.

Tei and coworkers [13] found the critical N dilution curve for the butterhead type (%Nc = 4.56 DW$^{-0.357}$) grown in Central Italy. In California (USA), Bottoms and coworkers [14] working on a large dataset of experimental and commercial field data of both crisphead and romaine lettuce types suggested the empirical linear regression %Nc = 4.20 − 2.8 DW to distinguish between N-deficient and N-sufficient conditions across the entire season, underlining the unsuitability of %Nc dilution curve proposed by Tei et al. [13] for the butterhead. Nevertheless, considering that the environment as well as the large variability in physio-morphological traits between lettuce typologies strongly affect the dry mass production and N uptake [15,16], a type-specific critical %N dilution curve may need to be locally calibrated [6,17] to tune N crop demand.

To the best of our knowledge, no such studies are available for modeling N crop nutrition and optimizing N fertilization of lettuce grown under Southern Italian/Mediterranean climatic conditions. Therefore, the main aims of the present paper were to define (i) the growth and productivity, (ii) the nitrogen use efficiency, (iii) the critical N dilution curve, and (iv) the N demand of the most widely cultivated lettuce typologies (butterhead and crisphead), grown at three N fertilization levels.

2. Materials and Methods

2.1. Field Experimental Site and Climatic Conditions

Two open field lettuce (*Lactuca sativa* L., var. *capitata*) trials were carried out in 2008 (exp. 1) and 2009 (exp. 2) on a commercial farm located in Foggia Province, Puglia Region (Italy) (latitude 41°46′ N, longitude 15°5′ E, 74 m above mean sea level). The site is within an area dominated by a Mediterranean climate with a mild winter and dry-and-warm summer; mean minimum and maximum temperatures are 10.8 ± 1.7 °C and 19.9 ± 2.2 °C, respectively, and the mean temperatures of the coldest (January) and hottest (August) months are 7.1 and 24.5, respectively. The average annual rainfall is 537 mm. The weather conditions during the two growing seasons are reported in Figure 1.

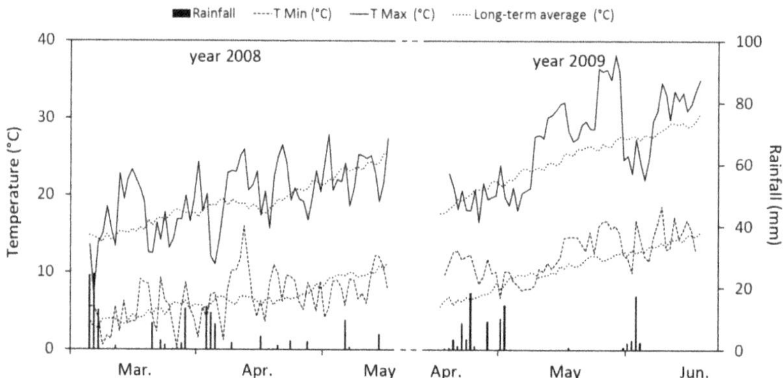

Figure 1. Climate data for the two lettuce trials (5 March–14 May 2008 and 16 April–12 June 2009).

The area falls within a nitrate vulnerable zone; therefore, it is subjected to the European Directive 109 91/676/EEC prescriptions.

Both trials were carried out in the same field and the soil characteristics were 24% clay, 34% silt, 42% sand, pH 7.52 (soil:water 1:2.5), 2.2% organic matter, 1.52‰ total N, 382 ppm NH$_4$OAc-extractable K, 24 ppm Olsen P, and 7% active carbonate.

Seedlings with 4–5 true-leaves were transplanted 30 cm apart in 30 cm spaced rows (11.1 plants m^{-2}) on 5 March 2008 (exp. 1) and 16 April 2009 (exp. 2). Both trials included three nitrogen fertilization rates: 0, 50, and 100 kg ha^{-1} (indicated as N_0, N_{50}, N_{100}) and two types of capitata lettuce: the butterhead (cv. Faustina—ISEA) and crisphead (cv. Silvinas—Rijk Zwaan) type. Each experiment was arranged in a split-plot design with four replications and with N rate as the main factor and cultivar as sub factor. The experimental plot unit included 6 rows and was 1.8 m wide and 5 m long (9 m^2).

In both years, the preceding crop was broccoli and crop residues were incorporated into the soil one month before the lettuce transplanting. In each growing cycle, the level of PK fertilization was adjusted to the plant nutritional requirements; a sprinkler irrigation system was used to satisfy the water requirements of the crops, which were assessed through the water balance method using the Penman–Monteith approach to estimate the reference evapotranspiration. A total of 1460 and 2010 m^3 ha^{-1} of irrigation water was supplied in the first and the second cycle, respectively.

Nitrogen fertilizer was applied in surface strips as NH_4NO_3 (34:0:0, N:P:K) (Yara International ASA, Oslo, Norway). Forty percent of the N was applied at transplanting and the remaining 60% at the 20th true-leaf stage (7 April 2008 and 11 May 2009 for exp. 1 and 2, respectively).

Starting from approximately 2 weeks after transplanting until harvest, 9 (exp. 1) and 5 (exp. 2) destructive samplings (picking three plants in the four central rows of each experimental plot) were carried out, with a one-week interval in the first year (14, 21, 28, 35, 42, 49, 56, 64, and 70 DAT (days after transplant)) and a two-week interval in the second (19, 26, 33, 40, and 57 DAT). At each sampling, shoot and root fresh (FW) and dry weight (DW), number of leaves, leaf area, and (only at 14, 28, 42, 49, 64, and 70 DAT in 2008; 19, 26, 33, 40, and 57 DAT in 2009) shoot N concentration were determined. Harvest was carried out on 13 May 2008 in exp. 1 (70 days after transplant—DAT) and 11 June 2009 in exp. 2 (57 DAT), respectively. Twenty lettuce heads per replication were randomly selected at the optimal stage for fresh consumption to determine fresh yield, chlorophylls, and nitrate concentration.

Leaf area was measured by LI-COR 3100 (LICOR, Lincoln, NE, USA). Root samples were washed through two sieves to remove soil and weighed after drying up water using filter papers.

For DW determination, plant material was dried in a ventilated oven at 65 °C until the achievement of constant weight. Dry mass concentration was calculated as DW/FW (g kg^{-1}). Dried shoot material was then milled through a 1.0 mm sieve (IKA Labortechnik, Staufen, Germany) and the nitrogen (N) concentration was determined using the Kjeldahl method (Kjeltec model 1035—Foss Tecator). Nitrates were extracted from 0.5 g of dried shoots with 50 mL solution containing 3.5 mM sodium carbonate (Na_2CO_3) and 1.0 mM sodium bicarbonate (Na_2HCO_3). Nitrates were measured by ion chromatography (Dionex^{TM} ICS 3000, Thermo Scientific, Waltham, MA, USA) with a conductivity detector, using the pre-column Dionex^{TM} IonPack AG23 and the Dionex^{TM} IonPack AS23 (4 mm × 250 mm, 5 μm) separation column, according to the method reported in Bonasia et al. [18]. Reduced N was calculated as the difference between total N content and N-nitrate content. The chlorophyll a and b concentrations were spectrophotometrically determined as reported by Conversa et al. [19].

Shoot dry weight accumulation, leaf area index (LAI), and N crop uptake during the crop cycles were plotted using days after transplant. At harvest, specific leaf area (SLA) was calculated as leaf dry weight/leaf area ratio (g m^{-2}); specific leaf nitrogen (SLN) was calculated as leaf N content/leaf area ratio (g m^{-2}); and specific leaf-reduced N (SLNred) was calculated as leaf reduced N content/leaf area ratio (g m^{-2}).

2.2. Nitrogen Use Efficiency Indices

The nitrogen use efficiency (NUE) and NUE components (expressed as kg kg^{-1}) were calculated following Conversa et al. [17]. Briefly, the NUE components were:

Partial factor productivity of applied N (PFP), which represents the kilogram of product (heads' fresh weight—HFW) harvested per kilogram of N fertilizer applied (NA), also called simply NUE, can be used as an index of total economic output relative to the use of all N sources (soil N and applied fertilizer N):

$$PFP = HFW/NA \ (kg\ kg^{-1}) \tag{1}$$

Agronomic efficiency of N (NUEa), which represents the kilogram of yield (in terms of heads dry weight) increase per kilogram of N fertilizer applied (NA), is calculated as:

$$\text{NUEa} = (\text{HDWf} - \text{HDWc})/\text{NA} \ (\text{kg kg}^{-1}) \tag{2}$$

where HDWf = heads dry weight of fertilized treatment, HDWc = heads dry weight of control treatment, and NA = dose of applied N.

Apparent N recovery efficiency (REC) by the crop, which represents the kilogram increase in N uptake per kilogram of N applied, is calculated as:

$$\text{REC} = (\text{TNf} - \text{TNc})/\text{NA} \ (\text{kg kg}^{-1}) \tag{3}$$

where TNf = total N uptake in aboveground biomass at maturity (kg ha^{-1}) when an amount NA is applied, and TNc is the corresponding total plant N uptake in aboveground biomass at maturity (kg ha^{-1}) when no N-fertilizer is applied.

Physiological efficiency of N (NUEp), which represents the kilogram of yield (in terms of heads dry weight) increase per kilogram increase in N uptake from fertilizer, is calculated as:

$$\text{NUEp} = (\text{HDWf} - \text{HDWc})/(\text{TNf} - \text{TNc}) \ (\text{kg kg}^{-1}) \tag{4}$$

2.3. N Critical Curve

The critical N concentration was determined following Justes and coworkers [20]. The critical N curve indicates the minimum N concentration observed at a given time among all N treatments that had given, to that date, the maximum amount of the aboveground DW. For each trial, sampling date, and lettuce type, the observed aboveground DWs, obtained with the different N rates, were compared with the corresponding total N concentrations (%N). The Students' two-tail *t*-test was used to test the hypothesis of means equality at $p \leq 0.1$. All the values identified as critical (%Nc, in g 100 g^{-1} DW) were related to the aboveground dry biomass (DW, Mg ha^{-1}) according to the equation of Lemaire and Salette [21]:

$$\%\text{Nc} = a \ \text{DW}^{-b} \tag{5}$$

where a represents the critical N concentration in the dry biomass when DW $\geq$ 0.9 Mg ha^{-1}, and b is a statistical parameter governing the slope of the relationship.

The fitting of the Equation (5) was performed for %N pooled data and for each lettuce type. Additionally, all the observed %N values were compared with the critical N concentrations predicted by the curves proposed for butterhead lettuce (%Nc = 4.56 DW$^{-0.357}$) by Tei and coworkers [13] and for crisphead and romaine lettuce by Bottoms and coworkers [14] to distinguish between N-deficient and N-sufficient conditions (%Nc = 4.20 − 2.8 DW).

2.4. Statistical Elaboration of Data

All the data were submitted to analysis of variance by using the GLM Procedure of SAS software [22]. Mean separations were based on Fisher's protected Least Significant Difference (LSD) test at the 0.05 probability level.

The Gauss–Newton method of the NLIN procedure of SAS software was used for the non-linear regression fittings of Equation (5) on %Nc against the DW accumulation. The evaluation of model accuracy was performed using adjusted R^2 (adjR^2), root mean square error (RMSE), and relative root mean square error (RRMSE). The last two indices were calculated as follows:

$$\text{RMSE} = \sqrt{\sum_{i=1}^{n} (S_i - O_i)^2 / n} \tag{6}$$

$$RRMSE = RMSE\ 100/\bar{O} \tag{7}$$

where S_i and O_i are simulated and observed values, respectively, and $\bar{O}$ is the observed mean value. RMSE describes the difference between model simulations and observations in the units of the variable. Its value close to zero indicates a perfect fit; however, a value less than half of the standard deviation of the observations may be considered low [23]. RRMSE provides a measure (%) of the relative difference between simulated and observed data. Adjusted R^2 and RMSE and RRMSE indices were also used to evaluate the goodness of prediction of plant critical N concentration by the N curve proposed by Tei et al. [13] and by the equation proposed by Bottoms et al. [14].

3. Results

3.1. Weather Conditions

Minimum and maximum averaged daily temperatures were 6.7 °C and 19.8 °C, respectively, for the 2008 cycle (5 March–14 May), while for the 2009 cycle (16 April–15 June) they were 12.2 °C and 26.5 °C, respectively. In 2009, both maximum and minimum temperatures were higher than the long-term averages (10.5/24.9 °C, respectively) and particularly during the period from the 4th to the 7th week after transplanting and in the last 10 days before harvest. Total rainfall was 157 and 99 mm in 2008 and 2009, respectively (Figure 1). The cumulated thermal time (700 versus 942 day degrees (°C)) and global radiation (1222 versus 1308 MJ m^{-2}) were lower in the first than in the second trial.

3.2. Yield, Dry Weight Accumulation, and Partitioning

Neither N fertilizer application nor genotype affected the shoot fresh weight (yield) (Table 1). At transplanting, the crisphead and butterhead lettuce had similar dry weight of seedlings (1.164 ± 0.089 and 1.110 ± 0.039 g m^{-2}, respectively). During both crop cycles, the butterhead typology (Figure 2A,C) exhibited a higher accumulation of shoot DW than the crisphead one (Figure 2B,D).

Table 1. Effect of N fertilization rate, lettuce genotype, and year of cultivation on crop yield and plant dry weight partitioning.

Treatments	Shoot Fresh Weight (Yield)	Shoot Dry Weight Content	Shoot Dry Weight Concentration	Root Dry Weight	Root/Shoot Ratio
	(Mg ha^{-1})	(g m^{-2})	(g kg^{-1})	(g m^{-2})	
N rate (kg ha^{-1}) (N)					
0	79.2 a [1]	267 b	32.1 b	32.1 a	0.11 a
50	84.8 a	304 a	34.3 ab	32.9 a	0.11 a
100	86.9 a	329 a	36.1 a	32.8 a	0.11 a
Genotype (G)					
Butterhead	86.6 a	350 a	38.7 a	49.0 a	0.14 a
Crisphead	80.7 a	251 b	29.6 b	16.1 b	0.08 b
Year (Y)					
2008	67.4 b	180 b	26.5 b	19.5 b	0.11 a
2009	99.9 a	420 a	41.8 a	51.4 a	0.11 a
Significance [2]					
N	ns	**	*	ns	ns
G	ns	**	***	***	***
Yr	***	***	***	***	ns
Yr*N	ns	ns	ns	ns	ns
Yr*G	ns	ns	ns	***	ns
N*G	ns	*	*	ns	*
N*G*Yr	ns	ns	ns	ns	ns

[1] Means in columns not sharing the same letters are significantly different according to Least Significant Difference (LSD) test ($p = 0.05$). [2] ns, *, **, and ***, not significant or significant at $p \leq 0.05$, $p \leq 0.01$, or $p \leq 0.001$, respectively.

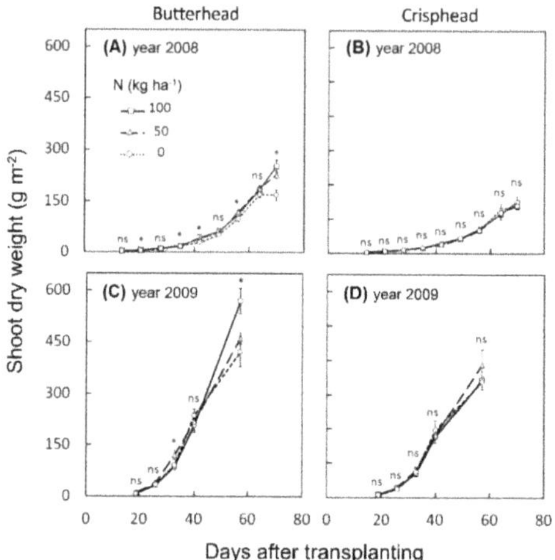

Figure 2. Shoot dry weight (DW) accumulation during the growing seasons (2008: **A,B**; 2009: **C,D**) of butterhead (**A,C**) and crisphead (**B,D**) lettuce as affected by N level. Vertical bars represent standard error (n = 4) with notes indicating a significant (*: p = 0.05) or not significant difference (ns) between N levels, according to the LSD test.

The DW accumulation was particularly evident three weeks before the harvest during the linear growth phase, with butterhead type also showing a positive response to N supply. Specifically, by increasing N fertilization rate at harvest, shoot DW rose from 295 to 410 g m^{-2} in the butterhead lettuce, while it was 251 g m^{-2}, on average among N rates, in the crisphead one (Figure 3A).

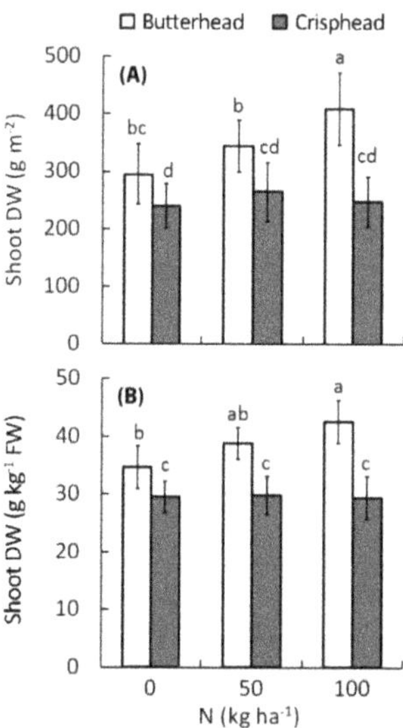

Figure 3. Shoot dry weight accumulation (**A**) and concentration (**B**) at harvest, in butterhead and crisphead plants as affected by N fertilization level. Vertical bars (± Standard Error (SE) of mean; n = 8] with different letters are significantly different according to the LSD test (p = 0.05).

Similarly, dry weight concentration was only enhanced in the butterhead lettuce by the dose of N fertilizer, and it was higher than the crisphead one (29.6 g kg^{-1} DW, on average) (Figure 3B). The root dry weight was also higher in butterhead than in crisphead lettuce (Table 1), especially in the 2009 trial. In the butterhead, a greater root:shoot DW ratio was also detected (Table 1), with the highest values in the unfertilized plants.

In both lettuce types, growth and yield were much lower (−32% and −57%, respectively) in first compared with the second experiment ($p \le 0.001$) (Table 1).

3.3. Shoot Morpho-Physiological Traits

Approaching the harvest, N_{50} and especially N_{100} treatment significantly improved leaf area index (LAI) in both lettuce types (Figure 4) as confirmed by the final values (Table 2). They were slightly higher in butterhead (Figure 4A,C) than in crisphead lettuce (Figure 4B,D).

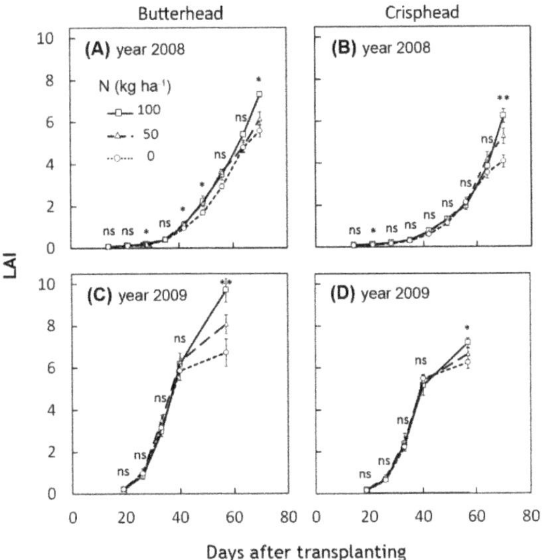

Figure 4. Changes in leaf area index (LAI) during the 2008 (**A**,**B**) and 2009 (**C**,**D**) crop cycles as affected by lettuce type (butterhead: **A**,**C**; crisphead: **B**,**D**) and N level. Bars indicate ± SE of mean ($n = 4$). The notes: ns, *, **, indicate differences not significant or significant at $p \le 0.05$, $p \le 0.01$, respectively.

N fertilization did not substantially modify the number of leaves, specific leaf area (SLA), specific leaf nitrogen (both SLNtot and SLNred), and the total chlorophyll content (on a leaf area basis) (Table 2). All these characteristics were affected by the genotype, with the butterhead type showing a greater leaves number ($p \le 0.001$), LAI ($p \le 0.001$), SLA ($p \le 0.05$), SLNtot ($p \le 0.05$), SLNred ($p \le 0.05$), and chlorophyll concentration ($p \le 0.05$) (Table 2). As for the yield and DW accumulation, they were higher in the 2009 than 2008 experiment ($p \le 0.001$) (Table 2).

Table 2. Effect of N fertilization rate, lettuce genotype and year of cultivation on head morpho-physiological characteristics.

Treatments	Leaves (no.)	LAI (m^2 m^{-2})	SLA [1] (g m^{-2})	SLN$_{tot}$ [2] (g m^{-2})	SLN$_{red}$ (g m^{-2})	Total Chlorophyll (μg cm^{-2})
N rate (kg ha^{-1}) (N)						
0	38.6 a [3]	5.62 c	45.9 a	1.51 a	1.38 a	18.6 ab
50	39.2 a	6.54 b	44.9 a	1.45 a	1.30 a	20.4 a
100	41.1 a	7.61 a	41.6 a	1.36 a	1.20 a	17.1 b
Genotype (G)						
Butterhead	48.5 a	7.26 a	46.8 a	1.53 a	1.39 a	22.3 a
Crisphead	30.8 b	5.92 b	41.5 b	1.35 b	1.21 b	15.1 b
Year (Yr)						
2008	35.8 b	5.78 b	31.6 b	1.17 b	1.13 b	14.5 b
2009	43.5 a	7.41 a	56.7 a	1.71 a	1.47 a	19.2 a
Significance [4]						
N	ns	**	ns	ns	ns	*
G	***	***	*	*	*	*
Yr	*	***	***	***	**	*
Yr*N	ns	ns	ns	ns	ns	ns
Yr*G	ns	ns	ns	ns	ns	ns
N*G	ns	ns	ns	ns	ns	ns
N*G*Yr	ns	ns	ns	ns	ns	ns

[1] SLA = Speficic Leaf Area. [2] SLN = Specific Leaf Nitrogen. [3] Means in columns not sharing the same letters are significantly different according to LSD test ($p = 0.05$). [4] ns, *, **, and ***, not significant or significant at $p \leq 0.05$, $p \leq 0.01$, or $p \leq 0.001$, respectively.

3.4. Crop N Uptake, Nitrogen, and Nitrate Shoot Concentration

The highest N removal rates occurred starting from 25 days before harvest in both years (Figure 5) during the phase of rapid growth (Figure 2), when 68–74% of the total N uptake was detected. The daily crop N uptake (averaged over the years) peaked in the most fertilized plants of butterhead lettuce (2.7, 3.0, and 3.7 kg ha^{-1} d^{-1} with N$_0$, N$_{50}$, and N$_{100}$, respectively), while in the crisphead one it averaged 2.4 kg ha^{-1} d^{-1}.

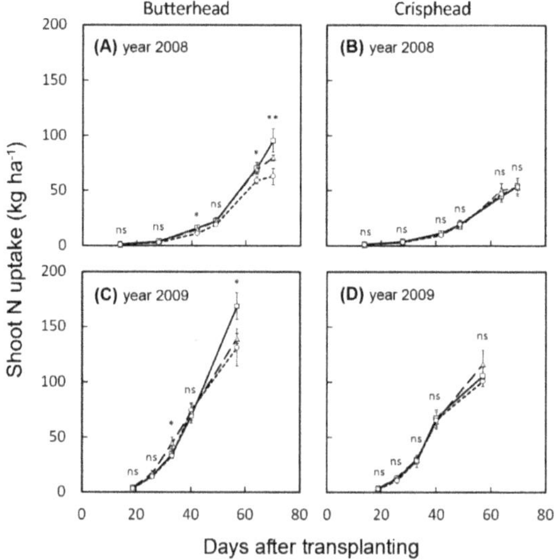

Figure 5. Changes in aboveground N uptake during the 2008 (**A** and **B**) and 2009 (**C** and **D**) crop cycles as affected by lettuce type and N level. Bars indicate ± SE of mean ($n = 4$). The notes: ns, *, **, indicate differences not significant or significant at $p \leq 0.05$, $p \leq 0.01$, respectively.

As a consequence, in the butterhead type the final N uptake rose from 97.0 in N_0 to 132.1 kg ha^{-1} in N_{100} treatment, while in the crisphead it was lower and almost unchanged between the three N levels (80.5 kg ha^{-1}, on average) (Table 3; Figure 6).

Table 3. Effect of N fertilization rate, lettuce genotype, and year of cultivation on N crop uptake, N tissue concentration, and nitrate content on a DW and a fresh weight (FW) basis.

Treatments	N Crop Uptake (kg ha^{-1})	Reduced N (g 100 g^{-1} N Uptake)	N Concentration (g kg^{-1} DW)	NO$_3$ (g kg^{-1} DW)	NO$_3$ (g kg^{-1} FW)
N rate (kg ha^{-1}) (N)					
0	87.0 b [1]	92.1 a	33.8 a	10.9 b	324 b
50	97.2 ab	91.5 b	33.2 a	11.9 ab	371 ab
100	106.0 a	90.4 b	33.6 a	13.3 a	424 a
Genotype (G)					
Butterhead	112.9 a	91.6 a	33.6 a	11.7 a	397 a
Crisphead	80.6 b	91.0 a	33.6 a	12.4 a	348 a
Year (Yr)					
2008	66.5 b	97.0 a	36.9 a	4.7 b	131 b
2009	127.0 a	85.7 b	30.3 b	19.2 a	615 a
Significance [2]					
N	***	***	**	***	***
G	*	*	ns	*	*
Yr	**	ns	ns	ns	ns
Yr*N	ns	ns	ns	ns	ns
Yr*G	ns	ns	ns	ns	ns
N * G	*	ns	ns	ns	ns
N *G*Yr	ns	ns	ns	ns	ns

[1] Means in columns not sharing the same letters are significantly different according to LSD test ($p = 0.05$). [2] ns, *, **, and ***, not significant or significant at $p \le 0.05$, $p \le 0.01$, or $p \le 0.001$, respectively.

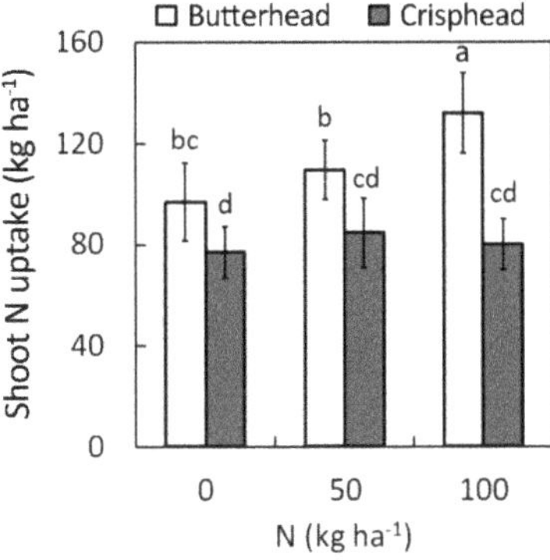

Figure 6. Effect of N fertilization rate and lettuce type on the shoot N uptake. Vertical bars (± SE of mean; *n* = 4) with different letters are significantly different according to the LSD test ($p = 0.05$).

On average, the fraction of reduced N over total N accumulated by the heads at the harvest decreased slightly in fertilized compared with unfertilized plants ($p \le 0.05$) (Table 3). Nitrogen fertilizer application did not affect final nitrogen concentration in lettuce tissues although it resulted in an increase ($p \le 0.05$) in nitrate concentration (both on a DW and a FW basis) passing from N_0 to N_{100}. Neither the concentration of total N and nitrate nor the incidence of reduced N on the total N taken up by the crop changed in the two lettuce types (Table 3). In 2009, crop N uptake was 48% higher than in the 2008 cycle ($p \le 0.001$), despite the total N concentration being lower ($p \le 0.01$). However, a greater nitrate concentration in lettuce tissues and a decrease in N-reduced percentage was observed (Table 3).

3.5. Nitrogen Fertilization Use Efficiency Indices

In both experiments, neither the agronomical N use efficiency (NUEa) nor its component, the apparent nitrogen recovery (REC), were affected by N fertilization and they were much greater in butterhead than in crisphead lettuce ($p \leq 0.05$) (Table 4). A significant interaction N*G was found for the second component of NUEa namely the physiological nitrogen use efficiency (NUEp). Irrespective of N fertilizer rate, lower NUEp values were detected in the crisphead type, while NUEp rose in the butterhead, particularly with the lower N rate (Figure 7). Partial factor productivity (PFP) was higher with the N_{50} than the N_{100} rate particularly in the 2009 cycle (Table 4), but it did not differ between lettuce types.

Table 4. Effect of year of cultivation, N fertilization rate, and lettuce genotype on N use efficiency indices.

Treatments	REC	NUEp	NUEa	PFP
	(kg kg^{-1})			
N rate (kg ha^{-1}) (N)				
50	0.20 a [1]	30.9 a	7.5 a	1696 a
100	0.19 a	21.8 a	6.2 a	869 b
Genotype (G)				
Butterhead	0.30 a	37.8 a	10.7 a	1326 a
Crisphead	0.09 b	14.8 b	2.9 b	1239 a
Year (Yr)				
2008	0.17 a	31.6 a	5.4 a	1033 b
2009	0.22 a	21.1 a	8.3 a	1532 a
Significance [2]				
N	ns	ns	ns	***
G	*	*	*	ns
Yr	ns	ns	ns	***
Yr*N	ns	ns	ns	*
Yr*G	ns	ns	ns	ns
N*G	ns	*	ns	ns
N*G*Yr	ns	ns	ns	ns

[1] Means in columns not sharing the same letters are significantly different according to LSD test ($p = 0.05$). [2] ns, * and ***, not significant or significant at $p \leq 0.05$ or $p \leq 0.001$, respectively.

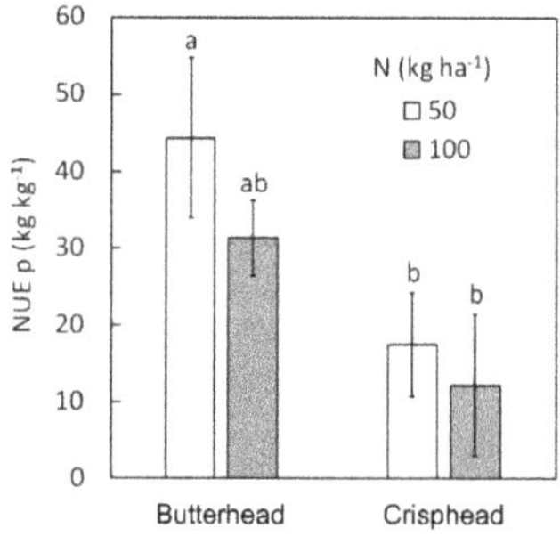

Figure 7. Physiological nitrogen use efficiency (NUEp) as affected by N levels and lettuce type. Vertical bars (± SE of mean; $n = 12$) with different letters are significantly different according to the LSD test ($p = 0.05$).

3.6. Critical Nitrogen Dilution Curves

The levels of critical N concentration (%Nc) were determined for each lettuce type during both cropping cycles following the procedure suggested by Justes et al. [18]. The critical dilution curves were applied for aboveground dry weight (DW) values ranging from 0.9 to 3.7 Mg ha^{-1} for the crisphead and from 0.9 to 5.7 Mg ha^{-1} for butterhead lettuce. This fraction of DW accumulation occurred in the last 25 days of the crop cycle (from about 40 to 63 DAT, averaged over both years). During this period, both growth (Figure 2) and LAI (Figure 4) exhibited a linear increase, suggesting that the competition for light among plants was already occurring with LAI values higher than 2. In the previous phase, N concentration rose from 2.7%, observed at 16 DAT, to 3.7%. The fitting of the power function %Nc = aDW^{-b} performed both on type-pooled %N data and on the type-specific ones showed a high goodness (Table 5). However, when the type-specific N critical values were considered, the fitting gave slight lower RMSE and RRMSE with adjR^2 showing the highest values especially in crisphead lettuce (Table 5). The type-specific power function for the butterhead (%N$_{BH}$) and crisphead (%N$_{CH}$) had a similar "a" coefficient, while the "b" parameter was statistically lower in butterhead than in crisphead lettuce (Table 5; Figure 8). In general, by evaluating the prediction performance of the N-critical functions proposed by Tei et al. [13] (%N$_{Tei}$) and by Bottoms et al. [14] (%N$_{Bottoms}$) against the %Nc observed data, the adjR^2 values were lower, and the RMSE and RRMSE higher than those obtained by type-specific modeling, particularly with the %N$_{Tei}$ modeling and for the butterhead type (Table 5, C–D).

Table 5. Indices of model performance evaluation, estimates and standard errors of parameters, after fitting the N power function (%N = a DW^{-b}) on the critical N concentration with the shoot dry weight. Our modeling fits were performed, considering the experimental %N data all together (A) or pooling them by lettuce type (B). The prediction performance of the Tei et al. [13] and Bottoms et al. [14] functions were also evaluated (C, D).

Model Evaluation	adj$R^{2\,(1)}$	RMSE	RRMSE	Function Parameters	
				a	b
(A) with all %N$_C$ data	0.835 **	0.11	3.17	3.778 ± 0.056	−0.152 ± 0.022
(B) by lettuce type					
Butterhead (%N$_{BH}$)	0.823 **	0.09	2.69	3.654 ± 0.067	−0.115 ± 0.024
Crisphead (%N$_{CH}$)	0.940 **	0.09	2.41	3.964 ± 0.145	−0.205 ± 0.041
Significance $^{(2)}$				ns	*
(C) using Tei et al. critical N					
Butterhead	0.806 **	0.31	8.45		
Crisphead	0.925 **	0.19	4.81		
(D) using Bottoms et al. equation					
Butterhead	0.789 *	0.18	5.01		
Crisphead	0.938 **	0.09	2.45		

$^{(1)}$ adjR^2 values with * and **, indicate that the ANOVA for the regression between the predicted and observed values is significant at $p \le 0.05$ or $p \le 0.01$, respectively. $^{(2)}$ ns and *, not significant or significant at $p \le 0.05$, respectively, between the lettuce types.

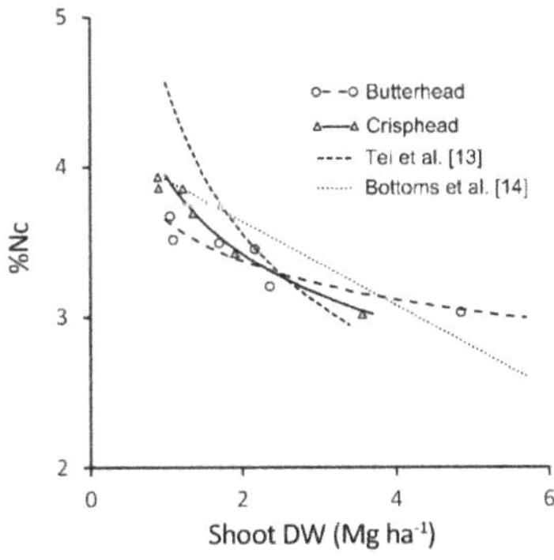

Figure 8. Plot of the critical dilution curves found for butterhead and crisphead lettuce over the 2008 and 2009 field trial data. The dilution N curves for lettuce as indicated by Tei et al. [13] and by Bottoms et al. [14] are also reported. Symbols: Observed data.

4. Discussion

4.1. Crop Growth and Production

Nitrogen effect on crop growth depends on the advantage in terms of carbon gain resulting from both improvement of light interception via leaf area growth and the enhancement of the photosynthesis rate at leaf level [24]. The latter is reported to be strictly linked to the specific leaf N (SLN) and chlorophyll content per unit leaf area [25,26].

In this research, the effect of N fertilization was related to lettuce typology, with a clear enhancement of growth in the butterhead and a negligible response in the crisphead (Figures 2 and 3A). By considering the potential effect of nitrogen on photosynthesis rate, it can be observed that in neither lettuce type did N fertilization affect the content either of the total (SLNtot) or the organic nitrogen (SLNred), which is directly related to the photosynthetic machinery [27]. Moreover, the SLA accounting for the amount of dry biomass accumulated per unit leaf area, remained unchanged in N-fertilized compared to unfertilized plants, and the chlorophyll content even showed a slight reduction when the largest N rate was applied (Table 2). Overall these results show that both in butterhead and crisphead lettuce, the photosynthesis rate remained fairly constant in N_0, N_{50}, and N_{100} leaves so it appears not to be involved in the growth response to N fertilization of the two lettuces. Consequently, it is reasonable to suppose that the observed difference in growth between fertilized butterhead and crisphead typology was mainly mediated by light interception level. Although the intercepted radiation has not been measured in this research, it is likely that leaf area expansion prompted by N fertilization (Figure 4; Table 2) resulted in an increase in light interception and so in growth, mainly in the butterhead type. Whereas, the lack of growth response in fertilized crisphead lettuce could have been caused by a very limited enhancement of the light interception due to the specific head shape of this type, where most of the leaves are in a hidden or occluded position with respect to the light. Regardless of the genotype features, the present data seem to be in agreement with results reported by other authors [3,28] who underline that N fertilization in lettuce affects leaf area, and hence the light interception, more than leaf photosynthesis.

Irrespective of N fertilization, butterhead lettuce showed higher efficiency in terms of dry biomass accumulation, as can also be inferred by comparing the N_0 plants (Figure 3A). In the butterhead type, the higher leaf number and expansion along with the greater SLN and level of chlorophylls may suggest a higher photosynthetic activity at the canopy level. Moreover, its greater root apparatus (higher root DW, higher root/shoot ratio) resulting in a larger acquisition of soil-N (Figure 6) contributes to explain the higher shoot dry weight accumulation as suggested by the findings of Kerbiriou et al. (2014) [29].

Crop productivity was not significantly enhanced by N fertilization, despite the positive evidence on butterhead growth (Figure 3A). However, in this type the better N crop nutrition improved the commercial quality due to the higher dry mass concentration of leaves (Figure 3B). In both years, the two lettuce types had a very similar yield (Table 1), since the higher water content (lower dry mass concentration) of the crisphead type compensated for its lower dry biomass accumulation (Figure 3B).

In the plant tissues, the fraction of total N content as organic form (reduced N) was very high even in fertilized plants (close to 91%, on average). Therefore, nitrate concentration, both on a dry (11.9 mg kg^{-1}) and fresh weight basis (373 mg kg^{-1}) (Table 3), was always far below the maximum limits imposed by the European Communities (EC) for lettuce grown in an open field and harvested from March to October (2.500 mg kg^{-1} of FW; EC Regulation No. 1258/2011) [30].

Better growth and yield performances were obtained in the second growing season, when temperatures were closer to the optimal values for lettuce [31] and the mean daily solar radiation was higher than in the 2008 cycle (22.5 versus 17.0 MJ m^{-2} day^{-1}) mainly due to the second trial being carried out later in the spring.

4.2. Plant N Uptake and Nitrogen Use Efficiency

The aboveground N uptake (Figure 5) closely followed the pattern of crop growth (Figure 2) showing that the highest N requirement of lettuce occurs in the last three weeks of the crop cycle as also reported by Sosa et al. [32] who observed 60% of the biomass and N accumulation in the 22 days before the harvest of lettuce. The nitrogen taken up by the butterhead plants was higher than that of the crisphead, particularly in the most fertilized ones (+65%) (Figure 6). However, in both lettuces the nitrate concentration detected by raising the N fertilization rate (Table 3) linkable to a decreasing trend in NUEp (Figure 7) can be considered very low. This suggests no large luxury N consumption even at the highest N rate.

The marginal increase in head dry biomass for each unit of nitrogen supplied (NUEa) was very low irrespective of N fertilizer rate (6.5 kg kg^{-1}, on average). It was clearly restricted by the low apparent N recovery (REC) with only 19% of the applied N recovered by the crop. REC remained unchanged with increasing N rate, suggesting that the highest N supply did not greatly exceed the crop N demand [33], also confirmed by the low increase in nitrate concentration in the N$_{100}$ plant tissues. The observed REC was comparable with that reported by Greenwood et al. [33] (~15%), and Bottoms et al. [14] (16%) for crisphead and romaine lettuce grown with much higher N rates (from 175 to 236 kg ha^{-1}) than the N fertilizer applied in this study. Di Gioia et al. [16] have reported a higher REC (32%) in lettuce for N rates ranging from 60 to 180 kg ha^{-1}, while Tei et al. [5,34] observed a decrease in REC from ~70 to ~35% when increasing the N rate from 50 to 200 kg ha^{-1}. The variability in the fraction of N-fertilizer taken up by lettuce could be explained by the changes in the contribution of sources of soil N other than the applied fertilizer (e.g., from organic matter mineralization). Bottoms et al. [14] found a very low REC (7%) with 150 kg ha^{-1} of N fertilization and a concentration of 20 mg kg^{-1} of native NO$_3$–N in the soil. The low REC values in our research could be due to the quite high N availability deriving both from soil organic matter and/or residues of the previous broccoli crop. It was confirmed by the high N uptake in unfertilized crop (87 kg ha^{-1}) which represented 80% of the N taken up by the fertilized ones. Vegetable crop residues are considered a potential major source of N for the subsequent crop as they often have a small harvesting index, with broccoli in particular leaving up to 180 kg ha^{-1} of N in the residues [35].

Furthermore, this study highlighted a clear difference in N-fertilizer use efficiency according to lettuce typology with noticeably low REC, NUEp, and so NUEa by the crisphead type (Table 4; Figure 8). All these results confirm the difference between the lettuce typologies in their ability to acquire/absorb N and in their use efficiency of the absorbed N for producing dry biomass due to the higher efficiency of both root and photosynthetic apparatus. Di Gioia et al. [16] have also reported differences in NUE indices between romaine (REC = 32%; NUEp = 22 kg^{-1} kg^{-1}; NUEa = 6.5 kg^{-1} kg^{-1}) and oak-leaf lettuce types (REC = 27%; NUEp = 15 kg^{-1} kg^{-1}; NUEa = 3.5 kg^{-1} kg^{-1}).

In our trials, the difference in climatic trend during the two trials did not affect NUEa, REC or NUEp (Table 4). However, the NUEp, averaged over years (27 kg kg^{-1}, Table 4), was much greater than that reported for fall–winter cycles in the same region (~14 kg^{-1} kg^{-1}; [16]), probably due to the higher global radiation in spring cycles compared with the fall–winter ones.

The efficiency with which N supply is converted into economic yield (PFP) is the most important index for growers because it integrates both indigenous and applied N uptake efficiency and a decrease with increasing fertilization rate is expected [36]. In these trials, PFP almost halved with 100 kg ha^{-1} of N (0.90 Mg kg^{-1}) compared with the lower N rate (1.70 Mg kg^{-1}) justifying the negligible response, in terms of fresh yield, to the increase in N fertilization observed in both cultivars.

4.3. Nitrogen Critical Curve and Critical Uptake

By considering the type-specific critical N dilution curve, the fitting was excellent, highlighting a faster reduction in plant N concentration during the cycle in the crisphead ($b = -0.205$) than in butterhead ($b = -0.115$) lettuce (Table 5; Figure 8). These results suggest that the butterhead type, consistently with its greater dry mass production (Figure 3) and N requirement (Figures 3 and 6), maintains its photosynthetically active compartment longer with higher N concentration than in the structural pool, as reported in [36].

The critical N dilution curve of Tei et al. [13] (%N_{Tei}), which was defined on a butterhead lettuce grown under sub-continental/continental climate, proved its limits in predicting the %Nc for this type, mainly due to the much higher values in the initial growth (Figure 8) confirming the results of Bottoms et al. [14]. Furthermore, although the %N_{Tei} covers the range of DW yield up to 3.4 Mg ha^{-1}, its trend over this threshold would underline a large underestimation of %Nc values compared to the %N_{BH} (Figure 8) due to the greater difference in *b*parameter (−0,357 versus −0.115). This result highlights a marked effect of the climate condition on the growth pattern of the buttered lettuce, confirming the need for the local re-calibration of the critical %N curve. A closer pattern was found by comparing the %N_{BH} and %N_{CH} with the empirical equation proposed by Bottoms et al. [13] (N‰ = 42-2.8 DW (Mg ha^{-1})) (%$N_{Bottoms}$) for romaine and crisphead lettuce grown under the Californian Mediterranean-like climate. In particular, for the crisphead lettuce, the %N_{CH} and %$N_{Bottoms}$ functions are mostly overlapping, with very similar adjR^2, RMSE, and RRMSE values of their predictions. For the butterhead type, the %$N_{Bottoms}$ equation gives a clear underestimation of N concentration, when DW is in the range between 4 and 6 Mg ha^{-1} (Figure 8), likely due to the differences in plant growth and physiology between butterhead and the lettuce typologies used by Bottoms et al.

In our case, based on the type-specific %Nc functions, the N demand ranged from 80 to 170 kg ha^{-1} of N for the butterhead, and from 50 to 110 kg ha^{-1} of N for the crisphead for sustaining a dry biomass production varying in the two seasons from 2.5 to 6.0 Mg ha^{-1} in the butterhead and from 1.4 to 4.0 Mg ha^{-1} in the crisphead lettuce (Figure 9A). Taking into consideration the fresh biomass production (yield) (Figure 9B), the N demand diverges more between the two types, with the butterhead type having a higher DW concentration than the crisphead.

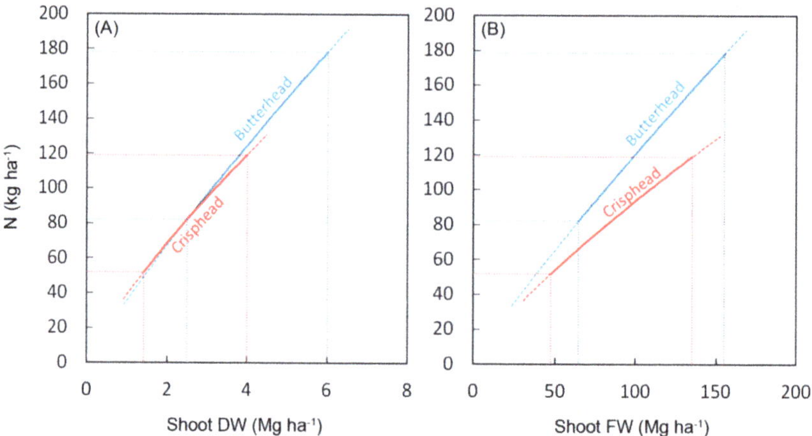

Figure 9. N demand (kg ha^{-1}) for butterhead and crisphead lettuce (**A**) as a function of dry biomass yield (in the range used for critical N curve validation—solid line) and (**B**) of the corresponding trend when using fresh yield as the independent variable.

5. Conclusions

The study provides evidence that growth and N uptake in lettuce are affected by genetic characteristics, with the butterhead type having a higher ability to uptake nitrogen and higher efficiency in using nitrogen for dry mass production than the crisphead one. The calibration of the specific N critical dilution curve performed for the two lettuce typologies may optimize their N nutrition, accounting for their own potential in dry mass production linked to the genetic characteristics and the interaction with climatic conditions.

Under the Mediterranean climate, the critical N dilution curves %Nc = 3.96 DW$^{-0.205}$ and %Nc = 3.65 DW$^{-0.115}$ are suggested for crisphead and butterhead lettuce, respectively. On an average, the optimal N uptake ranges from 80 kg ha^{-1} in crisphead to 125 kg ha^{-1} in butterhead lettuce to produce 2.5 and 4.3 Mg ha^{-1} of dry biomass, respectively, which correspond in terms of fresh biomass to an average yield of between 90 and 110 Mg ha^{-1} for both lettuce types.

Author Contributions: Conceptualization, G.C. and A.E.; Data curation, G.C. and A.E.; Formal analysis, G.C. and A.E.; Writing—original draft, G.C. and A.E.; Writing—review & editing, G.C. and A.E.

Funding: This research received no external funding.

Conflicts of Interest: The authors declare no conflict of interest.

References

1. FAOSTAT. Food and Agriculture Organization of the United Nations, Statistic Division. 2019. Available online: http://faostat3.fao.org/ (accessed on 5 August 2019).
2. Greenwood, D.J.; Cleaver, T.J.; Turner, M.K.; Hunt, J.; Niendorf, K.B.; Loquens, M.S. Comparison of the effects of nitrogen fertilizer on the yield, content, and quality of 21 different vegetable and agricultural crops. *J. Agric. Sci.* **1980**, *95*, 471–485. [CrossRef]
3. Broadley, M.R.; Escobar-Gutierrez, A.J.; Burns, A.; Burns, I.G. What are the effects of nitrogen deficiency on growth components of lettuce? *New Phytol.* **2000**, *147*, 519–526. [CrossRef]
4. Santamaria, P. Nitrate in vegetables: Toxicity, content, intake and EC regulation. *J. Sci. Food Agric.* **2006**, *86*, 10–17. [CrossRef]
5. Tei, F.; Benincasa, P.; Guiducci, M. Effect of nitrogen availability on growth and nitrogen uptake in lettuce. *Acta Hortic.* **2000**, *533*, 385–392. [CrossRef]
6. Elia, A.; Conversa, G. Agronomic and physiological responses of a tomato crop to nitrogen input. *Eur. J. Agron.* **2012**, *40*, 64–74. [CrossRef]
7. Elia, A.; Conversa, G. A decison support system (GesCoN) for managing fertigation in open field vegetable crops. Part I—Methodological approach and description of the software. *Front. Plant Sci.* **2015**, *6*, 319. [CrossRef] [PubMed]
8. Scholberg, J.; McNeal, B.L.; Jones, J.W.; Boote, K.J.; Stanley, C.D.; Obreza, T.A. Growth and canopy characteristics of field-grown tomato. *Agron. J.* **2000**, *92*, 152–159. [CrossRef]
9. Gastal, F.; Lemaire, G.; Durand, J.L.; Louarn, G. Quantifying crop responses to nitrogen and avenues to improve nitrogen-use efficiency. In *Crop Physiology—Applications for Genetic Improvement and Agronomy*, 2nd ed.; Sadras, V.O., Calderini, D.F., Eds.; Academic press/Elsevier: Burlington, MA, USA, 2015; pp. 161–206.
10. Greenwood, D.J.; Gastal, F.; Lemaire, G.; Draycott, A.; Millard, P.; Neeteson, J.J. Growth rate and %N of field grown crops: Theory and experiments. *Ann. Bot.* **1991**, *67*, 181–190. [CrossRef]
11. Lemaire, G.; Gastal, F. N uptake and distribution in plant canopies. In *Diagnosis of Nitrogen Status in Crops*; Lemaire, G., Ed.; Springer: Berlin, Germany, 1997; pp. 3–41.
12. Lemaire, G.; Jeuffroy, M.H.; Gastal, F. Diagnosis tool for plant and crop N status in vegetative stage. Theory and practices for crop N management. *Eur. J. Agron.* **2008**, *28*, 614–624. [CrossRef]
13. Tei, F.; Benincasa, P.; Guiducci, M. Critical nitrogen concentration in lettuce. *Acta Hortic.* **2003**, *627*, 187–194. [CrossRef]
14. Bottoms, T.G.; Smith, R.F.; Cahn, M.D.; Hartz, T.K. Nitrogen requirements and N status determination of lettuce. *HortScience* **2012**, *47*, 1768–1774. [CrossRef]
15. Conversa, G.; Santamaria, P.; Gonnella, M. Growth, yield, and mineral content of butterhead lettuce (*Lactuca sativa* var. capitata) grown in NFT. *Acta Hortic.* **2004**, *659*, 621–628. [CrossRef]
16. Di Gioia, F.; Gonnella, M.; Buono, V.; Ayala, O.; Santamaria, P. Agronomic, physiological and quality response of romaine and red oak-leaf lettuce to nitrogen input. *Ital. J. Agron.* **2017**, *12*, 47–58. [CrossRef]
17. Conversa, G.; Lazzizera, C.; Bonasia, A.; Elia, A. Growth, N uptake and N critical dilution curve in broccoli cultivars grown under Mediterranean conditions. *Sci. Hortic. (Amst.)* **2019**, *244*, 109–121. [CrossRef]

18. Bonasia, A.; Conversa, G.; Lazzizera, C.; Elia, A. Pre-harvest nitrogen and Azoxystrobin application enhances postharvest shelf-life in Butterhead lettuce. *Postharvest Biol. Technol.* **2013**, *85*, 67–76. [CrossRef]

19. Conversa, G.; Bonasia, A.; Lazzizera, C.; Elia, A. Bio-physical, physiological, and nutritional aspects of ready-to-use cima di rapa (*Brassica rapa* L. subsp. *sylvestris* L. Janch. var. *esculenta* Hort.) as affected by conventional and organic growing systems and storage time. *Sci. Hortic. (Amst.)* **2016**, *213*, 76–86. [CrossRef]

20. Justes, E.; Mary, B.; Meynard, J.M.; Machet, J.M.; Thelier-Huche, L. Determination of a critical nitrogen dilution curve for winter wheat crops. *Ann. Bot.* **1994**, *74*, 397–407. [CrossRef]

21. Lemaire, G.; Salette, J. Croissance estivale en matière sèche de peuplements de fétuque élevée (*Festuca arundinacea* Schreb) et de dactyle (*Dactylis glomerata* L.) dans l'ouest de la France: I. Etude en conditions de nutrition azotée et d'alimentation hydrique non limitantes. *Agronomie* **1984**, *7*, 381–389. [CrossRef]

22. SAS Institute. *SAS Version 8.02*; SAS Institute Inc.: Cary, NC, USA, 1999.

23. Moriasi, D.N.; Arnold, J.G.; Van Liew, M.W.; Binger, R.L.; Harmel, R.D.; Veith, T. Model evaluation guidelines for systematic quantification of accuracy in watershed simulations. *Trans. ASABE* **2007**, *50*, 885–900. [CrossRef]

24. Gastal, F.; Lemaire, G. N uptake and distribution in crops: An agronomical and ecophysiological perspective. *J. Exp. Bot.* **2002**, *53*, 789–799. [CrossRef]

25. Lemaire, G.; Oosterom, E.; van Sheehy, J.; Jeuffroy, M.H.; Massignam, A.; Rossato, L. Is crop N demand more closely related to dry matter accumulation or leaf area expansion during vegetative growth? *F. Crop. Res.* **2007**, *100*, 91–106. [CrossRef]

26. Pilbeam, D.J. The Utilization of Nitrogen by Plants: A Whole Plant Perspective. In *Annual Plant Reviews, Nitrogen Metabolism in Plants in the Post-Genomic Era*; Foyer, C.H., Zang, H., Eds.; Wiley-Blackwell: Chiechester, UK, 2011; Volume 42, pp. 305–352.

27. Evans, J.R. Developmental constraints on photosynthesis: Effects of light and nutrition. In *Photosythesis and the Environment*; Baker, N.R., Ed.; Kluwer: Dordrecht, The Netherlands, 1996; pp. 281–304.

28. De Pinheiro Henriques, A.R.; Marcelis, L.F.M. Regulation of growth at steady-state nitrogen nutrition in lettuce (*Lactuca sativa* L.): Interactive effects of nitrogen and irradiance. *Ann. Bot.* **2000**, *86*, 1073–1080. [CrossRef]

29. Kerbiriou, P.J.; Stomph, T.J.; Lammerts van Bueren, E.T.; Struik, P.C. Modelling concept of lettuce breeding for nutrient efficiency. *Euphytica* **2014**, *199*, 167–181. [CrossRef]

30. European Commission. Regulation of 2 December 2011 amending Regulation (EC) No. 1881/2006 as regards maximum levels for nitrates in foodstuffs, 1258/2011/EC. *Off. J. Eur. Union* **2011**, *L320*, 15–17.

31. Wittwer, S.H.; Honma, S. *Greenhouse Tomatoes, Lettuce and Cucumbers*; Michigan State University Press: Lansing, MI, USA, 1979; p. 225.

32. Sosa, A.; Padilla, J.; Ortiz, J.; Etchevers, J.D. Biomass Accumulation and its Relationship with the Demand and Concentration of Nitrogen, Phosphorus, and Potassium in Lettuce. *Commun. Soil Sci. Plant Anal.* **2012**, *43*, 121–133. [CrossRef]

33. Greenwood, D.J.; Burns, I.G.; Draycott, A.; Kubo, K. Apparent recovery of fertilizer N by vegetable crops. *Soil Sci. Plant Nutr.* **1989**, *35*, 367–381. [CrossRef]

34. Tei, F.; Benincasa, P.; Guiducci, M. Nitrogen fertilization of lettuce, processing tomato and sweet pepper: Yield, nitrogen uptake and the risk of nitrate leaching. *Acta Hortic.* **1999**, *506*, 61–67. [CrossRef]

35. De Neve, S. Organic matter mineralization as a source of nitrogen. In *Advances in Research on Fertilization Management of Vegetable Crops*; Tei, F., Nicola, S., Benincasa, P., Eds.; Springer: Cham, Switzerland, 2017; pp. 65–83.

36. Dobermann, A. Nitrogen use efficiency-state of the art. In Proceedings of the IFA International Workshop on Enhanced-Efficiency Fertilizers, Frankfurt, Germany, 28–30 June 2005; pp. 1–18.

Article

Sustainable and Profitable Nitrogen Fertilization Management of Potato

Anita Ierna [1],* and **Giovanni Mauromicale [2]**

1   Institute of BioEconomy, National Research Council (CNR-IBE) Via P. Gaifami 18, 95126 Catania, Italy
2   Department of Agriculture, Food and Environment (Di3A), University of Catania Via Valdisavoia 5, 95123 Catania, Italy; g.mauromicale@unict.it
*   Correspondence: anita.ierna@cnr.it; Tel.: +39-095-4783412

Received: 1 August 2019; Accepted: 17 September 2019; Published: 25 September 2019

Abstract: Nitrogen fertilization is indispensable to improving potato crop productivity, but there is a need to manage it suitably by looking at environmental sustainability. In a three-season experiment, we studied the effects of five nitrogen (N) fertilization rates: 0 (N0), 100 (N100), 200 (N200), 300 (N300) and 400 (N400) kg N ha^{-1} on crop N uptake, apparent nitrogen recovery efficiency (ANRE), tuber yield, nitrogen use efficiency (NUE), nitrogen uptake efficiency (NUpE), nitrogen utilization efficiency (NUtE) and agronomic nitrogen use efficiency (AgNUE) of five different potato cultivars: Daytona, Ninfa, Rubino, Sieglinde and Spunta. The economically optimum N fertilizer rates (EONFR) were also calculated. In seasons with high soil nitrogen availability for the crop (about 85 kg ha^{-1} of N), tuber yield increased only up to N100 and ANRE was about 50%; in seasons with medium (from 50 to 60 kg ha^{-1} of N) soil N availability, tuber yield increased up to N200 and ANRE was about 45%. Rubino and Sieglinde (early cultivars) responded for tuber yield only up to N100; Daytona, Ninfa, Spunta (late cultivars) up to N200, showing the highest values of NUE, NUpE, NUtE and AgNUE at N100. EONFR ranged from 176 to 268 kg ha^{-1} in relation to cultivar and season, but the reduction by 50% led to a tuber yield decrease of only around 16%. The adoption of cultivars characterized by high AgNUE at a low N rate and a soil nitrate test prior to planting, are effective tools to achieve a more sustainable and cost-effective nitrogen fertilization management.

Keywords: potato; nitrogen fertilization; environmental sustainability; cost-effective; nitrogen use efficiency; tuber yield; EONFR

---

1. Introduction

Potato is a very important crop in the Mediterranean basin, occupying an overall area of a little less than one million ha and producing 30 million tons of tubers [1]. In several countries such as Tunisia, Morocco, Egypt, Cyprus, Israel, Lebanon, Turkey, Spain and in southern Italy, potatoes are not grown in the usual cycle (spring–summer) owing to the high temperatures and considerable demand for irrigation water, but are largely grown in two offseason crops for early production: Winter–spring (planting from December to January and harvesting from March to early June) [2], and summer–autumn (planting in early September and harvesting from November to the end of January). Early potatoes, defined as "potatoes harvested before they are completely mature, marketed immediately after harvesting and whose skin can easily be removed without peeling" (United Nations Economic Commission for Europe of Geneva, Fresh Fruit and Vegetables-30/2001), are highly appreciated and are mainly exported to northern European countries, with considerable profit [3]. The substantial commercial value of the product and the intensive use of the land prompt farmers to supplement the potato crop with water and nutrients, which have undoubtedly been responsible for increased early potato yields in recent decades. As a consequence of low nitrogen (N) reserves and high mineralization potential in Mediterranean soils [4], N fertilization is considered indispensable to improve crop productivity. Indeed, N application has a substantial effect on the leaf area index (LAI) of potatoes by increasing both the rate of leaf expansion and the number of emerging leaves, and directly influences seasonal patterns of photon interception and crop production [5,6]. Because of the central role of this macronutrient in determining crop growth and yield capability, N fertilization of the early potato cultivation in the Mediterranean basin is excessive and often even irrational, with N rates higher than 600–700 kg ha^{-1} frequently being applied [7]. These rates are far greater than the usual crop N uptake, which for a tuber yield of about 20 t ha^{-1} is equal to about 100 kg ha^{-1} of N [7]. The excess nitrogen (N) not taken up by the crop remains in the soil profile and may be subject to losses by denitrification, volatilization, surface runoff and leaching to the groundwater, resulting in pollution of the environment [8]. This is favored by the high amounts of irrigation water applied, low efficiency of irrigation methods such as furrow or sprinkler [9] and by light-textured soils [10], common in early potato cultivation. The risk of pollution, as well as the fact that producing mineral N fertilizer is highly demanding in terms of fossil fuels, has increased the urgency for environmental care [11]. The Nitrate Directive 91/676/EEC [12] and the Water Framework Directive 2000/60/EC [13] are implementing a reduction in N supply to crops in Europe. The focus of agronomic research has therefore shifted from finding the optimum rate of input for maximizing tuber yield to how to make best use of the permitted maximum amount of the external supply of N [6]. Environmental losses of N from potato production systems are frequently high despite improvement in fertilizer N management practices. One approach to reducing environmental losses of N is to increase the nitrogen use efficiency (NUE) of the crop. Potato is characterized by a relatively low NUE ranging between 50% and 60% [14], due to it having a naturally shallow and poorly developed root system which is less efficient in taking up N than other crops such as wheat, maize or sugar beet [15]. On the other hand, with exaggerated N fertilization rates, the profit for the producer will also drop, as fertilizers are becoming more expensive [11]. Selection or identification of the N fertilizer rate is one of the most basic, yet most important decisions in managing N fertilizer [16]. For farmers, it has become more important to manage N fertilization in terms of providing a cost-effective yield, even if not necessarily the maximum possible yield, which can reduce environmental impact at the same time. As the N crop response is genotype-dependent [14], it would be useful to have this information on selected cultivars that may differ for biological, morphological and productive traits. With the exception of a few contributions [17–19], each investigating the effects of the nitrogen fertilization rate on the fate of N fertilizer, N uptake capacity and tuber yield in the Mediterranean environment, on one sole cultivar and for no more than two years, no attempts have focused to date on NUE and on defining economically optimum N fertilizer rates for early potato production. The goal of this work was, over a three-season period, (i) to evaluate the effects of different nitrogen fertilization rates on N uptake, tuber yield, nitrogen efficiency and ii) to determine the economically optimum N fertilizer rates in five different genotypes to achieve a more sustainable and cost-effective nitrogen fertilization management of early potato crops in a Mediterranean environment.

2. Materials and Methods

2.1. Site, Climate and Soil

Experiments were conducted in 2010, 2011 (season I and II, respectively) and 2014 (season III) at our experimental field, with a wheat pre-crop two years before, on the coastal plain, south of Siracusa (37°03′ N, 15°18′ E, 15 m a.s.l.), a typical area for potato cultivation in Sicily (South Italy). The climate is semi-arid Mediterranean, with mild winters, and commonly rainless springs. Frost occurrence is virtually unknown (only two events in 30 years). During the potato crop season for early production (from December–January to May), the mean maximum day temperatures and the mean minimum night temperatures of the 30-year period 1977–2006 were 15.4 and 7.1 °C in January, 16.2 and 7.6 °C in February, 17.7 and 8.8 °C in March, 20.2 and 10.9 °C in April, 24.3 and 14.4 °C in May, respectively. Rainfall over the same period averages about 180 mm (Figure 1). In the three seasons of the experiment, we used three adjoining plots in the same field. The soil, moderately deep, was classified as Calcixerollic Xerochrepts on the basis of the USDA Soil Taxonomy Classification [20]. At the start of the experiments, the soil characteristics analyzed in our laboratory were as follows: Sand (41%), silt (30%), clay (29%), limestone (4%), pH (7.9), organic matter (2.1%), total N (1.6‰), assimilable P_2O_5 (46 ppm), exchangeable K_2O (414 ppm).

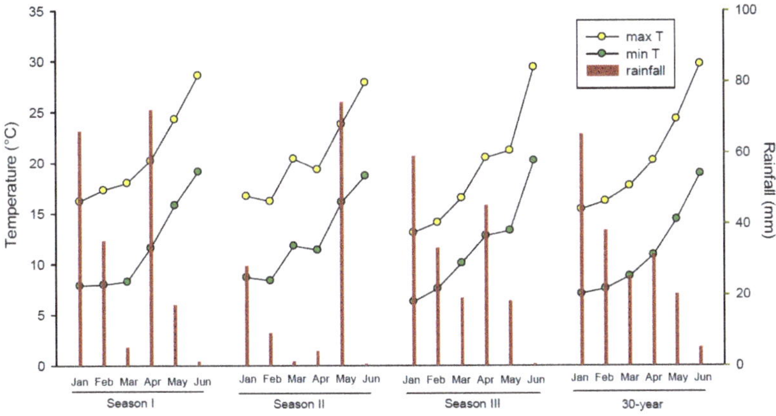

Figure 1. Average monthly maximum and minimum air temperatures and total monthly rainfall for the three seasons and 30-year period.

2.2. Experimental Design, Plant Material and Management Practices

The experiment (seasons I and II) was arranged in a randomized split-plot design with four replications including five nitrogen rates (0, 100, 200, 300, and 400 kg ha^{-1} referred after as N0, N100, N200, N300 and N400) as main plots and four cultivars of potato (*Solanum tuberosum* L.), e.g., Spunta, Sieglinde, Daytona, and Ninfa as sub-plots. In season III, the experiment was arranged in a split-plot design with four replications including five nitrogen rates (N0, N100, N200, N300 and N400) as main plots and two cultivars of potato (Rubino and Ninfa) as sub-plots. The five cultivars (Spunta, Sieglinde, Daytona, Ninfa and Rubino) utilized in this research differ for their morphological, biological, physiological and productive characteristics. Spunta and Sieglinde are widely cultivated in the Mediterranean region. Spunta is an early-medium ripening ware potato with long, regular, and very large tubers; plants produce few erect and vigorous stems and are well adapted to the Mediterranean climate, where they produce a high tuber yield [21]. Sieglinde is a firm flesh early cultivar with oblong, regular, and moderate-sized tubers; plants produce numerous stems of medium height, semi-erect and are moderately vigorous; they usually develop only limited biomass and deliver low tuber yield [2]. Daytona, Ninfa and Rubino are Italian cultivars bred within the Italian project "Breeding of Potato": Daytona was bred by Agenzia per la Sperimentazione Tecnologica e la Ricerca Agroambientale (ASTRA)—Innovation and Development (ex Mario Neri), Imola (Bologna), Ninfa and Rubino by CREA—Research Centre for Cereal and Industrial Crops, Bologna [22]. Daytona is a cultivar of medium to late maturity with short, oval, and regular tubers; stems are of medium size. Ninfa is a cultivar of medium to late maturity, with oblong, regular, and very large tubers; plants produce fairly tall and erect stems and provide marketable tuber yields superior to those of commercial cultivars frequently cultivated in southern Italy. Rubino is an early cultivar with oval and moderate-sized tubers; it was selected for earliness and suitability to early production. Whole virus-free seed-tubers were planted on February 3 (season I), on January 29 (season II) and on January 28 (season III). Plants emerged between 30 and 40 days after planting (DAP). In all experiments, the sub-plot size was 4.2 × 4.2 m, with 84 plants and consisted of six rows; tubers were planted at 0.3 intervals, in rows 0.7 m apart (equivalent to a planting density of 4.76 plants m^{-2}). In the three seasons, tillage consisted of a 40 cm depth ploughing followed by harrowing in October; at planting 100 kg ha^{-1} of P_2O_5 (as mineral superphosphate) and 150 kg ha^{-1} of K_2O (as potassium sulphate) were applied, whereas 50% of nitrogen (as ammonium nitrate) was supplied at complete crop emergence and the remaining 50% three weeks after as top dressing. Chlorpyrifos (30 kg ha^{-1}) was applied before planting; other standard crop management was applied, involving post-emergence weeding with linuron and pest control when needed. Crop water requirements were completely satisfied by drip irrigation, supplying 100% of crop maximum evapotranspiration, when the accumulated daily evaporation measured by class A pan evaporimeter reached 30–40 mm. Over the crop cycle, 197 (season I), 210 (season II) and 170 mm (season III) irrigation water were applied.

2.3. Data Collection and Calculations

2.3.1. SPAD Measurements

Leaf SPAD absorbance (correlated to chlorophyll content) was measured in the field using a portable Chl meter (SPAD 502, Minolta Camera, Osaka, Japan). Measurements were made on the distal leaflet of the youngest fully expanded leaf (usually the third or fourth leaf from the apex) between 11:00 and 13:00 (local solar time). Triplicate readings were taken from fully sun-exposed leaflets of 4 potato plants randomly sampled in four central rows of each sub-plot [23]. Between the 5th–6th leaf appearance and beginning of plant senescence, ten measurements were taken in season I, seven in season II and four in season III.

2.3.2. Plant Weight and Tuber Yield

When about 70% of leaves were dry (126, 121 and 120 DAP in season I, II and III, respectively), plants from central rows of each subplot were hand collected by removing an undisturbed soil sample. Plants were separated into aboveground biomass (stem + leaves), roots + stolons and tubers; roots, stolons and tubers were washed in gently running water. Tubers were classified in marketable (unitary weight > 20 g) and unmarketable (unitary weight < 20 g). All plant parts (marketable and unmarketable tubers, aboveground biomass and roots) were weighed separately to measure fresh weight. Marketable tubers were utilized to determine tuber yield. Three samples of about 50 g of all plant parts for each plot were oven-dried at 105 °C until constant weight and weighed to determine dry matter content.

2.3.3. Economically Optimum N Fertilizer Rate

To predict the economically optimum N fertilizer rates (EONFR), a quadratic equation model (SigmaPlot 11, Systat Software Inc.) described by Fontes et al. [24] and Belanger et al. [25] was utilized:

$$Y = b_0 + b_1N + b_2N^2 \tag{1}$$

where Y is the expected marketable fresh tuber yield expressed in kg ha^{-1}, N is applied fertilizer N expressed in kg ha^{-1}, and b_0, b_1 and b_2 are coefficients that are calculated from the experimental data. The EONFR, defined as the rate of N application where €1 of additional N fertilizer returned €1 of potatoes, was calculated as follows:

$$N_{op} = P - b_1/2b_2 \tag{2}$$

where N_{op} is the economically optimum application rate of fertilizer N expressed in kg ha^{-1}, P is the ratio of the cost of N fertilizer (€ 1.6 kg^{-1} N) to the selling price of potatoes (€ 0.25 kg^{-1} tuber), on the average of 2015 to 2016 [26], b_1 and b_2 as in Equation (1); this analysis assumes that fertilizer N costs are the only variable costs and that all other costs are fixed. According to Neeteson [27], EONFR was adjusted by considering the amount of available N in the soil at planting according to the following formula:

$$N_{op} = P - b_1/2b_2 - 0.7\,N_A \tag{3}$$

N_A represents the amount of available N in the 0–0.40 m soil layer at planting, which was the sum of mineral N (NO_3-N and NH_4-N) plus the N released through mineralization during the growing season, plus the N supplied through fertilization [28]. The initial mineral nitrogen content of the soil profile (0–0.40 m) was set at about 1% of total N, determined by the Kjeldahl method. N soil mineralization per month was calculated according to Gariglio et al. [29], from total N content corrected by an N mineralization factor as a function of soil temperature. On the basis of this procedure, the quantity of available mineral nitrogen in the soil for the crop cycle at planting was, regardless of cultivars, about 48 kg ha^{-1} in season I, 84 kg ha^{-1} in season II and 64 kg ha^{-1} in season III. In addition, four reduced rates (90, 75, 50 and 25%) of N_{op} were simulated and the relative yields decrease was calculated from the response quadratic curves.

2.3.4. Determination of Crop Nitrogen Content and Nitrogen Uptake

Nitrogen concentration in roots + stolons, marketable and unmarketable tubers and above-ground biomass was determined for each replication by dried materials collected at harvest, which was finely ground through a mill (IKA, Labortechnick, Staufen, Germany) with a 1.0 mm sieve. Nitrogen was determined by means of the Kjeldahl method (Kjeltec 2300 Auto Analyser; Foss-Tecator, Hilleroed, Denmark) [30]. The N content of each part of the plant was calculated as the product of the measured N concentration and dry weight (DW). Crop nitrogen uptake (CNU) was calculated as the sum of N contents of roots + stolons, marketable and unmarketable tubers and aboveground biomass.

2.3.5. Nitrogen Efficiency Indices

The efficiency of N fertilizer utilization was calculated using the following equation adapted from Vos [6]:

$$ANRE = (N_U - N_0)/N_F \ 100 \tag{4}$$

where ANRE is apparent nitrogen recovery efficiency expressed in %, N_U is the N uptake of the N-fertilized plot, N_0 is the uptake of the N-unfertilized plot (control), N_F is the amount of N applied by fertilization.

The inefficiency of N fertilizer utilization was calculated as follows:

$$NRI = 100 - ANRE \tag{5}$$

where NRI is nitrogen recovery inefficiency expressed in %.

Using the following equations adapted from van Bueren and Struik [8]:

$$NUE = Y_N/N_A \tag{6}$$

where NUE is nitrogen use efficiency expressed as kg ha^{-1} tuber DW kg ha^{-1} N, Y_N is marketable dry tuber yield, N_A as in Equation (3);

$$NUpE - N_U/N_A \tag{7}$$

where NUpE is nitrogen uptake efficiency expressed as kg ha^{-1}N kg^{-1} ha^{-1} N, N_U represents the amount of N uptake by the crop, N_A as in Equation (3);

$$NUtE = Y_N/N_U \tag{8}$$

where NUtE is nitrogen utilization efficiency expressed as kg ha^{-1} tuber DW kg ha^{-1} N, Y_N as in Equation (6), N_U as in Equation (7).

$$AgNUE = Y_N/N_F \tag{9}$$

where AgNUE is agronomic nitrogen use efficiency expressed as kg ha^{-1} tuber DW kg ha^{-1} N, Y_N as in Equation (6), N_F as in Equation (4).

2.4. Meteorological Data

Air temperature, relative humidity and rainfall were monitored during the experiments by a meteorological station (CR 21 data logger, Campbell Scientific, Inc., Utah, U.S.A.) sited at the experimental field. Measurements were made every 30 min.

2.5. Statistical Analysis

Data collected were first submitted to Bartlett's test to check the homoscedasticity, then analyzed using ANOVA [31]. A preliminary statistical analysis done for the same cultivars for season I and II showed a significant ($P = 0.001$) effect of interaction "season x nitrogen rate" for all parameters with the exception of the SPAD reading, indicating that the average response of 4 cultivars to the N rate was different for seasons I and II; the other preliminary statistical analysis made for Ninfa in all three seasons (I, II and III) showed a significant ($P = 0.001$) effect of interaction "season x nitrogen rate" for all parameters indicating that its response to N rate was different for the three seasons. Consequently, we analyzed each season's results separately, based on a factorial combination of "nitrogen rate × cultivar". Means were compared by a Least Significant Difference (LSD) test, when the F-test was significant. Table 1 shows the statistical significance from the analysis of variance for all studied variables separately for each season. CoStat Version 6.003 (CoHort Software, Monterey, CA, USA) was used. Polynomial effects up to the second degree were made where appropriate to define the linear or quadratic trend of N treatments and all studied parameters.

Table 1. Summary of statistical significance from analysis of variance for all studied variables: Crop nitrogen uptake (CNU), apparent nitrogen recovery efficiency (ANRE), nitrogen use efficiency (NUE), nitrogen uptake use efficiency (NUpE), nitrogen utilization use efficiency (NUtE), agronomical nitrogen use efficiency (AgNUE), in the three seasons; df indicates degree of freedom; **, *** indicate significant at $P \leq 0.01$, 0.001, respectively; NS = not significant.

Variable	Source of Variation	df	Season I	Season II	df	Season III
CNU	Nitrogen rate (N)	4	***	***	4	***
	cultivar (C)	3	***	***	1	**
	(N) × (C)	18	***	***	8	***
SPAD readings	Nitrogen rate (N)	4	***	***	4	***
	cultivar (C)	3	***	***	1	***
	(N) × (C)	18	***	***	8	**
ANRE	Nitrogen rate (N)	3	***	***	3	***
	cultivar (C)	3	***	***	1	***
	(N) × (C)	14	**	***	6	**
Tuber yield	Nitrogen rate (N)	4	***	***	4	***
	cultivar (C)	3	***	***	1	**
	(N) × (C)	18	***	**	8	***
NUE	Nitrogen rate (N)	4	***	***	4	***
	cultivar (C)	3	***	***	1	**
	(N) × (C)	18	***	**	8	***
NUpE	Nitrogen rate (N)	4	***	***	4	***
	cultivar (C)	3	***	***	1	NS
	(N) × (C)	18	***	**	8	NS
NUtE	Nitrogen rate (N)	4	***	***	4	**
	cultivar (C)	3	**	***	1	***
	(N) × (C)	18	NS	NS	8	***
AgNUE	Nitrogen rate (N)	3	***	***	3	***
	cultivar (C)	3	***	***	1	***
	(N) × (C)	14	**	**	6	***

2.6. Weather Conditions

The average monthly maximum and minimum temperatures from January to June were similar in the 3 seasons and to the 30-year (1977/2006) average, with the exception of March (Figure 1). In that month in season II, monthly maximum temperatures were 2.4 °C higher than in season I, 3.9 °C higher than season III and 2.7 °C higher than the 30-year average; minimum temperatures were higher by

3.5 °C compared to season I, by 1.7 °C compared to season III and by 3.0 °C compared to the 30-year average. The volume of rainfall from January to June and the distribution was similar in season I (196 mm) and season III (174 mm) and also with respect to the 30-year mean (184 mm); during season II rainfall was lower (116 mm) and was concentrated for about 65% in May, whereas it was absent in March and April (Figure 1).

3. Results

3.1. Crop Nitrogen Uptake, SPAD Readings, ANRE

In N0 plots, CNU, averaged over cultivars, was about 48 (season I), 84 (season II) and 64 (season III) kg ha^{-1} (Table 2). In N-fertilized plots, CNU increased linearly (all cultivars and seasons) and quadratically (all cultivars in season I and in season III) with the increase of the N rate (Table 2). Our results also indicate that CNU in relation to the N rate was cultivar-dependent. In fact, the highest increase in N uptake, increasing from N100 to N400, was found in Spunta (96 and 52 kg ha^{-1}, respectively in season I and II) and in Ninfa (104 and 91 kg ha^{-1} respectively in season I in season II); the lowest in Sieglinde (43 and 24 kg ha^{-1} respectively in season I in season II). In season III (in the same intervals) Rubino showed far less increases of N uptake (7 kg ha^{-1}) than Ninfa (36 kg ha^{-1}) (Table 2).

Table 2. Crop nitrogen uptake and SPAD readings as affected by "nitrogen rate x cultivar" interaction in the three seasons. Relationship tested by regression analysis, between N rate and responses of each variable and cultivar (L = linear, Q = quadratic; *, **, *** indicate significance at $P \leq 0.05$; $P < 0.01$; $P \leq 0.001$.

Season	N Rate	Crop Nitrogen Uptake (kg ha^{-1})				SPAD Readings (units)			
		Spunta	Sieglinde	Daytona	Ninfa	Spunta	Sieglinde	Daytona	Ninfa
I	N0	43.3	47.4	52.2	48.9	36.0	32.9	33.1	34.0
	N100	113.2	111.4	115.8	114.3	40.0	35.1	36.9	36.1
	N200	158.4	137.4	160.4	163.7	41.1	37.1	39.0	38.0
	N300	185.9	156.4	175.1	171.1	44.0	37.0	39.8	38.9
	N400	209.0	154.5	190.0	217.9	45.1	39.1	39.9	39.9
	L	***	***	***	***	***	***	***	
	Q	***	***	***	***			***	**
	LSD inter. ($P \leq 0.05$) **15.9**						**1.3**		
II	N0	74.9	83.5	90.9	88.1	38.6	35.7	36.0	33.3
	N100	116.0	131.3	156.1	133.0	39.3	38.6	40.0	36.7
	N200	107.7	115.3	162.8	145.1	41.2	39.3	40.6	38.9
	N300	166.2	126.8	177.0	159.0	43.0	40.0	41.3	40.0
	N400	168.0	154.9	201.3	198.8	44.6	39.0	40.7	40.7
	L	***	***	***	***	***	*	***	***
	Q						*	**	*
	LSD inter. ($P \leq 0.05$) **27.7**						**1.2**		
III				Rubino	Ninfa			Rubino	Ninfa
	N0			67.1	60.5			32.4	34.8
	N100			106.4	115.2			37.9	39.3
	N200			105.9	128.3			38.4	40.9
	N300			106.6	123.6			40.7	41.8
	N400			113.7	151.0			38.1	41.9
	L			***	***				*
	Q			**	***				
	LSD inter. ($P \leq 0.05$) **9.7**						**1.2**		

Chlorophyll meter readings, measured by SPAD-502, increased with increase of the nitrogen rate (Table 2). Generally, the major increases in SPAD units, rising from N100 to N400, were found in Spunta and Ninfa, and less so in Sieglinde and Rubino, confirming what was found for CNU.

ANRE, which expresses the proportion of N applied taken up by the plants, generally showed higher values in season I (50%) than in season II (28%) and season III (26%) (Table 3).

Table 3. Apparent nitrogen recovery efficiency (ANRE) and tuber yield as affected by "nitrogen rate x cultivar" interaction in the three seasons. Relationship tested by regression analysis, between N rate and responses of each variable and cultivar (L = linear, Q = quadratic; *, **, *** indicate significance at $P \le 0.05$; $P \le 0.01$; $P \le 0.001$.

Season	N Rate	ANRE (%)				Tuber Yield (t ha^{-1})			
		Spunta	Sieglinde	Daytona	Ninfa	Spunta	Sieglinde	Daytona	Ninfa
I	N0	-	-	-	-	14.3	14.5	19.4	20.5
	N100	80	64	64	65	35.9	31.1	35.9	37.0
	N200	62	45	54	57	46.7	35.3	43.5	47.4
	N300	51	36	41	41	48.7	38.1	41.1	46.7
	N400	44	27	34	42	48.5	35.1	39.0	46.2
	L	***	***	***	***	***	***	***	***
	Q					***	***	***	***
	LSD inter. ($P \le 0.05$) **6.0**					**5.0**			
II	N0	-	-	-	-	35.7	25.3	43.8	40.6
	N100	41	48	65	45	47.7	36.8	50.8	47.5
	N200	30	16	36	28	47.6	33.8	55.6	50.7
	N300	16	14	29	24	53.3	32.8	58.0	48.5
	N400	23	18	28	28	54.6	35.2	61.8	48.2
	L	*	**	***	**	***		**	
	Q	*	**	*	*				*
	LSD inter. ($P \le 0.05$) **11.2**					**5.4**			
III				Rubino	Ninfa			Rubino	Ninfa
	N0			-	-			19.7	16.3
	N100			39	55			28.7	32.9
	N200			19	34			30.9	43.0
	N300			13	21			25.0	40.1
	N400			12	23			28.3	33.6
	L			***	***			*	***
	Q			***	***			**	***
	LSD inter. ($P \le 0.05$) **3.0 3.7**								

As shown in Table 3, ANRE linearly (all cultivars and seasons) and quadratically (all cultivars in season II and in season III) decreased with the increasing nitrogen rate. In season I and II, the decrease in ANRE from N100 to N400 was more pronounced in Sieglinde (−58% and −62%, respectively in season I and II) than in Spunta (−45% and −44%, respectively), Daytona (−47% and −57%, respectively) and Ninfa (−35% and −38%, respectively); in season III the decrease was more pronounced in Rubino (−69%) than in Ninfa (−58%).

NRI (see Equation (4)), which represents the possible environmental impact, increased dramatically with the increasing N rate reaching N400, regardless of cultivars, and with values of about 63% in season I, 76% in season II and 83% in season III (data not shown).

3.2. Tuber Yield and Economically Optimum N Fertilizer Rate

In unfertilized N_0 plots, marketable tuber yield was, averaged over cultivars, 17.2 t ha^{-1} (season I), 36.3 t ha^{-1} (season II) and 18.0 t ha^{-1} (season III) (Table 3). In N fertilized plots, tuber yield increased linearly and quadratically in all cultivars (season I and III) with the increase of the N rate (Table 3). In season II the increase was linear in Spunta and Daytona, whereas it was quadratic in Ninfa (Table 3). Sieglinde (season I and II) and Rubino (season III), responded noticeably only up to N100, whereas Ninfa in the three seasons responded up to N200. Spunta and Daytona in season I responded significantly up to N200, whereas in season II were able to exploit higher doses of N fertilizers (300 kg ha^{-1}).

EONFR for each cultivar and season are reported in Table 4. For cultivar Ninfa, the sole cultivar to be used in all three seasons, EONFR changed substantially over the three seasons, showing a clear increasing trend (from 176 (season II), to 197 (season III), to 254 (season I) kg N ha^{-1}) with decreasing soil N mineral availability reserves (from 84 to 64, to 48 kg N ha^{-1}, respectively).

Table 4. Quadratic equations, the economically optimum nitrogen fertilizer rate (EONFR) and corresponding tuber yields in relation to cultivar and season.

Cultivar	Season	Quadratic Equation	R	EONFR (kg N ha^{-1})	Tuber Yield (t ha^{-1})
Spunta	I	$Y = 15094 + 230.9\,x - 0.374\,x^2$	0.995	268	50.1
Sieglinde	I	$Y = 15380 + 164.2\,x - 0.290\,x^2$	0.989	241	38.1
Daytona	I	$Y = 20157 + 179.2\,x - 0.337\,x^2$	0.986	227	43.5
Ninfa	I	$Y = 20897 + 189.9\,x - 0.322\,x^2$	0.993	254	48.3
Ninfa	II	$Y = 41031 + 72.8\,x - 0.141\,x^2$	0.963	176	49.4
Ninfa	III	$Y = 16363 + 210.9\,x - 0.423\,x^2$	0.995	197	41.5
Rubino	III	$Y = 21034 + 69.2\,x - 0.139\,x^2$	0.773	181	29.0

However, a decrease in EONFR led to a much less than proportional tuber yield decrease (Figure 2); applying 90% of EONFR, the yield decreased by only about 2%, applying 75% of EONFR, the yield decreased by only about 6%; applying 50% of EONFR, yield decreased by only about 6% in Ninfa in season II, 10% in Rubino in season III and about 19% on average in the other cultivars.

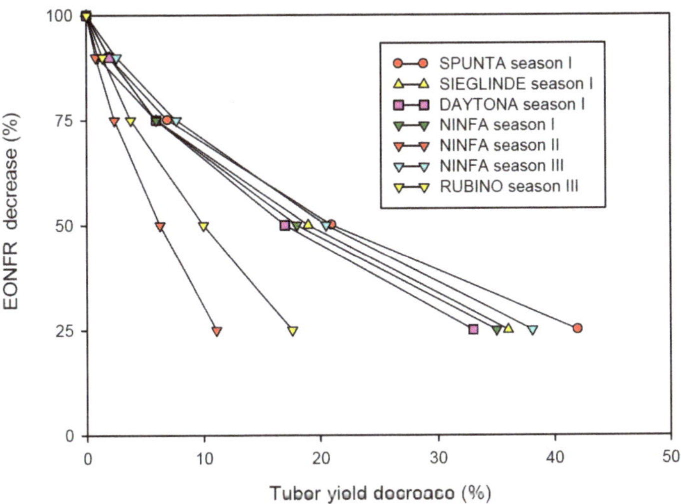

Figure 2. Tuber yield as affected by economically optimum nitrogen fertilization rate in relation to cultivar and season.

3.3. Nitrogen Efficiency Indices

The four nitrogen efficiency indices studied (NUE, NUpE, NUtE, AgNUE) tend to decline linearly and quadratically with increasing N application rates (Table 5; Table 6). The magnitude of this decline was generally genotype-dependent as demonstrated by the significance (nine cases out of 12) of the "nitrogen rate x cultivar" interaction. Sieglinde, compared to Daytona and Ninfa, showed a less evident decrease in NUE with increasing N rates up to N100 (season I) and up to N200 (season II); Rubino, in season III showed a more drastic decrease than Ninfa.

Table 5. NUE (nitrogen use efficiency) and NUpE (nitrogen uptake efficiency) as affected by "nitrogen rate x cultivar" interaction in the three seasons. Relationship tested by regression analysis, between N rate and responses of each variable and cultivar (L = linear, Q = quadratic; *, **, *** indicate significance at $P < 0.05$; $P < 0.01$; $P < 0.001$.

Season	N Rate	NUE (kg Tuber DW kg N^{-1})				NUpE (kg N kg N^{-1})			
		Spunta	Sieglinde	Daytona	Ninfa	Spunta	Sieglinde	Daytona	Ninfa
I	N0	68.2	83.0	112.3	101.8	0.64	0.91	1.00	0.94
	N100	52.5	45.2	57.4	50.3	0.74	0.73	0.76	0.75
	N200	36.4	30.4	41.1	37.0	0.63	0.55	0.64	0.65
	N300	25.3	22.8	25.8	23.4	0.53	0.44	0.50	0.49
	N400	20.9	16.5	20.7	18.6	0.46	0.34	0.42	0.48
	L	***	***	***	***	***	***	***	***
	Q	**	***	***	***	**		*	
	LSD inter. ($P \leq 0.05$)			**14.4**				**0.09**	
II	N0	78.0	63.2	111.5	102.6	0.81	0.90	0.98	0.95
	N100	49.6	44.8	62.4	55.9	0.60	0.68	0.81	0.56
	N200	32.2	25.7	42.6	39.6	0.37	0.39	0.56	0.50
	N300	26.8	18.7	30.6	28.0	0.42	0.32	0.45	0.40
	N400	22.4	15.1	24.3	25.1	0.34	0.31	0.41	0.40
	L	***	***	***	***	***	***	***	***
	Q	***	**	***	**	**	**	*	***
	LSD inter. ($P \leq 0.05$)			**12.7**				**0.1**	
III				Rubino	Ninfa			Rubino	Ninfa
	N0			51.5	49.8			1.00	0.93
	N100			28.7	36.9			0.63	0.67
	N200			17.8	25.7			0.39	0.47
	N300			11.3	16.3			0.29	0.34
	N400			10.1	12.3			0.24	0.32
	L			***	***			***	***
	Q			***				***	***
	LSD inter. ($P \leq 0.05$) **3.1**								

Table 6. NUtE (nitrogen utilization efficiency) and AgNUE (agronomical nitrogen use efficiency) as affected by the "nitrogen rate x cultivar" interaction in the three seasons. Relationship tested by regression analysis, between N rate and responses of each variable and cultivar (L = linear, Q = quadratic; *, **, *** indicate significance at $P < 0.05$; $P < 0.01$; $P < 0.001$.

Season	N Rate	NUtE (kg Tuber DW kg N^{-1})				AgNUE (kg Tuber DW kg N^{-1})			
		Spunta	Sieglinde	Daytona	Ninfa	Spunta	Sieglinde	Daytona	Ninfa
I	N0	81.5	70.4	87.3	82.1	-	-	-	-
	N100	64.7	56.6	69.2	62.0	73.3	63.1	80.1	70.2
	N200	55.1	53.6	61.5	54.1	43.7	36.3	49.2	44.2
	N300	46.4	49.7	50.4	47.1	28.7	25.8	29.1	26.4
	N400	44.2	47.0	47.9	37.5	23.0	18.1	22.8	20.3
	L	***	***	***	***	***	***	***	***
	Q	***			*	***	***	***	***
	LSD inter. ($P \leq 0.05$)					**6.4**			
II	N0	100.8	72.0	115.1	108.5	-	-	-	-
	N100	82.4	65.5	77.3	100.9	95.8	86.5	120.4	107.9
	N200	87.1	65.9	77.1	80.3	47.1	37.7	62.3	58.0
	N300	63.5	57.7	67.6	69.1	35.1	24.5	40.1	36.7
	N400	65.6	48.2	59.8	62.2	27.6	18.6	30.0	31.0
	L	**	*	***	***	***	***	***	***
	Q	***			*	***	***	***	***
	LSD inter. ($P \leq 0.05$)					**11.0**			
III				Rubino	Ninfa			Rubino	Ninfa
	N0			48.6	54.3			-	-
	N100			48.1	60.8			47.9	58.9
	N200			45.9	70.1			23.8	33.3
	N300			38.3	67.0			13.8	19.6
	N400			41.2	46.4			11.9	14.1
	L			*				***	***
	Q				***			***	***
	LSD inter. ($P \leq 0.05$) **8.2 3.4**								

Values of AgNUE reported in Table 6, show how in seasons I and II, Daytona proved to be the most efficient in the productive use of nitrogen supplied, while Sieglinde resulted in being the least effective; in season III, Rubino showed lower values of AgNUE than Ninfa at all N rates.

4. Discussion

Nitrogen fertilization has proved an effective means of improving tuber yield in the Mediterranean environment. Although yields responded positively to N application, there was no significant and consistent response of potato crops to varying nitrogen levels above 100 or to the maximum of 200 kg ha^{-1} of the N rate, depending on season and cultivar. In similar Mediterranean environments [7,18], potato yields increased with increasing nitrogen rate up to 120 kg ha^{-1}, but did not change further with higher rates. Even if increasing the nitrogen rate further does not lead to greater tuber yield, it did result in an increase of crop nitrogen uptake and chlorophyll meter readings, measured by SPAD-502, which are considered a promising tool to assess the N status of the potato crop [23,32]. The increase in crop nitrogen uptake was linear in agreement with Vos [6] and/or quadratic as highlighted by Darwish et al. [18] and Badr et al. [19]. In this research, values of crop nitrogen uptake were found to be lower than those in other Mediterranean environments at equal doses of N fertilizers applied [18,19]. This is mainly attributable to the fact that we distributed nitrogen top-dressed in the solid state (as is usually applied), whereas these researchers used fertigation, which is known to enhance N recovery and N use efficiency [17,18]. Increasing nitrogen rates resulted in a marked decrease of nitrogen recovery efficiency (ANRE), in agreement with other authors [18,19,33] and in a decrease of all nitrogen efficiency indices studied, confirming the trends reported by literature [8,16,18,19,24,34]. On average, passing from N100 to N200, the yield increased from 38.4 to 43.4 t ha^{-1}, while the ANRE decreased from 57% to 38%; by further augmenting the N fertilizer the yield remained constant, whereas the ANRE dropped drastically until reaching the maximum dose studied (N400) values of about 28%. This means that applying N400, which is very close to the conventional N fertilizer application dose, a significant amount of fertilizer, about 290 kg ha^{-1} of N, remained not up-taken by the crop and unused in the soil. Only a small part of the N given in excess carries over to the succeeding crops, whereas most of fertilizer N applied to potato is presumably lost over summer by volatilization (N_2O and NH_3) and in autumn, when rainfall exceeds evapotranspiration, by leaching of NO_3 and becomes a risk especially for groundwater and watercourses. ANRE values of about 60% found in N100 plots are in agreement with those found in semi-arid regions [18] and in temperate areas [33], while in more arid areas, such as Turkey and Jordan, values < 40% were found [10,17]. Greenwood and Drycott [34] attributed lower values of ANRE in potato crop, compared to other crops like cereals and grasses, to its lower root density, which causes some of the N fertilizer applied to potato to be remote from the roots for a considerable time before being absorbed. The theoretical economically optimum N fertilizer rates were quite high, ranging from 176 to 268 kg ha^{-1} of N in relation to cultivar and season. This is mainly due to the use of the quadratic model for the calculation of EONFR, which tends to overestimate them [35]. However, the quadratic model was chosen because it proved the most suitable for predicting EONFR because it minimizes the risks of potential economic losses in relation to the cost of the fertilizer and sell price of potatoes [25]. Nonetheless, considering that a decrease in EONFR led to a much less than proportional yield decrease; for example, applying 50% of EONFR, yield decreased by only about 16%, indicating how also from a cost viewpoint, it is possible to reduce currently excessive applications. The response of crop to N fertilization rates was season-dependent. Differences among seasons may largely be attributable to weather conditions, in particular to rainfall occurring before planting. Indeed, very high rainfall in the three months before planting in season I (with a peak of 388 mm in November) and in season III (peak of 350 mm in September) were recorded compared to season II, in which the rains did not exceed 70 mm monthly (data not shown). The high and concentrated rains in the autumn of season I and III probably favored NO_3 leaching, leaving less N availability for the crop in the soil (48 and 64 kg ha^{-1} in season I and III compared to 84 kg ha^{-1} of N in season II). This significantly affected production response of the crop. In unfertilized plots, tuber yields were in fact only about 17.0 t ha^{-1} in season I, 18.0 t ha^{-1} in season III compared to 36.0 t ha^{-1} in season II, in agreement with Greenwood and Draycott [34] and Rodrigues et al. [36], who found crop response to N rate depends on soil N availability at preplant. Moreover, in fertilized plots, tuber yield response to nitrogen fertilization was up to N200 in season I and III and only up to N100 in season II. In addition, the agronomic response to nitrogen seems to depend largely on soil nitrogen availability, as demonstrated by higher values in all cultivars of the agronomic use efficiency in season II, compared to season I and season III. The EONFR for cultivar Ninfa, the sole cultivar to be used in all three seasons, showed a clearly increasing trend with decreasing soil N mineral availability reserves. Therefore, in the Mediterranean environment, characterized by variability in the amount and distribution of autumn rains, nitrogen fertilization should be commensurate to rainfall. This also suggests the importance of carrying out the soil N concentration test before potato planting, already recommended in some cases to predict the fertilizer N rate in other crops [16]. Our results also indicate that crop response in relation to N rate was cultivar-dependent. Generally, the early cultivars Sieglinde (season I and II) and Rubino (season III), compared to the medium or late cultivars Spunta, Daytona and Ninfa, with an increasing N rate from N100 to N400 showed less increase in plant N uptake and SPAD units and a more pronounced decrease in nitrogen recovery efficiency. Furthermore, they showed less ability to use the soil N available (N residual + N fertilizers) for production of tuber dry matter (NUE), due to both their generally lower removal efficiency of available N (NUpE) and by less efficiency of N taken up to produce yield (NUtE). The lower values of NUpE of the early compared to the late cultivars may be due to the smaller size of the root system [37], whereas the lower NUtE values can be attributed to the shorter crop cycle, lower canopy size (lower % soil coverage), and lower photosynthesis activity [8]. Our previous research has shown how Sieglinde usually develops only limited biomass and delivers low tuber yield [2]. The specific literature reports that differences in NUE among potato cultivars are attributed to their different earliness [14,16,38]. Sieglinde and Rubino also proved less efficient in the productive use of nitrogen supplied than Spunta, Daytona and Ninfa, and responded for tuber yield markedly only up to N100, whereas Spunta, Daytona and Ninfa, were able to exploit higher doses of N fertilizers up to N300 in some cases. Van Bueren and Struik [8], studying a large set of cultivars, found that the majority of genotypes performing well under low N and showing a good response to N were late. Therefore, in consideration of the fact that crop nitrogen uptake generally grows only up to N100 in early cultivars and up to N400 in late cultivars, these last ones have luxury consumption of nitrogen fertilizers in plots supplied with higher N rate (> 200 kg ha^{-1}).

5. Conclusions

This experiment demonstrated that the potato crop, despite the variability between the seasons and the cultivars, benefited only from up to 100, and at most up to 200 kg ha^{-1}, of nitrogen, namely much lower rates than those usually supplied. The variability of the response to N supplied (100 or 200 kg ha^{-1}) found between the years seems to be due to variations in soil available N for the crop over the years. This suggests it is advisable to carry out the soil N concentration test before planting. The early cultivars like Rubino and Sieglinde responded well only up to 100 kg ha^{-1} of nitrogen, the late cultivars Spunta, Daytona and Ninfa up to 200 kg ha^{-1} of nitrogen, also showing, under low N, high agronomic use efficiency. Furthermore, the theoretical economically optimum N fertilizer rates, ranging from 176 to 268 kg ha^{-1} of N in relation to cultivar and season, could be halved without suffering any major yield reduction. Our results can be used to optimize and thus reduce nitrogen fertilization, thereby making savings for the farmer and ensuring a more environmentally-friendly crop in the Mediterranean.

Author Contributions: In this research article, authors contributions as the following: A.I.: data curation, formal analysis, investigation, methodology, writing-original draft, writing-review and editing; G.M.: writing-review and editing, visualization, supervision.

Funding: This research received no external funding.

Conflicts of Interest: The authors declare no conflicts of interest.

References

1. Faostat. *Food and Agriculture Data*; Food and Agricultural Organization: Rome, Italy, 2017; Available online: http:www//fao.org/faostat/en/#data/QC (accessed on 21 June 2018).
2. Ierna, A.; Mauromicale, G. Tuber yield and irrigation water productivity in early potatoes as affected by irrigation regime. *Agric. Water Manag.* **2012**, *115*, 276–284. [CrossRef]
3. Ierna, A. Tuber yield and quality characteristics of potatoes for off-season crops in a Mediterranean environment. *J. Sci. Food Agric.* **2010**, *90*, 85–90. [CrossRef] [PubMed]
4. Ierna, A.; Parisi, B. Crop growth and tuber yield of "early" potato crop under organic and conventional farming. *Sci. Hortic.* **2014**, *165*, 260–265. [CrossRef]
5. Vos, J.; Van der Putten, P.E.L. Effect of nitrogen supply on leaf growth, leaf nitrogen economy and photosynthetic capacity in potato. *Field Crop Res.* **1998**, *59*, 63–72. [CrossRef]
6. Vos, J. Nitrogen responses and nitrogen management in potato. *Potato Res.* **2009**, *52*, 305–317. [CrossRef]
7. Ierna, A.; Mauromicale, G. Potato growth, yield and water productivity response to different irrigation and fertilization regimes. *Agric. Water Manag.* **2018**, *201*, 21–26. [CrossRef]
8. Van Bueren, E.T.L.; Struik, P.C. Diverse concepts of breeding for nitrogen use efficiency. A review. *Agron. Sustain. Dev.* **2017**, *37*, 50. [CrossRef]
9. Foti, S.; Mauromicale, G.; Ierna, A. Influence of irrigation regimes on growth and yield of potato cv. Spunta. *Potato Res.* **1995**, *38*, 307–317. [CrossRef]
10. Halitligil, M.B.; Akin, A.; Ylbeyi, A. Nitrogen balance of nitrogen-15 applied as ammonium sulphate to irrigated potatoes in sandy textured soils. *Biol. Fertil. Soils* **2002**, *35*, 369–378.
11. Spiertz, J.H.J. Nitrogen, sustainable agriculture and food security: A review. *Agron. Sustain. Dev.* **2010**, *30*, 43–55. [CrossRef]
12. European Economic Community. Council Directive 91/676/EEC. 1991. Protection of Waters Against Pollution Caused by Nitrates from Agricultural Sources. Available online: https://www.eea.europa.eu/policy-documents/council-directive-91-676-eec/ (accessed on 21 December 2018).
13. European Commission. Directive 2000/60/EC. 2000. Framework for Community Action in the Field of Water Policy. Available online: http://www.ec.europa.eu/environment/water/water-framework/ (accessed on 13 December 2018).
14. Swain, E.Y.; Rempelos, L.; Orr, C.H.; Hall, G.; Chapman, R.; Almadni, M.; Stockdale, E.A.; Kidd, J.; Leifert, C.; Cooper, J.M. Optimizing nitrogen use efficiency in wheat and potatoes: Interactions between genotypes and agronomic practices. *Euphytica* **2014**, *199*, 119–136. [CrossRef]
15. Wishart, J.; George, T.S.; Brown, L.K.; Ramsay, G.; Bradshaw, J.E.; White, P.J.; Gregory, P.J. Measuring variation in potato roots in both field and glasshouse: The search for useful yield predictors and a simple screen for root traits. *Plant Soil* **2013**, *368*, 231–249. [CrossRef]
16. Zebarth, B.J.; Tai, G.; Tarn, R.; De Jong, H.; Milburn, P.H. Nitrogen use efficiency characteristics of commercial potato cultivars. *Can. J. Plant Sci.* **2004**, *84*, 589–598. [CrossRef]
17. Mohammad, M.J.; Zuraiqi, S.; Quasmeh, W.; Papadopoulos, I. Yield response and nitrogen utilization efficiency by drip-irrigated potato. *Nutr. Cycl. Agroecosyst.* **1999**, *54*, 243–249. [CrossRef]
18. Darwish, T.M.; Atallah, T.W.; Hajhasan, S.; Haidar, A. Nitrogen and water use efficiency of fertigated processing potato. *Agric. Water Manag.* **2006**, *85*, 95–104. [CrossRef]
19. Badr, M.A.; El-Tohamy, W.A.; Zaghloul, A.M. Yield and water use efficiency of potato grown under different irrigation and nitrogen levels in an arid region. *Agric. Water Manag.* **2012**, *110*, 9–15. [CrossRef]
20. USDA (U.S. Department of Agriculture). Soil Taxonomy: A basic system of soil classification for making and interpreting soil surveys. In *Soil Conservation Service, Agricultural Handbook*; Government Printing Office: Washington, DC, USA, 1975; p. 754.
21. Mauromicale, G.; Signorelli, P.; Ierna, A.; Foti, S. Effects of intraspecific competition on yield of early potato grown in Mediterranean environment. *Am. Potato J.* **2003**, *80*, 281–288. [CrossRef]
22. Habyarimana, E.; Parisi, B.; Mandolino, G. Genomic prediction for yields, processing and nutritional quality traits in cultivated potato (*Solanum tuberosum* L.). *Plant Breed.* **2017**, *136*, 245–252. [CrossRef]
23. Vos, J.; Bom, M. Hand-held chlorophyll meter: A promising tool to assess the nitrogen status of potato foliage. *Potato Res.* **1993**, *36*, 301–308. [CrossRef]

24. Fontes, P.C.R.; Braun, H.; Busato, C.; Cecon, P.R. Economic optimum nitrogen fertilization rates and nitrogen fertilization rate effects on tuber characteristics of potato cultivars. *Potato Res.* **2010**, *53*, 167–179. [CrossRef]

25. Belanger, G.; Walsh, J.R.; Richards, J.E.; Milburn, P.H.; Ziadi, N. Comparison of three statistical models describing potato yield response to nitrogen fertilizer. *Agron. J.* **2000**, *92*, 902–908. [CrossRef]

26. CCIAA (Camera di Commercio Industria Artigianato e Agricoltura) di Bologna [Chamber of Commerce, Industry, Crafts and Agriculture of Bologna] 2017. Available online: http://www.bo.camcom.gov.it (accessed on 29 December 2018).

27. Neeteson, J.J. Evaluation of the performance of three advisory methods for nitrogen fertilization of sugar beet and potatoes. *Neth. J. Agric. Sci.* **1989**, *37*, 143–155.

28. Moll, R.H.; Kamprath, E.J.; Jackson, W.A. Analysis and interpretation of factors which contribute to efficiency of nitrogen utilization. *Agron. J.* **1982**, *74*, 562–564. [CrossRef]

29. Gariglio, N.F.; Pilatti, R.A.; Baldi, B.L. Using nitrogen balance to calculate fertilization in strawberries. *HortTechnology* **2000**, *10*, 147–150. [CrossRef]

30. AOAC International. *Official Methods of Analysis*, 16th ed.; sec. 33.2.11, Method 991.20; Association of Official Analytical Chemists: Gaithersburg, MD, USA, 1995; Available online: http://www.aoac.org/ (accessed on 15 July 2018).

31. Snedecor, G.W.; Cochran, W.G. *Statistical Methods*, 8th ed.; The Iowa State University Press Publishing: New York, NY, USA, 1989.

32. Gianquinto, G.; Goffart, J.E.; Olivier, M.; Guarda, G.; Colauzzi, M.; Dalla Costa, L.; Delle Vedove, G.; Vos, J.; Mackerron, D.K.L. The use of hand-held chlorophyll meters as a tool to assess the nitrogen status and to guide nitrogen fertilization of potato crop. *Potato Res.* **2004**, *47*, 35–80. [CrossRef]

33. Saoud, A.A.; Van Cleemput, O.; Hofman, G. Uptake and balance of labelled fertilizer nitrogen by potatoes. *Fert. Res.* **1992**, *31*, 351–353. [CrossRef]

34. Greenwood, D.J.; Draycott, A. Recovery of fertilizer-N by diverse vegetable crops: Processes and models. In *Nitrogen Efficiency in Agricultural Soils*; Jenkinson, D.S., Smith, K.A., Eds.; Elsevier Applied Science: London, UK, 1988; pp. 46–61.

35. Cerrato, M.E.; Blackmer, A.M. Comparison of models for describing corn yield response to nitrogen fertilizer. *Agron. J.* **1990**, *82*, 138–143. [CrossRef]

36. Rodrigues, M.A.; Coutinho, J.; Martins, F.; Arrobas, M. Quantitative sidedress nitrogen recommendations for potatoes based upon crop nutritional indices. *Eur. J. Agron.* **2005**, *23*, 79–88. [CrossRef]

37. Iwama, K. Physiology of the potato: New insights into root system and repercussions. *Potato Res.* **2008**, *51*, 333–353. [CrossRef]

38. Ospina, C.A.; van Bueren, E.T.; Allefs, J.J.H.M.; Engel, B.; van der Putten, P.E.L.; van der Linden, C.G.; Struik, P.C. Diversity of crop development traits and nitrogen use efficiency among potato cultivars grown under contrasting nitrogen regimes. *Euphytica* **2014**, *199*, 13–29. [CrossRef]

 *agronomy*

Article

Planting Density and Fertilization Evidently Influence the Fiber Yield of Hemp (*Cannabis sativa* L.)

Gang Deng [1], Guanghui Du [1], Yang Yang [1], Yaning Bao [2] and Feihu Liu [1,*]

[1] School of Agriculture, Yunnan University, Kunming 650504, China
[2] College of Plant Science and Technology, Huazhong Agricultural University, Wuhan 430070, China
* Correspondence: xrhynu@hotmail.com; Tel.: +86-871-65031539; Fax: +86-871-6503-1539

Received: 16 June 2019; Accepted: 9 July 2019; Published: 11 July 2019

Abstract: Hemp is one of the most important green (i.e., environmentally sustainable) fibers. Planting density, nitrogen (N), phosphorus (P) and potassium (K) significantly affect the yield of hemp fiber. By optimizing the above main four cultivation factors is an important way to achieve sustainable development of high-fiber yield hemp crops. In this study, the effects of individual factors and factor × factor interactions on the yield of hemp fiber over two trial years were investigated by the central composite design with four factors, namely planting density, nitrogen application, phosphorus application, and potassium application rate. The influences of these four test factors on the yield of hemp fibers were in the order nitrogen fertilizer (X_2) > planting density (X_1) > potassium fertilizer (X_4) > phosphate fertilizer (X_3). To obtain yields of hemp with high-quality fiber greater than 2200 kg ha^{-1}, the optimal range of cultivation conditions were planting density 329,950–371,500 plants/ha, nitrogen application rate 251–273 kg ha^{-1}, phosphorus application rate 85–95 kg ha^{-1}, and potassium application rate 212–238 kg ha^{-1}. This study can provide important technical and theoretical support for the high-yield cultivation of hemp fiber into the future.

Keywords: planting density; fertilization; the central composite design; fiber yield; analog optimization

1. Introduction

Hemp (*Cannabis sativa* L.) is an ancient and eco-friendly cultivated crop that was first cultivated in China, and is currently used for the manufacture of clothes, household supplies, paper pulp, drugs, food, and recyclable composite materials and among others [1–3]; hemp is known to have been used to make more than 2500 products [4,5]. Hemp textile industries first began in Europe and Asia around 8000 BC [6]. In the middle of the 20th century, hemp was banned from cultivation by governments as an illegal drug crop. However, in recent years, governments and researchers became more interested in the cultivation of hemp, as one of the most important crops for green fiber, seed oil (rich in omega-3 and omega-6 in the right ratio) and domestic drugs uses [2,5]. The cultivation of a number of hemp cultivars with low (<0.3%) THC (tetrahydrocannabinol) concentration has been allowed, to the point where some European governments even provide agricultural subsidies for hemp cultivation [7]. Hemp can be grown with little or no chemical fertilizers, herbicides or pesticides and the crop is now cultivated all around the world [6].

The leading plant macronutrients, nitrogen (N), phosphorus (P) and potassium (K), are important components of plant amino acids, hormones, genetic materials (DNA and RNA) and other substances. Also, it involved in many life processes, such as plant growth, metabolism, cell structure, signal transduction, osmotic regulation, and response to stresses [8–10]. However, due to the low efficiency of utilization of crop fertilizers, especially NPK fertilizers, more than 50% of the nutrients applied to land as chemical fertilizers is wasted [11,12], which also leads to contamination of soil and water resources [13]. Therefore, it is important to determine how to effectively utilize fertilizers or to improve the nutrient-use efficiency of crops to achieve high fiber yields while protecting the environment. The characteristics of hemp are rapid growth, a well-developed root system and a high above-ground biomass (25,000 kg ha^{-1}) [14]. Its growth is sensitive to environmental factors, particularly fertilizers. Of these, the demand for nitrogen by fiber type hemp is greater than that for phosphate or potassium [15,16]. An appropriate planting density allows for efficient use of available resources, such as light, water and nutrients, by the crop, significantly increasing yield of hemp fibers [17,18]. In general, a high planting density is associated with the production of high-quality long fibers [19]. However, different varieties in different regions vary in their optimal planting densities, although an appropriate planting density for hemp fiber cultivation in China is 40–60 plant m^{-2} [20–23].

According to current research, both NPK application rates and planting density can influence the fiber yield of hemp. However, the extent of their respective impacts on the yield of hemp for the fiber industry, the pattern of the impacts, and the optimal cultivation methods are still unclear. In addition, due to the unscientific use of fertilizer in actual production, it not only causes waste of fertilizer, but also leads to an increase in production costs. Therefore, studying the effects of fertilizers and density on the fiber yield of hemp can effectively solve these problems. For example, this study can provide a suitable NPK ratio, optimal planting density, etc.

In order to identify the optimal agronomic conditions (including planting density, nitrogen, phosphorus, potassium) for high-yield cultivation of hemp fiber, the current study analyzed the extent of the effects of the individual factors N, P and K fertilizer application rates and planting density on the yield of hemp fibers, using the most important hemp variety in China, 'Yunma 1' as the test material. Varies optical agronomic methods can be obtained in this study, and it will provide important technical and theoretical support for the high-yield cultivation of hemp fiber into the future.

2. Materials and Methods

2.1. Materials

The experiment was carried out on the experimental farm of the Agricultural College of Yunnan University, Kunming, China, in 2016 and 2017, on sites with uniform soil fertility. The basic soil characteristics were pH 5.98, organic matter content 35.85 g kg^{-1}, total nitrogen 0.17%, total phosphorus 0.09%, total potassium 1.63%, available nitrogen 151.3 mg kg^{-1}, available phosphorus 44.04 mg kg^{-1}, and available potassium 239 mg kg^{-1}. The farm was unprecedented, with good irrigation and drainage conditions. Kunming lies at 25°01′ E, 102°41′ N, at an altitude of 1896 m, and it has a dry season from November to April, with an annual rainfall of 2016, 2017 were 1017 mm, 1049 mm, respectively, and a monthly average temperature of 2016, 2017 were 15.6 °C, 16 °C, 1049 mm, respectively. The test material was 'Yunma 1' (THC < 0.3%), a fiber hemp variety, seeds of which were provided by the Yunnan Academy of Science.

2.2. Methods

Four factors, planting density (X_1), nitrogen fertilizer rate (X_2), phosphate fertilizer rate (X_3), and potassium fertilizer rate (X_4), were tested in this study. Five levels were set for each factor. The experiment was conducted in each year by the central composite design with four factors and 36 combinations. The dimensions of each trial plot were 3.5 m × 2.6 m (area 9.1 m^2), with the 36 plots set out in a completely randomized arrangement, and all the 36 combinations with three replicates (total 108 plots).

The fertilizers applied in this study were urea (containing 46% N), calcium phosphate (containing 14% P_2O_5), and potassium chloride (containing 54% K_2O). Table 1 shows the variable factors and their levels. All phosphate fertilizer and potassium fertilizer were incorporated into the seedbed as base fertilizer, and nitrogen fertilizer was applied twice in March (60%, sowing), June (40%, rapid growth period). The seeds were sown by hand in early May, with an inter-row spacing of 40 cm, and a total of eight rows of hemp are planted in each plot. No chemical pesticides were used during the entire growth period. The hemp was harvested when mature (late September, 70–80% male plant flowering). Twenty plants in the middle of the plot (chosen from rows No. 3 to 5) were randomly selected in each plot. The rods and fibers in the above-ground biomass were separated using a special stripping machine. The fibers were dried (80 °C) and weighed. The yield of fibers (kg ha^{-1}) per plot was then calculated according to the effective number of plants in each plot.

Table 1. Central composite design of plant density and fertilizer dose.

Agronomic Variable	Alternative Gradient	Variable Design				
		−2	−1	0	1	2
Density (plants ha^{-1}) (X_1)	150 000	100,000	250,000	400,000	550,000	700,000
N (kg ha^{-1}) (X_2)	75	75	150	225	300	375
P (kg ha^{-1}) (X_3)	30	30	60	90	120	150
K (kg ha^{-1}) (X_4)	75	75	150	225	300	375

2.3. Data Analysis

The experimental design was based on the calculation principle of the combined design of the central composite design [24]. This study used the software processing system of the statistical software DPS 6.01 (Hangzhou Ruifeng Information Technology Co., Ltd., Hangzhou, China) [25], to establish the mathematical model of indicators such as fiber yield (dependent variables) and test factors (independent variables), and to statistically analyze the model. In this paper, *p*-value < 0.05 was used as a significant difference level, but the difference is a not significant difference.

3. Results

3.1. Establishment and Verification of the Fiber Yield Model

The fiber yield of hemp under different combinations ranged from 1701 to 3205 kg ha^{-1} (Table 2). The results were analyzed using the regression model. With the yield of hemp fibers as the target trait (Y), the regression model was established between the test factors (X_1, X_2, X_3 and X_4) and the target trait:

$$Y = 2907.75 - 112.75X_1 - 156.17X_2 + 1.08X_3 - 21.50X_4 - 196.02X_1{}^2 - 160.15X_2{}^2 - 123.65X_3{}^2 \\ - 144.15X_4{}^2 + 28.631X_2 - 2.25X_1X_3 - 65.88X_1X_4 - 70.75X_2X_3 + 46.63X_2X_4 - 9.00X_3X_4 \tag{1}$$

Through the analysis of variance (ANOVA) of fiber yield (Table 3), the best fitting models were determined by multiple linear regressions with backward elimination, whereby non-significant ($p > 0.05$) factors and interactions were removed from models. The determination coefficient for hemp fiber yield in this study was $r^2 = 0.699$, meaning that the models explained 70% of the variability in hemp fiber yield and were found to be adequate for the data. Analysis of variance (ANOVA) also showed that the regression models for hemp fiber yield were significant, and the models had no significant lack of fit (0.499, $p > 0.05$) (Table 2). In this way, well-fitting models for hemp fiber yield were established. Not all interaction parameters were significant ($p > 0.05$) (Table 3).

Table 2. Structure matrix and the production results from 2016 and 2017.

No	Density (X_1)	N (X_2)	P$_2$O$_5$ (X_3)	K$_2$O (X_4)	Mean Fiber Yield (kg ha^{-1})
1	1	1	1	1	2251
2	1	1	1	−1	2479
3	1	1	−1	1	2193
4	1	1	−1	−1	2821
5	1	−1	1	1	2051
6	1	−1	1	−1	2131
7	1	−1	−1	1	1708
8	1	−1	−1	−1	2239
9	−1	1	1	1	2387
10	−1	1	1	−1	2523
11	−1	1	−1	1	2824
12	−1	1	−1	−1	2399
13	−1	−1	1	1	2035
14	−1	−1	1	−1	2603
15	v	−1	−1	1	2102
16	−1	−1	−1	−1	2236
17	−2	0	0	0	2436
18	2	0	0	0	1701
19	0	−2	0	0	1968
20	0	2	0	0	2456
21	0	0	−2	0	2336
22	0	0	2	0	2380
23	0	0	0	−2	1935
24	0	0	0	2	2617
25	0	0	0	0	2833
26	0	0	0	0	3175
27	0	0	0	0	3048
28	0	0	0	0	2976
29	0	0	0	0	3294
30	0	0	0	0	3000
31	0	0	0	0	2849
32	0	0	0	0	2809
33	0	0	0	0	2936
34	0	0	0	0	3205
35	0	0	0	0	2579
36	0	0	0	0	2189

Table 3. Analysis of variance (ANOVA) of fiber yield of hemp.

Source	Sum of Squares	df	Mean Squares	Partial Correlation	F-Value	p-Value
X_1	305,101.49	1	305,101.49	−0.3761	3.4596	0.0770
X_2	585,312.64	1	585,312.64	0.4900	6.6369	0.0176 *
X_3	28.17	1	28.17	0.0039	0.0003	0.9859
X_4	11,094.00	1	11,094.00	−0.0772	0.1258	0.7264
X_1^2	122,9573.30	1	1,229,573.30	−0.6317	13.9422	0.0012 **
X_2	820,693.98	1	820,693.98	−0.5541	9.3059	0.0061 **
X_3^2	489,225.33	1	489,225.33	−0.4571	5.5474	0.0283 *
X_4^2	664,896.66	1	664,896.66	−0.5140	7.5393	0.0121 *
X_1X_2	13,110.25	1	13,110.25	0.0838	0.1487	0.7037
X_1X_3	81.00	1	81.00	−0.0066	0.0009	0.9761
X_1X_4	69,432.25	1	69,432.25	−0.1901	0.7873	0.3850
X_2X_3	80,089.00	1	80,089.00	−0.2036	0.9081	0.3515
X_2X_4	34,782.25	1	34,782.25	0.1358	0.3944	0.5368
X_3X_4	1296.00	1	1296.00	−0.0264	0.0147	0.9047
Regression	4,304,716.47	14	307,479.75	F2 = 3.48654		0.0100 **
Residual	1,851,999.75	21	88,190.46			
Lack of fit	865,925.50	10	86,592.55	F1 = 0.96597		0.4991
Pure error	986,074.250	11	89,643.11			
Total error	6,156,716.2222	35				

Note: * $p < 0.05$, ** $p < 0.01$. *df*, degree of freedom, X_1, planting density, X_2, nitrogen, X_3, phosphate, X_4, potassium.

3.2. Main-Effect Analysis of Factors

The sub-models of the relationship between hemp fiber yield and the main effects of planting density, nitrogen fertilizer, phosphorus fertilizer and potassium fertilizer were calculated by using the method of descending dimension as the mathematical model with the other factors of the fixed Equation # 11 at 0 levels (Equation (2)):

$$
\begin{aligned}
\text{Planting density: } Y_1 &= 2907.75 - 112.75X_1 - 196.02X_1^2 \\
\text{Nitrogen fertilizer: } Y_2 &= 2907.75 + 156.17X_2 - 160.15X_2^2 \\
\text{Phosphorus fertilizer: } Y_3 &= 2907.75 + 1.08X_3 - 123.65X_3^2 \\
\text{Potassium fertilizer: } Y_4 &= 2907.75 - 21.50X_4 - 144.15X_4
\end{aligned}
\tag{2}
$$

The above equation showed that the partial regression coefficients of planting density (X_1), nitrogen (X_2), phosphorus (X_3) and potassium application rates (X_4) were −112.75, 156.17, 1.08 and −21.5, respectively. As positive effects, increasing nitrogen and phosphorus would increase the yield of hemp fiber, whereas, as negative effects, increasing planting density and potassium as would reduce the yield of hemp fiber. According to the absolute value discriminant method of linear coefficients, the influence of each factor on fiber yield can be defined directly from the absolute value of the respective regression coefficient, as in the order nitrogen > planting density > potassium > phosphorus.

3.3. Analysis of Single Factor Effects

Figure 1 shows that, according to the sub-model (Equation (2)), the four test factors have a parabolic relationship with hemp fiber yield within the constraint range of −2 ≤ Xi ≤ 2. Fiber yield increased with increasing plant density level from −2 to 0.29, and then decreased, with a maximum fiber yield of 2923 kg ha^{-1} (at −0.29; 356,500 plants ha^{-1}). Fiber yield increased with increasing nitrogen level from −2 to 0.49, then decreased, with a maximum fiber yield of 2946 kg ha^{-1} (at 0.49; 262 kg ha^{-1}). Fiber yield increased with increasing phosphorus level from -2 to 0, and then decreased, with a maximum fiber yield of 2908 kg ha^{-1} (at 0; 90 kg ha^{-1}). Fiber yield increased with increasing potassium level from −2 to 0.07, and then decreased, with a maximum fiber yield of 2906 kg ha^{-1} (at 0.07; 230 kg ha^{-1}). Figure 1 illustrates that fiber yield decreased rapidly at planting density and potassium levels from 0–2, the decrease being faster than with nitrogen fertilizer and phosphate fertilizer, with planting density decreasing the fastest, and nitrogen decreasing at the slowest rate. Therefore, the fiber yield of hemp would not increase but would decrease rapidly once the plant density and potassium fertilizer levels increased beyond the optimal level.

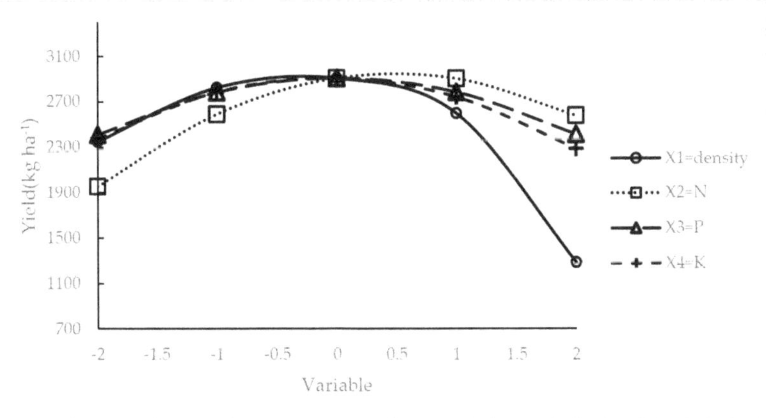

Figure 1. The effect of the factors (X_1 = density, X_2 = N, X_3 = P, X_4 = K) on mean fiber yield for 2016 and 2017.

3.4. Analysis of the Marginal Yield Effect of Single Factors

Marginal yield is the increase in yield for each additional unit of a variable factors. The rate of change caused by each factor could be calculated by the first-order partial derivative (dY/dX_i) of the fiber yield (Y) in response to a particular factor (X_i). It was further analyzed to obtain the reasonable collocation of the levels of different factors when a certain factor was the main control factor, and the highest yield under different conditions. The sub-model (Equation (3)) of the marginal yield effect of a single factor was calculated from the first derivative in the sub-model (Equation (2)):

$$
\begin{aligned}
\text{Planting density: } dY/dX_1 &= -112.75 - 392.04X_1 \\
\text{Nitrogen fertilizer: } dY/dX_2 &= 156.17 - 320.3X_2 \\
\text{Phosphorus fertilizer: } dY/dX_3 &= 1.08 - 247.3X_3 \\
\text{Potassium fertilizer: } dY/dX_4 &= -21.50 - 288.3X_4
\end{aligned}
\tag{3}
$$

According to the corresponding marginal models with different horizontal values of each factor, the planting density showed the greatest change in the marginal effect changes at different levels of each factor, while nitrogen and potassium showed less change, and phosphorus showed the least change. The marginal effects of all four factors, from 0 level to the highest level, were negative, which indicated that the increase in NPK application rate and planting density led to reduced hemp fiber yield, which also indicated that reduction in fertilizer application or planting density could increase the fiber yield of hemp. However, combining the fiber yield regression model, it was clear that there were certain interaction effects among the factors, but that the difference in the interaction effect between each factor was not significant. The influence of each factor on the increase of fiber yield at different levels varied, and the order of their effects on fiber yield increase was: $X_2 > X_1 > X_4 > X_3$ at the −2 and −1 levels; the order of their effects on fiber yield reduction was $X_1 > X_4 > X_3 > X_2$ at the 1 and 2 levels (Figure 2).

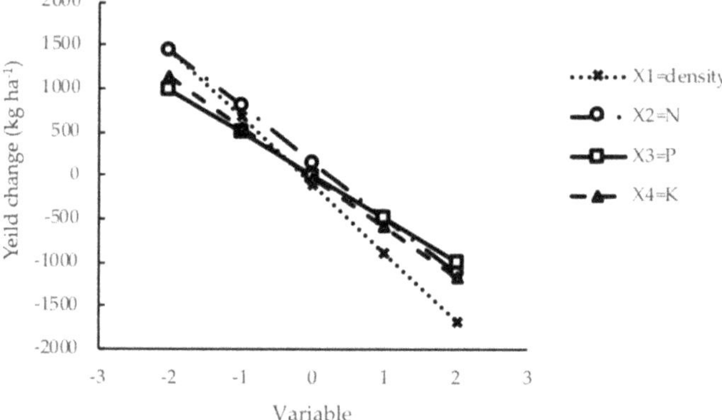

Figure 2. Marginal effects of increases in experimental factors (X_1 = density, X_2 = N, X_3 = P, X_4 = K) on boundary fiber hemp production.

3.5. Optimization of Agronomic Methods Plan

Within the constraint range of $-2 \leq X_i \leq 2$, 134 sets of combinations with the yield of hemp fiber greater than 2200 kg ha^{-1} were selected and further analyzed for frequency (Table 4). Table 4 shows that there were many routes by which to obtain high fiber yield according to various agronomic methods as shown by the combination plans with high yields of hemp fiber but were mainly concentrated in the horizontal −1 to +1 level range.

Table 4. Frequency of special X_i value of hemp fiber yield beyond 2200 kg ha^{-1}.

Factor		Density (X_1)		N (X_2)		P (X_3)		K (X_4)	
		Degree	Frequency (%)	Degree	Frequency (%)	Degree	Frequency (%)	Degree	Frequency (%)
Variable design	−2	10	7.46	0	0	8	5.97	8	5.97
	−1	44	32.84	19	14.18	36	26.87	36	26.87
	0	60	44.78	48	35.82	46	34.33	46	34.33
	1	20	14.93	48	35.82	36	26.87	36	26.87
	2	0	0	19	14.18	8	5.97	8	5.97
Weighted mean		−0.328		0.500		0		0	
Standard error		0.0710		0.0780		0.0870		0.0870	
95% Confidence interval		−0.467~−0.190		0.347~0.653		−0.171~0.171		−0.171~0.171	
Optimal range		329,950~371,500 plant ha^{-1}		251.03~273.98 kg ha^{-1}		84.87~95.13 kg ha^{-1}		212.18~237.83 kg ha^{-1}	

Under similar experimental conditions, in order to obtain high yields of raw hemp fiber greater than 2200 kg ha^{-1}, the relatively optimal combination plan of cultivation involved a planting density of 329,950–37,1500 plants ha^{-1}, a nitrogen application rate of 251–273 kg ha^{-1}, a phosphorus application rate of 85–95 kg ha^{-1}, and a potassium application rate of 212–238 kg ha^{-1}.

4. Discussion

Hemp has attracted much attention from the market and from researchers due to its multiple uses, therefore, improving the production of hemp fiber through research will help promote the promotion and competitiveness of hemp products [26,27]. Many studies have focused on the effects of N fertilizer application rate and planting density on the growth and fiber yield of hemp [15,18,28], but the current study is the first to provide a comprehensive analysis of the effects and interactions of NPK fertilizer application rates and planting density on hemp fiber yield. Because hemp for the fiber industry has the characteristics of high biomass and rapid growth, a large amount of fertilizer is required during the growth period. It has been reported that the amount of NPK applied and hemp fiber yield per unit area were positively correlated [29]. Nitrogen fertilizer in the current study exhibited the greatest influence on the fiber yield of hemp, and its contribution to yield at high nitrogen level was higher than that of the other three factors; at a moderate nitrogen level (level 0; 225 kg ha^{-1}), however, nitrogen contributed less to fiber yield than did the other three factors. When nitrogen level reached 262 kg ha^{-1}, the fiber yield was maximal, with yield response to nitrogen application rate increasing more at −2 to 0 levels than that at the 0 to 2 levels. The results were similar to those reported by Struck et al. [14]. The current study also found that the fiber yield at the high nitrogen level 2 was still higher than that at the lowest nitrogen level −2, while the fiber yield at the highest level of the other treatments was lower than that at the lowest level, an observation which demonstrated the importance of optimizing nitrogen fertilizer application to achieving the goal of high hemp fiber yield.

This present study revealed that increasing either phosphorus or nitrogen application rates exhibited a positive effect on hemp fiber yield. However, the effect of phosphorus on hemp fiber yield was smaller, the increase was not significant, and the overall change curve was relatively flat, findings which were similar to those reported by Vera et al. [30,31]. Meanwhile, the current study found that increasing either potassium application rate or planting density exhibited a negative effect on hemp fiber yield, results which differed from those of previous studies that showed increased hemp fiber yield in response to increased potassium, but were in line with the results of Finnan and Burke's research [32], which concluded that there was no significant correlation between hemp fiber yield and soil potassium levels. The demand for potassium by hemp may be lower than expected. Despite there being high potassium uptake by hemp under high-potassium conditions, the extra uptake of potassium had no significant effect on fiber yield increase of hemp, which was considered to be luxury uptake [32].

Planting density directly affects the structure of the hemp population, and thus affects the fiber yield. According to previous studies, it has found that, when the density reached a certain level, hemp fiber yield decreased due to a self-thinning effect [28]. In this study, it was found that increasing the planting density had a negative effect on the yield of hemp. The fiber yield level was lower than that achieved by other factors at planting density levels above 0, and the decrease was the greatest. Therefore, it is not appropriate to increase the planting density in hemp production, as, once the planting density exceeded a certain range, fiber yield per area was significantly reduced. The present study demonstrated that the optimal planting density was 32–37 plants m^{-2}.

In order to obtain high fiber yield of hemp under similar conditions to those experienced in the present study, this study optimized the agronomic methods, and showed the relatively optimal combination plan of cultivation methods which could reach high fiber yields of greater than 2200 kg ha^{-1}, namely planting density of 329,950–371,500 plants/ha, a nitrogen application rate of 251–273 kg ha^{-1}, a phosphorus application rate of 85–95 kg ha^{-1}, and potassium application rate of 212–238 kg ha^{-1}, with an approximate N:P:K fertilizer application ratio (relative to the soil NPK levels described in Section 2.1) of 5:2:4. This present study can provide important guidance for optimizing the agronomic conditions for hemp cultivation for fiber.

5. Conclusions

The four tested factors effects on the fiber yield of hemp was shown in this study to be in the order nitrogen fertilizer rate (X_2) > planting density (X_1) > potassium fertilizer rate (X_4) > phosphate fertilizer rate (X_3). The study also revealed that increasing the amount of N, P, or K applied or the planting density could lead to fiber yield reductions. This study suggested that the relatively optimal combination plan of cultivation to obtain hemp fiber yield greater than 2200 kg ha^{-1} involved a planting density of 329,950–371,500 plants ha^{-1}, a nitrogen application rate of 251–273 kg ha^{-1}, a phosphorus application rate of 85–95 kg ha^{-1}, and a potassium application of 212–238 kg ha^{-1}, with an approximate N: P: K fertilizer.

Author Contributions: Conceptualization, F.L; methodology, G.D. (Gang Deng); resources, Y.Y.; data curation, G.D. (Guanghui Du); writing—original draft preparation, G.D. (Gang Deng); writing-review & editing, F.L., Y.B.; funding acquisition, F.L.

Funding: This research was supported by a grant from the Natural Science Foundation of Yunnan (2016FB068 and 2017FD060), and China Agriculture Research System (CARS-16-E15).

Conflicts of Interest: The authors declare no conflict of interest.

References

1.   Blade, S.F.; Gaudiel, R.G.; Kerr, N. *Low-THC Hemp Research in the Black and Brown Soil Zones of Alberta, Canada*; ASHS Press: Alexandria, VA, USA, 1999.

2.   Decorte, T. The case for small-scale domestic cannabis cultivation. *Int. J. Drug Policy* **2010**, *21*, 271–275. [CrossRef] [PubMed]

3.   Zhou, Y.K.; Zhang, J.; Zhang, J.C. Structure and properties of hemp xyloid stem based viscose fiber. *J. Text. Res.* **2008**, *29*, 4.

4.   Johnson, R. *Hemp as an Agricultural Commodity*; Library of Congress Washington DC Congressional Research Service: Washington, DC, USA, 2014.

5.   Liu, F.H.; Hu, H.R.; Du, G.H.; Deng, G.; Yang, Y. Ethnobotanical research on origin, cultivation, distribution and utilization of hemp (*Cannabis sativa* L.) in China. *Indian J. Tradit. Knowl.* **2017**, *16*, 235–242.

6.   Hemphouse. Hemp History. Available online: http://www.hemphousemaui.com/hemp-history/ (accessed on 8 June 2019).

7.   Forapani, S.; Carboni, A.; Paoletti, C.; Cristiana, V.M.; Ranalli, N.P.; Mandolino, G. Comparison of hemp varieties using random amplified polymorphic DNA markers. *Crop Sci.* **2001**, *41*, 1682–1689. [CrossRef]

8. Maathuis, F.J.; Ichida, A.M.; Sanders, D.; Schroeder, J.I. Roles of higher plant K$^+$ channels. *Plant Physiol.* **1997**, *114*, 1141–1149. [CrossRef] [PubMed]

9. Krouk, G.; Lacombe, B.; Bielach, A.; Perrine-Walker, F.; Malinska, K.; Mounier, E.; Hoyerova, K.; Tillard, P.; Leon, S.; Ljung, K.; et al. Nitrate-regulated auxin transport by NRT1.1 defines a mechanism for nutrient sensing in plants. *Dev. Cell* **2010**, *18*, 927–937. [CrossRef]

10. Chevalier, F.; Rossignol, M. Proteomic analysis of *Arabidopsis thaliana* ecotypes with contrasted root architecture in response to phosphate deficiency. *J. Plant Physiol.* **2011**, *168*, 1885–1890. [CrossRef]

11. Lynch, J.P. Root phenes for enhanced soil exploration and phosphorus acquisition: Tools for future crops. *Plant Physiol.* **2011**, *156*, 1041–1049. [CrossRef]

12. Xu, G.H.; Fan, X.R.; Miller, A.J. Plant nitrogen assimilation and use efficiency. *Annu. Rev. Plant Biol.* **2012**, *63*, 153–182. [CrossRef]

13. Zimmermann, R.; Bauermann, U. Effects of growing site and nitrogen fertilization on biomass production and lignin content of linseed (*Linum usitatissimum* L.). *J. Sci. Food Agric.* **2006**, *86*, 415–419. [CrossRef]

14. Struik, P.C.; Amaducci, S.; Bullard, M.J.; Stutterheim, N.C.; Venturi, G.; Cromack, H.T.H. Agronomy of fibre hemp (*Cannabis sativa* L.) in Europe. *Ind. Crop Prod.* **2000**, *11*, 107–118. [CrossRef]

15. Papastylianou, P.; Kakabouki, I.; Travlos, I. Effect of nitrogen fertilization on growth and yield of industrial hemp (*Cannabis sativa* L.). *Not. Bot. Horti Agrobot.* **2018**, *46*, 197–201. [CrossRef]

16. Liu, H.; Zhang, Y.Y.; Hu, H.R.; Du, G.H.; Yang, Y.; Liu, F.H. Effects of combined application of N, P, K fertilizers on accumulation and distribution of dry matter and nutrients in industrial hemp (*Cannabis sativa* L.). *Plant Fiber Sci. China* **2015**, *37*, 6.

17. Schafer, T.A.; Honermeier, B. Effect of sowing date and plant density on the cell morphology of hemp (*Cannabis sativa* L.). *Ind. Crop Prod.* **2006**, *23*, 88–98. [CrossRef]

18. Tang, K.; Struik, P.C.; Yin, X.; Calzolari, D.; Musio, S.; Thouminot, C.; Bjelkova, M.; Strarnkale, V.; Magagnini, G.; Amaducci, S. A comprehensive study of planting density and nitrogen fertilization effect on dual-purpose hemp (*Cannabis sativa* L.) cultivation. *Ind. Crop Prod.* **2017**, *107*, 427–438. [CrossRef]

19. Westerhuis, W.; Amaducci, S.; Struik, P.C.; Zatta, A.; van Dam, J.E.G.; Stomph, T.J. Sowing density and harvest time affect fibre content in hemp (*Cannabis sativa*) through their effects on stem weight. *Ann. Appl. Biol.* **2009**, *155*, 225–244. [CrossRef]

20. Sun, A.G.; Sun, Y.X.; Liang, D.P.; Zhu, H.B. Study on the method and density of hemp. *Plant Fiber Sci. China* **1981**, *2*, 7.

21. Lv, Y.M.; Yang, L.; Hu, W.Q. Standardized planting mode of hemp in Lu'an. *Anhui Agric. Sci. Bull.* **2011**, *17*, 3.

22. Hu, X.L.; Yang, M.; Xu, Y.P.; Guo, M.B.; Chen, Y.; Deng, W.P.; Wu, J.X.; Zeng, M.; Guo, H.Y. Effect of planting density on yield and agronomic characters of hemp variety Yunma no. 1. *Plant Fiber Sci. China* **2009**, *31*, 3.

23. Jiang, G.C.H.; Chen, X.W.; Yu, J.; Sun, T. Effects of fertilization pattern and planting density on yield of hemp. *J. Hunan Agric. Univ.* **2018**, *44*, 4.

24. Box, G.E.P.; Hunter, W.G.; Hunter, J.S. *Statistics for Experiments*; John Willey and Sons: New York, NY, USA, 1978.

25. Tang, Q.Y.; Zhang, C.X. Data Processing System (DPS) software with experimental design, statistical analysis and data mining developed for use in entomological research. *Insect Sci.* **2013**, *20*, 254–260. [CrossRef] [PubMed]

26. Aubin, M.P.; Seguin, P.; Vanasse, A.; Lalonde, O.; Tremblay, G.F.; Mustafa, A.F.; Charron, J.B. Evaluation of eleven industrial hemp cultivars grown in Eastern Canada. *Agron. J.* **2016**, *108*, 1972–1980. [CrossRef]

27. Faux, A.M.; Draye, X.; Lambert, R.; d'Andrimont, R.; Raulier, P.; Bertin, P. The relationship of stem and seed yields to flowering phenology and sex expression in monoecious hemp (*Cannabis sativa* L.). *Eur. J. Agron.* **2013**, *47*, 11–22. [CrossRef]

28. van der Werf, H.M.G.; Wijlhuizen, M.; de Schutter, J.A.A. Plant density and self-thinning affect yield and quality of fibre hemp (*Cannabis sativa* L.). *Field Crop Res.* **1995**, *40*, 11. [CrossRef]

29. Yang, Y.; Zhou, H.Y.; Du, G.H.; Li, D.W.; Liu, G.H.; Liu, F.H. Preliminary study of relationship between population structure and ontogenesis in hemp. *Plant Fiber Sci. China* **2014**, *36*, 4.

30. Vera, C.L.; Malhi, S.S.; Phelps, S.M.; May, W.E.; Johnson, E.N. N, P, and S fertilization effects on industrial hemp in Saskatchewan. *Can. J. Plant Sci.* **2010**, *90*, 179–184. [CrossRef]

Agronomy **2019**, *9*, 368

31. Vera, C.L.; Malhi, S.S.; Raney, J.P.; Wang, Z.H. The effect of N and P fertilization on growth, seed yield and quality of industrial hemp in the parkland region of Saskatchewan. *Can. J. Plant Sci.* **2004**, *84*, 939–947. [CrossRef]
32. Finnan, J.; Burke, B. Potassium fertilization of hemp (*Cannabis sativa*). *Ind. Crop Prod.* **2013**, *41*, 419–422. [CrossRef]

MDPI

St. Alban-Anlage 66

4052 Basel

Switzerland

Tel. +41 61 683 77 34

Fax +41 61 302 89 18

www.mdpi.com

Agronomy Editorial Office

E-mail: agronomy@mdpi.com

www.mdpi.com/journal/agronomy